THE OXFORD HANDBOOK OF

PHILOSOPHY OF PHYSICS

THE OXFORD HANDBOOK OF

PHILOSOPHY OF PHYSICS

Edited by ROBERT BATTERMAN

OXFORD
UNIVERSITY PRESS

Oxford University Press is a department of the University of Oxford.
It furthers the University's objective of excellence in research, scholarship,
and education by publishing worldwide.

Oxford   New York
Auckland   Cape Town   Dar es Salaam   Hong Kong   Karachi
Kuala Lumpur   Madrid   Melbourne   Mexico City   Nairobi
New Delhi   Shanghai   Taipei   Toronto

With offices in
Argentina   Austria   Brazil   Chile   Czech Republic   France   Greece
Guatemala   Hungary   Italy   Japan   Poland   Portugal   Singapore
South Korea   Switzerland   Thailand   Turkey   Ukraine   Vietnam

Oxford is a registered trade mark of Oxford University Press
in the UK and certain other countries.

Published in the United States of America by
Oxford University Press
198 Madison Avenue, New York, NY 10016

© Oxford University Press 2013

First issued as an Oxford University Press paperback, 2017.

All rights reserved. No part of this publication may be reproduced, stored in a
retrieval system, or transmitted, in any form or by any means, without the prior
permission in writing of Oxford University Press, or as expressly permitted by law,
by license, or under terms agreed with the appropriate reproduction rights organization.
Inquiries concerning reproduction outside the scope of the above should be sent to the Rights
Department, Oxford University Press, at the address above.

You must not circulate this work in any other form
and you must impose this same condition on any acquirer.

Library of Congress Cataloging-in-Publication Data
The Oxford handbook of philosophy of physics / edited by
Robert Batterman.
p. cm.
ISBN: 978-0-19-539204-3 (hardcover : alk. paper); 978-0-19-085518-5 (paperback : alk. paper)
1. Physics—Philosophy.   I. Batterman, Robert W.   II. Title: Handbook of philosophy of physics.
QC6.O925   2012
530.1—dc23   2012010291

Contents

	Contributors	*vii*
	Introduction *Robert Batterman*	1
1.	For a Philosophy of Hydrodynamics *Olivier Darrigol*	12
2.	What Is "Classical Mechanics" Anyway? *Mark Wilson*	43
3.	Causation in Classical Mechanics *Sheldon R. Smith*	107
4.	Theories of Matter: Infinities and Renormalization *Leo P. Kadanoff*	141
5.	Turn and Face the Strange . . . Ch-ch-changes: Philosophical Questions Raised by Phase Transitions *Tarun Menon and Craig Callender*	189
6.	Effective Field Theories *Jonathan Bain*	224
7.	The Tyranny of Scales *Robert Batterman*	255
8.	Symmetry *Sorin Bangu*	287
9.	Symmetry and Equivalence *Gordon Belot*	318
10.	Indistinguishability *Simon Saunders*	340

11. Unification in Physics — Margaret Morrison — 381

12. Measurement and Classical Regime in Quantum Mechanics — Guido Bacciagaluppi — 416

13. The Everett Interpretation — David Wallace — 460

14. Unitary Equivalence and Physical Equivalence — Laura Ruetsche — 489

15. Substantivalist and Relationalist Approaches to Spacetime — Oliver Pooley — 522

16. Global Spacetime Structure — John Byron Manchak — 587

17. Philosophy of Cosmology — Chris Smeenk — 607

Index — 653

Contributors

Guido Bacciagaluppi is Reader in Philosophy at the University of Aberdeen. His field of research is the philosophy of physics, in particular the philosophy of quantum theory. He also works on the history of quantum theory and has published a book on the 1927 Solvay conference (together with A. Valentini). He also has interests in the foundations of probability and in issues of time symmetry and asymmetry.

Jonathan Bain is Associate Professor of Philosophy of Science at the Polytechnic Institute of New York University. His research interests include philosophy of spacetime, scientific realism, and philosophy of quantum field theory.

Sorin Bangu is Associate Professor of Philosophy at the University of Bergen, Norway. He received his Ph.D. from the University of Toronto and has previously been a postdoctoral fellow at the University of Western Ontario and a fixed-term lecturer at the University of Cambridge, Department of History and Philosophy of Science. His main interests are in philosophy of science (especially philosophy of physics, mathematics, and probability) and later Wittgenstein. He has published extensively in these areas and has recently completed a book manuscript on the metaphysical and epistemological issues arising from the applicability of mathematics to science.

Robert Batterman is Professor of Philosophy at the University of Pittsburgh. He is a Fellow of the Royal Society of Canada. He is the author of *The devil in the details: Asymptotic reasoning in explanation, reduction, and emergence* (Oxford, 2002). His work in philosophy of physics focuses primarily upon the area of condensed matter broadly construed. His research interests include the foundations of statistical physics, dynamical systems and chaos, asymptotic reasoning, mathematical idealizations, the philosophy of applied mathematics, explanation, reduction, and emergence.

Gordon Belot is Professor of Philosophy at the University of Michigan. He has published a number of articles on philosophy of physics and related areas—and one small book, *Geometric possibility* (Oxford, 2011).

Craig Callender is Professor of Philosophy at the University of California, San Diego. He has written widely in philosophy of science, metaphysics, and philosophy of physics. He is the editor of *Physics meets philosophy at the Planck length* (with

Huggett) and the *Oxford handbook of the philosophy of time*. He is currently working on a book monograph on the relationship between physical time and time as we experience it.

Olivier Darrigol is a CNRS research director in the SPHERE/Rehseis research team in Paris. He investigates the history of physics, mostly nineteenth and twentieth century, with a strong interest in related philosophical questions. He is the author of several books including *From c-numbers to q-numbers: The classical analogy in the history of quantum theory* (Berkeley: University of California Press, 1992), *Electrodynamics from Ampère to Einstein* (Oxford: Oxford University Press, 2000), *Worlds of flow: A history of hydrodynamics from the Bernoullis to Prandtl* (Oxford: Oxford University Press, 2005), and *A history of optics from Greek antiquity to the nineteenth century* (Oxford: Oxford University Press, 2012).

Leo P. Kadanoff is a theoretical physicist and applied mathematician who has contributed widely to research in the properties of matter, the development of urban areas, statistical models of physical systems, and the development of chaos in simple mechanical and fluid systems. His best-known contribution was in the development of the concepts of "scale invariance" and "universality" as they are applied to phase transitions. More recently, he has been involved in the understanding of singularities in fluid flow.

John Byron Manchak is an Assistant Professor of Philosophy at the University of Washington. His primary research interests are in philosophy of physics and philosophy of science. His research has focused on foundational issues in general relativity.

Tarun Menon is a graduate student in Philosophy at the University of California, San Diego. His research interests are in the philosophy of physics and metaphysics, particularly time, probability, and the foundations of statistical mechanics. He is also interested in formal epistemology and the cognitive structure of science.

Margaret Morrison is Professor of Philosophy at the University of Toronto. She is the author of several articles on various aspects of philosophy of science including physics and biology. She is also the author of *Unifying scientific theories: Physical concepts and mathematical structures* (Cambridge, 2000) and the editor (with Mary Morgan) of *Models as mediators: Essays on the philosophy of natural and social science* (Cambridge, 1999).

Oliver Pooley is University Lecturer in the Faculty of Philosophy at the University of Oxford and a Fellow and Tutor at Oriel College, Oxford. He works in the philosophy of physics and in metaphysics. Much of his research focuses on the nature of space, time, and spacetime.

Laura Ruetsche is Professor of Philosophy at the University of Michigan. Her *Interpreting quantum theories: The art of the possible* (Oxford, 2011) aims to articulate questions about the foundations of quantum field theories whose answers might hold interest for philosophy more broadly construed.

Simon Saunders is Professor in the Philosophy of Physics and Fellow of Linacre College at the University of Oxford. He has worked in the foundations of quantum field theory, quantum mechanics, symmetries, thermodynamics, and statistical mechanics and in the philosophy of time and spacetime. He was an early proponent of the view of branching in the Everett interpretation as an "effective" process based on decoherence. He is co-editor (with Jonathan Barrett, Adrian Kent, and David Wallace) of *Many worlds? Everett, quantum theory, and reality* (OUP 2010).

Chris Smeenk is Associate Professor of Philosophy at the University of Western Ontario. His research interests are history and philosophy of physics, and seventeenth-century natural philosophy.

Sheldon R. Smith is Professor of Philosophy at UCLA. He has written articles on the philosophy of classical mechanics, the relationship between causation and laws, the philosophy of applied mathematics, and Kant's philosophy of science.

David Wallace studied physics at Oxford University before moving into philosophy of physics. He is now Tutorial Fellow in Philosophy of Science at Balliol College, Oxford, and university lecturer in Philosophy at Oxford University. His research interests include the interpretation of quantum mechanics and the philosophical and conceptual problems of quantum field theory, symmetry, and statistical physics.

Mark Wilson is Professor of Philosophy at the University of Pittsburgh, a Fellow of the Center for Philosophy of Science, and a Fellow at the American Academy of Arts and Sciences. His main research investigates the manner in which physical and mathematical concerns become entangled with issues characteristic of metaphysics and philosophy of language. He is the author of *Wandering significance: An essay on conceptual behavior* (Oxford, 2006). He is currently writing a book on explanatory structure. He is also interested in the historical dimensions of this interchange; in this vein, he has written on Descartes, Frege, Duhem, and Wittgenstein. He also supervises the North American Traditions Series for Rounder Records.

THE OXFORD HANDBOOK OF

PHILOSOPHY OF PHYSICS

INTRODUCTION

ROBERT BATTERMAN

When I was in graduate school in the 1980s, philosophy of physics was focused primarily on two dominant reasonably self-contained theories: Orthodox nonrelativisitic quantum mechanics and relativistic spacetime theories. Of course, there were a few papers published on certain questions in other fields of physics such as statistical mechanics and its relation to thermodynamics. These latter, however, primarily targeted the extent to which the reductive relations between the two theories could be considered a straightforward implementation of the orthodox strategy outlined by Ernest Nagel.

Philosophical questions about the measurement problem, the question of the possibility of hidden variables, and the nature of quantum locality dominated the philosophy of physics literature on the quantum side. Questions about relationalism vs. substantivalism, the causal and temporal structure of the world, as well as issues about underdetermination of theories dominated the literature on the spacetime side. Some worries about determinism vs. indeterminism crossed the divide between these theories and played a significant role in shaping the development of the field. (Here I am thinking of Earman's *A Primer on Determinism* (1986) as a particular driving force.)

These issues still receive considerable attention from philosophers of physics. But many philosophers have shifted their attention to other questions related to quantum mechanics and to spacetime theories. In particular, there has been considerable work on understanding quantum field theory, particularly from the point of view of algebraic or axiomatic formulations. New attention has also been given to philosophical issues surrounding quantum information theory and quantum computing. And there has, naturally, been considerable interest in understanding the relations between quantum theory and relativity theory. Questions about the possibility of unifying these two fundamental theories arise. Relatedly, there has been a focus on understanding gauge invariance and symmetries.

However, I believe philosophy of physics has evolved even further, and this belief prompts the publication of this volume. Recently, many philosophers have focused their attentions on theories that, for the most part, were largely ignored in the past. As noted above, the relationship between thermodynamics and statistical mechanics—once thought to be a paradigm instance of unproblematic theory reduction—is now a hotly debated topic. Philosophers and physicists have long implicitly or explicitly adopted a reductionist methodological bent. Yet, over the years this methodological slant has been questioned dramatically. Attention has been focused on the explanatory and descriptive roles of "nonfundamental," *phenomenological* theories. In large part because of this shift of focus, "old" theories such as classical mechanics, once deemed to be of little philosophical interest, have increasingly become the focus of deep methodological investigations.

Furthermore, some philosophers have become more interested in less "fundamental" contemporary physics. For instance, there are deep questions that arise in condensed matter theory. These questions have interesting and important implications for the nature of models, idealizations, and explanation in physics. For example, model systems, such as the Ising model, play important computational and conceptual roles in understanding how there can be phase transitions with specific characteristics. And, the use of the thermodynamic limit is an idealization that (some have argued) plays an essential, ineliminable role in understanding and explaining the observed universality of critical phenomena. These specific issues are discussed in several of the chapters in this volume.

In the United States during the 1970s and 1980s, there was a great debate between particle physicists who pushed for funding of high-energy particle accelerators and solid-state or condensed-matter theorists for whom the siphoning off of so much government funding to "fundamental" physics was unacceptable. A famous paper championing the latter position is Philip Anderson's "More Is Different" (1972). Not only was this a debate over funding, but it raised issues about exactly what should count as "fundamental" physics. While historians of physics have focused considerable attention on this public debate, philosophers of physics have really only recently begun to engage with the conceptual implications of the possibility that condensed matter theory is in some sense just as fundamental as high-energy particle physics.

This collection aims to do two things. First, it tries to provide an overview of many of the topics that currently engage philosophers of physics. And second, it focuses attention on some theories that by orthodox 1980s standards would not have been considered fundamental. It strives to survey some of these new issues and the problems that have become a focus of attention in recent years. Additionally, it aims to provide up-to-date discussions of the deep problems that dominated the field in the past.

In the first chapter, "For a Philosophy of Hydrodynamics," Olivier Darrigol focuses attention on lessons that can be learned from the historical development of fluid mechanics. He notes that hydrodynamics has probably received the least attention of any physical theory from philosophers of physics. Hydrodynamics is not a "fundamental" theory along the lines of quantum mechanics and relativity theory, and its basic formulation has not evolved much for two centuries. These facts, together with a lack of detailed historical studies of hydrodynamics, have kept the theory off the radar.[1] Darrigol provides an account of the development of hydrodynamics as a complex theory—one that is not fully captured by the basic Navier-Stokes equations. For the theory to be applicable, particularly for it to play an explanatory role, a host of techniques—idealizations, modeling strategies, and empirically determined data must come into play. This discussion shows clearly how intricate, sophisticated, and modern the theory of hydrodynamics actually is. Darrigol draws a number of lessons about the structures of phenomenological theories from his detailed discussion, focusing particularly on what he calls the "modular structure" of hydrodynamics.

Continuing the discussion of "old"—but by no means dead or eliminated—theories, Mark Wilson takes on the formidable task of trying to say exactly what is the nature of classical mechanics. A common initial reaction to this topic is to dismiss it: "Surely we all know what classical mechanics is! Just look at any textbook." But as Wilson shows in "What Is 'Classical Mechanics' Anyway?", this dismissive attitude is misleading on a number of important levels. Classical mechanics is like a five-legged stool on a very uneven floor. It shifts dramatically from one foundational perspective to another depending upon the problem at hand, which in turn is often a function of the scale length at which the phenomenon is investigated. In the context of planetary motions, billiards, and simplified ideal gases in boxes, the point-particle interpretation of classical mechanics will most likely provide an appropriate theoretical setting. However, as soon as one tries to provide a more realistic description of what goes on inside actual billiard ball collisions, one must consider the fact that the balls will deform and build up internal stresses upon collision. In such situations, the point-particle foundation will fail and one will need to shift to an alternative foundation, provided by classical continuum mechanics. Yet a third potential foundation for classical mechanics can be found within so-called analytic mechanics, in which the notion of a rigid body becomes central. Here constraint forces (such as the connections that allow a ball to roll, rather than skid, down an inclined plane) play a crucial role. Forces of this type are not wholly consistent with the suppositions central to either the point-particle or continuum points of view. A major lesson from Wilson's discussion is that classical mechanics should best be thought of as constituted by various foundational methodologies that do not fit

---

[1] Darrigol's recent *Worlds of Flow* fills this lacuna providing an exceptional discussion of the history (Darrigol 2005).

particularly well with one another. This goes against current orthodoxy that a theory must be seen as a formally axiomatizable consistent structure. On the contrary, to properly employ classical mechanics for descriptive and explanatory purposes, one pushes a foundational methodology appropriate at one scale of investigation to its limiting utility, after which one shifts to a different set of classical modeling tools in order to capture the physics active at a lower size scale. Wilson argues that a good deal of philosophical confusion has arisen from failing to recognize the complicated scale-dependent structures of classical physics.

Sheldon Smith's contribution adds to our understanding of a particular aspect of classical physics. In "Causation in Classical Mechanics," he addresses skeptical arguments initiated by Bertrand Russell to the effect that causation is not a fundamental feature of the world. In the context of classical physics, one way of making this claim more precise is to argue that there is no reason to privilege retarded over advanced Green's functions for a system. Green's functions, crudely, describe the effect of an instantaneous, localized disturbance that acts upon the system. It seems that the laws of motion for electromagnetism or for the behavior of a harmonic oscillator do not distinguish between retarded (presumably "causal") and advanced (presumably "acausal") solutions. If there is to be room for a principle of causality in classical physics, then it looks like we need to find extra-nomological reasons to privilege the retarded solutions. Smith surveys a wide range of attempts to answer the causal skeptic in the contexts of the use of Green's functions and the imposition of (Sommerfeld) radiation conditions, among other attempts. The upshot is that it is remarkably difficult to find justification within physical theory for the maxim that causes precede their effects.

The next chapter, by Leo Kadanoff, focuses on condensed matter physics. In particular, Kadanoff discusses progress in physically understanding the fact that matter can abruptly change its qualitative state as it undergoes a phase transition. An everyday example occurs with the boiling water in a teakettle. As the temperature increases, the water changes from its liquid phase to its vapor phase in the form of steam. Mathematically, such transitions are described by an important concept called an order parameter. In a first-order phase transition, such as the liquid vapor transition, the order parameter changes discontinuously. Certain phase transitions, however, are continuous in the sense that the discontinuity in the behavior of the order parameter approaches zero at some specific critical value of the relevant parameters such as temperature and pressure. For a long time there were theoretical attempts to understand the physics involved in such continuous transitions that failed to adequately represent the actual behavior of the order parameter as it approached its critical value. The development of the *renormalization group* in the 1970s remedied this situation. Kadanoff played a pivotal role in the conceptual development of renormalization group theory. In this chapter, he focuses on these developments (particularly, the improvement upon early mean field theories) and on a deeply interesting feature he calls the "extended singularity theorem." This is the idea that sharp, qualitatively distinct, changes in phase involve the presence of a mathematical singularity. This singularity typically emerges in the limit in which the system size becomes infinite. The understanding of the behavior of systems at and near phase transitions requires radically different conceptual apparatuses. It involves a synthesis between standard statistical mechanical uses of probabilities and concepts from dynamical systems theory—particularly, the topological conceptions of basins of attraction and fixed points of a dynamical transformation.

The discussion of the renormalization group and phase transitions continues as Tarun Menon and Craig Callender examine several philosophical questions raised by phase transitions. Their chapter, "Turn and Face the Strange …Ch-ch-changes," focuses on the question of whether phase transitions are to be understood as genuinely emergent phenomena. The term "emergent" is much abused and confused in both the philosophical and physics literatures and so Menon and Callender provide a kind of road map to several concepts that have been invoked in the increasing number of papers on emergence and phase transitions. In particular, they discuss conceptions of reduction and corresponding notions of emergence: conceptual novelty, explanatory irreducibility, and ontological irreducibility. Their goal is to establish that for any reasonable senses of reducibility and emergence, phase transitions are not emergent phenomena, and they do not present problems for those of a reductionist explanatory bent. In a sense, their discussion can be seen as challenging the importance of the extended singularity theorem mentioned above. Menon and Callender also consider some recent work in physics that attempts to provide well-defined notions of phase transition for finite systems. Their contribution serves to highlight the controversial and evolving nature of our philosophical understanding of phase transitions, emergence, and reductionism.

Jonathan Bain's contribution on "Effective Field Theories" looks at several physical and methodological consequences of the fact that some theories at low-energy scales are effectively independent of, or decoupled from, theories describing systems at higher energies. Sometimes we know what the high-energy theory looks like and can follow a recipe for constructing low-energy effective theories by systematically eliminating high-energy interactions that are essentially "unobservable" at the lower energies. But, at other times, we simply do not know the correct high-energy theory, yet nonetheless, we still can have effective low-energy theories. Broadly construed, hydrodynamics is an example of the latter type of effective theory, if we consider it as a nineteenth century theory constructed before we knew about the atomic constitution of matter. Bain's focus is on effective theories in quantum field theory and condensed matter physics. His discussion concentrates on the intertheoretic relations between low-energy effective theories and their high-energy counterparts. Given the effective independence of the former from the latter, should one think of this relation as autonomous or emergent? Bain contends that an answer to this question is quite subtle and depends upon the type of renormalization scheme employed in constructing the effective theory.

My own contribution to the volume concerns a general problem in physical theorizing. This is the problem of relating theories or models of systems that appear at widely separated scales. Of course, the renormalization group theory (discussed by Kadanoff, Menon and Callender, and Bain in this volume) is one instance of bridging across scales. But more generally, we may try to address the relations between finite statistical theories at atomic and nanoscales and continuum theories that apply at scales 10+ orders of magnitude higher. One can ask, for example, why the Navier-Cauchy equations for isotropic elastic solids work so well to describe the bending behavior of steel beams at the macroscale. At the microscale the lattice structure of iron and carbon atoms looks nothing like the homogeneous macroscale theory. Nevertheless, the latter theory is remarkably robust and safe. The chapter discusses strategies for upscaling from theories or models at small scales to those at higher scales. It examines the philosophical consequences of having to consider, in one's modeling practice, structures that appear at scales intermediate between the micro and the macro.

There has been considerable debate about the nature of symmetries in physical theories. Recent focus on gauge symmetries has led philosophers to a deeper understanding of the role of local invariances in electromagnetism, particle physics, and the hunt for the Higgs' particle. Sorin Bangu provides a broad and comprehensive survey of concepts of symmetry and invariance in his contribution to this volume. One of the most seductive features of symmetry considerations comes out of Wigner's suggestion that one might be able to understand, explain, or ground laws of nature by appeal to a kind of superprinciple expressing symmetries and invariances that constrain laws to have the forms that they do. On this conception symmetries are, perhaps, ontologically and epistemically prior to laws of nature. This raises deep questions for further research on the relationship between formal mathematical structures and our physical understanding of the world.

Gordon Belot also considers issues of symmetry and invariance. His contribution explores the connections between being a symmetry of a theory—a map that leaves invariant certain structures that encode the laws of the theory—and what it is for solutions to a theory to be *physically* equivalent. It is fairly commonplace for philosophers to adopt the idea that, in effect, these two notions coincide. And if they do, then we have tight connection between a purely formal conception of the symmetries of a theory and a methodological/interpretive conception of what it is for two solutions to represent the same physical state of affairs. Belot notes that in the context of spacetime theories there seem to be well-established arguments supporting this tight connection between symmetries and physical equivalence. However, he explores the difficulties in attempting to generalize this connection in contexts that include classical dynamical theories. Belot examines different ways one might make precise the notion of the symmetries of a classical theory and shows that they do not comport well with reasonable conceptions of physical equivalence. The challenge to the reader is then to find appropriate, nontrivial notions of symmetries for classical theories that will respect reasonable notions of physical equivalence.

Yet another type of symmetry, permutation symmetry, is the subject of the chapter by Simon Saunders, entitled "Indistinguishability." He focuses on the proper understanding of particle indistinguishability in classical statistical mechanics and in quantum theory. In the classical case, Gibbs had already (prior to quantum mechanics) recognized a need to treat particles, at least sometimes, as indistinguishable. This is related to the infamous *Gibbs paradox* that Saunders discusses in detail. The concept of "indistinguishability" had meanwhile entered physics in a completely new way, involving a new kind of statistics. This came with the derivation of Planck's spectral distribution, in which Planck's quantum of action h first entered physics. Common wisdom has long held that particle indistinguishability is strictly a quantum concept, inapplicable to the classical realm; and that classical statistical mechanics is anyway only the classical limit of a quantum theory. This fits with the standard view of the explanation of quantum statistics (Bose-Einstein or Fermi-Dirac statistics): departures from classical (Maxwell-Boltzmann) statistics are explained by particle indistinguishability. With this Saunders takes issue. He shows how it is possible to treat the statistical mechanical statistics for classical particles as invariant under permutation symmetry in exactly the same way that it is treated in the quantum case. He argues that the conception of permutation symmetry deserves a place alongside all the other symmetries and invariances of physical theories. Specifically, he argues that the concept of indistinguishable, permutation invariant, *classical* particles is coherent and reasonable contrary to many claims found in the literature.

Margaret Morrison's topic is "Unification in Physics." She argues that there are a number of distinct senses of unification in physics, each of which has different implications for how we view unified theories and phenomena. On the one hand, there is a type of unification that is achieved via reductionist programs. Here a paradigm example is the unification provided by Maxwellian electrodynamics. Maxwell's emphasis on mechanical models in his early work involved the introduction of the displacement current, which was necessary for a field theoretic representation of the phenomena. These models also enabled him to identify the luminiferous aether with the medium of transmission of electromagnetic phenomena. Two aethers were essentially reduced to one. When these models were abandoned in his later derivation of the field equations, the displacement current provided the unifying parameter or theoretical quantity that allowed for the identification of electromagnetic and optical phenomena within the framework of a single field theoretic account. This type of unification was analogous to Newton's unification of the motions of the planets and terrestrial trajectories under the same (gravitational) theoretical framework. However, not all cases of unification are of this type. Morrison discusses the example of the electroweak theory in some detail, arguing that this unificatory success represents a kind of synthetic, rather than reductive, unity. The electroweak theory also involves a unifying parameter, namely, the "Weinberg angle." However, the unity achieved through gauge symmetry is a synthesis of structure, rather than of substance, as exemplified by the reductive cases. Finally, in calling attention to the difficulties with the Standard Model more generally, Morrison notes that yet a different kind of unification is achieved in the framework of effective field theory. This provides another vantage point from which to understand the importance of the renormalization group. Morrison argues for a third type of unification in terms of the universality classes, one that focuses on unification of phenomena but should be understood independently of the type of micro-reduction characteristic of unified field theory approaches.

As noted earlier, there continues to be significant research on foundational problems in quantum mechanics. Guido Bacciagaluppi's chapter provides an up-to-date discussion of work on two distinct problems in the foundations of quantum mechanics that are typically conflated in the literature. These are the problem of the classical regime and the measurement problem. Both problems arise from deep issues involving entanglement and the failure of an ignorance interpretation of reduced quantum states. Bacciagaluppi provides a contemporary and thorough introduction to these issues. The problem of the classical regime is that of providing a quantum mechanical explanation or account of the success of classical physics at the macroscale. It is, in essence, a problem of intertheoretic relations. Contemporary work has concentrated on the role of environmental decoherence in the emergence of classical kinetics and dynamics. Bacciagaluppi argues that the success of appeals to decoherence to solve this problem will depend upon one's interpretation of quantum mechanics. He surveys an ontologically minimalist instrumental interpretation and a standard, ontologically more robust or realistic interpretation.

The measurement problem is the distinct problem of deriving the collapse postulate and the Born rule from the first principles (Schrödinger evolution) of the quantum theory. In examining the measurement problem, Bacciagaluppi provides a detailed presentation of a modern, realistic theory of measurement that goes beyond the usual idealized discussions of spin measurements using Stern-Gerlach magnets. This discussion generalizes the usual collapse postulate and the Born rule to take into account the fact that real measurements are unsharp. It does so by employing the apparatus of positive operator value (POV) measures and observables. The upshot is that the measurement problem remains a real worry for someone who wants to maintain a standard, reasonably orthodox interpretation of quantum theory. Perhaps Everett theories, GRW-like spontaneous collapse theories, and so on are required for a solution.

The Everett, or Many Worlds, interpretation of quantum mechanics is the subject of David Wallace's chapter. It is well known that the linearity of quantum mechanics leads, via the principle of superposition, to the possibility that macroscopic objects such as cats can be found in bizarre states—superpositions of being alive and being dead. Wallace argues that a proper understanding of what quantum mechanics actually says will enable us to understand such bizarre situations in a way that does not involve changing the physics (e.g., as in Bohmian hidden variable mechanics or GRW spontaneous collapse theories). Neither, he claims, does it involve changing one's philosophy by, for example, providing an operationalist interpretation that imposes some special status to the observer or to what counts as measurement, along the lines of Bohr. Such interpretations are at odds with our understanding of, say, the role of the observer in the rest of science. Wallace argues for a straightforward, fully realist interpretation of the bare mathematical formalism of quantum mechanics and claims that this interpretation will make sense of superposed cats, and so on, without changing the theory and without changing our overall view of science. The straightforward realist interpretation that is to do all of this work is the Everett interpretation. Prima facie, this claim is itself bizarre: after all, the Everett interpretation has us multiplying worlds or universes upon measurements. Nevertheless, Wallace makes a strong case that an understanding of superposition as a description of multiplicity, rather than of the indefiniteness of states, is exactly what is needed. Furthermore, that is exactly what the Everett interpretation (and no other) provides. The bulk of Wallace's contribution examines various problems that have been raised for the Everett interpretation. In particular, he focuses on (1) the problem of providing a preferred basis—what actually justifies our understanding of superposition in terms of multiplicity of worlds, and (2) the probability problem—how to understand the probabilistic nature of quantum mechanics if one has only the fully deterministic dynamics provided by the Schrödinger equation. He argues that the contemporary understanding of the Everett interpretation has the resources to address these issues.

Laura Ruetsche's chapter "Unitary Equivalence and Physical Equivalence" investigates a question of deep physical and philosophical importance: The demand for criteria establishing the physical equivalence of two formulations of a physical theory. In "ordinary" quantum mechanics the received view is that two quantum theories are physically equivalent just in case they are unitarily equivalent. Any pair of theories purporting, say, to describe two entangled spin 1/2 systems are really just one and the same because of the Jordan and Wigner theorem showing that a theory that represents the canonical anticommutation relations for a system of n spins is unique up to unitary equivalence. A similar theorem due to Stone and von Neumann guarantees an analogous result for any Hilbert space representation of the canonical commutation relations for a Hamiltonian system. What are the consequences of the breakdown of unitary equivalence for those quantum systems for which these theorems fail to hold? Such systems include the infinite systems studied in quantum field theory, quantum statistical mechanics, and even simpler infinite systems like an infinite one-dimensional chain of quantum spins. She calls these theories collectively QM_∞. The plethora of unitarily inequivalent representations in these infinite cases demands that we revisit our assumptions about physical equivalence and the nature of quantum theories. Ruetsche examines various competing suggestions, or competing principles that may guide the investigation into this problem.

The next chapter, by Oliver Pooley, provides an up-to-date, comprehensive discussion of substantivalist and relationalist approaches to spacetime. Crudely, this is a debate about the ontology of our theories of space and spacetime. The substantivalists hold that among the fundamental objects of the world is spacetime itself. Relationists, to the contrary, deny that propositions about spacetime are ultimately to be understood in terms of claims about material objects and possible spatiotemporal relations that may obtain between them. Pooley presents a historical introduction, as well as a detailed discussion of the current landscape in the literature. Specifically, he considers recent relationist, neo-Machian proposals by Barbour, as well as dynamical approaches favored by Brown, and Pooley and Brown, that aim to provide a reductive account of the spacetime symmetries in terms of the dynamical symmetries of laws governing the behavior of matter. In addition, Pooley provides a current assessment of the impact of the so-called Hole Argument against substantivalism.

In "Global Spacetime Structure" John Manchak examines the qualitative, primarily topological and causal, aspects of general relativity. He provides an abstract classification of various local and global spacetime properties. In the global causal context he explicitly defines a set of causal conditions that form a strict hierarchy of possible casual properties of spacetime. The strongest is the condition of global hyperbolicity, which implies others including causality and chronology. Another set of global properties of spacetime concerns in what sense a spacetime can be said to possess singularities. Here he focuses on the notion of geodesic incompleteness. Manchak then takes up philosophical questions concerning the physical reasonableness of these various spacetime properties. In a local context, being a solution to Einstein's Field Equation is typically taken to be physically reasonable. But, global properties concerning the existence and nature of singularities and the possibility of time travel lead to open questions of philosophical interest that are currently being investigated.

Last, but not least, Chris Smeenk's contribution concerns philosophical issues raised in contemporary work on cosmology. A common view is that cosmology requires a distinctive methodology because the universe-as-a-whole is a unique object. Restrictions on observational access to the universe due to the finite speed of light pose severe challenges to establishing global properties of the universe. How can we know that the local generalizations we take to be lawful in our limited region can be extended in a global fashion? Here, of course, there is overlap with the discussions of the previous chapter. Successes of the so-called Standard Model for cosmology include big-bang nucleosynthesis and the understanding of the cosmic background radiation, among others. Challenges to the Standard Model result from growing evidence that if it is correct, then *most* of the matter and energy present in the universe is not what we would consider ordinary. Instead, there apparently needs to be dark matter and dark energy. Smeenk provides an overview of recent hypotheses about dark matter and energy, and relates these discussions to philosophical debates about underdetermination. A different kind of problem arises in assessing theories regarding the very early universe. These theories are often motivated by the idea that the initial state required by the Standard Model is highly improbable. This deficiency can be addressed by introducing a dynamical phase of evolution, such as inflationary cosmology, that alleviates this need for a special initial state. Smeenk notes that assessing this response to fine-tuning is connected with debates about explanation and foundational discussions regarding time's arrow. One very important aspect of recent work in cosmology is the appeal to anthropic reasoning to help explain features of the early universe. A second recent development, often related to anthropic considerations, is the multiverse hypothesis—the existence of causally isolated pocket universes. This chapter brings these fascinating issues to the fore and raises a number of philosophical questions about the nature of explanation and confirmation appropriate for cosmology.

It is my hope that readers of this volume will gain a sense of the wide variety of issues that constitute the general field of philosophy of physics. The focus of the field has expanded tremendously over the last thirty years. New problems have come up, and old problems have been refocused and refined. It is indeed my pleasure to thank all of the authors for their contributions. In addition, I would like to thank Peter Ohlin from Oxford University Press. A number of others contributed to this project in various ways. I am particularly indebted to Gordon Belot, Julia Bursten, Nicolas Fillion, Laura Ruetsche, Chris Smeenk, and Mark Wilson for invaluable advice and support.

References

P. W. Anderson. More is different. *Science*, 177(4047):393–396, 1972.

Olivier Darrigol. *Worlds of Flow: A History of Hydrodynamics from the Bernoullis to Prandtl*. Oxford University Press, Oxford, 2005.

John Earman. *A Primer on Determinism*. Reidel, Dordrecht, 1986.

CHAPTER 1

FOR A PHILOSOPHY OF HYDRODYNAMICS

OLIVIER DARRIGOL

Among the major theories of physics, hydrodynamics is probably the one that has received the least attention from philosophers of science. Until recently, three circumstances easily explained this neglect. First, there was very little historical literature on which philosophers could rely. Second, philosophers tended to focus on fundamental theories such as relativity theory and quantum theory and to neglect more phenomenological theories. Third, they harbored a neo-Hempelian concept of explanation following which the foundations of a theory implicitly contain all its explanatory apparatus.[1] Even Thomas Kuhn, who brought the "normal" phases of science to the fore, restricted conceptual innovation to the revolutionary phases.[2] Since the fundamental equations of hydrodynamics have remained essentially the same for about two centuries, this view reduces the development of this theory to a matter of technical prowess in solving the equations.

In recent years these three circumstances have lost much of their weight. We now have fairly detailed histories of hydrodynamics.[3] The superiority of fundamental theories over lower scale or phenomenological theories has been multiply challenged, both within science and in the philosophy of science.[4] And there has been a growing awareness that explanation mostly resides in devices that are not contained in the bare foundations of a theory. For example, Mary Morgan and

---

I thank Robert Batterman for his useful comments on a draft of this essay.

[1] For a criticism of the Hempelian view, cf. Heidelberger 2006, 49–50.
[2] Kuhn 1961.
[3] Darrigol 2005, hereinafter abbreviated as *WF*; Eckert 2005.
[4] Cf., e.g., Cartwright 1983, 1999; Cat 1998.

Margaret Morrison have emphasized the role of models as mediators between theory and experiment; Jeffry Ramsey has argued the conceptual significance of approximations and "transformation reductions"; Robert Batterman has made explanation depend on strategies for the elimination of irrelevant details in the foundations; Paul Humphreys has placed computability at the center of his assessment of the nature and value of scientific knowledge. Eric Winsberg has shown the importance of extratheoretical considerations in judging the validity of numerical simulations based on the fundamental equations. Already in 1983, Ian Hacking and C. W. F. Everitt, who were more in touch with the actual practice of physicists than average philosophers, introduced "theory articulation" or "calculation" as an essential "semantic bridge between theory and observation."[5]

Granting that theory articulation is as philosophically important as the building of foundations, hydrodynamics becomes a topic of exceptional philosophical interest largely because of the huge time span between the establishment of its foundations and its successful application to some of the most pressing engineering problems. This delay is an indirect proof of the creativity needed to expand the explanatory power of theories. It enables us to observe a rich sample of the devices through which explanatory expansion may occur. Margaret Morrison, Michael Heidelberg, and Moritz Epple have recently given philosophical studies of two of these devices: Ludwig Prandtl's boundary-layer theory and his wing theory. The present essay is conducted in the same spirit.[6]

The first section gives a few historical examples of the means by which hydrodynamics became applicable to a growing number of concrete situations. The second provides a tentative classification of these means. The third contains a definition of physical theories that includes their evolving explanatory apparatus. Special emphasis is given to a "modular structure" of theories that makes them more amenable to tests, comparisons, communication, and construction.[7]

1. Some History

In the mid-eighteenth century, Jean le Rond d'Alembert and Leonhard Euler formulated the general laws of motion of a nonviscous fluid. In Euler's form, calling v the velocity of the fluid, P its pressure, ρ its density, and f an impressed force

---

[5] Morrison and Morgan 1999; Ramsey 1993, 1995; Batterman 2002; Humphreys 2004; Winsberg 1999; Hacking 1983, 215. Kuhn earlier applied the word "articulation" to the paradigms of normal science. In 1974, Hilary Putnam noted "in passing" a pervasive but neglected schema for scientific problems, "schema III," in which the fundamental laws of the theory and some auxiliary statements are known but the factual consequences are unknown (Putnam 1974, 261–62).

[6] Morrison 1999; Heidelberger 2006; Epple 2002.

[7] The foundations of hydrodynamics, though historically stable, are not devoid of philosophical interest. As Clifford Truesdell pointed out long ago, some of its basic concepts, such as the concept of internal stress, are indeed problematic (Truesdell 1968; Darrigol 2007). The relation of these foundations to general mechanics and to statistical mechanics (for instance, the kinetic theory of gases) is another philosophically interesting topic (Yamalidou 1998). For the sake of homogeneity, I confine this essay to post-foundational developments.

density, these laws are given by the equation of motion

$$\rho\left(\frac{\partial \mathbf{v}}{\partial t} + (\mathbf{v}\cdot\nabla)\mathbf{v}\right) = \mathbf{f} - \nabla P,$$

the equation of continuity,

$$\nabla\cdot\rho\mathbf{v} + \frac{\partial\rho}{\partial t} = 0,$$

and the boundary condition that the fluid velocity next to the walls of a rigid container should be parallel to these walls. If the fluid has a free surface at which it touches another fluid, the boundary conditions (later provided by Lagrange) are the equality of the pressures of the two fluids, and the condition that a particle of the surface of one fluid should remain on its surface.[8]

Euler's derivation of the equation of fluid motion assumes the pressure between two contiguous fluid parts to be perpendicular to the separating surface, as is the case in hydrostatics. In 1822 Claude Louis Navier implicitly dropped this assumption by comparing the internal fluid forces with the molecular forces of his general theory of elasticity. The resulting equation of motion is the Navier-Stokes equation

$$\rho\left[\frac{\partial \mathbf{v}}{\partial t} + (\mathbf{v}\cdot\nabla)\mathbf{v}\right] = \mathbf{f} - \nabla P + \mu\Delta\mathbf{v},$$

which involves the viscosity μ. This equation was reinvented several times. There was much hesitation on the proper boundary conditions, although in 1845 George Gabriel Stokes correctly argued for a vanishing relative velocity of the fluid next to rigid bodies.[9]

From a mathematical point of view, the most evident goal of the theory is to integrate the equations of motion for any given initial conditions and boundary conditions. There are at least three reasons not to confine fluid mechanics to this goal:

1. In the case of a compressible fluid, the system of equations is not complete because one needs the relation between pressure and density. This relation implies thermodynamic considerations, and therefore forces us to leave the narrow context of fluid mechanics.
2. It is generally impossible to solve the equations by analytical means because of their nonlinear character. Moreover, the few restricted cases in which this is possible may have little or no resemblance with actual flow because of instabilities. Nowadays, numerical integration is often possible and is, indeed, sufficient for some engineering problems. This leads us to the third caveat.
3. The answer to most physical questions regarding fluid behavior is not to be found in the solution of specific boundary-value problems. Rather, the

---

[8] Euler 1755. Cf. Truesdell 1954; WF, chap. 1.
[9] Navier 1822; Stokes [1845] 1849. Cf. WF, chap. 3.

physicist is often interested in generic properties of classes of solutions. In mathematical terms, we need to have a handle on the structure of the space of solutions.

What do physicists do when the solution of boundary problems no longer serves their interests? In order to answer this question, we will consult some of the historical evolution of hydrodynamics.

1.1 Bernoulli's Law

From a practical point of view, the main result that Euler could derive from his new hydrodynamics was the law

$$P = \rho \mathbf{g} \cdot \mathbf{r} - \tfrac{1}{2}\rho v^2 + \text{constant}$$

relating the pressure P, the position $\mathbf{r}$, and the velocity $\mathbf{v}$ for the steady motions of an incompressible fluid that admit a velocity potential ($\mathbf{g}$ is the acceleration of gravity). This achievement may seem meager for the following reasons:[10] the law had already been derived by Daniel Bernoulli in the 1730s as an application of the conservation of live force (energy) to steady, parallel-slice, incompressible fluid motion; the law requires a narrow specialization of the theory; one aspect of this specialization, the existence of a velocity potential, is (or was) physically obscure (its original purpose was to simplify the equations of motion and to permit their integration); under this specialization, the law is a straightforward mathematical consequence of Euler's equations.

From these remarks, one might be tempted to judge that Bernoulli's law adds nothing significant to the fundamental equations of hydrodynamics. Yet the practice of physicists and engineers suggests the contrary: This law is used in many circumstances, surely more often than Euler's equations themselves. There are several good reasons for this:

(1) Bernoulli's law relates easily accessible parameters of fluid motion in a simple manner, without any reference to the subtleties of the underlying dynamics;
(2) it is related to the general principle of energy conservation, which bridges hydrodynamics with mechanics;
(3) it provides the basis for the hydraulicians' language of pressure head, velocity head, and gravity head; and
(4) this language is still used when the law is violated.

Although this last point may seem paradoxical, it illustrates a highly important mode of concept formation in the post-foundational life of a theory: the solutions of the general theory are characterized with reference to the solutions of a more workable specialization of this theory. The concepts engendered by the specialization

---

[10] Euler 1755; Bernoulli 1738.

thus enrich the language of the general theory. They are useful as long as the law is valid in parts of the investigated system and as long as the loci of its violations are sufficiently understood. In typical hydraulic systems, there are regular pipes and reservoirs in which the law applies with a known correction (viscous or boundary-layer retardation in pipes) and there are phenomenologically or theoretically known "losses of head" when some accidents, such as pipe-to-pipe connections or sudden enlargements of the section of a pipe, occur.

1.2 Surface Waves

Historically, the second successful application of Euler's equations was to the problem of water waves. In this case, specialization is also necessary: the fluid is taken to be incompressible and a velocity potential is assumed. Moreover, some approximations must be introduced to circumvent the nonlinearity of the equations. In a memoir of 1781, Joseph Louis Lagrange originally assumed waves of small amplitude and of length much larger than the depth of the water. In the mid-1810s, Siméon Denis Poisson and Augustin Cauchy did without the latter approximation. The resulting differential equation for the deformation of the water surface is linear, and it admits sine-wave solutions whose propagation velocity depends on the wavelength. At this (first-order) approximation, one may use an autonomous language of sine waves that is no longer reminiscent of the underlying fluid dynamics and that is equally applicable to other kinds of linear waves. All one needs to know is how to combine (superpose) various sine waves in order to accommodate given initial shapes or perturbations of the water surface. We here encounter a second case of bridging of hydrodynamics with other theories: the introduction of concepts that apply to similar modes of motion in different theories (optics, hydrodynamics, acoustics ...).[11]

This is not to say that all linear wave problems are understood once we know the dispersion law (how the velocity of a sine wave depends on its wavelength). Historically, much effort was needed to understand the structure of a superposition of sine waves. Employing strictly mathematical methods, Poisson and Cauchy only succeeded in describing the wave created by a stone thrown into a pond. John Scott Russell (in 1844) and William Froude (in 1873) later observed that the front of a group of waves traveled at a smaller velocity than individual waves in the group. In 1876, Stokes gave the modern theoretical explanation in terms of phase and group velocity. Ten years later, William Thomson (Lord Kelvin) determined the form of ship waves by a clever application of these concepts. On the physical side of his deduction, he relied on the optical "principle of interference." On the mathematical side, he invented the method of the stationary phase, which is now commonly used in various domains of physics. Again, we have a case of concepts and tools generated in a region of a given theory but ultimately applied to regions

---

[11] Lagrange 1781; Poisson 1816; Cauchy [1815] 1827. Cf. *WF*, 35–47.

of many other theories (by region, I mean a restriction of the theory to a limited class of systems and boundary conditions). These concepts were partly derived by a mathematical process of specialization and approximation, partly by observation, partly by analogy with other domains of physics.[12]

Similar remarks apply to the case of nonlinear waves. George Biddell Airy and Stokes tamed nonlinear periodic waves by successive approximations to the fundamental equations, with applications to ocean waves and river tides. This was a mostly mathematical process of a cumbersome but fairly automatic nature. In contrast, Scott Russell observed solitary waves (isolated swells) of invariable shape long before theorists admitted their possibility. When Joseph Boussinesq and Lord Rayleigh at last deduced such waves from theory, it became clear that qualitative results (such as the deformation of traveling waves) derived by considering separately a small-depth (nondispersive) approximation and a small-amplitude (linear) approximation, no longer obtained when the depth and amplitude were both large. The compensation of the dispersive and nonlinear causes of deformation for waves of a properly selected shape is a mechanism which, again, applies to many other domains of physics.[13]

1.3 Vortex Motion

Early fluid mechanics usually assumed the existence of a velocity potential because it greatly simplified the fundamental equations and also because Lagrange had shown that it resulted from the equations of motion for a large class of boundary conditions (motion started from rest and caused by moving solids). Another reason, emphasized by British fluid theorists, was the fact that the velocity potential of an incompressible fluid obeys the same differential equation (Laplace's equation) as the gravitational, electric, and magnetic potentials. This formal analogy was a constant source of inspiration for Stokes, Thomson, and James Clerk Maxwell. It permitted an intuitive demonstration of some basic theorems of the abstract "potential theory," and it provided fluid-mechanical analogs of electrostatic, electrokinetic, and magnetostatic phenomena.

The general case in which no velocity potential exists was judged intractable until 1858 when Hermann Helmholtz discovered a few remarkable theorems that pushed this case to the forefront of the theory. As Cauchy and Stokes had earlier proved, the infinitesimal evolution of a fluid element can be regarded as the superposition of three kinds of motion: a translation of the center of gravity of the element, a dilation of the element along three mutually orthogonal axes, and a rotation. Formally, the rotation per unit time is half the vector $\omega = \nabla \times v$, which has the components $\partial v_z/\partial y - \partial v_y/\partial z$ etc. This vector, now called *vorticity*, vanishes if and only if there exists a velocity potential (in a connected domain). This

---

[12] Stokes 1876; Thomson 1887b. Cf. *WF*, 85–100.
[13] Airy 1845; Stokes 1847; Russell 1839; Boussinesq 1871; Rayleigh 1876a. Cf. *WF*, 69–84. Similar comments could be made about the compression waves studied by Euler and Lagrange.

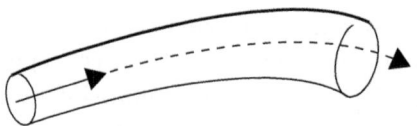

Figure 1.1 A portion of a vortex filament. The product of the vorticity (indicated by the arrows) by the normal section of the filament is a constant along the filament. It is also invariable during the motion of the fluid.

kinematic analysis of infinitesimal fluid motion is part of the conceptual furniture of modern fluid mechanics. Maxwell used it to develop the physico-mathematical concepts of curl and divergence that apply to any field theory. Helmholtz reinvented it to interpret the non-existence of the velocity potential and the vector $\omega = \nabla \times v$ geometrically.[14]

Helmholtz extended the geometrical interpretation to the "vorticity equation,"

$$\frac{\partial \omega}{\partial t} + (v \cdot \nabla)\omega = (\omega \cdot \nabla)v,$$

which derives from Euler's equations when the fluid is incompressible. For this purpose, he defined *vortex filaments* as thin bundles of lines everywhere tangent to the vorticity, and the *intensity of a filament* as the product of a normal section of this filament by the value of the vorticity in the section (see figure 1.1). He then showed that the intensity of a filament was a constant along a filament and that the vorticity equation was equivalent to the statement that the vortex filaments moved together with the fluid without altering their intensity. This theorem implies that the distribution of vorticity in a perfect liquid is in a sense invariant: it travels together with the fluid without any alteration.[15]

In this light, Helmholtz argued that the vorticity field (as today's physicists say) better represented arbitrary flows than the velocity field: its invariant properties completely determine the rotational component of the flow, while the irrotational component is ruled by the theorems of potential theory. With the help of an electromagnetic analogy, Helmholtz then determined the velocity fields associated with simple distributions of vorticity: straight vortex lines, vortex sheets, and vortex rings. He also calculated the interactions of vortices and verified his predictions experimentally.

The vortex sheets played an important role in Helmholtz's later writings. They are mathematically equivalent to a finite slide of fluid over fluid, and they should occur, according to Helmholtz, whenever a fluid is forced to pass the edge of an immersed body. As an illustration of this process, Helmholtz gave the formation of smoke jets when he blew the smoke of a cigar through his lips. Through ingenious reasoning, he proved the instability of the discontinuity surfaces or vortex sheets: any small bump on them must roll up spirally. This mechanism, now

---

[14] Helmholtz 1858. Cf. WF, 149.
[15] Cf. WF, 148–58.

called Helmholtz-Kelvin instability, plays an important role in many hydraulic and meteorological phenomena, as Helmholtz himself foresaw.[16]

Helmholtz not only meant to improve the applicability of hydrodynamics but also to equip this theory with a new mode of description for fluid motion in which vortices and discontinuity were the leading structural features. The enormous success of this project in the later history of hydrodynamics is somewhat paradoxical, because Helmholtz's theorems only hold in the unrealistic case of a perfect liquid. The physicists' use of the vorticity concept in much more general situations is comparable to the hydraulicians' use of the concept of hydraulic head in situations in which Bernoulli's theorem does not apply. In some cases of vortex motion, the effects of compressibility and viscosity can be shown to be negligible. In all cases, one can take Helmholtz's theorems as a reference and correct them through terms derived from the Navier-Stokes equation, as Vilhelm Bjerknes did in the late nineteenth century. As for the vortex sheets, we will see in a moment that in the early twentieth century Ludwig Prandtl used them to approximately describe important aspects of fluid resistance at high Reynolds number (low viscosity).[17]

In the historical examples discussed so far, it became increasingly difficult to produce the needed new conceptual apparatus. The degree of difficulty can be taken to be proportional to the time elapsed between the invention of Euler's equations and the introduction of this apparatus. For example, Bernoulli's law was easiest to derive, as it only requires a simple integration. But pure mathematics did not suffice to discover the laws of wave propagation on a water surface. Some intuition of interference processes (borrowed from optics), and also a few experimental observations (groups of waves, solitary waves), were instrumental. The discovery of the laws of vortex motion was even more difficult. A century elapsed from the time when d'Alembert and Euler gave the vorticity equation to the time when Helmholtz interpreted it through his theorem. Experiments or observations did not by themselves suggest this interpretation, though Helmholtz's efforts were, in fact, part of a project for improving the theoretical understanding of organ pipes. Helmholtz's success primarily depended on his ability to combine various heuristic devices including algebraic manipulation in the style of Lagrange, geometric visualization in the style of Thomson and Maxwell, and a focus on invariant quantities as exemplified in his own work on energy conservation.

1.4 Instabilities

Exact solutions of Euler's or Navier's equations under given boundary conditions may differ widely from the flow observed in a concrete realization of these conditions. For instance, the flow of water in a pipe of rapidly increasing diameter never has the smooth, laminar character of exact steady solutions of the Navier-Stokes

---

[16] Helmholtz 1868, 1888. Cf. *WF*, 159–71.
[17] Bjerknes 1898; Prandtl 1905.

equation in this case. As Stokes already suspected in the 1840s, this discrepancy has to do with the instability of the exact steady solutions: any small perturbation of these solutions will induce wide departures from the original motion. Consequently, the knowledge of exact solutions of the fundamental equations or (more realistically) the knowledge of some features of these solutions under given boundary conditions is not sufficient for the prediction of observed flows. One must also determine whether these solutions or features are stable.[18]

In principle this question can be mathematically decided, by examining how a slightly perturbed solution of the equations evolves in time. As we saw, a first success in this direction was Helmholtz's prediction of the spiral rolling up of a bump on a discontinuity surface. Later in the century, Lord Rayleigh and Lord Kelvin treated the more difficult problem of the stability of plane parallel flow. Their results were only partial (Rayleigh's inflection theorem in the nonviscous case), or wrong (Kelvin's prediction of stability for the plane Poiseuille flow). Most of these questions exceeded the mathematical capacity of nineteenth-century theorists, and some of them have remained unresolved to this day. The efforts of Rayleigh and others nonetheless yielded a general method and language of perturbative stability analysis. Rayleigh linearized the equation of evolution of the perturbation, and sought plane-wave solutions. These solutions are "proper modes" whose oscillatory or growing character depends on the real or imaginary character of the frequency. This proper-mode analysis of stability goes beyond hydrodynamics: it originated in Lagrange's celestial mechanics and it can be found in many other parts of physics.[19]

As the mathematical discussion of stability was nearly as difficult as the finding of exact solutions of the fundamental equations, the most important results in this domain were reached by empirical means. Plausibly, the observed instability of jets motivated Helmholtz's derivation of the instability of discontinuity surfaces. Certainly, Tyndall's observations of this kind motivated Rayleigh's calculations for parallel flow. Most important, Gotthilf Hagen (1839) and Osborne Reynolds (1883) discovered that pipe flow, for a given diameter and a given viscosity, suddenly changed its character from laminar to turbulent when the velocity passed a certain critical value. The sharpness of this transition was a surprise to all theorists. From Reynolds to the present, attempts to mathematically determine the critical velocity (or Reynolds number) in cylindrical pipes have failed. This is a question of academic interest only, because unpredictable entrance effects (the way the fluid is introduced into the pipe), not the inherent instability in a pipe of infinite length, usually determine the transition.[20]

In the twentieth century, significant progress has been made in understanding the transition from laminar to turbulent flow. In the first half of the century, Ludwig Prandtl, Walter Tollmien, Werner Heisenberg, and Chia Chiao Lin proved the

---

[18] Stokes 1843. Cf. WF, 184–87.
[19] Rayleigh 1880; Thomson 1887a. Cf. WF, 208–18; Drazin and Reid 1981.
[20] Tyndall 1867; Hagen 1839; Reynolds 1883. Cf. WF, 243–63.

instability of the plane Poiseuille flow and unveiled the spatial periodicity of the mechanism of this instability.[21] In the second half of the century, developments in the theory of dynamical systems at the intersection between pure mathematics, meteorology, and hydrodynamics permitted a detailed qualitative understanding of the transition to turbulence, with intermediate oscillatory regimes, bifurcations, and strange attractors.[22] It remains true that most of the practical applications of hydrodynamics only require a rough knowledge of the conditions under which turbulence occurs. The source of this knowledge is partly theoretical and partly empirical. There is no easy way to gather it from the fundamental equations. In most cases, the best that can be done is to repeat Reynolds's rough argument that the full vorticity equation has two terms, a viscous term that tends to damp any eddying motion, and an inertial term which preserves the global amount of vorticity. The laminar or turbulent character of the motion depends on the ratio of these two terms, whose order of magnitude is given by the Reynolds number.

1.5 Turbulence

The state of motion that follows the turbulent transition is even more difficult to analyze than the transition itself. Casual observation of turbulent flow reveals its chaotic and multi-scale character. The detailed description of any motion of this kind seems to require a huge amount of information, much more than is humanly accessible (without computers at least). As Reynolds pondered, we are here facing a situation similar to that of the kinetic theory of gases: the effective degrees of freedom are too numerous to be handled by a human calculator. Unfortunately, turbulent motion is more often encountered in nature and in manmade hydraulic devices than laminar motion. Engineers and physicists have had to invent ways of coping with this difficulty.[23]

One strategy is to design the hydraulic or aeronautic artifacts so that turbulence does not occur. When turbulence cannot be avoided, one may adopt a purely empirical approach and seek relations between measured quantities of interest. For instance, nineteenth-century engineers gave empirical laws for the retardation (loss of head) in hydraulic pipes. A second approach is to find rules allowing the transfer of the results of measurements done at one scale to another scale. Stokes, Helmholtz, and Froude pioneered this approach in the contexts of pendulum damping, balloon steering, and ship resistance, respectively. They derived the needed scaling rules from the scaling symmetries of the Navier-Stokes equation or of the underlying dynamical principles. This is an example of a hybrid approach, founded partly on the fundamental equations, and partly on measurements of theoretically unpredictable properties.[24]

---

[21] Cf. Eckert 2008.
[22] Cf. Franceschelli 2007.
[23] Reynolds 1895. Cf. WF, 259–60.
[24] Stokes 1850; Helmholtz 1873; Froude [1868] 1957; 1874. Cf. WF, 256–58, 278–79.

In a third approach, one may completely ignore the foundations of fluid mechanics and cook up a model based on a grossly simplified picture of the flow. An important example is the laws for open channel flow discovered in the 1830s and 1840s by a few French Polytechnique-trained engineers: Jean Baptiste Bélanger, Jean Victor Poncelet, and Gaspard Coriolis. They assumed the flow to occur through parallel slices that rubbed against the bottom of the channel according to a phenomenological friction law, and they applied momentum or energy balance to each slice.[25]

In the 1840s Adhémar Barré de Saint-Venant emphasized the "tumultuous" character of the fluid motion in open channel flow and suggested a distinction between the large-scale average motion of the fluid and the smaller-scale tumultuous motion. The main effect of the latter motion on the former, Saint-Venant argued, was to enhance momentum exchange between successive (large-scale) fluid layers. Based on this intuition, he replaced the viscosity in the Navier-Stokes equation with an effective viscosity that depended on various macroscopic circumstances such as the distance from a wall. In the 1870s, Boussinesq solved the resulting equation for open channels of simple section and thus obtained laws that resembled Bélanger's and Coriolis's laws, with different interpretations of the relevant parameters.[26]

In 1895, Reynolds relied on analogy with the kinetic theory of gases to develop an explicitly statistical approach to turbulent flow. In the spirit of Maxwell's kinetic-molecular derivation of the Navier-Stokes equation, he derived a large-scale equation of fluid motion by averaging over the small-scale motions governed by the Navier-Stokes equation. Reynolds's equation depends on the "Reynolds stress," which describes the turbulent exchange between successive macro layers of the fluid. Like Saint-Venant's effective viscosity, the Reynolds stress cannot be determined without further assumptions concerning the turbulent fluctuation around the large-scale motion. There have been many attempts to fill this gap in the twentieth century. The most useful ones were Kármán's and Prandtl's derivations of the logarithmic velocity profile of a turbulent boundary layer. The assumptions made in (improved) versions of these derivations are simple and natural (uniform stress, matching between the turbulent layer and a laminar sublayer next to the wall), and the resulting profile fits experiments extremely well (much better than earlier phenomenological laws). The logarithmic profile is the basis of every modern engineering calculation of retardation in pipes or open channels.[27]

Despite powerful studies by Geoffrey Taylor, Andrey Nikolaevich Kolmogorov, and many others, the precise manner in which turbulence distributes energy between different scales of fluid motion remains a mystery.[28] There is no doubt, however, that the general idea of describing turbulent flow statistically has been fruitful since its first intimations by Saint-Venant, Boussinesq, and Reynolds. In the case

---

[25] Cf. *WF*, 221–28.
[26] Saint-Venant 1843; Boussinesq 1877. Cf. *WF*, 229–38.
[27] Reynolds 1895; Kármán 1830; Prandtl 1831 Cf. Eckert 2005, chap. 5; *WF*, 259–62, 297–301.
[28] Cf. Farge and Guyon 1999; Frisch 1995.

of turbulent fluid mechanics, as in statistical mechanics, a new conceptual structure emerges at the macroscale of description. Similar questions can be raised in both cases concerning the nature of the reduction or emergence. Does the microscale theory truly imply the macroscale structure? Is this structure uniquely defined? Can this structure be used without further reference to the microscale? Are there singular situations in which the reduction fails? The answers to these questions tend to be more positive in the case of statistical mechanics than in the case of the statistical theory of turbulence, because the relevant statistics are better known in the former than in the latter case.

1.6 Boundary Layers

From a practical point of view, two outstanding problems of fluid mechanics are fluid resistance and fluid retardation. Fluid resistance is the decelerating force experienced by a rigid body moving through a fluid. Fluid retardation is the fall of pressure or loss of head experienced by a fluid during its travel along pipes or channels. The two problems are related, since they both involve the mutual action of a fluid and an immersed solid. In 1768, d'Alembert challenged "the sagacious geometers" with the paradox that resistance vanished for a perfect liquid in his new hydrodynamics. There were various strategies to circumvent this theoretical failure. Some engineers determined by purely empirical means how the resistance depended on the velocity and shape of the immersed body. Others retreated to Isaac Newton's naïve theory by the impact of fluid particles on the front of the body, although some consequences of this theory (such as the irrelevance of the shape of the end of the body) had already been refuted. In the mid-nineteenth century, Saint-Venant, Poncelet, and Stokes traced resistance to viscosity and the production of eddies. With the damping of pendulums in mind, Stokes successfully determined the resistance of small spheres and cylinders by finding solutions to the linearized Navier-Stokes equation. For most practical problems, the larger size of the immersed body and the smallness of the viscosities of air and water imply that the nonlinear term of this equation cannot be neglected (the Reynolds number is too high). Stokes had nothing to say in such cases beyond the qualitative idea of dissipation by the production of eddies.[29]

In the ideal case of vanishing viscosity, the proof of d'Alembert's paradox implicitly assumes the continuity of the fluid motion. However, Helmholtz's study of vortex motion implies that finite slip of fluid over fluid is perfectly compatible with Euler's equations. Around 1870, Kirchhoff and Rayleigh realized that Helmholtz's discontinuity surfaces yielded a finite resistance for an immersed plate. According to Helmholtz, a tubular discontinuity surface is indeed produced at the sharp edges of the plate. The water behind the plate and within this surface is stagnant, so that its pressure vanishes (when measured in reference to its uniform value at

---

[29] D'Alembert 1768; Saint-Venant 1843; Poncelet 1839; Stokes 1850. Cf. *WF*, 135–39, 265–67, 270–73.

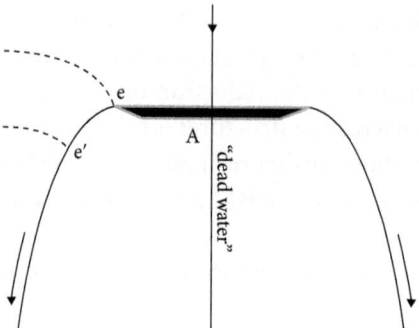

Figure 1.2 Discontinuity surface (ee′) formed when a downward flow encounters the disk A. From Thomson (1894, 220).

infinite distances from the plate) (see figure 1.2). Since the pressure at the front of the plate is positive, there is a finite resistance, which Kirchhoff and Rayleigh determined by analytical means. The result roughly agreed with the measured resistance.[30]

In the case of ships, the resistance problem is complicated by the fact that ships are not supposed to be completely immersed. Consequently, wave formation at the water surface is a significant contribution to the resistance. The leading nineteenth-century experts on this question, William John Macquorn Rankine and William Froude, distinguished three causes of resistance: wave resistance, skin resistance, and eddy resistance. Skin resistance corresponds to some sort of friction of the water when it travels along the hull. Eddy resistance corresponds to the formation of eddies at the stern of the ship; it is usually avoided by proper profiling of the hull. Rankine and Froude traced skin resistance to the formation of an eddying fluid layer next to the hull. They derived this notion from the observation that the flow of water around the ship, when seen from the deck, appears to be smooth everywhere expect for a narrow tumultuous layer next to the hull and for the wake. Rankine assumed the validity of Euler's equations in the smooth part of the flow and solved it to determine the hull shapes that minimized wave formation. Froude gave a fairly detailed description of the mechanism of retardation in the eddying layer, although he was not able to draw quantitative conclusions. In the end, Froude measured skin friction on plates, total resistance on small-scale ship models, and then used separate scaling laws for skin and wave resistance in order to determine the resistance of a prospective ship hull.[31]

In sum, Rankine and Froude distinguished two different regions of flow amenable to different theoretical or semi-empirical treatments and combined the resulting insights to determine the total resistance. Froude thus obtained the first quantitative successes in the problem of fluid resistance at a high Reynolds number.

---

[30] Kirchhoff 1869; Rayleigh 1876b.
[31] Rankine 1858, 1865, 1870; Froude 1874, 1877. Cf. Wright 1983; *WF*, 273–82.

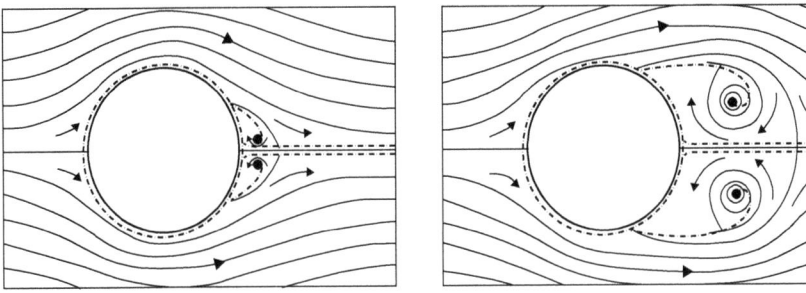

Figure 1.3 Formation of a discontinuity surface behind a cylinder. From Prandtl (1905, 579–80).

Although his and Rankine's considerations appealed to higher theory in several manners, they also required considerable empirical input.

The next and most famous progress in the high Reynolds-number resistance problem occurred in Göttingen, under the leadership of Ludwig Prandtl. Impressed by the qualitative success of Helmholtz's surfaces of discontinuity, Prandtl assumed that the solution of the Navier-Stokes equation for high-Reynolds flow around a body somewhat resembled a solution of Euler's equation (with strictly vanishing viscosity). In the latter solution, the fluid slides along the surface of the body, whereas for a viscous fluid the relative velocity of the fluid must vanish at the surface of the body. Consequently, for the real flow Prandtl assumed a thin (invisible) layer of intense shear that imitated the finite slide of the Eulerian solution. He also assumed that in some cases this layer could shoot off the surface of the body to mimic a Helmholtzian surface of discontinuity (with its characteristic instability resulting in an eddying trail). This is the so-called separation process. Outside the boundary layer, Prandtl naturally applied Euler's equations. Within the boundary layer, the intense shear allowed him to use an approximation of the Navier-Stokes equation that could be integrated to determine the evolution of the velocity profile along the body. For sufficiently curved bodies, Prandtl found that at some point the flow was inverted in the part of the boundary layer closest to the body. He interpreted this point as the separation point from which a (quasi) discontinuity surface was formed. In the case of a flat or little curved surface (for which separation does not occur), he determined the resistance by integration of the sheer stress along the surface of the body. He illustrated the separation process through experiments done with a tank and a paddle-wheel machine (figure 1.3).[32]

Comparison with Froude's earlier concept of eddying layer leads to the following remarks. Unlike Froude, Prandtl was able to determine theoretically and precisely the flow within the boundary layer. This determination requires a previous solution of the Eulerian flow problem around the body, because the evolution of the boundary layer depends on the pressure at its confines. Conversely, this evolution may

---

[32] Prandtl 1905. Cf. Eckert 2005, chap. 2; Heidelberger 2006; WF, 283–89.

induce separation, which necessarily affects the Eulerian part of the flow. Prandtl himself emphasized this interaction between the Eulerian flow and the boundary layer. Whereas Froude had no interest in separation (which ship builders systematically avoided), Prandtl had a precise criterion for its occurrence. Whereas Froude could only measure the sheer stress of the boundary layer, Prandtl could determine it theoretically.

So far the comparison seems to favor Prandtl. In reality, in many cases including ship resistance, the boundary layer has an internal turbulence that is not taken into account in Prandtl's original theory. In 1913, Prandtl's former student Heinrich Blasius found that beyond a certain critical Reynolds number, the edgewise resistance of a plate obeyed Froude's empirical law and not Prandtl's theoretical law. Prandtl explained that the profile of a laminar boundary layer could become unstable and thus lead to a turbulent boundary layer à la Froude. He used this notion to explain the bizarre reduction of the resistance of spheres that Gustave Eiffel had observed at a certain critical velocity: turbulence in a boundary layer, Prandtl explained, delays the separation process and thus sharply decreases the resistance. Paradoxically, it is when the boundary layer is turbulent that the global flow mostly resembles the smooth Eulerian flow.[33]

As the boundary layers around airplane wings are turbulent, Prandtl needed to know the sheer stress along such layers in order to determine the drag of the wings. He originally relied on plate resistance measurements, as Froude had done in the past. As was already mentioned, it became possible to calculate this stress in the 1830s when Kármán and Prandtl discovered the logarithmic velocity profile of turbulent layers.

It is now time to reflect on the relation that boundary-layer theory has to the foundational theory of Navier-Stokes. Prandtl's idea (if we believe his own plausible account) has its theoretical origin in the idea of using solutions to Euler's equations as a guide for solving the Navier-Stokes equation at a high Reynolds number. This is only a heuristic, because Prandtl had no mathematical proof that the low-viscosity limit of a solution of the Navier-Stokes equation is a solution of Euler's equation. Yet the motion imagined by Prandtl, with its Eulerian, high-sheer, and stagnant regions, clearly is an approximate solution of the Navier-Stokes equation. What is missing is a proof of the uniqueness of this solution (under given boundary conditions), as well as a general proof of its existence for any shape of the immersed body. With this concession, the boundary-layer theory can legitimately be regarded as an approximation of the Navier-Stokes theory.

An interesting feature of the boundary-layer theory is its use of different approximate equations in different regions of the flow. Our discussion of Bernoulli's law showed that this law is often used regionally (i.e., in laminar regions of the flow) with head losses localized in turbulent regions. Boundary-layer theory similarly introduces different regions of flow, although it does so in a more interactive manner. Each region is described through computable solutions of appropriate equations

---

[33] Prandtl 1914. Cf. Eckert 2005; WF, 293–94.

of motion, and the precise conditions for the matching of the regional solutions are known (continuity of pressure, stress, and velocity). These matching conditions imply causal relations between features of the two regions: for instance, the pressure distribution in the Eulerian region determines the evolution of the velocity profile in the boundary layer, and in the case of separated flow, the position of the separating surface affects the Eulerian region.[34]

In qualitative applications, Prandtl's theory may be restricted to the general ideas of a boundary layer, a free fluid, and their interaction sometimes leading to separation. In quantitative engineering applications, this picture must be supplemented with a law for the evolution of the sheer stress along a boundary layer (laminar or turbulent), and with quantitative criteria for separation and for the transition between laminar and turbulent layer. Granted that this supplementary information is available, the theory can be used without reference to the Navier-Stokes theory. The gain in predictive efficiency is enormous, as verified by the immense success of Prandtl's theory in engineering applications. Yet one should not forget that much of the supplementary information comes from the intimate connection between the boundary-layer theory and the Navier-Stokes theory. In fact the legitimacy of the whole picture depends upon this intimate connection. The boundary-layer theory, unlike the early French models of open channel flow, is not an ad hoc model that owes its simplicity to counterfactual assumptions. It is a legitimate articulation of the Navier-Stokes theory.

2. Explanatory Progress

The above examples make clear that in the course of its history, hydrodynamics has acquired a sophisticated explanatory apparatus without which it would remain merely a "paper" theory. The explanatory apparatus is presented in various chapters in modern textbooks. We will now reflect on the ways this apparatus was obtained, on its components, and on its functions.

2.1 The Sources of Explanatory Progress

In some cases, explanation was improved through blind mathematical methods. For instance, a simple integration yielded Bernoulli's law (after proper specialization), the symmetries of the Navier-Stokes equation yielded scaling laws, and standard approximation procedures yielded the theory of waves of small amplitude. Despite the relatively easy and automatic way in which these results were obtained, they considerably improved the explanatory power of the theory by directly relating quantities of physical interest.

---

[34] Heidelberger 2006 rightly insists on this causal structure of the boundary-layer theory.

In other cases, more intra- or intertheoretical heuristics was needed. Kinematic analysis of the vorticity equation led to Helmholtz's vortex theorems; asymptotic reasoning led to Prandtl's notions of laminar boundary layer and separation; scaling and matching arguments led to the logarithmic velocity profile of turbulent boundary layers. These heuristics required an unusual amount of creativity; they involved intuitions bound to personal styles of thinking. Such intuitions are tentative and may lead to erroneous guesses. For instance, the great Kelvin erred in his stability analysis of parallel flow. A rigorous check of the compatibility of the conclusions with the fundamental equations is always needed.[35]

In still other cases, observations or experiments suggested new concepts such as group velocity, solitary waves, the stability or instability of laminar flow, and turbulent boundary layers. The very fact that pure theory was historically unable to lead to these concepts (and sometimes even resisted their introduction) shows the vanity of regarding them as implicit consequences of the fundamental equations. They nevertheless belong to fundamental hydrodynamics inasmuch as their compatibility with the fundamental equations can be verified a posteriori.

Lastly, the impossibility of solving the fundamental equation and the evident complexity of observed flows sometimes forced engineers and even physicists to arbitrarily and drastically simplify aspects of the flow. This happened for instance in early models of open channel flow. These models cannot be strictly regarded as parts of fundamental hydrodynamics, since some of their assumptions contradict both observed and theoretical properties of the flow. Yet their success suggests a looser sort of relation with the Navier-Stokes theory. In the case of open-channel flow, the models can be reinterpreted as re-parametrizations of the true equations for the approximate, large-scale motion derived from turbulent solutions of the Navier-Stokes equations.

In every case, the theoretical developments occurred with specific applications in mind: some kind of flow frequently observed in nature needed to be explained or the functioning of some instruments or devices needed to be understood. Purely mathematical methods broadly applied to general flow were of little avail. Insight was gained as a result of investigation directed at concrete and restricted goals. This is why the heroes of nineteenth-century and early twentieth-century fluid mechanics were either mathematically fluent engineers or physicists who had a foot in the engineering world.

2.2 The Components of Explanation

A first alley toward better explanation involves the restriction of the scope of a theory. The Navier-Stokes equations, regarded as the general foundation of hydrodynamics, can be specialized in various ways. There are homogeneous specializations or idealizations in which the restricted choice of parameters and kinds of systems

---

[35] On misleading intuitions in fluid mechanics, cf. Birkhoff 1950.

(boundary conditions) leads to more tractable integration problems or successful statistical approaches. Typical examples are irrotational Eulerian flow, low Reynolds-number flow, and fully turbulent flow. There are also heterogeneous specializations in which the restrictions on parameters and systems lead to flows that have different regions, each of which depends upon a specific simplification of the Navier-Stokes equations. This is the case for the high-Reynolds resistance problem and the airplane wing problem according to Prandtl. As was already mentioned, success here requires proper matching between the different regions.

Another explanatory resource is the identification of invariant structures of a flow belonging to a given class. The most impressive example of this sort is Helmholtz's demonstration of the conservation of vortex filaments. As the mind tends to focus on invariant aspects of our environment, the identification of new invariants often shape our descriptive language. As Helmholtz predicted, this has, in fact, happened in fluid mechanics: the vorticity field is now often preferred to the velocity field as a description of flow.

Third, instead of seeking structure in a given solution, we may attend to the structure of the space of solutions of the fundamental equation when the boundary conditions vary. For instance, we may ask whether laminar solutions are typical, whether small perturbations lead to different sorts of solutions: this is the issue of stability. We may also ask whether some classes of solution share common large-scale features, as we do in the statistical theories of turbulence. And, we may ask whether some properties or laws are generic in some regime of flow: this is the issue of universality, which we briefly touched with the logarithmic profile of turbulent boundary layers.

Lastly, explanation and understanding may come from linking hydrodynamics to other theories. We have encountered a few examples of this kind: potential theory, wave interference, group velocity, solitary waves, field kinematics, and proper-mode analysis of stability. In half of these cases, concepts of hydrodynamic origin were brought to bear on other theories and not vice versa. The cross-theoretical sharing of concepts nonetheless remains a token of their explanatory value.

2.3 A Pragmatic Definition of Explanation

As was stated above, the goal of fluid mechanics cannot be reduced to finding integrals of the fundamental equations that satisfy given boundary conditions. This is usually impossible by analytical means, and modern numerical means require a different simulation for each choice in the infinite variety of boundary conditions. As Batterman, Ramsey, and Heidelberger have argued, bare foundations do not answer the questions that truly interest physicists and engineers. Practitioners want to be able to characterize a physical situation by a humanly accessible number of physical parameters and to possess a picture of the situation that enables them to derive relations between these parameters in a reasonable amount of time. In other words, they need a concept of explanation that integrates our human capacity at representing and intervening. As Batterman emphasizes, this requires means for eliminating irrelevant details in our description of systems. This also implies the elaboration of a descriptive language, the concepts of which directly refer to controllable features of the system.[36]

With this pragmatic definition of explanation, it becomes clear that the earlier described developments of hydrodynamics served the purpose of increasing the explanatory power of the theory. Homogeneous specializations do so by offering adequate concepts and methods for certain kinds of flow. Heterogeneous specializations do so by combining the former specializations to describe flows that occur in problems of great practical import. The identification of invariant structures for certain classes of motion improves the economy of the representation. Attention to structure in the space of solutions enables us to decide to what extent smaller details of the motion affect the features of practical interest, and to what extent their effect can be smoothed out by some averaging process. Intertheoretical links produce familiar concepts that can indifferently be used in various domains of physics.

In this light, the practice of physics has more similarity with engineering than is usually assumed. The remark is not uncommon in recent writings in the philosophy of science. For instance, Ramsey revives J. J. Thomson's old characterization of theories as tools for solving physics or engineering problems; Epple compares the formation of Prandtl's wing theory to an engineering process combining multiple theoretical and experimental resources. In these scholars' view, the engineer only differs from the physicist by (usually) not participating in the invention of the theories and by his more systematic appeal to extra-theoretical components. Physicists and engineers not only share the goal of efficient intervention, they also share some of the means.[37]

Ramsey and Heidelberger insist that the articulation of theories implies the formation of new, adequate concepts. One could even argue that the bare Navier-Stokes theory has no physical concepts. It harbors only mathematical concepts such as the velocity field that correspond to an ideal description of the flow, ignoring molecular structure and presuming indefinite resolution. A concept, in the etymological sense of the word (*concipio* in Latin, or *begreifen* in German), is a mental means to grasp some concrete object or situation. Hydraulic head, vortices, wave groups, solitary waves, the laminar-turbulent transition, boundary layers, separation, etc. are concepts in this practical sense. The detailed velocity field or the various terms of the Navier-Stokes equation are not. What Thomas Kuhn once belittled as the "mopping up" of theories in the normal phases of science truly is concept formation.[38]

---

[36] Batterman 2002; Ramsey 1992, 1993, 1995; Heidelberger 2006.
[37] Ramsey 1995, 16; Epple 2002.
[38] Kuhn 1962, 24. Hilary Putnam similarly criticized another of Kuhn's characterizations of normal science: "The term 'puzzle solving' is unfortunately trivializing; searching for explanations of phenomena and for ways to harness nature is too important a part of human life to be demeaned" (Putnam 1974, 261).

3. Theories and Modules

3.1 Defining Physical Theories

Once we recognize the cognitive impotence of the bare foundations of a theory, we need a general definition of "theory" that is not limited to the fundamental equations and a few naïve rules of application. The definition must allow for evolving components, since the cognitive efficiency of any good theory always increases in time. It must include explanatory devices and it must allow the intertheoretical connectivity found in mature theories. The following is a sketch of such an enriched definition.[39]

A physical theory is a mathematical construct including:

(a) a *symbolic universe* in which systems, states, transformations, and evolutions are defined by means of various magnitudes based on Cartesian powers of $\mathbf{R}$ (or $\mathbf{C}$) and on derived functional spaces.
(b) *theoretical laws* that restrict the behavior of systems in the symbolic universe.
(c) *interpretive schemes* that relate the symbolic universe to idealized experiments.
(d) methods of *approximation* and considerations of *stability* that enable us to derive and judge the consequences that the theoretical laws have on the interpretive schemes.

The symbolic universe and the theoretical laws are permanently given. They correspond to the "family of models" of the semantic view of physical theories. In the case of hydrodynamics, the symbolic universe consists in the velocity, pressure, and density fields for each fluid of the system, in the boundaries of rigid bodies that may or may not move, and in force densities such as gravity. The theoretical laws are the Navier-Stokes equations, boundary conditions, and (for compressible fluids) a relation between density and pressure that may involve modular coupling with thermodynamics (we will return to this point).

In the semantic view of theories, the empirical content of a theory is defined by an isomorphism between parts of the symbolic universe and empirical data; although the means by which this isomorphism is determined are usually left in the dark. The notion of an interpretive scheme is intended to fill part of this gap. By definition *an interpretive scheme consists in a given system of the symbolic universe together with a list of characteristic quantities that satisfy the three following properties.*(1) *They are selected among or derived from the (symbolic) quantities that define the state of this system.* (2) *At least for some of them, ideal measuring procedures*

---

[39] For a discussion of this definition and a comparison with the definition of Sneedian structuralists, cf. Darrigol 2008, 198–203.

are known. (3) *The laws of the symbolic universe imply relations of a functional or a statistical nature among them.* More specifically, interpretive schemes are blueprints of conceivable experiments whose outcomes depend only on relations between a finite set of mutually related quantities, a sufficient number of which are measurable. In some cases, the intended experiments may be designed to determine some theoretical parameters from the measured quantities. In other cases, the theoretical parameters are given, and theoretical relations between the measured quantities are verified. In all cases, the interpretive schemes do not contain rigid linguistic connections between theoretical terms and physical quantities; their concrete implementation is analogical, historical, and subject to revisions.[40]

The introduction of interpretive schemes implies a selection of systems and quantities from the infinite variety of elements in the symbolic universe of the theory. This selection can evolve dramatically with the number and nature of the imagined applications of the theory. The two main classes of interpretive schemes of early hydrodynamics were the pierced vessel, in which the efflux of water is related to the height of the water surface; and the resistance scheme in which a solid body immersed in a stream of water experiences a force related to the velocity of the stream. Another interesting scheme, Bernoulli's pipe of variable section, implied pressure measurement through vertical columns of water. A sample of later schemes includes the determination of the velocity of surface waves as a function of depth and wavelength, the visualized motion of vortices as a function of their relative configuration, the visualized lines of flow around an immersed body as a function of the asymptotic velocity, the drag and lift of a wing as a function of asymptotic velocity and angle of attack. Some schemes were reactions to well-identified practical problems and others to some new theoretical development. In the latter category, we may cite the determination of the separation point for the flow around an immersed sphere, the measurement of instability thresholds, Prandtl's aspiration of the boundary layer to prevent separation, and the post-theoretical visualization of laminar boundary layers.

For an interpretive scheme to serve its purpose as an experimental blueprint, a few conditions must be met: one must know how to realize concretely the system picked in the symbolic universe; one must know how to implement the ideal measuring procedures; one must be able to compute the relations between measured quantities and theoretical parameters; and one must know something about the stability of these relations. Point (d) of my general definition of theories is meant to meet these two last requirements. In this regard, the reader may consult the growing literature regarding the philosophy of approximation, numerical analysis, and stability. The following discussion is restricted to aspects of the working of interpretive schemes that have to do with the *modular structure* of theories.[41]

---

[40] This is a considerable weakening of the logical-empiricist strictures on the meanings of theoretical terms.

[41] A more detailed discussion is given in Darrigol 2008. Interpretive schemes supplemented with the requirement of computability are similar to Humphreys's "computational templates." According to Humphreys 2004, it is at the level of computational templates that questions about theoretical representation, empirical fitness,

3.2 Modules

By definition, a *module* is a component of a theory which is itself a theory, with a different domain of application. Our ability to apply a theory crucially depends on integrated modules. First, there are *defining modules* that serve to define some of the quantities in the symbolic universe. In the case of hydrodynamics, the list of these modules includes a Euclidian geometrical module that defines the spatial relations of the systems; a mechanical module that defines external force densities, external pressures, and the motion of immersed bodies; a thermodynamic module that defines relations between fluid density, pressure, and temperature (sometimes also heat transfer). These modular definitions enable us to transfer already known measuring procedures into the interpretive schemes of hydrodynamics. In the case of compressible fluids, they are essential to the completeness of the theory: no prediction can be made without knowing how the density varies according to the thermal properties of the system.

Second, there are *idealizing modules* obtained by simplifying the symbolic universe and retaining similar interpretive schemes (of course, the functional relations between schematic quantities are different). In the case of hydrodynamics, the most important modules of this kind are the theory of incompressible fluids, the theory of inviscid fluids, and the theory of incompressible inviscid fluids. Incompressibility enables us to ignore the coupling of hydrodynamics with thermodynamics. Inviscidity eliminates one term in the Navier-Stokes equations and yields Euler's simpler equations. The usefulness of these idealizations comes from the relative smallness of the compressibility of water and from the smallness of the viscosities of air and water.

Third, there are *specializing modules* that are exact substitutes of the theory for subclasses of schemes that meet certain conditions. For instance, Lagrange's theory of irrotational incompressible fluid motion can replace Euler's theory for schemes in which the fluid motion is started from rest by the motion of walls or immersed bodies; Helmholtz's theory of vortex motion can replace the incompressible specialization of Euler's theory for schemes based on the vortex structure.

Idealizing and specializing modules are not by themselves sufficient to design effective interpretive schemes. We also need *approximating modules* that can be seen as limits of the theory for a given subclass of systems when a parameter of this class or a parameter of the symbolic universe (or a combination of both kinds of parameters) takes extreme but still finite values (the limit may involve statistical considerations). Hydrodynamic examples of modules of this kind concern the small-depth and small-amplitude limits of surface wave schemes, the high Reynolds-number limit of fluid resistance or fluid retardation schemes (boundary-layer theory), and the low Reynolds-number limit of these schemes (creeping flow). In most cases, it is only at

---

realism, and so on must be discussed; knowledge "in principle" must be subordinated to knowledge "in practice," which involves the available technologies of measurement and calculation.

the level of approximating modules that the functional relations between schematic quantities can be effectively computed.[42]

There is a last kind of modules, the *reducing modules*, that has more to do with the foundations of the theory than with its applications. These are theories diverted from their original domain of application in order to build the whole symbolic universe of another theory.[43] This is what happens, for instance, when the mechanics of a system of interacting mass points is used in Clerk Maxwell's manner as a molecular-kinetic-theoretical foundation for the Navier-Stokes equation.[44] There is a difference between saying that T is a reducing module of T′ and saying that T′ is an approximating module of T: in the latter case, the schemes of T′ are a subclass of those of T, whereas in the former case the schemes of T have nothing to do with the schemes of T′ (they lose their empirical realizability in the reducing process). In the case of reducing modules, the theory T is necessarily known before the reduction is done and the theory T′ may even be invented through the reduction, as was the case with Maxwell's theory of electrodynamics. With approximating modules, the theory T may or may not precede the theory T′. Whereas the Navier-Stokes theory preceded its boundary-layer module, Euler's hydrodynamics postdated its narrow-vase module à la Bernoulli. Maxwell's electrodynamics postdated its quasi-stationary module and wave optics postdated its rays-optics module.

Modules, qua theories, can have submodules. For instance, the incompressible idealizing module of the Navier-Stokes theory has an inviscid specializing module. More interestingly, the boundary-layer theory, as an approximating module, relies on defining modules that are idealizing, specializing, or approximating modules of the Navier-Stokes theory. These defining modules respectively correspond to inviscid fluid motion (in the "free fluid"), discontinuity surfaces (in the case of separation), and the boundary-layer equation. This means that a module of a theory can also be a submodule of another module of the same theory (see figure 1.4). It also means that the same theory can be a module of different theories. More evident examples of multiply inserted modules are Euclidian geometry and Newtonian mechanics, which are defining modules of all the main theories of classical physics.

The modular structure varies as the theory develops. The defining modules are there, by necessity, from beginning to end. Reducing modules may occur at any stage of the life of the theory: at its birth, in its middle age, or even at its death. An instance of the last case occurred when the electromagnetic theory of light replaced elastic-solid theories of light. Specializing and approximating modules are gradually introduced, for the sake of mathematical simplification and efficient application. The status of a module may vary. For instance, a defining module

---

[42] Approximating modules correspond to what Jeffry Ramsey calls transformation reduction (Ramsey 1993, 1995).

[43] In this case, being a module of another theory does not imply a sort of inclusion; but it remains true that a module of a theory serves this theory.

[44] More exactly, the low-density gas specialization of the Navier-Stokes theory is an approximating module of the kinetic theory of gases, of which the mechanics of a set of interacting molecules is a reducing module.

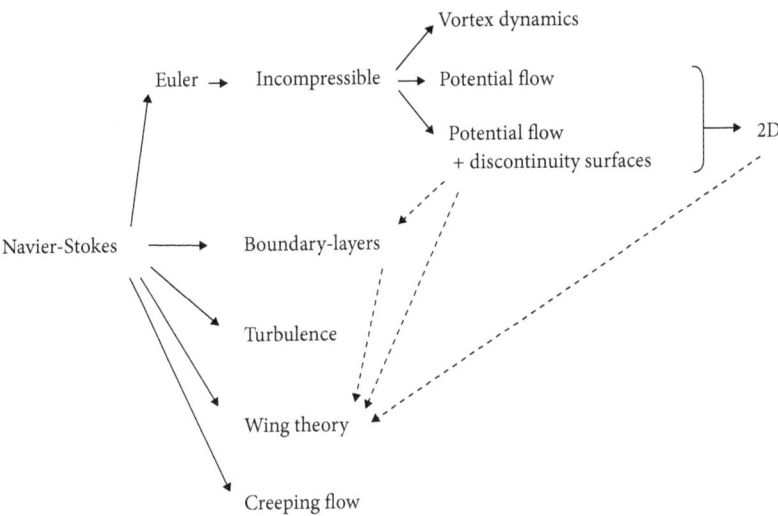

Figure 1.4 Some of the modular structure of modern hydrodynamics. The solid arrows correspond to specializing or approximating modules, the dotted arrows to defining or idealizing modules.

may become a reducing module or vice versa. In the course of the history of electrodynamics, mechanics was successively a defining module (Coulomb, Ampère, Neumann, Weber), a reducing module (Thomson, Maxwell), and again a defining module (Hertz). This variability of the status of modules is the reason why I have introduced a fairly wide spectrum of modular interrelations.

As I have argued elsewhere, modules play an essential role in the application, construction, comparison, and communication of theories.[45] In the case of hydrodynamics, the role of modules in permitting efficient applications of the theory is most evident. They yield conceptual structures that are better adapted to concrete problem situations than the bare Navier-Stokes equation. They instruct us about the choice of accessible, causally interrelated aspects of fluid motion and they tell us how to measure them. Through a nesting hierarchy of modules, we can capitalize on our concrete knowledge of the schemes of the most basic modules to imagine and control the complex experimental environment through which the predictions of higher-level theories are tested.

The constructive role of modules is evident in the case of defining and reducing modules. Idealizing, specializing, and approximating modules also help theory construction when they are known before the projective theory. They may play an instrumental role in theoretical unification or in the rejection of a tentative unification. And they may provide a "correspondence principle" for guiding the design of the symbolic universe of a new theory, as was the case when Bohr and

[45] Darrigol 2008.

Heisenberg appealed to classical electrodynamics in the construction of quantum theory.

The comparison of two theories requires shared interpretive schemes whose concrete realization is not tied to either of these theories. This is possible if all the schematic quantities can be defined by means of shared modules. For example, the predictions of various nineteenth-century theories of electrodynamics could be compared thanks to the sharing of electrostatic, electrokinetic, and magnetostatic modules. Shared modules are also essential for the communication between different subcultures of physics and other communities of scientists and engineers who use physics in their work. These shared modules enable someone to use results of a theory whose foundation he ignores or even rejects. They permit the sharing of apparatus whose functioning depends only on lower-level modules. Lastly, modular structure is essential to the teaching of theories. A typical textbook is organized by chapters that correspond to modules of the theory. Thus, the student can connect the new theory to other theories with which he is already familiar, he can get a grasp on how to apply the theory in concrete situations, and he can learn techniques that transcend the domain of this theory.

3.3 Models and Modules

In recent philosophy of science, there has been a strong emphasis on models as mental constructs that differ both from full-fledged theory and from narrow empirical induction. Mary Morgan and Margaret Morrison regard models as mediating instruments between theory and phenomena. In their view, models are partially autonomous from theory: some of their components have extratheoretical origins. The models help to shape theories as much as they rely on theory. They are more directly relevant to the empirical world than theories, at the price of a more limited scope. For all these reasons, Morgan and Morrison insist that models are not theories.

Yet (physics) models fit my definition of theories, since they necessarily have a symbolic universe, internal laws, and interpretive schemes. In my view, they differ from other theories only by having a smaller scope or less structural unity. This difference is largely a matter of degree and convention. The partial autonomy of models from more fundamental theories results from the modular character of their interconnection with these theories. Typically, fundamental theories are defining or reducing modules of models; or else models are approximating modules of a more fundamental theory.[46] The relation between models and theories is just a particular case of the modular relation between two theories. It therefore implies the same sort of mutual fitness without fusion. There is no need to sharply discriminate models

---

[46] In conformity with the physicists' usage, Morrison and Morgan also call "models" what I call a "reducing module." For instance, Maxwell's mechanical model of 1862 for the electromagnetic field is a model in this sense. This kind of model widely differs from ad hoc models for limited classes of phenomena.

from theories once the modular structure of theories is taken into account. It is sufficient to recognize that some theories are more fundamental than others.

We may now revisit Prandtl's boundary-layer theory, which has received more attention from philosophers of science than any other aspect of hydrodynamics. The reason for this interest, no doubt, is the glaring cognitive superiority of Prandtl's theory compared to any earlier approach to the high Reynolds-number resistance problem. Margaret Morrison calls Prandtl's theory a model and insists on its extratheoretical origins in conformity with her general views on models. In her opinion, Prandtl's concept of boundary layer originated in an inductive inference from the flow patterns that Prandtl observed with his water mill and tank. Michael Heidelberger denies this reconstruction and favors an account in terms of theoretical heuristics. As he correctly remarks, laminar boundary layers could not be seen in Prandtl's tank, and Prandtl himself cited asymptotic reasoning as the true source of this concept. However, the scenario imagined by Morrison is frequently encountered in the history of hydrodynamics. For instance, Rankine and Froude's concept of eddying boundary layer did result from casual observation of the flow around a ship hull.[47]

Despite his disagreement with Morrison over the origins of Prandtl's theory, Heidelberger continues to call it a model. Presumably, he means to indicate that Prandtl's theoretical heuristics implied more creative guessing than would be needed in a mere deduction from the Navier-Stokes theory would engender, and that it created a new efficient, and fairly autonomous, conceptual structure. Prandtl himself did not call his theory a model. The reasons are not difficult to guess. The word was then used in Göttingen as a way to characterize semi-concrete theories that saved the phenomena without pretending to reach the true causes. In contrast, Prandtl's boundary-layer theory was meant to represent the true flow around bodies at a high Reynolds number; it did not imply any counterfactual hypothesis; and it was demonstrably compatible with the Navier-Stokes equation. In my terminology, Prandtl's theory was an approximating module of the Navier-Stokes theory. In conformity with the physicists' parlance, I would rather reserve the word "model" for theories that imply conscious simplifications of the system under consideration, for instance, the early nineteenth-century "models" of open channel flow.[48]

These terminological subtleties matter inasmuch as an overly generous use of the word "model" implies a neglect of the modular structure of theories, which I regard as pervasive and essential. Morrison's and Heidelberger's insights into the function of what they prefer to call models are nevertheless important. They both emphasize the impotence of bare fundamental theories and the need to supplement them with conceptual structures that somehow mediate between theory and experiment. And they both understand that unification, in the context of a fundamental theory, remains a desideratum. In a witty allusion to Nancy Cartwright's criticism

---

[47] Morrison 1999, 53–60; Heidelberger 2006, 60–62.
[48] For instance, Walther Ritz (1903, 3) called his vibrating-square theory of series spectra a "model" (his quotation marks).

of fundamental theories, Heidelberger claims that the Navier-Stokes theory "does not even lie about the world." At the same time, he understands that the boundary-layer theory, which so much improves the explanatory power of hydrodynamics, is an approximation of the Navier-Stokes theory. In my view, the moral is that the Navier-Stokes theory, or any other of the great theories of physics, should not be considered independently of its ever-increasing modular structure. Although the result of this evolution can never fulfill the dream of a transparent and automatic application of the fundamental equations to every conceivable situation, it has the organic unity and efficiency that we need in order to understand and control some of the physical world.[49]

REFERENCES

Airy, George Biddell (1845). Tides and waves. In *Encyclopedia Metropolitana*, 5: 291–396.
Batterman, Robert (2002). *The devil in the details: Asymptotic reasoning in explanation, reduction, and emergence*. Oxford: Oxford University Press.
Bernoulli, Daniel (1738). *Hydrodynamica, sive de viribus et motibus fluidorum commentarii*. Strasbourg: J. R. Dulsecker.
Birkhoff, Garrett (1950). *Hydrodynamics: A study in logic, fact, and similitude*. Princeton: Princeton University Press.
Bjerknes, Vilhelm (1898). Über einen hydrodynamischen Fundamentalsatz und seine Anwendung besonders auf die Mechanik der Atmosphäre und des Weltmeeres. Kongliga Svenska, Vetenskaps-Akademiens, *Handlingar*, 31: 3–38.
Boussinesq, Joseph (1871). Théorie de l'intumescence liquide appelée *onde solitaire* ou *de translation*, se propageant dans un canal rectangulaire. Académie des Sciences, *Comptes rendus hebdomadaires des séances*, 72: 755–59.
——. (1877). Essai sur la théorie des eaux courantes. Académie des Sciences de l'Institut de France, *Mémoires présentés par divers savants*, 23: 1–680.
Cartwright, Nancy (1983). *How the laws of physics lie*. Oxford: Oxford University Press.
——. (1999). *The dappled world: A study of the boundaries of science*. Cambridge: Cambridge University Press.
Cat, Jordi (1998). The physicists' debates on unification in physics at the end of the 20th century. *Historical Studies in the Physical and Biological Sciences* 28: 253–99.
——. (2005). Modeling cracks and cracking models: Structure, mechanisms, boundary conditions, constraints, in inconsistencies and the proper domain of natural laws. *Synthese* 146: 447–87.
Cauchy, Augustin (1827). Théorie de la propagation des ondes à la surface d'un fluide pesant d'une profondeur indéfinie. Académie des Sciences de l'Institut de France, *Mémoires présentés par divers savants*, 1: 1–123.
D'Alembert, Jean le Rond (1768). Paradoxe proposé aux géomètres sur la résistance des fluides. In *Opuscules mathématiques*, vol. 5, 34th memoir, 132–38. Paris: David.

---

[49] Cartwright 1983; Heidelberger 2006, 64.

Darrigol, Olivier (2005). *Worlds of flow: A history of hydrodynamics from the Bernoullis to Prandtl.* Oxford: Oxford University Press.

———. (2007). On the necessary truth of the laws of classical mechanics. *Studies in the History and Philosophy of Modern Physics* 38: 757–800.

———. (2008). The modular structure of physical theories. *Synthese* 162: 195–223.

Drazin, Philip, and William Reid (1981). *Hydrodynamic stability.* Cambridge: Cambridge University Press.

Eckert, Michael (2005). *The dawn of fluid dynamics: A discipline between science and technology.* Berlin: Wiley.

———. (2008). Turbulenz: ein problemhistorischer Abriss. *NTM* 16: 39–71.

Epple, Moritz (2002). Präzision versus Exaktheit: Konfligierende Ideale der angewandten mathematischen Forschung. Das Beispiel der Tragflügeltheorie. *Berichte zur Wissenschaftsgeschichte* 25: 171–93.

Euler, Leonhard (1755) [printed in 1757]. Principes généraux du mouvement des fluides. Académie Royale des Sciences et des Belles-Lettres de Berlin, *Mémoires*, 11: 274–315.

Farge, Marie, and Etienne Guyon (1999). A philosophical and historical journey through mixing and fully-developed turbulence. In *Mixing: Chaos and turbulence*, ed. Hugues Chaté et al., 11–36. New York: Kluwer Academic/Plenum Publishers.

Franceschelli, Sara (2007). Construction de signification physique pour la transition vers la turbulence. In *Chaos et systèmes dynamiques: Eléments pour une épistémologie*, ed. S. Franceschelli, M. Paty, and T. Roque, 213–37. Paris: Hermann.

Frisch, Uriel (1995). *Turbulence: The legacy of A. N. Kolmogorov.* Cambridge: Cambridge University Press.

Froude, William [1868] 1955. Observations and suggestions on the subject of determining by experiment the resistance of ships, Memorandum sent to E. J. Reed, Chief Constructor of the Navy, dated December 1868. In *The papers of William Froude*, ed. A. D. Duckworth, 120–27. London: Institution of Naval Architects.

———. (1874). Reports to the Lords Commissioners of the Admiralty on experiments for the determination of the frictional resistance of water on a surface, under various conditions, performed at Chelston Cross, under the authority of their Lordships. British Association for the Advancement of Science, *Reports*, 249–55.

———. (1877). The fundamental principles of the resistance of ships. Royal Institution, *Proceedings*, 8: 188–213.

Hacking, Ian (1983). *Representing and intervening: Introductory topics in the philosophy of natural science.* Cambridge: Cambridge University Press.

Hagen, Gotthilf (1839). Über die Bewegung des Wassers in engen cylindrischen Röhren. *Annalen der Physik* 46: 423–42.

Heidelberger, Michael (2006). Applying models in fluid dynamics. *International Studies in the Philosophy of Science* 20: 49–67.

Helmholtz, Hermann (1858). Über Integrale der hydrodynamischen Gleichungen, welche den Wirbelbewegungen entsprechen. *Journal für die reine und angewandte Mathematik* 55: 25–55.

———. (1868). Über diskontinuirliche Flüssigkeitsbewegungen. Akademie der Wissenschaften zu Berlin, mathematisch-physikalische Klasse, *Sitzungsberichte*, 215–28.

———. (1873). Über ein Theorem, geometrisch ähnliche Bewegungen flüssiger Körper betreffend, nebst Anwendung auf das Problem, Luftballons zu lenken. Königliche Akademie der Wissenschaften zu Berlin, *Monatsberichte*, 501–14.

———. (1888). Über atmospherische Bewegungen I. Akademie der Wissenschaften zu Berlin, mathematisch-physikalische Klasse, *Sitzungsberichte*, 652.

Humphreys, Paul (2004). *Extending ourselves: Computational science, empiricism, and scientific method*. Oxford: Oxford University Press.

Kármán, Theodore von (1930). Mechanische Ähnlichkeit und Turbulenz. Gesellschaft der Wissenschaften zu Göttingen, mathematisch-physikalische Klasse, *Nachrichten*, 58–76.

Kirchhoff, Gustav (1869). Zur Theorie freier Flüssigkeitsstrahlen. *Journal für die reine und angewandte Mathematik* 70: 289–98.

Kragh, Helge (2002). The vortex atom: A Victorian theory of everything. *Centaurus* 44: 32–114.

Kuhn, Thomas (1961). The function of measurement in modern physical science. *Isis* 52: 161–93.

———. (1962). *The structure of scientific revolutions*. Chicago: University of Chicago Press.

Lagrange, Joseph Louis (1781). Mémoire sur la théorie du mouvement des fluides. Académie Royale des Sciences et des Belles-Lettres de Berlin, *Nouveaux mémoires*. Also in *Oeuvres* (1869), 4: 695–750.

Morrison, Margaret (1999). Models as autonomous agents. In *Models as mediators: Perspectives on natural and social science*, ed. Mary Morgan and Margaret Morrison, 38–65. Cambridge: Cambridge University Press.

———. (2000). *Unifying scientific theories: Physical concepts and mathematical structures*. Cambridge: Cambridge University Press.

Morrison, Margaret, and Mary Morgan (1999). Models as mediating instruments. In *Models as mediators: Perspectives on natural and social science*, ed. Mary Morgan and Margaret Morrison, 10–37. Cambridge: Cambridge University Press.

Navier, Claude Louis (1822). Sur les lois du mouvement des fluides, en ayant égard à l'adhésion des molécules [read on 18 March 1822]. *Annales de chimie et de physique* 19 (1821) [in fact 1822]: 244–60.

Poisson, Siméon Denis (1816). Mémoire sur la théorie des ondes. Académie Royale des Sciences, *Mémoires*, 1: 71–186 (read on 2 October and 18 December 1815, published in 1818).

Poncelet, Jean Victor (1839). *Introduction à la mécanique industrielle, physique ou expérimentale*. 2d ed. Paris.

Prandtl, Ludwig (1905). Über Flüssigkeitsbewegung bei sehr kleiner Reibung. In III. internationaler Mathematiker-Kongress in Heidelberg vom 8. bis 13. August 1904, ed. A. Krazer, *Verhandlungen* (Leipzig), 484–91. Also in *Gesammelte Abhandlungen* 2: 575–84.

———. (1914). Der Luftwiderstand von Kugeln. Gesellschaft der Wissenschaften zu Göttingen, mathematisch-physikalische Klasse, *Nachrichten*. Also in *Gesammelte Abhandlungen* 2: 597–608.

———. (1931). On the role of turbulence in technical hydrodynamics. World Engineering Congress in Kyoto, *Proceedings*. Also in *Gesammelte Abhandlungen* 2: 798–811.

Putnam, Hilary (1974). The "corroboration" of theories. In *The philosophy of Karl Popper*, ed. P. A. Schilpp, vol. 2, La Salle. Also in H. Putnam, *Philosophical papers*, vol. 1: *Mathematics, matter and method* (Cambridge University Press, 1975), 250–69.

Ramsey, Jeffry (1992). Towards an expanded epistemology for approximations. In *PSA 1992*: Proceedings of the 1992 Biennial Meeting of the Philosophy of Science

Association, ed. K. Okruhlik, A. Fine, and M. Forbes, 1: 154–64. East Lansing, MI: Philosophy of Science Association.

———. (1993). When reduction leads to construction: Design considerations in scientific methodology. *International Studies in the Philosophy of Science* 7: 239–51.

———. (1995). Construction by reduction. *Philosophy of Science.* 62: 1–20.

Rankine, William John Macquorn (1858). Resistance of ships, Letter to the editors of 26 August 1858. *Philosophical Magazine* 16: 238–39.

———. (1865). On plane water-lines in two dimensions. Royal Society of London, *Philosophical Transactions*, 154: 369–91.

———. (1870). On stream-line surfaces. Royal Institution of Naval Architects, *Transactions*, 11: 175–81.

Rayleigh, Lord (William Strutt) (1876a). On waves. *Philosophical Magazine.* Also in *Scientific Papers* 1: 251–71.

———. (1876b). On the resistance of fluids. *Philosophical Magazine* 11: 430–41.

———. (1880). On the stability, or instability, of certain fluid motions. London Mathematical Society, *Proceedings*, 11: 57–70.

Reynolds, Osborne (1883). An experimental investigation of the circumstances which determine whether the motion of water shall be direct or sinuous, and of the law of resistance in parallel channels. Royal Society of London, *Philosophical Transactions*, 174: 935–82.

———. (1895). On the dynamical theory of incompressible viscous fluids and the determination of the criterion. Royal Society of London, *Philosophical Transactions*, 186: 123–64.

Ritz, Walther (1903). Zur Theorie der Serienspektren. *Annalen der Physik.* Also in *Oeuvres* (Paris, 1911), 1–77.

Russell, John Scott (1839). Experimental researches into the laws of certain hydrodynamical phenomena that accompany the motion of floating bodies, and have not previously been reduced into conformity with the known laws of the resistance of fluids. Royal Society of Edinburgh, *Transactions*, 14: 47–109.

Saint-Venant, Adhémar Barré de (1843). Note à joindre au mémoire sur la dynamique des fluides, présenté le 14 avril 1834. Académie des Sciences, *Comptes-rendus hebdomadaires des séances*, 17: 1240–43.

Stokes, George Gabriel (1843). On some cases of fluid motion. Cambridge Philosophical Society, *Transactions*, 8: 105–37.

———. (1847). On the theory of oscillatory waves. Cambridge Philosophical Society, *Transactions*. Also in *Mathematical and Physical Papers*, 1: 197–225.

———. (1849) [read in 1845]. On the theory of the internal friction of fluids in motion, and of the equilibrium and motion of elastic solids. Cambridge Philosophical Society, *Transactions*, 8: 287–319.

———. (1850). On the effect of the internal friction of fluids on the motion of pendulums. Cambridge Philosophical Society, *Transactions*. Also in *Mathematical and Physical Papers*, 3: 1–141.

———. [1876]. Smith prize examination papers for 2 Feb. 1876. In *Mathematical and Physical Papers*, 5: 362.

Thomson, William (1887a). Rectilinear motion of viscous fluid between two parallel planes. *Philosophical Magazine* 24: 188–96.

———. (1887b). On ship waves [lecture delivered at the "Conversazione" in the Science and Art Museum, Edinburgh, on 3 Aug. 1887]. Institution of Mechanical Engineers, *Minutes of Proceedings*, 409–34.

———. (1894). On the doctrine of discontinuity of fluid motion, in connection with the resistance against a solid moving through a fluid. *Nature* 50: 524–25, 549, 573–75, 597–98.

Truesdell, Clifford (1954). Rational fluid mechanics, 1657–1765. In Euler, *Opera Omnia*, ser. 2, 12: ix–cxxv. Lausanne: Orell Füssli.

———. (1968). The creation and unfolding of the concept of stress. In *Essays in the history of mechanics*, 184–238. Berlin: Springer.

Tyndall, John (1867). On the action of sonorous vibrations on gaseous and liquid jets. *Philosophical Magazine* 33: 375–91.

Winsberg, Eric (1999). Sanctioning models: The epistemology of simulation. *Science in Context* 12: 275–92.

Wright, Thomas (1983). Ship hydrodynamics 1770–1880. Ph.D. dissertation (Science Museum, South Kensington, London).

Yamalidou, Maria (1998). Molecular ideas in hydrodynamics. *Annals of Science* 55: 369–400.

CHAPTER 2

WHAT IS "CLASSICAL MECHANICS" ANYWAY?

MARK WILSON

1. Preliminary Considerations

One of the prominent sources of unhelpful folklore within philosophy is the historical controversy whose proper intricacies have been underappreciated. Misunderstood problems beget mistaken "morals" that can lead philosophical thinking astray for long epochs thereafter. This has occurred, to an extent that few philosophers recognize, with respect to the so-called "foundations of classical mechanics." As matters are commonly represented within modern college primers, "classical physics" appears to be a transparent subject matter firmly founded upon Newton's venerable laws of motion. But this placid appearance is deceptive. Any purchaser of an old home is familiar with parlor walls that seem sound except for a few imperfections that "only require a little spackle and paint." When those innocent dimples are opened up, the ancient gerry-rigged structure comes tumbling down and our hapless fix-it man finds himself confronted with months of dusty reconstruction. So it is with our subject, whose basic concepts can seem so "clear and distinct" on first acquaintance that unwary thinkers have mistaken them for a priori verities. But the true lesson of "classical mechanics" for philosophy should be exactly the opposite: the conceptual matters that initially strike us as simple and pellucid often unwind into hidden complexities when probed more adequately.[1]

---

[1] This is an extract (skillfully edited by Julia Bursten) of a longer survey to appear in a collection of essays entitled *Physics Avoidance*. I would like to thank Julia Bursten and Bob Batterman for their helpful advice.

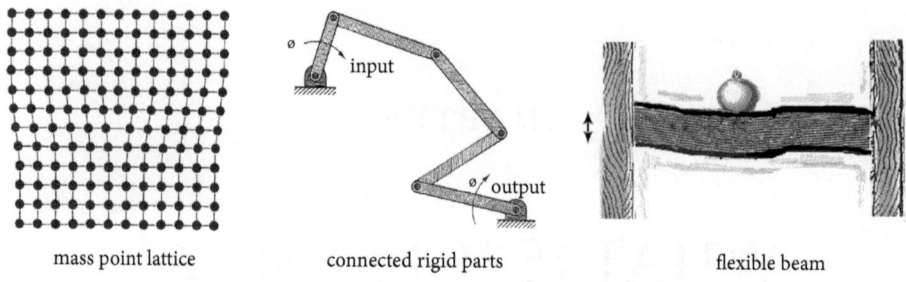

mass point lattice   connected rigid parts   flexible beam

Figure 2.1

Matters have been rendered more confusing by the fact that a conceptually simple *surrogate* for classical doctrine is readily available, even though its formally articulated doctrines skirt most of the tricky conceptual problems encountered within classical tradition. The tenets of this simple theory comprise the themes that we shall investigate under the heading of "point-mass mechanics." Within this approach the term *point mass* designates an isolated, zero-dimensional point that carries concentrated mass, charge, and so on. In contrast, there are two other sorts of "fundamental objects" with which a "classical mechanics" can be potentially concerned: *rigid bodies*, understood as extended solids whose points never alter their relative distances to one another and *flexible bodies* such as fluids or solids that are completely malleable at every size scale (figure 2.1).

Commonly, the latter are also called *continua*, a practice we shall adopt here. Of course, any of these entities can be joined together in larger combinations, as when individual rods are assembled into a *mechanism* or one flexible body is embedded within another as a *composite* (e.g., a jelly doughnut).[2]

Mathematicians commonly label our continua as *fields* due to their distributed character. We will generally avoid this terminology and will not discuss classical electrodynamics at all. In the sequel, I shall employ the phrase *material point* to designate a zero-dimension region within a continuously distributed body (either in its interior or along some bounding surface). In contrast to our point masses, material points are connected with one another quite densely and (usually) do not carry finite values of mass or impressed force (they, instead, only display mass and charge *densities* that sum to genuine masses and densities over regions of an adequate measure). The phrase *analytic mechanics* will serve as a generic title for the sundry formalisms that deal with connected systems of rigid bodies.

As just noted, the "conceptually simple surrogate" for classical doctrine that most commonly dominates philosophical discussions of "Newtonian mechanics" comprises a set of prescriptions that make coherent sense only with respect to isolated *point masses* that never come into contact with one another. We shall discuss

---

[2] In textbooks, *ontologically mixed circumstances* (a point mass sliding upon a rigid plane) often appear. Usually these need to be viewed as degenerations of dimensionally consistent schemes (i.e., a ball sliding on a plane or a free mass floating above a lattice of strongly attracting masses).

the specific features of these doctrines in section 3. From a point-mass perch, any appeal to rigid bodies or continua merely represents a convenient means of discussing large swarms of point masses held together through cohesive bonding at short scale lengths.

The deceptive simplicity available to the point-mass approach traces largely to the fact that, within its frame, matter can exist only in the form of isolated singularities, thereby sidestepping the substantial mathematical concerns that arise when extended objects *come in contact* with one another (on rare occasions, point masses can collide with one another, but these contacts only occur at fleeting moments that can usually be handled through appeal to conservation principles). As a result, point masses act upon one another only through *action-at-a-distance forces*,[3] but higher dimensional objects require direct *contact forces* as well. As we will learn, getting action-at-a-distance forces and contact forces to work in tandem is a non-trivial affair, but it becomes a conceptual obligation that vanishes from view if we are allowed to restrict our fundamental ontology to point masses alone.

However, there is a wide range of subtle reasons why it can easily *look as if* a specific classical author embraces the point-mass viewpoint. As we will observe in section 3, Newton's celebrated laws of motion are difficult to parse coherently unless terms like "body" are interpreted in a punctiform manner. A host of significant *mathematical complexities* attach to the notion of "material point" as it appears within continuum physics (i.e., as a point-sized region within a continuous body), and these are sometimes bypassed by confusing embedded continuum points with the simple isolated singularities of the point-mass treatment. We shall survey several of these shifts in the pages to follow. From a formal point of view, it is important to distinguish between the *ordinary differential equations* (ODEs) pertinent to point masses and analytic mechanics and the trickier *partial differential equations* (PDEs) required in continuum modeling.[4]

The fact that the real world proves *quantum mechanical* within its small-scale behaviors occasions confusion as well. Although particles like electrons appear to be "point-like" in their scattering behaviors, they also "fill" larger effective volumes courtesy of the uncertainty relations. In many cases, one obtains the requisite Schrödinger equation for a system of particles (which is a PDE describing a field spread out within a high dimensional space) by "quantizing" a parallel set of ODEs for a classical point-mass system.[5] But this mathematical linkage does not entail

---

[3] If a mathematical treatment happens to make two point masses coincide, that occurrence is generally viewed as a *blowup* (= breakdown of the formalism) rather than a true contact. It is often possible to push one's treatment through such blowups through appeal to sundry conservation laws and the rationale for these popular procedures will be scrutinized in section 3.

[4] Modern investigations have shown that true ODEs and PDEs are usually the resultants of foundational principles that require more sophisticated mathematical constructions for their proper expression (integro-differential equations; variational principles, weak solutions, etc.). We shall briefly survey some of the reasons for these complications when we discuss continua in section 4 (although such concerns can even affect point-mass mechanics as well). For the most part, the simple rule "ODEs = point masses or rigid bodies; PDEs = continua" remains a valuable guide to basic mathematical character.

[5] Often internal variables such as spin are tolerated in these ODEs, even though they lack clear counterparts within true classical tradition.

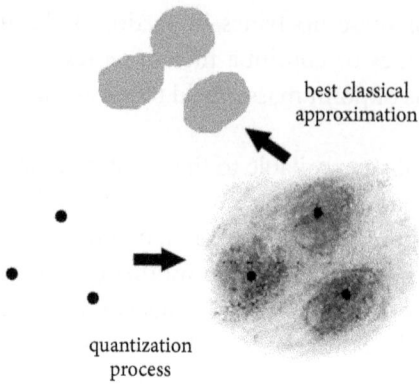

Figure 2.2

that nature behaves much like any classical point-mass system at a small size scale (figure 2.2).

Quite the contrary, constructing a classical system that can approximate the "effective volumes" of quantum clouds accurately at the size scale of so-called "molecular modeling" often requires classical blobs of extended size and flexibility. Most scientists working in the final epoch when classical mechanics could plausibly claim to govern the world in its entirety, namely the late nineteenth century, rejected the point-mass viewpoint as empirically inadequate for the bloblike characteristics of real-life atoms and molecules.

Nonetheless, there are convenient mathematical associations between the ODEs for classical point-mass models and the Schrödinger equation, so many contemporary physicists and philosophers of physics are familiar with the point-mass formalism alone. However, scholars hoping to extract methodological morals from the struggles over "matter," "atoms," and "force" that occurred toward the end of the nineteenth century will be misled if they study point-masses only, for it misses the conceptual complexities at the heart of the historical disputes. Viewed retrospectively, the degree to which the technical arcana of classical mechanics have impacted the development of scientifically attuned philosophy over the past several centuries is quite striking, even if this influence is not always recognized by modern readers. In this review, we shall sketch some of the chief ways in which the subtleties of classical mechanics have impacted philosophy.

There are two major arenas in which these effects have arisen. First, many of our greatest historical thinkers (Newton, Leibniz, Kant, Duhem, and others) directly struggled with the problems of classical matter, and their developed philosophies often prove intimately entangled with the specific foundational pathways they chose to follow.[6] Such portions of our philosophical heritage are often

---

[6] The abstract ruminations of *The Critique of Pure Reason*, for example, appear to have derived in part from the nitty-gritty worries about flexible matter that we shall review later. We look forward to Michael Friedman's big book on these issues.

misunderstood nowadays simply because the true contours of the physical problems our forebears faced have been forgotten. Second, as a result of these struggles, the great philosopher-scientists formulated a wide range of philosophical attitudes including anti-realism and instrumentalism as a response to the technical oddities they confronted. The twentieth-century logical empiricists who came later—after the chief focus of academic physics had shifted to quantum theory and relativity—were influenced by those older philosophical conclusions without adequate appreciation of the concrete issues that prompted them. Unfortunately, many philosophers have continued to hew to these old presumptions as if they represented firm verities, illustrating Darwin's celebrated aperccu: "False facts are highly injurious to the progress of science, for they often endure long; but false views, if supported by some evidence, do little harm, for everyone takes a salutary pleasure in proving their falseness."[7] A large folklore of "false facts" concerning classical mechanics continues to bend contemporary philosophy along unprofitable contours even today.

It is not the chief intent of this essay to pursue these satellite philosophical concerns with any vigilance, but to instead concentrate upon the key tensions that render classical doctrine hard to capture in the first place. Nonetheless, I hope that our prolegomena on larger themes suggests that significant points of general philosophical edification still lodge within the cracks of mechanics' hoary edifice.

2. Axiomatic Presentation

It will serve as a convenient benchmark for our investigations to recall that David Hilbert placed the rigorization of mechanics on his celebrated 1899 list of problems that mathematicians should address in the century to come (it is his sixth problem). He wrote, "The investigations on the foundations of geometry suggest the problem: To treat in the same manner, by means of axioms, those physical sciences in which mathematics plays an important part; in the first rank are the theory of probabilities and mechanics."[8] Indeed, Hilbert's own work in geometry and elsewhere comprised a chief inspiration for the logical empiricist program. Following this lead, we will serially examine the prospects for meeting Hilbert's challenge based upon the three foundational choices identified in section 1: point masses, rigid bodies, and continua.

Since this essay will conclude that Hilbert's objectives cannot be completely satisfied with respect to classical mechanics in the manner anticipated, let me first distance this evaluation from a popular viewpoint with which it might be otherwise

---

[7] Charles Darwin, *The Descent of Man and Selection in Relation to Sex*, Part II (New York: American Dome, 1902), 780.
[8] David Hilbert, "Mathematical Problems," in *Mathematical Developments Arising from Hilbert Problems*, ed. Felix Browder (Providence, RI: American Mathematical Society, 1976), 14.

confused. Many recent philosophers have responded to the axiomatic expectations of the logical empiricist school by concluding that science cannot be usefully studied in a formal manner at all. "Real life physics represents an ongoing practice," they claim, "and any attempt to capture its free-spirited antics within the rigid net of mathematical formalization represents an intrinsic distortion." But this is not what I shall claim, for I reject such a point of view entirely. Writing idly of "practices" in the loose manner of such authors offers little prospect for either appreciating or correctly identifying the concrete conceptual difficulties to be documented in this essay. Indeed, it was precisely through careful formal studies in Hilbert's manner that twentieth-century practitioners eventually reached a much sharper understanding of the fundamental requirements of continuum mechanics than was available in 1899. Indeed, Hilbert's own lectures in 1905 and the pioneering efforts of his student, Georg Hamel, comprised early landmarks along this long and tortuous development.[9] The only anti-Hilbertian moral we will extract from our examination is that a descriptive regime can often address large-scale objects more successfully if its underpinnings are structured in an overall "theory facade" manner somewhat at odds with standard axiomatic expectations. In every other way, I completely endorse the motivating intent of Hilbert's sixth problem.

We cannot appreciate the old puzzles of classical matter in their historical dimensions unless we keep the mathematical difficulties of continua firmly in mind. Scientists planning bridges or studying the musical qualities of violins in early eras did not have the luxury of waiting until the twentieth century to gather the tools they properly require. They simply had to cobble by with the mathematics they had on hand, even at the price of rather dodgy justifications. For example, due to the lack of clearly articulated PDE equations, Leibniz and his school could not deal directly with the three-dimensional complexities of a shaking beam straight on; they were forced to dissect the problem as illustrated into a connected sequence of one-dimensional tasks locally governed by ODEs (figure 2.3).

Newton followed a similar procedure in investigating how rotation affects the earth's shape: he began his treatment with a one-dimensional "canal" through the planet's interior.[10] Even today, most textbook problems adopt similar reductive stratagems: witness the standard treatment of the vibrating string.

Studying physics within these reduced, lower-dimensional settings can be very misleading from a "foundational" point of view (encouraging one to, e.g., think of stress as simply a kind of force). However, it is unlikely that classical physics could have staggered its way to an adequate treatment of continua without relying upon a broad array of results for systems that, from a foundational point of view, cannot represent their proper conceptual ingredients.

Finally, to appreciate the historical debates over classical physics in a proper context, we must disentangle the term "foundations" from certain absolutist demands

---

[9] Georg Hamel, *Theoretische Mechanik: Eine einheitliche Einführung in die gesamte Mechanik* (Berlin: Springer Verlag, 1949).
[10] Isaac Newton, *Principia*, vol. 1 (Berkeley: University of California Press, 1966), 349.

WHAT IS "CLASSICAL MECHANICS" ANYWAY? 49

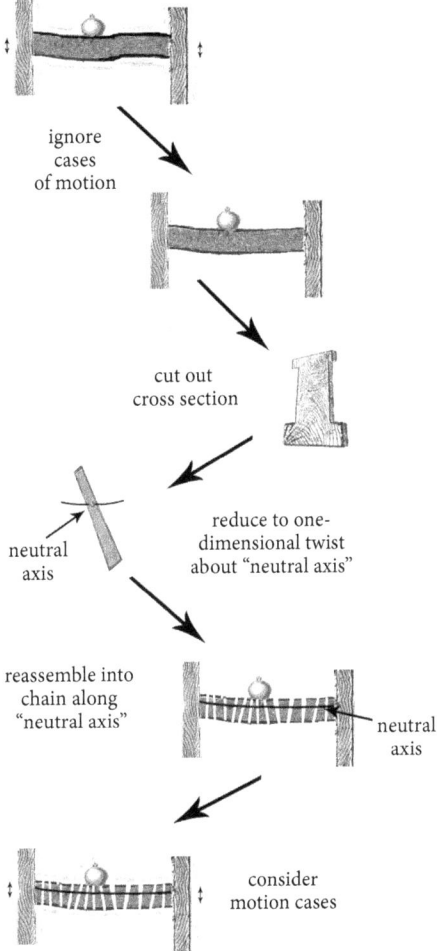

Figure 2.3

that contemporary philosophers are inclined to make. If we mark out clear axiomatic "foundations" for point masses, say, have we thereby selected an absolute *bottom layer of entities* from which any other object or system considered within a classical frame should be constructed? Many contemporary philosophers almost instinctively answer "yes," but the more prevalent historical assumption would have rejected "ultimate foundations" for classical mechanics in that vein. Indeed, calls for axiomatization per se need not inherently favor any unique choice of "ideology and ontology" in an absolutist manner, for one may instead believe that different selections of base entities and primitive terms may prove better suited for different agendas. Indeed, nineteenth-century mathematicians influenced by Julius Plücker maintained that traditional Euclidean geometry lacks any privileged basic ontology—there is no special reason to regard *points* as the subject's primitive objects rather than lines or circles. Indeed, a chief objective of traditional "foundational" work within geometry was interested in learning how the subject appears when it is dissected into alternative choices of elementary forms (points, lines, circles, etc.), under the assumption that each dissection into "primitives" offers fresh insights into the structural relationships that interlace the subject. Hilbert may have approached his sixth-problem axiomatization project with similarly tolerant expectations.

Most of the great scientists of Hilbert's time tacitly recognized that descriptive success in reliable modeling invariably relies upon some tacit *choice of scale length*. Matter generally reveals a hierarchy of qualities depending on how closely one inspects its structural details (it is traditional to designate this depth of focus by a "characteristic scale length" ΔL). For example, on an observational scale ΔL^O, well-made steel obeys simple isotropic rules for stretch and compression under normal loads (figure 2.4).

But closer inspection reveals that this macroscopic uniformity and toughness represents the resultant of a carefully engineered randomness at the level of the crystalline grain ΔL^G making up the material (such a scale length is sometimes dubbed the "mesoscopic level"). Considered at this lowered ΔL^G length, each component granule will stretch and compress in a more complicated manner than the bulk steel, but their randomized orientations supply the larger body with its simple behavior

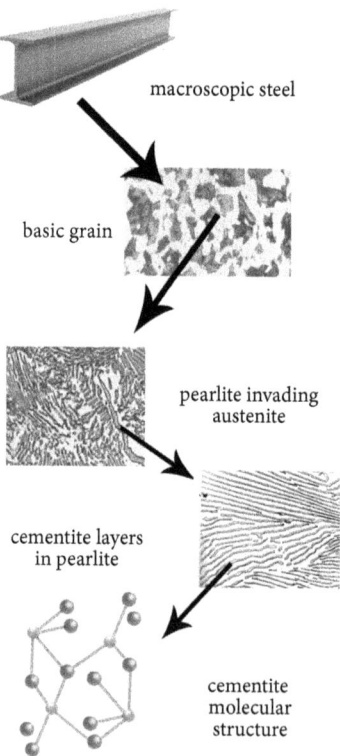

Figure 2.4

at the macroscopic level (so-called "homogenization theory" concerns itself with the details of how this ΔL^G scale to ΔL^O scale process operates). Lowering our focus to the molecular lattice ΔL^L composing the grain, we find that its capacity to *transmit dislocations* supplies the true underpinnings of the admirable toughness witnessed in the bulk steel at the much longer characteristic length ΔL^O. If we attempt to capture these various scale-dependent behaviors individually utilizing classical modeling techniques alone (as we can, to a remarkable degree of success), we will generally find ourselves selecting different ontological base units according to the implicit scale length we have selected. In such a mode, civil engineers usually model a steel beam upon a ΔL^O scale as a single *flexible body* of considerable homogeneity, whereas technicians interested in steel manufacture typically concern themselves with the thermodynamics of structural formation at the ΔL^G level. As such, the latter often adopt an ontology of *rigid crystalline forms* bound together into a complex material matrix. Initial efforts in modeling materials at the ΔL^L scale often employ *point-mass atoms* bound together in an irregular grid. But a more refined approach to these same lattice "atoms" will instead assign them flexible shapes—at the cost of considerable computational complexity. And so the modeling shifts proceed, each alteration in characteristic scale length commonly favoring a different "ontology" in its modeling material.

Here is a useful way to think about the relationships between scale sizes. In presuming that the point masses within a rigid part retain their comparative distances, we are actually pursuing a rough-hewn stratagem for profitable variable reduction, in the sense that we are attempting to evade consideration of the huge class of descriptive parameters needed to fully fix the position and velocity of every point mass within its surrounding rigid-body cloud. By treating the cloud as a united whole, we can track its *dominant behaviors* with a simple choice of six descriptive parameters (three to locate its center of mass; three to mark its angles of rotation around that center). But in tracking these values, we are only attending to the dominant behavior of the cloud because any normal collection of point masses will need to jiggle in very complex ways as they move forward. So our six rigid-body coordinates count as an effective set of *reduced variables* for our complicated point-mass swarm. Modern mathematicians like to picture such reductions as consisting of the trajectories etched upon a smallish "reduced manifold" sitting inside some much larger dynamic space. Our point-mass swarm (which is symbolized within a standard high dimensional "phase space" as the movements of a single dot) will wander throughout the larger space in an exceedingly complicated way, but it may fly fairly close (for certain portions of its journey at least) to a smaller "reduced variable" manifold, as illustrated (figure 2.5).

If so, we can gauge its complex movements with reasonable accuracy by simply tracking its shadow upon the surface of the reduced manifold. Such reduced-variable techniques have been long employed within celestial mechanics and it remains the hope of modern modelers in, for example, hydrodynamics that some allied set of reduced quantities might be found to simplify the refractive complexities within those topics.

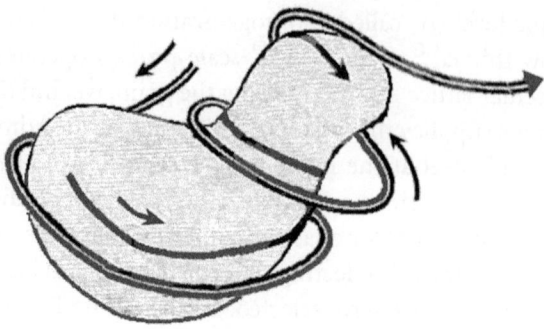

Figure 2.5

Speculative philosophers such as Leibniz opined that this alteration of ontological units would continue forever as one descends to smaller scales. More cautious observers have merely observed that experiment had not established any clear choice of lowest scale unit for classical mechanics. In this regard, it should be recalled that the evidence for fundamental particles only became overwhelming at the very end of the classical period, in the guise of Rutherford's experiments on radioactive scattering and the like. Once quantum mechanics enters our descriptive arena, its percepts increasingly dominate at smaller scale lengths and we eventually fall beyond the resources of classical modeling tools altogether.

Unfortunately, the various crossover points at which classical treatments lose their accuracy do not favor any uniform choice of fundamental classical entity. Sometimes point-mass treatments supply the most convenient form of lowest-scale classical modeling, but more often continua or rigid bodies provide better modeling accuracy. So while quantum mechanics may select certain entities as physically "bottom level," it does not follow that classical mechanics will do the same when considered upon its own merits. Accordingly, Hilbert's sixth-problem formalization project should not be saddled with the burden of satisfying a contemporary philosopher's expectations with respect to bottom-level ontology. What we will want to investigate carefully, as part of our "foundationalist" enterprise, is the degree to which principles applicable on a higher scale level ΔL^* relate to those applicable at the lower length ΔL. I call such transfers of doctrine across size scales *lifts*, and I employ "lift" in the elevator sense: one can go both up and down in a hoist.

Hilbert's own articulation stresses the importance of understanding these lifts more centrally than the simpler task of formalizing our three starting perspectives. He wrote:

> Boltzmann's work on the principles of mechanics suggests the problem of developing mathematically the limiting processes, there merely indicated, that lead from the atomistic view to the laws of motion of continua. Conversely, one might try to derive the laws of motion of rigid bodies by a limiting from a system of axioms depending upon the idea of continuously

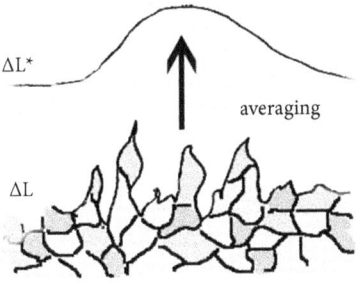

Figure 2.6

varying conditions of matter filling all space continuously, these conditions being defined by parameters. For the question of equivalence of different systems of axioms is always of great theoretical interest.[11]

Here Hilbert calls our attention to the various relationships between scale length that have been intensely studied in recent times under the general headings of "homogenization" and "degeneration."[12] He observes that the vague invocation of "limits" rarely provides an adequately precise diagnosis of the relationships involved, an observation that modern investigations heartily underscore. Observe that Hilbert's final sentence suggests that he did not anticipate that any of his suggested starting points would prove *fundamental* in the bottom-layer sense just canvassed. According to the applicational task at hand, different modes of ontological dissection (e.g., flexible continua or Boltzmannian swarms of rigid bodies) may possess their descriptive utilities in the same manner in which alternative decompositions of geometry into "primitive elements" prove fruitful. Even so, Hilbert insists that we must guard against erroneously lifting physical doctrines from one decompositional program to another without adequate precaution (figure 2.6).

In standard textbook practice, these lifts usually appear as dubious "derivations" of, for example, rules of continua considered at a ΔL^* scale level on the basis of rigid body swarms at a ΔL scale. As we will later see in detail, such improper doctrinal transfers are common in practice and sometimes serve as the source of substantial conceptual confusion.[13]

---

[11] Hilbert, "Mathematical Problems," 15.

[12] I do not have the space to survey such modern studies here, which attempt to, for example, recover the tenets of rigid body mechanics from continuum principles by allowing certain material parameters to become infinitely stiff (thus "degeneration"). Generally the results are quite complex, with corrective modeling factors emerging in the manner of Prantdl's boundary layer equations. Sometimes efforts are made to weld our different foundational approaches into unity through employing tools like Stieltjes-Lesbeque integration. More generally, a "homogenization" recipe *smears out* the detailed processes occurring across a wide region ΔW in an "averaging" kind of way, whereas "degeneration" instead concentrates the processes within ΔW onto a spatially singular support like a surface (the Riemann-Hugoniot approach to shock waves provides a classic exemplar).

[13] After a sufficient range of mechanical considerations has been surveyed in later sections, we shall be able to sketch a more favorable view of the useful offices that standard textbook lifts provide. I should also add that we shall generally consider our "ΔL to ΔL^* lifts" in two simultaneous modes: (1) as a modeling shift

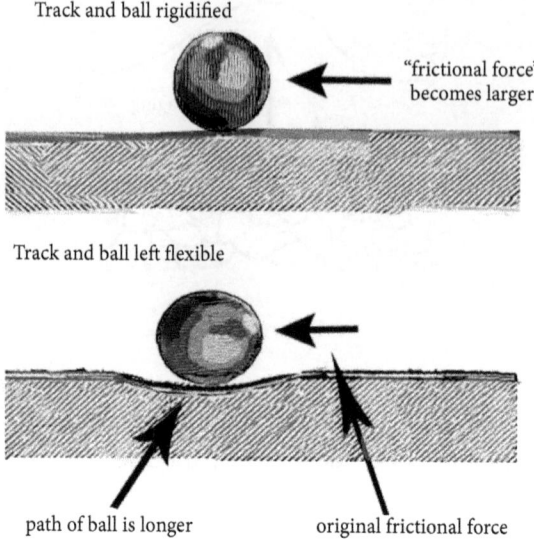

Figure 2.7

Consider a simple example of the problems that can arise in such shifts from ΔL to ΔL^*. The term *force* has a notorious tendency to alter its exact significance as characteristic scale lengths are adjusted. At a macroscopic level, the "rolling friction" that slows a ball upon a rigid track is a simple Newton-style force opposing the onward motion. But at a lower scale length, the seemingly "rigid" tracks are not so firm after all: they elongate under the weight of the sphere to a nontrivial degree. So part of the work required to move our ball against friction consists in the fact that it must *travel further* than is apparent. But when we consider the "forces" on our ball at a macrolevel, we instinctively treat the track length as fixed and allocate the effects of its actual elongation to a portion of the "force of rolling friction" budget (figure 2.7). A similar phenomenon occurs with the "viscosity" of a fluid.

When such adjustments in reference occur, one cannot legitimately lift a doctrine about "forces" applicable on scale level ΔL to scale level ΔL^*, for "force" does not mean quite the same thing in the two applications. Of course, if these innocent drifts were the only kinds of problematic lift to which mechanical practice was liable, serious conceptual debates would not have arisen in the subject. But these humble illustrations supply a preliminary sense of the problems we must watch for.

The properties we ascribe to a system with respect to an upper-scale length ΔL^* ("rolling on a rigid track") usually represent *averages* (or some allied form of homogenization or degeneration) over the more elaborate behaviors we will witness

from one *finite scale length* to another (e.g., from ΔL^G to ΔL^O in our steel bar example) and (2) as a mathematical shift from a *lower dimensional object* (a point mass or line) to a higher dimensional gizmo such as a three-dimensional blob. Properly speaking, these represent distinct projects, although, in historical and applicational practice, they blur together.

at a finer scale of resolution ΔL ("stretching the molecular lattice"). Obtaining a workable scheme of physical description tailored to ΔL* usually requires that a fair amount of fine detail gets *frozen over* in our modelings. In other words, we generally hope to capture only the *dominant behaviors* of our real-life system within in our ΔL* treatment and anticipate that we will sometimes need to open up the suppressed degrees of freedom whenever the complexities of the lower scale begin to intrude upon the patterns normally predominant at the coarser scale ΔL*.

Generically, the use of a smaller set of quantities to capture system behaviors dominant upon a higher scale length ΔL* is called a *reduced variable treatment*. There are a large number of ways in which these reduced-variable models can arise. For example, a reasonable policy of *homogenization* might adjust its descriptive terms from those suited to a ΔL^G assembly of iron grains to a smoothed-over steel bar described as continuous at the ΔL^O level.[14] But a quite different exemplar of reduced-variable "freezing" can be witnessed in Newton's celebrated treatment of the planets. At the scale lengths appropriate to celestial mechanics, one can ignore the complexities attendant upon the earth's shape and size by modeling it as a simple point mass. Rather than smearing out the properties of the planets over wider regions (as occurs in homogenization), we instead concentrate their extended traits upon much smaller supports.

Such policies of compressing complex expanses into singularities (or other lower-dimensional structures such as one-dimensional strings) are sometimes called *degenerations* (a term I regard as preferable to the misleading phrase *idealization*). Plainly, when very detailed astrophysical calculations are wanted, one must open up those internal complexities and treat the earth as a continuum subject apt to distort under rotational effects. However, there are many forms of reduced-variable lift that involve a mixture of the two policies or other sorts of tactic altogether.

Some of the anti-atomism advocated by late nineteenth-century scientists such as Duhem and Mach traces not to some obtuse dismissal of lower scale structure per se, but to the widely shared assumption that, in any application, modelers must invariably engage in such "freezing to a scale level" procedures. Their primary disagreement with other mechanists of their era concerns the format that should be regarded as the optimal embodiment of "classical principle" within such a scale-sensitive setting. Specifically, Duhem and Mach maintained that "basic physics," as an organizational enterprise, should develop tools that will prove maximally useful *at any chosen scale length*. This requirement almost automatically favors a "thermomechanical" approach of the sort described in the discussion of flexible bodies in section 5. Their opponents, such as Ludwig Boltzmann, generally favored the simplest base ontology that could plausibly support the more complex forms of

---

[14] Strictly speaking, a lift to continuous variables from an ODE-style treatment involving a large number of discrete variables at the ΔL level should not be called a "reduced variable" treatment, as we actually *increased* the number of degrees of freedom under the lift (normally, a true "reduced variable" treatment will supply a ΔL* level manifold lying near to some submanifold contained within the ΔL phase space). However, the descriptive advantages of a lift to continuous variables often resembles those supplied within a true "reduced variable" treatment, so in the sequel I will often consider both forms of lift under a common heading.

mechanics in a ΔL to ΔL^* manner (they often employed point masses or connected rigid bodies as their base level ingredients). In these respects, we might observe that Duhem and Mach's strictures better suit the methodological percepts of empiricists such as David Hume, who opined that any postulation of lower-scale structure must be based upon "laws" directly verifiable at the laboratory level.

Prima facie, we might reasonably expect that it should prove possible to formalize any of our three basic ontologies independently of one another, placing them on their own bottoms, as it were. Thus Hilbert probably anticipated that we should be able to frame distinct axiomatic encapsulations for point masses, rigid bodies and flexible bodies and then proceed to investigate how ably such formalisms relate to one another under ΔL to ΔL^* lifts. However, a somewhat surprising obstacle impedes such projects, whose various ramifications will comprise the bulk of this essay. They collectively trace to the simple consideration that if we attempt to frame general principles applicable to a higher ΔL^* scale length based upon behaviors operative on a lower scale length ΔL, we will find that our ΔL^* level principles generally *display gaps, holes, or gross inaccuracies* in special circumstances.

The general explanation for such upper-scale gaps is quite straightforward: a useful selection of "reduced variables" at the ΔL^* level will focus upon behaviors that *dominate* at that size scale. But, invariably, there will be special ΔL-level arrangements where the effects suppressed in our ΔL^* treatment obtain equal or greater importance than the usual dominant behaviors. I shall sometimes call such shifts "escape hatches," for they provide ladders that allow us to evade the inferential instructions of a formalism that no longer serves its empirical purposes. But such practices create a formal difficulty for axiomatization projects in Hilbert's vein because the domain of interest frequently becomes re-ontologized under the scale shift. But axiomatic presentations rarely include provisos for ontology shifts. Instead, we anticipate that their formal tenets will supply behavioral principles applicable to its ontology in all circumstances, even if, in real-life practice, we would normally escape such descriptive straitjackets in favor of some revised treatment operating at a lower length scale ΔL.

In short, conventional axiomatized theories are expected to supply principles that can govern even the bad spots within their ranges of empirical coverage. Such formal expectations lead many philosophers to further suppose that "classical mechanics" must completely specify the behaviors tolerated *within its own parochial range of possible worlds*, in spite of the fact that we would never apply such modelings to real-world dominions of a strongly quantum mechanical or relativistic character. But such dogmas presume that some fairly complete axiomatization of overall "classical mechanics" is available, a thesis we shall critically examine in this essay.

Let us now ask ourselves a commonsensical question. Considered from a practical point of view, is it really wise or meritorious to fill out a formalism in a manner that carries with it no discernible empirical merit? Mightn't it be better to deliberately leave our stocks of physical principle somewhat incomplete, allowing its very holes to signal when we should look for suitable ΔL^* to ΔL escape hatches? Indeed, explicit indications in the mathematics of when modeling problems begin should

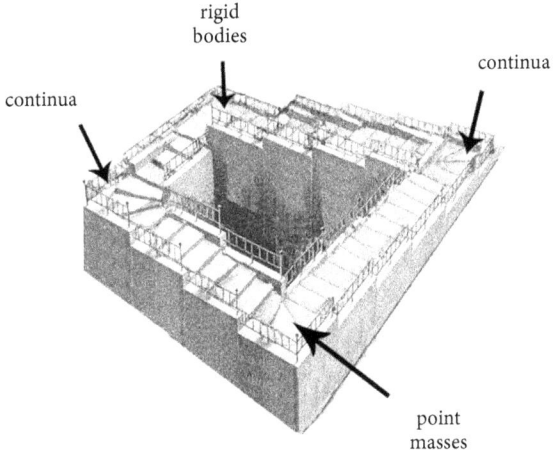

Figure 2.8

be greatly cultivated, for we surely want to avoid the fate of the computers who cheerfully compute worthless data simply because no one has told them to stop.[15] Training in mechanics generally inculcates considerable skill in knowing when one should adventitiously shift from one modeling framework to another. So it is sometimes unwise to push a formalism's axiomatized coverage beyond the limits of its real-life modeling effectiveness.

This point of view suggests that we might look upon the inherited compendium of descriptive lore we call "classical mechanics" as a series of descriptive patches (corresponding to our three basic choices of fundamental objects) linked together at their descriptive bad spots by various ΔL^* to ΔL escape hatches. However, whenever manifolds are constructed through sewing together local patches in this way, twisted topologies can potentially emerge in the final result (Klein bottles and Möbius strips provide classic illustrations of the phenomenon). In these respects, nature shows little favoritism as to which of our three basic ontologies of classical objects should be viewed as "fundamental" from an applicational point of view.

If we attempt to understand "classical physics" as a *conceptual system closed unto itself*, we thereby obtain a structure like one of those impossible Escher etchings: local plates connected by staircases that never stabilize upon a lowest landing (figure 2.8).

But such topographical oddities do not indicate that "classical physics" has not served its descriptive purposes perfectly well. As long as the salient escape routes are clearly marked, our Escherish edifice serves a base frame upon which a wide range of interconnected forms of reduced-variable modeling techniques can be conveniently located (I sometimes call structures of this sort *theory facades*). By operating with

---

[15] In many statistical problems, the population under review is artificially increased to an infinite size, simply so that the applicable mathematics will supply crisp answers to the questions we commonly ask. Left to its own devices, *mathematics is rather stupid* in a literal-minded kind of way and finds it very difficult to answer questions in a "well, almost all of the time" vein, which is often the best that can be achieved with respect to a finite population. But if the same community is modeled as infinite, we can often fool the mathematics into supplying us with the brisk replies we desire.

a proper regard for the requisite level shifts, we can thereby assemble the most fruitful terminology yet devised for dealing with the complex physical world about us at nonmicroscopic scale lengths: the shared language of "classical physics." The twisted topology within its connection manifold merely reflects the "exit from bad patches" considerations that allow the scheme to cover extremely wide swatches of application with great efficiency.

The historical triumph of "classical mechanics" as a descriptive enterprise would have never occurred had the subject not lightly skipped over the many problematic transitions of the sort we shall survey. Historically, the price of a vigorous conceptual enlargement is often a lingering residue of confusion that can occasionally blossom into full paradox when suitably nurtured. And such has been the career of classical mechanics: full of predictive glories but comingled with mystifying transitions that have led some of our greatest philosophical minds down the garden path to strange assessments of our descriptive position within nature. In the sequel, we consider three basic descriptive patches handed down to us in our classical legacy and examine the typical confusions that arise when one shifts from one framework to another without noticing.

3. Point-Mass Mechanics

Let us first consider the point-mass formalism of classical mechanics, suggested by Newton's familiar formulation of the fundamental laws of motion. To begin, it is worth noting that substantive foundational issues immediately arise if we scrutinize these laws with a critical eye. In their original form, these principles are hard to interpret with any exactitude due to the ambiguous manner in which Newton employs his terms. Here they are in Motte's translation:

> Law I: Every body persists in its state of being at rest or of moving uniformly straight forward, except insofar as it is compelled to change its state by force impressed.
> Law II: The alteration of motion is ever proportional to the motive force impressed; and is made in the direction of the right line in which that force is impressed.
> Law III: To every action there is always opposed an equal reaction: or the mutual actions of two bodies upon each other are always equal, and directed to contrary parts.[16]

Look carefully at Law I. If a "body" represents an isolated point mass, then the phrase "moves uniformly straight forward" is not ambiguous. But what is the parallel intent if a *rotating rigid object* can be selected as a body? Or a packet within a compressible

---

[16] Isaac Newton, *op. cit.* 13–14.

fluid? One cannot demand that every point within a freely moving boomerang must translate rectilinearly—at best, they can rotate around some *representative center* within the full projectile (such as its center of mass). Allied interpretational problems affect Newton's remaining laws as does the question of precisely where the "impressed forces" are supposed to act. Indeed, the three laws can be readily interpreted only if "body" is read as "isolated point mass" throughout. However, this was neither Newton's intent nor that of the many subsequent writers who have cited the three laws with approval, as illustrated by Peter Tait and Lord Kelvin in their celebrated *Treatise on Natural Philosophy*:

> We cannot do better, at all events in commencing, than follow Newton somewhat closely. Indeed, the introduction to the *Principia* contains in a most lucid form the general foundations of Dynamics. The definitiones and *Axiomata sive Leges Motus*, there laid down, require only a few amplifications and additional illustrations, suggested by subsequent developments, to suit them to the present state of science, and to make a much better introduction to dynamics than we find in even some of the best modern treatises.[17]

But such claims are misleading. Why have Newton's laws been allowed to stand so long in such a confusing form? Newton himself proved somewhat wobbly with respect to precise content of his own first law, in that he offers as an illustration the fact that a rotating hoop will continue in its angular movements if not acted upon by "outside forces": "A spinning hoop, which has parts that by their cohesion continually draw one another back from rectilinear motions, does not cease to rotate, except insofar as it is retarded by the air."[18] Plainly a tacit appeal to some *generalized* inertia principle is implicated: the activities of the wholly "internal" forces within a rigid body should not affect its overall rotation. Newton, of course, knew that this same claim will not hold for a flexible object such as the earth or a falling cat. The "rigidity" of the hoop somehow underpins a lift that converts an inertial principle relevant to isolated point masses into a requirement upon composite objects operating at the scale size of a hoop. But shouldn't Newton have properly attended to the constitutive modeling assumptions that render the internal constitution of a rigid ring different from a cat or a flexible earth? Yes—as we have already observed, such forms of ΔL to ΔL^* scale lift pose the same kinds of justificatory problems as arise when we shift from a ΔL-level swarm of interacting molecules to their "averaged" statistical mechanics at level ΔL^*.

Modern scholarship generally credits the standard modern reading of Newton's second law to Euler, who introduces the expectation that "$F = ma$" supplies the central framework upon which suitable sets of ODE modeling equations for

---

[17] William Thomson (Lord Kelvin) and P. G. Tait, *Treatise on Natural Philosophy*, Vol. 1, (retitled as *Principles of Mechanics and Dynamics*) (New York: Dover, 1962), 219.

[18] Newton, *Principia Mathematica*, 416.

point-mass modeling can be assembled. This recipe of Euler's unfolds as follows: Choose a target system S to model in a point-mass mode. Count the number of masses one needs in S. For each $i \in S$, write down the following framework for constructing a well-posed set of (vectorial) ordinary differential equations:

$$m_i d^2 q_i / dt^2 = \sum f_i(j, t).$$

In this formula, m_i is the mass of the particle numbered as i, $q_i(t)$ is its vector location at time t and the various $f_i(j, t)$ will supply the strengths of specific forces applicable to particle i that have their origins in a particle j (for $i \neq j$). But this merely lays down the *basic scaffolding* we will need for a properly completed equational set. In particular, we must yet specify the sundry $f_i(j,t)$ in concrete ways that can lead to a set of equations that are uniquely solvable (at least most of the time) with respect to an arbitrary set of initial conditions (supplementary "data" that provides $q_i(t_0)$ and $dq_i(t)/dt|_{t_0}$ information for each particle $i \in S$ and some particular time t_0). To achieve such formal closure within a completed recipe, each $f_i(j,t)$ must represent a *special force law* that links the strength of the $f_i(j,t)$ forces somehow to the q-locations of particles i and j (as we will see, Newton's third law puts sharp restrictions on the nature of this dependency). The basic prototype for a "special force law" of this ilk is Newton's law of gravitation:

$$f_i^G(j, t) = (G m(i) m(j)) / |q_i(t) - q_j(t)|^2,$$

but allied principles are needed to govern all other applicable forces (such as those responsible for the cohesion and repulsion of matter). Each additional law is expected to carry in its wake its own range of *material constants*, such as the *charges* $c(i)$ that show up in the static form of Coulomb's law[19]:

$$f_i^C(j, t) = (K c(i) c(j)) / |q_i(t) - q_j(t)|^2$$

(where such $c(i)$ can sometimes carry negative values). We are said to have supplied a *constitutive modeling*[20] for the system S in point-mass terms once we have specified S's full complement of particles i and the values of the applicable "force law" constants $m(i)$, $c(i)$, and so on (the list of constants then tells us how many $f_i(j,t)$ terms are "turned on" within S).

One of the most frustrating aspects of the classical point-mass tradition is that it never fully resolved what these special "force laws" (besides gravitation) should be. Modern molecular modelers frequently utilize sundry mixtures of sixth and twelfth power principles (e.g., the familiar Lennard-Jones potential) between point masses to simulate the molecular interactions within a gas, but no one maintains

---

[19] In many circumstances, it is natural to borrow Coulomb's law from Maxwellian electrodynamics, but, strictly speaking, this rule only suits static circumstances. Accommodating dynamic circumstances within a "classical physics" frame, we must normally introduce a foreign element (the electromagnetic field) that carries us beyond the limits of our point-mass framework. Indeed, no one has yet figured out a wholly satisfactory way to amalgamate classical point masses with such a field.

[20] Sometimes this phrase is tacitly restricted by further requirements on the locations $q(i,t)$: it seems strange to say that we have supplied a "constitutive modeling" for a cuckoo clock if we are willing to consider that "modeling" in a condition where its component masses are scattered across the wide universe!

that such rough rules enjoy any canonical status within classical mechanics. This incompleteness traces to the fact that nature has indicated no special preference for *classical* principles governing, say, small-scale cohesion and repulsion because it has decided to let matter behave in a strongly quantum mechanical manner in such close quarters. So molecular modelers are left with a rather diffuse collection of principles that might possibly model close range interaction ably, with the final choice being decided by what appears to work. Indeed, textbooks frequently sidestep the need to fill in the special-force-law holes in an Euler's recipe modeling through *various forms of evasion*, such as the appeals to rigid-body constraints. Typically, such diversionary appeals tacitly shift us into ontological realms natural to continua and rigid bodies, which approach the problems of cohesion in an inherently different manner than currently contemplated.

The absence of enough special force laws in the point-mass setting engenders another familiar difficulty for point-mass mechanics that is usually "solved" by shifting the underlying ontological framework. Suppose we construct a constitutive modeling for the solar system, where we treat the sun and planets as point masses and the only special force we turn on is gravitation. The resulting Euler's-recipe equational set will be "formally well-posed" in the sense that we are supplied the right number of equations to potentially possess unique solutions given initial conditions.

But that is merely a "formal" guarantee in the sense that it tells us that we are somewhere in the ballpark of getting the solutions we wanted. It does not completely assure us that the solutions really exist. And we have good grounds for worrying about this. Rocket designers appreciate the fact that one can supply a projectile with a significant increase in kinetic energy by slingshoting it through the strong gravitational field of a planet (the technique was used several times to generate enough boost to propel the Cassini space probe to Saturn) (figure 2.9).

Since point masses have no size, and because we have not included any sort of repulsive force in our equational set, can a particle possibly extract enough energy

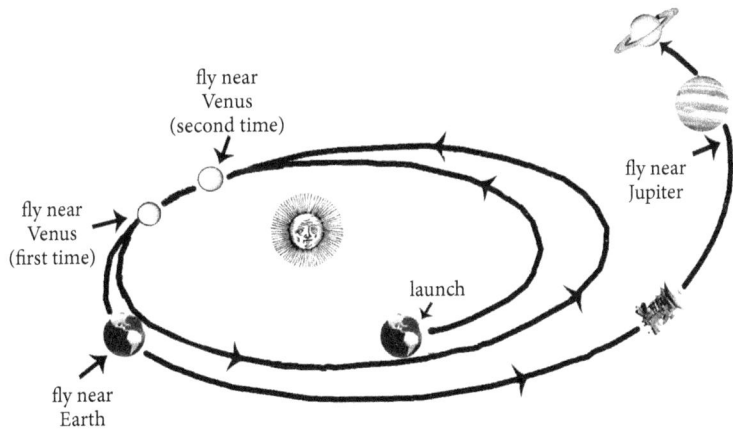

Figure 2.9

from its planetary near approaches to produce an *infinite velocity boost* within a finite span of time? This was a famous mathematical question that was settled in the affirmative by Zhihong Xia[21] in the 1980s. As such, the velocity blowup indicates that the viability of our point-mass modeling has self-destructed of its own accord. Rather than searching for repulsive special force laws that might inhibit the effect, physicists typically brush the problem aside, "Oh, you've just neglected the finite size of real planets." That is true, but they have thereby escaped to the dominions of rigid body or continuum mechanics in making such appeals.

The full battery of special force laws that point-mass mechanics requires is further skirted by the common practice of presuming that, however the true missing laws precisely operate, their net effects can be linearized or otherwise simply approximated as long as their activities are not strong (the telltale symptom of this ploy are terms in equations such as a linear "$Wq(x,t)$" whose special-force-law origins are left hazy). Many of the descriptive successes of nineteenth-century physics were prosecuted under the guidance of such approximatizing assumptions. It is to be expected that such modelings will frequently self-generate holes in their descriptive coverage due to the fact that their solution sets can evolve into situations where the presumptive *ansatz* that "the activities of the unspecified force law can be approximated by $Wq(x,t)$" must plainly fail. It is striking that if one inspects the stock point-mass modelings provided in popular textbooks, very few of them completely satisfy the provisos of Euler's modeling recipe and instead invoke some tactic for special-force-law avoidance. Such *methodologies of avoidance* can be prudent in practice, through preventing the merits of a modeling from being held hostage to the delicate specifics of an unproven special force law. Absent any definitive resolution of what its full complement of special force laws $f_i(j,t)$ should comprise, our Eulerian recipe employs "$F = ma$" as a skeletal frame upon which a formally closed differential equation modeling might be eventually assembled as soon as adequate skin and clothes can be found for the task. Certainly, one cannot coherently discuss issues such as whether classical mechanics is deterministic until these issues of special force laws have gotten fleshed out in some fuller manner.[22]

Special force laws represent the natural point-mass analog to the constitutive laws of modern continuum mechanics. In both cases, we must learn to watch out for physics avoidance stratagems that bypass some of the expected ingredients in the relevant recipe. As we have noted, such constitutive-modeling evasions frequently take the form of mixed scale-level lifts in which the descriptive vocabulary natural to a higher scale ΔL^* becomes invoked in a manner that allows the modelers to evade the nontrivial constitutive modeling concerns they would otherwise need to confront had they remained resolutely at the original modeling scale ΔL.

---

[21] Z. Xia, "The Existence of Non-collision Singularities in Newtonian Systems," *Annals of Mathematics* 135 (1992), 411–468.

[22] For a more detailed discussion of these issues, see Mark Wilson, "Determinism: The Mystery of the Missing Physics," *British Journal for the Philosophy of Science* (2009), 173–193. I might add that the common technique of *dropping dimensions* (e.g., confining point masses to a plane with no specification of the forces that keep them there) should be considered as a further variety of "Euler's recipe avoiding" policy (such moves should be scrutinized with a close methodological eye whenever they are invoked).

There are several widely discussed aspects of Newton's laws of motion that merit quick remark. Regarding his second law, implicit within our Eulerian recipe is the assumption that sound modeling equations can be set up for every system S based upon their Cartesian locations $q_i(t)$ within "absolute space" or, more minimally, with respect to some choice of inertial frame. Newton himself, insofar as I can tell, never quite made such a claim, for he often set up his equations using what are often called "natural coordinates"—quantities that possess a palpable physical significance within the target system itself. For instance, in the case of a bead sliding along a curved wire, the arc length along the wire qualifies as a natural coordinate, whereas the bead's location within an externally defined frame does not. Within celestial mechanics, this distinction enjoys little purchase, but the issue becomes pertinent when material constraints such as "moving along a rigid wire" come into play.

Regarding Newton's third law, its original formulation (as the so-called principle of "action = reaction") seems patently hazy and has been historically subject to substantially divergent interpretations. In a modern point-mass reading, it is usually regarded as placing various strong restrictions upon the special force laws we are allowed to employ in Euler's recipe:

> (3a) All forces arise between pairs of particles and have their source in one of the pair.
> (3b) These forces are directed along the line between the masses (the forces are "central") and opposite in magnitude ("balanced").
> (3c) The strength of these forces depends only upon the spatial separation between the bodies and not, say, upon their relative velocities.

In other words, if a special force law claims that mass j exerts a specific force $f_i(j,t)$ upon mass i, then j must exert a reciprocating force $f_j(i,t)$ upon i equal in magnitude to $f_i(j,t)$ but reversed in direction (observe that only action-at-a-distance forces are relevant within a point-mass setting, so $f_i(j,t)$ acts at i's position, whereas $f_j(i,t)$ acts at j's position). Although Newton's own law of gravitation suits these requirements, it is unclear that he would have accepted the (3a-c) supplements in the strength stated. Requirement (3c), for example, stands in apparent conflict with most varieties of frictional force because their strength generally depends upon the rate $(dq(x)/dt)$ whereby bodies slip past one another. Requirement (3b) seems to rule out sheering forces, such as arise when one layer of water slips over another, or the sideways force that a charged particle feels near a magnetic pole (note, however, that some of these situations only pertain to extended objects in contact along an interface and may not directly concern us now). One of the chief reasons for making such strong restrictions on forces is that they are required to establish vital tenets like Galilean relativity, balance of angular momentum, and the conservation of energy within a point-mass frame (Newton did not maintain energy conservation himself).

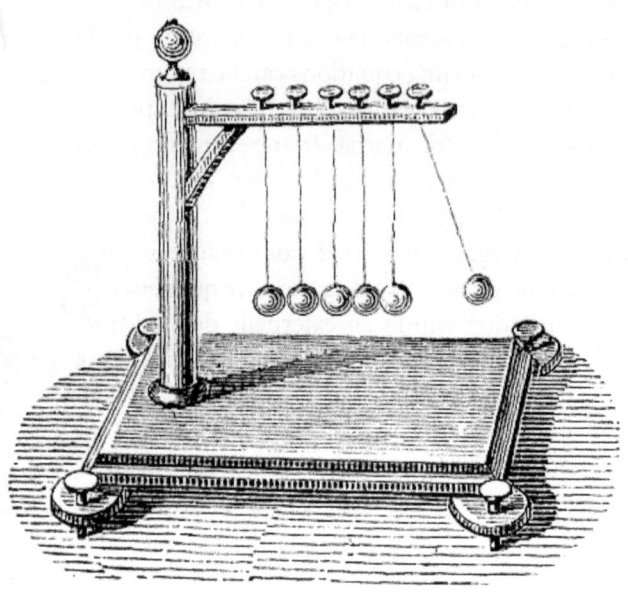

Figure 2.10

This is because the underlying notion of potential energy requires some restriction akin to (3c).

Partially due to its vaguely expressed contours, Newton's third law often serves as a significant site of substantial lifts within mechanics. Let's look at a typical example in the context of a familiar scientific toy: a line of steel ball pendulums lying adjacent to one another (figure 2.10).

If the ensemble is struck by a falling ball b_R to the right, it will come to rest and the ball b_L at the left end will fly off. But as soon as gravity pulls b_L into collision with the group, b_L will halt and b_R will fly off again, to return, more or less, to its original state. And back and forth the knocking oscillations will go, until friction eventually brings the ensemble to rest. And it is natural to conceptualize this situation in this manner. The originally falling b_R externally exerts an impactive force upon its first member of the adjacent ensemble b_{R-1}, which then imparts a congruent internal force upon its nearest neighbor b_{R-2} and so on across the array until we reach b_L. Since b_L lacks any leftward neighbor upon which to exert a leftward force, it is forced to convert that potential into its own kinetic motion, which will be of the same magnitude as b_R originally possessed, under the presumption that the masses m_L and m_R are identical. Expressed in ersatz third-law-style jargon, we can say: ball b_R originally supplies an impressed *external* force upon its neighbors, which then excites a spectrum of internal forces in direct contact with another. Because of the third law, these internal forces will exactly cancel each other out in terms of any work they can perform on the ensemble, hence the central packet of balls will

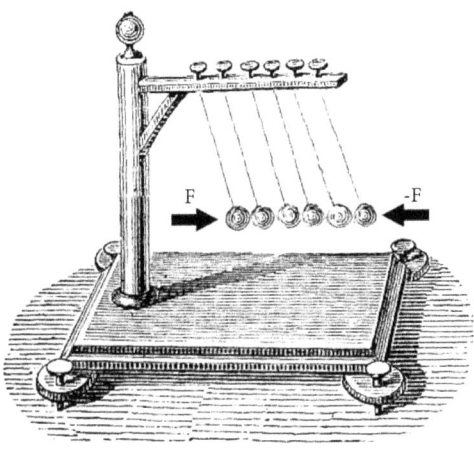

Figure 2.11

display no visible movement. But ball b_L lacks a balancing left-hand neighbor, so it is forced to convert the impressed force upon it into its own kinetic movement.

Often related reasoning is presented in a somewhat more elaborate guise. Rather than allowing b_R to fall against the group, let us simply push against the entire group at ball b_R with an applied force of magnitude F (figure 2.11).

What countervailing force should we apply to b_L to maintain the whole group in *equilibrium*? −F, obviously. Let us now conceptualize b_L's so-called "inertial reaction" $m_L^2 x / dt^2$ as a kind of "force" (until recent times, it was quite common to employ the term "force" in this wider manner). Returning to our original "b_R supplying an external force to the group" case, we can codify our prediction in the guise: b_L will develop an inertial reaction force exactly equal in magnitude and direction to −F. In this format, the reasoning of our previous paragraph can be extended: mechanical systems always maintain a kind of equilibrium, wherein certain members will counter any unbalanced forces upon them by forming the requisite inertial reactions. All of this reasoning is well and good in a certain sense, except that (1) its notion of "force balance" has nothing to do with Newton's third law *as we have interpreted it* and (2) reasoning of this type properly requires the realm of rigid bodies for its firm support and can only be regarded as a rough approximative lift within the strict context of point-mass mechanics.

To see what has gone wrong, let us replace our array of pendulum balls with a lattice line of legitimate point masses. To reconstruct a point-mass substitute for the pendulum-like behavior of the balls, we need (1) some special force law $F_{rep}(x_i, x_j)$ to generate repulsive force that point i will exert upon point j under close approach and (2) some outside source of attractive force $F_{att}(x_i)$ to hold each point mass i within a neighborhood of its lattice rest position. Now apply a force F to the lattice point p_R. What does our third law, *as heretofore interpreted*, demand? Only that $F_{rep}(x_i, x_j) = -F_{rep}(x_j, x_i)$ and that the unspecified sources of F and $F_{att}(x_i)$ should feel reciprocal forces upon themselves. There is absolutely no requirement that the *summed* forces upon our sundry lattice points **i** will "perform no work" upon them. In fact, this will generally be false: the initial blow will send waves of compression and expansion through the lattice, at each stage of which small amounts of work will be exerted on each **i**. It is only if F_{rep} and F_{att} forces of a very stiff character are posited that we will witness a lattice behavior similar to our pendulum ball expectations. In other circumstances; the blow at p_R might induce negligible transmissive effects at p_L (e.g., we might see a point-mass simulacrum for a line of pendulums composed of *putty*).

All of these specific requirements upon F_{rep} and F_{att} fall under our earlier heading of "constitutive modeling conditions." How did we manage to overlook such *constitutive concerns* in our original reasoning about our pendulums? The answer is that we inadvertently punned on the term "force balance," thereby lifting a local point-mass requirement on F_{rep} and F_{att} to the level of visible "balls." The shift is facilitated by the innocent-looking invocation of a distinction between "external" and "internal" forces, where it appears as if all of the "internal forces" in the pack possess their required "third law" correlates, while the leftward force on b_R lacks a match, a lapse that b_R can only rectify through its "inertial reaction."

There is a celebrated passage in Thomson and Tait that explicitly interprets Newton's third law in this "lifted" manner:

> [I]f we consider any one material point of a system, its reaction against acceleration must be equal and opposite to the resultant of the forces which that point experiences, whether by the actions of other parts of the system upon it, or by the influence of matter not belonging to the system. In other words, it must be in equilibrium with those forces. Hence, by the principle of superposition of forces in equilibrium, all the forces acting upon the system form, with the reactions against acceleration, an equilibrating set of forces upon the whole system. This is the celebrated principle first explicitly stated, and very usefully applied, by d'Alembert in 1742, and still known by his name.[23]

But if we do that, we abandon some of the original specifics that permit a ready pathway from Newton's three laws *as we interpreted them* to the conservation of energy and the like. What did Newton himself intend by his "third law"? His examples suggest drifts in his own thinking, sometimes straying close to those of Thomson and Tait.

As indicated earlier, most physicists had firmly abandoned the point-mass approach by 1850 or so, only to be revived in the twentieth century as offering the easiest pedagogical bridge to quantum theory. Why did this happen? A number of salient considerations can be extracted from the wonderful articles that James Clerk Maxwell composed for the celebrated ninth edition of the *Encyclopedia*

---

[23] Thomson and Tait, *Treatise on Natural Philosophy*, 1: 248.

Britannica.[24] Many of his concerns trace to the simple fact that natural materials vibrate in the manner that spectroscopy indicates and can transmit waves. But attempts to construct point-mass lattices capable of imitating the experimentally determined behaviors usually proved disappointing, whereas models constructed upon continuum or rigid body principles did much better. For example, in the 1820s Claude-Louis Navier had developed a celebrated point-mass model for elastic materials leading to substances whose macroscopic behaviors are characterized entirely by their Young's modulus. Working from general principles in a top-down, continuum mechanics mode, Cauchy instead concluded that isotropic elastic materials require *two* independent constants (Poisson's ratio in addition to Young's modulus) to fix their behaviors rather than Navier's solitary value. These issues were of great scientific moment because the varieties of wave that can travel through an elastic material are intimately linked to these constants. After a long period of controversy, Cauchy's "multi-constant" predictions were eventually confirmed by experiment. By the end of the century, it was widely presumed that nature was composed of continua of some sort, with its apparent point-like "particles" comprising whirlpool-like structures within an underlying continuous medium.[25]

Cauchy did not fully appreciate the methodological advantages of the approach he initiated (he sometimes worked in Navier's bottom-up mode as well), but later writers such as Green and Stokes strongly emphasized the merits of the top-down approach, which eventually became the core construction within modern continuum mechanics (in a manner we shall survey in section 5). To this day, their top-down techniques generally supply more reliable models with respect to the materials of macroscopic experience. In fact, many of the celebrated philosophical percepts developed by writers such as Pierre Duhem and Ernst Mach in the late nineteenth century trace, in part, to their appreciation of the descriptive superiority of the top-down methods. More recently, the rise of swift computers has rendered the project of working directly with point-mass swarms in a bottom-up manner a more viable enterprise, but the results obtained are generally more suggestive than accurate. Shortly after Cauchy's work, Poisson was able to reproduce the "two constants" predictions from a molecular model composed of attracting spheroids rather than point masses. Likewise, one obtains better results within molecular simulation today by working with swarms of extended bodies rather than points, although the computational costs are much higher. But, from a foundational point of view, these modeling adjustments transport us into the realms of *rigid body mechanics*, which we shall canvass in the next section.

Some folks, however, become so smitten with point masses that they strive mightily to found "classical mechanics" upon that basis, no matter how physically

---

[24] Cf. the entries "Constitution of Bodies," "Atom," and "Attraction" in J. C. Maxwell, *Collected Scientific Papers*, ed. Ivan Niven (New York: Dover, 1952). Maxwell also worried that point-mass swarms could not remain structurally stable when vigorously shaken or explain the fact that the world's wide variety of materials only displays a very limited palette of spectra. Reint de Boer, *Theory of Porous Media* (New York: Springer, 2000) provides a good capsule summary of these developments.

[25] J. S. Rowlinson, *Cohesion* (Cambridge: Cambridge University Press, 2002), 110–126 (this book is an excellent introduction to the fascinating conceptual problems that attach to "cohesion" generally).

implausible the constructions they employ may appear. Thus, we might theoretically piggyback upon Poisson's "two constant" results by collecting large swarms of point masses into mock spheroids held together by strange attractive forces. But such assemblies bear no relationship to any structures present in real-life materials (whereas Poisson's spheroids often do). I am not sure what one gains from vain reductive enterprises like this.

A logical observation is pertinent as well. When one strives to explain why modeling principles $\mathscr{P}^*$ work well at scale level ΔL^* based upon the principles $\mathscr{P}$ operative at ΔL, one is further obliged to explicate why the $\mathscr{P}^*$ principles operate *over the full range* that they do. It is often easy to construct specific "toy models" at a ΔL level that will implement the desired $\mathscr{P}^*$ behaviors at the ΔL^* scale, but one little skirmish does not win a war. At best, one has merely built what the Victorians called a $\mathscr{P}$ - principle *analogy* to the $\mathscr{P}^*$ events. To be sure, the construction ensures that some of $\mathscr{P}^*$'s ontological claims are technically compatible with $\mathscr{P}$, but this signifies comparatively little if the supportive "analogies" require such elaborate contrivances on a ΔL scale that they cannot serve as general underpinnings for the higher scale behaviors.

I would have presumed that this observation was so obvious that it is scarcely worth drawing, but several times in the past year I have heard philosophers proudly declare that they have "derived the Navier-Stokes equations" (or the like) upon a more elementary basis, when, in fact, they had merely concocted a weak and contrived analogy to such a system (by such standards, one can probably "found" the same equations upon *The Pickwick Papers*). There are many loose claims afloat within the philosophical world as to how the various branches of physics allegedly "reduce" to each other; readers should approach most of these with a wary eye.

There is a final issue we should survey before returning to our main themes. As I have explicated our Eulerian recipe, it fails as a modeling scheme as soon as quantities like acceleration lose their required features. But this is exactly what happens if, for example, a point mass runs into another point mass or into one of the hard-shell barriers discussed earlier. From a strict point-mass perspective, one should not tolerate acceleration-destroying interactions. But fulfilling this ambition in a plausible manner is not easy (and we must furthermore tame the additional blowup problems that emerge in the Xia construction mentioned above). In real-life modeling practice, "impactive" encounters between point masses are usually addressed through ad hoc remedies that temporarily relax our Euler's recipe requirements, rather than searching for elusive special force laws. In fact, Newton's own approach to billiard collisions implements this basic "turn off the laws temporarily" stratagem. He surrounds the center of each ball with a crisp finite boundary (so that the central mass point is credited with a "hard shell potential," although utilizing that vocabulary is quite anachronistic in application to Newton) (figure 2.12).

Whenever these radii contact one another (we shall only worry about the head-on collision case), Newton abandons the requirement that the "a" in "$F = ma$"

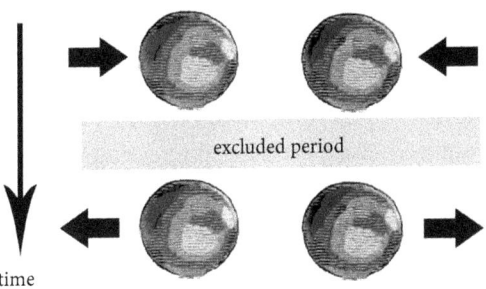

Figure 2.12

must make sense and shifts his focus to the two balls' incoming stores of linear momentum and kinetic energy (as we now dub them), together with a purely empirical factor called a *coefficient of restitution* (it governs how much the total kinetic energy budget will diminish post-collision). In effect, this treatment blocks out the crucial interval of time Δt where "F = ma" fails to make sense and glues together the incoming and outgoing events exterior to Δt through a mixture of conservation principles[26] (conservation of linear momentum) and raw empirics (coefficients of restitution extracted from experiment). Formally, tactics that patch over problematic intervals or regions in this manner are frequently called *matched asymptotics*.

It is now time to extract the central morals of our discussion from the underbrush of specifics. The main descriptive holes within the point-mass approach trace to the absence of the special force laws that would be needed to complete its Eulerian recipe for ODE model construction. It is hard to repair these lapses with any assurance because the missing laws concern the nature of close range cohesive forces and nature offers few robust indications as to how a classical point-mass modeler should tackle such phenomena. In consequence, our underlying recipe lacks many of the ingredients it would require before it could ratify, upon a purist point-mass basis, the many non-punctiform modeling techniques that practitioners regularly employ at higher ΔL^* scale lengths. In pedagogical practice, these lapses in constitutive modeling are frequently disguised by covert lifts to alternative approaches better suited to the ΔL^* level techniques. But conceptual complications within those rival schemes encourage frequent retreats back to the stolid redoubt of point masses, where the conceptual setting—if not the livin' itself—is easy. The result is an intellectual landscape pockmarked with easy lifts and quick escapes that can seem quite perplexing if your physics instructor assures you that everything you see is rigorously wrought and intellectually beyond reproach.

---

[26] If no kinetic energy is lost to heat (a so-called "purely elastic collision"), then we possess enough "conservation laws" (energy and linear momentum) to guide two colliding point masses uniquely through a collision (as every elementary college text demonstrates). But these principles alone are not adequate to three-way collisions, energetic losses, or to more oblique modes of scattering.

4. Rigid Body Mechanics

Let us now investigate the foundational prospects for a physics resting squarely upon a basic ontology composed of rigid bodies interacting through contact. There are a number of somewhat different treatments available in this arena, falling under the generic heading of "analytical mechanics." Our plan is no longer to analyze such bodies as swarms of point masses or to allow them any internal flexibility. Accordingly, when our ensembles of rigid parts flex, it must be through the internal realignment of completely stiff components, maintained as a coherent collection through an admixture of action-at-a-distance attractions and directly contacting linkages such as hinges, pins, wires, and so on.[27] The rigidity of any part is mathematically expressed by the fact that its current location can be completely fixed by six numbers: three Cartesian coordinates to locate a representative point within the body, and three angles to indicate how the figure has rotated about that point. Any connection between such parts is usually called a *constraint* and is expressed with constraint equations that interrelate the coordinates of the sundry parts.[28] Two useful paradigms that I shall often cite for the systems under review are (i) a bead sliding frictionlessly along a rigid wire and (ii) the sewing machine mechanism illustrated (figure 2.13).

One does not expect a normal lattice of point masses to remain completely rigid when disturbed—the gentlest attempt to move them en masse is likely to send little waves of disturbance across the swarm. But if we can safely assume that substantial hunks of a mechanism remain approximately rigid in their gross movements, we can potentially ignore a huge amount of internal complexity within the device. Consider our sewing device, whose bottom eccentric link is turned by a motor. If we were forced to model these arrangements explicitly as a swarm of strongly attracting point masses, we would need to painstakingly plot how the binding forces allow the input movement to gradually transmit itself from one little piece to another across the mechanism. This story must surely involve very complex processes in light of the branching causal pathways that initiate at the motor. As observed earlier, it is scarcely evident that orthodox point-mass mechanics contains enough internal resources to provide an adequate simulacrum of the expected behaviors. But once we are assured that the component pieces will remain nearly rigid throughout all of the device's ordeals, high school geometry can compute exactly how much the needle will wiggle as the drum at the bottom gets turned through an angle θ. Admittedly, this is not a trivial high-school calculation, but its demands are vastly simpler than computing how the whole point-mass swarm will behave under the same conditions. In other words, we can employ our upper scale ΔL^* knowledge that our sewing machine

---

[27] Constraint relationships are sometimes maintained through factors external to the device (such as the pressures of an ambient fluid or the gravitational attraction that binds a cam to its follower), in which case the device is said to be *force closed*. Descartes, for example, essentially dissected the universe into component mechanisms, but they were usually held together through force closure rather than internal pinning.

[28] Such contacts are further classified as "higher or lower pairs" according to the contacting geometry they implement.

WHAT IS "CLASSICAL MECHANICS" ANYWAY?

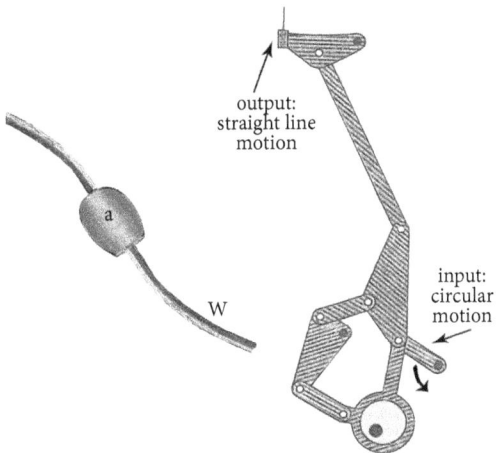

Figure 2.13

parts stay rigid and obey their connective constraints to avoid the very complicated mechanical relationships that hold among the device's component masses at the ΔL level.

Or, at least, that is how our target mechanism would appear *from a point-mass perspective*. But in the present section, we wish to consider "foundations" for classical mechanics in which notions like "rigid body" and "pinned constraint" comprise the *mechanical primitives* of the subject and are not introduced as convenient approximations to complex point-mass underpinnings.

This perspective supplies analytical mechanics with a huge computational advantage over point-mass-based modelings. Thomson and Tait articulate these virtues as follows: "[T]he forces which produce, or tend to produce, [the actions] may be left out of consideration. Thus we are enabled to investigate the action of machinery supposed to consist of separate portions whose form and dimension are unalterable."[29] Earlier we noted that it is hard to model, from a point-mass perspective, the simplest forms of *redirection of thrust*, as occurs when a plug slides along a curved track. In our sewing machine, the redirection is of a far cleverer design, but proceeds according to the same analytical mechanical principles. To formulate doctrines of this type correctly, we generally need to capture the system's current configuration in terms of *generalized coordinates*, rather than the Cartesian coordinates that are central to the point-mass reading of the second law. Often, good generalized coordinates use natural coordinates of the kind mentioned earlier: quantitative measures of displacement that are closely correlated with the system's available motions. For a plug sliding in a slot, its placement in terms of *arc length along the slot* represents the single natural coordinate we require to fix the plug's position, whereas its three locations expressed

---

[29] Thomson and Tait, *Treatise on Natural Philosophy*, Vol. 2, §441, 2.

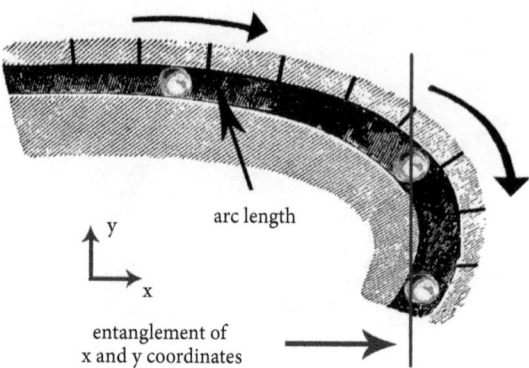

Figure 2.14

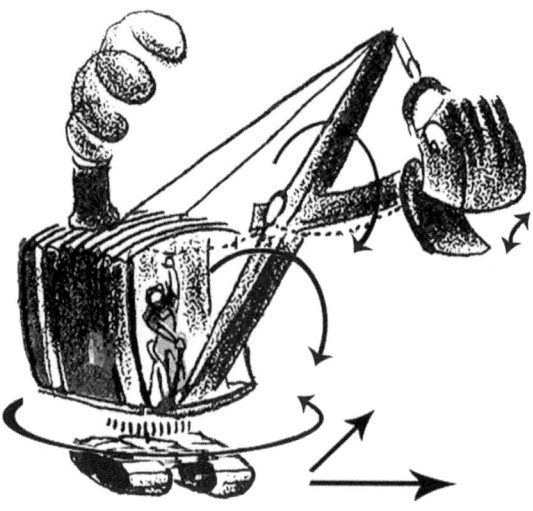

Figure 2.15

on Cartesian axes do not relate to the motion in any internally "natural" way (figure 2.14).

A second vital feature of the coordinates usually employed in analytical mechanics is that they are *independently variable* with respect to each other: any specific coordinate can be altered without necessarily disturbing the others. With a steam shovel, for example, we will want to employ the five decompositional movements illustrated to fix its configuration, rather than the regular Cartesian coordinates of its many parts. The latter are descriptively entangled in a manner that prevents us from applying the usual forms of virtual-work reasoning (figure 2.15).

Much of the practical success of analytic mechanics traces to the fact that suitable independent coordinates for a complex system can often be divined simply through experimentally determining how it wiggles under manipulation and the directions

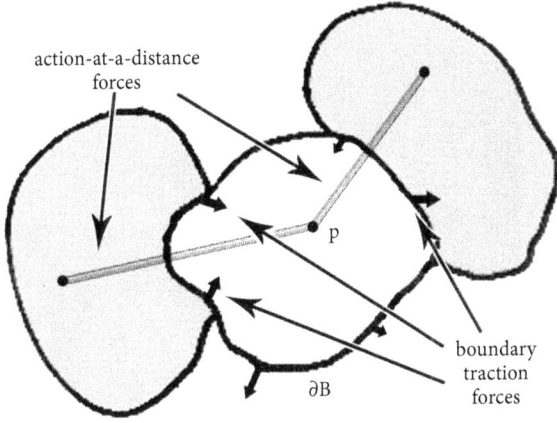

Figure 2.16

in which input thrust travels across its interior. Such data represents raw higher-scale information about our system's dominant behaviors.

As indicated earlier, as soon as material bodies genuinely fill finite volumes, a new type of "force" quietly enters the scene. This force eventually becomes a secret source of significant tensions within mechanical thinking. Since point masses are inherently zero-dimensional in nature, they can be provided with a surrogate for normal "size" only by erecting rough "effective volumes" through a battery of strong, short-range repulsive forces. But if our fundamental objects possess true sizes, then the *contact forces* will arise upon the interface between two contacting bodies. These new items can no longer qualify as *action-at-a-distance forces* simply because no distance separates the embedded points where the transmission occurs (figure 2.16).[30]

There are two grades of contact forces with which we must eventually deal. The first are the *boundary forces* applied along the outside surface ∂B[31] of a body B, such as a loaded weight passively resting on top of a four bar mechanism or a hammer blow applied somewhere (the first is traditionally called a *static load* or a *dead load* and the second a *dynamic loading*) (figure 2.17).

These are the only contact forces at issue within rigid body mechanics. But when we turn to flexible continua, a second grade of interior contact forces emerge in the guise of the *traction forces* that appear across the boundary of (almost) any internal surface S that we might mark out within the larger body B. Such internal surfaces S are commonly called *free body diagrams* or *Eulerian cuts* (figure 2.18).

Each such S will bristle with an array of *traction forces* that point either inward or outward at each surface point—it is then presumed in third-law fashion that the material outside S will push or get pulled in the opposite direction at that same place. The most familiar exemplar of such traction vectors are the normal

---

[30] Indeed, it is not evident to which body the contact point "belongs" (one needs to beware of making simplistic assumptions about "how points belong to bodies" in such circumstances).
[31] It is common to designate the external closure of a body B with the notation "∂B."

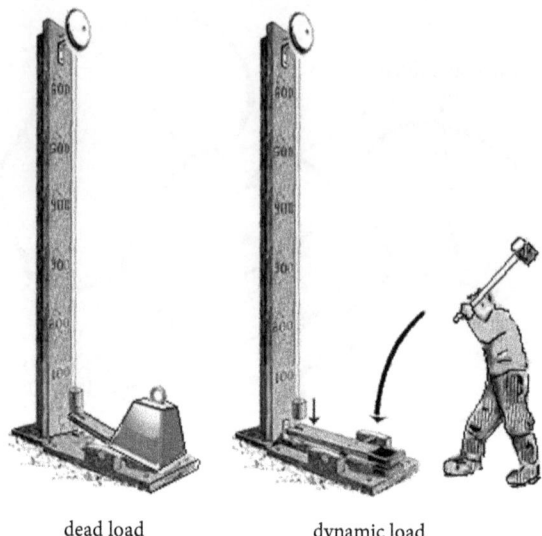

dead load   dynamic load

Figure 2.17

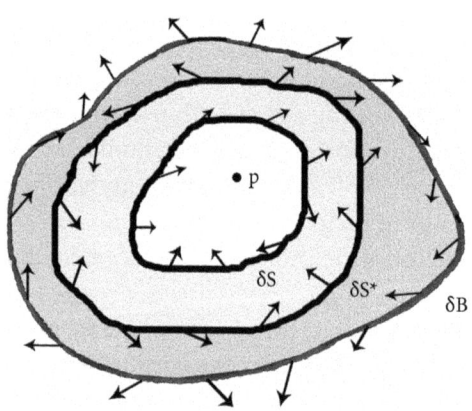

Figure 2.18

pressures acting within a nonviscous fluid, but the complicated internal pushes and pulls operative within other flexible bodies mandate the introduction of the more general notion of *stress*.

But once the interior of a body is claimed to be completely rigid, as we shall assume throughout the present section, then this interior grade of contact force becomes ill-defined, as do allied notions such as internal pressure. So we will not worry about how to deal with such internal tractions now and will concentrate upon the surface forces that appear along the boundaries between contacting bodies.

It is common for elementary textbooks to vaguely claim that all forms of contact force really represent short-range cohesive forces between separate particles. This contention might be true insofar as contact forces represent *classical distillations*

WHAT IS "CLASSICAL MECHANICS" ANYWAY?

of quantum processes of roughly that character, but such asseverations can prove very misleading insofar as our text appears to be concerned with classical processes exclusively, where it is not evident that plausible short-range cohesive forces of a classical point-mass character can ground, on a ΔL basis, the standard rigid body or continuum behaviors witnessed upon a ΔL^* scale. In fact, such offhanded appeals to short-range forces often disguise the fact that fundamentally new issues about how forces operate mathematically appear on the scene as soon as contact forces are tolerated. Often their resolution requires that we reinterpret Newton's laws in a significantly altered manner or turn to other forms of "foundational principle" altogether. In the face of these conceptual challenges, hazy appeals to fictitious short-range forces between point centers at a lower scale length ΔL merely serve as convenient escape hatches that allow authors to evade addressing these foundational issues squarely (these evasions become particularly troubling in the context of continua, as we shall later see). To be sure, the texts eventually stagger their way to the requisite ΔL^* level equations, but only along pathways that are apt to confuse a critical student.

One of these difficulties—which arises even with the exterior boundary forces of rigid body mechanics—traces to the simple fact that there is a *disparity in dimension* between contact forces that act upon *bounding surfaces* ∂B and forces such as gravity that act upon the localized *points* inside B. As soon as our attention shifted to extended bodies, we should have properly stopped calling such items "forces" at all and instead considered "force *densities*" of dimensionally incompatible grades. The motives for these adjustments trace to the usual difficulties of making sense of continuously distributed quantities that date to the time of Zeno.

Suppose we have a target with a bull's-eye and two archers: skilled Marian and inept Robin (figure 2.19). What are Marian's and Robin's respective probabilities for hitting the exact center of the target **c**? Answer: most likely zero in both cases, because if the "hit **c** exactly" answers were credited with any finite amount ε, then (under

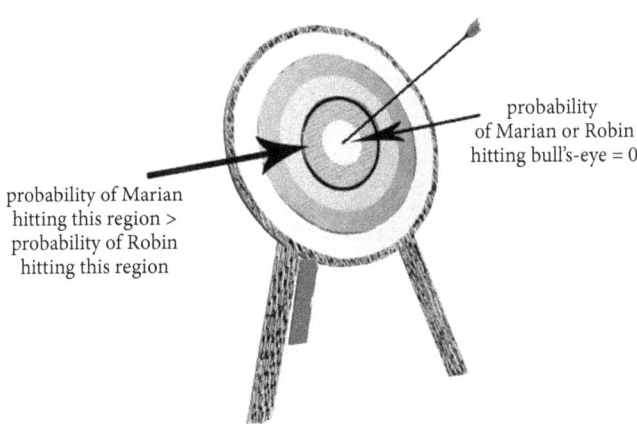

Figure 2.19

the assumption that points near to c should be credited with probabilities close to ε) the summed probabilities of hitting any finite region of the target will become infinite (due to the infinity of individual points contained in such a region). But if Marian's and Robin's probabilities for hitting the target at any individual point are always zero, shouldn't it follow that their summed probabilities of hitting any finite region A also need to be identical (viz., zero), rendering them equally lousy marksmen? Obviously not, but the task of straightening out these riddles is the business of the modern theory of measure. This theory addresses our problem by crediting Marian and Robin with different probability *densities* with respect to the individual points in the target. To extract a proper *probability* from a density, one must "add up" (integrate over) these densities over sufficiently large areas. Based upon their different densities, the true *probability differences* between Marian and Robin's skills will show up only after sufficiently large expanses of target come into consideration. Getting all of this to work out correctly requires very careful mathematical preparation. Plainly, we need to adopt similar policies with respect to our new "forces": considered at a point-length scale only the *force densities* can be nonzero—true forces should not emerge until we have integrated these local densities over larger regions.

The awkward tension that segregates surface forces from body forces such as gravity stems from the fact that, considered properly as densities, their respective quantities must be *dimensionally inharmonious*. Why? In the case of the tractions pulling and pushing upon a boundary ∂B, we expect to reach genuine resultant forces after we have integrated over finite stretches ∂S of the exterior surface ∂B. But with gravitation, we must integrate over *volumes* V of points within B itself (not just along stretches of ∂B) before we can assemble forces of comparable strength from gravitational attraction (figure 2.20).

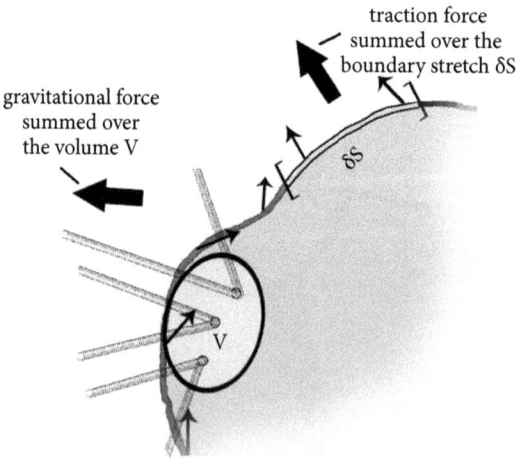

Figure 2.20

Considered from the point of view of the normal volume measure on $\mathcal{B}$, any surface piece $\partial \mathcal{S}$ will qualify as "of 0 measure," so we cannot use this same measure in dealing with contact forces. In sum: genuine forces can be assembled from much smaller sets of points in the case of a contact force than in the case of gravity, for they reach the level of a "finite resultant force" more quickly in the former case.

It is only after these surface and volume resultants (the fat arrows in the diagram) have been obtained that we will possess genuine forces—not densities—that can be *meaningfully combined*. This dimensional disparity of our densities is not merely an awkward mathematical issue, for the fact that surface forces inherently overwhelm body forces within small regions plays a vital role in determining the logical character of vital notions like "stress." We shall discuss these features in section 5.

Let us return to the problem of combining body and surface forces, now construed as densities. We find that two basic gizmos are needed to fulfill the roles that "total force" serves within point-mass mechanics. We first require a dimensionally correct analog for the notion of total force, which we now compute as the vector resultant of two density integrations $\int_S \mathbf{f}_s \, ds$ and $\int_V \mathbf{f}_b \, dv$ (where $\mathbf{f}_s$ and $\mathbf{f}_b$ are the surface force and body force densities, respectively). Observe that these two integrations transpire over the requisite regions: S for outer surface and V for interior volume. In so doing, we are summing a large number of force densities that act in different locales, unlike in the point-mass case where forces all act in the same place. But in composing our new notion of total force, we simply ignore these differences in point of application.

Using these new notions, we obtain an analog of Newton's Second Law suitable to isolated rigid bodies: $(\int_V \rho \, dv) \, d^2 r_i/dt^2 = \int_S \mathbf{f}_s \, ds + \int_V \mathbf{f}_b \, dv$, where $\int_V \rho \, dv$ is the summation of the mass density over the entire rigid body $\mathcal{B}$. But with which point in $\mathcal{B}$ should the location $\mathbf{r}$ in the term $(\int_V \rho \, dv) \, d^2 r_i/dt^2$ be computed? It does not really matter: every point will display the same linear acceleration in any direction we look. Some writers link $\mathbf{r}$ to the center of mass of the body, but there is no especial reason for doing so (especially when the center of mass is often not located inside $\mathcal{B}$ at all, as in a doughnut).

Although the points in $\mathcal{B}$ accelerate in the same way, they certainly do not have the same velocities. An additional notion is needed to describe the velocity relations, or the turning of a rigid body. This new notion is called the *torque* τ (or *turning moment*) of the summed force densities acting upon $\mathcal{B}$. Once again, this summation needs to be broken into two integrals that separately average the lever arm contributions of the surface and body forces with respect to some center A within $\mathcal{B}$ (it does not matter where, although certain centroids can make the calculations easier) (figure 2.21). More exactly, $\tau = \int_S (\mathbf{f}_s \times \mathbf{r}) ds + \int_V (\mathbf{f}_b \times \mathbf{r}) \, dv$, where $\mathbf{r}$ represents distance to the chosen reference point A. Quite different distributions of force density across a rigid body can move it in identical ways as long as their averaged total force and averaged torque about A are the same.

To complete our scheme, we must now quantify how our rigid body creates an inertial resistance to an applied torque as well. Here we need to compute how far

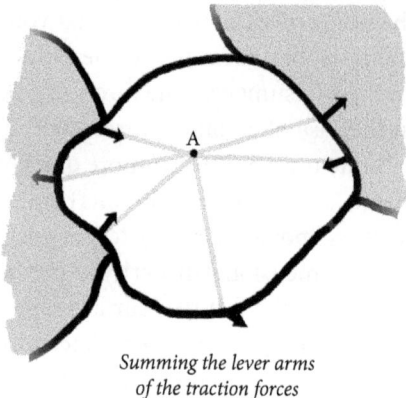

Summing the lever arms of the traction forces

Figure 2.21

away from A the mass density ρ within $\mathcal{B}$ tends to lie on average (viz. $\int_V (\rho \cdot r^2)\, dv$). This new quantity I is called $\mathcal{B}$'s "moment of inertia" around A. Using it, we can express "Euler's Second Law of Motion"[32] for torques as $I\, d^2\theta/dt^2 = \int_S (f_s \times r)\, ds + \int_V (f_b \times r)\, dv$, where $d^2\theta/dt^2$ is the angular acceleration of $\mathcal{B}$.

Working within a point-mass framework, Euler's second law is provable from Newton's second law in conjunction with the third law restrictions on action-at-a-distance forces. But this dependence no longer holds as soon as the forces tolerated multiply into new varieties. In particular, Euler's second law is required as an independent postulate to show that stress tensors must be symmetric within a continuum physics setting. Unjustified lifts from point-mass mechanics often disguise this crucial fact in many texts.

However, our two Eulerian principles alone do not tell us how *hinged assemblies* of rigid bodies should act, which is our main objective in this section. A general answer to this question was supplied by Lagrange, who elevated some of the reasonings we have already canvassed into a general framing principle. Specifically, Lagrange maintained that, in any system of rigid parts characterized by n sites of impressed force, either (i) the device remains in equilibrium and the total *virtual work* associated with all impressed forces vanishes or (ii) the device moves with exactly the requisite inertial reactions at the n sites to compensate for the virtual

---

[32] In this context, "Euler's First Law" is often viewed as simply "Newton's Second Law" in application to rigid bodies. Credit for regarding the "$F = ma$" scheme as a framework upon which "recipes" for differential equations for both forms of mechanics can be built is historically due to Euler, not Newton. As we shall see, the analogous recipe for continua relies upon a formula traditionally called "Cauchy's Law," which many writers regard as yet "another version of $F = ma$" (although it actually employs the tricky notion of *stress* that Cauchy originated). The similarities of these three "recipe" formulas support the strong "family resemblance" character of "classical mechanics." Terminology issues become more confusing within the context of continua, in which analogs of Euler's two laws are also applied to the *sub-bodies* in the interior of container blobs. In such contexts, these analogs are often dubbed the "balance principles" for momentum and angular momentum. In the context of rigid bodies, once specific values for moments of inertia et al. have been computed with respect to such entities, these values remain the same, allowing the import of Euler's principles to be expressed as equations of ODE type. Within flexible bodies, in contrast such values fluctuate as they flex and so PDEs are required to capture the requisite relationships.

work imbalance. Traditionally, consideration (i) is dubbed the *principle of virtual work* and (ii) is called *d'Alembert's principle*. Combined into one formula, we obtain *Lagrange's principle*:

$$\sum F_i \delta q_i = \sum m_i \, d^2 q_i / dt^2 \qquad (1)$$

where δq_i represents a "virtual adjustment" in the coordinate value q_i, leading to a measure of the work that the applied force F_i would supply if it could be prolonged through that distance.[33]

Lagrange's formula is particularly useful if we have employed *independent* coordinates as our q_i, for then we can write down a formula that expresses how work supplied at, for example, site q_1 gets transmitted across the mechanism to any other site *on the assumption that the remaining sites can stay fixed* in the process. Working out these rules for each pair of sites provides a collection of equations that can completely fix how our hinged rigid body will move. The formulas familiar from analytic mechanics that are couched in terms of "Lagrangians" or "Hamiltonians" simply represent these new equations rewritten reliant upon certain further assumptions about the nature of the forces at issue (viz., their derivability from potentials).

Many interesting geometrical problems are closely connected to the generalized-coordinate representations of rigid-body mechanics. The configuration spaces of our earlier point-mass swarms are comparatively uninteresting from a mathematical point of view. But if one considers the *mobility space* of a complicated mechanism like our crane, as determined by its varying generalized coordinates, we obtain a quite novel structure, largely because its coordinates are angle-like in character: they return to their starting values after a 360° rotation. Mathematically, we obtain such mobility spaces by cutting out all of the "can't be visited" regions from a regular Cartesian 3n space and gluing together the remaining edges according to the pathways of angle-like returns. The resulting substructures can prove very complicated geometrically and comprise a topic of great mathematical interest far beyond the limits of the *kinematics of mechanisms* (which is the traditional name for the study of machine mobility).

From a point-mass vantage, we are plainly skipping over a huge amount of internal structure. Let us examine a small piece of our crane from a punctiform point of view (figure 2.22).

---

[33] Although I have quoted Lagrange's principle in its standard textbook form, it conceals a subtle ambiguity; specifically as to whether the "r" cited is a true position coordinate or rather represents something "generalized" like an angle. If the latter (which is usually what is needed), then the corresponding "mass" terms "m" must be read as moments of inertia, etc. Presumably, we require some instruction in how these "generalized inertial terms" are to be found. Such unnoticed shifts are often sites of significant "lifts" (and sometimes outright errors, which are common in this branch of mechanics).

The restriction to "virtual variations" is necessary because the *mechanical advantages* of most mechanisms continuously adjust as they move through their cycles. This means that inputted forces F_1, F_2, F_3 on our crane will not be able to balance quite the same output force F_4 when the machine stands in a different configuration. But the "instantaneous work" performed by the input forces will always equal the "instantaneous work" expended at the outputs, which is the key idea that we need to capture in our "virtual work" formula for static situations.

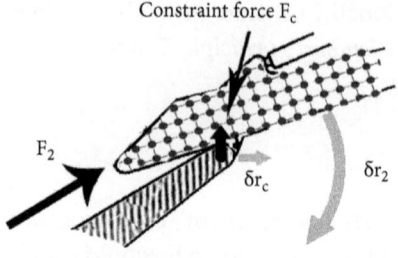

Figure 2.22

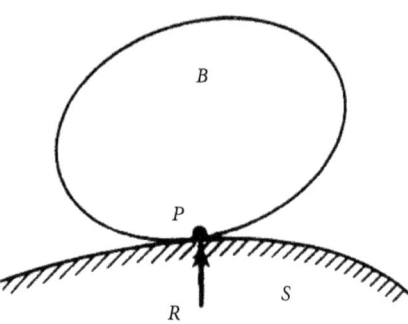

Figure 2.23

Clearly, strong cohesive forces F_{ij} will be required to lock point i into a lattice with point mass j and some kind of binding force F_C will also be needed to keep our piece fixed to its pin. All mention of these has vanished from Lagrange's principle. Why are we allowed to ignore these extra forces? Textbooks commonly argue as follows: (1) "The net work of the cohesive forces vanishes because they occur in internal-force pairs where $F_{ij} = -F_{ji}$. Since their virtual displacements will be the same, their virtual work contributions will cancel each other out." Or: (2) "The constraint force F_c does no work because its action is orthogonal to the path of virtual movement δr_c."

Here is Donald Greenwood's version of this last argument, presented in the course of "justifying" Lagrange's principle from a point-mass standpoint: "[C]onsider a body B which slides without friction on a fixed surface S. . . . The constraint force is normal to the surface at the tangent point P, but any virtual displacement of P involves sliding in the tangent plane at that point. Hence no work is done by the constraint force R in a virtual displacement" (figure 2.23).[34]

But in our point-mass frame, all forces are supposed to act from one point to another along the line connecting them. But our constraint force R looks as if it

---

[34] Donald T. Greenwood, *Classical Dynamics* (New York: Dover, 1997), 16–18. I do not intend these remarks to be as critical as they may presently seem. Eventually, we come to see Greenwood's "proofs" as functioning, not as derivations proper, but as "Task C" indicators of profitable ways to avoid ΔL constitutive assumptions through the exploitation of knowledge of a material's ΔL^* behaviors (specifically, its apparent "rigidities").

WHAT IS "CLASSICAL MECHANICS" ANYWAY?

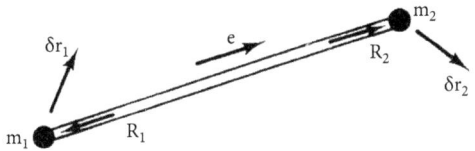

Figure 2.24

starts and ends in exactly the same spot P, which was not permitted under our old reading of the third law. Plainly, some new kind of "force" has been smuggled into Greenwood's text, without adequate prior warning.

On virtually the same page Greenwood argues for Lagrange's principle in a different setting as follows (figure 2.24):

> Assume that two particles are connected by a rigid, massless rod.... Because of Newton's third law, the forces exerted by the rod on the particles m_1 and m_2 are equal, opposite and collinear. Hence $R_2 = -R_1$ ... as shown. Furthermore, since the rod is rigid, the displacement components in the direction of the rod must be equal or $e.\delta r_1 = e.\delta r_2$ [where e is a unit vector pointing in the direction of the rod]. Therefore the virtual work of the constraint forces is zero: $\delta W = R_1.\delta r_1 + R_2.\delta r_2 = 0$.[35]

But by what right can we insert a "rigid, massless rod" in our system and still maintain that "Newton's third law" equates $R_2 = -R_1$? It is not as if the two masses are directly exerting forces on one another, as our earlier reading of the third law expects. Indeed, suppose that the intervening rod is curved, rather than straight. We still want our reasoning to hold, yet plainly $R_2 \neq -R_1$ (the vectors point in quite different directions).

Passages like these trade upon unnoticed elisions between the foundational sense of "isolated particle" and looser policies of talking of "representative points" within more extended bodies. By exploiting our alleged freedom to place representative points where we wish, Greenwood allows point masses to sometimes sit on top of one another or locate themselves at the far ends of ghostly, massless rods. Through simple appeals of this character, we find ourselves miraculously lifted to the characteristic scale level of objects far above the realm of the component particles (atoms, molecules, the tiny crystals in iron) that we originally sought to model as point masses.

Observe that a doctrine that is essentially philosophical in nature ("scientists idealize their targets through selecting representative points") has been tacitly employed as a cover for a missing stretch of substantial mathematics ("how do point-mass foundations support the principles of analytical mechanics?"). As we observed, one of Hilbert's stated objectives in his sixth problem was to study these

---

[35] Ibid., p. 16.

lifts in a more rigorous spirit, although he did not observe that stock textbook arguments like Greenwood's involve moves of this character. We also discussed the general manner in which *constitutive modeling considerations* get mysteriously bypassed in arguments of this character. But in the situations Greenwood discusses, we have somehow persuaded ourselves that we can derive salient predictions based upon "general laws" pertinent to point masses alone, without needing to supply any constitutive modeling that can explain why systems would behave differently if they had been composed of nonrigid parts.

Thus, standard "derivations" of analytical mechanics doctrines from point-mass foundations are rarely cogent, if scrutinized from the point of view of Hilbertian rigor. But none of our considerations show that an alternative set of foundational principles cannot be coherently framed that accepts rigid bodies as its primitive objects, possibly in conjunction with point masses as well. In fact, the best modern writers on mechanics recognize that pretending that analytical mechanics can be adequately founded upon point-mass foundations is simply a mistake. Cornelius Lanczos comments:

> Those scientists who claim that analytical mechanics is nothing but a mathematically different formulation of the laws of Newton must assume that Lagrange's principle is deducible from the Newtonian laws of motion. The author is unable to see how this can be done. Certainly the third law of motion, "action equals reaction," is not wide enough to replace Lagrange's principle.[36]

He particularly has in mind some of the third-law ambiguities discussed in section 2.

In criticizing derivations like Greenwood's for their lack of rigor, we should never forget that the modeling techniques they are intended to justify are of vital importance to working physics. For the import of virtual-work schemes in practice is that they allow us to avoid working through an awful lot of difficult physics that runs the risk of introducing large errors into our calculations with little gain in predictive power. We have already discussed the advantages of working with data drawn from a range of scale sizes. If we already know how the principal patterns of thrust transmission operate within our crane at a large scale size ΔL^*, why not exploit that information to reduce the complexity of our modeling, even at the cost of a certain degree of approximation with respect to the point masses that comprise it at a scale ΔL? The essential genius of virtual work and the other techniques of analytical mechanics lies in their ability to combine data types in this manner.

Analytical mechanics, if stoutly set on its own feet axiomatically, should appear an odd choice for serving as a baseline ontology for classical mechanics due to the tremendous number of descriptive holes it contains. This section has devoted its attention to analytic mechanics' prospects as a foundational enterprise largely because the subject commonly serves as a favored point of refuge when one

---

[36] Cornelius Lanczos, *The Variational Principles of Mechanics* (New York: Dover, 1986), 70.

encounters conceptual difficulties in pursuing our other basic ontologies. In fact, the position of rigid bodies within classical physics is much like that of the disreputable uncle who possesses most of the money in the family: you do not fully admire his character but you appreciate all of the good things he can buy for you.

We have already examined several ways in which point-mass mechanics commonly appeals to rigid-body notions as escape hatches when it finds itself in descriptive hot water. Thus, we invoke "massless rods" to hide the fact that we do not know the special force laws that bind two point-masses at a constant distance. Or we enlarge our point-mass planets to become finite spheroids to avoid Xia-type blowups. Or, like Poisson, we correct the one-elastic-constant deficiencies of a material modeling by replacing the point centers with rigid ellipsoids. But these doors of conceptual escape swing both ways, for analytical mechanics commonly evokes the resources of its ontological rivals to sustain its own reasonableness.

5. Continuum Mechanics

If we could mark out the salient differences clearly and poll most of the prominent classical physicists of the past with respect to their favored choice of foundational object, the majority would undoubtedly select continuous, flexible bodies. Let us now try to articulate principles capable of governing the behaviors of continua coherently. We immediately confront a more difficult form of the surface/volume force coordination problem than we confronted in the case of rigid bodies. In rigid-body mechanics, the relevant traction forces operate only along the exterior surfaces ∂B of the rigid body B under consideration. But inside a flexible body, we can carve out infinitely many internal surfaces ∂S able to support their own arrays of traction vectors as well (figure 2.25).

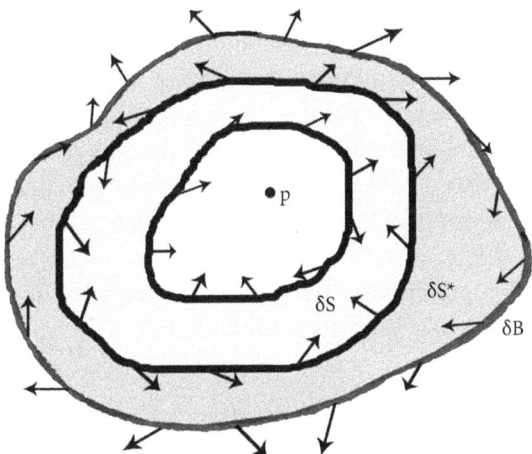

Figure 2.25

Furthermore, the tractions on each different ∂S will generally differ from one another and from the exterior tractions applied along the outer boundary ∂B. Indeed, we anticipate that as we push and pull upon ∂B in different ways, these exterior modifications will make themselves felt at an interior point q through progressively altering the traction forces upon all of the interior surfaces ∂S that surround q. Moreover, this process of inward transmission will require some time to complete: the tractions on ∂S^* must alter before the tractions on ∂S can change. Inside a truly rigid body, however, such inner tractions and waves no longer make sense, for essentially the same reasons that the notion of "absolute pressure" becomes problematic when a fluid is assumed to be incompressible. Usually, notions of "rigid body" are regarded as incompatible with a continuum physics point of view.

In the previous section, we summed the surface forces around the outer boundary ∂B and the volume forces inside B employing two integrations whose results we then added to get a resultant *applied force*. We then learned that we should also compute a combined *torque* in a similar manner. With those two ingredients in place, Euler's two laws of motion could tell us how our rigid body would respond. But in that context we only had to contend with the body forces and traction vectors around the outer boundary ∂B. How should we address the vast army of *differing* ∂S's that have now entered our stage in the entourage of flexible bodies? If we naïvely compute resultant forces and torques from these, we can obtain substantially different answers according to the inner surface ∂S chosen. This is a surface/body force coordination problem of considerably greater subtlety than we addressed earlier.

The eventual solution invokes the notions of "stress" and "strain." Before proceeding further, a few words of warning are in order concerning these innocent-looking words. Most philosophers interested in physics have already run across those words—in the guise, say, of their close cousin, the "stress-energy tensor" of general relativity—without reflecting sufficiently on their conceptual oddities ("stress" is not "just a form of force" and "strain" is not "just a form of shape change"). Historically, it was not until the end of the nineteenth century that the true novelty of these constructions was adequately recognized.[37] Some of this confusion traced to the carelessness about "points" that "representative point" talk encourages. So let us reiterate that in dealing with interior points like q, we are not longer considering the isolated points of mass point physics: our new points come densely surrounded by infinitely many neighbors, situated as close to them as one might like. And they should not to be identified with points in the ambient container space; our *material points* (their most common name) wander through that background space in a trackable way.

It is best, at this preliminary stage of our discussion, to conceptualize our material points q, not as bare geometrical entities, but as *decorated points* or *physical*

---

[37] Cf. Clifford Truesdell, "The Creation and Unfolding of the Concept of Stress," in *Essays in the History of Mechanics* (Berlin: Springer-Verlag, 1968:184-238). One needs to be wary of framing one's conception of these notions from one-dimensional continua such as strings or lamina, for in such reduced contexts "stress" does appear like a simple force density. In the main text, I am trying to bring forth the funny kind of three-dimensional structuring that is inherent in the notion of a "tensor."

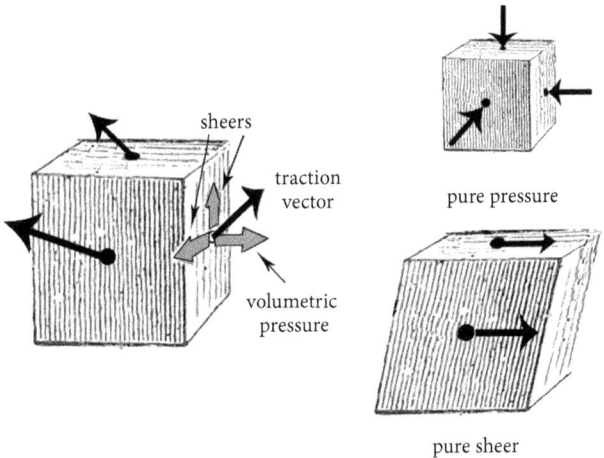

Figure 2.26

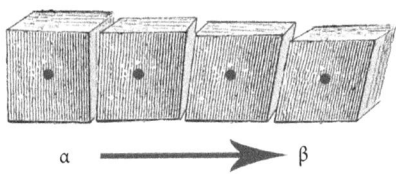

Figure 2.27

infinitesimals that have temporarily parked themselves at various spatial points p (which are not "decorated"). In particular, if **q** forms part of a solid material like iron and the material in **q**'s immediate neighborhood is fully relaxed, we should picture its "decorated" condition as an infinitesimal little cube about the central point p (figure 2.26).

But if **q** is subject to stress, infinitesimal traction vectors should appear on each of its faces, one to a side and pointing inward or outward in any direction desired. In response to these tractions, our little boxes will adjust their decoration, by adjusting their infinitesimal *volumes* or shifting shape in a *shearing pattern* (or displaying combinations of these two basic alterations). In fact, the descriptive purpose of a *stress tensor* is to capture the local pushes and pulls of the traction vectors on **q**, while the usual (Cauchy-Green) *strain tensor* captures **q**'s degree of distortion from its cubic relaxed state. Considering our material over a broader scale, we realize that the material points **q*** found at locations near to **q** must be decorated in a manner very close to, but not identical with, that of **q**—otherwise, the material would display fissures (this relationship among nearby infinitesimals is called *compatibility*[38]) (figure 2.27).

---

[38] This proviso is enforced within a PDE modeling through the Saint-Venant compatibility equations.

Because all of these modes of decoration occur on an infinitesimal scale, stresses and strains behave like the *densities* introduced earlier in connection with *mass* and *force*: such measures do not sum to become legitimate masses and forces until we consider finite volumes of material. But the rules whereby localized stresses and strains eventually sum to produce finite characteristics of the material stuff to which they belong are more complicated than the procedures used for simple densities.

I have highlighted the odd, decorated-point aspects of the material points of continuum mechanics to help readers properly appreciate the conceptual novelties that lie before us—we are no longer considering the familiar isolated points of the easy-to-comprehend point-mass framework. And although we shall eventually appeal to various limiting procedures to persuade our stresses and strains to work together harmoniously, readers should not presume that the conceptual difficulties of continuum physics are merely matters of "explaining away infinitesimals" in the familiar δ/ε fashion of elementary calculus courses. No: deep questions of *physical principle* are central to our concerns; we are not simply striving to make hygienic sense of infinitesimal points. Our foundational difficulties might be fairly dubbed the *problem of the physical infinitesimal,* but our problems mainly belong to physics and are considerably more substantive than the comparable *problem of the mathematical infinitesimal* from freshman calculus. To be sure, philosophers have sometimes presumed otherwise, but only as a result of underestimating the pertinent physical concerns. It strikes me that many of the deepest worries about matter in our philosophical heritage trace, in one way or another, to our "problem of the physical infinitesimal."

With these warnings not to underestimate stress and strain in hand, let us now turn to the foundational ailments for which they will eventually provide part of the cure. Let us recall the rather complicated physical situation that pertains at the level of the complete blob B of material to which q belongs (figure 2.28).

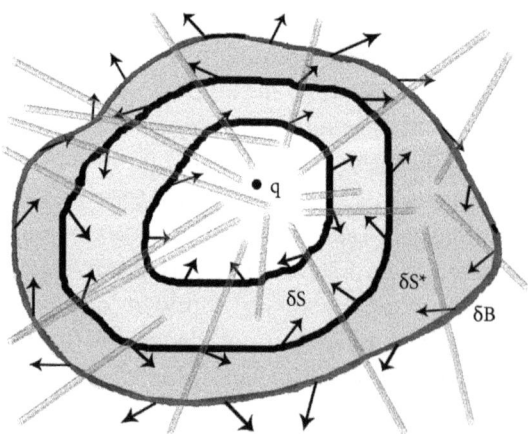

Figure 2.28

Its interior will be pulled upon by gravity g and other action-at-a-distance forces of that ilk. But $\mathcal{B}$ will also be affected by the various twists and pulls that we exert directly upon its exterior surface $\partial\mathcal{B}$ as "contact forces." If the material inside $\mathcal{B}$ is perfectly rigid, the basic problem of coordinating these two classes of "force" could be resolved fairly easily by computing resultant torques and applying Euler's two laws of motion for rigid bodies. But this simple policy works only because the material is *rigid*: every point q^* inside $\mathcal{B}$ must display the same linear acceleration and the whole will rotate in exactly the same way no matter from which reference point its torque is gauged. However, if the matter inside $\mathcal{B}$ is not perfectly stiff (let $\mathcal{B}$ be a blob of jelly or water), then the response behavior immediately around q will usually look quite different from the corresponding behaviors around q^* (neither the local accelerations nor rotations will be the same). And it is these local differences within flexible bodies that allow them to carry interior *waves*, which rigid bodies cannot support. When we twist and pull upon the external surface of a flexible body $\mathcal{B}$, we generate a lot of internal tractions, for the effects of our surface manipulations generate compression waves whose effects eventually reach q by progressively altering the local tractions upon a contracting collection of surfaces $S_1, S_2, S_3, \ldots$ surrounding q. It is these internal surfaces and their shifting arrays of traction vectors that greatly complicate our earlier surface/body force coordination problem in the case of flexible bodies.

That difficulty, the reader will remember, traces to the dimensional disparity generated by the fact that contact tractions represent surface "forces" (properly force densities) in the sense they must attach to some shell of surface ∂S surrounding a point q before they can be coherently integrated, whereas body "forces" (again, densities) such as gravitation apply directly to simple points q and need to be integrated over *volumes*. In the rigid-body case, only the outer layer of exterior traction pulls needed to coordinate with its interior points q, but in flexible bodies we are confronted with a host of additional shells ∂S to coordinate, appearing as the interior cuts whose characteristics are continually altered by the waves that pass through them.

Plainly this represents a fairly complicated physical problematic. It is often remarked that physics is simpler in the small, indicating that uncomplicated laws of material behavior can be elegantly formulated only at the infinitesimal level. Will this methodology help us here? Consider the material at a point q, where it will display a local mass density ρ and allied characteristics like charge. It will be pulled upon directly by gravitation and other possible long-distance "body forces," which can be summed to supply a local resultant vector g^*. We can presume our material will react to its full schedule of local pushes and pulls by manifesting an acceleration D^2q/Dt^2 (the capital D's signify the *material derivative*, which is explained in every textbook on continuum mechanics). Unfortunately, the compression waves passing through q will also affect its full schedule of local pushes and pulls and it is these that make our force-coordination problem so difficult. It is easy to understand how a passing wave will affect a shell of surface ∂S: run a tangent plane through any point on ∂S and see which way the traction T supplied by the wave locally

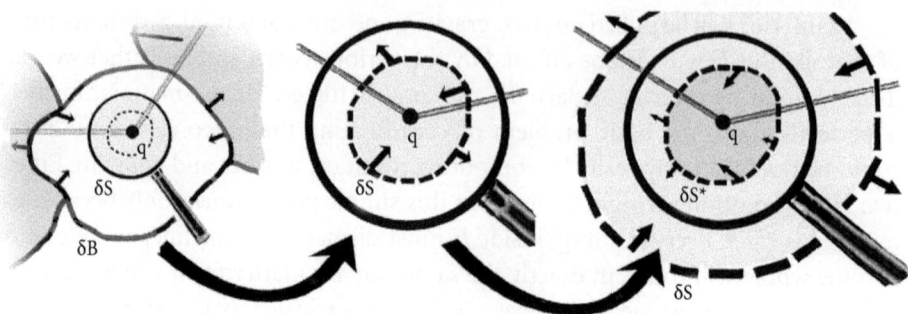

Figure 2.29

points across the plane. So to understand how the compression waves will affect q, we should set up a little shell around q and compute the traction vectors on ∂S created by the passing waves. All we need to do, it would seem, is to compute how the resulting surface "force" summation F* should compare to the body force summation g* acting at q. But wait a minute: no part of ∂S is actually located at q and, in fact, we can easily carve out a smaller cut ∂S^* inside ∂S whose integrated tractions may differ considerably from those on ∂S itself (why? because ∂S^* is affected by different wave movements than ∂S) (figure 2.29). And we can draw an even smaller cut ∂S^{**} inside ∂S^* where the same phenomenon reappears. And so on, ad infinitum.

In short, we have gone smaller in our physics, but nothing has become simpler! The regress traces, of course, to the fundamental *scale invariance* of homogeneous classical continua. Whatever characteristic length ΔL we choose, volumes of such materials will always behave exactly alike in terms of the principles they obey (of course, one can also consider *composite continua* where various sectors obey different rules, but these raise further difficulties, which we shall discuss later).

Somehow we must arrest this regress of unprofitable descent if we hope to get anywhere in continuum physics. But how can we do this? One cannot blithely say, "Oh, just take a 'limit' as you shrink to q," for it is not at all apparent what should happen to our traction forces when the cuts on which they live shrink to nothingness at q itself. (1) Will the result be merely a simple *pressure*, which operates to expand or contract our element in terms of its volume? (2) Can such local "pressures" pull differently in different directions? (3) Can the directionalities of our tractions lean sideways in a manner that can shear an infinitesimal blob S without altering its volume? (4) If so, will they act differently upon different planes around S? (5) How differently? (6) If so, how much latitude can they display with respect to these variations? (7) Will turning torques also leave a residual infinitesimal turning moment within S?

The standard (although not invariable) answers to these questions are: (1) no; (2) no; (3) no; (4) yes; (5) yes; (6) they must interrelate in the manner of a 3D vector space; (7) no. But few of these should seem entirely obvious. Internal pressures,

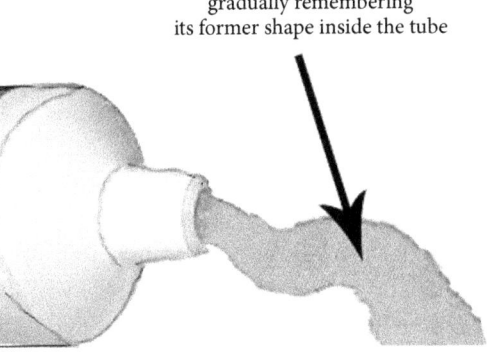

Figure 2.30

for example, can vary considerably across a fluid—mightn't these longer range inequalities deposit an unbalanced pulling upon our small blob S as a local residue? Prior to the time of Cauchy, the greatest practitioners of mechanics answered "no" to (3), often on the basis of the way in which they correctly answered "no" to (7).[39] Although the conventional textbook response to (7) is "no," there are well-developed theories of directed media that address this question differently. The fact that it is hard to augur intuitively how *infinitesimal portions* of a continuous medium should behave helps explain why the old controversy between rari- and multi-constant theories of elasticity took so long to resolve.

Such questions concern only the static responses of materials. Once dynamics come into play, an even wider range of difficult questions emerge. Can our infinitesimal elements retain long-term "memories" of their previous history? Certainly, *macroscopic* media often behave in this way: two identical looking paper clips made of the same material may respond differently to bending pressures because clip A has been flexed many times in the past but clip B has not. Can an infinitesimal blob S display allied memories as well, or must such processes emerge due to complicated interactions between finite portions of a composite system? Likewise, might our "infinitesimals" display "delayed memories" in the sense that a blob S might respond to altered conditions in a non-immediate manner? Again, toothpaste acts like this: it gradually "remembers" its shape back in the tube and tardily reverts back to it (figure 2.30).

Such questions lay behind the twentieth-century revival of interest in the "foundations" of classical continuum mechanics: scientists who confronted with new industrial substances needed *guidance* as to how such complicated materials might be reliably modeled.

---

[39] One can witness some of this struggle in Kant's *Metaphysical Foundations of Natural Science* (Cambridge: Cambridge University Press, 2004) where he is plainly aware that some source of sheer is needed to make sense of conventional "solidity," but cannot find a way to incorporate such a quantity into his descriptive framework.

Ultimately, the answers to all of these questions depend upon physics because, insofar as mathematics is alone concerned, such issues can be resolved in many different ways. That is why our "problem of the physical infinitesimal" (which is equivalent to answering our questions coherently) is not mainly an issue in δ/ε rigorization.

In sum, we are confronted with a serious conceptual regress: the complicated behaviors of these materials never seem to become simpler no matter how small the portions we consider. How can we halt this unhelpful descent into what Leibniz called "the labyrinth of the continuum"? I will first sketch two traditional answers and then the modern view. The first of these claims that at some minute scale length ΔL, the volumes S around q will "stiffen" enough that we will see a simpler physics there. We will not want our infinitesimal S to become totally rigid, lest we never recover any flexibility in the larger bodies B to which it belongs, but perhaps a small S might move like a *little mechanism*, so that some of the techniques of the previous section become applicable. For example, a standard weighted beam can be assigned a small-scale mechanical element that eventuates in the stock Bernoulli-Euler equation for such structures (figure 2.31).

In this situation, our element is allowed to turn about its centroid, as well as move up and down in a plane, although a series of springs sets up an internal resistance to turning. In addition, gravity acts in the center of the element according to the weight W it bears. In this situation, Euler's rules for torque play a role in the derivations.

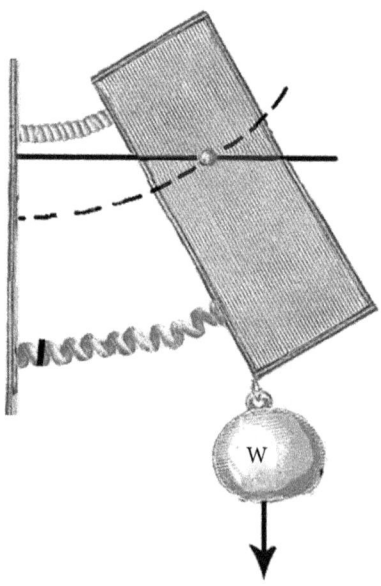

Figure 2.31

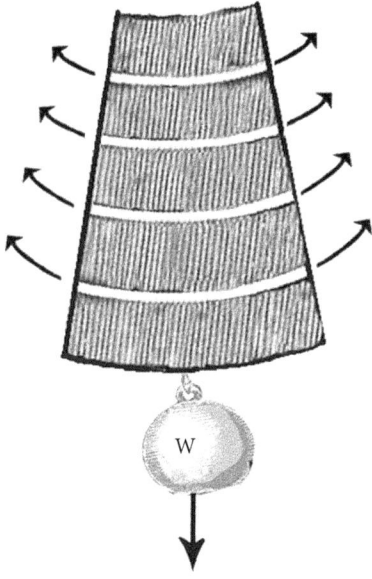

Figure 2.32

Except in early works,[40] it is fairly rare to see presumed "mechanical elements" decked out in blocks and springs quite like this. But there are several alternate modes of presentation that can achieve comparable results by invoking the controlled-virtual-work behaviors that we briefly discussed in the previous section (figure 2.32).

Thus, we might portray our Bernoulli-Euler element as illustrated, where we have an element that is intrinsically flexible, but which responds to contact tractions only at specific sites. As stated before, such restrictions represent a diagnosis of how applied thrusts are expected to transmit themselves through the element. It is evident that we get our required "simplification in the small" through locating these sites of controlled thrust; otherwise, we would simply be looking at a small section S of the original blob B we began with, displaying exactly the same behavioral complexities as where we started.

Modern books in engineering—at least, the sophisticated ones—no longer follow these old policies, which trade upon *rigid-body mechanics* as an intermediary. From a practical point of view, such presentations leave us rather confused as to which behaviors are possible—and which are not possible—within a continuous material. A common model for a drumhead, in effect, constrains its movements in the mode of the blocks-and-cords construction illustrated in figure 2.33.

Its little elements have been linked together in such a way that they can only move up and down, but not horizontally. Translating those limitations into wave terms, this means that such membranes can transmit only *transverse waves* (like

---

[40] I have patterned my first Bernoulli-Euler "element" after a diagram that Leibniz provides for a loaded beam. Cf. Clifford Truesdell, *The Rational Mechanics of Flexible or Elastic Bodies 1638–1788*, (editor's introduction to Euler, *Opera Omnia* II, vol. 12) (Lausanne: 1954).

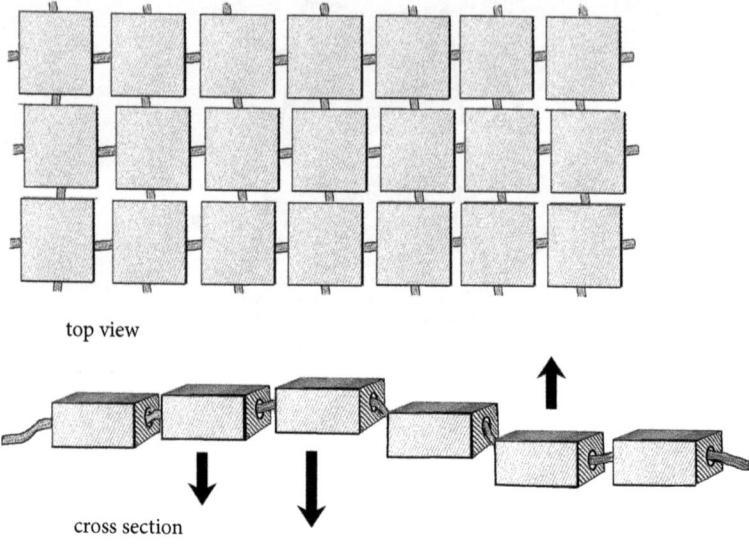

top view

cross section

Figure 2.33

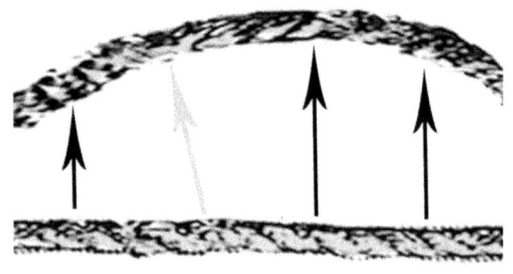

each section of string
remains above its rest position

Figure 2.34

surface water waves) but not *compression waves* (like sound). But are such materials really possible, except in coarse approximation? This is the kind of *inductive guidance* with respect to the behavioral capabilities of materials that we would like continuum mechanics to provide.

The strangeness of our drumhead's hypothetical capacities can be made quite vivid if we consider its one-dimensional analog, an oddity that lies concealed within the basic equation for a vibrating string discussed in every classical physics primer (figure 2.34).

In its derivation we tacitly posit that, in its stretching each section of string "remembers" its rest position well enough to remain constantly above it, never veering left or right in the manner of the gray arrow. How can a dumb piece of string achieve this remarkable feat? In the drumhead case, we surreptitiously employed the rigidity of the blocks to enforce the vertical-only movements, inserting cords to allow each element to become effectively longer as it does so. But our string lacks any comparable enforcement mechanism of this kind. Should we conclude that no continuous material can truly behave as our textbook model prescribes or simply that it is unlikely, except in crude approximation?

In fact, nearly all of the standard continuum models studied in undergraduate primers contain some hidden dimension of unlikely behavior of this ilk: they continually ask beams to bend in a plane, say, but to not bulge outward as they do so. But see if you can find a real material that will be so obliging.

On the other hand, real materials do display odd abilities to "remember" their earlier states. If we attempt to find an infinitesimal mechanism-like element that duplicates these capacities, we are likely to require strange, Rube Goldberg-like devices.

So, at base, our "problem of the physical infinitesimal" is one of delineating, with some measure of confidence, the full range of infinitesimal behaviors that can be legitimately expected of the points **q** within a continuous body. The great twentieth-century investigations into the foundations of continuum mechanics led by Clifford Truesdell and his school decided that traditional approaches of the character we have surveyed had jumbled together three basic tasks that should be kept distinct:[41] (A) to establish the local existence of stress, strain, rate of deformation, and allied tensors within a continuous body; (B) to supply "constitutive relationships" that capture why a material like iron differs so greatly from putty or water; and (C) to exploit empirical determinations of the dominant patterns of thrust propagation within a medium to render the results of tasks A and B more mathematically tractable. According to this modern reassessment, the policies pursued by the great nineteenth-century masters of continuum physics (Kelvin, Stokes, and others) had mixed *approximative considerations* properly reserved for task C together with the general theoretical principles required for Tasks A and B. Such blurring made it impossible to answer our "what range of infinitesimal behaviors are possible?" question with any confidence.

To get a better sense of what is at issue here, let us return to our old problem of how to combine the *traction vectors* acting upon a surrounding shell ∂S modelings with the body forces (including accelerations) that act inside S. In particular, let us carve out a finite internal volume S of a body B with an imaginary Eulerian cut. As before, sum (= integrate) all of these actors as resultant forces F^* and torques τ^* over S or ∂S according to need, just as we did with rigid bodies. But where inside S do F^* and torques τ^* act? What *representative point* should be appropriate for the finite volume S? In the case of rigid bodies, the answer did not matter because of the rigidity, but lumps of putty will act quite differently according to where F^*

---

[41] Clifford Truesdell, *A First Course in Rational Continuum Mechanics* (San Diego, CA: Academic Press, 1991) and C. Truesdell and R. A. Toupin, "Classical Continuum Physics," in S. Flugge, ed., *Handbook of Physics*, Vol. 3/i (Berlin: Springer, 1960:226-376). Morton Gurtin, Eliot Fried, and Lallit Anand, *The Mechanics and Thermodynamics of Continua* (Cambridge: Cambridge University Press, 2010) is also recommended and up to date. I should also indicate that many researchers outside of Truesdell's school contributed to the new understandings we shall outline, without fully embracing the "purism" characteristic of the latter's approach.

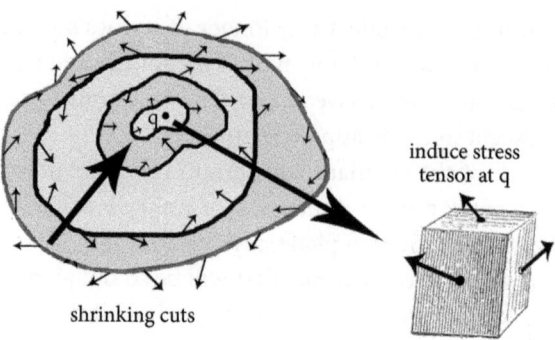

shrinking cuts

induce stress tensor at q

Figure 2.35

is placed. Once we establish how S as a whole behaves, we might be able to assign it some reasonable representative centers (its center of gravity, perhaps) but, right now, such centers move around inside S considerably according to how the blob is affected by the outside forces. It is at this stage that traditionalist approaches invoke rigidification or little mechanisms within S's that are sufficiently firm to allow our F^* and τ^* to work upon them in a more determinant manner. But to gain this firmness, the traditionalists invoke *constraints* and other modeling restrictions that our modernists regard as approximative and wish consigned to the "simplify the mathematics" purposes characteristic of task C's portfolio.

Accordingly, the modern approach advises us to overlook these "how do we halt the regress" concerns for the moment and assures us that we can nonetheless regard Euler's two basic laws of motion (or *balance principles*, as they are usually called in this context) as fully applicable to (almost) any Eulerian cut S. This is a rather abstract claim to accept, due to the fact that we possess little concrete sense of where or how F^* and τ^* will operate upon S. "Have patience," our modernists advise, "we'll trap it eventually." Crudely speaking, the proposal is that if we continue shrinking S to ever smaller dimensions, in the final limit, we will recover those infinitesimal cubes C we considered earlier (figure 2.35).

In fact, these C's are so small that they no longer qualify as Eulerian S's at all (the S's possess finite volumes, whereas our C's comprise "decorated points"). Due to their minute character, such C's will possess one traction vector on each face and only one (summed) body force vector and acceleration inside. Furthermore, the tractions upon opposing faces must be diametrically opposed lest our C cube find itself subject to an infinitesimal turning moment. The net effect of these forces is to make C either alter its volume or sheer, or some combination of the two. Once we know what happens here, then we can determine what happens in cuts with larger volumes S by simply integrating all of the infinitesimals C's that comprise it.

These procedures probably sound obscure (or even mystical) due to the fact that I have framed the proposal in the language of infinitesimals. So let us purge those notions from my presentation using tensorial objects instead (the basic technique for

WHAT IS "CLASSICAL MECHANICS" ANYWAY?

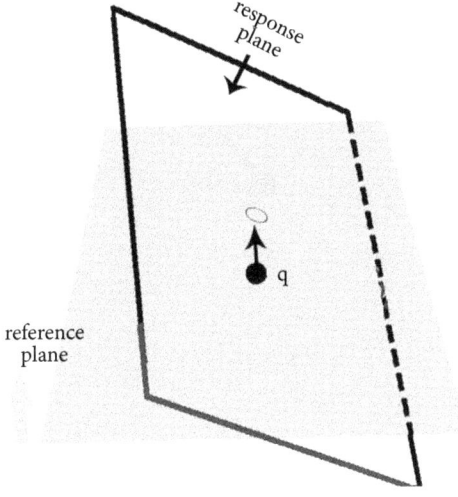

Figure 2.36

doing so is rather abstract but beautiful). To do this we must understand how *stress* and *strain tensors* function. I will begin with the latter, conventionally designated by ε. Take a point q inside a finite blob S and run an oriented reference plane through it (the orientation is supplied by the little gray arrow) (figure 2.36).

Our strain tensor intuitively provides, in the guise of a matrix of nine numbers, how much the corresponding face (or "response plane") of an infinitesimal cube at q has expanded or contracted (according to whether the center of the response plane has moved outward or inward from the reference plane) and also the degree to which the response plane has become tilted with respect to that original orientation (obviously it can tilt in both x and y directions). In other words, a *strain tensor* is a gizmo that maps planes through points q to new planes (this is part of its proper definition). Employing this strain tensor information about the response planes through q, we can, in effect, reconstruct our original strained infinitesimal C by calculating the dilation (= compression or expansion) and reorientation experienced by various choices of reference frame as we run them through q. Now there needs to be a gradualist coherence among our answers for we want our reconstructed "infinitesimal" to turn out to be a skewed cube and not, say, a skewed dodecahedron. We enforce this coherence among our answers through the standard "vector space" qualities demanded of any tensor. The upshot of all of this is that the strain tensor attaching to q can be fairly characterized as "the ghost of a vanishing shape"—the technique captures the data that we need to have installed at q in a manner that explains why we are intuitively inclined to picture q's strained state as an infinitesimal cube *with sides*.[42]

---

[42] Philosophers new to the peculiar world of continuum physics parlance should prepare themselves for phraseology such as "dimensionless point cube" (J. D. Reddy, *An Introduction to Continuum Mechanics* (Cambridge: Cambridge University Press, 2008), 126—an excellent book, by the way).

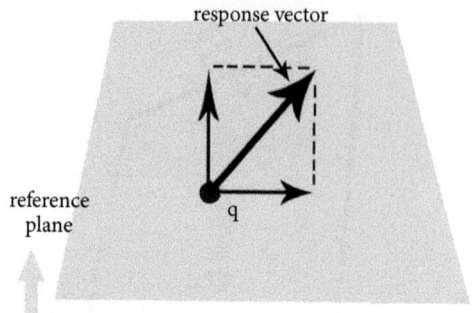

Figure 2.37

We employ the same techniques to make sense of the *stress tensor* σ at q, except that σ now places a tilted traction vector F upon our reference plane (figure 2.37).

The component of F that runs normal to the reference plane represents the pressure (compressive or dilatory) that strives to alter the volume of $\mathcal{S}$; the planar component of F captures its sheering capacities. Because we normally do not want to deposit any unbalanced torques on S, we require the F on the other side of our cube (= a reference plane with a reversed orientation) to be equal and opposite in magnitude. Operationally, this requires that the matrix of numbers corresponding to σ must be symmetric, with only six independent values.

In any case, our ε and σ tensors provide the basic information we require within our infinitesimal cubes,[43] while eschewing any talk of infinitesimals per se. I hope it is evident that, while the tensorial method for eschewing infinitesimals is quite clever, most of the entangled difficulties within our physical infinitesimal packet have been left untouched, for they largely concern the question of the local traits that need to be deposited at q for continuum mechanics to work coherently. Once those physical issues have been resolved, any "infinitesimal" proposal can be easily reworked into a collection of tensors or allied objects.

Using this language, the result of enforcing Euler's two laws of motion upon (almost) every cut $\mathcal{S}$ we can carve out of a body $\mathcal{B}$ tells us that stress and strain tensors will be locally defined at (most) points q inside $\mathcal{B}$ and will, furthermore, obey Cauchy's celebrated law of motion:

$$\rho D^2 q/Dt^2 = \text{div}\sigma + \rho g.$$

---

[43] A. N. Whitehead did some foundational work in mechanics at the turn of the twentieth century and his "method of extensive abstraction" was later popularized by Bertrand Russell in *Our Knowledge of the External World* (London: Routledge, 2009). I am not sure how Whitehead understood his construction (which shrinks in on points through decreasing volumes), but Russell plainly regarded the technique entirely as a logical procedure for "defining away points." Russell's misunderstanding of the underlying physical problematic continues to reverberate within the halls of analytic philosophy. For a survey, see Mark Wilson, "Beware of the Blob," in Dean Zimmerman, ed., *Oxford Studies in Metaphysics* (Oxford: Oxford University Press, 2008).

A *subtle point*: when we combine our stress and strain information, should our resultant vectors situate themselves on the reference or the response planes? This matter becomes important in nonlinear elasticity and requires the careful delineation of different stress tensors ("Piola-Kirchhoff" versus "Cauchy") that one finds in modern textbooks.

WHAT IS "CLASSICAL MECHANICS" ANYWAY?

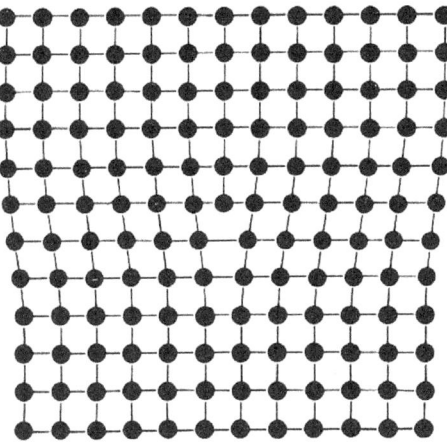

Figure 2.38

Here the divergence operator (div) evaluates how the stress field varies in the vicinity of q and provides us with a vectorial assessment of where the greatest changes in σ lie.[44] This provides us with a density vector that can be meaningfully summed with the body force densities that act at q. Observe that Cauchy's principle looks very much like Newton's second law as it appeared within our point-mass setting and many authors identify it as such (although that can be only regarded as a rather diffuse "family resemblance" claim, in that we are plainly dealing with a considerably more sophisticated construction now).

Indeed, it is a mistake—although many elementary textbooks encourage the opposite point of view—to assimilate notions like *stress* too glibly to more straightforward notions like *force* (I devoted a fair amount of space to their proper mathematical nature for this reason). Thus, many writers will assure their readers that stresses "reflect the short range forces within a material," which is true in some loose "stresses reflect information about such arrangements" sense (in a fashion that encourages us to conceptualize the underlying material in molecular terms). But there is no ready recipe that converts these molecular short-range forces into the numerical values that belong to the stresses assigned to points q within a continuum modeling of the situation.

Perhaps this last point can be clarified with a specific example. The short-range forces active within most real materials rarely bind them into perfect lattices, but tolerate the irregularities known as dislocations (figure 2.38). Large numbers of

---

[44] ρg, it will be recalled, captures the summed body forces acting upon q. In following this standard representation, we are tacitly ignoring the third law demands that persuaded us to distinguish $V(q)$ from $V^*(q^n)$ earlier (the mathematics of continua is rough enough without fussing about that!). It is important to realize that the accelerative term behaves mathematically very much like g and is often called an "inertial force" as a result (some of the third law ambiguities surveyed earlier trace to this drift in the significance of "force"). And an important symmetry with respect to constitutive equations is relevant as well: materials (usually) respond to an applied schedule of accelerations by exactly the same rules as they react to a comparable array of genuine forces (this requirement is called "material frame indifference" or "objectivity").

these lattice defects can emerge at scale lengths that need to be treated as short-range and can affect the macroscopic qualities of a material in significant ways. How, in a classical continuum modeling of our material, should its dislocational properties be registered? A lot of recent work in extended continuum mechanics (I will discuss some of this later) has supplied a variety of answers to this question. In some of these schemes, the dislocations are not captured in the material's strain tensor at all, but within other mathematical constructions attributed to the point q (e.g., to a torsion within the underlying manifold on which q lives). Such a torsion can be recognized as the "short-range forces" within the material just as ably as does its conventional strain, but follows a different coding scheme.

In truth, when we casually parse stresses as short-range forces, we are tacitly making a lift from continuum mechanics into a different conceptual arena within which the tricky notion of *stress* can be "rationalized" through a rough alignment with a more readily understandable form of material structure. Such lifts (which are a common occurrence in continuum mechanics) are fully in accord with the theory-facade character of "classical mechanics" overall, but they can obscure the fact that, considered in their own terms, *tensor fields* are novel mathematical constructions with their own spectrum of characteristics (mathematicians did not isolate the notion clearly until the end of the nineteenth century). Indeed, it is precisely these special qualities that allow modernists to halt our "labyrinth of the continuum" regress in a novel way: they claim that the traction vectors around shrinking S's will deposit a localized residue on q in the form of a stress tensor. With the help of a simple divergence computation, we can then extract a vector to add to the body force and acceleration in a mathematically coherent manner.

If we survey the conceptual framework just sketched, we realize that none of the fundamental principles employed directly concern points q, but instead talk, in sometimes very abstract ways, about how *finite volumes* S behave. Thus Euler's two laws of motion hold only of finite "cuts" S extracted from a body B; they do not make sense for individual points q. Conservation of mass, likewise, concerns how finite blobs S relate to the reference manifold. And so on. Mathematically, such principles need to be expressed by *integral differential equations*, not as localized differential equations per se. Cauchy's law, to be sure, is of the latter class but it has been derived from fundamental integral principles; it has not been posited as basic.

Within the point-mass setting of section 3, particular materials were credited with behavioral individualities through choosing the number of particles present within the system and assigning them material constants (mass, charge, etc.). These constants then turned on an appropriate set of special force laws in modeling, such as Newton's law of universal gravitation. Modeling specifications of this character we called "constitutive assumptions," secretly borrowing terminology from a continuum context. When we have successfully assembled a closed equational system by these procedures, we say that we have thereby followed "Euler's recipe". Although this portrait of modeling techniques within physics is both simple and appealing, in point of brute fact one regularly finds practitioners evading the recipe's dictates through appeal to ΔL^* level constraints and allied modes of physics avoidance. Indeed, those methodological intrusions have become so pervasive in practice that most point-mass modelers appear to forget that they have any obligations to track down a full set of special force laws at all. But the appeals that typically displace special force laws within such contexts scarcely seem lawlike in their own right: "X is a rigid rod" does not sound much like a law of nature. In that sense at least, it is misleading to insist that such physicists and engineers are seeking to find the laws to which nature conforms.

Within our current continuum-mechanics program, we are not allowed to invoke constraints in setting up our fundamental modeling equations. But what is the present analog to our former "special force laws"? The stress/strain constitutive assumptions we have just examined. "But there are zillions of these," we might protest. "Don't workers in continuum mechanics attempt to reduce their multitude to a smaller collection?" The answer is no, they don't; they merely try to sort the possibilities into general classes, so that the simplest forms of stress/strain behaviors can be studied first. In other words, they supply a *taxonomy* of possible constitutive behaviors, but no reductive listing of special force laws is ever offered. Indeed, in typical modeling practice, the constitutive principles assigned to a material are generally determined through *direct experimentation* on large hunks of the material in the laboratory. In this manner, in a Cauchy-recipe modeling, its core constitutive equations reflect a *projection* of behaviors witnessed experimentally at a large ΔL^* length scale down to an infinitesimal scale.

Let us finally turn to our Task C. According to the modern program under review, we should not invoke constraints of any sort in setting up our basic constitutive modelings. But this methodological prohibition is commonly violated within traditionalist presentations of continuum mechanics. Consider again the illustrated infinitesimal element for a Bernoulli-Euler beam (figure 2.39). Note that the applicable pushes and pulls upon the element are assumed to *balance along fibers* running across the material. This assumption represents a constraint on permissible behavior, of the same general character as we examined in the previous section. According to Cauchy's modeling recipe, we should properly supply constitutive equations of a Hooke's law ilk able to insure that stresses will be largely conveyed across the element in this fashion. Great—but see if you can fill out a matrix of coefficients that will do this. The sad truth is that this task is not at all easy—in fact, we have canvassed this same problem already, in the humble form of the vibrating string. The constraint critical to the simple "derivations" found in most college textbooks maintains that string elements forever hover infallibly above their original rest positions. But try to

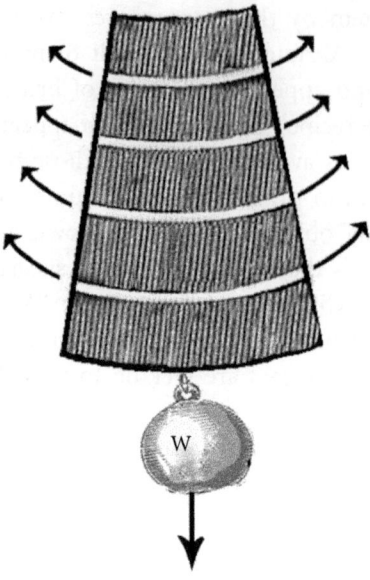

Figure 2.39

find a set of stress/strain relationships that can simulate this behavior approximately within a three-dimensional material. This is again a daunting task.[45]

Plainly, adhering to the foundational clarity demanded within our modern approach places ghastly burdens upon beginners in continuum physics, for the path to the one-dimensional wave equation becomes strewn with knotty mathematical thorns. Such considerations provide a rationale for instituting a new, approximative division of continuum mechanics that investigates how our strict Task A/Task B modeling requirements can be profitably circumvented through the wise exploitation of ΔL^* level constraint information. Such work frames a third Task C for continuum mechanics: develop mixed-level modeling techniques that relate to the strict constitutive-modeling requirements of "foundational" continuum mechanics in the same manner as the evasive techniques of analytical mechanics relate to Euler's recipe.[46] From this point of view, we can anticipate that the characteristic emphases of analytical mechanics will make a strong reappearance within practical continuum mechanics, for the simple reason that the former practices the approximative art of exploiting ΔL^* scale constraints to isolate the pathways of *dominating activity* within a complex ΔL-level medium. So it is not surprising that the lore of old-fashioned continuum mechanics appears riddled with innumerable

---

[45] For a vivid illustration of the divergence between traditional methods and the approved "modern" approach, see Stuart S. Antman, "The Equations for the Large Vibration of Strings," *American Mathematical Monthly* 87 (1980). Drops in dimension through appeal to symmetries usually act in the manner of constraints.

[46] As we have seen, traditional modelers commonly appealed to little mechanisms as a means of introducing Task C simplifications into their modelings, so that analytical mechanics serves as a convenient house of refuge for continuum mechanics as well.

lifts into rigid-body mechanics, for the demanding requirements of Cauchy's recipe needed to be relaxed before the equations that support the great eighteenth- and nineteenth-century advances in wave motion, and so on, could emerge into central focus.

Continuum modeling could have never gotten on its feet historically without the temporary assistance of rigidified infinitesimals and "little mechanisms." If d'Alembert, the author of the first PDE for a vibrating string, had felt obliged to deal with matrix equations containing 21 independent constants, continuum technique would have been abandoned as stillborn at birth. All of this merely underscores the lessons we have noted with respect to the secret contribution of lifts with respect to classical mechanics' triumphant hegemony.

But the specific constraint-assisted lifts that helped traditional continuum modelers on their way had the curious effect of encouraging themes within the philosophy of science that continue to reverberate strongly even to this day. They trace to the following factors. In order to block the "never simplifying" regress created by flexible materials that behave identically on all size scales, traditional modelers assumed that small portions of a material behave like little mechanisms. In doing so, they tacitly credited the lower scale lengths of a material with *characteristics that they did not believe they really possess*. It then appears that we cannot set up coherent "foundations" for flexible bodies without injecting *patent descriptive fictions* to arrest an otherwise vicious regress. And so the thesis emerges that physics cannot begin its descriptive tasks until it has first indulged in a preliminary degree of *essential idealization*: smallish portions of materials must be credited with patently incorrect characteristics. After we reach a completed modeling, we can throw away the idealized ladder we have climbed, for our final equations will describe materials that behave identically at every scale. But on route there, we must accept, in the physicist J. H. Poynting's phrase, a fictive "scaffolding from without."[47]

The assumption that some form of essential idealization must be invoked to arrest continuum physics' "labyrinth of the continuum" problem has played an important, if often unacknowledged, role in shaping the doctrines of the philosophers who pondered the problems of classical matter carefully: Leibniz, Kant, Duhem, Hertz, Mach, and others.[48] Its enduring legacy is the lingering presumption that *intentional misdescription* represents a commonplace activity in scientific activity. In retrospect, however, this philosophical thesis seems to have engendered by the lifts required to link Task A/Task B modeling demands with the more relaxed standards required in practical work of a Task C cast.

In the naïve form we articulated, our Task A approach to mechanics presumes that Euler's two laws hold true, in an abstract manner, for any contracting sequence of cuts $S, S', S, \ldots$ surrounding a target point q. And this presumes that their

---

[47] J. H. Poynting, "1899 Presidential Address to the Mathematical and Physical Section of the British Association (Dover)," *British Association Report* (1899), 615–624.
[48] As a case in point, a key document within the rise of "anti-realism" is Karl Pearson's once influential *The Grammar of Science* (London: Thoemmes Continuum, 1992), which is very explicit in its continuum mechanics roots, commingled with a variety of neo-Kantian themes.

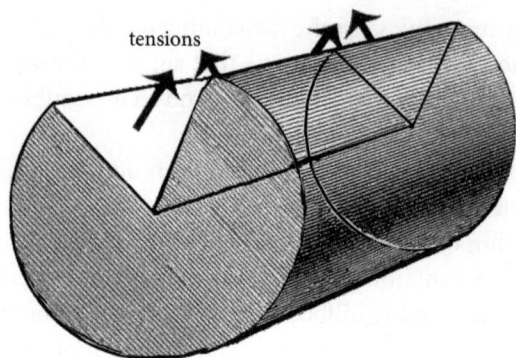

Figure 2.40

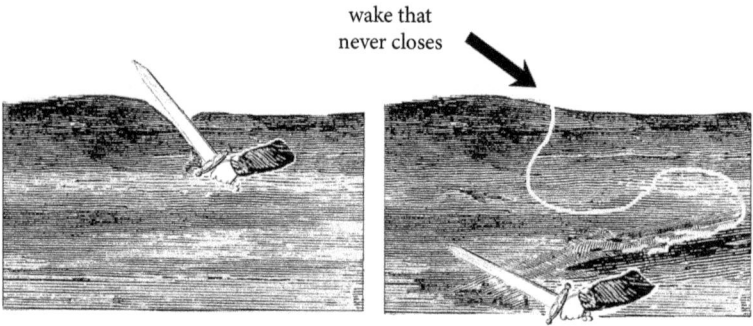

Figure 2.41

respective perimeters ∂S, $\partial S'$, $\partial S, \ldots$ can carry full complements of traction vectors. But this demand is too strong, partially because some ∂S are too irregular to bear such measures, but also because such requirements need to fail when ∂S cuts through a portion of shock wave surface (some irregularity must prevent the contracting cuts S, S', $S, \ldots$ from installing stress and strain tensors upon these problematic points).

In point of fact, the canonical modelings of traditional mechanics have long tolerated funny spots, namely singularities, upon their boundaries. For example, take a notched rod and pull upon its two open faces with a uniform tension (figure 2.40). The result is an infinite twist along the base of the cut. Modern treatments of continua employ rather fancy tools from functional analysis, such as trace operators, to bring the inner and boundary descriptions of continuous bodies into better mathematical accord. Very subtle considerations with respect to energy storage typically lie in the background of such interior/boundary "harmonizations."

Here is an example of this phenomenon that I regard as particularly telling. Pull a knife through some water. The knife draws the top layer of the water with it (figure 2.41). Intuitively, we expect that, after a certain period of mixing, the waters on the two sides of the cut will soon fuse together. But according to the story that

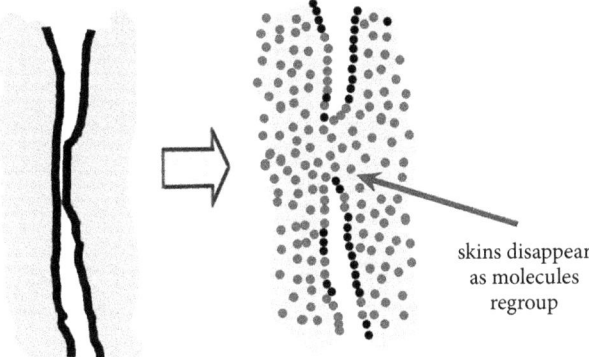

Figure 2.42

the PDEs of continuum mechanics tell, this wound can never heal, for differential equations cannot alter the topologies of the flows they track. But these descriptive limitations entail that, without some significant alteration, the mathematical framework of orthodox continuum mechanics can model neither the fusion nor the fracture of ordinary materials (which is why the subject traditionally confines its attention to noncomposite blobs in circumstances where they are unlikely to suffer fracture or fission.

To anyone who has not scrutinized the standard lifts of mechanical tradition in the critical manner of this essay, this claim will seem outrageous: "Of course, classical mechanics can readily handle mixing: the molecules from each side of the cut rapidly intermingle until it becomes impossible to determine where the dividing boundary had been" (figure 2.42). Yes, but observe that in this rationalization we have escaped into a reontologized ΔL domain governed by point-mass mechanics or something similar. It is "classical mechanics" all right, but it is not the same continuum mechanics with which we started. Due to these readily available lifts, one can learn a substantial amount of fluid mechanics without realizing that one's PDE tools are limited in this way.[49]

Although I have here discussed such issues in rather formal terms, many of the great historical philosophers of matter (e.g., Locke, Leibniz, Kant) commented upon the fact that the everyday processes of cohesion and disassociation appear very mysterious from a mechanical point of view. Only the point-mass approach handles such topics with any satisfaction. Yet it is unable to equip materials with the characteristics they need when they are not about the business of breaking or fusing. The only route to a satisfactory coverage of common forms of everyday material behavior is to weld together a classical mechanics from

---

[49] The discussion in Richard E. Meyer, *An Introduction to Mathematical Fluid Dynamics* (New York: Dover, 2007) brought home the point to me.

different descriptive platforms assessable to one another along suitable escape-hatch ladders.

6. Conclusion

In sum, if we go searching for the "foundational core" of classical physics practice in Hilbert's manner, we are likely to feel as if we have become trapped in a novel by Kafka, with particular branches of a vast bureaucracy claiming greater authorities than they truly possess and, when challenged, shunting us off to other departments that assist us no further in our quest. And the most maddening aspect of these unsettled convolutions is that the resulting interconnections appear, when evaluated from the perspective of brute pragmatics, as exceptionally well plotted in their organizational architecture, for the intricate interwebbing we call "classical mechanics" comprises as *effective a grouping of descriptive tools* as man has yet assembled, at least for the purposes of managing the macroscopic aspects of the universe before us with well-tuned efficiency.

In the final analysis, our investigations provide us with a richer understanding of why "family resemblance" structures often possess great pragmatic utility. The crucial point to observe is that the frequent lifts that populate the pages of college textbooks do not function as the "derivations" their authors suppose them to be, but instead provide Task C-style guidelines for how difficult modeling problems can be evaded through the exploitation of data (e.g., rigidity or principal directions of thrust propagation) extracted from observation along a mixture of scale lengths.[50] So while we have been critical of such textbook lifts when evaluated from a Hilbertian point of view, these same passages perform a crucial pedagogical purpose in directing a modeler's efforts to locally effective results. In the final analysis, it is the astounding success of these well-tuned models with respect to the macroscopic world that insure that "classical physics," as an important intellectual activity, will probably remain with us forever. So while it is important to recognize, from a methodological point of view, that the routes whereby standard textbook prose stitches the fabric of "classical mechanics" into a well-engineered facade rarely comprise "derivations" in a proper sense, the good offices they perform for us should not be devalued in rendering that judgment. I trust that many readers had the uneasy sense, when we criticized worthy textbooks earlier for failing to satisfy Hilbertian standards of rigor, that somehow our target authors were "doing the right thing" in their presentations regardless. Yes, but such passages serve a *different organizational purpose* than we have been led to expect.

---

[50] In the applications considered here, only two characteristic scale lengths are generally relevant, but Batterman's essay in this volume surveys some of the exciting recent work that promises a capacity to intermingle data extracted from a wider array of scale sizes.

Although we cannot properly explore the possibilities here, deeper answers are still wanted as to why the characteristic ingredients of "classical mechanics" bind together into a facade as effectively as they do. Although Wittgensteinians sometimes claim otherwise, our remarkable capacities to sort human faces into "family" groups wants explaining: the brain must perform some rough form of statistical analysis over facial features when it computes its groupings, although the psychological mechanisms involved do not appear to be well understood at present. Just so: the strong feelings of "family resemblance" with which every student of classical physics is familiar merit probing in the same vein. Tait invokes the phenomenon well:

> [A]ll who have even a slight acquaintance with the subject know that the laws of motion, and the law of gravitation, contain absolutely all of Physical Astronomy, in the sense in which that term is commonly employed: viz., the investigation of the motions and mutual perturbations of a number of masses (usually treated as mere points, or at least as rigid bodies) forming any system whatever of sun, planets, and satellites. But, as soon as physical science points out that we must take account of the plasticity and elasticity of each mass of such a system, the amount of liquid on its surface, . . . [etc.], the simplicity of the data of the mathematical problem is gone; and physical astronomy, except in its grander outlines, becomes as much confused as any other branch of science.[51]

Here Tait expresses his conviction that point-mass physics best encapsulates the elusive "central core" to classical mechanics, although he realizes that this "core" must be dressed within the confusing garments of flexible bodies before reliable empirical results can be obtained. But what is the true nature of this "central core"? I believe that any reasonable answer must come from a deeper understanding of how our classical descriptive tools *sit on top of quantum mechanics*: the ways in which we usefully track macroscopic "work" and "energy" at the ΔL^* level must somehow trace to the ΔL-importance of correspondent notions within the quantum domain.

However such issues resolve themselves, classical mechanics, as studied here, offers many valuable lessons to philosophy as a whole: in particular, that well-wrought conceptual structures can be assembled as facades tied together through "look across size scales" linkages. But to praise a family-resemblance fabric in this manner is not to deny that its organizational patterns can be accorded rational underpinnings. On the contrary, we should scrutinize lifts and escape hatches within a facade with formal care so that their operative strategies of physics avoidance become accurately identified and their empirical outreach accordingly improved. As a prerequisite to those diagnostic endeavors, we must first recognize that the

---

[51] P. G. Tait, *Heat* (London: MacMillan, 1895), 9–10.

"derivations" provided in elementary textbooks rarely satisfy Hilbertian demands on rigor but instead fulfill the "look across scale sizes" offices that allow the basic terminology of classical mechanics to cover wide swatches of macroscopic experience with an admirable efficiency. In these respects, current work in continuum mechanics provides an excellent paragon of how a useful base scheme can be profitably extended to wider applications once its conceptual supports have become viewed without methodological illusion.[52]

---

[52] The claim that everyday classificatory words operate along organizational principles similar to those surveyed here comprises the chief argumentative burden of my *Wandering Significance: An Essay on Conceptual Behavior* (Oxford: Oxford University Press, 2006).

CHAPTER 3

CAUSATION IN CLASSICAL MECHANICS

SHELDON R. SMITH

Before the nineteenth century, it was common to think that much of our understanding of the physical world was organized around the concept of cause and general "causal principles." According to David Hume, "All reasonings concerning matters of fact [roughly, non-tautologous truths] seem to be founded on the relation of *Cause and Effect*" (Hume 1748, 16). Later, for Immanuel Kant, the category of cause was one of the pure categories of the understanding which the mind uses to structure its experiences and without which comprehension of a coherent world would be impossible. By the late nineteenth century, however, it became common among physicists and like-minded philosophers to assert that, at bottom, the concept *cause* was not particularly important for understanding the physical world because careful study of the physical world had revealed causation to be absent from it.

The backdrop in philosophy for most contemporary discussions of such skepticism about causation and its import is Bertrand Russell, who claimed that considerations of causation play no role in theorizing in advanced sciences, especially

---

I would like to thank the following for comments on an earlier draft: John Norton, an audience at the University of South Carolina (especially Michael Dickson, Matthew Kizner, and Michael Stoeltzner), an audience at UC San Diego (especially Craig Callender, Ioan Muntean, and Chris Wüthrich), and the Southern California Philosophy of Physics Reading group (Jeff Barrett, Craig Callender, Sam Fletcher, Jason Hoelscher-Obermaier, David Malament, Casey McCoy, Tarun Menon, and Jim Weatherall). Special thanks to Sam Fletcher for calling many typos and other mistakes to my attention. I would also like to thank Stephen Parrott for a very helpful email correspondence, Mark Wilson for conversations on these topics over many years, and Bob Batterman both for past encouragement and for his patience in giving comments on this essay.

physics.¹ Although Russell's discussion is complex and involves a number of distinct considerations, two of the main points seem to be the following:²

1. It is impossible to unequivocally identify "causes" and "effects" within the fundamental equations of physics, but physics gets along just fine in spite of that fact.³ This shows that causation is not a fundamental feature of the world.
2. There are no general causal principles that place restrictions on physical behavior that are not otherwise there.⁴

With regard to the latter point, Russell claims, "The law of causality, I believe, like much that passes muster among philosophers, is a relic of a bygone age, surviving, like the monarchy, only because it is erroneously assumed to do no harm" (Russell 1981, 132). Russell's explicit target—the "law of causality"—seems to be "same cause, same effect," but the general tenor of his discussion suggests that he would also reject maxims like "everything has a cause" and "the cause comes before the effect."

These days, many think that Russell's skepticism is misguided.⁵ For, it is noted that physicists frequently invoke considerations of "causality" when evaluating equations, including candidates for fundamental equations of classical physics. Although "causality" can mean different things (e.g., determinism and restrictions on the velocity of causal propagation),⁶ I shall focus in this essay on the maxim, "the cause precedes the effect," and on some places within classical physics where appeal to it allegedly enters.⁷ (My discussion will be limited to issues that arise already in a flat spacetime without closed causal curves.⁸) Here is a brief list of the places where it is claimed that a "Principle of Causality" applies:⁹

---

[1] Three recent papers that are sympathetic to Russell are Norton (2007a), Norton (2007b), and Hitchcock (2007). See, more generally, the papers in Price and Corry (2007).
[2] Whether this is actually an accurate interpretation of Russell is an open question. It does, I believe, represent a view that has become associated with his name.
[3] Russell never puts the point this way, but he does claim, "In the motions of mutually gravitating bodies, there is nothing that can be called a cause, and nothing that can be called an effect; there is merely a formula" (Russell 1981, 141).
[4] My formulation of this claim owes to Norton since this is how he often expresses his own skepticism about causal principles (Norton 2007a).
[5] Among those who argue this via examination of physical theory are Steiner (1986), Cartwright (1983, 1989), and Frisch (2005, 2009a).
[6] For interesting discussion of the status of determinism in classical physics, see Earman (1986, 2007), Norton (2006, 2007a), and Wilson (1989). For discussion of restrictions on the velocity of propagation, see Earman (1987) and Maudlin (2002). Obviously, in a relativistic context, if there is causation over spacelike separation, then there is backward causation in some frames of reference unless one "reinterprets" the direction of causation in such frames. I will limit myself, however, to cases of backward causation that are within or on the light cone.
[7] Mathias Frisch prefers to frame the principle as "the cause does not come after the effect." This way of stating the principle seems to have as its sole motivation the desire to maintain that nothing is amiss with equations like Newton's Second Law, $F = ma$, even though the force does cause the acceleration but does not come before it, since they are simultaneous. However, because I will not discuss alleged cases of simultaneous causation, I will stick with the more standard wording, since nothing will depend upon the difference.
[8] For discussion of "causality restrictions" in curved spacetimes brought about by the existence of closed-causal curves in some models of General Relativity, see Hawking and Ellis (1973) and Earman (1995).
[9] Frisch has appealed to nearly all of these items from physics in support of the importance of causality considerations within physics (Frisch 2005, 2009a).

1. Advanced Green's functions—to be described later—are often declared by physicists to be unphysical because they are "acausal." They suggest that the effect comes before the cause.
2. In radiation theory, the Sommerfeld radiation condition is used to rule out waves collapsing in on a source from infinity. Intuitively, these are waves that are caused by the source but are caused "into the past." So, we invoke the Sommerfeld radiation condition so as to adhere to causality.
3. Some equations of motion—for example, the Abraham-Lorentz Equation and the related Lorentz-Dirac Equation—are dismissed by physicists because they violate causality.
4. In dispersion theory, dispersion relations are shown to follow for systems which are causal. Since one invokes causality early on in the theory of dispersion, it might be thought that causality is an important, fundamental principle in physics.

I shall discuss each of these items more or less in turn with an eye toward examining what role—if any—assumptions of causality play. A complete weighing of the role of causality is out of the question in an essay of this scope. But, I hope to give at least a sense of the sorts of arguments that have been presented.

1. Identifying Causes and Effects: Advanced and Retarded Green's Functions

Although Russell's skepticism is not obviously coming from these quarters, one major driving force for causal skepticism of the sort that he advocates is the denial that there is any reason to privilege the so-called "retarded" Green's function for a system over the "advanced" Green's function. Green's functions will be described in detail below, but roughly a Green's function describes the effect of an instantaneous impulse acting on a system. The retarded Green's function describes the effect as coming after the cause, whereas the advanced Green's function describes the effect as preceding the cause.

As a sociological matter, it is certainly not uncommon for physicists to announce the privilege of the retarded Green's function or related notions like the retarded potentials. The following quotations will give some sense of the ubiquity of such claims:

- "Although the advanced potentials are entirely consistent with Maxwell's equations, they violate the most sacred tenet in all of physics: the principle of causality." (Griffiths 1981, 399)[10]

---

[10] Frisch, someone who thinks that causality does play a substantive role in theorizing, points to this passage among others (Frisch 2009a).

- "In the usual theory one omits the advanced solutions [or advanced Green's functions] as [their] existence would be in violation of our ordinary concept of causality." (Weigel 1986, 194)
- "The advanced potentials are mathematically allowed solutions, but they conflict with the basic physical concept of causality." (Johnson 1965, 21)
- "For the time being, at least, we discard the advanced solution as unphysical." (Vanderline 2004, 272)

I shall begin with the very basic example of an undamped harmonic oscillator, since much of the problematic already arises in this simple context. An example of a harmonic oscillator would be a block attached to a spring having a linear restoring force with no damping present. The equation for a simple harmonic oscillator is just a concrete instance of Newton's Second Law, $F = ma$:

$$\ddot{x}(t) + \omega_0^2 x(t) = f(t), \tag{1}$$

where dots represent time derivatives and ω_0 is taken to be constant. In (1), $f(t)$ is an "inhomogeneous term," a term that does not involve the dependent variable (in this case x). Such terms are typically thought to represent a cause acting on the system from outside. There are several reasons for this thought: insofar as the inhomogeneous term does not involve the dependent variable, it is not representing something that is a function of the internal state of the system being modeled. But since it makes a difference to the evolution of the system, it represents an outside influence on it. Another consideration is that a nonzero inhomogeneous term will increase (or decrease as the case may be) the energy of the oscillator, but there should not be any such energy change in a system that is "closed" (in the sense of being noninteracting with its environment).

Perhaps, we have found a rationale for thinking of $f(t)$ as a cause of changes to the system. Insofar as this is so, the first claim frequently made by causal skeptics—that one cannot identify causes in physics—appears to be overblown. However, a consequence sometimes drawn from this skeptical claim might still be thought to be correct: Nothing terribly important hinges upon the identification of $f(t)$ as a cause for the purposes of doing mathematical physics; one can solve the equation—given some $f(t)$ and suitable conditions leading to uniqueness—without thinking of $f(t)$ as representing a cause, and such a solution would give the entire evolutionary history of the oscillator. Since one derives the entire trajectory of the system without having to identify $f(t)$ as a cause of anything, it is not so obvious that the identification of causes is particularly important for physics. I shall, however, leave the question of what import this identification has to one side. For, even bracketing that issue, we still have not seen that we can locate causes *and effects*. What is the effect of the cause represented by the inhomogeneous term, $f(t)$?

Since such a term can be rather complex depending upon what function $f(t)$ is, it is easiest to think in terms of a "point-source," a source that acts only at a single instant of time. This will be fruitful in this case, since we have a linear equation, and one will be able to use this simpler inhomogeneous term to compose more complex ones. Such an instantaneous source is represented by the Dirac delta function. A solution to (1) (along with other conditions) when there is a Dirac delta function source is known as a Green's function.[11] Thus, the Green's function problem is based around the following equation, which tells us how a harmonic oscillator responds to a delta function kick or gives the effect of such a kick:

$$\ddot{g}(t) + \omega_0^2 g(t) = \delta(t). \tag{2}$$

Any $g(t)$ that solves the Green's function problem is a Green's function.

The easiest path to a Green's function is via Fourier transforms.[12] The Fourier transform of a function, $f(t)$, may be defined as follows:

$$\frac{1}{\sqrt{2\pi}} \int_{-\infty}^{\infty} f(t) e^{i\omega t} dt. \tag{3}$$

Thinking for a moment of (1), we can Fourier transform each term of the equation:

$$\frac{1}{\sqrt{2\pi}} \int_{-\infty}^{\infty} \ddot{x}(t) e^{i\omega t} dt + \frac{1}{\sqrt{2\pi}} \int_{-\infty}^{\infty} \omega_0^2 x(t) e^{i\omega t} dt = \frac{1}{\sqrt{2\pi}} \int_{-\infty}^{\infty} f(t) e^{i\omega t} dt. \tag{4}$$

Using the fact that the Fourier transform of a derivative is $-i\omega$ times the Fourier transform of the undifferentiated function, we get

$$(-i\omega)^2 \hat{x}(\omega) + \omega_0^2 \hat{x}(\omega) = \hat{f}(\omega), \tag{5}$$

where $\hat{x}$ indicates the Fourier transform of x. Simple algebra gets us to

$$\hat{x}(\omega) = \frac{\hat{f}(\omega)}{\omega_0^2 - \omega^2}. \tag{6}$$

One hopes to arrive at the solution to equation (1) via inverse Fourier transform from (6):

$$x(t) = \frac{1}{\sqrt{2\pi}} \int_{-\infty}^{\infty} \frac{\hat{f}(\omega) e^{-i\omega t}}{\omega_0^2 - \omega^2} d\omega. \tag{7}$$

Let us return to the Green's function problem where $f(t) = \delta(t)$. Up to a factor of $\frac{1}{\sqrt{2\pi}}$, the Fourier transform of $\delta(t)$ has the value that $e^{i\omega t}$ takes at $t = 0$.[13] By Euler's identity, this is just the value that $\cos \omega t + i \sin \omega t$ takes at $t = 0$, and this is just 1. So, for the Fourier transform of the Green's function, we derive

$$\hat{g}(\omega) = \frac{1}{\sqrt{2\pi}} \cdot \frac{1}{\omega_0^2 - \omega^2}. \tag{8}$$

---

[11] Typically, a Green's function will be a "weak solution," which means that it is not sufficiently differentiable to be a solution in the classical sense.
[12] Some of what follows requires the theory of distributions to be made rigorous.
[13] For a proof of this using the theory of distributions, see Richards and Youn (1990, 67).

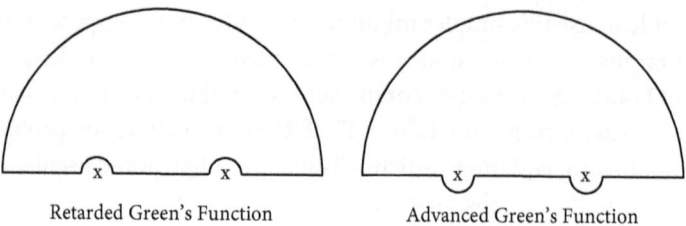

Figure 3.1 Contours for harmonic oscillator Green's functions.

To get "the" Green's function, one just needs to take the inverse Fourier transform:

$$g(t) = \frac{1}{\sqrt{2\pi}} \int_{-\infty}^{\infty} \hat{g}(\omega) e^{-i\omega t} d\omega. \tag{9}$$

The problem with the standard interpretation of this integral is that $\hat{g}(\omega)$ has poles on the real axis (at $\omega_0 = \pm\omega$). To make sense of the integral one may choose a contour that goes around the poles (like the contours in figure 3.1 where "x" represents a pole of $\hat{g}(\omega)$). One then takes some limit so that the contour encompasses the entire real axis (except for the dimples around the poles) with the contribution from the large semi-circle going to zero. (More details about this will follow when the damped oscillator is discussed.) Depending upon what contour one takes, one gets either the retarded Green's function, g_{ret}, or the advanced Green's function, g_{adv}, or some linear combination of them. Those functions are as follows (Butkov 1968, 282):

$$g_{ret}(t) = \begin{cases} 0 & t < 0 \\ \frac{1}{\omega_0} \sin \omega_0 t & t > 0 \end{cases} \tag{10}$$

$$g_{adv}(t) = \begin{cases} -\frac{1}{\omega_0} \sin \omega_0 t & t < 0 \\ 0 & t > 0 \end{cases} \tag{11}$$

The first arises from taking the integral along a contour including the real axis but with dimples that go over the poles (as in the left of figure 3.1). The second arises from a contour with dimples that go under the poles (as in the right of figure 3.1).[14] A linear combination of the two arises from taking a contour that goes over one pole and under the other.

Insofar as these Green's functions represent the response of the system to a Dirac delta function cause, it seems reasonable to think that they represent the effect of such a cause for a harmonic oscillator system. Unfortunately, however, we have not found a unique effect. We have, rather, two[15] different evolutions associated with the Dirac delta function cause. The retarded Green's function suggests that the Dirac delta function kick causes harmonic oscillations *after* the kick is applied

---

[14] As will be clear from the discussion of the damped oscillator to follow, the contours depicted in figure 3.1 represent the Green's function only for $t < 0$.
[15] In fact, again, there are more than two, since other contours of integration result in linear combinations of those two.

to the system; the advanced Green's function suggests that similar oscillations are caused *before* the application of the kick. A causal skeptic can note here that we have failed to identify the effect of the delta function kick, but it is not clear that we are the worse for it.

The situation is analogous for Maxwell's equations describing the behavior of the electromagnetic field in a vacuum. After some mathematical work to decouple the equations, one gets the wave equation for the electric field (and an analogous equation for the magnetic field):

$$\nabla^2 E - \mu_0 \epsilon_0 \frac{\partial^2 E}{\partial t^2} = f(t). \tag{12}$$

If one solves for "the" Green's function for the wave equation via such Fourier transform means, one finds that what one gets depends upon a contour since, as in the case of the harmonic oscillator, the Fourier transform of the Green's function has poles on the real axis (Griffel 1981, 74; Barton 1989, 406). The retarded Green's function for the wave equation suggests that if the field is subject to a Dirac delta function kick, a spherical wave spreads out at the speed of light into the future. The advanced Green's function, on the other hand, suggests that such a wave collapses in on the source at the speed of light. Unless one can find some reason to privilege one of these representations of the effect of a Dirac delta function kick, we have failed to determine the effect of such a kick.

1.1 Privileging the Retarded Green's Function

Equations (1) and (12) are insufficient to uniquely determine the effect of the inhomogeneous term. Nevertheless, there are two standard approaches to selecting the retarded Green's function as—in some sense—preferred. They apply similarly in the oscillator and wave equation cases. The first is to think in terms of a different differential equation for which the Fourier transform procedure sketched above gives a unique Green's function. The second is to add additional constraints to the differential equation for the Green's function but without modifying the differential equation. Let me consider them in turn.

1.1.1 Adding Damping

One standard "physical motivation" for a particular contour of integration is that any real, macroscopic oscillator will have damping (Butkov 1968, 281). For example, for a block attached to a spring, there would be internal damping in the spring but also from the ambient air. Since such damping was not included in the equation, there is missing physics in the undamped oscillator equation. To remedy this, one turns to the damped oscillator equation. If we add dissipation (of a simple sort), we get

$$\ddot{x}(t) + \gamma \dot{x}(t) + \omega_0^2 x(t) = f(t), \tag{13}$$

where $\gamma > 0$ and constant. As in the undamped case, we Fourier transform everything to get

$$(-i\omega)^2 \hat{x}(\omega) - \gamma \omega i \hat{x}(\omega) + \omega_0^2 \hat{x}(\omega) = \hat{f}(\omega). \tag{14}$$

Solving for the Fourier transform of $x(t)$, we get

$$\hat{x}(\omega) = \frac{\hat{f}(\omega)}{(\omega_0^2 - \omega^2) - \gamma \omega i}. \tag{15}$$

Thinking in terms of a delta function forcing, we arrive at the Fourier transform of the Green's function for the case as before:

$$\hat{g}(\omega) = \frac{1}{\sqrt{2\pi}} \cdot \frac{1}{(\omega_0^2 - \omega^2) - \gamma \omega i}. \tag{16}$$

The Fourier transform of the Green's function for this case has poles in the lower half of the complex plane, not on the real axis.[16] In fact, adding damping, however small, to the undamped oscillator will bring the poles of the Fourier transform of the Green's function down into the lower part of the complex plane. Because of this, one no longer has the issue of what contour to take around the poles lying on the real axis, since the poles no longer lie on the real axis.

From the inverse Fourier transform, one gets a Green's function for the damped oscillator that is retarded in the sense that it is zero before the application of the delta function kick. To give a picture of how one arrives at this result, one takes the inverse Fourier transform of the Green's function

$$g(t) = \frac{1}{2\pi} \int_{-\infty}^{\infty} \frac{e^{-i\omega t}}{(\omega_0^2 - \omega^2) - \gamma \omega i} d\omega \tag{17}$$

via contour integration. One starts by evaluating the integral over a contour from $-R$ to R on the real axis along with a semi-circle either over or under that interval and then taking the limit as R goes to infinity. (See figure 3.2.) If one closes the contour downward via a semi-circle in the lower half of the complex plane, it can be made to include the poles of the integrand and, thus, by the Residue Theorem the integral will be $-2\pi i$ times the sum of the residues of those poles (since the curve is oriented clockwise). One then shows that as R goes to infinity, the integral along the semi-circle will go to zero when $t > 0$. So, in the limit, the integral over the contour is just the integral on the real axis, since the contribution from the semi-circle goes to zero, but the integral is nonzero, since there are poles within the contour. This will not, however, represent $g(t)$ for $t < 0$ for, in that case, the integral over the chosen semi-circle would not vanish; in fact, it becomes unbounded.[17] However, if, instead, one closes the contour of integration with a semi-circle in the upper half of the complex plane, one does get a representation of $g(t)$ for $t < 0$ since, in that

---

[16] For discussion of their placement, see Butkov (1968, 277).
[17] This is not hard to see. If we think of ω as a complex variable, $\omega_r + i\omega_i$, the $e^{-i\omega t}$ in the numerator of the inverse Fourier transform is $e^{-it(\omega_r + i\omega_i)} = e^{-i\omega_r t + \omega_i t} = e^{-i\omega_r t} e^{\omega_i t}$. When $t < 0$ and $\omega_i < 0$, the second exponent is positive and will grow as ω_i gets smaller in the limit.

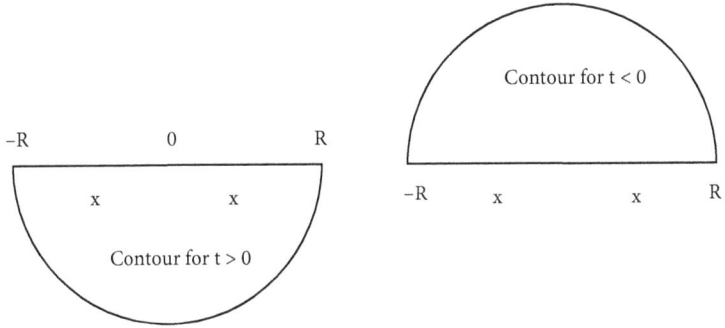

Figure 3.2 Contours for the damped oscillator.

case, the contribution of the semi-circle does vanish as one takes the limit as R goes to infinity. However, since the integrand is analytic within that contour, one gets that the value of the integral is zero. Thus, $g(t)$ turns out to be 0 when $t < 0$. As such, one arrives via these means at a retarded Green's function. (For details of the derivation, see Butkov (1968, 277–80); see also Wallace (1984, 157–160).)

The motivation, then, for thinking that the retarded Green's function for the undamped oscillator truly reveals the causal direction of the system is the following: the equation for the undamped oscillator really is just an approximate model of a system for which the damping is quite small; in reality, there are no macroscopic, undamped oscillators. So, we know that the system we are modeling is actually damped. As such, the correct Green's function is the retarded one, since that is the only one that arises via this procedure for the damped oscillator. So, the retarded Green's function correctly represents the causal directionality of the system. We just started with an approximate equation, the one for the undamped oscillator. But, had we used a better equation, the one for the undamped oscillator, we would have gotten the causal directionality of the system right.

Note that on this approach the retarded Green's function for the oscillator is not privileged because it is the only one that obeys the maxim, "the cause is prior to the effect." Rather, it turns out that the cause is prior to the effect because there is damping in any real macroscopic oscillator. So, it does not look as if there is any brute appeal to a "causality maxim" here. Rather, all of the work in privileging the retarded Green's function follows from the actual presence of damping. So, roughly, the constraint is: "If there is phenomenological damping, add damping to the model." Once we invoke that, we do not have to invoke anything like "the cause is prior to the effect." So, this is prima facie a case where one is not imposing "causality" but is imposing something else.

Moreover, even if this were the right way to think of the privilege of the retarded Green's function in the oscillator case, it is not clear that it is an appropriate way to think of all cases of interest. In a vacuum, the classical electromagnetic field is taken genuinely to be described by the undamped wave equation (12) even though the poles of the Fourier transform of the Green's function for it lie along the real axis.

In this case too, if one adds damping, one ends up with a retarded Green's function. But, if we do not think that the undamped equation is just an approximation to a better equation involving damping, why would we think that we have learned something about the causal directionality of the electromagnetic field in a vacuum from considering damped systems?

One possible response is that we learn from *other* systems what the appropriate causal directionality is for *this* system because from these other systems we learn about causal directionality *tout court*, causal directionality in nature as a whole rather than in this or that system. Since waves in a material medium are damped, we learn that retarded Green's functions properly represent causal directionality *tout court*. If it were the case that some materials were anti-damped and, thus, give rise to advanced Green's functions via the Fourier transform procedure, we would not be able to assign a unique causal directionality *tout court*.[18] Rather, we would have to talk of the causal directionality of this or that system. But, if there are no anti-damped materials, we are able to assign a causal directionality *tout court*. As such, we are able to assign a causal directionality for waves in a vacuum, and that directionality is properly represented by the retarded Green's function.

At this point, however, a causal skeptic will want to know what reason we have for thinking that there is causal directionality *tout court*. Why doesn't the fact that we did not find unambiguous causal directionality in the wave equation show that there is not? The claim that there is causal directionality *tout court* is motivated by the claim that for all systems that *have* a privileged causal direction, it is the same direction. Even if that is true it does not change the fact that there are some systems that do not have one. On the other hand, there are some grounds for thinking that the equations that do not reveal a privileged causal direction should be our guide to whether there is causal directionality *tout court*. For, typically, the most fundamental equations of classical physics do not have the time asymmetry that appears in equations with damping. Thus, one might think that the undamped equations are a better guide to causal directionality in nature than the damped equations. However, they show the absence of causal directionality. As such, we still have not been able to uniquely identify effects.

Of course, we have found causal directionality in the damped oscillator equation. This suggests that it is not true that it is impossible to identify causes and effects within some of the equations of physics. But, if the equations that do not have causal directionality are more fundamental than the ones that do, one does not get the impression that causal directionality is a fundamental feature of the world as is sometimes thought.

1.1.2 "Causality": The Initial Value Problem

There is another approach to selecting the retarded Green's function that is more commonly associated with invoking a maxim of "causality." Here, one does not look

---

[18] It is not obvious that there could not be anti-damped materials.

to a modified equation of motion, so no new physics is added. Rather, one adds supplementary data that will select out the retarded Green's function as unique. In particular, one posits that the Green's function satisfies the following initial condition: For $t < 0$

$$g(t) = 0. \tag{18}$$

It is evident that this rules out the advanced Green's function and also any nontrivial linear combination of the advanced and retarded Green's functions. Often, this condition is referred to as a causality condition. However, one might like a motivation for imposing this condition. A causal skeptic will not be too impressed with the claim that we impose it so that the cause precedes the effect, since he is skeptical about both the truth of that claim and of its importance. As we shall see, a motivation for such initial conditions can be found, but once we see what it is we may note that it has little to do with causation per se and that certain elements of the buildup to it are optional.

Let us start by thinking generally about the initial value problem. The general solution to an initial value problem for an inhomogeneous equation like (1) is the sum of the general solution to the associated homogeneous equation—i.e., (1) with $f(t)$ set to zero—and a particular solution to (1). Thus, the general solution to (1) is as follows:

$$x(t) = x_h(t) + x_p(t) \tag{19}$$

where $x_h(t)$ is the general solution to the homogeneous equation and $x_p(t)$ is a particular solution to the inhomogeneous equation. To get a unique solution, however, one typically assigns both the amplitude $x(t)$ and the velocity $\dot{x}(t)$ at a particular time. Suppose one assigns such initial conditions at a time t_0, before the inhomogeneous term turns on. In this case, a simple way to write the solution is to build the initial conditions into the solution to the homogeneous equation in the sense that $x_h(t_0) = x(t_0)$ and $\dot{x}_h(t_0) = \dot{x}(t_0)$. Because it satisfies the initial conditions, let us call that part of the solution $x_i(t)$. The solution to the relevant inhomogeneous problem may now be written in terms of the Green's function as follows:

$$x(t) = x_i(t) + \int_{-\infty}^{\infty} g(t - t') f(t') dt'. \tag{20}$$

But, for this to work out, one needs some further conditions upon $g(t - t')$ so that one does not end up contradicting one's initial conditions. What one does not want is for $\int_{-\infty}^{\infty} g(t - t') f(t') dt'$ to have either a nonzero total value or a nonzero velocity at the initial moment, t_0. One can intuitively see that if one were to add the advanced Green's function arising from a source later than t_0 to $x_i(t)$, one will have contradicted one's initial conditions. In more detail: How could the integral in (20) acquire a nonzero value at t_0? In the setup, $f(t)$ is zero until after t_0 only becoming nonzero at some later time t_1;[19] let us suppose that it remains nonzero until t_2. The integral can only acquire a nonzero value at t_0 if $g(t_0 - t')$ and $f(t')$ are nonzero together (and are so on more than a set of Lebesgue measure zero). If $g(t_0 - t')$ is nonzero on more than a set of Lebesgue measure zero in the interval where t' ranges from t_1 to t_2, then there risks being a contribution to the integral. However, all of the times in that interval are later than t_0. So, $g(t_0 - t')$ has a negative argument along that interval. If $g(t - t')$ takes only the value zero for negative arguments, then the integral cannot have a nonzero value at t_0. So, one will want to think of the Green's function as satisfying the initial condition (18) in addition to (2). These conditions leave only the retarded Green's function standing. Thus, the desired solution to the problem can be written as follows:

$$x(t) = x_i(t) + \int_{-\infty}^{\infty} g_{ret}(t - t') f(t') dt'. \tag{21}$$

Many view imposing condition (18) on the Green's function as imposing "causality." However, the motivation for (18) given above has nothing to do with wanting to maintain the truth of any causal dictum like "the cause precedes the effect." Instead, it had to do with not wanting to violate the principle of noncontradiction: once we have assigned initial conditions that represent the state of the oscillator at t_0 and we have built those initial conditions into the solution to the homogeneous equation, the solution to the inhomogeneous equation that we add to it cannot be such as to contradict them. It is a bit odd to call this approach the imposition of "causality" when the real driving force in the argument is the principle of noncontradiction and not anything having to do with causation.

As a matter of fact, quite often, physicists are considerably more lax about allowing advanced Green's functions to play a role when there is no possibility of contradicting specified state values. For example, advanced Green's functions are often allowed to play a role in the derivation of an equation of motion for a charged point-particle that senses its own field—an equation that includes a particle's "self-interaction" or "radiation reaction." Assuming that a charged particle with a nonzero radius, a, gives rise to a purely retarded field as it moves in one dimension, the force on it from its own field is (Feynman et al. 1989, 2: 28–6)

$$F_{ret} = \alpha \frac{e^2}{ac^2} \ddot{x} - \frac{2e^2}{3c^3} \dddot{x} + \lambda \frac{e^2 a}{c^4} \ddddot{x} + \ldots, \tag{22}$$

where α and λ depend upon the shape of the particle and the charge distribution. Like the last term written explicitly, all later terms go to zero as a goes to zero. However, the first term blows up in that limit. So, we cannot by this means define the self-force on a point-particle. Essentially the same problem with the limit would face us if we had, instead, assumed that the accelerated charge gives rise to an

---

[19] If one envisions cases where there is forcing before t_0, one still has a rationale for taking $f(t)$ to be zero before that time. Either the forcing before that time has no effect on the state at t_0 or its effect is already fully taken into account in $x_i(t)$.

advanced field, but in that case, we would get

$$F_{adv} = \alpha \frac{e^2}{ac^2}\ddot{x} + \frac{2e^2}{3c^3}\dddot{x} + \lambda \frac{e^2 a}{c^4}\ddddot{x} + \ldots \quad (23)$$

for the self-force.

Since Dirac, it has been common to derive an equation of motion for a point-electron by assuming that the interaction of the electron with itself is a combination of the advanced and retarded potentials, specifically, one-half of the difference of the advanced potential and the retarded potential. If we assume this, the first term which is problematic in the limit goes away. Since the later terms all go to zero in the limit, we are left with

$$F_{self} = \frac{2e^2}{3c^3}\dddot{x} = mt_r\dddot{x} \quad (24)$$

as the self-force on a point-charge (where $t_r = \frac{2e^2}{3c^3 m}$ is the time it takes for light to cross the classical electron radius). The equation of motion for a charged point-particle, known in the nonrelativistic case as the Abraham-Lorentz equation, is

$$m(\ddot{x} - t_r\dddot{x}) = F_{ext}. \quad (25)$$

(See Jackson 1975 for further details.)

We will return to this equation later, but for now it is worth noting that the assumptions in the derivation above are not always rejected as violating causality. For example, Feynman rehearses Dirac's derivation, and insofar as he is dismissive of it, it is because it contains an "arbitrary" assumption not because of any violation of "causality" (Feynman et al. 1989, 2-28-5). At any rate, when one derives this equation of motion, there are no imposed initial conditions, since one is not solving an initial value problem. As such, there is no fear of contradicting one's initial conditions. And, when this is so, it is much less common to balk at the inclusion of the advanced potentials even though, presumably, the causal interpretation of them as involving backward causation would remain. This provides evidence that when physicists reject the advanced potentials as *acausal* they are not abhorring the cause coming after the effect per se. Rather, they are thinking in terms of an initial value problem where the initial conditions are already built into the solution to the homogeneous equation and the retarded Green's function must be used on pain of contradiction.

Further fuel for the causal skeptic comes from realizing that various elements within the motivation for representing motion in terms of the retarded Green's function are optional. In particular, one does not need to build the initial values into the solution to the homogeneous equation. Rather, among other things, one can also solve a "final value problem" where the state of the system is assigned at a time *after* the inhomogeneous term has stopped acting. In this case, the very same solution will be represented as

$$x(t) = x_f(t) + \int_{-\infty}^{\infty} g_{adv}(t - t')f(t')dt', \quad (26)$$

where $x_f(t)$ is a solution to the homogeneous equation, which is determined by a "final condition" after the inhomogeneous term becomes zero.[20] In this case, so as not to contradict the "final condition" one needs to use the advanced Green's function. But here, except in special cases, $x_f(t)$ will not satisfy the initial conditions used in the previous representation. Only the sum $x_f(t) + \int_{-\infty}^{\infty} g_{adv}(t-t')f(t')dt'$ will satisfy those initial conditions. But, one can represent the exact same solution via the retarded Green's function and we are, thus, left without any sense of what it gets wrong.

Of course, some might claim that the advanced Green's function represents the causal directionality in the system incorrectly, since it depicts the system as if the cause comes after the effect. This is just to baldly assert that the retarded Green's function correctly represents causation, since it is the one that adheres to the maxim, "the cause precedes the effect." However, we have not thereby gained any insight into why we take the maxim to be true and substantive. Why not, on the contrary, think that the maxim is (or might be) false or of indeterminate truth value? A causal skeptic will continue to wonder here what the grounds are for thinking that the advanced Green's function incorrectly represents the effect of an impulsive force. Moreover, one will wonder why one needs to enter this causal morass in the first place: one could capture the entire motion of the system via the advanced Green's function and a suitable solution to the homogeneous equation. So, it is not clear what one will have missed if one did not get into the business of trying to privilege the retarded Green's function in the first place.

1.2 The Wave Equation and Spatial Propagation

The drama above involving the Green's function of the harmonic oscillator carries over rather straightforwardly to the wave equation. But, the wave equation contains additional complexities because it is a partial differential equation and, as such, involves spatial propagation in addition to mere temporal evolution. For now, consider the homogeneous wave equation:

$$\nabla^2 \psi - \frac{1}{c^2} \frac{\partial^2 \psi}{\partial t^2} = 0. \tag{27}$$

Solving this in all of infinite, unbounded space involves a pure initial value problem or pure Cauchy problem. Since it is second order in time, the wave equation needs two initial conditions:

$$\psi(\mathbf{x}, t_0) = \psi_1(\mathbf{x}) \tag{28}$$

and

$$\frac{\partial \psi(\mathbf{x}, t_0)}{\partial t} = \psi_2(\mathbf{x}), \tag{29}$$

---

[20] This is just another way of writing the general solution to (1) that uses a different particular solution to it (Hilgevoord 1960, 10).

where x represents all three spatial variables. For the wave equation in all of space, these conditions are sufficient to yield a unique solution (assuming sufficient differentiability of the initial data[21]). As such, with the wave equation it is not necessary to impose a "radiation condition" at infinity to rule out solutions that involve spherical waves collapsing down onto a point. Rather, a specification of the de facto initial conditions will be sufficient to rule out waves collapsing at a point if, in fact, there are none. Nature does not need extra constraints so as to keep "waves from coming in from infinity."

1.3 The Wave Equation in a Bounded Spatial Domain

There are, of course, other sorts of problems than a pure initial value problem for the wave equation. If one thinks in terms of a bounded spatial domain, one needs to supply boundary conditions, as well as initial conditions. There are a variety of different types of boundary conditions applicable to the wave equation, which I will not discuss in detail here. For each type of them, one specifies some aspect of the behavior of the field on the spatial boundary. In a bounded spatial domain, the Green's function will be required to solve homogeneous boundary conditions, meaning that some aspect of its behavior on the boundary—either its value or the value of its derivative—is set to zero. One can use such a Green's function to give a rather interesting and useful decomposition of a solution to an initial and boundary-value problem. A solution to the wave equation in bounded regions can be decomposed as follows (Barton 1989, 245):

$$\psi(\mathbf{x}, t) = \psi_f(\mathbf{x}, t) + \psi_b(\mathbf{x}, t) + \psi_i(\mathbf{x}, t) \tag{30}$$

$$\psi_f(\mathbf{x}, t) = \int_{t_0}^{t} dt' \int_V dV' g(\mathbf{x}, t, \mathbf{x}', t') f(\mathbf{x}', t') \tag{31}$$

$$\psi_b(\mathbf{x}, t) = \int_{t_0}^{t} dt' \int_S dS' [g(\mathbf{x}, t, \mathbf{x}', t') \partial_n' \psi_S(\mathbf{x}', t') - (\partial_n' g(\mathbf{x}, t, \mathbf{x}', t')) \psi_S(\mathbf{x}', t')] \tag{32}$$

$$\psi_i(\mathbf{x}, t) = \psi_{i1}(\mathbf{x}, t) + \psi_{i2}(\mathbf{x}, t) \tag{33}$$

$$\psi_{i1}(\mathbf{x}, t) = \frac{\partial}{\partial t} \int_V dV' \frac{1}{c^2} g(\mathbf{x}, t, \mathbf{x}', t_0) \psi_1(\mathbf{x}') \tag{34}$$

$$\psi_{i2}(\mathbf{x}, t) = \int_V dV' \frac{1}{c^2} g(\mathbf{x}, t, \mathbf{x}', t_0) \psi_2(\mathbf{x}'), \tag{35}$$

where $f(\mathbf{x}, t)$ is an inhomogeneous term, $g(\mathbf{x}, t, \mathbf{x}', t')$ is a retarded Green's function, t_0 is the initial time, and ψ_S represents ψ on the spatial boundary surface S. Gabriel

---

[21] Failing such differentiability, one may turn to a generalized wave equation, but I do not discuss the details of that here.

Barton calls this rather quaintly the "magic rule" for solving the wave equation, since it allows one to arrive at the solution via quadrature when the Green's function is known.

One might think of these functions as follows: $\psi_f(x, t)$ gives the dependence on the inhomogeneous term (i.e., the forcing term); it is a solution of the inhomogeneous equation with homogeneous boundary conditions and homogeneous initial conditions; $\psi_b(x, t)$ gives the dependence on the boundary conditions; it is a solution of the homogeneous equation with inhomogeneous boundary conditions and homogeneous initial conditions; $\psi_i(x, t)$ gives the dependence on the initial conditions; it is a solution of the homogeneous equation with homogeneous boundary conditions.

The Green's function in the magic rule above is the retarded Green's function. However, a retrodictive "magic rule" involving the advanced Green's function also exists. So, why don't we typically represent the solution using the advanced Green's function? Barton claims the following:

> The reason why less attention is paid to retrodiction than to prediction is that in general the requisite input information about the future is not available, and that one seldom needs to construct the past (even though the requisite data are known, referring as they do to the present). (Barton 1989, 257)

However, advanced Green's functions are not "unphysical" in the sense of being useless for the accurate description of physical processes. In fact, in final value problems, one has to use them in the retrodictive magic rule.

A causal skeptic can also note the following: there is room to quibble over whether these functions represent the way the solution breaks up into contributions from the various physical aspects of the system. For example, I described $\psi_i(x, t)$ as giving the dependence of the solution on the initial conditions. But, it is not obvious that this is the right way to think about it. To simplify, suppose that the problem being solved involves homogeneous boundary conditions of the "Dirichlet variety" (i.e., the specification that $\psi(x, t)$ vanishes on the boundary) and that there is no forcing. In this case, $\psi_i(x, t)$ will be the only contribution to the system. Because it involves such homogeneous boundary conditions, waves will reflect off of the boundary, and such reflection will be present in $\psi_i(x, t)$. As such, the statement of dependence given above suggests that a wave that is present in the initial conditions but reflects off of the boundary is caused by the initial conditions and not by the boundary. But, one could plausibly reason in either of the following two ways: (1) With homogeneous boundary conditions, the boundary is not really doing anything; it is not pumping energy into the system. There is no source present in that sense. The energy that is there in the reflected wave can simply be traced back to the initial conditions. So, it is caused by them and not by the boundary. Thus, the reflecting homogeneous boundary conditions really do represent what the initial conditions cause correctly: the reflected wave is due to the initial conditions. Or:

(2) Were the boundary not there, there would not have been any reflection of waves present in the initial data. So, the reflected waves that have been attributed to the initial conditions are attributed incorrectly. Rather, they are due to the boundary rather than (or, perhaps, in addition to) the initial conditions, since it is the boundary that reflects the waves.

Insofar as it is difficult to assess which of these two patterns of reasoning is "right," there is some sense in which our "concept of cause" does not push us to a natural "effect" of the initial conditions. But, all of this provides a new reason to wonder whether there is a uniquely correct "causal" decomposition of the field. This reason remains even if we have decided that the retarded Green's function correctly represents the temporal direction of causation. Clearly, there is no unique mathematical decomposition of the field, but it is not even obvious that there is a unique decomposition that most nearly reflects our thinking about causality. But, for all that, we do not seem to be any worse off in terms of understanding electromagnetism or other fields of physics where the wave equation plays a role.

2. Where and Why Does One Need a Radiation Condition?

For the wave equation in all of space, the initial conditions suffice to determine a unique solution. In a bounded domain, the "magic rule" gives the solution directly from the applied data involving the forcing term, initial conditions, and boundary conditions. In neither case does one need to invoke an additional radiation condition so as to select a unique solution. There are, however, contexts in which a radiation condition is invoked so as to ensure uniqueness. The most famous such condition is applied to the Helmholtz equation (which will be derived below in several ways) and is called the Sommerfeld radiation condition:

$$\lim_{r \to \infty} r(\frac{\partial \psi}{\partial r} - ik\psi) = 0 \qquad (36)$$

uniformly in all directions. What this condition does is rule out solutions that involve "incoming waves," waves that originate from infinity and propagate inward toward a source.

Here, I explore where and why such a condition is needed. Let us start with a simple problem of radiation:[22] we imagine a channel starting at $z = 0$ but having an infinite length along the positive z axis. (See figure 3.3.) We will imagine radiation being pumped into the channel via a time-dependent boundary condition. Our main equation is, of course, the wave equation (27) in which we shall assume

---

[22] The details and setup of this problem are from the excellent Snider (2006, 554), which should be consulted for additional insight.

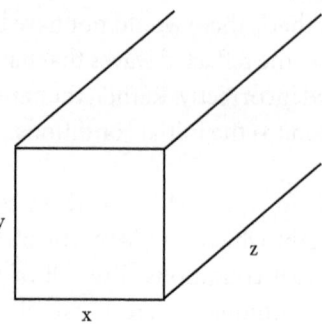

Figure 3.3 Waveguide.

that $c = 1$. In addition, the time-dependent boundary condition along the $z = 0$ boundary is

$$\psi(x, y, 0, t) = f(x, y, t). \tag{37}$$

We also need boundary conditions along the other sides of the cavity. For simplicity, we can assume that it is a perfectly reflecting cavity. This is ensured by requiring that

$$\psi(\mathbf{x}, t) = 0 \tag{38}$$

along all of those other sides. Moreover, we assume the following initial conditions at $t_0 = 0$:

$$\psi(\mathbf{x}, t_0) = 0 \tag{39}$$

and

$$\frac{\partial \psi(\mathbf{x}, t_0)}{\partial t} = 0. \tag{40}$$

Lastly, we assume a "finiteness condition" at $z = \infty$.[23] This finiteness condition, in itself, does not have the content of the Sommerfeld radiation condition, since "incoming waves" can be finite at infinity. In theory, the conditions imposed here are enough to yield a unique solution to the problem without any sort of radiation condition, but that does not mean that we can easily find the solution that fits the given data. One typically employs (as we did above to find Green's functions) a "transform" technique of some kind: one solves the problem in the transformed format and then transforms back—via an inverse transformation—to the original setting.

---

[23] One might wonder why one assumes this. The reason here is that (below) we are trying to solve the wave equation via the Helmholtz equation and it requires the elimination of certain solutions that are unbounded at spatial infinity. Such solutions clearly will not represent the solution to the wave equation we are seeking here.

2.1 The Laplace Transform Technique

One path to the solution is via the Laplace transform. The Laplace transform of $\psi(\mathbf{x}, t)$ is as follows:

$$\hat{\psi}(\mathbf{x}, s) = \int_0^\infty e^{-st} \psi(\mathbf{x}, t) dt, \tag{41}$$

where in this subsection $\hat{\psi}$ indicates the Laplace transform of ψ rather than its Fourier transform. To solve the wave equation, we want to Laplace transform each term of it. First, the second time derivative transforms as follows:

$$\int_0^\infty e^{-st} \frac{\partial^2 \psi(\mathbf{x}, t)}{\partial t^2} dt = s^2 \hat{\psi}(\mathbf{x}, s) - s\psi(\mathbf{x}, t_0) - \frac{\partial \psi(\mathbf{x}, t_0)}{\partial t}. \tag{42}$$

Next, one transforms the second derivatives of the spatial variables. I show what this amounts to for one spatial variable only, but it obviously works equally for the others:

$$\int_0^\infty e^{-st} \frac{\partial^2 \psi(\mathbf{x}, t)}{\partial x^2} dt = \frac{\partial^2}{\partial x^2} \int_0^\infty e^{-st} \psi(\mathbf{x}, t) = \frac{\partial^2 \hat{\psi}(\mathbf{x}, s)}{\partial x^2}. \tag{43}$$

Once everything is Laplace transformed, the wave equation ends up just being

$$\nabla^2 \hat{\psi}(\mathbf{x}, s) = s^2 \hat{\psi}(\mathbf{x}, s) - s\psi(\mathbf{x}, t_0) - \frac{\partial \psi(\mathbf{x}, t_0)}{\partial t}. \tag{44}$$

But, because of the quiescent initial conditions that we assumed (we are trying to solve the original problem), those last two terms are just zero. So, one ends up with

$$\nabla^2 \hat{\psi}(\mathbf{x}, s) = s^2 \hat{\psi}(\mathbf{x}, s), \tag{45}$$

which is known as the "Helmholtz equation" or the "reduced wave equation."[24]

To solve the original problem for the wave equation, one wants to solve the Helmholtz equation with the Laplace-transformed boundary conditions and then take the inverse transform so as to get back to the desired $\psi(\mathbf{x}, t)$, which has some time-dependence.[25] When one does this (which is not trivial), one sees that $\psi(\mathbf{x}, t)$ remains zero until the source (i.e., the z-boundary behavior) starts. After the source is turned on, there is a wave that travels in the outward z-direction but not one that travels in the inward z-direction. But, as this simply falls out of the imposed initial and boundary conditions no "radiation" or "causal condition" needs to be applied in this context so as to rule out an incoming wave. In the next section, we shall see when it does have to be applied to it.

---

[24] To make matters rather confusing and misleading, I know of some texts that refer to it as the wave equation in spite of the fact that it is not identical with the wave equation above. In part, that is because it does not involve the time variable. In essence, that has been partially transformed away and then further eliminated by the imposition of initial conditions.

[25] One may find the solution to these problems in Snider (2006, 555).

2.2 The Fourier Transform Technique

Let us now look at the Fourier transform treatment of the case. The Fourier transform is

$$\hat{\psi}(\mathbf{x},\omega) = \frac{1}{\sqrt{2\pi}} \int_{-\infty}^{\infty} \psi(\mathbf{x},t) e^{i\omega t} d\omega, \qquad (46)$$

where $\hat{\psi}$ is (once again) the Fourier transform of ψ with respect to t. When we take the Fourier transform of the wave equation, we again arrive at the Helmholtz equation, but this time for the Fourier transform:

$$\nabla^2 \hat{\psi}(\mathbf{x},\omega) = -\omega^2 \hat{\psi}(\mathbf{x},\omega). \qquad (47)$$

We again solve the Helmholtz equation. After discarding solutions that do not fit the boundary conditions and including a time-dependence, we get solutions of the form (Snider 2006, 558)

$$\psi(\mathbf{x},t) = A e^{i[\sqrt{\omega^2 - (\frac{m\pi}{x_1})^2 - (\frac{n\pi}{y_1})^2} z + \omega t]} + B e^{-i[\sqrt{\omega^2 - (\frac{m\pi}{x_1})^2 - (\frac{n\pi}{y_1})^2} z - \omega t]}, \qquad (48)$$

where x_1 represents the rightmost boundary in the x-direction, y_1 represents the uppermost boundary in the y-direction, and m and n are the numbers of "wave guide modes." The first summand of this solution represents an incoming wave that needs to be eliminated by the Sommerfeld radiation condition. So, here is where such a radiation condition is needed.

Why do we have an incoming wave in this case but not when we used the Laplace transform? Snider provides a clear answer:

> [W]hat is the story behind the incoming wave? And why didn't it appear in the Laplace Transform?
>
> This is best understood by recalling that the Fourier description is tailored to represent a system for all time, from minus infinity to plus infinity. Its "initial conditions" correspond to the system's status at $t = -\infty$ ...Now since the waveguide extends from $z = 0$ to $z = \infty$, if there were some disturbance in the tube at "$t = -\infty$" then by any *finite* time t its outgoing components would have propagated past every finite point z, but its incoming components would keep arriving (there being no damping mechanism). This possibility has to be accommodated by the Fourier description ...
>
> In the Laplace description we prescribed *quiescent* initial conditions throughout the waveguide at $t = 0$. This had the effect of zeroing out such "built-in" waves, and no indeterminacy occurred in the computations. (Snider 2006, 558–559)

In the case of the Laplace transform derivation of the Helmholtz equation, one has the occasion to apply the initial conditions for the problem on the route to the Helmholtz equation. In the Fourier transform derivation, one does not. Thus, one needs some other way to impose the correct initial conditions for the problem. This is, in effect, what the Sommerfeld radiation condition does. However, in essence, the Sommerfeld radiation condition has not been invoked so as to adhere to some general principle of causality. Rather, it has been invoked so as to get the correct solution to the initial value problem that one is solving, a solution that does not have incoming waves.

2.3 Time Harmonic Waves

In addition to the two paths to it given above, the Helmholtz equation results from the wave equation via separation of variables. If we assume that a solution to the wave equation $\psi(\mathbf{x}, t)$ is such that $\psi(\mathbf{x}, t) = \psi_{\mathbf{x}}(\mathbf{x})\psi_t(t)$, we end up with two functions: $\psi_{\mathbf{x}}(\mathbf{x})$, which can be shown to be a solution to the Helmholtz equation and $\psi_t(t)$, which gives a harmonic time-dependence (Zachmanoglou and Thoe 1986, 267). There are time-harmonic solutions to the wave equation that contain only "outgoing waves" but others that contain "incoming waves." The latter solutions are eliminated by the imposition of the Sommerfeld radiation condition.[26]

One thing that is odd about such time-harmonic solutions to the wave equation is that throughout all of space, waves are present. One can show that a solution to the Helmholtz equation that is twice differentiable (i.e., a classical solution to the Helmholtz equation) is analytic (Colton and Kress 1998, 18). So, if a solution to it vanishes in an open subset, it vanishes everywhere. Thus, there are no spatial regions that waves have not reached in these time-harmonic solutions to the wave equation.

This might make one wonder how such time-harmonic solutions to the wave equation relate to solutions arising from certain initial value problems for the wave equation. For, in many initial value problems, one starts with a field that is nonzero only in a bounded region of space. So what is the relation between these two types of solutions? Here is a reasonable suggestion from J. J. Stoker's classic book *Water Waves*:[27]

> A point of view which seems to the author reasonable is that the difficulty [in selecting sensible radiation conditions in certain cases where it is unclear what conditions should apply] arises because the problem of determining simple harmonic

---

[26] Given the time-dependence $\psi_t(t) = e^{-i\omega t}$, a spherically symmetric outgoing solution is $\psi_{\mathbf{x}}(\mathbf{x}) = \frac{e^{ikr}}{r}$. The analogous incoming solution is $\frac{e^{-ikr}}{r}$. The sum of those two is a standing wave. See Barton (1989, 336) for a discussion of the various types of solutions. The Sommerfeld radiation condition rules out standing waves in addition to incoming waves.

[27] This point of view is further elaborated along with proofs in Stoker (1956).

> *motions is an unnatural problem in mechanics.* One should in principle rather formulate and solve an initial value problem by assuming the medium to be originally at rest everywhere outside a sufficiently large sphere, say, and also assume that the periodic disturbances are applied at the initial instant and then maintained with a fixed frequency. As the time goes to infinity the solution of the initial value problem will tend to the desired steady state solution without the necessity to impose any but boundedness conditions at infinity.
>
> The steady state problem is unnatural—in the author's view, at least—because a hypothesis is made about the motion that holds for all time, while Newtonian mechanics is basically concerned with the prediction—in a unique way, furthermore—of the motion of a mechanical system from given initial conditions. Of course, in mechanics of continua that are unbounded it is necessary to impose conditions at ∞ not derivable directly from Newton's laws, but for the initial value problem it should suffice to impose only boundedness conditions at infinity. (Stoker 1957, 175)

Essentially, one thinks of the radiation condition as selecting that solution to the Helmholtz equation that corresponds to the infinite time limit of an initial value problem involving the wave equation where there is fixed-frequency periodic forcing from the initial moment and the field either is initially absent or it starts out confined to a bounded region. The problem of finding the solution to the Helmholtz equation corresponding with a certain initial value problem of the forced wave equation in the infinite time limit is sometimes called the "Principle of Limiting Amplitude." (Tikhonov and Samarskii 1990, 573–575). From thinking in these terms, one can see why one wants to eliminate certain solutions from the Helmholtz equation by using the radiation condition: when one applies a specific harmonic time-dependence to them, they do not correspond with the infinite time limit of the initial value problem to the wave equation that one is solving. So, in essence, what is motivating their dismissal is that they violate the long-time behavior associated with the imposed initial conditions and forcing. This is easy to lose sight of since, being elliptic and not involving the time variable, the Helmholtz equation does not accept initial conditions. But, in the end, what one is doing is getting one's solution to the Helmholtz equation (with supplemental harmonic time-dependence) to correspond with the behavior resulting from the given initial conditions attached to the wave equation in the infinite time limit.

In these cases of time-harmonic wave motions, the radiation condition is not justified by appeals to maxims like "the cause precedes the effect." Rather, the project initiated by Stoker is to get away from such vague appeals. In pursuit of a project that is similar to Stoker's, Wilcox (1959, 133) starts with the following claim:

> Nearly fifty years have passed since Sommerfeld introduced his radiation condition. During this period it has become customary to use the condition in formulating and solving the boundary value problems associated with the diffraction of time-harmonic waves. The radiation condition is satisfactory from the mathematical viewpoint in that it leads to boundary value problems having unique solutions. However, the physical reasons usually advanced for adopting it, rather than some other condition, are far from convincing. Our purpose here is to provide a more satisfying foundation for use of Sommerfeld's condition by deriving it from other facts concerning wave propagation that are both mathematically demonstrable and evident to physical intuition.

Wilcox shows that, among other radiation conditions, Sommerfeld's radiation condition is a consequence of a certain property of the solution to an initial and boundary-value problem[28] for the wave equation in the infinite time limit. Moreover, as Stoker notes, in many cases when one is dealing with the Helmholtz equation, the path to the right radiation conditions to impose comes from thinking in these terms. But, this suggests that the precise mathematical content of such radiation conditions is being driven by solutions to certain initial and boundary-value problems. A slogan like "the cause precedes the effect" will not get one to such mathematical content and is, thus, comparatively worthless. Moreover, in well-posed initial and boundary-value problems involving the wave equation, the solution is determined without a causality condition. Thus, such a causality condition does not seem to be of fundamental importance.

3. Backward Causation in Point-Particle Electrodynamics

Even if the radiation condition is not justified by a brute imposition of the maxim "the cause precedes the effect," there are other interesting cases in which a causality principle is frequently claimed to be invoked. Above I noted Dirac's derivation of an equation of motion (25) for a charged point-particle with self-interaction. That equation does raise worries among some physicists though some seem to take it to be an acceptable classical equation of motion.[29] One feature of the equation that one is not totally accustomed to is that it is of the third order, whereas standard classical

---

[28] Wilcox is thinking in terms of what is known as an "exterior boundary-value problem": Conditions are given on the boundary of some bounded volume such as a sphere and one solves for the state of the field external to that surface.

[29] For a summary of issues surrounding (25), see Erber (1961).

equations of motion for point-particles are second order. So, one is confronted by a different sort of beast than is usual.

Many of the worries about (25) surround the fact that it allows both runaway solutions (that is, solutions such that the acceleration grows continually even in the absence of external forces) and pre-accelerations (that is, accelerations that happen in advance of a force being applied). Runaway solutions are not in evidence in nature. So, one would like to rule them out. The general solution to (25) is (Levine et al. 1977, 75)

$$m\ddot{x}(t) = e^{\frac{t}{t_r}}[m\ddot{x}(0) - \frac{1}{t_r}\int_0^t e^{\frac{-t'}{t_r}} F_{ext}(t')dt']. \qquad (49)$$

If one selects the initial acceleration to be

$$m\ddot{x}(0) = \frac{1}{t_r}\int_0^\infty e^{\frac{-t'}{t_r}} F_{ext}(t')dt', \qquad (50)$$

it results formally in the acceleration at temporal infinity being zero, and for runaway solutions the acceleration at temporal infinity is not zero. So, one might motivate imposing this initial acceleration by the desire to rule out runaway solutions. By imposing this condition, one arrives at a new equation, sometimes called the "nonlocal" (in time) equation (Levine et al. 1977, 75; Jackson 1975, 797),

$$m\ddot{x}(t) = \int_0^\infty e^{-s} F_{ext}(t + t_r s) ds. \qquad (51)$$

This equation, however, obviously involves pre-acceleration insofar as the acceleration at t depends upon an integral involving values of the force at future times. Thus, a particle can start to accelerate due to forces on it in the future.

Such pre-acceleration is sometimes declared "unphysical." But, this does not give us much of a sense of what is wrong with it. Often "unphysical" just means *defying our antecedent expectations as to what ought to happen.*[30] There are many reasons that we would not have expected such pre-acceleration. Nothing in classical mechanics would have led us to expect this, since it is not found in the standard Newtonian equation of motion. But, sometimes such pre-acceleration is declared unphysical because it represents a violation of causality. Some of Mathias Frisch's claims that causality requirements enter physical theory refer to this equation and its dismissal by physicists. For example, he claims,

> [A] causal interpretation of Dirac's theory [of charged point-particles] also seems to be at the root of the feeling of unease that many physicists have toward the theory. For the causal structure of the theory violates several requirements we would like to place on causal theories. If nothing more were at issue than questions of determination, the nonlocal character of the equation of motion ought not to be

---

[30] See Norton (2006) for a more detailed discussion of various notions attached to "unphysical."

> troubling. That is, physicists themselves appear to be guided
> by causal considerations in their assessment of the theory.
> (Frisch 2005, 99)

Presumably, the idea here is that a "causal interpretation" involves the following ideas: (1) Forces cause accelerations. (2) Insofar as a particle can accelerate even though it only has forces on it in the future, one can see from the nonlocal equation (51) that later forces are causing the acceleration. (3) But, that involves pernicious backward causation, and that explains some of the unease toward the equation.

Even if that does capture how some physicists are thinking when they reject the equation, not all physicists reject it on these grounds. And, even if they all did, that alone is merely a sociological matter which does not give us much of a feel for whether they are warranted. Reasons as to why such a causality violation should be particularly troublesome are not typically given. In some cases, it is merely claimed that the equation "violates our ordinary conception of causality." That might be true as a psychological matter and it might explain some unease, but it gives one no sense as to why one ought to feel unease here. Perhaps our ordinary conception is simply naive. Obviously, physics has tended to show us that our ordinary conceptions (of space, of time, of the behavior of the microscopic, etc.) are not particularly respected by nature.

If the dominant worry about (51) were only that it involves backward causation but nothing concrete could be said about why that is a genuine worry (other than, perhaps, psychological facts about us), then no one who shares Russell's causal skepticism needs to be persuaded that there is a legitimate constraint that ought to rule out the equation. Moreover, one could note that the derivation of the Abraham-Lorentz equation that was given above that ultimately led to (51) assumed backward causation when it assumed that the field associated with the charge is a combination of the retarded and advanced potentials. Presumably, if backward causation is a cause for complaint, one never should have been willing to assume a premise involving it in the first place. One ought to have been antecedently dismissive of the initial steps of the derivation on these grounds. However, as we have seen, Feynman was not. We have yet to see a reason to be. Of course, maybe the backward causation that appears in the nonlocal equation (51) is somehow worse than the backward causation openly assumed in the derivation: perhaps not all backward causation is equally bad. But, if something can be said about what makes some backward causation worse than others, then it might be the case that it is not really backward causation per se that is the problem. Rather, it is the feature of the derived equation that makes the backward causation in it be particularly pernicious.

There are, to be sure, other derivations of (25) that work via energy-momentum conservation and that assume that the field associated with an accelerating charge is fully retarded. (For the two routes, see Poisson (1999). See also the derivation in Parrott (1987, 136–141).) Someone who prefers derivations that start along those lines can reject the backward causation of the resulting equation in good faith since, at least, he did not openly and willingly assume it in the derivation. However, one still lacks any feel for why backward causation is a real source for complaint. And, perhaps, now it will be even harder to say why one ought to reject the resulting equation on grounds of causality violation: even if one thinks that violation of our ordinary concept of causality creates some presumption (however weak) of falsity, certainly a derivation from such relatively more secure ideas as conservation of energy might be thought to defeat that presumption.[31] So, it is not clear how strong the grounds are for rejecting (25) or (51) on the basis of the maxim "the cause precedes the effect" alone.

This is not to say that there are not troubling circumstances surrounding (25) and its relativistic variant, the Lorentz-Dirac equation (Rohrlich 1965, 145):

$$ma^\mu = F^\mu_{ext} + \frac{2e^2}{3c^3}\left(\frac{da^\mu}{d\tau} - \frac{1}{c^2}a^\lambda a_\lambda v^\mu\right). \tag{52}$$

This equation also has pre-acceleration solutions and runaway solutions. Moreover, the Lorentz-Dirac equation suggests highly counterintuitive behavior that, as far as we know, does not appear in the world. In particular, as Eliezer originally proved (Parrott 1987, 198), there are solutions to the Lorentz-Dirac equation according to which a negatively charged particle heading toward a positively charged one will not collide with it, but will ultimately be turned away with the negative charge accelerating away from the positive charge in runaway fashion. As noted in Parrott (1987), this behavior is certainly not what one would expect from two oppositely charged particles, which would be expected to collide since opposite charges attract.[32] Perhaps, then, there are grounds for thinking that this equation is not an appropriate classical equation of motion, but it is not clear that worry over backward causation is or should be a major driving force in its rejection, even if one chooses to reject it as the right classical equation—which not all physicists do.

4. From Causality to Dispersion Relations

As fuel for his anti-Russellian stance, Frisch has recently brought up dispersion theory as an arena in which causality plays a role.[33] The cornerstone of dispersion theory is the derivation of "dispersion relations" which express the real part of some function of a complex variable in terms of its imaginary part and vice versa. Here is how such derivations can go: suppose we start with a Green's function that allows

---

[31] One will be able to find something to complain about in nearly every derivation of the equation. Discussion of how rigorous derivations of the equation really are is ongoing as are attempts to derive it more rigorously. But, it is not clear why our ordinary conception of causality ought to be thought to be more secure than the steps in any particular derivation.

[32] Eliezer-type results have been reinforced in Parrott (1993) and Parrott (2005). Even some who disagree with Parrott's earlier analysis concede that there are problems surrounding Eliezer's theorem that await resolution (Comay 1993, 1131–1132).

[33] For an exchange about causality in dispersion theory, see Frisch (2009a), Norton (2009), and Frisch (2009b). Some of this exchange is reviewed below.

us to write the state of a system as follows:

$$x(t) = \int_{-\infty}^{\infty} g(t-t') f(t') dt'. \tag{53}$$

We may think of $x(t)$ as the effect (or "output") and $f(t)$ as the "cause" (or "input"). Next, it is typical to invoke "causality," that the effect does not start before the cause. To ensure this, it is required that

$$g(t-t') = 0 \tag{54}$$

for $t < t'$. Because of (54), the Fourier transform of $g(\tau)$ (where $\tau = t - t'$) can be written as

$$\hat{g}(\omega) = \frac{1}{\sqrt{2\pi}} \int_0^{\infty} g(\tau) e^{i\omega\tau} d\tau. \tag{55}$$

The fact that this integral extends only over the positive reals ensures that $\hat{g}(\omega)$ has an analytic continuation into the upper half of the complex plane (i.e., when the imaginary part of ω is greater than zero).[34]

Once that is established, one starts via consideration of the following Cauchy integral

$$\int_C \frac{\hat{g}(\omega')}{\omega' - \omega} d\omega' \tag{56}$$

over a contour C like the one that appears on the right of figure 3.2. Provided that $\hat{g}(\omega')$ is analytic inside and on the contour, Cauchy's integral formula assures us (Saff and Snider 2003, 495–496) that when ω (in the denominator) lies on the contour,[35] the following holds:

$$P \int_C \frac{\hat{g}(\omega')}{\omega' - \omega} d\omega' = i\pi \hat{g}(\omega) \tag{57}$$

where P indicates a Cauchy principle value integral. We now assume that the integral along the semi-circle in the upper part of the complex plane vanishes as R goes to infinity,[36] and we take that limit. We end up with

$$\hat{g}(\omega) = \frac{1}{i\pi} P \int_{-\infty}^{\infty} \frac{\hat{g}(\omega')}{\omega' - \omega} d\omega'. \tag{58}$$

---

[34] See Fetter and Walecka (1980, 315–6), and Greiner (1998, 396). The rough way to think of the reason for this is as follows: if the Fourier transform converges for real values of ω, it converges better for values in the upper part of the complex plane. The integral

$$\hat{g}(\omega_r + i\omega_i) = \frac{1}{\sqrt{2\pi}} \int_0^{\infty} g(\tau) e^{i\tau(\omega_r + i\omega_i)} d\tau = \frac{1}{\sqrt{2\pi}} \int_0^{\infty} g(\tau) e^{i\tau\omega_r} e^{-\tau\omega_i} d\tau \tag{59}$$

now contains the term $e^{-\tau\omega_i}$, which keeps the function regular when $\omega_i > 0$.

[35] Since we later take the limit as R goes to infinity, the contour ultimately encompasses the entire real axis. Since the frequency, ω, will be real, it will, thus, lie on the contour.

[36] If this is not the case, one may use the "method of subtraction" to get around that but I do not discuss that here.

From here, we think in terms of the real and imaginary parts of $\hat{g}(\omega) = \hat{g}_r(\omega) + i\hat{g}_i(\omega)$:

$$\hat{g}_r(\omega) + i\hat{g}_i(\omega) = \frac{1}{i\pi} P \int_{-\infty}^{\infty} \frac{\hat{g}_r(\omega') + i\hat{g}_i(\omega')}{\omega' - \omega} d\omega'$$

$$= \frac{1}{i\pi} P \int_{-\infty}^{\infty} \frac{\hat{g}_r(\omega')}{\omega' - \omega} d\omega' + \frac{1}{i\pi} P \int_{-\infty}^{\infty} \frac{i\hat{g}_i(\omega')}{\omega' - \omega} d\omega'$$

$$= \frac{1}{i\pi} P \int_{-\infty}^{\infty} \frac{\hat{g}_r(\omega')}{\omega' - \omega} d\omega' + \frac{1}{\pi} P \int_{-\infty}^{\infty} \frac{\hat{g}_i(\omega')}{\omega' - \omega} d\omega'.$$

Since the second summand in the last line is real, $\hat{g}_r(\omega)$ can be equated to it. The first summand is imaginary so $i\hat{g}_i(\omega)$ can be equated to it. One concludes that the real and imaginary parts of $\hat{g}(\omega)$ are Hilbert transform pairs. That is,

$$\hat{g}_r(\omega) = \frac{1}{\pi} P \int_{-\infty}^{\infty} \frac{\hat{g}_i(\omega')}{\omega' - \omega} d\omega' \tag{60}$$

$$\hat{g}_i(\omega) = -\frac{1}{\pi} P \int_{-\infty}^{\infty} \frac{\hat{g}_r(\omega')}{\omega' - \omega} d\omega'. \tag{61}$$

These relations are known as dispersion relations (or Kramer's-Kronig relations). In the theory of waves in a dispersive medium, among the relevant functions of a complex variable is the relative permittivity (or dielectric constant), ϵ, whose imaginary part gives the absorptive properties of the medium while its real part gives the dispersive properties. Since the major driving engine of this derivation of dispersion relations is causality, it is typical to claim that causality implies such dispersion relations. But, of course, other assumptions went into the derivation as well. In particular, it was assumed that $\hat{g}(\omega')$ had no poles inside or on the contour of integration.

A simple illustration of a system for which such dispersion relations hold is that of the damped oscillator. In this case, the real and imaginary parts of (16) are as follows:

$$\hat{g}_r(\omega) = \frac{1}{\sqrt{2\pi}} \cdot \frac{\omega_0^2 - \omega^2}{(\omega_0^2 - \omega^2)^2 + \gamma^2 \omega^2} \tag{62}$$

and

$$\hat{g}_i(\omega) = \frac{1}{\sqrt{2\pi}} \cdot \frac{\gamma \omega}{(\omega_0^2 - \omega^2)^2 + \gamma^2 \omega^2}. \tag{63}$$

Those are, in fact, Hilbert transform pairs.

After this ground-setting, I want to suggest (following to some degree Norton (2009)) that, at a minimum, it is not clear that causality is a fundamental principle that plays the role in dispersion theory that Frisch believes. Both Norton and Frisch focus largely on the approach in Jackson's standard text. Part of what is at issue is whether Jackson invokes causality in his derivation of dispersion relations.

CAUSATION IN CLASSICAL MECHANICS

In the derivation of Kramers-Kronig relations, Jackson starts (as is typical) with an equation like (Jackson 1975, 307)

$$D(x,t) = E(x,t) + \int_{-\infty}^{\infty} g(\tau)E(x,t-\tau)d\tau, \tag{64}$$

where D is the displacement describing charges in materials and E is the electric field. At this point, Frisch sees "causality" as being invoked so as to limit the values of E that matter to D. He claims,

> The next step in the derivation is to impose an additional constraint that is generally identified as a causality condition. The condition is, as John Toll ([1956]) puts it, 'no output can occur before the input.' More precisely, we demand that the output field at time t is fully determined by the input field at all times prior to t. (Frisch 2009a, 463).

Frisch sees the need for the causality condition as arising from a larger part of the interpretation of these equations: We do not take them just as functional dependencies pace Russell. Rather "we interpret them causally"; we interpret E as the cause of D. This is supposed to explain why Jackson follows (64) with

$$D(x,t) = E(x,t) + \int_{0}^{\infty} g(\tau)E(x,t-\tau)d\tau. \tag{65}$$

However, as Norton notes, this is not actually how Jackson's derivation goes. Rather, Jackson starts out assuming that $g(\tau)$ is the inverse Fourier transform of $\epsilon(\omega) - 1$ (i.e., of the relative permittivity minus 1). He had previously derived that $\epsilon(\omega) - 1$ is $\frac{\omega_p^2}{\omega_0^2 - \omega^2 - i\gamma\omega}$. From there, Jackson then derives that $g(\tau)$ is a retarded Green's function as we did above for the damped harmonic oscillator. In fact, one can see that the equation for $\epsilon(\omega) - 1$ is relevantly similar to the Fourier transform of the damped oscillator Green's function. Jackson never postulates "causality" in this derivation. Of course, some assumptions went into the derivation of $\epsilon(\omega) - 1$. Jackson derives the formula for it from an assumption about a "phenomenological damping force" acting on the electrons of the medium in which electromagnetic waves are propagating (Jackson 1975, 285). But, Jackson has not concluded here that the Green's function must be retarded because E causes D. Rather, that the Green's function is retarded follows from a (partially phenomenological) model of the medium. So, it does not seem like the "causal interpretation" of the relation between D and E plays any role in Jackson's actual derivation.

Of course, as (Frisch 2009b) notes, Jackson assumes a model for the medium that is not time-reversal invariant (since it includes damping). However, this is not the same as invoking causality as a restriction on the model. That is something that Jackson does not (initially) do. Rather, Jackson adds a phenomenological damping term. So, he invokes the condition noted above in the discussion of the damped oscillator in section 1.1.1: If there is phenomenological damping, add damping to the model. Once one has done that, causality (for that model) follows and is not needed as an independent condition.

The appearance of phenomenological equations that are not time-reversal invariant is, of course, something about which much has be written that cannot even be approached with any rigor in an essay of this scope.[37] Moreover, many investigations into the source of irreversibility take one outside of classical physics to arenas of physics that are more appropriate to the microscopic and also to grand cosmological considerations. As such, in an essay of this scope, I can only note that there is controversy over what, exactly, the source of the irreversibility in Jackson's model is.

Leaving Jackson's actual derivation aside, Frisch claims that it is not so much that causality is invoked in any particular derivation where a particular model of a material has been given. Rather, causality is invoked in some more general sense in the derivation of Kramers-Kronig relations. He claims,

> Of course, once we have specified a particular model for the dielectric constant ϵ, the causality condition provides no additional constraint on that particular model. Rather the condition provides a general constraint on any physically legitimate model of ϵ and as such has content going beyond what is contained in any finite list of such models. (Frisch 2009a, 467)

As applied to derivations of dispersion relations, the idea seems to be that we are able to derive dispersion relations without any particular model in mind: "a derivation of the dispersion relations that begins with [the causality condition] allows us to ignore the details of the medium in question and its detailed interaction with the field" (Frisch 2009a, 468) But, for some models, Hilbert transform dispersion relations do not follow. For metals, $\epsilon(\omega)$ has a pole at $\omega = 0$ (Landau and Lifshitz 1960, 260). So, one has to take a contour around this point in a derivation of Hilbert-transform-like relations as above. Because of this, the second of the dispersion relations is modified by an addition of $4\pi\sigma/\omega$, where σ is the conductivity. So, the model matters somewhat to what one gets out of these derivations and cannot be completely ignored.

Moreover, we are left with little sense of the status of the causality principle that is invoked. Just as it is unclear what the source of the damping is in Jackson's model, it is unclear what the source of the irreversibility that results in causality in material media more generally is. Typically, appeal is made to "initial conditions" involving a low entropy past. But, it is not clear that that has anything directly to do with causality. Initial conditions are just a specification of the state (or sometimes a range of possible states) of the universe at a time. As such, they will not be expected

---

[37] Time-reversal invariance does not hold in fundamental physics. However, the failure of time-reversal invariance in the decay of neutral K mesons is not thought to be responsible for the sort of damping that makes Jackson's model viable nor for thermodynamical behavior more generally.

to say anything about what causes what or even what can cause what. At a minimum, someone of Russell's bent would want to hear more about why it is suspected that causality has a fundamental status if it results from special initial conditions.

On the other hand, Frisch does not seem particularly wedded to the idea that causality is fundamental. His main point is that it plays an ineliminable role in macroscopic electrodynamics. He claims, " Even if the asymmetric causal constraint were ultimately in some sense reducible, it remains part of a genuinely scientific theory and within certain contexts is explanatorily indispensable." (Frisch 2009b, 491). As I have described Russell's claims above, this is consistent with Russell's view, since I limit the claim that causation is not found in physics to the fundamental equations of physics. To be fair, it is not clear whether this limitation is found in Russell. But, that is a question of Russell scholarship, not a claim that will take us closer to understanding the status of causality in classical physics. So, I will not pursue it further.

As for whether invoking causality is indispensable, more would need to be said about its indispensability and the scope of it. For linear systems, one need not invoke a claim that the system in question obeys causality to derive dispersion relations. One could invoke instead that the system is "passive," where a system is passive if it absorbs energy but does not create it. One can show that a system which is passive and linear is also causal in the sense of (54) (Zemanian 1965, 300–303; Nussenzveig 1972, 391–392). Passivity is a notion whose relation to causality is not particularly transparent antecedently. Zemanian, for example, claims that the connection between passivity and causality is a "remarkable fact" (Zemanian 1965, 302). Prima facie passivity is a distinct property that can be invoked in the derivation of dispersion relations within the linear theory. As such, in a certain sense causality is not indispensable within the linear theory that Frisch discusses since one can invoke passivity instead.

5. Closing Thoughts

It would be difficult to summarize the outcome of this essay with respect to whether the imposition of causality in the intended sense is important. Of course, space limitations render any such discussion grossly incomplete anyway. It seems to me, however, that Russell's position (at least as I have rendered it) looks somewhat better than is sometimes suggested. We have, at least, failed to find any clear sense in which the retarded Green's function for a system like the undamped wave equation is privileged. Moreover, we have seen where a radiation condition (often thought to impose causality) is needed and where it is not. This has given us some sense that causality is a derivative condition used to incorporate certain initial conditions in cases where there has been no occasion to implement them. Moreover, we have seen that there are grounds to dismiss the Abraham-Lorentz equation, but most of them remain even if one did not have scruples about backward causation per se. And, it is not clear how seriously one should take the worry that the equation involves a causality violation. Lastly, we have seen that the assumption of causality only gets one so far in the derivation of Hilbert-transform dispersion relations. Rather, assumptions as to the material constitution of the medium are needed as well, since one needs to know whether the Fourier transform of the Green's function has poles along the real axis. But, when one has a model of the material medium such as Jackson gives, causality can be derived rather than postulated. Moreover, within the linear theory one could invoke something other than causality in the derivation of dispersion relations. As such, Russell might argue that causality is not such an ineliminable, fundamental principle in linear dispersion theory.

Even if I have not convinced the reader that Russell's skepticism is more warranted than sometimes supposed, I will be content to have framed the issues in a useful way.

References

Barton, G. (1989). *Elements of Green's functions and propagation: potentials, diffusion, and waves.* Oxford: Oxford University Press.

Born, M., and Wolf, E. (1999). *Principles of optics.* 7th ed. Cambridge: Cambridge University Press.

Butkov, E. (1968). *Mathematical physics.* Reading, MA: Addison-Wesley.

Cartwright, N. (1983). *How the laws of physics lie.* Oxford: Clarendon Press.

—— (1989). *Nature's capacities and their measurement.* Oxford: Clarendon Press.

Colton, D., and Kress, R. (1998). *Inverse acoustic and electromagnetic scattering theory.* 2nd ed., New York: Springer-Verlag.

Comay, E. (1993). Remarks on the physical meaning of the Lorentz-Dirac equation. *Foundations of Physics* 23(8): 1121–1136.

DuChateau, P., and Zachman, D. (2002). *Applied partial differential equations.* New York: Dover Publications.

Earman, J. (1986). *A primer on determinism.* Dordrecht: D. Reidel Publishing Company.

—— (1987). Locality, nonlocality, and action at a distance: A skeptical review of some philosophical dogmas. In Kargon, R., and Achinstein, P., editors, *Kelvin's Baltimore lectures and modern theoretical physics*, 449–490. Cambridge, MA: MIT Press.

—— (1995). *Bangs, crunches, whimpers and shrieks: Singularities and acausalities in relativistic spacetimes.* New York: Oxford University Press.

—— (2007). Aspects of determinism in modern physics. In Butterfield, J., and Earman, J., editors, *Handbook of the philosophy of physics, Part A*, 1369–1434. Amsterdam: North-Holland Press.

Erber, T. (1961). The classical theory of radiation reaction. *Fortschritte der Physik* 9: 343–392.

Fetter, A., and Walecka, J. D. (1980). *Theoretical mechanics of particles and continua.* New York: McGraw-Hill, Inc.

Feynman, R., Leighton, R., and Sands, M. (1989). *The Feynman lectures on physics.* Redwood City, CA: Addison-Wesley.

Frisch, M. (2005). *Inconsistency, asymmetry, and non-locality: A philosophical investigation of classical electrodynamics.* Oxford: Oxford University Press.

——— (2009a). "The most sacred tenet"? Causal reasoning in physics. *British Journal for the Philosophy of Science* 60(3): 459–474.

——— (2009b). Causality and dispersion: A reply to John Norton. *British Journal for the Philosophy of Science* 60(3): 487–495.

Greiner, W. (1998). *Classical electrodynamics.* New York: Springer-Verlag.

Griffel, D. H. (1981). *Applied functional analysis.* New York: Halsted Press.

Griffiths, D. (1981). *Introduction to electrodynamics.* 2d ed., Englewood Cliffs, N.J: Prentice Hall.

Hawking, S., and Ellis, G. (1973). *The large scale structure of space and time.* Cambridge: Cambridge University Press.

Hilgevoord, J. (1960). *Dispersion relations and causal description.* Amsterdam: North-Holland Publishing Company.

Hitchcock, C. (2007). What Russell got right. In Price and Corry (2007), *Causation, physics, and the constitution of reality: Russell's republic revisited,* 45–65. Oxford: Clarendon Press.

Hume, D. ([1748] 1977). *An enquiry concerning human understanding.* Indianapolis: Hackett Publishing Company.

Jackson, J. D. (1975). *Classical electrodynamics.* New York; John Wiley and Sons, Inc.

Johnson, C. (1965). *Field and wave electrodynamics.* New York: McGraw-Hill.

Landau, L., and Lifshitz, E. (1960). *Electrodynamics of continuous media.* Reading, MA: Addison-Wesley Publishing Company, Inc.

Levine, H., Moniz, E., and Sharp, D. (1977). Motion of extended charges in classical electrodynamics. *American Journal of Physics* 45(1): 75–78.

Maudlin, T. (2002). *Quantum non-locality and relativity.* Oxford: Blackwell Publishers.

Norton, J. (2006). The dome: A simple violation of determinism in Newtonian mechanics. *Philosophy of Science Assoc. 20th Biennial Meeting (Vancouver): PSA 2006 Symposia.*

——— (2007a). Causation as folk science. In Price and Corry (2007), *Causation, physics, and the constitution of reality: Russell's republic revisited,* 11–44. Oxford: Clarendon Press.

——— (2007b). Do the causal principles of modern physics contradict causal anti-fundamentalism? In Machamer, P. K., and Wolters, G., editors, *Thinking about causes: From Greek philosophy to modern physics,* 222–234. Pittsburgh: University of Pittsburgh Press.

——— (2009). Is there an independent principle of causality in physics? *British Journal for the Philosophy of Science* 60(3): 475–486.

Nussenzveig, H. M. (1972). *Causality and dispersion relations.* New York: Academic Press.

Parrott, S. (1987). *Relativistic electrodynamics and differential geometry.* New York: Springer-Verlag.

Parrott, S. (1993). Unphysical and physical(?) solutions of the Lorentz-Dirac equation. *Foundations of Physics* 23(8): 1093–1119.

——— (2005). Variant forms of Eliezer's theorem. *arXiv:math-ph/0505042v1,* 1–10.

Poisson, E. (1999). An introduction to the Lorentz-Dirac equation. *arXiv:gr-qc/9912045v1,* 1–14.

Price, H. and Corry, R., eds. (2007). *Causation, physics, and the constitution of reality: Russell's republic revisited.* Oxford: Clarendon Press.

Richards, J. I. and Youn, K. K. (1990). *Theory of distributions: A nontechnical introduction.* Cambridge: Cambridge University Press.

Rohrlich, F. (1965). *Classical charged particles.* Reading, MA: Addison-Wesley.

Russell, B. (1981). On the notion of cause. In *Mysticism and Logic,* chapter 9, 132–151. Totowa, NJ: Barnes and Noble Books.

Saff, E. and Snider, A. D. (2003). *Fundamentals of complex analysis.* 3rd ed., Englewood Cliffs, NJ: Prentice-Hall.

Snider, A. (2006). *Partial differential equations: Sources and solutions.* New York: Dover Publications, Inc.

Steiner, M. (1986). Events and causality. *Journal of Philosophy* 83(5): 249–264.

Stoker, J. J. (1956). On radiation conditions. *Communications on Pure and Applied Mathematics* 9: 577–595.

———(1957). *Water waves: The mathematical theory with applications.* New York: Interscience Publishers, Inc.

Tikhonov, A., and Samarskii, A. A. (1990). *Equations of mathematical physics.* New York: Dover Publications, Inc.

Vanderline, J. (2004). *Classical electromagnetic theory.* Dordrecht: Kluwer Academic Publishers.

Wallace, P. R. (1984). *Mathematical analysis of physical problems.* New York: Dover Publications, Inc.

Weigel, F. (1986). *Introduction to path-integral methods in physics and polymer science.* Singapore: World Scientific.

Wilcox, C. (1959). Spherical means and radiation conditions. *Archive for Rational Mechanics and Analysis* 3(1): 133–148.

Wilson, M. (1989). Critical notice: John Earman's *A primer on determinism. Philosophy of Science* 56: 502–532.

Zachmanoglou, E., and Thoe, D. W. (1986). *Introduction to partial differential equations with applications.* New York: Dover Publications, Inc.

Zemanian, A. (1965). *Distribution theory and transform analysis.* New York: Dover Publications, Inc.

CHAPTER 4

THEORIES OF MATTER
INFINITIES AND RENORMALIZATION

LEO P. KADANOFF

1. Introduction

1.1 The Discovery and Invention of Materials

From the "stone age" through the present day humankind has made use of the materials available to us in the earth and equally to materials we could manufacture for further use. Jared Diamond (1997), for example, in his book *Guns, Germs and Steel*, has pointed out how crucial metalworking was to the spread of European power.

However, it is only in relatively recent times that we were able to bring scientific understanding of the inner workings of materials to human benefit. The latter half of the nineteenth century brought the beginning of two major theoretical advances to the science of materials, advances that would deepen and grow into the twentieth century so that today we can boast of a fundamental understanding of the main properties of many materials. These advances are a theory of statistical physics developed initially by Rudolf Clausius (Brush 1976), J. C. Maxwell (Garber, Bush, and Everett 1986), and Ludwig Boltzmann (Brush 1976, 1983) and an understanding of the different phases of matter based in part upon an understanding of the changes from one phase[1] to another. These changes are called phase transitions and our

---

Copies of this paper have been published on arXiv and on the author's website: jfi.uchicago.edu\ ~ leop\.

[1] The word "phase" is interesting. According to the *Oxford Dictionary of Word Histories* (and the *Oxford English Dictionary*) it entered English language in the nineteenth century to describes the phases of the moon. The

understanding of them is based upon the work of Thomas Andrews (1869), Johannes van der Waals (Levelt-Sengers 1976), and Maxwell. It is the purpose of this essay to develop a description of the development of these basic ideas from the 1870s to the last quarter of the twentieth century.

Much of our first principles understanding of materials is based upon the fact that they are composed of many, many atoms, electrons, and molecules. This means that we cannot hope or wish to follow any motion of their individual constituents, but that instead we must describe their average or typical properties through some sort of statistical treatment. Further, though we should believe that the properties of these materials are built upon the properties of the constituents, we should also recognize that the properties of a huge number of constituents, all working together, might be quite different from the behaviors we might infer from thinking about only a few of these particles at once. P. W. Anderson has emphasized the difference by using the phrase "More Is Different" (1972; Ong and Bhatt 2001).

There are two surprising differences that have dominated the study of materials: irreversibility and the existence of sharply distinct thermodynamic phases. First, irreversibility: even though the basic laws of both classical mechanics and quantum mechanics are unchanged under a time reversal transformation, any appropriately statistical treatment of systems containing many degrees of freedom will not show such an invariance. Instead, such systems tend to flow irreversibly toward an apparently unchanging state called *statistical equilibrium*. This flow was recognized by Clausius in his definition of entropy, detailed by Boltzmann in his gas dynamic definition of the Boltzmann equation, and used by Gibbs (Rukeyser 1942) in his definitions of thermodynamics and statistical mechanics. It is a behavior that is best recognized as a property of a limiting case, either of having the number of degrees of freedom become infinite or having an infinite observation time. Time reversal asymmetry is not displayed by any finite system described over any finite period of time.[2]

This essay is concerned with thermodynamic equilibrium resulting from this irreversible flow. It focuses upon another property of matter that, as we shall see, only fully emerges when we consider the limiting behavior as an equilibrium material becomes infinitely large. This property is the propensity of matter to arrange itself into different kinds of structures that are amazingly diverse and beautiful. These structures are called *thermodynamic phases*. Figure 4.1 illustrates three of the many thermodynamic phases formed by water. Water has many different solid phases. Other fluids form liquid crystals, in which we can see macroscopic manifestations of the shapes of the molecules forming the crystals. The alignment of atomic spins or electronic orbits can produce diverse magnetic materials, including ferromagnets, with their substantial magnetic fields, and also many other more subtle forms of magnetic ordering. Our economic infrastructure is, in large measure,

---

Oxford English Dictionary lists a very early use in J. Willard Gibbs's writings about thermodynamics as the "phases of matter." Apparently Gibbs then extended the meaning to get "extension in phase" that then got further extended into the modern usages "phase transition" and "phase space."

[2] Particle physics does show a very weak time reversal asymmetry discovered by James W. Cronin and Val L. Fitch, but this asymmetry is immaterial for all mundane phenomena.

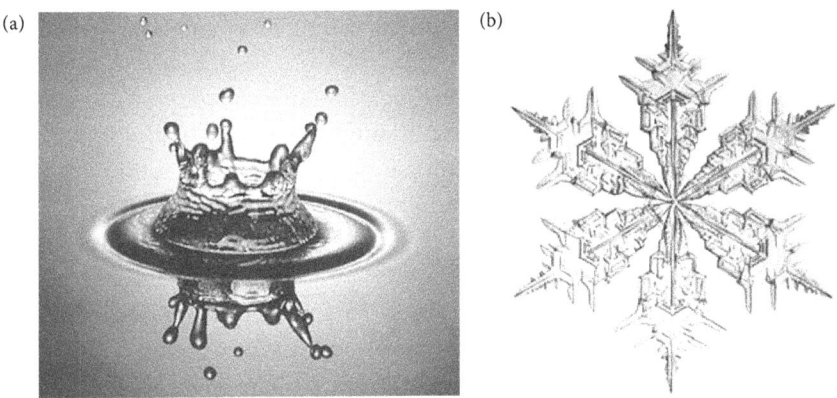

Figure 4.1 Splash and snowflake. This picture is intended to illustrate the qualitative differences between the fluid and solid phases of water. On the left is liquid water, splashing up against its vapor phase. Its fluidity is evident. On the right is a crystal of ice in the form of a snowflake. Note the delicate but rigid structure, with its symmetry under the particular rotations that are multiples of sixty degrees.

based upon the various phase-dependent capabilities of materials to carry electrical currents: from the refusal of insulators, to the flexibility of semiconductors, to the substantial carrying capacity of conductors, to the weird resistance-free behavior of superconductors. This flow, and other strange properties of superconductors, are manifestations of the subtle behavior of quantum systems, usually only seen in microscopic properties, but here manifested by these materials on the everyday scales of centimeters and inches. I could go on and on. The point is that humankind has, in part, understood these different manifestations of matter, manifestations that go under the name "thermodynamic phases." Scientific work has produced at least a partial understanding of how the different phases change into one another. This chapter is a brief description of the ideas contained in the science of such things.

As is the case with irreversibility, the differences among solid, liquid, and gas; the distinctions among magnetic materials and between them and nonmagnetic materials; and the differences between normal materials and superfluids are all best understood as distinctions that apply in the limit in which the number of molecules is infinite. For any finite body, these distinctions are blurry with the different cases merging into one another. Only in the infinite limit can the sharp distinction be maintained. Of course, our usual samples of everyday materials contain a huge number of molecules, so the blur in the distinction between different phases is most often too fine for us to discern. However, if one is to set up a theory of these materials, it is helpful to respect the difference between finite and infinite.

1.2 Different Phases; Different Properties

Three of the phases of water are illustrated in figure 4.1, which depicts a snowflake (a crystal of a solid phase of water) and a splashing of the liquid phase. A third phase of water, the low-density phase familiar as water vapor, or steam, exists in the empty-looking region above and around the splashing liquid.

Different phases of matter have qualitatively different properties. As you see, ice forms beautiful crystal structures. So do other solids, each with its own characteristic shape and form. Each crystal picks out particular spatial directions for its crystal axes. That selection occurs because of forces produced by the interactions of the microscopic constituents of the crystals. Crystalline materials are formed at relatively low temperatures. At such temperatures, microscopic forces tend to line up neighboring molecules and thereby produce strong correlations between the orientations of close neighbors. Such correlations extend through the entire material, with each molecule being lined up by several neighbors surrounding it, thereby producing an ordering in the orientation of the molecules that can extend over a distance a billion times larger than the distance between neighboring atoms. This orientational order thus becomes visible in the macroscopic structure of the crystal, as in figure 4.1, forming a macroscopic manifestation of the effects of microscopic interactions.

The same materials behave differently at higher temperature. Many melt and form a liquid. The long-range orientational order disappears. The material gains the ability to flow. It loses its special directions and gains the full rotational symmetry of ordinary space.

Many of the macroscopic manifestations of matter can be characterized as having *broken symmetries*. Phases other than simple vapors or liquids break one or more of the characteristic symmetry properties obeyed by the microscopic interactions of the constituents forming these phases. Thus, a snowflake's outline changes as it is rotated despite the fact that the molecules forming it can, when isolated, freely rotate.

1.2.1 Broken Symmetries and Order Parameters

A previous publication (Kadanoff 2009) describes the development of the theory of phase transitions up to and including the year 1937. In that year, Lev Landau (Landau 1937) put together a theoretical framework that generalized previously existing mean field theories of phase transitions. In Landau's approach, a phase transition manifests itself in the breaking of a mathematical symmetry. This breaking is, in turn, reflected in the behavior of an *order parameter* describing both the magnitude and nature of the broken symmetry. Two different phases placed in contact are seen to be distinguished by having different values of the order parameter. For example, in a ferromagnet the order parameter is the vector magnetization of the material. Since the spins generate magnetic fields, this alignment is seen as a large time-independent magnetic field vector, pointing in some particular direction. The order parameter in this example is then the vector describing the orientation and size of the material's magnetization and its resulting magnetic field. Note that the *orientation* of the order parameter describes the way in which the symmetry is broken, while the *magnitude* of this parameter describes how large the symmetry breaking might be.

Other order parameters describe other situations. A familiar order parameter characterizes the difference between the liquid and the vapor phases of water. This parameter is simply the mass density, the mass per unit volume, minus the value of that density at the critical point. This parameter takes on positive values in the liquid phase and negative ones in the vapor phase. These phases are clearly exhibited in processes of condensation, or boiling, or when the two phases stand in contact with one another. In superfluid examples, ones in which there is particle flow without dissipation, a finite fraction of the particles are described by a single, complex wave function. The order parameter is that wave function.

The alignment of atomic spins or electronic orbits can produce diverse magnetic materials, including ferromagnets, with their substantial magnetic fields, and also many subtler forms of magnetic ordering. A crystal in which the magnetization can point in any direction in the three-dimensional space conventionally labeled by the X, Y, and Z axes is termed an "XYZ" ferromagnet. Another kind of ferromagnet is called an "XY" system, and is one in which internal forces within the ferromagnet permit the magnetization to point in any direction in a plane. The next logical possibility is one in which there a few possible axes of magnetization, and the magnetization can point either parallel or antiparallel to one of these axes. A simple model describing this situation is termed an Ising model, named after the physicist Ernst Ising (1925), who studied it in conjunction with his adviser Wilhelm Lenz (Brush 1967; Niss 2005) (see section 2).

1.2.2 Dynamics and Equilibrium

We have already mentioned the propensity of condensed matter systems to approach an unchanging state called thermodynamic equilibrium. This approach can be either extremely fast or extremely slow, depending upon the situation. When an electrical current in a metal is set in motion by an applied voltage it can take as little as a millionth of a billionth of a second for the current to reach close to its full value. On the other hand, impurities may take years to diffuse through the entire volume of a metal. Because of this very broad range of timescales, and because of the wide variety of mechanisms for time-dependence, dynamics is a very complex subject.

Equilibrium is much simpler. The equilibrium state of a simple material is characterized by a very few parameters describing the thermodynamic environment around the material. To define the thermodynamic state of a container filled with water, one needs to know its temperature, the volume of the container, and the applied pressure. One can then directly calculate the mass of the water in the volume using the data from what is called the *equation of state*.

By the time our story begins, in the early twentieth century, *kinetic theory* will be well-developed as a description of common gases and liquids. That theory includes the statement that molecules in these fluids are in rapid motion, with a kinetic energy proportional to the temperature. *Thermodynamics* provides a more broadly applicable theory, brought close to its present state by J. Willard Gibbs in 1878 (Gibbs 1961; Rukeyser 1942 pp. 55–371). That discipline provides relations among the properties of many-particle systems based upon conservation of energy and the requirement that the system always develops in the direction of thermodynamic equilibrium. Thermodynamic calculations are based upon thermodynamic functions, for example, the Helmholtz free energy used in this essay. This function depends upon the material's temperature, volume, and the number of particles of various types within it. One can calculate all kinds of other properties by calculating derivatives of the free energy with respect to its variables. The free energy has the further important property that the actual equilibrium configuration of the system will produce the minimum possible value of the free energy. Any other configuration at that temperature, in that volume, with those constituents will necessarily have a higher free energy.

To understand phase transitions, one has to go beyond thermodynamics, which concerns itself with relations among macroscopic properties of materials, to the subject of *statistical mechanics* that defines the probabilities for observing various phenomena in a material in thermodynamic equilibrium, starting from a microscopic description of the behavior of the materials' constituents. Statistical mechanics further differs from thermodynamics in that the latter treats only summed or gross properties of matter, while the former looks to the individual constituents and asks about the relative probabilities for their different configurations and motions.

To use statistical mechanics, in principle, all one needs to know is a function called the *Hamiltonian*, which gives the system's energy as a function of the coordinates and momenta of the particles.[3] In practice, for many-particle systems, the actual calculations are sufficiently hard so that in large measure they only became possible after World War II.

Statistical mechanics is classically defined by using a *phase space* given by the momenta and coordinates of the particles in the system. The state of a particular classical system is then defined by giving a point in that space. The basics of statistical mechanics were put forward by Boltzmann (Cercignani 1998 p. 8, ch. 7; Uffink 2005; Gallavotti 2008) and then clearly stated by Gibbs (1902) in essentially the same form as it is used today. It describes the probability for finding a classical system in a particular configuration, c, in the phase space. The probability of finding the system in a small volume of phase space, $d\Omega$, around configuration c is given by

$$e^{-(\mathcal{H}(c)-F)/T} \, d\Omega. \tag{1}$$

Here, $\mathcal{H}(c)$ is the energy in this configuration, T is the absolute temperature in energy units, and F is a constant, which turns out to be the free energy. Since the total probability of all configurations of the system must add up to unity, one can determine the free energy by

$$e^{-F/T} = \int d\Omega \, e^{-(\mathcal{H}(c))/T} \tag{2}$$

where the integral covers all of phase space.

---

[3] This function is named after William Rowan Hamilton who described how to formulate classical mechanics using this Hamiltonian function.

This probability formulation, the basis of all of statistical mechanics, is known as the Maxwell-Boltzmann distribution, to most physicists and chemists, or the Gibbs measure, to most mathematical scientists.

This probability of Eq. (1) describes a collection of identical materials arrayed in different configurations. Gibbs calls the "ensemble" of configurations thus assembled a "canonical ensemble[4]".

1.2.3 Phase Transitions

A treatment of phase transitions may properly start with the 1869 experimental studies of Andrews (1869), who investigated the phase diagram of carbon dioxide and thereby discovered the qualitative properties of the liquid–gas phase transition. His results, as illuminated by the theoretical work of van der Waals and Maxwell (Levelt-Sengers 1976), are shown in figure 4.2. This plot gives the pressure as it depends upon volume in a container with a fixed number of particles. Each curve shows the behavior of the pressure at a given value of the temperature. The two curves at the bottom show results quite familiar from our experience with water. At high pressures one has a liquid and the liquid is squeezed into a relatively small volume. Since the density of the liquid is the (fixed) number of particles divided by a varying volume, this high-pressure region is one of high density. The curve moves downward showing a reduced pressure as the liquid is allowed to expand. At a sufficiently low pressure, the liquid starts to boil and thereby further reduce its density until a sufficiently low volume is reached so that it has attained the density of vapor. The boiling is what is called a *first-order phase transition*. The boiling occurs at a constant pressure and has liquid and vapor in contact with one another. Then after the vapor density is reached, additional expansion produces a further reduced density. At somewhat higher temperatures this same scenario is followed, on a higher curve, except that the region of boiling and its connected jump in density is smaller. Andrews's big discovery was that at a sufficiently high temperature, the jump in density disappears and the fluid goes from high pressure to low without a phase transition. This disappearance of the first-order phase transition occurs at what is called a *critical point*. The disappearance itself is called a *continuous phase transition*.

1.2.4 The First Mean Field Theory

In his thesis of 1873, Johannes van der Waals put together an approximate theory of the behavior of liquids using arguments based in ideas of the existence of molecules. The very existence of molecules was an idea then current, but certainly not proven. Van der Waals started from the known relation between the pressure and the volume of a perfect gas, that is, one that has no interactions between the molecules. Expressed in modern form, the relation is

$$p = T N/V. \tag{3}$$

---

[4] The word "canonical" seems to be a somewhat old-fashioned usage for something set to a given order or rule. The *Oxford English Dictionary* traces it back to Chaucer.

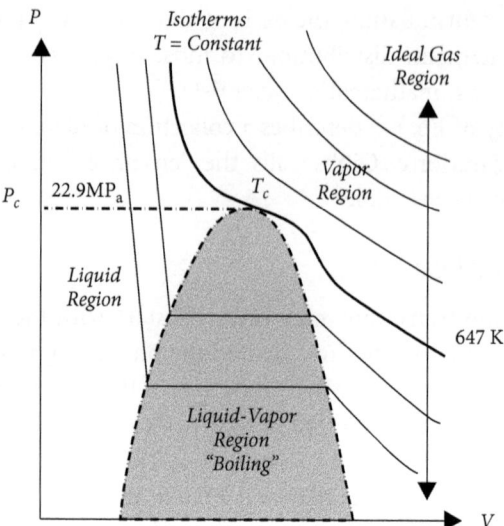

Figure 4.2 Cartoon PVT diagram for water. Each curve describes how the pressure depends upon volume for a fixed temperature. Note the figures for critical temperature and pressure on this diagram. They apply to water. The corresponding figures for carbon dioxide are 31.1° C and 73 atmospheres = 7.2 megapascals. These values are more easily accessible to experiments than the ones for water.

Here, p is the pressure, V is the volume of the container, N is the number of molecules within it, and T is the temperature expressed in energy units.[5] This *equation of state* relates the pressure, temperature, and density of a gas in the dilute-gas region in which we may presume that interactions among the atoms are quite unimportant. It says that the pressure is proportional to the temperature, T, and to the density of particles, N/V. This result is inferred by ascribing an average kinetic energy to each molecule proportional to T and then calculating the transfer of momentum per unit area to the walls. The pressure is this transfer per unit time. Of course, Eq. (3) does not allow for any phase transitions.

Two corrections to this law were introduced by van der Waals to estimate how the interactions among the molecules would affect the properties of the fluid.

First, he argued that the molecules could not approach each other too closely because of an inferred short-ranged repulsive interaction among the molecules. This effect should reduce the volume available to the molecules by an amount proportional to the number of molecules in the system. For this reason, he replaced V in Eq. (3) by the available volume, $V - Nb$, where b would be the excluded volume around each molecule of the gas.

The second effect is more subtle. The pressure, p, is a force per unit area produced by the molecules hitting the walls of the container. However, van der

---

[5] I use energy units in order to write fewer symbols. It is more conventional to write, instead of T, kT, where k is the Boltzmann constant.

Waals inferred that there was an attractive interaction pulling each molecule toward its neighbors. This attraction is the fundamental reason why a drop of liquid can hold together and form an almost spherical shape. As a molecule moves toward a wall, it is pulled back and slowed by the molecules left behind. Because of this reduced speed, it imparts less momentum to the walls than it would otherwise. The equation of state contains the pressure as measured at the wall, p. This pressure is the one produced inside the liquid, $NT/(V-Nb)$, minus the correction term coming from the interaction between the molecules near the walls. That correction term is proportional to the density of molecules squared. In symbols Van der Waals' corrected expression for the pressure is thus

$$p = NkT/(V-Nb) - a(N/V)^2. \tag{4}$$

Here, a and b are parameters that are different for different fluids and N/V is the density of molecules.

Eq. (4) is the widely used van der Waals equation of state for a fluid. Because it takes into account average forces among particles, we describe it and similar equations as the result of a *mean field theory*. This equation of state can be used to calculate the particle density, $\rho = N/V$, as a function of temperature and pressure. It is a cubic equation for ρ and thus has at most three solutions.

1.2.5 Maxwell's Improvement

The equation of state proposed by van der Waals is plotted in figure 4.3. Each curved line shows the dependence of pressure on volume. This equation of state has a major defect: it shows no boiling region. Worse yet, it contains regions in which,

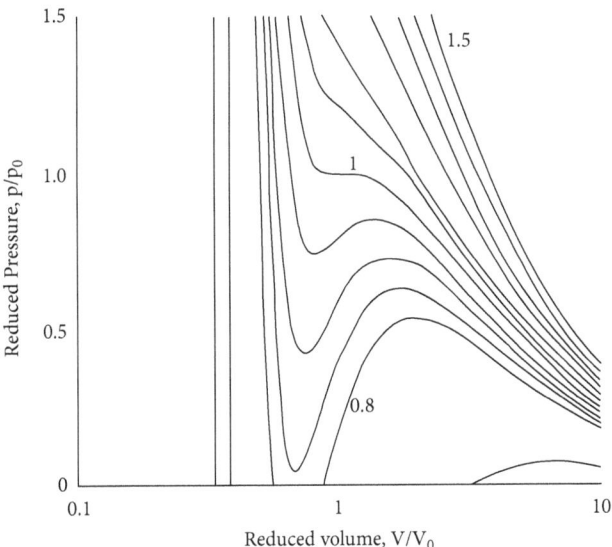

Figure 4.3 PVT curves predicted by the theory set up by van der Waals. The fluid is mechanically unstable whenever the pressure increases as the volume increases.

at fixed temperature and numbers of particles, the pressure increases as the volume increases. This situation is unstable. If the fluid finds itself in a region with this kind of behavior, the forces within it will cause it to separate into two regions, one at a high density the other at a lower one. In fact, exactly this kind of separation does happen in the boiling process in which a lower-density vapor is in contact with a high-density liquid.

The instability just described is termed a mechanical instability. It can be triggered by a fluctuation in which a piece of fluid acquires a density slightly different from that of the surrounding fluid elements. J. C. Maxwell (1874; 1875) in 1874 and 1875 recognized this instability and also the somewhat bigger region of thermodynamic instability against larger fluctuations. Maxwell identified this bigger region of instability with boiling and drew a phase diagram like that in figure 4.2. Note that this figure has a completely flat portion of the constant temperature lines to represent the predicted boiling of Maxwell's theory. We shall hear more of this Maxwell construction in section 4.

Maxwell's result gives a qualitative picture of the jump in density between the two phases over a quite wide range of temperatures. For the purposes of this essay, however, the most important region is the one near the critical point in which the jump is small. According to the theory, as the jump in density, $\rho = N/V$, goes to zero, it shows a behavior

$$\rho_{liquid} - \rho_{gas} = \text{constant} \times (T_c - T)^\beta, \tag{5}$$

where T_c is the critical temperature and β has the value one half. Andrews's data does fit a form like this, however with a value of the exponent, β, much closer to one third than one half. Later on, this discrepancy will become quite important.

Despite the known discrepancy between mean field theory and experiment in the region of the critical point, few scientists focused upon this issue in the years in which mean field theory was first being developed. There was no theory or model that yielded Eq. (5) with any power different from one half, so there was no focus for anyone's discontent. Thomas Kuhn (1962) has argued that an old point of view will continue on despite evidence to the contrary if there is no replacement theory.

Following soon after van der Waals, many other scientists developed mean field theories, applying them to many different kinds of phase transitions. All these theories have an essential similarity. They focus upon some property of the many-particle system that breaks some sort of global symmetry.[6] In mean field calculations, the ordering in one part of the system induces ordering in neighboring regions until, after some time, ordering is spread through the entire system. Thus mean field theory calculations are always descriptive of symmetry breaking and the induced

---

[6] In fact, the liquid–gas case is one of the most subtle of the phase transitions since the symmetry between the two phases, gas and liquid, is only an approximate one. In magnets and most other cases the symmetry is essentially exact, before is it broken by the phase transition.

correlations that carry the symmetry breaking through the material. These calculations are then most relevant and immediately useful for the description of the jumps that occur in first-order phase transitions.

1.3 Fluctuations

In equilibrium, the material has a behavior that, in a gross examination, looks time-independent. Hence, many of the phenomena involved may be described by using time-averages of various quantities. This averaging is the basis of mean-field-theory techniques. However, a more detailed look shows *fluctuations*, that is, time dependence, in everything. These fluctuations will call for additional calculational techniques beyond mean field theory, which will be realized with the renormalization group methods described in section 6. Here, I describe two important examples of fluctuations that arise near phase transitions.

1.3.1 Fluctuations I: Boiling

In the process of ordering, typically a material will display large amounts of disorder. For example, as the pressure is reduced at the liquid–gas coexistence line, a liquid turns into a vapor by an often-violent process of boiling. The boiling produces bubbles of low-density vapor in the midst of the higher-density liquid. Thus the fluid, which is quite homogeneous away from its phase transition, shows a rapidly fluctuating density in its boiling region. As every cook knows, one can reduce the violence of the fluctuations by making the boiling less rapid. Nonetheless, it remains true that the fluid shows an instability in the direction of fluctuations in its region of boiling in the phase diagram.

1.3.2 Fluctuations II: Critical Opalescence

A process, not entirely dissimilar to boiling, occurs in the equilibrium fluid near its critical point. Observers have long noticed that, as we move close to the liquid–gas critical point, the fluid, hitherto clear and transparent, turns milky. This phenomenon, called *critical opalescence*, was studied by Marian Smoluchowski (1908) and Albert Einstein (1910; Pais 1983, p. 100). Both recognized that critical opalescence was caused by the scattering of light from fluctuations in the fluid's density. They pointed out that the total amount of light scattering was proportional to the compressibility, the derivative of the density with respect to pressure.[7] They also noted that the large amount of scattering near the critical point was indicative of anomalously large fluctuations in that region of parameters. In this way, they provided a substantial explanation of critical opalescence.[8]

---

[7] As pointed out to me by Hans van Leeuwen, the opalescence is very considerably enhanced by the difficulty of bringing the near-critical system to equilibrium. The out-of-equilibrium system tends to have anomalously large droplets analogous to those produced by boiling. These droplets then produce the observed turbidity.

[8] Einstein then used the explanation of this physical effect to provide one of his several suggested ways of measuring Avogadro's number, the number of molecules in a mole of material.

1.4 Ornstein and Zernike

Leonard Ornstein and Frederik Zernike (Ornstein and Zernike 1914) subsequently derived a more detailed theory of critical opalescence. In modern terms, one would say that the scattering is produced by small regions, droplets, of materials of the two different phases in the near-critical fluid. The regions would become bigger as the critical point was approached, with the correlations extending over a spatial distance called the correlation or coherence length, ξ. Ornstein and Zernike saw this length diverge on the line of coexistence between the two phases of liquid–gas phase transition as the critical point was approached in the form

$$\xi = \text{constant} \times 1/|T - T_c|^\nu \quad \text{with} \quad \nu = 1/2, \qquad (6)$$

with $T - T_c$ being the temperature deviation from criticality. Thus, the correlation length does go to infinity as the critical point is approached. As we shall see below in section 2.2, this result is crucially important to the overall understanding of the critical point.

1.5 Outline of Essay

The next section defines the Ising model, a simple and basic model for phase transitions. It then uses that model to describe the *extended singularity theorem*, which describes the relationship between phase transitions, mathematical singularities in thermodynamic functions, and correlated fluctuations. Section 3 then defines mean field theories as one way of approaching the theory of phase transitions. The next section describes the 1937 Landau theory as the pinnacle of mean field theory descriptions. But Landau's work also starts the replacement of mean field theory by fluctuation-dominated approaches. In that same year, a conference in Amsterdam exhibited the confusion caused by the conflict between mean field theory and the extended singularity theorem. The long series of studies that indicated a need for supplementing mean field theory is described in section 5. Section 6 describes the development of a new phenomenology to understand fluctuations in phase transitions. Kenneth Wilson transformed that phenomenology into a theory by adding ideas described in section 7. The concepts that grew out of that revolution are discussed in the final section.

2. The Ising Model

2.1 Definition

The Ising model is a conceptually simple representation of a system that can potentially show ferromagnetic behavior. Its name comes from the physicist Ernst Ising

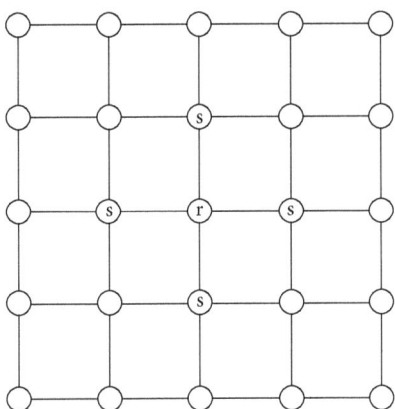

Figure 4.4 Lattice for two-dimensional Ising model. The spins are in the circles. The couplings, K, are the lines. A particular site is labeled with an "r." Its nearest neighbors are shown with an "s."

(1925), who studied it[9] in conjunction with his adviser Wilhelm Lenz (Brush 1967). Real ferromagnets involve atomic spins placed upon a lattice. The elucidation of their properties requires a difficult study via the band theory of solids. The Ising model is a shortcut that catches the main qualitative features of the ferromagnet. It puts a spin variable upon each site, labeled by **r**, of a simple lattice. (See figure 4.4.) Each spin variable, σ_r, takes on values plus or minus one to represent the possible directions that might be taken by a particular component of a real spin upon a real atom.

The sum over configurations is a sum over all these possible values. The Hamiltonian for the system is the simplest representation of the fact that neighboring spins interact with a dimensionless coupling strength, K, and a dimensionless coupling to an external magnetic field, h. The Hamiltonian is given by

$$-H/T = K \sum_{nn} \sigma_r \sigma_s + h \sum_r \sigma_r, \qquad (7)$$

where the first sum is over all pairs of nearest neighboring sites, and the second is over all sites. The actual coupling between neighboring spins, with dimensions of an energy, is often called J. Then $K = -J/T$. In turn, h is proportional to the magnetic moment of the given spin times the applied magnetic field, all divided by the temperature. Since lower values of the energy have a larger statistical likelihood, the two terms is Eq. (7) reflect respectively a tendency of spins to line up with each other and also a tendency for them to line up with an external magnetic field.

---

[9] Much of the historical material in this work is taken from the excellent book on critical phenomena by Cyril Domb (1966).

2.2 The Extended Singularity Theorem

Particle spin is certainly a quantum mechanical concept. There is no simple correspondence between this concept and anything in classical mechanics. Nonetheless, the concept of spin fits smoothly and easily into the Boltzmann-Gibbs formulation of statistical mechanics. When spins are present, the statistical sum in Eq. (2) includes a quantum summation over each spin-direction. In the Ising case, when the only variable is σ, standing for the z-component of the spin, that summation operation is simply a sum over the two possible values plus one and minus one of each spin at each lattice site.

For the Ising model, the integral over configurations in Eq. (2) is replaced by a sum over the possible spin values, thus making the result have particularly simple mathematical properties. Take the logarithm of that equation and find

$$-F/T = \ln[\sum_c e^{-\mathcal{H}(c)/T}], \tag{8}$$

On the right-hand side of this equation one finds a simple sum of exponentials. This is a sum of positive terms, and it gives a result that is a smooth function of the parameters in each exponential, specifically the dimensionless spin-coupling, K, and the dimensionless magnetic field h. A logarithm of a smooth function is itself a smooth function. Therefore, it follows directly that the free energy, F, is a smooth function of h and K.

The reader will notice that this smoothness seems to contradict our definition of a phase transition, the statement that a phase transition is a singularity, that is, failure of smoothness, in some thermodynamic quantity.

This seeming contradiction is the key to understanding phase transitions. No sum of a finite number of smooth terms can be singular. However, for large systems, the number of terms is the sum grows quite rapidly with the size of the system. When the system is infinite, the number of terms is infinite. Then singularities can arise. Thus, all singularities, and hence all phase transitions, are consequences of the influence of some kind of infinity. Among the likely possibilities are infinite numbers of particles, infinite volumes, or —more rarely— infinitely strong interactions. Real condensed matter systems often have large numbers of particles. A cubic centimeter of air contains perhaps 10^{20} particles. When the numbers are this large, the systems most often behave almost as if they had an infinity of particles.

I am going to give a name to the idea that phase transitions only occur when the condensed matter system exhibits the effect of some singularity extended over the entire spatial extent of the system. Usually the infinity arises because some effect is propagated over the entire condensed system, that is, over a potentially unbounded distance. I am going to call this result the "extended singularity theorem," despite

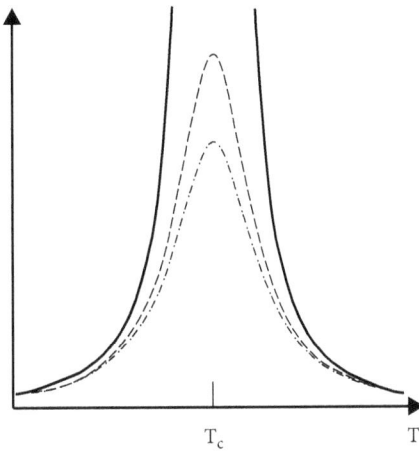

Figure 4.5 Cartoon view of a singularity in a phase transition. The *magnetic susceptibility*, the derivative of the magnetization with respect to the magnetic field, is plotted against temperature for different values of N. The thick solid curve is shows the susceptibility in an infinite system. The dashed curves apply to systems with finite numbers of particles, with the higher line being the larger number of particles. The compressibility of the liquid–gas phase transition also shows this behavior.

the fact that the argument is rather too vague to be a real theorem. It is instead a slightly imprecise mathematical property of real phase transitions.[10]

This theorem is only partially informative. It tells us to look for a source of the singularity, but not exactly what we should seek. In the important and usual case in which the phase transition is produced by the infinite size of the system, the theorem tells us that any theory of the phase transition should look to things that happen in the far reaches of the system. What things? How big are they? How should one look for them? Will they dominate the behavior near the phase transition or be tiny? The theorem is uninformative on all these points.

Sometimes it is very hard to see the result of the theorem. In an Ising or liquid–gas phase transition there is a singularity in the regions just touching the coexisting phases (Adreev 1964; Fisher 1978). This singularity is very weak. One must use indirect methods to observe or analyze it. Conversely, near critical points, singularities are very easy to observe and measure. For example, in a ferromagnet, the derivative of the magnetization with respect to the applied magnetic field is infinite at the critical point. (See figure 4.5.)

By looking at simulations of finite-sized Ising systems one can see how the infinite size of the system enters the susceptibility. Figure 4.5 is a set of plots of

---

[10] Imprecision can often be used to distinguish between the mathematician and the physicist. The former tries to be precise; the latter sometimes uses vague statements that can then be extended to cover more cases. However, in precisely defined situations, for example the situation defined by the Ising model, the extended similarity "theorem" is actually a theorem (Isakov 1984).

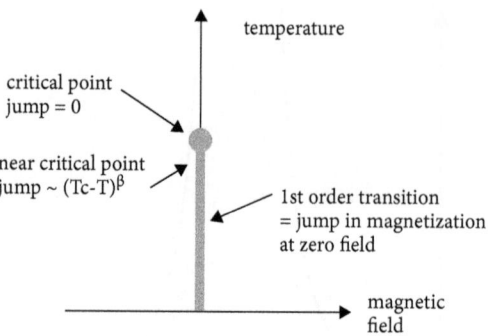

Figure 4.6 Phase diagram for ferromagnet and Ising model. The jump in magnetization occurs at zero magnetic field. In this representation, the jump region has been reduced to a line running from zero temperature up to the critical point.

susceptibility versus temperature in an Ising system with a vanishingly small positive magnetic field. The different plots show what happens as the number of particles increases toward infinity. As you can see, the finite N curves are smooth, but the infinite-N curve goes to infinity. This infinity is the singularity. It does not exist for any finite value of N. However, as N gets larger, the finite-N result approaches the infinite-N curve. When we look at a natural system, we tend to see phase transitions that look very sharp indeed, but are actually slightly rounded. However, a conceptual understanding of phase transitions requires that we consider the limiting, infinite-N, case.

Now we can see the importance of the Ornstein-Zernike infinity in the correlation length. This last infinity accompanies and causes the infinity in the susceptibility, and both of these then require an infinite system for their realization.

Figure 4.6 is the phase diagram of the Ising model. The x axis is the magnetic field; the y axis is the temperature. This phase diagram applies when the lattice is infinite in two or more dimensions. There is no phase transition for lower dimensionality.

3. More Is the Same

This section describes mean field theory, which forms the basis of much of modern many-particle physics and field theory. So far, we said that an infinite statistical system sometimes has a phase transition, involving a discontinuous jump in a quantity called the order parameter. But we have given no indication of how big the jump might be, nor of how the system might produce it. Mean field theory provides a partial and approximate answer to that question.

We begin with the statistical mechanics of one spin in a magnetic field. Then, we extend this one-spin discussion to describe how many spins work together to produce ferromagnetism.

3.1 One Spin

A single spin in a magnetic field can be described by a simplification of the Ising Hamiltonian of Eq. (7), $-H/T = h\sigma$. As before, σ is a component of the spin in the direction of the magnetic field. This quantum variable takes on two values ± 1, so that probability distribution of Eq. (1) gives the average value of the spin as

$$<\sigma> = \sum_{\sigma=\pm 1} \sigma\, e^{h\sigma - F} = \tanh h. \qquad (9)$$

(In general, we write the statistical average of any quantity, q, as $<q>$.)

3.2 Many Spins; Mean Fields

The very simple result, Eq. (9), appears again when one follows Pierre Curie (1895) and Pierre Weiss (1907) in their development of a mean field theory of ferromagnetism. Translated to the Ising case, their theory would ask us to concentrate our attention upon one Ising variable, say the one at **r**. We would then notice that this one spin would see a field with the value

$$h(\mathbf{r}) + K \sum_{s \text{ nn to } \mathbf{r}} \sigma_s, \qquad (10)$$

where $h(\mathbf{r})$ is the dimensionless magnetic field at **r** and the sum covers all the spins with positions, **s**, sitting at nearest neighbor sites to **r**. To get the mean field result, replace the actual values of all the other spins, but not the one at **r**, by their average values and find, by the same calculation that gave Eq. (9), a result in which the average is once more

$$<\sigma_{\mathbf{r}}> = \tanh h^{\text{eff}}(\mathbf{r}) \qquad (11a)$$

but now the actual field is replaced by an effective field

$$h^{\text{eff}}(\mathbf{r}) = h(\mathbf{r}) + K \sum_{s \text{ nn to } \mathbf{r}} <\sigma_s>. \qquad (11b)$$

3.3 Mean Field Results

Given this calculation of basic equations for the local magnetization, $<\sigma_{\mathbf{r}}>$, we can go on to find many different aspects of the behavior of this mean field magnet. We notice that when h is independent of position, the equation for $<\sigma>$ has a critical point, that is, an ambiguity in its solution, at zero magnetic field and $Kz = 1$. When we expand the equations around that critical point, we can find that the magnetization obeys a cubic equation like that of the van der Waals theory (or equally the Landau theory as described below in section 4.1. Thus, there is a full and complete correspondence between the van der Waals theory of the liquid–gas transition and the ferromagnetic mean field theory near its critical point, as one can see by comparing figure 4.2 with figure 4.6.

A brief calculation shows that in mean field theory the magnetic susceptibility, the derivative of the magnetization with respect to magnetic field, diverges as $1/|T - T_c|$ near the critical point. The analogy just mentioned between the liquid–gas system and the magnetic phase transition makes this magnetic susceptibility the direct analog of the compressibility. Please recall that the compressibility has an infinity that was used by Einstein to explain critical opalescence. (See section 1.3.2).

On the other hand, Ornstein and Zernike calculated the fluid analog of the more disaggregated quantity

$$g(\mathbf{r},\mathbf{s}) = \frac{\partial <\sigma_\mathbf{r}>}{\partial h(\mathbf{s})} \qquad (12)$$

called the spin correlation function. A sum over all lattice sites, s, of this correlation function will give the susceptibility. Then g can be evaluated from the equations of mean field theory ((Kadanoff 2000), p. 232) as

$$g(\mathbf{r},\mathbf{s}) = z\, a\, \frac{e^{-|\mathbf{r}-\mathbf{s}|/\xi}}{4\pi|\mathbf{r}-\mathbf{s}|} \qquad (13)$$

in the simplest case: three dimensions, $h = 0$, t small but greater than zero, and separation distance large compared to the lattice spacing, a. In Eq. (13), ξ is the correlation length that describes the range of influence of a change in magnetic field. Its value, given by Eq. (6), shows that the correlation length diverges as criticality is approached. This behavior is an expected consequence of the extended singularity theorem, which asks for infinite ranges of influence at phase transitions.

We previously argued that the extended singularity theorem called for fluctuations extending over large distances. Indeed that call is precisely answered by Eq. (13) and Eq. (6). A theorem of statistical mechanics relates the correlation function to spin fluctuations via

$$g(\mathbf{r},\mathbf{s}) = <(\sigma_\mathbf{r} - <\sigma_\mathbf{r}>)(\sigma_\mathbf{s} - <\sigma_\mathbf{s}>)>. \qquad (14)$$

The right-hand side of this expression relates g to the deviations of the spins at r and s from their averages values. According to Eq. (13), these fluctuations have correlations that persist over distances comparable to the length, ξ, which can then go to infinity as criticality is approached.

Note the scaling of the spin correlation function. For relatively small values of the distance, the correlation function in Eq. (13) has a form in which g varies as one divided by the distance. It is conventional to describe this correlation function, varying as $|\mathbf{r}-\mathbf{s}|^{-2x}$, by saying that there are two local quantities contributing to the correlation and then saying that each scales as distance to the power x. Therefore, in this case, the index going with the order parameter is $x = 1/2$.

3.4 Representing Critical Behavior by Power Laws

The reader will, no doubt, have noticed the appearance of "power laws" in the description of behavior near critical points. In these laws, some critical property is written as a power of a quantity that might become very large or very small, as for example, magnetization = constant $\times\ t^\beta$. So far, we have seen laws like this in the behavior of the order parameter (Eq. (5)), the correlation length (Eq. (6)), the magnetic susceptibility (figure 4.5), and the correlation function (Eq. (13)). Why does this power function appear repeatedly?

All of this singular behavior is rooted in the fact that phase transitions produce a variation over a tremendous range of length scales. For example, the basic interactions driving most phase transitions occur on a length scale described by the distance between atoms or molecules, that is some fraction of a nanometer (10^{-9} meters). On the other hand, we observe and work with materials on a characteristic length scale of centimeters (10^{-2} meters). The crucial issue in phase transitions is how the material interpolates phenomena over this tremendous length scale. The answer is roughly speaking that all the physical quantities mentioned follow the changes in the length scale.

As we shall see in section 7.2, in renormalization calculations, the changes of the length scale in turn follow from multiplicative laws. To get to a tremendous change in length scale, ℓ, one puts many small steps $\ell_1, \ell_2, \ell_3, \cdots$ into the renormalization calculation and the big change is produced by the multiplication of these factors,

$$\ell = \ell_1 \times \ell_2 \times \ell_3 \times \cdots.$$

This kind of behavior is explicitly built into renormalization calculations.[11]

Scale transformation is a symmetry operation. It describes an underlying symmetry of nature in which every scale—kilometer, centimeter, nanometer—is equally good for describing nature's basic laws. Whenever a physical phenomenon reflects a symmetry operation, observed physical quantities must transform under symmetry operations as mathematical representations of that symmetry. That is why we use scalars, vectors, and tensors to describe quantities that obey, say, the usual rotational symmetry. The same thing works for scale transformations. Here, power laws reflect the symmetries built into multiplication operations. The physical quantities behave as powers, ℓ^x, where x can be rational or irrational, positive or negative, or indeed zero. In the last case, the limiting behavior is that of a logarithm instead of a power, as is actually obtained in the heat capacity of the Onsager solution (Onsager 1944) of the two-dimensional Ising model.

The wide range of length scales also applies in particle physics where the basic scales for interactions may be vastly different from the scale at which observations

---

[11] Calculations of the effects of scale changes are much more implicit in mean field theories than in renormalization theories. In both cases we are treating variation over a huge range of scales, and power laws are a likely way of describing this huge range of variation. However, because the mean field theories deal less directly with scale transformations they do not get the relation between the renormalization scalings of fluctuations and free energy quite right.

are performed. Thus, in particle physics it is also true that renormalization and scaling have to interpolate behaviors over very large length scales.

Whenever one has a power law, say $<\sigma> = (-t)^\beta$, one has a power, here β. This power is called a "critical exponent" or a "critical index." During the many years in which critical behavior has been a subject of scientific study, many human-years of scientific effort have been devoted to the accurate determination of these indices. Sometimes scientists complained that this effort was misplaced. After all, there is little insight to be obtained from the statement that β (the index that describes the jump in the liquid–gas phase transition) has the value 0.31 versus 0.35 or 0.125 or 0.5. But these various values can be obtained from theories that give a direct calculation of critical quantities or related them one to another. The calculations or relations come from ideas with considerable intellectual content. Finding the index-values then gave an opportunity to check the theory and see whether the underlying ideas were sound. Thus, the small industry of evaluating critical indices supports the basic effort devoted to understanding critical phenomena.

4. The Year 1937: A Revolution Begins

4.1 Landau's Generalization

Lev Landau followed van der Waals, Pierre Curie, and Ehrenfest in noticing a deep connection among different phase transition problems (Daugherty 2007). Landau translated this observation into a mathematical theory in a novel and interesting way. Starting from the recognition that, in the neighborhood of a critical point, each phase transition was a manifestation of a broken symmetry, he used the order parameter to describe the nature and the extent of symmetry breaking (Landau 1937).

Landau generalized the work of others by writing the free energy as an integral over all space of an appropriate function of the order parameter. The dependence upon r indicates that the order parameter is considered to be a function of position within the system. In the simplest case, described above, the phase transition is one in which the order parameter, say the magnetization, changes sign.[12] In that case, the appropriate free energy takes the form

$$-F/T = \int d\mathbf{r} \left[Ah(\mathbf{r})\Psi(\mathbf{r}) + B\,\Psi(\mathbf{r})^2 + C\,\Psi(\mathbf{r})^4 + D\,[\nabla\Psi(\mathbf{r})]^2 + \cdots \right] \quad (15)$$

where $A, B, C, \ldots$ are parameters that describe the particular material and $\Psi(\mathbf{r})$ is the order parameter at spatial position $\mathbf{r}$. In recognition of the delicacy of the critical point, each term goes to zero more rapidly than $\Psi(\mathbf{r})^2$ as criticality is approached.

---

[12] The symmetry of the phase transition is reflected in the nature of the order parameter, whether it be a simple number (the case discussed here), a complex number (superconductivity and superfluidity), a vector (magnetism), or something else.

The next step is to use the well-known rule of thermodynamics that the free energy is minimized by the achieved value of every possible macroscopic thermodynamic variable within the system. Landau took the magnetization density at each point to be a thermodynamic variable that could be used to minimize the free energy. Using the calculus of variations one then gets an equation for the order parameter:

$$0 = h(\mathbf{r}) + 2(B/A)\,\Psi(\mathbf{r}) + 4(C/A)\,\Psi(\mathbf{r})^3 - 2(D/A)\,\nabla^2\Psi(\mathbf{r}). \tag{16}$$

One would get a result of precisely this form by applying the mean field theory magnetization equation near the critical point. The B-term is identified by this comparison as being proportional to the temperature deviation from criticality, $B = -At/2$.

In some sense, of course, Landau's critical point theory is nothing new. All his results are contained within the earlier theories of the individual phase transitions. However, in another sense his work was a very big step forward. By using a single formulation that could encompass all critical phenomena with a given symmetry type, he pointed out the close similarity among different phase transition problems. And indeed in the modern classification of phase transition problems (Kadanoff et al. 1967), the two main elements of the classification scheme are the symmetry of the order parameter and the dimension of the space. Landau got the first one right but not, at least in this variational formulation, the second classifying feature. On the other hand, Landau's inclusion of the space gradients that then brought together the theory's space dependence and its thermodynamic behavior also seems, from a present-day perspective, to be right on.

4.2 Summary of Mean Field Theories

As already mentioned, Landau's 1937 result provides a kind of mean field theory that agrees in all essential ways with the results of the main previous workers. The only difference is that Landau produced a specialized theory intended to apply mostly to the region near the critical point. From the point of view of the discussion that will follow the main points of his theory are:

- Universality. The Landau theory gives an equation for the order parameter as a function of the thermodynamic parameters (e.g., t and h) that is universal: it only depends upon the kind of symmetry reflected in the ordering.
- Symmetry. A first-order phase transition is often, but not always, a reflection of a change in the basic symmetry of the condensed system.
- Interactions. This symmetry change is usually caused by local interactions among the basic constituents of the system.
- Scaling. The results depend upon simple ratios of the thermodynamic parameters raised to powers. For example, in the ferromagnetic transition all physical quantities depend upon the ratio t^3/h^2. In subsequent theories, the restriction to simple powers will disappear.

- Order parameter jump. At the first-order phase transition, there is a discontinuous jump in the order parameter. As the critical point is approached, the jump goes to zero with critical index $\beta = 1/2$ as in Eq. (5).
- Correlation length. The correlation length goes to infinity at criticality as in Eq. (6) with an index $\nu = 1/2$.

As we shall see, for many purposes, the mean field theories have been replaced by a *renormalization group* theory of phase transitions. The qualitative properties of mean field theory, like universality and scaling, have been retained. On the other hand, all the quantitative properties of the theory, for example, the values of the critical indices, have been replaced.

4.3 Away from Corresponding States—Toward Universality

Landau's calculation represented the high-water mark of the class of theories described as "mean field theories." He showed that all of them could be covered by the same basic calculational method. They differed in the symmetries of the order parameter, and different symmetries could give different outcomes. However, within one kind of symmetry the result was always the same. This uniform outcome was very pleasing for many students of the subject, particularly so for the physicists involved. We physicists especially like mathematically based generalizations and Landau had developed an elegant generalization, which simplified a complex subject.

However, Landau's uniformity was different from the theoretical idea of uniformity that had come before him. Earlier work had been based upon the idea that different fluids have an almost identical relation expressing the dependence of their pressure upon temperature and density. This idea is called the "principle of corresponding states."

This principle of corresponding states had broad support among the scientists working on phase transitions. Starting with van der Waals, continuing with the work of Einstein (Pais 1983, p. 57), George Uhlenbeck, and E. A. Guggenheim (1945), work on phase transitions was inspired by the aim and hope that the phase diagrams of all fluids would be essentially alike. However, Landau's work marked a new beginning. His method would apply only near a critical point and his version of corresponding states could be expected to apply only in this region. So Landau deepened the theory but implicitly also narrowed its domain of application to a relatively small region of the phase diagram. This was the start of a new point of view, which we shall see develop in the rest of this essay. The new point of view would come with a new vocabulary so that instead of corresponding states people would begin to use the word "universality" (Kadanoff 1990).

4.4 Statistical Confusion: A Meeting in the Netherlands

The extended singularity theorem (see section 2) presents both an opportunity and a challenge for understanding phase transitions. The theorem is self-evident in the case of the Ising model with its simple sums and exponentials. It is less obviously true for the statistical mechanics of the liquid–gas phase transition since, in this case, the calculation of the free energy includes integrals and also unbounded potentials. However, the theorem remains true for that transition. Thus, the theorem would then demand an infinite system for a sharp liquid–gas phase transition. On the other hand, the van der Waals mean field argument would, for example, give a sharp phase transition in small systems. This contradiction might serve as a confusing element in the development of a theory of phase transitions.

The contradiction was as old as the first definitions of statistical mechanics and phase transitions, but was apparently not discussed for many years. It might well, however, have come up at a 1937 meeting held in Amsterdam to celebrate the centenary of van der Waals's birth. Hendrik Kramers, George Uhlenbeck, and Peter Debye were all present at that occasion. According to Uhlenbeck (1978), at that meeting Kramers pointed out that the sharp singularity of a phase transition could only occur in a system with some infinity built in and, for that, that an infinite system is required. Then the van der Waals theory's prediction of a phase transition in a finite system could be viewed as a grave failure of mean field theory and maybe even of statistical mechanics. E. G. D. Cohen described material by Uhlenbeck (Cohen n.d.): "Apparently the audience at this van der Waals memorial meeting in 1937, could not agree on the above question, whether the partition function could or could not explain a sharp phase transition. So the chairman of the session, Kramers, put it to a vote." The proposition was "Can statistical mechanics describe the liquid state?" The meeting is said to have split 50–50, with Debye (!) voting no!

Clearly half the people at that meeting were wrong. Seventy plus years later one can see the right answer, in close analogy to our understanding of irreversibility. Infinite size is required for a sharp phase transition, but a large system can very well approximate the behavior of the infinite system. Finite size slightly rounds off and modifies the sharp corners shown in the plot of figure 4.2. Conventional statistical mechanics, following the path begun by Andrews, van der Waals, and Maxwell, can describe quite well what happens in the liquid region, especially if one stays away from the critical region and from boiling. In fact, there are theories, including one by John Weeks, David Chandler, and Hans Anderson (1917), that do a good job of describing the liquid region of the fluid.

However, the extended singularity theorem does have its effects. There are indeed singularities near the first-order transition (Andreev 1964; Fisher 1967). These singularities are very weak in the Ising and liquid–gas transition, but will be stronger when an unbroken symmetry remains after the symmetry breaking of the first-order transition. (We see such a residual symmetry in the Heisenberg model of ferromagnetism.) Also, there are quite strong singularities in the neighborhood of the critical point, not correctly described by mean field theory. All these singularities are consequences of fluctuations, which are not included in the mean field approach.

However, the Amsterdam meeting was quite right, in my view, to be disquieted by the applicability of statistical mechanics. But they focused upon the wrong part of the phase diagram. The liquid region is described correctly by statistical mechanics.

But this theory does not work well in the two-phase, "boiling" region of figure 4.2. Here the fluctuations entirely dominate and the system sloshes between the two phases. The behavior of the interface that separates the phases is determined by delicate effects of dynamics and previous history and by hydrodynamic effects including gravity, surface tension, and the behavior of droplets. Hence, the direct application of statistical mechanics is fraught with difficulty precisely in the midst of the phase transition. Thus, the extended singularity theorem suggests that a new theory is required to treat all the fluctuations appearing near singularities.

5. Beyond (or Beside) Mean Field Theory

Weaknesses of mean field theory began to become apparent to the scientific community immediately after Landau's statement of the theory in its generalized form. This section will describe the process of displacement of mean field theory, at least for behavior near the critical point, which we might say began in 1937 and culminated in Kenneth Wilson's enunciation of a replacement theory in 1971 (Wilson 1971).

Landau's theory provided a standard and a model for theories of general phenomena in condensed matter physics. Looking at Landau's result one might conclude that a theory should be as general and elegant as the phenomena it explained. The mean field theories that arose before (and after) Landau's work were partial and incomplete, in that each referred to a particular type of system. That was certainly necessary in that the details of the phase diagrams were different for different kinds of systems, but somewhat similar for different materials of the same general kind. Landau's magisterial work swept all these difficulties under the rug and for that reason could not apply to the whole phase diagram of any given substance. Thus, if Landau were to be correct he would most likely be so in the region near criticality. Certainly his theory is based upon an order parameter expansion that only is plausible in the critical region. However, precisely in this region, as we shall outline below, both theoretical and experimental facts contradicted his theory.

5.1 Experimental Facts

The ghosts of Andrews and van der Waals might have whispered to Landau that a theory that predicts $\beta = 1/2$ near criticality cannot be correct. In addition, a much larger body of early work on fluids had pointed to this conclusion. These early data, developed and published by J. E. Verschaffelt (1900) and summarized by Levelt-Sengers (1976), touched almost every aspect of the critical behavior of fluids. Verschaffelt particularly stresses the incompatibility of the data with mean field theory.

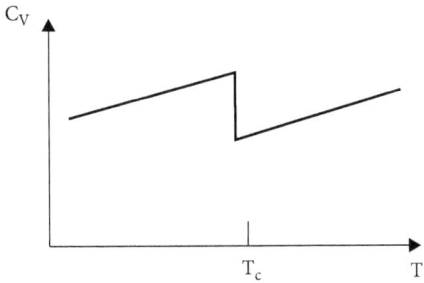

Figure 4.7 Cartoon sketch of heat capacity in the neighborhood of critical temperature as predicted by mean field theory. The heat capacity is higher below T_c because there is an additional temperature dependence in the free energy in this region produced by a term proportional to the square of the order parameter.

These same experimental facts appear once more in the 1945 work of E. A. Guggenheim (1945), who compared data for a wide variety of fluids. He says, "The principle of corresponding states may safely be regarded as the most useful by-product of van der Waals' equation of state. While this [van der Waals] equation of state is recognized to be of little or no value, the principle of corresponding states as correctly applied is extremely useful and remarkably accurate." He examined data for seven fluids on the line of the liquid–gas phase transition and fit the data to a power law with $\beta = 1/3$, rather than the mean field value $\beta = 1/2$. The latter value clearly does not work; the former fits reasonably well. Thus "corresponding states" receives support in this region, but not mean field theory per se.

But neither Guggenheim nor Heike Kamerlingh Onnes (Levelt-Segers 1976) before him was ready to receive information suggesting that behavior in the critical region was special, so that the former rejected mean field theory while the latter accepted it with reservations as to its quantitative accuracy.

Later, near-critical data on heat capacity, the derivative of average energy with respect to temperature, became available. Mean field theory predicts a discontinuity in the constant volume heat capacity as in figure 4.7. L. F. Kellers (1960) looked at the normal fluid to superfluid transition in helium-4 (see figure 4.8.) The data on this phase transition seemed to support the view that the heat capacity diverges weakly, perhaps as a logarithm of $|T - T_c|$, as criticality is approached. Similar heat capacity curves were observed by Alexander Voronel' (Bagatskii, Voronel', and Gusak 1963; Voronel' et al. 1964) in the liquid–gas transition of classical gases and in further work in helium (Moldover and Little 1965).

5.2 Theoretical Facts

As we have seen, experimental evidence suggested that mean field theory was incorrect in the critical region. A further strong argument in this direction came from Lars Onsager's exact solution (1944) of the two-dimensional Ising model, followed

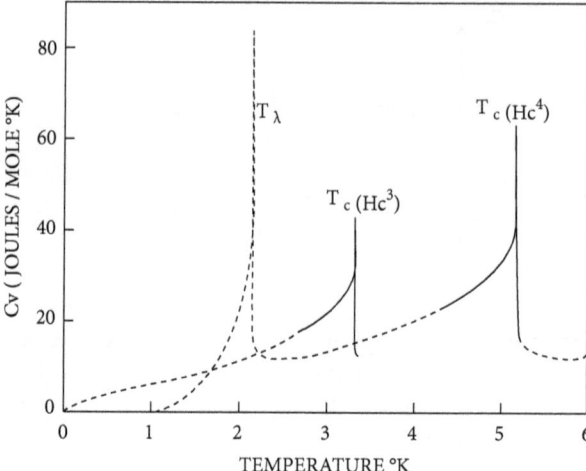

Figure 4.8 Heat capacity as measured. This picture, the work of Moldover and Little (1965), shows measured heat capacities for the normal–superfluid transition of helium-4, labeled as T_λ, and the liquid–vapor transition of helium-3 and helium-4. Note that all three heat capacities seem to spike at the critical point, in contrast to the prediction of mean field theory.

by C. N. Yang's calculation (1952) of the zero-field magnetization for that model. Onsager's result for the heat capacity diverged as the logarithm of $T - T_c$ as did the experimental observations, as shown, for example, in figure 4.8, but did not resemble the discontinuity of mean field theory.[13]

Yang's results, for which $\beta = 1/8$, also disagreed with mean field theory, which has $\beta = 1/2$. The Onsager solution implies a correlation length with $\nu = 1$, which is not the mean field value $\nu = 1/2$; see Eq. (6).

The most systematic theoretical discrediting of mean field theory came from the series expansion work of the King's College (London) school, under the leadership of Cyril Domb, Martin Sykes, and—after a time—Michael Fisher (Domb 1996; Niss 2005). Recall that the Ising model is a simplified model that can be used to describe magnetic transitions. It is described by a strength of the coupling between neighboring spins proportional to a coupling constant J. The statistical mechanics of the model is defined by the ratio of coupling to temperature, specifically $K = -J/T$. One can get considerable information about the behavior of these models by doing expansions of quantities like the magnetization and the heat capacity in power series in K, for high temperatures, and e^{-K} for low temperatures. The group at King's developed and used methods for doing such expansions and then analyzing them to obtain approximate values of critical indices like β and ν. The resulting index-values

---

[13] Onsager's results looked different from those depicted in figure 4.8 in that they showed much more symmetry between the high-temperature region and the low-temperature region. This difference reflects the fact that two-dimensional critical phenomena are markedly different in detail from three-dimensional critical phenomena. Further, subsequent work has indicated that none of the heat capacity singularities shown in figure 4.8 are actually logarithmic in character. They are all power law singularities.

in two dimensions agreed very well with values derived from the Onsager solution. In three dimensions, models on different lattices gave index values roughly agreeing with experiment on liquids and magnetic materials, but differing substantially from predictions of mean field theory. This work provided a powerful argument indicating that mean field theory was wrong, at least near the critical point. It also played a very important role in focusing attention upon that region.

Another reason for doubting mean field theory, ironically enough, came from Landau himself. In 1941, Andrei Kolmogorov developed a theory of turbulence based upon concepts similar to the ones used in mean field theory, in particular the idea of a typical velocity scale for velocity differences over a distance r (Kolmogorov 1941). These differences would, in his theory, have a characteristic size that would be a power of r. Landau criticized Kolmogorov's theory saying that it did not take into account fluctuations (Frisch 1995), whereupon Kolmogorov modified the theory to make it substantially less similar to mean field theory (1962).

5.3 Spatial Structures

The spatial structure of mean field theory does not agree with the theorem that phase transitions can only occur in infinite systems. Mean field theory is based on the alignment of order parameter values at neighboring sites, so that particles will order if neighboring particles are ordered also. Any collection of coupled spins can have a mean field theory phase transition. Thus, two spins and a bond are quite sufficient to produce a phase transition in a mean field argument like that in section 3.2. On the other hand, the extended singularity theorem insists that the occurrence of a phase transition requires some sort of infinity, most often the existence of an infinite number of interacting parts within the system.

As we shall see, what is wrong with mean field theory is that in the critical region the effect of the average behavior of the order parameter can be completely swamped by fluctuations in this quantity. In 1959 and 1960, A. P. Levanyuk and Vitaly Ginzburg described a criterion that one could use to determine whether the behavior near a phase transition was dominated by average values or by fluctuations (Levanyuk 1959; Ginzburg 1960). For example, when applied to critical behavior of the type seen in the simplest version of the Ising model, this criterion indicates that fluctuations dominate in the critical region whenever the dimension is less than or equal to four. Hence, mean field theory is wrong(47) for all the usual critical phenomena in systems with dimension smaller than or equal to four.[14] Conversely, this criterion suggests that mean field theory gives the leading behavior above four dimensions.

---

[14] There are exceptions. Mean field theory works quite well whenever the forces are sufficiently long-ranged so that many different particles will interact directly with any given particle. By this criterion mean field theory works well for the usual superconducting materials studied up through the 1980s(7, 8), except extremely close to the critical point. However, mean field theory does not work for the newer "high-temperature superconductors," a class discovered in 1986 by Georg Bednorz and Alexander Müller(9).

6. New Foci; New Ideas

6.1 Bureau of Standards Conference

So far, the field of phase transitions had lived up perfectly to Thomas Kuhn's (1962) view of the conservatism of science. Before World War II, the only theory of phase transitions was mean field theory. No theory or model yielded Eq. (5) with any value of β different from one half. There was no focus for anyone's discontent. For this reason, the mean-field-theory point of view continued on, despite evidence to the contrary, until a set of events occurred that would move the field in a new direction. One crucial event was the conference on critical phenomena held at the US National Bureau of Standards in 1965 (NBS 1965). The late Melville Green was the moving spirit behind this meeting. The point of this conference was that behavior near the critical point formed a separate body of science that might be studied on its own merits, independent of the rest of the phase diagram. In the years just before the conference, enough work (Domb and Miedema 1964; Fisher 1967; Heller 1967) had been done so that the conference could serve as an inauguration of a new field. We have mentioned the experimental studies of Kellers and of the Voronel' group. At roughly the same time important theoretical work was done by Alexander Patashinskii and Valery Pokrovsky (Patashinskii and Pokrovsky 1964), Benjamin Widom (1964; 1965a and b) and myself (1966), which would form a basis for a new synthesis. The experimental and theoretical situation just after the meeting was summarized in reviews (Fisher 1967; Heller 1967; Kadanoff et al. 1967). This section begins by reporting on those new ideas and then describes their culmination in the work of Kenneth G. Wilson (1971).

6.2 Correlation Function Calculations

For many years the Landau group had been using field theory to describe the critical point. Two young theoreticians, Patashinskii and Pokrovsky, focused their attention upon the correlated fluctuations of order parameters at many different points in space. Their result was simple but powerful. Consider the result of calculating the average of the product of m local order parameter operators at m different positions, r_m, in a system at the critical point. (All differences between positions of the operators should be large in comparison to the distance between neighboring sites or the range of forces.) Compare this average with the same correlation function calculated at the positions $\ell \times r_m$. All that has been done is to change the length-scale on which the correlations have been defined. Patashinskii and Pokrovsky then argued that this change in scale was an invariance of the system so that the two correlation functions will have precisely the same structure (Patashinskii and Pokrovsky 1964), and differ by a factor ℓ^{-mx}. A similar rule, with a different index holds, for other kinds of fluctuating quantities near the critical point (Patashinskii and Pokrovsky

1978). These authors succeeded in getting the right general structure of the correlations. In building upon the early work of Widom, these authors succeeded in constructing most of the elements of the two-index scaling theory of critical phenomena. Patashinskii and Pokrovsky's work pointed the way toward future field theoretic calculations of correlation behavior.

In parallel, I calculated (1966) the long-distance form of the spin correlation function for the two-dimensional Ising model by making use of the Onsager solution. This was the part of a long series of calculations that would give insight into the structure of that model (McCoy and Wu 1973). Those insights would be quite crucial in establishing the fundamental theory of behavior at the critical point.

6.3 Widom Scaling

Benjamin Widom (1964; 1965 a and b; see figure 4.9) developed a phenomenological theory of the thermodynamics near critical points. (He studied the liquid–gas transition, but here his results will be stated in the language of the magnetic transition, in which the temperature deviation from criticality is t and the symmetry of the ordering is broken by a magnetic field, h.) If t is zero, the average order parameter, $<\sigma>$, was experimentally seen to be proportional to $h^{1/\delta}$ where δ is a critical index known to be close to 4.4 in three dimensions (Widom and Rice 1955) and 15 in two dimensions (Kadanoff et al. 1967). As discussed above, if $h=0$ and $t<0$, then $<\sigma>$ is proportional to $\pm(-t)^\beta$. He then said that, near the critical point, no one of these three quantities has a natural size, but instead each one should be measured against the size of the others. This led him to suggest (1965) a general formula for the magnetization near the critical point that could fit both limiting forms, specifically

$$<\sigma> = h^{1/\delta} g(h^{1/\delta}/t^\beta), \qquad (17)$$

where g is a function that would have to be determined experimentally.

In this way Widom got very concrete and precise results from his initial requirement that each small quantity, $<\sigma>, h, t$, and so on be measured against another small quantity. He was able to predict the index-value to describe how every thermodynamic quantity would go to zero or infinity at criticality. All these critical indices would then be determined from just the two indices, β and δ. (See Sengers and Shanks (2009) for a comparison of the results of this theory with experiment.)

These results were published in a paper in the *Journal of Chemical Physics* (1965). In an adjacent paper, Widom also got a scaling relation (1965) for the surface tension, the free energy of the boundary between liquid and vapor, by relating it to the coherence length. To get this, think of an interface covered by many structures produced by the critical behavior. One might expect the characteristic size and spacing of these structures to be a correlation length and that each such structure would bring in an extra free energy of order T. Therefore one expects that the entire interface would produce an extra free energy per unit area of order of T times the number of structures that could be placed to fill a unit area, ξ^{-2}. In view of

Figure 4.9 Benjamin Widom, left, and Michael Fisher, right. Widom is a Chemistry Professor at Cornell. Fisher has been at King's College (London), Cornell, and the University of Maryland.

Eq. (6), which ascribes a critical index ν to ξ, we would have a surface tension that varies as $t^{2\nu}$. This result derived by Widom must have pleased him very much, since it shed light on a difficulty that went back to van der Waals. The theory of the latter gave a critical index for the surface tension of 1.5, while van der Waals's and later experiments gave results in the range 1.22 to 1.27. Widom's new theory offered a hope that this old discrepancy between theory and experiment could soon be resolved.

An estimation essentially similar to the one Widom used for the surface tension indicates that the singular term in the bulk free energy has an expected behavior like $t^{3\nu}$. This result follows from the idea that the free energy in a three-dimensional system would have its singular part determined by excitations with an energy of order T. These excitations would have a size equal to the correlation length and a density equal to the inverse cube of the correlation length. Thus, the surface tension and the free energy density provide a bridge between an understanding of the correlation length and an understanding of thermodynamic properties. This bridge is not like anything contained in the mean field theories. In fact these relations, termed hyperscaling relations, are the most characteristic feature of the renormalization theory that will soon arrive on the scene.

6.4 Less Is the Same: Block Transforms and Scaling

Widom supplied much of the answer to questions about the thermodynamics of the critical point. I then supplied a part of the strategy for deriving the answer (1966).

I will now describe the method in a bit of detail, since the calculation provides some insight into the structure of the solution.

Imagine calculating the free energy of an Ising model near its critical temperature based upon the interactions incorporated in Ising's Hamiltonian function for the problem. The result will depend upon the number of lattice sites, the temperature deviation from criticality and the dimensionless magnetic field. Next imagine redoing the calculation using a new set of variables constructed by splitting the system into cells containing several spins and then using new spin variables, each intended to summarize the situation in a block containing several old spin variables. (See figure 4.10.) To make that happen one can, for example, pick the new variables to have the same direction as the sum of the old spin variables in the block and the same magnitude as each of the old variables. The change could then be represented by saying that the distance between nearest neighboring lattice sites would change from its old value, a, to a new and larger value, a'. (See figure 4.10, in which the lattice constant has grown by a factor $\ell = 3$.) In symbols, the change is given by

$$a' = \ell a. \tag{18a}$$

One can then do an approximate calculation and set up a new "effective" free energy calculation that will give the same answer as the old calculation based upon an approximate "effective" Hamiltonian making use of the new variables. Near the critical point, one could argue on the basis of universality[15] that the new Hamiltonian could be written to have the same structure as the old one. However, near criticality, the new parameters in the effective Hamiltonian, the number of lattice sites, the temperature deviation from criticality, and the dimensionless magnetic field all are proportional to the corresponding old parameters. This change can be represented by writing

$$N' = (\ell)^{-d} N \tag{18b}$$

$$h' = (\ell)^{y_h} h \tag{18c}$$

$$t' = (\ell)^{y_t} t. \tag{18d}$$

In the first of these statements, Eq. (18b), N is the number of lattice sites and d is the dimension of the lattice. The equation simply describes how the number of sites depends upon the spacing between lattice sites.

The other two equations are far, far less simple. Eq. (18c) says that the new situation has a symmetry breaking field of the same sign as the previous one. That would be a reflection of the fact that both situations would have the same kind of ordering. The coefficient, $(\ell)^{y_h}$, might be derived after some sort of statistical mechanical analysis of the situation. It is, as it stands, just a number defined by the

---

[15] I used Eq. (18), but I did NOT make an explicit argument based upon universality in my paper in which I first applied this block transformation. My discussion would have been much stronger had I the wisdom to do so. But wisdom often comes after the fact.

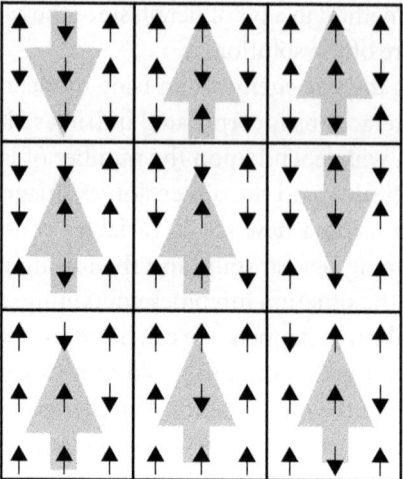

Figure 4.10 Making blocks. In this illustration a two-dimensional Ising model containing 81 spins is broken into blocks, each containing 9 spins. Each one of those blocks is assigned a new spin with a direction set by the average of the old ones. We imagine the model is reanalyzed in terms of the new spin variables.

result of that calculation and one that might depend upon the exact way in which we chose to define the new spin variable.

The equation for the new value of the new deviation from criticality, $t = K_c - K$, could be described in similar terms. It is reasonable to assume that if the original system is at its critical point, so is the new description obtained after the block transformation. Further it is reasonable to argue that the transformation should engender no singularities, thus requiring that a new temperature-deviation from criticality would have a linear dependence upon the old deviation. So the remaining point is to calculate the coefficient in the linear relation and express it in the special manner given in Eq. (18d).

7. The Wilson Revolution

7.1 Physical Space; Fourier Space

Before entering into Wilson's construction of the renormalization group theory, I should touch upon a point of technique.

The proportionalities in Eq. (18d) and Eq. (18c) are representations of scaling, and the coefficients in the linear relations define the scaling relations among the variables. Note that here scaling is viewed as a change in the effective values of the thermodynamic parameter produced by a change in the length scale at which the system is analyzed. The length scale must be irrelevant to the determination of the eventual answer and must drop out of the final result for the free energy. It is this dropping out that gives the empirical relations proposed by Widom. These scaling relations then give a theory with all the empirical content of Widom's work (1965a), but backed by the outlines of a conceptual and calculational scheme.

This theoretical work of Kadanoff (1966); Patashinskii and Pokrovsky (1964); and Widom (1965a) was well-received. The review paper of (Kadanoff et al. 1967) was particularly aimed at seeing whether the new phenomenology agreed with the experimental data. It reviewed most of the recent experiments but missed large numbers of the older ones that are included in Domb (1966) and Levelt–Sengers (1976). All of this activity validated the consideration of the critical region as an appropriate subject of study and led to a spate of experimental and numerical work, but hardly any further theoretical accomplishments until the work of Wilson (1971).

There are two traditional ways of setting up a Hamiltonian or free energy that will then provide a microscopic description of the system. One way is in coordinate space, the real XYZ space in which you and I live. This setup is the one we used for the Ising model, the Landau theory, and for the description of the previous subsection. It is relatively easy to visualize and the most effective method for problems in low dimensions, specifically for phase transitions in two dimensions.

The other method employs Fourier transforms. It represents every variable in terms of its Fourier transform. For example, the order parameter field of the Landau theory has a transform

$$\psi(\mathbf{k}) = \int d\mathbf{r}\, \Psi(\mathbf{r}) e^{-i\mathbf{k}\cdot\mathbf{r}}. \tag{19}$$

The integral covers a space of dimensionality d. Using $\psi(\mathbf{k})$ as our basic statistical variable the Landau free energy may be written as

$$-F/T = \int \frac{d\mathbf{k}}{(2\pi)^d}\Big[Ah(-\mathbf{k})\psi(\mathbf{k}) + B\,|\psi(\mathbf{k})|^2 - D\,k^2|\psi(\mathbf{k})|^2$$
$$+ C \int \frac{d\mathbf{m}}{(2\pi)^d} \int \frac{d\mathbf{n}}{(2\pi)^d} \int \psi(-\mathbf{k}-\mathbf{m}-\mathbf{n})\psi(\mathbf{m})\psi(\mathbf{n})\psi(\mathbf{k})\Big]. \tag{20}$$

This form in Eq. (20) is used to reach beyond mean field theory and take into account possible fluctuations in the local variables that describe the system. To do this, one uses F/T as a kind of a kind of Hamiltonian for phase transition problems. In this use, the k-space is divided into small pieces and $\psi(\mathbf{k})$ is taken to be an integration variable in each piece. In this context the expression in Eq. (20) is called the Landau-Ginzburg-Wilson free energy.

The k-shell integration just described is easily performed if the free energy includes only linear and quadratic terms in the variable, $\psi(\mathbf{k})$. The fourth-order term provides a problem, one that can be attacked by using the renormalization method. The term involving k^2 ensures that the contribution to the integral for the highest values of k will be small and relatively easily controlled. So, one successively integrates over shells in k-space, starting from the highest values of $|\mathbf{k}|$, and working downward. As each integral is done, one stops and regroups terms to bring everything back close to the form of the original Landau-Ginzburg-Wilson free energy. As one does this, the coefficients multiplying the various terms change.

The k-space method is particularly appropriate for higher dimensions, going down to roughly three dimensions. It is the usual method of choice in particle physics. In statistical physics, Wilson and Fisher (1972) have done a very convincing calculation in which they analyze the behavior near four dimensions by assuming that the fourth-order term is quite small. (See the discussion of e-expansion in section 7.3.1 below.)

Both real-space and k-space methods have added considerably to our understanding of phase transitions. I use the former to describe the concept of renormalization, since I find it more natural to think about phenomena in real space rather than Fourier space. In particle physics, however, our basic conceptualization is based upon, naturally enough, particles. These are best followed in k-space, since the k labels the momentum of particles. So the two different formulations are complementary, with the best applications to problems in different dimensionalities and indeed to different fields of science.

The extended singularity theorem, of course, applies equally in both the real-space and the Fourier-space formulations. In real-space, in order to have the potential for generating singularities, and thereby phase transitions, the system must be infinite in two or more dimensions. In Fourier-space, the corresponding statement is that two or more components of the k-vector must extend to infinity. The remaining requirement in either formulation is that the renormalization must lead to a nontrivial fixed point, one with infinitely large values of some of the couplings.

7.2 Wilson's Contribution

Around 1970, these concepts were extended and combined with previous ideas from particle physics (Gell-Mann and Low 1954; Stueckelberg and Peterman 1953) to produce a complete and beautiful theory of critical point behavior, the renormalization group theory of Kenneth G. Wilson 1971. (See figure 4.11.) The basic idea of reducing the number of degrees of freedom, described in Kadanoff 1966, was extended and completed.

Wilson, in essence, converted a phenomenology into a calculational method by introducing ideas not present in the earlier phenomenological treatment (Kadanoff 1966):

- Instead of using a few numbers, for example, t, h, to define the parameters multiplying a few coupling terms, he extended the list of possible couplings to include all the kinds of terms that might be found in the Hamiltonian of the system. Thereby it became automatically true that the renormalization would maintain the different coupling terms, but only change the size of the parameters which multiplied them.

Figure 4.11 Kenneth G. Wilson at California Tech where he did a Ph.D. thesis under Murray Gell-Mann, a major contributor to early work on renormalization in particle physics. This was followed by a Junior Fellowship at Harvard, a year's stay at CERN, and then an academic appointment at Cornell. The renormalization group work was done while Wilson was at Cornell.

- Wilson considers indefinitely repeated transformations, as in the earlier particle physics work. Each transformation increases the size of the length scale. In concept, then, the transformation would eventually reach out for information about the parts of the system that are infinitely far away. In this way, the infinite spatial extent of the system became part of the calculation. The idea that behaviors at the far reaches of the system would determine the thermodynamic singularities were thence included in the calculation.
- Furthermore, Wilson added the new idea that a phase transition would occur when the transformations brought the coupling to a *fixed point*. That is, after repeated transformations, the couplings all would settle down to a behavior in which further renormalization transformation would leave them unchanged.
- Finally, at the fixed point, the correlation length would be required to be unchanged by renormalization transformations. The transformation multiplies the length scale by a factor that depends upon the details of the transformation. Wilson noted that there are two ways that the correlation length might be unchanged. For transformations related to a continuous transition, the correlation length is infinite, thence reflecting the infinite-range correlation. For transformations related to first-order transitions, the correlation length is zero, reflecting the local interactions driving the transition.

A very important corollary to the use of repeated transforms is the idea of *running coupling constants*. As the length scale changes, so do the values of the different parameters describing the system. In the earlier field theoretical work (Gell-Mann and Low 1954; Stueckelberg and Peterman 1953), the important parameters were the charge, masses, and couplings of the "elementary" particles described by the theory. The parameters to be varied were specified at the beginning and were, in no sense, the outcome of the renormalization calculation. The change in length scale then changed these prespecified parameters from the "bare" values appearing in the basic Hamiltonian to renormalized values that might be observed by experiments examining a larger scale.

The use of renormalized or "effective" couplings was current not only in particle theory but also in the quasiparticle theories that are pervasive in condensed matter physics (Anderson 1997). In these theories one deals with particles that interact strongly with one another. Nonetheless, one treats them using the same Hamiltonian formalism that one would use for noninteracting particles. The only difference from free particles is that the Hamiltonian is allowed to have a position and momentum dependence that reflects the changes produced by the interactions. In this work, the quantities to be renormalized are prespecified. In contrast, Wilson's renormalization calculation determines what is to be renormalized as a part of the calculation.

7.3 Building upon the Revolution

This renormalization theory provided a basis for the development of new methods that could be used for building an understanding of critical phenomena and additional subjects as diverse as particle physics, the development of chaos, the behavior of computer programs, as well as dynamical behavior in condensed matter physics. It provided a framework into which one could fit a variety of different theories and physical problems. There was a tremendous flowering of new work following upon Wilson's.

7.3.1 The ϵ-expansion

But first the renormalization method had to gain acceptance. The most substantial step in that direction came from the ϵ-expansion of Wilson and Fisher (1972). Here ϵ means dimension minus four. This calculational method focuses upon the dependence of physical quantities upon dimension. It uses renormalization transforms near four dimensions, where mean field theory is almost, but not quite, correct.[16] The idea of using the dimension of the system as a continuously variable parameter seems a bit strange at first sight. However, in the momentum-space representation of statistical ensembles, each term in a perturbation expansion can be evaluated for all integer values of the dimension and then the analysis can be continued to all values of the dimension, including noninteger values.

When applied near four dimensions this method allows an almost exact analysis of the fixed point behavior. Near the fixed point, the nontrivial terms in the

---

[16] The idea of variable dimensionality is also used in particle physics under the name dimensional regularization. One of the earliest applications in particle physics was in the work of 't Hooft and Veltman (1972a and b) proving that the gauge theory of strong interactions was renormalizable.

free energy, like the term proportional to C in Eq. (20), go to zero as the dimension approaches four. Because of this simplification, the method gives quite accurate results for critical behavior near four dimensions. Further, it provides a series expansion that gives useful answers for many different models in three dimensions. The close correspondence of theory and experiment helped to convince people that both the variable-dimension method and the renormalization method were valid. The way had been opened for an explosion of new calculations and new understandings.

7.4 Different Kinds of Fixed Points

Wilson's theory gives three different kinds of fixed point corresponding to three qualitatively different points in phase diagrams. For the *weak coupling fixed point*, couplings can go to zero and the correlation length goes to zero. The symmetry represented by the order parameter will remain unbroken. This kind of fixed point describes all areas of the phase diagram that do not touch a phase transition. For the *strong coupling fixed point*, some couplings will go to infinity and the correlation length goes to zero. Here the basic symmetry represented by the order parameter gets broken by at least one nonzero coupling that violates that symmetry. This kind of fixed point describes all areas of the phase diagram that touch a first-order phase transition. For the *critical fixed point* couplings remain finite, the symmetry remains unbroken, and a correlation length goes to infinity.

Critical fixed points may be classified by their dimension and by the symmetry of their order parameter. The combination of the Landau theory and the ϵ-expansion gave the first steps in that direction. The later calculations of critical behavior were then fit into this scheme.

8. New Concepts

8.1 Different Scalings: Relevant, Irrelevant, Marginal

Since the Wilsonian point of view generated the renormalization of many different couplings, it became important to keep track of the different ways in which the couplings in the free energy would change as the length scale changes. This work starts with an eigenvalue analysis. One takes linear combinations of couplings and arranges the combinations so that, after a renormalization, every combination reproduces itself except for a multiplicative factor. In other words, this approach makes every linear combination of couplings obey an equation like the ones in Eq. (18), so that the combination, s, obeys

$$s' = \ell^{y_s} s \tag{21}$$

The different combinations are then classified according to the values of the index, y_s, which may be complex. There are three possibilities (Domb, Green, and Lebowitz 2001; see F. Wegner, Vol. 6, p. 8):

- Relevant, real part of y_s greater than zero. These are the couplings like t and h that grow larger as the length scale is increased. Each of these will, as they grow, push one away from the critical point. In order to reach the critical point, one must adjust the initial Hamiltonian so that these quantities are zero.
- Irrelevant, real part of y_s less than zero. These couplings will get smaller and smaller as the length scale is increased so that, as one reaches the largest length scales, they will have effectively disappeared
- Marginal, real part of y_s equal to zero.

The last case is rare. Let us put it aside for a moment and argue as if only the first two existed.

8.2 Universality Classes

To study critical phenomena based upon renormalization transformations, one sets all the relevant combinations of couplings to zero and then does a sufficient number of successive renormalizations so that all the irrelevant combinations have effectively disappeared. We thus end up with a unique *fixed point* independent of the value of all of the irrelevant couplings. The act of renormalization is a sort of focusing in which many different irrelevant couplings fade away and we end up at a single fixed point representing a whole multidimensional continuum of different possible Hamiltonians. These Hamiltonians form what is called a *universality class*. Each Hamiltonian in its class has exactly the same critical point behavior, with not only the same critical indices but also the same long-ranged correlation functions, and the same singular part of the free energy function.

The identity among different problems is not just a theoretical artifact. The Ising model, single axis ferromagnets, and the liquid–gas phase transition all show identical critical properties (Lee and Yang 1952a and b). The theory makes these critical properties vary with dimension, and experiments bear out the predicted universality in the two observable cases: $d = 2$ and $d = 3$. As another example, XY ferromagnets have a two component order parameter, with the same symmetry properties as superfluids, with their complex order parameter.

This universality-class idea has been applied to many different problems beyond critical phenomena.[17] Whenever two systems show an unexpected or deeply rooted identity of behavior they are said to be in the same universality class.

---

[17] I must admit to a certain pride connected with universality. The 1967 review paper (Kadanoff et al.) in which I participated was organized about universality classes. I borrowed the word "universality" from the conversation of Sasha Polyakov and Sasha Migdal, who were apparently translating a usage common in the Landau group. I then imported this usage into the English language (1990). Alternatively, one might argue

There are, of course, many different universality classes corresponding to different dimensionalities, different symmetries of the order parameter, and to different stability properties of the fixed points.

Before leaving this subject, focus once more on the possibility of a marginal behavior. In the marginal case, we have a coupling that does not vary under renormalization. That kind of coupling can produce critical properties that vary continuously as some parameter is varied. For example, a pair of coupled Ising models living in the same space show a marginal behavior of this kind (Kadanoff and Wegner 1973). A different marginal behavior is shown by the XY model in two dimensions (Hadzibabic 2006; Kosterlitz and Thouless 1973).

8.3 New Kinds of Answers

In one sense the renormalization group is rather different from anything that had come before in statistical physics, and by extension in other parts of physics as well. Previous work in statistical physics had emphasized finding the properties of *problems* defined by statistical sums, each sum being based upon a probability distribution defined by particular values of coupling constants like K and h. Such sums would be called *solutions* to the problems in question. In the renormalization group work the emphasis is on connecting problems by saying that different problems could have identical solutions. The method involved finding different values of couplings that would then give identical free energies and other properties. These set of couplings would then form a representation of a universality class. All the interactions that flow into a given fixed point in the course of an infinite number of renormalizations belong to the universality class of that fixed point.

A universality class would give a solution, in the old sense, if one finds within the class a set of couplings so simple that the solution is obvious. This is what happens when the running couplings produce infinitely weak interactions, thereby producing a weak coupling fixed point. A strong coupling fixed point might also be trivial if no important symmetry remains after the order parameter takes on a nonzero value. However, a first-order phase transition might produce a nontrivial situation with quite a bit of remaining symmetry. In that case, further analysis is necessary before one can get anything like a solution in the old sense of the word.

Finally, a critical fixed point is not really a "solution" in the old sense. It gives us values of critical indices and describes scaling behavior, which can then be used to infer many of the qualitative properties of a solution. But many of the details of the old-sense solution may not be available from a knowledge of the fixed point alone.

that universality was a product of many different authors, including Robert Griffiths (1975), as well as the entire King's College school (Domb 1996).

8.4 Flows and Flow Diagrams

As already mentioned, a renormalization operation differs from the calculations performed within the statistical mechanics of Boltzmann and Gibbs. In statistical mechanics you start with a statistical ensemble, usually defined with a Hamiltonian, and use that ensemble to calculate an average. In a renormalization operation, you start with a statistical ensemble, usually defined by a Hamiltonian, and you calculate another ensemble, often described by a Hamiltonian containing different couplings. In one case the calculation is, in brief, *ensemble generates averages*; in the other, the calculation is *ensemble generates ensemble*. This is quite a substantial difference.

The part of mathematics that goes with standard statistical mechanics is probability theory. One part of the mathematics that goes with renormalization is called "dynamical systems theory" and describes how things change under transformations. The concepts of a fixed point and of a basin of attraction belong to dynamical systems theory rather than probability theory.

Dynamical systems theory is often used to describe continuous changes, as, for example, the changes in a mechanical system as its state changes in time. For the purposes of this section, I will speak as if all renormalization transformations were continuous changes produced by an infinitesimal increase in a basic length. Thus, the transformation will be $a \to a + d\ell$. Then, every coupling also undergoes an infinitesimal change $K \to K + dK$. In particle physics this kind of approach has the name of the Callen-Symanzik equation (Callan 1970; Symanzik 1970, 1971).

The simplest kinds of flow pictures look at a single coupling constant, K, and how that coupling changes under renormalization. In the one-dimensional Ising model, depicted in figure 4.12, each renormalization makes the coupling weaker. Thus the coupling flows toward the weak coupling fixed point at $K = 0$. In contrast, the flow in figure 4.13 describes the two-dimensional Ising model. The flow is zero at the critical fixed point. To the left of the critic fixed point, all couplings flow toward the weak coupling fixed point at $K = 0$; to the right, all flows go toward the strong coupling point at $K = \infty$. This diagram describes a system with a single critical point, but a total of three fixed points.

We have come a long way from the starting point set by Boltzmann and Gibbs. Solutions to problems in statistical mechanics have here been described in terms of renormalization group flows, universality classes, and types of fixed points. This new language has become important in statistical physics and has been extended to applications well beyond the situations described here. This language, derived from statistical mechanics, has become even more pervasive in particle physics, where

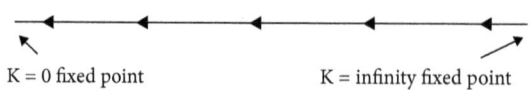

Figure 4.12 Flow diagram for one-dimensional Ising model. Renormalization weakens the coupling and pushes it toward a weak coupling fixed point at $K = 0$.

THEORIES OF MATTER

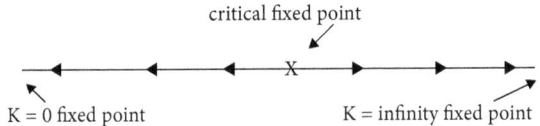

Figure 4.13 Flow diagram for two-dimensional Ising models. There is a criticial fixed point at $K = K_c$. For initial couplings weaker than this value, renormalizations weaken the coupling and push it toward a weak coupling fixed point at $K = 0$. Conversely, if the initial couplings are stronger than K_c, renormalizations produce a flow toward a strong coupling fixed point, describing a ferromagnetic state.

coupling constants run everywhere. A calculational method is more than a way of putting symbols on paper. It provides a way of looking at, and conceptualizing, nature.

8.5 The Renormalization Group Is Not a Group

Although the renormalization operation is usually described as a part of a group, block transformations actually produce a semigroup. A group is a set of operations with three characteristics:

- Two operations in the group, taken in succession, produce another group operation.
- The group contains an element called the identity, which has the effect of changing nothing whatsoever.
- For each operation in the group there is, as part of the group, an inverse operation, so that when you successively perform the operation and its inverse, that pair of operations produces the identity element.

A semigroup lacks the third characteristic. Once you have performed a group operation you cannot necessarily undo that operation.

The reason that renormalization produces a semigroup is that a block transformation (see section 6.4) loses information. After the transformation, the system contains fewer lattice sites and so can hold less "information." Some irrelevant couplings, which could be seen before the transformation have simply disappeared. (These couplings have the index value $y = -\infty$. In addition, there are other kinds of couplings, called "redundant," that do not affect the free energy and so disappear without a trace in the course of a renormalization.) Both kinds of coupling make renormalization a semigroup operation.

This characteristic is important because it eliminates the possibility of finding the small-scale Hamiltonian of the system by looking at large-scale phenomena. Before the use of renormalization methods scientists often thought that a sufficiently accurate and detailed study of a system, albeit a study conducted on a large length scale, could determine all the basic laws governing the system, down to the smallest scale. In practice, the disentanglement of microscopic laws has always proved to be hard. But, in principle, it was always assumed to be possible. However, the renormalization group theory says that information will disappear in the process of changing length-scales. Even ordinary irrelevant operators have effects that disappear with exponential rapidity in the course of a renormalization transformation. These do not produce, in principle, a disappearance of information but they make it well-nigh impossible to reconstruct a small-scale Hamiltonian from large-scale data.

8.6 A Calculational Method Defines Many Worlds

Wilson, in essence, converted a slightly vague phenomenological theory into a well-defined calculational method. So are they all: classical mechanics, quantum mechanics, statistical mechanics, field theory,...., all calculational methods. But they are also each complete descriptions of some "sub-universe." As such, they each engender their own world and their own philosophy. As you can see from the set of ideas outlined in this section, the renormalization group built its own set of philosophical perspectives, which then displayed statistical physics and condensed matter physics in a new way.

However, there is an additional sense in which the renormalization group defines its own worlds. Each fixed point has its own basin of attraction defined by its very large set of irrelevant couplings. This basin of attraction is the region in the space of possible Hamiltonians that will eventually flow into our particular fixed point. Within this basin, the flow of the scaling variables play a crucial role. When two or more of these variables interfere, we can see non-linear effects. Since we expect that our listing of variables is a complete list, we expect that two variables acting together will produce an effect that we can describe as a summed effect of the variables on our previous list. In this way, we get a kind of multiplication table in which the product of any two variables is a sum of the others with specified coefficients. Such a multiplication table is what the mathematicians call an algebra. This algebra defines what is happening in the phase transition.

The algebras that have actually been studied are a little deeper than the one just described. They are produced not just by the couplings, but by the specifications of the couplings in local regions of the system. Therefore the algebras combine the properties of space with the properties of the particular fixed point. They have been most richly studied in two dimensions (Belavin, Polyakov, and Zamolodchikov 1984; Friedan, Qiu, and Shenker 1984) in which the spatial part common to all these algebras is called the Virasoro algebra (Virasoro 1970). This approach also plays an important role in string theory.

Each fixed point has its own unique algebra (Kadanoff 1969; Wilson 1969), called a short distance expansion or an operator product expansion, that describes the structure of the local correlations determining the fixed point behavior.

8.7 Extended Singularities Revisited

The renormalization group has an entirely different spatial structure from that of mean field theory. The difference can best be seen by comparing the Ising model mean field theory of Section (3) with the block spin formulation of Section (6.4).

In the mean field formulation, the value of an average magnetization at point **r** is determined, first of all, by the values of the magnetization at points connected by bonds to the initial point. These are then determined, in turn, by magnetizations at points connected to these new points by bonds. This extension process might continue indefinitely or terminate after only finding a finite number of spins. In either case, the theory may or may not predict a phase transition. Mean field theory does not have the right spatial structure for the correct prediction of phase transitions.

In contrast, the blocking procedure of the renormalization group determines the couplings in a given region, in the first analysis, by the effects of couplings in a region of size ℓ larger. The blocking then reaches out in geometric progression to regions each expanded by a factor of ℓ.

Of course, the block transformation reaches out more quickly and effectively than do the steps of the mean field calculation. But that is not the main difference. The mean field theory can have a pseudo-phase transition determined by just a few couplings. On the other hand, if the block transformation ever reaches out and sees no more couplings in the usual approximation schemes (Niemeijer and Leeuwen 1973) that will signal the system that a weak coupling situation has been encountered and will cascade back to produce a weak coupling phase. Hence the blocking approach has the potential of using the right fact about the spatial topology to determine the possibility of a phase transition.

By this argument the extended singularity theorem suggests that phase transitions are triggered by a very elegant mathematical juxtaposition put before us by Nature. On one hand, the phase transition is connected with a symmetry operation built into the microscopic couplings of the system. For example, the ferromagnetic based upon the breaking of a symmetry in the possible direction of spins. On the other hand, the phase transitions also make use of the extended topology of a system that extends over an effectively infinite region of space. This coupling of microscopic with macroscopic has an unexpected and quite breathtaking beauty.

Acknowledgment

Some of the material in this review was first prepared for a talk I gave at the Royal Netherlands Academy of Arts and Sciences in 2006. Still more of the material appeared in a talk at the 2009 Seven Pines meeting on the Philosophy of Physics under the title "More Is the Same, Less Is the Same, Too; Mean Field Theories and

Renormalization." These talks have appeared on the authors' website (2009) since then. This Seven Pines meeting was generously sponsored by Lee Gohlike. The paper also incorporates material from "More Is the Same" published in *J. Stat. Phys.* in 2009.

This work was supported in part by the University of Chicago MRSEC program under NSF grant number DMR0213745. It was completed during visits to the Perimeter Institute, which is supported by the Government of Canada through Industry Canada and by the Province of Ontario through the Ministry of Research and Innovation, and by the present NSF DMR-MRSEC grant number 0820054.

I had useful discussions related to this essay with Tom Witten, E. G. D. Cohen, Gloria Lubkin, Ilya Gruzberg, Gene Mazenko, Hans van Leeuwen, Wendy Zhang, Franz Wegner, Roy Glauber, Yitzhak Rabin, Gerard 't Hooft, Sidney Nagel, and Subir Sachdev. I owe particular thanks to Michael Fisher who, as he has done many times, helped me understand this interesting subject.

REFERENCES

H. C. Andersen, J. D. Weeks, and D. Chandler. Relationship between the hard-sphere fluid and fluids with realistic repulsive forces. *Phys. Rev. A*, 4:1597–1607, 1917.

P. W. Anderson. More is different. *Science*, 177:393–396, 1972.

P. W. Anderson. *Concepts in Solids*. World Scientific, Singapore, 1997.

A. F. Andreev. Singularity of thermodynamic quantities at a first order phase transition point (Singularity of thermodynamic potential for liquid at boiling point). *Sov. Phys. JETP*, 18:1415, 1964.

T. Andrews. On the continuity of the gaseous and liquid states of matter. *Phil. Trans. Roy. Soc.*, 159:575–590, 1869.

M. I. Bagatskii, A. V. Voronel', and V. G. Gusak. Measurement of the specific heat C_V of Argon in the immediate vicinity of the critical point. *JETP*, 16:517, 1963.

J. Bardeen, L. N. Cooper, and J. R. Schrieffer. Microscopic theory of superconductivity. *Phys. Rev.*, 106:162–164, 1957.

J. Bardeen, L. N. Cooper, and J. R. Schrieffer. Theory of superconductivity. *Phys. Rev.*, 108:1175, 1957.

J. G. Bednorz and K. A. Müller. Possible high Tc superconductivity in the Ba-La-Cu-O system. *Z. Physik*, B 64:189–193, 1986.

A. A. Belavin, A. M. Polyakov, and A. B. Zamolodchikov. Infinite conformal symmetry in two-dimensional quantum field theory. *Nucl. Phys.*, B241:333–380, 1984.

Stephen G. Brush. History of the Lenz-Ising model. *Rev. Mod. Phys*, 39, 1967.

Stephen G. Brush. *Statistical Physics and the Atomic Theory of Matter from Boyle and Newton to Landau and Onsager*. Princeton University Press, NJ, 3rd edition, 1983.

Steven G. Brush. *The Kind of Motion We Call Heat: A History of the Kinetic Theory of Gases in the 19th Century*. North Holland, 1976.

C. G. Callan, Jr. Broken scale invariance in scalar field theory. *Phys. Rev.*, D 2:1541–1547, 1970.

Carlo Cercignani. *Ludwig Boltzmann: The Man Who Trusted Atoms.* Oxford University Press, Oxford and New York, 1998.

E. G. D. Cohen. private communication.

P. Curie. Lois expérimentales du magnétisme. Propriétés magnétiques des corps à diverses températures [thesis reprint]. *Ann. Chem. Phys.*, 5:289, 1895.

Debra Daugherty. Elaborating the crystal concept: Scientific modeling and ordered states of matter, Ph.D. thesis, University of Chicago, 2007.

J. Diamond. *Guns, Germs, and Steel: The Fates of Human Societies.* W. W. Norton and Company, New York, New York, 1997.

Cyril Domb. *The Critical Point.* Taylor and Francis, 1996.

Cyril Domb and A. R. Miedema. *Progress in Low Temperature Physics*, 4:296, 1964.

Cyril Domb, Melville S. Green, and Joel Louis Lebowitz, editors. *Phase Transitions and Critical Phenomena*, volume 20. Academic Press, 2001.

M. E. Fisher. The theory of equilibrium critical phenomena. *Reports on Progress in Physics*, XXX, part II:615, 1967.

M. E. Fisher. Lee-Yang edge singularity and ϕ^3 field theory. *Phys. Rev. Lett.*, 40:1610, 1978.

James W. Cronin and Val L. Fitch http://www.nobelprize.org/nobel_prizes/physics/laureates/1980/.

Daniel Friedan, Zongan Qiu, and Stephen Shenker. Conformal invariance, unitarity, and critical exponents in two dimensions. *Phys. Rev. Lett.*, 52:1575, 1984.

U. Frisch. *Turbulence: The Legacy of A. N. Kolmogorov.* Cambridge University Press, Cambridge, 1995.

G. Gallavotti, W. L. Reiter, and J. Yngvason editors, Entropy, nonequilibrium, chaos and infinitesmals. in *Boltzmann's Legacy. European Mathematical Society.* Zurich, 2008, pages 39–60.

Elizabeth Garber, Stephen G. Brush, and C. W. F. Everett, editors. *Maxwell on Molecules and Gases.* The Massachusetts Institute of Technology, 1986.

Murray Gell-Mann and F. E. Low. Quantum electrodynamics at small distances. *Phys. Rev.*, 95:1300–1312, 1954.

J. Willard Gibbs. *Elementary Principles of Statistical Mechanics.* Charles Scribners Sons, New York, 1902.

J. Willard Gibbs. *Scientific Papers of J Willard Gibbs.* Dover, New York, 1961.

V. L. Ginzburg. Some remarks on phase transitions of the second kind and the microscopic theory of ferroelectric materials. *Soviet Phys.–Solid State*, 2:1824, 1960.

Robert B. Griffiths. Phase diagrams and higher-order critical points. *Phys. Rev. B*, 12(1):345–355, 1975.

E. A. Guggenheim. The Principle of Corresponding States. *J. Chem. Phys.*, 13:253, 1945.

Z. Hadzibabic et al. Berezinskii-Kosterlitz-Thouless crossover in a trapped atomic gas. *Nature*, 441:1118, 2006.

P. Heller. Experimental investigations of critical phenomena. *Reports on Progress in Physics*, XXX, part II:731, 1967.

S. N. Isakov. Nonanalytic features of the first order phase transition in the Ising model. *Communications in Mathematical Physics*, 95:427–443, 1984.

E. Ising. "Beitrag zur Theorie des Ferromagnetismus [Contribution to the theory of ferromagnetism]." *Z. Phys.*, 31:253, 1925.

Leo P. Kadanoff. Scaling laws for Ising models near T_c. *Physics*, 2:263, 1966.

Leo P. Kadanoff. Spin-spin correlation in the two-dimensional Ising model. *Nuovo Cimento*, 44:276, 1966.

Leo P. Kadanoff. Operator algebra and the determination of critical indices. *Phys. Rev. Lett.*, 23:1430, 1969.

Leo P. Kadanoff. Scaling and universality in statistical physics. *Physica A*, 163:1–14, 1990.

Leo P. Kadanoff. *Statistical Physics: Statics, Dynamics and Renormalization*. World Scientific, Singapore, 2000.

Leo P. Kadanoff. web page: http://jfi.uchicago.edu/leop/Chicago, 2009.

Leo P. Kadanoff. More is the same; mean field theory and phase transitions. *J. Stat Phys.*, 137:777–797, 2009.

Leo P. Kadanoff and Franz Wegner. Some critical properties of the eight-vertex model. *Phys. Rev.*, B 4:3989–3993, 1973.

L. P. Kadanoff, W. Gotze, D. Hamblen, R. Hecht, E. A. S. Lewis, V. V. Palciauskas, M. Rayl, J. Swift, D. Aspnes, and J. W. Kane. Static phenomena near critical points: Theory and experiment. *Rev. Mod. Phys.*, 39:395, 1967.

L.F. Kellers. Ph.D. thesis, Duke University, unpublished, 1960.

Andrey Nikolaevich Kolmogorov. The local structure of turbulence in incompressible viscous fluid for very large Reynolds numbers. *Proc. USSR Acad. Sci.*, 30:299–303, 1941.

Andrey Nikolaevich Kolmogorov. A refinement of a previous hypothesis concerning the local structure of turbulence in a viscous incompressible fluid at high Reynolds numbers. *J. Fluid Mech.*, 13:82–85, 1962.

J. M. Kosterlitz and D. J. Thouless. Ordering, metastability and phase transitions in two-dimensional systems. *J. Phys. C: Solid State Phys.*, 6:1181–1203, 1973.

Thomas S. Kuhn. *The Structure of Scientific Revolutions*. University of Chicago Press, Chicago, 1962.

L. D. Landau. Zur Theorie der Phasenumwandlungen. I *Phys. Z. Sow*, 11:26–47 and Zur Theorie der Phasenumwandlungen. II 545–555, 1937.

T. D. Lee and C. N. Yang. Statistical theory of equations of state and phase transitions: I. Theory of condensation. *Phys. Rev.*, 87:404–409, 1952a.

T. D. Lee and C. N. Yang. Statistical theory of equations of state and phase transitions: II. Lattice gas and Ising model. *Phys. Rev.*, 87:4010–419, 1952b.

A. P. Levanyuk. *Zh. Eksp. Teor. Fiz.*, 36:810, 1959.

J. M. H. Levelt-Sengers. Critical exponents at the turn of the century. *Physica*, 82-A:319, 1976.

J. C. Maxwell. *Nature*, 10:477, 1874.

J. C. Maxwell. *Nature*, 11:418, 1875.

B. M. McCoy and T. T. Wu. *The Two-Dimensional Ising Model*. Harvard University Press, 1973.

M. R. Moldover and W. A. Little. Ultraviolet behavior of non-abeilan gauge theories. *Phys. Rev. Lett.*, 15:54–56, 1965.

National Bureau of Standards. *Critical Phenomena. Proceedings of a Conference Held in Washington, D.C., April 1965*. United States Department of Commerce, National Bureau of Standards, Washington, 1965.

Th. Niemeijer and J. M. van Leeuwen. Wilson Theory for Spin Systems on a Triangular Lattice. *Phys. Rev. Letts.*, 31:1411, 1973.

Martin Niss. *Phenomena, Models, and Understanding, The Lenz-Ising Model and Critical Phenomena 1920-1971*. Springer, Berlin/Heidelberg, 2005.

Nai-Phuan Ong and Ravin Bhatt, editors. *More Is Different: Fifty Years of Condensed Matter Physics*. Princeton Series in Physics, Princeton, NJ, 2001.

L. Onsager. Crystal Statistics. I. A Two-Dimensional Model with an Order-Disorder Transition *Phys. Rev.*, 65:117, 1944.

L. S. Ornstein and F. Zernike. Accidental deviations of density and opalescence at the critical point of a single substance. *Proc. Acad. Sci. Amsterdam*, 17:793, 1914.

Abraham Pais. *Subtle Is the Lord ...* Oxford University Press, paperback, 1983.

A. Z. Patashinskii and V. L. Pokrovsky. Second order phase transition in a Bose liquid. *Soviet Phys. JETP*, 19:667, 1964.

A. Z. Patashinskii and V. L. Pokrovsky. *Fluctuation Theory of Phase Transitions*. Amsterdam, Elsevier, 1978.

Muriel Rukeyser. *Willard Gibbs: American Genius*. Ox Bow Press, Woodbridge, CT, 1942.

Jan V. Sengers and Joseph G. Shanks. Experimental critical-exponent values for fluids. *J. Stat. Phys.*, 137:857–877, 2009.

E. C. G. Stueckelberg and A. Peterman. Normalization of constants in the quantum theory. *Helv. Phys. Acta*, 26:499, 1953.

K. Symanzik. Small-distance behaviour in field theory and power counting. *Commun. Math. Phys.*, 18:227, 1970.

K. Symanzik. Small-distance-behaviour analysis and Wilson expansions. *Commun. Math. Phys.*, 23:49, 1971.

G. 't Hooft. Renormalizable Lagrangians for massive Yang-Mills fields. *Nucl. Phys. B*, 35:167, 1971.

G. 't Hooft and M. Veltman. Regularization and renormalizatioin of gauge fields. *Nucl. Phys. B*, 44:189, 1972.

G. 't Hooft and M. Veltman. Combinatorics of gauge fields *Nucl. Phys. B*, 50:318, 1972.

Jos Uffink. Rereading Ludwig Boltzmann. In L. Valdes and D. Westerstahl, editors, *Methodology and Philosophy of Science. Proceedings of the Twelfth International Congress*, pages 537–555. King's College Publications, London, 2005.

George E. Uhlenbeck. Some historical and critical remarks about the theory of phase transitions. In Shigeji Fujita, editor, *Science of Matter. Festschrift in Honor of Professor Ta-You Wu*. Gordon and Breach Science Publishers, New York, 1978.

J. D. van der Waals. Over de Continuiteit van den Gas- en Vloeistoftoestand, Leiden, The Netherlands, 1873.

J. E. Verschaffelt. *Commun. Phys. Lab. Leiden*, (55), 1900.

M. A. Virasoro. Subsidiary conditions and ghosts in dual-resonance models. *Phys. Rev*, D1:2933–2936, 1970.

A. V. Voronel', Yu. Chaskin, V. Popov, and V. G. Simpkin. *JETP*, 18:568, 1964.

P. Weiss. *J. Phys.*, 6:661, 1907.

B. Widom. *J. Chem. Phys.* 41:1633, 1964

B. Widom. Surface Tension and Molecular Correlations near the Critical Point. *J. Chem. Phys.*, 43:3892, 1965a.

B. Widom. Equation of State in the Neighborhood of the Critical Point. *J. Chem. Phys.*, 43:3898, 1965b.

B. Widom and O. K. Rice. Critical Isotherm and the Equation of State of Liquid-Vapor Systems. *J. Chem. Phys.*, 23:1250, 1955.

K. Wilson. Non-Lagrangian Models of Current Algebra *Phys. Rev.*, 179:1499, 1969.

K. G. Wilson. Renormalization group and critical phenomena. I. Renormalization group and the Kadanoff scaling picture. *Phys. Rev.*, B4:3174–3183, 1971.

K. G. Wilson and M. E. Fisher. Critical exponents in 3.99 dimensions. *Phys. Rev. Lett.*, 28:240–243, 1972.

C. N. Yang. The Spontaneous Magnetization of a Two-Dimensional Ising Model. *Phys. Rev.*, 85:808, 1952.

CHAPTER 5

TURN AND FACE THE STRANGE... CH-CH-CHANGES
PHILOSOPHICAL QUESTIONS RAISED BY PHASE TRANSITIONS

TARUN MENON AND
CRAIG CALLENDER

Phase transitions are abrupt changes in the macroscopic properties of a system. Examples of the phenomenon are familiar: freezing, condensation, magnetization. Often these transitions are particularly dramatic, as when solid objects composed of the silvery metal gallium vanish into puddles when picked up (the temperature of the hand is just enough to raise gallium's temperature past its melting point). Characterized generally, one finds them inside and outside of physics, in systems as diverse as neutron stars, DNA helices, financial markets, and traffic. In the past half-century, the study of phase transitions and critical phenomena has been a central preoccupation of the statistical physics community. In fact, it is now a truly interdisciplinary area of research. Phase transitions manifest at many different scales and in all sorts of systems, so they are of interest to atomic physicists, materials engineers, astronomers, biologists, sociologists, and economists. However, philosophical attention to the foundational issues involved has thus far been limited.

Our thanks to Daniel Arovas, Robert Batterman, Jeremy Butterfield, Sarang Gopalakrishnan, John Norton, and Larry Sklar for helpful comments on an earlier draft of this chapter.

This is unfortunate because the theory of phase transitions is unusual in many ways and offers a novel perspective that could enrich a number of debates in the philosophy of science. In particular, questions about reduction, emergence, explanation, and approximation all arise in a particularly stark manner when considering this phenomenon. Here we will focus on these questions as they relate to the most studied type of phase transition, namely, transitions between different equilibrium phases in thermodynamics. These are sudden changes between one stable thermodynamic state of matter and another while one smoothly varies a parameter. A paradigmatic example is the change in water from liquid to gas as the temperature is raised or the pressure is reduced.

In the small philosophical commentary on this topic, such changes have provoked many surprising claims. Many have claimed that phase transitions cannot be reduced to statistical mechanics, that they are truly emergent phenomena. The argument for this conclusion hangs on one's understanding of the infinite idealization invoked in the statistical mechanical treatment of phase transitions. In this chapter we will focus on puzzles associated with this idealization. Is infinite idealization necessary for the explanation of phase transitions? If so, does it show that phase transitions are, in some sense, emergent phenomena? If so, what precisely is that sense? Questions of this sort provide a concrete basis for the exploration of philosophical approaches to reduction and idealization, and they also bear on the ongoing scientific study of these systems.

1. The Physics of Phase Transitions

Phase transitions raise interesting questions about intertheoretic relationships because they are studied from three distinct theoretical perspectives. Thermodynamics provides a macroscopic, phenomenological characterization of the phenomenon. Statistical mechanics attempts to ground the thermodynamic treatment by explaining how this macroscopic behavior arises out of the interaction of microscopic degrees of freedom. This project has led to the employment of renormalization group theory, a tool first developed in the context of particle physics for studying the behavior of systems under transformations of scale. While renormalization group theory is usually placed under the broad rubric of statistical mechanics, the methods employed are importantly different from the traditional tools of statistical mechanics. Rather than a probability distribution over an ensemble of configurations of a single system, the primary theoretical device of renormalization group theory is the flow generated by the scaling transformation on a space of Hamiltonians representing distinct physical systems. In this section we describe how these three approaches treat the phenomenon of phase transitions, with special attention to the employment of the infinite particle idealization.

1.1 Thermodynamic Treatment

The thermodynamic treatment of phases and phase transitions began in the nineteenth century. Experiments by Andrews, Clausius, Clapeyron, and many others provided data that would lead to developed theories of phase transitions and critical phenomena. Gradually it was recognized that at certain values of temperature and pressure a substance can exist in more than one thermodynamic phase (e.g., solid, liquid), while at other values there can be a change in phase but no coexistence of phases.

For instance, as pressure is reduced or temperature is raised, liquid water transitions to its gaseous phase. At the boundary between these phases, both liquid and gaseous states can coexist; the thermodynamic parameters of the system do not pick out a unique equilibrium phase. In fact, at the triple point of water (temperature 273.16 K and pressure 611.73 Pa), all three phases—solid, liquid, and gas—can coexist. The transitions at these phase boundaries are marked by a discontinuity in the density of water. As the pressure is reduced at a fixed temperature, the equilibrium state of water switches abruptly from a high-density liquid phase to a low-density gaseous phase. This is an example of a *first-order* phase transition. As the temperature is increased past the critical temperature of 647 K, water enters a new phase. In this regime, there are no longer macroscopically distinct liquid and gas phases, but a homogenous supercritical fluid that exhibits properties associated with both liquids and gasses. Changing the pressure leads to a continuous change in the density of the fluid; there are no phase boundaries. This supercritical phase allows a transition from liquid to gas that does not involve any discontinuity in thermodynamic observables: raise the temperature of the liquid past the critical temperature, reduce the pressure below the critical pressure (22 MPa for water), then cool the fluid back to below the critical temperature. This path takes the system from liquid to gas without crossing a phase boundary. The transition of a system past its critical point to the supercritical phase is a *continuous* phase transition.

Mathematically, phase transitions are represented by nonanalyticities or singularities in a thermodynamic potential. A singularity is a point at which the potential is not infinitely differentiable, so at a phase transition some derivative of the thermodynamic potential changes discontinuously. A classification scheme due to Ehrenfest provides the resources to distinguish between first- and second-order transitions in this formalism. A first-order phase transition involves a discontinuity in the first derivative of a thermodynamic potential. In the liquid–gas first-order transition, the volume of the system, a first derivative of the thermodynamic potential known as the Gibbs free energy, changes discontinuously. For a second-order phase transition the first derivatives of the potentials are continuous, but there is a discontinuity in a second derivative of a thermodynamic potential. At the liquid–gas critical point, we see a discontinuity in the compressibility of the fluid, which is a first derivative of volume and hence a second derivative of the Gibbs free energy. Ehrenfest's scheme extends naturally to allow for higher-order phase transitions as well. An n-th order transition would be one whose n-th derivative is discontinuous. Contemporary statistical mechanics retains the category of first-order phase transitions (sometimes referred to as abrupt transitions), but all other types of non-analyticities in thermodynamic potentials are grouped together as continuous phase transitions.

Continuous phase transitions are often referred to as order–disorder transitions. There is usually some symmetry in the supercritical phase that is broken when we cross below the critical point. This broken symmetry allows for the material to be ordered in various ways, corresponding to different phases. A stark example of the transition between order and disorder is the transition in magnetic materials, such as iron, between paramagnetism and ferromagnetism. At room temperature, a piece of iron is permanently magnetized when exposed to an external magnetic field. In the presence of a field, the minimum energy configuration is the one with the largest possible net magnetic moment reinforcing the field, so the individual dipoles within the iron align to maximize the net moment. This configuration remains stable even when the external field is removed. Materials with this propensity for induced permanent magnetization are called *ferromagnetic*. If the temperature is raised above 1043 K, the ferromagnetic properties of iron vanish. The iron is now *paramagnetic*; it can no longer sustain induced magnetization when the external field is removed. In the stable configuration, there is no correlation between the alignments of neighboring dipoles. In the paramagnetic phase, no direction is picked out as special after the magnetic field is switched off. The material exhibits spatial symmetry. In the ferromagnetic phase, this symmetry is broken. The dipoles line up in a particular spatial direction even after the field is removed. The order represented by this alignment does not survive the transition past criticality.

A simple way to understand this transition between order and disorder is in terms of the minimization of the Helmholtz free energy of the system:

$$F = E - TS. \tag{1}$$

Here E is the energy of the system, T is the temperature, and S is the entropy. The stable configuration minimizes free energy. At low temperatures, the energy term dominates, and the low-energy configuration with dipoles aligned is favored. At high temperatures, the entropy term dominates, and we get the high-entropy configuration with uncorrelated dipole moments. The change in magnetic behavior is explicable as a shift in the balance of power in the battle between the ordering tendency due to minimization of energy and the disordering tendency due to maximization of entropy. As indicated, the paramagnetic–ferromagnetic transition is continuous, not first order. All first derivatives of the free energy are continuous, but second derivatives (such as the magnetic susceptibility $\chi = \frac{\partial^2 F}{\partial H^2}$, where H is the magnetization) are not.

The transition from order to disorder is also represented, following Landau, as the vanishing of an *order parameter*. In the case under consideration, this parameter is the net magnetization M of the system. Below the critical point, you have different phases with distinct values of the order parameter. If we simplify our model of the magnetic material so that the induced magnetization of the dipoles is only along one spatial axis (as in the Ising model), then at each temperature below criticality the order parameter can take two values, related by a change of sign. The magnetization vanishes as we approach the critical point and remains zero in the supercritical phase, corresponding to a disappearance of distinct phases.

The vanishing of the order parameter close to the critical temperature T_c is characterized by a power law:

$$M \propto (-t)^\beta \qquad (2)$$

where t is the reduced temperature $(T - T_c)/T_c$. The exponent β characterizes the rate at which the magnetization falls off as the critical temperature is approached. It is an example of a *critical exponent*, one of many that appear in power laws close to the critical point. The experimental and theoretical study of critical exponents has been crucial to recent developments in the theory of phase transitions.

1.2 Statistical Mechanical Treatment

Statistical mechanics is the theory that applies probability theory to the microscopic degrees of freedom of a system in order to explain its macroscopic behavior. The tools of statistical mechanics have been extremely successful in explaining a number of thermodynamic phenomena, but it turned out to be particularly difficult to apply the theory to the study of phase transitions. There were two significant obstacles to the development of a successful statistical mechanical treatment of phase transitions: one experimental and one conceptual.

The experimental obstacle had to do with the failure of mean field theory. This was the dominant approach to the statistical mechanics of phase transitions up to the middle of the twentieth century. The theory is best explicated by considering the Ising model, which represents a system as a lattice of sites, each of which can be in two different states. The states will be referred to as spin up and spin down, in analogy with magnetic systems. However, Ising models have been successfully applied to a number of different systems, including the liquid-gas system near its critical point. The Hamiltonian for the Ising model involves a contribution by an external term, corresponding to the external magnetic field for magnetic systems, and internal coupling terms. The only coupling is between neighboring spins on the lattice. It is energetically favorable for neighboring spins to align with one another and with the external field. This model is supposed to represent the way in which local interactions can produce the kinds of long-range correlations that characterize a thermodynamic phase.

In statistical mechanics, all thermodynamic functions are determined by the canonical partition function. The coupling terms in the Hamiltonian make the calculation of the partition function for the Ising model mathematically difficult. To make this calculation tractable, we approximate the contribution of a particular lattice site to the energy of the system by supposing that all its neighbors have a spin equal to the ensemble average. This approximation ignores fluctuations of spins from their mean values. The fluctuations become less relevant as the number of neighbors of a particular lattice site increases, so the mean field approximation works better the higher the dimensionality of the system under consideration. Once the partition function is calculated using this approximation, there is an elegant method due to Landau for determining the critical exponents. Unfortunately, Landau's method gives results that conflict with experiment. For instance, the mean field value for the critical exponent β is 0.5, but observation suggests the actual value is about 0.32. The approximation fails close to the critical point of a magnetic system. In fact, this failure is predicted by Landau theory itself. The theory tells us that as we approach the critical point, the *correlation length* diverges. This is the typical distance over which fluctuations in the microscopic degrees of freedom are correlated. As this length scale increases, fluctuations become more relevant, and the mean field approximation, which ignores fluctuations, weakens. Mean field theory cannot fully describe continuous phase transitions because of this failure near criticality. Another approach is needed for a full statistical mechanical treatment of the phenomenon.

As mentioned, there was also a deeper conceptual obstacle to a statistical mechanics of phase transitions. If one adopts the definition of phase transitions employed by thermodynamics, then *phase transitions in statistical mechanics do not seem possible*. The impossibility claim can be explained very easily. As mentioned above, thermodynamic functions are determined by the partition function. For instance, the Helmholtz free energy is given by:

$$F = -kT \ln Z \qquad (3)$$

where k is Boltzmann's constant, T is the temperature of the system, and Z is the canonical partition function:

$$Z = \sum_n \exp\left(-\frac{E_n}{kT}\right) \qquad (4)$$

with E_n labeling the different possible mechanical energies of the system. Recall the definition of a phase transition according to thermodynamics:

(Def 1) An equilibrium phase transition is a nonanalyticity in the free energy.

Depending on the context, one might choose a nonanalyticity in a different thermodynamic potential; however, that freedom will not affect matters here.

As natural as it is, Def 1 makes a phase transition seem unattainable in statistical mechanics. The reason is that each of the exponential functions in (4) is analytic, the partition function is just a sum of exponentials, and the free energy essentially is just the logarithm of this sum. Since a sum of analytic functions is itself analytic and the logarithm of an analytic function itself analytic, the Helmholtz free energy, expressed in terms of the logarithm of the partition function, will also be analytic. Hence, there will be no phase transitions as defined by Def 1. Since the same reasoning can be applied to any thermodynamic function that is an analytic function of the canonical partition function modifications of Def 1 to other thermodynamic functions will not work either. (For a rigorous proof of the above claims, see Griffiths (1972).)

In the standard lore of the field, this problem was resolved when Onsager in 1944 demonstrated for the first time the existence of a phase transition from nothing but the partition function. He did this rigorously for the two-dimensional Ising model with no external magnetic field. How did Onsager manage the impossible? He worked in the thermodynamic limit of the system. This is a limit where the number of particles in the system N and the volume of the system V go to infinity while the density $\rho = N/V$ is held fixed. Letting N go to infinity is the crucial trick in getting around the "impossibility" claim. The claim depends on the sum of exponentials in (4) being finite. Any finite sum of analytic functions will be analytic. Once this restriction is removed, however, it is possible to find nonanalyticities in the free energy. The apparent lesson is that statistical mechanics can describe phase transitions, but only in infinite particle systems.

It is common to visualize what is going on in terms of the Yang-Lee theorem. The free energy is a logarithm of the partition function, so it will exhibit a singularity where the partition function goes to zero. But the partition function is a polynomial of finite degree with all positive coefficients, so it has no real positive roots. Instead the roots are imaginary and the zeros of the partition function must be plotted on the complex plane. The Yang-Lee theorem, for a two-dimensional Ising model, says that these zeros sit on the unit sphere in the complex plane. As the number of particles increases, the zeros become denser on the unit sphere until at the thermodynamic limit they intersect the positive real axis. Since a real zero of the partition function is only possible in this limit, it is only in this limit that we can have a phase transition (understood as in Def 1).

An alternative definition of phase transitions is sometimes used, one proposed by Lebowitz (1999). A phase transition occurs, on this definition, just in case the Gibbs measure (a generalization of the canonical ensemble) is nonunique for the system. This corresponds to a coexistence of distinct phases and therefore a phase transition. Using this alternative definition, however, will not change philosophical matters. The Gibbs measure can only be nonunique in the thermodynamic limit, just as Def 1 can only be satisfied in the thermodynamic limit. That said, this way of looking at the issue perhaps makes it easier to see the similarities between the foundational issues raised by phase transitions and those raised by spontaneous symmetry breaking.

1.3 Renormalization Group Theory

We mentioned in the previous section that mean field theory fails near the critical point for certain systems because it neglects the importance of fluctuations in this regime. Dealing with this strongly correlated regime required the introduction of a new method of analysis, imported from particle physics. This is the renormalization group method. While mean field theory hews to tools and forms of explanation that are orthodox in statistical mechanics, such as determining aggregate behavior by taking ensemble averages, renormalization group theory introduced a somewhat alien approach with tools more akin to those of dynamical systems theory than statistical mechanics.

To explain the method, we return to our stalwart Ising model. Suppose we coarse-grain a 2-D Ising model by replacing 3×3 blocks of spins with a single spin pointing in the same direction as the majority in the original block. This gives us a new Ising system with a longer distance between lattice sites, and possibly a different coupling strength. You could look at this coarse-graining procedure as a transformation in the Hamiltonian describing the system. Since the Hamiltonian is characterized by the coupling strength, we can also describe the coarse-graining as a transformation in the coupling parameter. Let K be the coupling strength of the original system and R be the relevant transformation. The new coupling strength is $K' = RK$. This coarse-graining procedure could be iterated, producing a sequence of coupling parameters, each related to the previous one by the transformation R. The transformation defines a flow on parameter space.

How does this help us ascertain the critical behavior of a system? If you look at an Ising system at its critical point, you will see clusters of correlated spins of all sizes. This is a manifestation of the diverging correlation length. Now squint, blurring out the smaller clusters. The new blurry system that you see will have the same general structure as the old one. You will still see clusters of all sizes. This sort of scale invariance is characteristic of critical behavior. The system has no characteristic length scale. Coarse-graining produces a new system that is statistically identical to the old one. At this point, the Hamiltonian of the system remains the same under indefinite coarse-graining, so it must be a fixed point in parameter space (i.e., a point K_f such that $K_f = RK_f$). The nontrivial (viz., not $K = 0$ or $K = \infty$) fixed points of the flow characterize the Hamiltonian of the system at the critical point, the point at which correlation length diverges and there is no characteristic scale for the system. The critical exponents can be calculated by series expansions near the critical point. Critical exponents predicted by renormalization group methods agree with experiment much more than the predictions of mean field theory.

The same approach can be applied to systems with more complicated Hamiltonians involving a number of different parameters. Some of these parameters will be relevant, which means they get bigger as the system is rescaled. If a system has a nonzero value for some relevant parameter, then it will not settle at a nontrivial fixed point upon rescaling, since rescaling will amplify the relevant parameter and therefore change the couplings in the system. At criticality, then, the relevant parameters must be zero. An example of a relevant parameter for the Ising system is the reduced temperature t. If $t = 0$, the system can flow to a nontrivial fixed point. However, if t is perturbed from zero, the system will flow away from this critical fixed point toward a trivial fixed point. So a continuous transition only takes place when $t = 0$, which is at the critical temperature. Other parameters might turn out to be irrelevant at large scales. They will get smaller and smaller with successive coarse-grainings, effectively disappearing at macroscopic scales. This elimination of microscopic degrees of freedom means that the renormalization group transformation can be irreversible (which would, strictly speaking, make it a semi-group rather than a group), and there can be attractors in parameter space. These are fixed points into which a number of microscopically distinct systems flow. This is the basis of *universality*, the shared critical behavior of quite different sorts of systems. If the systems share a fixed point their critical exponents will be the same, even if their microscopic Hamiltonians are distinct. The differences in the Hamiltonians are in irrelevant degrees of freedom that do not affect the macroscopic critical behavior of the system. Systems that flow to the same nontrivial fixed point are said to belong to the same universality class. The liquid–gas transition in water is in the same universality class as the paramagnetism–ferromagnetism transition. They have the same critical exponents, despite the evident differences between the systems.

The difference between relevant and irrelevant parameters can be conceptualized geometrically. In parameter space, if we restrict ourselves to the hypersurface on which all relevant parameters are zero, so that the differences between systems on this hypersurface are purely due to irrelevant parameters, then all points on the hypersurface will flow to a single fixed point. Perturb the system so that it is even slightly off the hypersurface, however, and the flow will take it to a different fixed point.

It is significant that the fixed point only appears when the system has no characteristic length scale. This is why the infinite particle limit is crucial to the renormalization group approach. If the number of particles is finite, then there will be a characteristic length scale set by the size of the system. Coarse-graining beyond this length will no longer give us statistically identical systems. The possibility of invariance under indefinite coarse-graining requires an infinite system. The requirement for the thermodynamic limit in renormalization group theory can be perspicuously connected to the motivation for this limit in the standard statistical mechanical story. The correlation length of a system near its critical point can be characterized in terms of some second derivative of a thermodynamic potential. For instance, in a magnetic system the range of correlations between parts of the system is proportional to the susceptibility, a second derivative of the free energy. On the thermodynamic treatment, the susceptibility diverges as we approach the critical point, and according to the statistical mechanical treatment this is impossible unless we are in the thermodynamic limit. This means the correlation length cannot diverge, as is required for renormalization group methods to work, unless the system is infinite.

2. The Emergence of Phase Transitions?

All of the above should sound a little troubling. After all, the systems we are interested in, the systems in which we see phase transitions every day, are not infinite systems. Yet the physics of phase transitions seems to make crucial appeal to the infinitude of the systems modeled. It appears that, according to both statistical mechanics and renormalization group theory, phase transitions cannot occur in finite systems.

Additionally, the explanation of the universal behavior of systems near their critical point seems to require the infinite idealization. Considerations of this sort have led many authors to say that phase transitions are genuinely emergent phenomena, suggesting that statistical mechanics cannot provide a full reductive account of phase transitions in finite systems. The eminent statistical mechanic Lebowitz says phase transitions are "paradigms of emergent behavior" (Lebowitz, 1999, S346) and the philosopher Liu says they are "truly emergent properties" (Liu, 1999, S92).

Needless to say, if this claim is correct, phase transitions present a challenge to philosophers with a reductionist bent. The extent of this challenge depends on how we interpret the claim of emergence. The concept of "emergence" is notoriously slippery, interpreted differently by different authors. We will consider a number of different arguments for phase transitions being emergent, corresponding to varying conceptions of emergence. What these arguments have in common is that they all involve a rejection of what Andrew Melnyk has called "reductionism in the core sense" (Melnyk, 2003, 83). This is the intuitive conception of reduction that underlies various more precise philosophical accounts of reduction. A theory T_h reduces to a lower-level theory T_l if all the nomic claims made by T_h can be explained using only the resources of T_l and necessary truths.

This conception is deliberately vague, allowing for various precisifications depending on one's theory of explanation and how one delineates the explanatory resources available to a particular theory. One possible precisification is Ernest Nagel's account of reduction (Nagel, 1979), which says that T_l reduces T_h if and only if the laws of the latter can be deduced from the laws of the former in conjunction with appropriate bridge laws. In this account the core sense of reduction has been filled out with a logical empiricist theory of explanation according to which the explanatory resources of a theory are the deductive consequences of its lawlike statements. It is important to recognize that reductionists are committed to this account of reduction only insofar as endorse such a theory of explanation. The proper motivation for Nagel's theory lies in the extent to which it successfully captures the core sense of reduction.

In this chapter we do not endorse any particular account of reduction. Instead we consider three broad ways in which the explanatory connection between a higher-level theory and a lower-level theory may break down, and examine the extent to which these explanatory breakdowns are manifested in the case of phase transitions. Whether we have a genuine explanatory failure in a particular instance will depend on the details of our account of explanation. Often, the reductionist may be able to avoid a counterexample by simply reconceiving what counts as an adequate explanation.[1] However, certain instances will be regarded as explanatory failures under a wide variety of plausible accounts of explanation, perhaps even under all

---

[1] As an example, consider multiple realization, often presented as a failure of reduction. However, it is only a failure if we believe that a lower-level explanation of the higher-level law must be unified (i.e., the explanation must be the same for every instance of the higher-level law). If we are willing to allow for disunified explanation, then we may indeed have a genuine lower-level explanation of the higher-level law, preserving the core sense of reduction.

plausible accounts of explanation. The weaker the assumptions about explanation required for the counterexample to work, the stronger the case for emergentism. We can arrange our examples of purported explanatory failure into a hierarchy based on the constraints placed on an account of explanation.

At the bottom of this hierarchy (at least for the purposes of this chapter) is *conceptual novelty*. This is the sort of "irreducibility" involved when there is some natural kind in the higher-level theory that cannot be equated to a single natural kind in the lower-level theory. It may be the case that the phenomena that constitute the higher-level kind can be individually explained by the lower-level theory, but the theory does not unite them as a single kind. Conceptual novelty involves a failure of type–type reduction, but need not involve a failure of token–token reduction. In the case of phase transitions, it has been suggested that although one can provide a perfectly adequate explanation of individual transitions using statistical mechanics, the theory does not distinguish these phenomena as a separate kind. For instance, from the perspective of statistical mechanics, the transition from ice to water in a finite system as we cross 273.16 K is not qualitatively different from the transition from cold ice to slightly warmer ice as we cross 260 K, at least if something like the standard story is correct. The only difference is that the thermodynamic potentials change a lot more rapidly in the former situation than in the latter, but they are still analytic, so this is merely a difference of degree, not a difference of kind.

There are two tacks one can take in response to this observation. The first is that this is a case where statistical mechanics corrects thermodynamics. Just as it showed us that the second law is not in principle exceptionless, it shows us that rigorous separation of phases, the only phenomenon worthy of the name "phase transition," is only possible in infinite systems. This view of the emergence of phase transitions is expressed by Kadanoff when he says that "in some sense phase transitions are not exactly embedded in the finite world but, rather, are products of the human imagination" (Kadanoff 2009, 778). Thermodynamics classifies a set of empirical phenomena as phase transitions, involving a qualitatively distinct type of change in the system. Statistical mechanics reveals that these phenomena have been misclassified. They are not genuinely qualitatively distinct and should not be treated as a separate natural kind. This response does not appear to pose much of a threat to reductionism. It may be true that thermodynamics has not been reduced to statistical mechanics in a strict Nagelian sense, but this seems like much too restrictive a conception of reduction. There are many paradigmatic cases of scientific reduction where the reducing theory explains a corrected version of the reduced theory, not the theory in its original form. This correction may often involve dissolving inappropriate distinctions. If this is all there is to the challenge of conceptual novelty, it is not much of a challenge.

However, one might want to resist this eliminativism and reject the notion that thermodynamics has misclassified phenomena. Perhaps the right thing to say is that at the thermodynamic level of description it is indeed appropriate to have a distinct kind corresponding to phase transitions in finite systems. But the appropriateness of this kind is invisible at the statistical mechanical level of description, since statistical mechanics does not have the resources to construct such a class. This is a more substantive challenge to reductionism, akin to cases of multiple realizability. As an analogy, consider that from the perspective of our molecular theory there is no natural kind (or indeed finite disjunction of kinds) corresponding to the category "can opener." It seems implausible that we will be able to delineate the class of can openers using only the resources of our microscopic theory. Yet we do not take this to mean that our microscopic theory corrects our macroscopic theory, demonstrating that can openers do not exist as a separate kind. Can openers do exist. They are an appropriate theoretical kind at a certain level of description. Similarly, the fact that statistical mechanics does not have the resources to delineate the class of finite particle phase transitions need not lead us to conclude that this classification is bogus.

How might the reductionist respond to conceptual novelty of this sort? One response would be to develop a sense of explanation that makes reduction compatible with multiple realization. Even though statistical mechanics does not group phase transitions together the way that thermodynamics does, it is still able to fully explain what goes on in individual instances of phase transition. Perhaps the existence of individual explanations in every case constitutes an adequate explanation of the nomic pattern described by thermodynamics. If this is the case, the core sense of reduction is satisfied. One does not need to look at phase transitions to notice that any claim about the reduction of thermodynamics to statistical mechanics must be based on a conception of reduction that is compatible with multiple realizability. Temperature, that most basic of thermodynamic properties, is not (the claims of numerous philosophers notwithstanding) simply "mean molecular kinetic energy." It is a multiply realizable functional kind. If our notion of reduction precludes the existence of such properties, then the project of reducing thermodynamics cannot even get off the ground.

To us, this seems like the correct response to claims of emergence based on the conceptual novelty of phase transitions. If this is all it takes for emergence, then practically every thermodynamic property is emergent. Perhaps the emergentist is willing to bite this bullet, but we think it is more plausible that the argument from conceptual novelty to emergence relies on a much too restrictive conception of scientific explanation. It is, however, worth noting another line of response. It may be the case that a class of finite particle phase transitions can be constructed within statistical mechanics that overlaps somewhat (but not completely) with the thermodynamic classification. This would be a case of statistical mechanics correcting thermodynamics, but not by eliminating the phenomenon of phase transitions in finite systems. Instead, statistical mechanics would redefine phase transitions in a manner that preserves our judgments about a number of empirical instances of the phenomenon. If such a redefinition could be engineered, phase transitions would not be conceptually novel to thermodynamics. The prospects for this strategy are discussed in section 3.1.

Let us suppose our conception of reduction is broad enough that mere conceptual novelty does not indicate a failure of reduction. We accept with equanimity that under certain conditions it might be appropriate to model phenomena using a conceptual vocabulary distinct from that of our reducing theory. For instance, at a sufficiently coarse-grained level of description a certain set of thermodynamic transformations is fruitfully modeled as exhibiting singular behavior, and appropriately grouped together into a separate natural kind. However, one might not think that a fully reductive explanation has been given unless one can explain using the resources of the reducing theory why this model is so effective under those conditions. Why does modeling a finite particle phase transition as nonanalytic work so well at the thermodynamic level of description if finite systems cannot exhibit non-analyticities at the statistical mechanical level of description? If we cannot give such an explanation, we have another potential variety of emergence: *explanatory irreducibility*.

To give an idea of the kind of story we are looking for, consider the infinite idealization involved in explaining the extensivity of certain thermodynamic properties. Many thermodynamic properties are extensive, such as the entropy, internal energy, volume, and free energy. What this means is that if we divide a system into macroscopic parts, the values of those properties behave in an additive way. Loosely put, if we double the size of the system (that is, double internal energy, particle number, volume), then we double that system's extensive properties (e.g., the entropy).[2] Intensive properties, by contrast, do not scale this way; for example, if we double the size of a system, we do not double the pressure. Extensivity and intensivity are features usefully employed by phenomenological thermodynamics. However, when we look at a system microscopically, we quickly see that no finite system is ever strictly extensive or intensive. Correlations exist between the particles in one part of a system and another part. If we want to reproduce the thermodynamic distinction exactly, we are stymied: no matter how large the system, if it is finite, surface effects contribute to the partition function, which will mean that systems' energies and entropies cannot be neatly halved. For instance, if we define the entropy as a function over the joint probability distributions involved (as with the Gibbs entropy), we see that the entropy is extensive only when the two subsystems are probabilistically independent of one another. The only place we can reproduce the sharp distinction is by going to the thermodynamic limit. There we can define a variable f as extensive if f goes to infinity as we approach the thermodynamic limit while f/V is constant in the limit, where V is the volume of the system.[3] Strictly speaking, only in infinite systems are entropy, energy, and so on, truly extensive.

Does this fact imply that there is a great mystery about extensivity, that extensivity is truly emergent, that thermodynamics does not reduce to finite N statistical mechanics? We suggest that on any reasonably uncontentious way of defining these terms, the answer is no. We know exactly what is happening here. Just as the second

---

[2] Strictly speaking, additivity and extensivity are different properties; see Touchette (2002). Since they overlap for many real systems, they are commonly run together; however, it is a mistake to do so in general, for some quantities scale with particle number N (and hence are extensive), yet are not additive.

[3] Some textbooks even go in the other direction, namely, defining the thermodynamic limit as that state wherein entropy and energy are extensive.

law of thermodynamics is no longer strict when we go to the microlevel, neither is the concept of extensivity. The notion of extensivity is an idealization, but it is one approximated well by finite particle statistical mechanics. For boxes of length l containing particles interacting via short-range forces, the surface effects scale as l^2 and the volume as l^3. Surface effects become less and less important as the system gets larger. Beings restricted to macroscopic physics would do well to call upon the extensive/intensive distinction, since in most cases the impact of surface effects would be well beyond the precision of measurements made by such beings. Here we see that extensivity in finite systems is conceptually novel to thermodynamics. It does not exist in statistical mechanics. However, leaving the story there is unsatisfactory. We need a further account, from a statistical mechanical perspective, of why this new concept works so well in thermodynamics. And indeed such a story is forthcoming. It relies crucially on the fact that the resolution of our measurements is limited, but this in itself does not, or at least should not, derail the reductionist project. As long as we have a story that explains why beings with such limitations could fruitfully describe sufficiently large systems as extensive—a story in terms of the components of the system and their organization, and how relevant quantities scale as the system gets larger—we do not have a genuine challenge to reductionism in the core sense.

The question is whether a similar sort of explanation is available to account for the efficacy of the infinite idealization involved in the statistical mechanical analysis of phase transitions. If there is not, we would have a case for emergence. There would be something about the systems under consideration that could not be accounted for reductively, namely, the fact that their behavior at a phase transition can, under certain conditions, be adequately modeled as the behavior of an infinite system. This feature of finite systems is crucial to understanding their behavior at phase transitions, so if it cannot be explained it would be legitimate to say that phase transitions are emergent. In section 3.2 we examine the possibility of a reductive explanation of the efficacy of the infinite idealization.

Modeling the behavior of particular systems is not the only function of the infinite idealization in the study of phase transitions. The idealization plays a central role in the renormalization group explanation for universal behavior at the critical point. As we have discussed above, universal behavior is accounted for by the presence of stable fixed points in the space of Hamiltonians, each of which is the terminus of a number of different renormalization flow trajectories. This sort of explanation raises special problems that do not arise when we consider the sort of infinite idealization involved in the assumption of extensivity. There we have a property that, as it turns out, can only be approximated by finite systems. It is only actually instantiated in infinite systems. However, the property itself can be characterized without recourse to the infinite idealization. We could in principle construct an explanation of why a finite thermodynamic system approximates extensive behavior without any appeal to the infinite idealization. The idealization gives us a model of a genuinely extensive system, but it is not essential to an understanding of why it is useful to treat macroscopic finite systems as extensive.

It appears that the situation is different when we consider the renormalization group explanation of universality. There, the infinite idealization plays a different role. Talking about how a particular large finite system approximates the behavior of an infinite system will not be helpful, because universality is not about the behavior of individual systems, finite or infinite. It is a characteristic of classes of systems. The renormalization group method explains why physical systems separate into distinct universality classes, and it explains this in terms of certain structural features of the space of systems, the fixed points of the renormalization flow. It is the existence of these features, and their connection to the phenomenon of universality, that requires the infinite idealization. We might be able to give an account of why a particular large finite system approaches very close to a fixed point as it is rescaled, approximating the behavior of an infinite system, but this will not tell us why this behavior matters. In order to see the connection between approaching a fixed point and exhibiting universal behavior, we need the infinite idealization. This argument is made in Batterman (2011). We address it in section 4.

In a case of explanatory irreducibility the higher-level theory models a particular phenomenon in a conceptually novel manner, and the efficacy of this model cannot be explained by the lower-level theory. However, this does not preclude the possibility that the phenomenon can be modeled within the lower-level theory in a different way. There may be aspects of the phenomenon (such as, say, its macroscopic similarity to other phenomena) that cannot be captured by the descriptive resources of the theory, but the phenomenon itself can be described by the theory. Consider, for instance, the relationship between neuroscience and folk psychology. It might be argued that the latter is explanatorily irreducible to the former. Perhaps there is no viable neuroscientific account of why the reasons explanations common in folk psychology are successful, but a materialist about the mind could maintain that this is merely because the neuroscientific theory operates at too fine a scale to discern the patterns that ground this sort of explanation. In every token instance covered by the folk psychological explanation, there is nothing relevant going on that is not captured by neuroscience. It is just that the way neuroscience describes what is going on is not conducive to the construction or justification of reasons explanations. The patterns that the neuroscientific description fails to see are nonetheless wholly generated by processes describable using neuroscience.

A substance dualist, however, would argue that there is an even deeper failure of reduction going on here. The phenomena and processes described by neuroscience are by themselves inadequate to even generate the kinds of patterns that characterize reasons explanations. This is because the lower-level theory does not have the resources to describe a crucial element of the ontological furniture of the situation, the mind or the soul. Here we have more than a mere case of explanatory irreducibility. We may call cases like this, where the lower-level theory cannot even fully describe a phenomenon that can be modeled by the higher-level theory, examples of *ontological irreducibility*.

This is probably the sense in which the British emergentists conceived of emergence (see McLaughlin (1992) for an illuminating analysis of this school of thought).

With reference to phase transitions, this view is perhaps most starkly expressed in Batterman (2005). Batterman argues that the discontinuity in the thermodynamic potential at a phase transition is not an artifact of a particular mathematical representation of the physical phenomenon but is a feature of the physical phenomenon itself. He says, "My contention is that thermodynamics is correct to characterize phase transitions as real physical discontinuities and it is correct to represent them mathematically as singularities" (ibid., 234). If there are genuine discontinuities in physical systems, it seems we could not represent them accurately using only continuous mathematical functions. So, since the statistical mechanics of finite systems does not give us discontinuities, it is incapable of fully describing this physical phenomenon. We can only approach an explanation of the phenomenon by working in the infinite limit. The idealization is a manifestation of the inability of the theory to fully describe the phenomenon of phase transitions in finite systems. We discuss these ideas further in section 3.3.

In the remainder of this chapter, we discuss the status of these three notions of emergence—conceptual novelty, explanatory irreducibility, and ontological irreducibility—as they apply to both the standard statistical mechanical notion of phase transitions and the treatment of critical phenomena by the renormalization group. These topics are treated separately because, as discussed above, the renormalization group introduces new issues bearing on the topic of emergence and reduction that go beyond issues involving infinite idealization in traditional statistical mechanics.

3. The Infinite Idealization in Statistical Mechanics

In the previous section, we discussed three ways in which the relationship between statistical mechanics and thermodynamics might be nonreductive. There is a hierarchy to these different senses of emergence set by the varying strengths of the assumptions about explanation required in order for them to represent a genuine failure of the core sense of reduction. Conceptual novelty is the weakest notion of emergence, explanatory irreducibility is stronger, and ontological irreducibility is stronger still. In this section, we discuss the case that can be made for phase transitions exemplifying each of these notions of emergence. We conclude that in the domain of ordinary statistical mechanics (excluding the renormalization group), the case for phase transitions being either ontologically or explanatorily irreducible is weak. The case for phase transitions being conceptually novel is stronger, but even here there are questions that can be raised.

3.1 Conceptual Novelty

A natural kind in a higher-level theory is conceptually novel if there is no kind in any potential reducing theory that captures the same set of phenomena. Are thermodynamic phase transitions conceptually novel? That is, does the kind 'phase transition' have a natural counterpart kind in statistical mechanics? If we restrict ourselves to finite N systems, it is commonly believed that there is not a kind in statistical mechanics corresponding to phase transitions and that one can only find such a kind in infinite N statistical mechanics. *We believe, to the contrary, that no theory, infinite or finite, statistical mechanical or mechanical, possesses a natural kind that perfectly overlaps with the thermodynamic natural kind.* Yet if one relaxes the demand of perfect overlap, then there are kinds—even in finite N statistical mechanics—that overlap in interesting and explanatorily powerful ways with thermodynamic phase transitions. Strictly speaking, thermodynamic phase transitions are conceptually novel; more loosely speaking, they are not.

To begin, one might wonder in what sense "phase transition" is a kind even in thermodynamics. After all, there are ambiguities in the way we define phases. Is glass a supercooled liquid or a solid? It depends on which criteria one uses and no set seems obviously superior. Be that as it may, the notion of a transition is relatively clear in thermodynamics, and it is defined, as above, as a discontinuity in one of the thermodynamic potentials. Let's stick with this.

Now, is the kind picked out by Def 1 the counterpart of the thermodynamic definition? Despite many claims that it is, Def 1's extension is clearly very different than that given by thermodynamics. To mention the most glaring difference—and on which, more later—there are many systems that do not have well-defined thermodynamic limits. Do they not have phase transitions? One can define words as one likes, but the point is that there are many systems that suffer abrupt macroscopic changes, changes that thermodynamics would count as phase transitions, but which do not have thermodynamic limits. Systems with very long-range interactions are prominent examples. But in fact the conditions on the existence of a thermodynamic limit are numerous and stringent, so in some sense most systems do not have thermodynamic limits. A strong case can be made that Def 1, as a result, provides at best sufficient conditions for a phase transition, and not necessary conditions.

How does finite N statistical mechanics fare? The conventional wisdom is that finite N statistical mechanics lacks the resources to have counterparts of thermodynamics phase transitions. However, we believe that people often assent to this claim too quickly. One of the more interesting developments in statistical mechanics of late has been challenges to ordinary statistical mechanics from the realms of the very large and the very small. These are regimes that test the applicability of normal Boltzmann-Gibbs equilibrium statistical mechanics. The issues arise from the success of statistical mechanical techniques in new areas. In cosmology, statistical mechanics is used not only to explain the inner workings of stars but also to explain the statistical distribution of galaxies, clusters, and more. In these cases, the force of interest is of course the gravitational force, one that is not screened at short distances like the Coulomb force. Systems like this do not have a well-defined thermodynamic limit, often are not approximately extensive, suffer negative heat capacities, and more (see Callender (2011) for discussion). There has also been an extension of statistical mechanical techniques to the realm of the small. Sodium clusters obey a solidlike to liquidlike "phase transition," Bose-Einstein condensation occurs, and much more. These atomic clusters have been surprisingly amenable to statistical mechanical treatment, yet they too do not satisfy the conditions for the application of the thermodynamic limit. Physically, one way to think about what is happening here is that in small systems a much higher proportion of the particles reside on the surface, so surface effects play a substantial role in the physics. As a result, these systems also raise issues about extensivity, negative specific heats, and much more.[4]

These systems are relevant to our concerns here for a very simple reason: they appear to have phase transitions, yet lack a well-defined thermodynamic limit, so Def 1 seems inadequate. Orthogonal to our philosophical worries about reduction, there are also purely physical motivations for better understanding thermodynamic phase transitions from the perspective of finite statistical mechanics. Naturally, some physicists appear motivated by both issues, the conceptual and the physical:

> Conceptually, the necessity of the thermodynamic limit is an objectionable feature: first, the number of degrees of freedom in real systems, although possibly large, is finite, and, second, for systems with long-range interactions, the thermodynamic limit may even be not well defined. These observations indicate that the theoretical description of phase transitions, although very successful in certain aspects, may not be completely satisfactory. (Kastner 2008, 168)

As a result of this motivation, there are already several proposals for finite-particle accounts of phase transitions. These are sometimes called *smooth phase transitions*. The research is ongoing, but what exists already provides evidence of the existence of thermodynamic phase transitions in finite systems. There are many different schemes, but we will concentrate on the two most well known.

3.1.1 Back-Bending

Inspired in part by van der Waals theory and its S-shaped bends, this theory has been developed by Wales and Berry (1994), Gross and Votyakov (2000) and Chomaz, Gulminelli, and Duflot (2001). Unlike in the traditional theory of phase transitions, here the authors work with the microcanonical ensemble, not the canonical ensemble. The general idea is that the signatures of phase transitions of different orders are read off from the curvature of the microcanonical entropy, $S = k_b \ln \Omega(E)$, where $\Omega(E)$ is the microcanonical partition function. In particular, if written in terms of the associated caloric curve, $T(E) = 1/\partial_E \ln[\Omega(E)]$, we can understand a first-order transition as a "back-bending" curve, where for a given value of $T(E)$ one can have

---

[4] For the thermodynamic limit to exist, two conditions on the potential in the Hamiltonian must be satisfied, one on large distances, one on small distances. These extensions can be viewed as challenges in either length scale. In another sense, however, one can view both types of systems as unified together as "small" systems. If we define a system as "small" if its spatial extension is less than the range of its dominant interaction, then even galactic clusters are small.

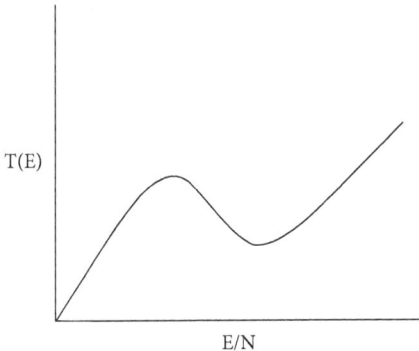

Figure 5.1 Back-bending of the caloric curve.

more than one set of values for E/N (see figure 5.1). For our illustrative purposes, we will use this as our definition:

(Def 2) A first-order phase transition occurs when there is "back-bending" in the microcanonical caloric curve.

Def 2 is equivalent to the entropy being convex or the heat capacity being negative for certain values. As expected, back-bending can be seen in finite-N systems. So with Def 2 we have an alternative criterion of phase transitions that nicely characterizes phase transitions even in systems that do not have thermodynamic limits. We hasten to add that the theory is not exhausted by a simple definition. Rather, the hope—which has to some extent been realized—is that it and its generalizations can predict and explain both continuous phase transitions and also phase transitions in systems lacking a thermodynamic limit.

Def 2 is rather striking when one realizes that it is equivalent to a region of negative heat capacities appearing. The reader familiar with the van Hove theorem may be alarmed, for that theorem forbids back-bending in the thermodynamic limit. Since our concerns are about the finite case, this in itself is not troubling. But if one hopes that this definition goes over to the infinite N definition in the thermodynamic limit, where ensemble equivalence holds for many systems, this might be a problem: the canonical ensemble can never have negative heat capacity, whereas the microcanonical one can, and yet they are equivalent for "normal" short-range systems in the thermodynamic limit. Does "ensemble equivalence" in the infinite limit squeeze out these negative heat capacities? No, for one must remember that ensemble equivalence holds, where it does, only when systems are not undergoing phase transitions. This is a point originally made by Gibbs (1902). And indeed, ensemble inequivalence can be used as a marker of phase transitions. What is happening is that the microcanonical ensemble has structure that the canonical ensemble cannot see; the regions of back-bending (or convex entropy, or negative heat capacity) are missed by the canonical ensemble. Yet since the canonical ensemble is equivalent to the microcanonical—if at all—only when no phase transition obtains, there is no opportunity for conflict with "equivalence" results.

This remark provides a clue to the relation between Def 1 and Def 2 and a way of thinking about the first as a subspecies of the second. When there is back-bending there is ensemble inequivalence. From the perspective of the canonical ensemble for an infinite system, this is where a nonanalyticity appears in the thermodynamic limit. It can "see" the phase transition in that case; but when finite it is blind to this structure. Def 2 can then be seen as more general, since it triggers the nonanalyticity seen in infinite systems and captured by Def 1 but also applies to finite systems.

Many more interesting facts have recently been unearthed about the relationships among back-bending, nonconcave entropies, negative heat capacity, ensemble inequivalence, phase transitions, and nonextensivity. We refer the reader to Touchette and Ellis (2005) for discussion and references. For rigorous connections between Def 1 and Def 2, see Touchette (2006).

3.1.2 Distribution of Zeros

This approach grows directly out of the Yang-Lee picture. The Yang-Lee theorem is about the distribution of zeros of the grand canonical ensemble's partition function in the complex plane. A critical point is encountered when this distribution "pinches" the real axis, and this can only occur when the number of zeros is infinite. Fisher and later Grossmann then provided an elaborate classification of phase transitions in terms of the distribution of zeros of the canonical partition function in the complex temperature plane. Interested in Bose-Einstein condensation, nuclear fragmentation and other "phase transitions" in small systems, a group of physicists at the University of Oldenburg sought to extend this approach to the finite case (see Borrmann, Mülken, and Harting 2000). For our purposes, we can define their phase transitions as:

> (Def 3) A phase transition occurs when the zeros of the canonical partition function align perpendicularly to the real temperature axis and the density scales with the number of particles.

The distribution of zeros of a partition function contains a lot of information. The idea behind this approach is to extract three parameters (α, γ, τ_1) from the partition function that tell us about this distribution: τ_1 is a function of the number of zeros in the complex temperature plane, and it is positive for finite systems; γ is the crossing angle between the real axis and the line of zeros; and α is determined from the approximate density of zeros on that line. What happens as we approach a phase transition is that the distribution of zeros in the complex temperature plane "line up" and gradually gets denser and straighter as N increases.[5]

We stress that, as with the previous group, the physicists involved do not offer a stray definition but rather a comprehensive theory of phase transitions in small

---

[5] A small movie of this occurring for small magnetic clusters is available at http://smallsystems.isn-oldenburg.de/movie.gif.

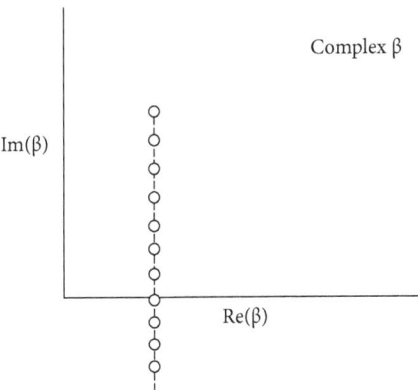

Figure 5.2 Distribution of zeros in the complex inverse temperature ($\beta = 1/kT$) plane.

systems. In particular, the Oldenburg group can use this theory to not only predict whether there is a phase transition but also to identify the correct order of the transition. Their classification excels when treating Bose-Einstein condensation, as it reproduces the space dimension and particle number dependence of the transition order.

Like the approach using Def 2, the present approach works for both finite and infinite systems. For finite systems, τ_1 is always positive and we look for cases where $\alpha = \gamma$: these correspond to first-order transitions in finite systems. More complicated relations between α and γ correspond to higher-order transitions. For infinite systems, phase transitions of the first-order occur when $\alpha = \gamma = \tau_1 = 0$ and for higher-order when $\alpha > 0$. So the scheme includes the Def 1 case as a subspecies. One can then view Def 3—or more accurately, the whole classification scheme associated with (α,γ,τ_1)—as a wider, more general definition of phase transitions, one including small systems, with Def 1 as a special case when the thermodynamic limit is legitimate.

What is the relationship between Def 2 and Def 3? It turns out that they are almost equivalent. Indeed, if one ignores a class of systems that may turn out to be unphysical, they are demonstrably equivalent; see Touchette (2006).[6] The rich schemes of which these definitions form a part may not be equivalent, but on the question of what counts as a phase transition they will largely agree.

As a result of the work on finite-N definitions—and while duly recognizing that it is very much ongoing—it seems to us that statistical mechanics is hardly at a loss to describe phase transitions in finite systems. The situation instead seems to us to be more subtle. No definition in statistical mechanics, infinite or finite, exactly reproduces the extension picked out by thermodynamics with the kind "phase transition." What one judges the best definition then hangs on what

---

[6] This chapter shows that yet another definition, one based on a bimodality of the energy distribution, is almost equivalent to Def 3. However, the bimodality definition is equivalent to Def 2, so the demonstration links Def 2 and Def 3.

extension one wants to preserve. If focusing on thermodynamic systems possessing thermodynamic limits, then Def 1 is fine. Then the kind "phase transition" is conceptually emergent relative to finite-N statistical mechanics. But if impressed by long-range systems, small systems, nonextensive systems, and "solidlike-to-liquidlike" mesoscopic transitions, then one of the finite-N definitions is necessary. Relative to these definitions, the kind "phase transition" is not conceptually novel. If one wants a comprehensive definition, for finite and infinite, then the schemes described provide the best bet. Probably none of the definitions provide necessary and sufficient conditions for a phase transition that overlaps perfectly with thermodynamic phase transitions. That, however, is okay, for thermodynamics itself does not neatly characterize all the ways in which macrostates can change in an "abrupt" way.

In any case, we do not believe that conceptual novelty by itself poses a major threat to reductionism. After all even a (too) strict Nagelian notion of reduction can accommodate conceptual novelty (as long as the novel higher-level kind is expressible as a finite disjunction of lower-level kinds). Conceptual novelty is only a problem when you do not have explanatory reducibility of the conceptually novel kind, a question to which we now turn.

3.2 Explanatory Irreducibility

Explanatory irreducibility occurs, we said, when the explanation of a higher-level phenomenon requires a conceptual novelty, yet the reducing theory does not have the resources to explain why the conceptual novelty is warranted.[7] Where phase transitions are especially interesting, philosophically, lies in the fact that, at first glance, they seem to be a real-life and prominent instance of explanatory irreducibility. To arrive at this claim, let us suppose that the finite-N definitions surveyed above are theoretically inadequate. Assume that Def 1 is employed in the best explanation of the phenomena. Then we have already seen that no finite-N statistical mechanics can suffer phase transitions so understood. If the "reducing theory" is finite-N

---

[7] There are some potential connections between "explanatory irreducibility" and notions in the literature on idealization. In particular, depending upon how one understands Galilean idealization, it is possible that a conceptual novelty is explanatorily irreducible just in case it is not a "harmless" Galilean idealization. Coined by McMullin, a Galilean idealization in a scientific model is a deliberate distortion of the target system that simplifies, unifies or generally makes more useful or applicable the model. Crucially, a Galilean idealization is also one that allows for controlled "de-idealization." In other words, it allows for adding realism to the model (at the expense of simplicity or usefulness, to be sure) so that one can see that the distortions are justified by convenience and are not ad hoc. Idealizations like this are sometimes dubbed "controllable" idealizations and are widely viewed as harmless. What to make of such non-Galilean idealizations is an ongoing project in philosophy of science. One prominent idea—see, e.g., Cartwright (1983) or Strevens (2009)—is that the model may faithfully represent the significant causal relationships involved in the real system. The departure from reality need not then accompany a corresponding lack of faith in the deliverances of the model. It is possible that we could understand the standard explanation of phase transitions as a distortion that nonetheless successfully represents the causal relationships of the system. Perhaps the thermodynamic limit is legitimatized by the fact that surface effects are not a difference-maker (in the sense of Strevens) in the systems of interest. We will leave this line of thought to others to develop.

statistical mechanics, then we potentially have a case of explanatory irreducibility. But should the reducing theory be restricted to finite-N theory?

One quick way out of difficulty would be to include the thermodynamic limit as part of the reducing theory. However, this would be a cheat. The thermodynamic limit is, we believe, the production of another phenomenological theory, not a piece of the reducing theory. The ontology of the classical reducing theory is supposed to be finite-N classical mechanics. Such a theory has surface effects, fluctuations, and more, but the thermodynamic limit squashes these out. More importantly, the ontology of the system in the thermodynamic limit is not the classical mechanics of billiard balls and the like. A quick and interesting way to see this point is to note that the thermodynamic limit is mathematically equivalent to the continuum limit (Compagner 1989). The continuum limit is one wherein the size and number of particles is decreased without bound in a finite-sized volume. When thermodynamics emerges from this limit, it is emerging from a theory describing continuous matter, not atomistic matter. New light is shed on all that is regained in the thermodynamic limit if we see it as regained in the continuum limit. For here we do not see properties emerging from an atomic microworld behaving thermodynamically, but rather properties emerging from a continuum, a realm well "above" the atomic. For this reason, with respect to the reduction of thermodynamics to statistical mechanics, we do not see proofs that thermodynamic properties emerge in the thermodynamic limit as cases whereby thermodynamic properties are reduced to mechanical properties.

If this is right, then we have a potential case of explanatory irreducibility. The best explanation of the phenomenon of phase transitions contains an idealization whose efficacy cannot be explained from the perspective of finite-N theory. So are phase transitions actually explanatorily irreducible? The answer hangs on whether de-idealization can be achieved within finite-N statistical mechanics. We believe that it can be. We have already hinted at one possibility. If one could show that one or more of the finite-N definitions approximate in a controlled way Def 1, then we could view Def 1 as "really" talking about one of the other definitions. Indeed, this seems very much a live possibility with either Def 2 or Def 3 above. However, suppose we believe that this is not possible. Is there any other way of de-idealizing the standard treatment of phase transitions? We believe that there is, and both Butterfield (2011) and Kadanoff (2009) point toward the right diagnosis.

Before getting to that, however, notice that the actual practice of the science more or less guarantees that some finite-N approximation must be available. In recent years there has been an efflorescence of computational models of statistical mechanical phenomena (see Krauth 2006). Since we cannot simulate an infinite system, these models give an inkling of how we might approximate the divergences associated with critical behavior in a finite system. Consider, for instance, the Monte Carlo implementation of the Ising model (see, for instance, Wolff (1989)). The Monte Carlo method involves picking some probabilistic algorithm for propagating fluctuations in the lattice configuration of an Ising system as time evolves. Each run of the simulation is a random walk through the space of configurations, and we study the statistical properties of ensembles of these walks.

It might be argued that the system size in these simulations is effectively infinite, since the lattice is usually implemented with periodic boundary conditions. However, this periodicity should be interpreted merely as a computational tool, not as a simulation of infinite system size. The algorithm is supposed to study the manner in which fluctuations propagate through the lattice, but the model will only work if the correlation length is less than the periodicity of the system. If fluctuations propagate over scales larger than the periodicity, we will have a conflict between the propagation of fluctuations and the constraints set by the periodicity of boundary conditions. So the periodic boundary conditions should be interpreted as setting an effective system size. The model is only useful as long as the correlation length remains below this characteristic length scale. Unfortunately, the periodic boundary conditions also mean that the model is not accurate at the critical point, only close to it. As the correlation length approaches system size in a real system, surface effects become relevant, and the simulation neglects these effects.

Nonetheless, the Monte Carlo method does allow us to see how Ising systems approach critical behavior near the critical point. For instance, models exhibit the increase of correlation length as the critical point is approached and the associated slow-down of equilibriation (due to the increased length over which fluctuations propagate). As we construct larger and larger systems, the model is precise closer and closer to the critical point, and we can see the correlation length get larger. We can also model the nonequilibrium phenomenon of avalanches, where the order parameter of the system changes in a series of sharp jumps as the external parameter in the Hamiltonian is varied. As an example, the magnetization of a magnetic material exhibits avalanches as the external field is tuned. The avalanches are due to the way in which fluctuations of clusters of spins trigger further fluctuations. At the critical point, we get avalanches of all sizes. Again, the approach to this behavior can be studied by examining how the distribution of avalanches changes as the system approaches the critical point. These are just some examples of how finite models can be constructed to examine the behavior of a system arbitrarily close to the critical point. These models fail sufficiently close to criticality because they do not adequately deal with boundary effects. However, they do give an indication of how the behavior of large finite systems can be seen as smoothly approximating the behavior of infinite systems.

We now turn to a more explicit attempt to understand the idealization. Butterfield (2011, § 3.3 and § 7) thinks the treatment of phase transitions does not occasion any great mystery. We agree and reproduce his mathematical analogy (with slight modifications) to illustrate the point. Consider a sequence of real functions g_N, where N ranges over the natural numbers. For each value of N, the function $g_N(x)$ is continuous. It is equal to -1 when x is less than or equal to $-1/N$, increases linearly with slope N when x is between $-1/N$ and $1/N$, and then stays at 1 when x is greater than or equal to $1/N$. The slope of the segment connecting the two constant segments of the function gets steeper and steeper as N increases.

While every member of this sequence of functions is continuous, the limit of the sequence $g_\infty(x)$ is discontinuous at $x = 0$. Now consider another sequence of real functions of x, f_N. These are two-valued functions, defined as follows:

$$f_N(x) = \begin{cases} 0, & g_N(x) \text{ is continuous at } x \\ 1, & g_N(x) \text{ is discontinuous at } x. \end{cases}$$

Given these definitions, $f_N(x)$ is the constant zero function for all N. If we just look at the sequence of functions, we would expect the limit of the sequence f_N as $N \to \infty$ to also be constant. However, if we construct $f_\infty(x)$ from $g_\infty(x)$ using the above definition, we will not get a constant function. The function will be discontinuous; it will take on the value 1 at $x = 0$. If one focuses only on f_N without paying attention to how it is generated from g_N, the behavior in the limit will seem mysterious and inexplicable given the behavior at finite N.

Imagine that we represent a physical property in a model in terms of $f_N(x)$ taking on the value 1, where N is a measure of the size of the physical system. This property can only be exemplified in the infinite-N limit, of course. And if we restricted ourselves to considering f_N when trying to explain the property, we would be at a loss. No matter how big N gets, as long as it is finite there is no notion of being nearer or further away from the property obtaining. We might conclude that the property is emergent in the infinite limit, since we cannot "de-idealize" as we did in the case of extensivity and show how a finite system approximates this property. However, this is only because we are not paying attention to the $g_N(x)$. Realizing the relationship between f_N and g_N allows us to account for the behavior of f_N in the infinite limit from a finite system perspective, since there is a clear sense in which the functions g_N approach discontinuity as N approaches infinity.

We might put the point as follows. Suppose we have a theory of some physical property that utilizes the predicates g, N, and x. Suppose further that we are particularly interested in the rapid increase in $g_N(x)$ around $x = 0$ when N is large. Rather than analyze $g_N(x)$ for particular finite values of N, it might make sense from a computational perspective to work with the infinite idealization $g_\infty(x)$, where the relevant behavior is stark and localized at $x = 0$. We may introduce a new "kind" represented by the predicate f that picks out the phenomenon of interest in the infinite limit. This kind is conceptually novel to the $\{g, N, x\}$ framework. Indeed, one can imagine a whole theory written in terms of f, without reference to g. Using such a theory it could be difficult to see how f is approximated by some function of finite-N. Because f is two-valued, the property it represents will appear to just pop into existence in the infinite limit without being approximated in any way by large finite systems. Restricted to f (and hence $g_\infty(x)$), one would not have the resources present to explain how f emerges from the shape of g when N is finite.

This is precisely what happens in phase transitions. As Butterfield shows, the example of f and g translates nicely into the treatment of phase transitions. The magnetization in an Ising model behaves like $g_N(x)$, where N is the number of particles and x is the applied field. For finite systems, the transition of the system between the two phases of magnetization occurs continuously as the applied field goes from negative to positive. In the infinite case, the transition is discontinuous. The sequence of functions f_N isolate one aspect of the behavior of the functions g_N—whether or not they are continuous. If we just focus on this property, it might seem like there is entirely novel behavior in the infinite particle case. The shape of $f_\infty(x)$ around $x = 0$ is not in any sense approximated or approached by f_N as N gets large. If it is the case that large finite systems can be successfully modeled as infinite systems, this might seem to be a sign of explanatory irreducibility. The success of the infinite particle idealization cannot be explained because the infinite particle function is not the limit of the finite particle function sequence f_N. The illusion of explanatory irreducibility is dispelled when we realize that any explanation involving f_∞ can be rephrased in terms of g_∞, and the latter function does not display inexplicably novel behavior. It is in fact the limit of the finite particle functions g_N. As N increases, g_N approaches g_∞ in a well-defined sense. At a sufficiently large but finite system size N_0, the resolution of our measuring instruments will not be fine-grained enough to distinguish between $g_{N_0}(x)$ and $g_\infty(x)$. We have an explanation, much like the one we have for extensivity, of the efficacy of the infinite idealization.

Recognizing that the predicate f only picks out part of the information conveyed by the predicate g dissolves the mystery. The new predicate is useful when we are working with the idealization, but it makes de-idealization a more involved process. To see the connection between a phase transition defined via Def 1 and real finite systems, one must first "undo" the conceptual innovation and write the theory as a limit of nascent functions. At that point one can then see that the idealization is an innocent simplification and extrapolation of what happens to certain physical curves when N grows large.[8]

3.3 Ontological Irreducibility

Ontological irreducibility involves a very strong failure of reduction, and if any phenomenon deserves to be called emergent, it is one whose description is ontologically irreducible to any theory of its parts. Batterman argues that phase transitions are emergent in this sense (Batterman 2005). It is not just that we do not know of an adequate statistical mechanical account of them, we cannot construct such an account. Phase transitions, according to this view, are cases of genuine physical discontinuities. The discontinuity is there in nature itself. The thermodynamic representation of these phenomena as mathematical singularities is quite natural on this view. It is hard to see how else to best represent them. However, canonical statistical mechanics does not allow for mathematical singularities in thermodynamic functions of finite systems, so it does not have the resources to adequately represent these physical discontinuities. If the density of a finite quantity of water does as a

---

[8] Thanks to Jim Weatherall for kick-starting our thinking of phase transitions as delta functions that can be approximated by analytic functions and to Jeremy Butterfield for kindly letting us use an advance copy of his 2011 article.

matter fact change discontinuously at a phase transition, then it seems that statistical mechanics is incapable of describing this phenomenon, so the thermodynamics of phase transitions is genuinely ontologically irreducible.

Why think phase transitions are physically discontinuous? Batterman appeals to the qualitative distinction between the phases of fluids and magnets. Yet describing the distinction between the phases as "qualitative" is potentially misleading. It is true that the different phases of certain systems appear macroscopically distinct to us. A liquid certainly seems very different from a gas. However, from a thermodynamic perspective the difference is quantitative. Phases are distinguished based on the magnitudes of certain thermodynamic parameters. The mere existence of distinct states of the system exhibiting these different magnitudes does not suggest that there is any discontinuity in the transition between the systems. This is a point about the mathematical representation, but the lesson extends to the physical phenomenon. While it is true that the phases of a system are macroscopically distinct, this is not sufficient to establish that the physical transition from one of these phases to the other as gross constraints are altered involves a physical discontinuity.

In order to see whether there really is a discontinuity that is appropriately modeled as a singularity we need to understand the dynamics of the change of phase. So we take a closer look at what happens at a first-order phase transition. Consider the standard representation of an isotherm on the liquid-gas P-V diagram at a phase transition (figure 5.3).

The two black dots are coexistence points. At these points the pressure on the system is the same, but the system separates into two distinct phases: low-volume liquid and high-volume gas. The two coexistence points are connected by a horizontal tie-line or Maxwell plateau. On this plateau, the system exists as a two-phase mixture. It is here that the dynamics of interest takes place. However, the representation above is too coarse-grained to provide a full description of the behavior of the system at transition. This representation certainly involves a mathematical

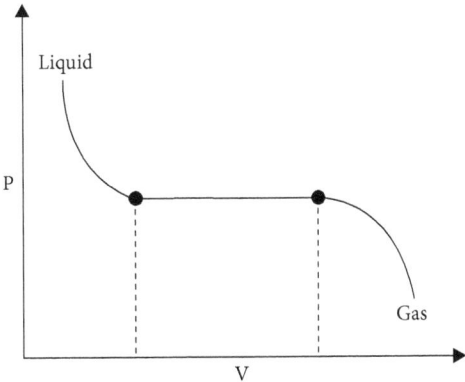

Figure 5.3 P-V diagram for a liquid–gas system at a phase transition.

singularity: as the pressure is reduced, the volume of the system changes discontinuously. But a closer look at how the transition takes place demonstrates that this is just an artifact of the representation, and not an accurate picture of what is going on at the transition. The *P-V* diagram ignores fluctuations, but fluctuations are crucial to the transition between phases. The process by which this takes place is nucleation. When we increase the pressure of a gas above the coexistence point it does not instantaneously switch to a liquid phase. It continues in its gaseous phase, but this supersaturated vapor is meta-stable. Thermal fluctuations cause droplets of liquid to nucleate within the gaseous phase. In this regime, the liquid phase is energetically favored, and this encourages the expansion of the droplet. However, surface effects at the gas–liquid interface impede the expansion. When the droplet is small, surface effects predominate, preventing the liquid phase from spreading, but if there is a fluctuation large enough to push the droplet over a critical radius, the free energy advantage dominates and the liquid phase can spread through the entire system. A full account of the gas–liquid transition will involve a description of the process of nucleation, a nonequilibrium phenomenon that is not represented on the equilibrium *P-V* diagram in figure 5.3.

Perhaps the nucleation of droplets from zero radius could be seen as an example of a physical discontinuity. However, an analysis of this process is not beyond the reach of finite particle statistical mechanics. We can study the nucleation of a new phase using the Ising model. As the external field crosses zero, simulations of the model show that initially local clusters of spins flip. Some of these clusters are too small, so they shrink back to zero, but once there is a large enough cluster—a critical droplet—the flipping spreads across the entire system and the new phase takes over. All of this is observable in a simple finite particle Ising system, so the phenomenon of nucleation can be described by statistical mechanics without having to invoke the thermodynamic limit. If it is the case that physical discontinuities cannot be accurately described by statistical mechanics, then we have good reason for believing there are no such discontinuities in the process of phase transition.

Even if we grant that phase transitions involve a physical discontinuity and can only be accurately represented by a mathematical singularity, the ontological irreducibility of the phenomenon does not follow. Very recently it has been shown that the microcanonical entropy, unlike the canonical free energy, can be nonanalytic for finite systems. And indeed, a research program has sprung up based on this discovery that tries to link singularities of the microcanonical entropy to thermodynamic phase transitions (Franzosi, Pettini, and Spinelli 2000, Kastner 2008). That program demonstrates that nonanalyticities in the entropy are associated with a change in the topology of configuration space. Consider the subset of configuration space M_v that contains all points for which the potential energy per particle is lower than v. As v is varied, this subset changes, and at some critical values of v the topological properties of the subset change. This topology change is marked by a change in the Euler characteristic. For finite systems, there is a nonanalyticity in the entropy wherever there is a topology change. For infinite systems there is a continuum of points at which the topology changes, so a straightforward identification of phase transitions with topology change is inappropriate.[9] Nevertheless, it is widely believed that there is some connection between these finite nonanalyticities and thermodynamic phase transitions.

This is a fledgling research program and there are still a number of open questions. It is unclear what topological criteria will be necessary and sufficient to define phase transitions, if any such criteria can be found. What is important for our purposes is that it is clear that the microcanonical ensemble does exhibit singularities even in the finite particle case and that there is a plausible research program attempting to understand phase transitions in terms of these singularities. As such, it is certainly premature to declare that phase transitions are ontologically irreducible even if they involve genuine physical discontinuities. Statistical mechanics might well have the resources to adequately represent these discontinuities without having to advert to the thermodynamic limit.

4. The Infinite Idealization in the Renormalization Group

We have argued that there is good reason to think the use of the infinite limit in the statistical mechanical description of phase transitions does not show that the phenomenon is either ontologically or explanatorily irreducible. Here we examine whether similar claims can be made about the way the infinite idealization is used in renormalization group theory. While this theory is usually included under the broad rubric of statistical mechanics, there are significant differences between renormalization group methods and the methods characteristic of statistical mechanics. Statistical mechanics allows us to calculate the statistical properties of a system by analyzing an ensemble of similar systems. Renormalization group methods enter when correlations within a system extend over scales long enough to make straightforward ensemble methods impractical (see Kadanoff (2010a) for more on this distinction). The properties of the system are calculated not from a single ensemble but from the way in which the ensemble changes upon rescaling. In statistical mechanics, the infinite idealization is important for the effect it has on a single ensemble (allowing nonanalyticities, for instance). In renormalization group theory, the infinite idealization is important because it allows unlimited rescaling as we move from ensemble to ensemble. The apparent difference in the use of the idealization suggests the possibility of significant philosophical distinctions. It will not do to blithely extend our conclusions about statistical mechanics to cover renormalization group theory.

---

[9] The problem with identifying these singularities with phase transitions in thermodynamics is that as N grows the order of the phase transition also increases, roughly as $N/2$. These transitions are far weaker than the ones encountered in thermodynamics, and in any case, unobservable in real noisy data unless N is really small.

We distinguish two different types of explanation that utilize the renormalization group framework. The first is an explanation of the critical behavior of particular systems, and the second is the universal behavior of classes of systems. The first type of explanation does not raise any fundamentally new issues that we did not already consider in our discussion of the explanatory reducibility of phase transitions in statistical mechanics. The second type of explanation does raise significant new issues, since we move from the examination of phenomena in particular systems to phenomena characterizing classes of systems. Batterman (2011) argues that the renormalization group explanation of universality is a case of explanatory irreducibility. While we might be able to tell a complex microphysical story that explains why a particular finite system exhibits certain critical behavior (the first type of explanation), we cannot account for the fact that many microscopically distinct systems exhibit identical critical behavior (the second type of explanation) without using the infinite idealization.

We begin with a brief discussion the first type of explanation: the renormalization group applied to the critical behavior of individual systems. We know from theory and experiment that there are large-scale correlations near the critical point and that mean field theory does not work in these conditions. We need a method that can handle systems with long correlation lengths, and this is exactly the purpose that the renormalization group method serves. We idealize the correlation length of the system as infinite so that it flows to a fixed point under rescaling and then calculate its critical exponent by examining the behavior of the trajectory near the fixed point.

This raises the question of why a system with a large correlation length can be successfully represented as a system with an infinite correlation length. If we have no explanation of the success of this idealization, we have a case of explanatory irreducibility. However, when we are focusing on the behavior of a particular system, any irreducibility in the renormalization group theory is inherited from orthodox statistical mechanics. The justification of the infinite correlation length idealization will coincide with the justification for the infinite system size idealization. Why does the renormalization group method need the infinite limit? Because it relies on the divergence of the correlation length at the critical point, which is impossible in a finite system. Why does the correlation length diverge? Because it is related to the susceptibility, which is a second derivative of the free energy and diverges. Why does the susceptibility diverge? Because there is a nonanalyticity in the free energy. Explaining why (or whether) this nonanalyticity exists takes us back to the statistical mechanical definition of phase transitions. If statistical mechanics can explain phase transitions reductively, then the renormalization group does not pose an additional philosophical problem *when we focus on its application to particular systems*. It is true that the system must be idealized in order to employ renormalization group theory, but that idealization can be justified outside renormalization group theory.

The more interesting case is the second type of explanation, the explanation of universality. Without the renormalization group method, we might examine the behavior of individual finite system and discover that a number of such systems, though microscopically distinct, exhibit strikingly similar macroscopic behavior near criticality. However, this would not tell us why we should *expect* this macroscopic similarity, and so it is not really a satisfactory explanation of universality. The renormalization group method gives us a genuine explanation: when the correlation length diverges, there is no characteristic length scale. If the relevant parameters for the system vanish, as they do at criticality, the system will flow to a fixed point under repeated rescaling. Fixed points can function as attractors, leading to similar critical behavior for a number of different systems.

If the system size is finite, the system will not flow to a fixed point. We might be able to show that a number of distinct large finite systems flow to points in system space that are very close to each other, but once again all that we have done is *revealed* the universality of critical (or near-critical) behavior. We have not *explained* it. There is a generic reason to expect distinct infinite systems to flow to stable fixed points, but without mentioning fixed points there does not seem to be a generic reason to expect distinct finite systems to flow to points that are near each other. So it seems that fixed points play an indispensable role in the explanation of universal behavior. We cannot "de-idealize" and remove reference to fixed points in the explanation, the way we can for nonanalyticities in particular systems. Think back to Butterfield's example described in section 3.2. In that example, the apparent explanatory irreducibility of the behavior of f_∞ was resisted by rephrasing our explanations in terms of g_∞, a function whose behavior in the limit is not novel. In the case of the renormalization group, it seems that this move is unavailable to us. Fixed points are a novel feature that only appear in the infinite limit. There does not seem to be a clear sense in which the renormalization flow of finite systems can approximate a fixed point. A point is either a fixed point for the flow or it is not; it cannot be "almost" a fixed point. And unlike Butterfield's example, there does not seem to be a way of rephrasing the explanation of universality in terms that are approximated by large finite systems.

So there is a strong prima facie case that universality is explanatorily irreducible. However, we do not believe that the case stands up to scrutiny. To see how it fails, we begin by showing that we can explain why finite systems exhibit universal behavior near criticality. However, this explanation does require the full resources of the renormalization group method, including fixed points. So it is not an explanation of the sort that we were contemplating above, one that does away with reference to fixed points. We will argue that this should not actually trouble the reductionist, but first we present the explanation.

Consider an Ising system extending over a finite length. When the system is rescaled, the separation between the nodes on the lattice increases. Since we are keeping the system size fixed, this means the number of nodes will decrease. So unlike the infinite system case, for a finite system the number of nodes is a parameter that is affected by rescaling. If the number of nodes is N, we can now think of $1/N$ as a relevant parameter (as defined in section 1.3). When we restrict ourselves to the infinite case, we are considering a particular hypersurface of this new parameter space where $1/N$ is set to 0. However, since $1/N$ is a relevant parameter, perturbing the system off this hypersurface (i.e., switching from the infinite to a finite system) will take the system away from the critical fixed point. This should be cause for concern. It seems there is no hope for an explanatory reduction. If even a slight perturbation off the $1/N = 0$ hypersurface changes the critical behavior, how can we think of finite systems as approximating the behavior of infinite systems? As Kadanoff says, "if the block transformation ever reaches out and sees no more couplings in the usual approximation schemes . . . that will signal the system that a weak coupling situation has been encountered and will cascade back to produce a weak coupling phase [a trivial fixed point with $K = 0$]" (Kadanoff 2010b, 47).

However, all is not lost. The difference between the behavior of finite and infinite systems depends on the correlation length. When the correlation length is very small relative to the system size, the finite system behaves much like the infinite system. The values of thermodynamic observables will not differ substantially from their values for an infinite system. The behavior of the finite system will only exhibit a qualitative distinction when the correlation length becomes comparable to the system size. This phenomenon is known as finite size crossover (see Cardy (1996), ch. 4) for a full mathematical treatment). It is a manifestation of the fact that the behavior of the system is sensitive to the large-scale geometry of the system only when the correlation length is large enough to be comparable to the system size. The crossover is controlled by the reduced temperature. As long as this parameter is above a certain value (given by an inverse power of the system size), the correlation length will be small enough that no distinction between finite and infinite systems will be measurable. It is only below the crossover temperature that finite-size effects become significant and the system flows away from the critical point. For a large system, the crossover temperature will be very small, and its difference from the critical temperature $t = 0$ may be within experimental error. So for a sufficiently large system, it is plausible that the infinite size approximation will work all the way to criticality. Renormalization group theory itself predicts this. A similar point is made in Butterfield (2011).

Crossover theory also provides tools for estimating the changes to critical behavior that come from changing the geometry of the system by limiting its size. Adding system size as a parameter gives us a new scaling function for the susceptibility, a description of how the susceptibility changes with changes in relevant parameters. As described above, this scaling function gives a behavior for the susceptibility similar to the infinite limit as long as the ratio of correlation length to system size is low. It also allows us to predict the behavior of the susceptibility when this ratio becomes close to one. The susceptibility of a finite system will not diverge; it will have a smooth peak. The height of the peak of susceptibility scales as a positive power of size of the system. So for a large system, the susceptibility will be large but not infinite. In addition, the location of the peak shifts, and this shift scales as an inverse power of the size. This means that for a large system the difference between the critical temperature (the temperature at the critical fixed point of the infinite system) and the temperature at which it attains maximum susceptibility is very small. So for a macroscopic system, crossover theory explains why it is a good approximation to treat the susceptibility as diverging at the critical point.

The point of this discussion is that we can tell an explanatory story about the circumstances under which particular large finite systems can be treated like infinite systems. If the crossover temperature is sufficiently small, then limitations of our measurement procedures might make it difficult or even impossible to distinguish that the system does not flow to the critical point. However, this explanatory story does make reference to fixed points in system space. So the worry is that it is not a fully reductive account. We may have explained why individual finite systems can be successfully idealized as flowing to the critical fixed point, but have we accounted for the existence of the critical fixed point? We are taking for granted in our explanation the topological structure of system space, a topological structure that is to a large extent determined by the behavior of *infinite* systems.[10]

This is true, but does it lead to explanatory irreducibility? Is it illicit to include the topological structure of system space among the explanatory resources of our lower-level theory? It would be if this structure involved an idealization whose efficacy could not be accounted for within the lower-level theory. Isn't an irreducible infinite idealization involved in the postulation of a renormalization flow with fixed points? It is not. As we have seen, the renormalization flow can be defined for all systems, finite and infinite alike, since $1/N$ can be introduced as a relevant parameter. Fixed points will appear on the hypersurface where $1/N = 0$. There is no infinite idealization involved here. Of course, we are talking about infinite systems and how they behave under the renormalization flow, but this should not be problematic from a reductive point of view. The problem would arise if we model finite systems as infinite systems without explanation. But at this stage, when we are setting up the space and determining its topological characteristics, we are not modeling particular systems. Insofar as finite systems are represented in our description of the space, they are represented as finite systems, and infinite systems are represented as infinite systems.

So the topological structure of the space can be described without problematic infinite idealization. When we try to explain the universality of critical behavior in finite systems, we do have to employ the infinite idealization, but as we have seen, this idealization is not irreducible if we can use the topological structure of system space in our reductive explanation. We can de-idealize for particular systems and see why they can be treated as if they flow to the critical point. Understanding the behavior of infinite systems is crucial to explaining the behavior of finite systems, since we only get the fixed points by examining the behavior of infinite systems, but this in itself does not imply emergence. We agree with Batterman (2011) that mathematical singularities in the renormalization group method are information sources, not information sinks. We disagree with his contention that the renormalization group explanation requires the infinite idealization and is thus emergent. It requires consideration of the behavior of infinite systems, but it does not require

---

[10] Our thanks to Robert Batterman for pushing us on this point.

us to idealize any finite system as an infinite system. Any actual infinite idealizations in a renormalization group explanation can be de-idealized using finite-size crossover theory. Locating fixed points does not require an infinite idealization, it just requires that our microscopic theory can talk about infinite systems, and indeed it can.

5. Conclusion

Phase transitions are an important instance of putatively emergent behavior. Unlike many things claimed emergent by philosophers (e.g., tables and chairs), the alleged emergence of phase transitions stems from both philosophical and scientific arguments. Here we have focused on the case for emergence built from physics. We have found that when one clarifies concepts and digs into the details, with respect to standard textbook statistical mechanics, phase transitions are best thought of as conceptually novel, but not ontologically or explanatorily irreducible. And if one goes past textbook statistical mechanics, then an argument can be made that they are not even conceptually novel. In the case of renormalization group theory, consideration of infinite systems and their singular behavior provides a central theoretical tool, but this is compatible with an explanatory reduction. Phase transitions may be "emergent" in some sense of this protean term, but not in a sense that is incompatible with the reductionist project broadly construed.

References

Batterman, R. W. (2005). Critical phenomena and breaking drops: Infinite idealizations in physics. *Studies in History and Philosophy of Modern Physics* 36: 225–244.
———. (2011). Emergence, singularities, and symmetry breaking. *Foundations of Physics* 41(6): 1031–1050.
Borrmann, P., Mülken, O., and Harting, J. (2000). Classification of phase transitions in small systems. *Physical Review Letters* 84: 3511–3514.
Butterfield, J. (2011). Less is different: emergence and reduction reconciled. *Foundations of Physics* 41(6): 1065–1135.
Callender, C. (2011). Hot and heavy matters in the foundations of statistical mechanics. *Foundations of Physics* 41 (6): 960–981
Cardy, J. (1996). *Scaling and renormalization in statistical physics*. Cambridge: Cambridge University Press.
Cartwright, N. (1983). *How the laws of physics lie*. Oxford: Clarendon Press.

Chomaz, P., Gulminelli, F. and Duflot, V. (2001). Topology of event distributions as a generalized definition of phase transitions in finite systems. *Physical Review E* 64: 046114.

Compagner, A. (1989). Thermodynamics as the continuum limit of statistical mechanics. *American Journal of Physics* 57: 106–117.

Franzosi, R., Pettini, M. and Spinelli, L. (2000). Topology and phase transitions: Paradigmatic evidence. *Physical Review Letters* 84: 2774–2777.

Gibbs, J. W. (1902). *Elementary principles in statistical mechanics.* New York: Charles Scribner's Sons.

Griffiths, R. (1972). *Phase transitions and critical phenomena*, Vol. 1, ed. C. Domb and M. Green. London: Academic Press.

Gross, D. H. E. and Votyakov, E. V. (2000). Phase transitions in "small" systems. *The European Physical Journal B — Condensed Matter and Complex Systems* 15: 115–126.

Kadanoff, L. (2009). More is the same; mean field theory and phase transitions. *Journal of Statistical Physics* 137: 777–797.

———. (2010a). Relating theories via renormalization. Available at http://jfi.uchicago.edu/~leop/AboutPapers/RenormalizationV4.0.pdf.

———. (2010b). Theories of matter: Infinities and renormalization. (arXiv:1002.2985)

Kastner, M. (2008). Phase transitions and configuration space topology. *Reviews of Modern Physics* 80: 167–187.

Krauth, W. (2006). *Statistical mechanics: Algorithms and computations.* Oxford: Oxford University Press.

Lebowitz, J. L. (1999). Statistical mechanics: A selective review of two central issues. *Reviews of Modern Physics* 71: S346–S347.

Liu, C. (1999). Explaining the emergence of cooperative phenomena. *Philosophy of Science* 66: S92–S106.

McLaughlin, B. P. (1992). The rise and fall of British emergentism. In *Emergence or Reduction?*, ed. A. Beckermann, H. Flohr, and J. Kim, 49–93. Berlin: Walter de Gruyter.

Melnyk, A. (2003). *A physicalist manifesto: Thoroughly modern materialism.* Cambridge: Cambridge University Press.

Nagel, E. (1979). *The structure of science: Problems in the logic of scientific explanation.* Indianapolis: Hackett.

Strevens, M. (2009). *Depth: An account of scientific explanation.* Harvard: Harvard University Press.

Touchette, H. (2002). When is a quantity additive, and when is it extensive? *Physica A: Statistical Mechanics and Its Applications* 305: 84–88.

———. (2006). Comment on "First-order phase transition: Equivalence between bimodalities and the Yang-Lee theorem." *Physica A: Statistical Mechanics and Its Applications* 359: 375–379.

Touchette, H. and Ellis, R. S. (2005). Nonequivalent ensembles and metastability. In *Complexity, Metastability and Nonextensivity*, ed. C. Beck, G. Benedek, A. Rapisarda, and C. Tsallis. 81–88. World Scientific.

Wales, D. J. and Berry, R. S. (1994). Coexistence in finite systems. *Physical Review Letters*, 73: 2875–2878.

U. Wolff (1989). Collective monte carlo updating for spin systems. *Physical Review Letters* 62: 361–364.

CHAPTER 6

EFFECTIVE FIELD THEORIES

JONATHAN BAIN

1. Introduction

An effective field theory (EFT) of a physical system is a theory of the dynamics of the system at energies small compared to a given cutoff. For some systems, low-energy states with respect to this cutoff are effectively independent of ("decoupled from") states at high energies. Hence one may study the low-energy sector of the theory without the need for a detailed description of the high-energy sector. Systems that admit EFTs appear in both relativistic quantum field theories (RQFTs) and condensed matter physics. When an underlying high-energy theory is known, an effective theory may be obtained in a "top-down" approach by a process in which high-energy effects are systematically eliminated. When an underlying high-energy theory is not known, it may still be possible to obtain an EFT by a "bottom-up" approach in which symmetry and "naturalness" constraints are imposed on candidate Lagrangians. In both cases, the intertheoretic relation between the EFT and its (possibly hypothetical) high-energy theory is complicated, and, arguably, cannot be described in terms of traditional accounts of reduction. This has suggested to some authors that the EFT intertheoretic relation (and/or the phenomena associated with it) should be described in terms of a notion of emergence. Other authors have described the process of constructing an EFT as one in which idealizations are made in order to produce a computationally tractable, yet inherently approximate, theory that is empirically equivalent (to a given range of accuracy) to a typically computationally more complex, but complete, high-energy theory. One such claim is that, in the context of RQFTS, the set of possible worlds associated with an EFT are ones in which space is discrete and finite.

This essay reviews effective field theory techniques, focusing on the intertheoretic relation that links an EFT with its (possibly hypothetical) high-energy theory. The goal is to contribute to discussions on how EFTs can be interpreted, and, in particular, to investigate the extent to which a notion of emergence is viable in such interpretations. Section 2 sets the stage by reviewing the general steps in the construction of an EFT in the top-down and bottom-up approaches. Section 3 then reviews the extent to which an EFT can be said to be empirically equivalent to its high-energy theory: typical EFTs are nonrenormalizable and thus break down at high energies; however, this has not stopped physicists from using them to derive predictions for low-energy phenomena. Section 4 indicates the extent to which the explicit form of an EFT depends on the type of renormalization scheme one employs to handle divergent integrals that can arise when one uses the EFT to calculate the values of observable quantities. It is argued that the choice of renormalization scheme is irrelevant for calculating such values. However, to the extent that this choice determines the explicit form of the EFT, arguably, it has nontrivial consequences when it comes to the question of how the EFT can be interpreted. These consequences are investigated in section 5. Finally, section 6 takes up the task of assessing the extent to which the intertheoretic relation between an EFT and its high-energy theory can be described in terms of emergence.

2. The Nature of EFTs

The construction of an EFT follows one of two general procedures, top-down and bottom-up, depending on whether a high-energy theory is known. Both procedures are based on an expansion of the effective action (which formally represents the EFT) in terms of a (possibly infinite) sum of local operators, constrained by symmetry and "naturalness" considerations. They differ on how the effective action is obtained: the top-down approach obtains it by eliminating degrees of freedom from the action of the high-energy theory, whereas the bottom-up approach constructs it from scratch.

2.1 Top-Down

The top-down approach starts with a known theory and then systematically eliminates degrees of freedom associated with energies above some characteristic energy scale E_0. The practical goal is to obtain a low-energy theory that allows one to more easily calculate the values of observable quantities associated with energies below E_0 than in the original theory. Intuitively, calculations in such a low-energy "effective" theory have fewer parameters to deal with (namely, all those parameters associated with high-energy degrees of freedom) and thus are simpler than calculations in the original theory. However, the construction of a low-energy effective theory that accomplishes this is not just a matter of simply ignoring the high-energy degrees of freedom, for they may be intimately tangled up with the low-energy degrees of freedom in nontrivial ways. (As we will see below, one way to distinguish a "renormalizable" theory from a "nonrenormalizable" theory is that in the former, the high-energy degrees of freedom are independent of the low-energy degrees of freedom, whereas in the latter, they are not.) One method of disentangling the high-energy and low-energy degrees of freedom was pioneered by Wilson and others in the 1970s. In the following, I will refer to it as the Wilsonian approach to EFTs. It typically involves two steps: (I) The high-energy degrees of freedom are identified and integrated out of the action. These high-energy degrees of freedom are referred to as the high momenta, or "heavy," fields. The result of this integration is an *effective* action that describes nonlocal interactions between the low-energy degrees of freedom (the low momenta, or "light," fields). (II) To obtain a local effective action (i.e., one that describes local interactions between low-energy degrees of freedom), the effective action from Step I is expanded in terms of local operators. The following describes these steps in slightly more technical detail.[1]

(I) Given a field theory described by an action S and possessing a characteristic energy scale E_0, suppose we are interested in the physics at a lower scale $E \ll E_0$. First choose a cutoff Λ at or slightly below E_0 and divide the fields ϕ into high and low momenta parts with respect to Λ: $\phi = \phi_H + \phi_L$, where ϕ_H have momenta $k > \Lambda$ and ϕ_L have momenta $k < \Lambda$. Now integrate out the high momenta fields. In the path integral formalism, one does the integral over the ϕ_H. Schematically

$$\int \mathcal{D}\phi_L \int \mathcal{D}\phi_H e^{iS[\phi_H, \phi_L]} = \int \mathcal{D}\phi_L e^{iS_\Lambda[\phi_L]}, \qquad (1)$$

where the Wilsonian effective action is given by $e^{iS_\Lambda[\phi_L]} = \int \mathcal{D}\phi_H e^{iS[\phi_L, \phi_H]}$. The effective Lagrangian density $\mathcal{L}_{eff}$ is thus given by $S_\Lambda[\phi_L] = \int d^D x \mathcal{L}_{eff}[\phi_L]$, where D is the dimension of the spacetime.

(II) Typically, the integration over the heavy (i.e., high momenta) fields will result in a nonlocal effective action (i.e., one in which terms occur that consist of operators and derivatives that are not all evaluated at the same spacetime point). In practice, this is addressed by expanding the effective action in a set of local operators:

$$S_\Lambda = S_0(\Lambda, g^*) + \sum_i \int d^D x g^i \mathcal{O}_i, \qquad (2)$$

where the sum runs over all local operators $\mathcal{O}_i$ allowed by the symmetries of the initial theory, and the g_i are coupling constants. Assuming weak

---

[1] This exposition is based on Polchinski (1993), and Campbell-Smith and Mavromatos (1998). See also Burgess (2004, 2007), Dobado et al. (1997), Manohar (1997), Pich (1998), Rothstein (2004), and Schakel (2008).

coupling, the expansion point S_0 may be taken to be the free action of the initial theory, so that $g^* = 0$.

To see how the effective action relates to the initial high-energy action, one can perform a dimensional analysis on the operators that appear in (2). This allows one to obtain information about their behavior as the cutoff Λ is lowered. This analysis involves three steps:

(i) Choose units in which the action is dimensionless ($\hbar = 1$, $c = 1$). In such units, energy has dimension $+1$ while length has dimension -1. The free action can now be used to determine units for the field operators.[2] This then determines units for the coupling constants, and subsequently, for terms in the expansion (2). For instance, if an operator $\mathcal{O}_i$ has been determined to have units E^{δ_i} (thus dimension δ_i), then its coupling constant g_i has units $E^{D-\delta_i}$, and the magnitude of the ith term is $\int d^D x \, \mathcal{O}_i \sim E^{\delta_i - D}$.

(ii) To make the cutoff dependence of the terms explicit, one can define dimensionless coupling constants by $\lambda_i = \Lambda^{\delta_i - D} g_i$. The order of the ith term in (2) is then

$$\lambda_i (E/\Lambda)^{\delta_i - D}. \qquad (3)$$

(iii) The terms in the expansion (2) can now be classified into three types:

- *Irrelevant*: $\delta_i > D$. This type of term falls at low energies as $E \to 0$. Such terms are suppressed by powers of E/Λ.
- *Relevant*: $\delta_i < D$. This type of term grows at low energies as $E \to 0$.
- *Marginal*: $\delta_i = D$. This type of term is constant and equally important at low and high energies (insofar as quantum effects can modify its scaling behavior toward either relevancy or irrelevancy).

This dimensional analysis indicates that, in cases of physical relevance, there will only be a finite number of relevant and marginal terms in (2).[3] In such cases, the low-energy EFT will only depend on the underlying high-energy theory through a finite number of parameters. It is typical in the literature on EFTs to elevate these considerations to a principle. Polchinski (1993, 6) articulates such a principle in the following way:

> The low energy physics depends on the short distance theory only through the relevant and marginal couplings, and

---

[2] Consider, for instance, a scalar field theory with free action $S = (1/2) \int d^D x (\partial_\mu \phi)^2$. This action contains D powers of the spacetime coordinate from $d^D x$ (with total energy units E^{-D}), and -2 powers from the two occurrences of the spacetime derivative $\partial_\mu \equiv \partial/\partial x^\mu$ (with total units E^2). Thus, in order for the action to be dimensionless (with units E^0), the field ϕ must have units E^y satisfying $E^{-D} E^2 E^y E^y = E^0$, and thus dimension $y = D/2 - 1$.

[3] Consider, again, scalar field theory. From note 2, the dimension of a scalar field ϕ is given by $D/2 - 1$; hence, in general, an operator $\mathcal{O}_i$ constructed from M ϕ's and N derivatives will have dimension $\delta_i = M(D/2 - 1) + N$. For $D \geq 3$, there are only a finite number of ways in which $\delta_i < D$ and $\delta_i = D$.

possibly through some leading irrelevant couplings if one measures small enough effects.

Note that arbitrarily many irrelevant terms can occur in (2), but they are suppressed at low energies by powers of E/Λ. Moreover, the cutoff can be used as a regulator for any divergences associated with these terms in calculations of the values of observable quantities. Thus, "even though the [effective] Lagrangian may contain arbitrarily many terms, and so potentially arbitrarily many coupling constants, it is nonetheless predictive so long as its predictions are only made for low-energy processes, for which $E/\Lambda \ll 1$" (Burgess 2004, 17). A little more will be said on this matter in section 3 below.

Finally, in addition to symmetry considerations, a further constraint is typically applied to the expansion (2) of the effective action: one assumes that the dimensionless coefficients λ_i are of order 1. This is associated with a hypothesis of "naturalness," insofar as it rules out the presence of very large or very small numbers, relative to the cutoff, in the expansion.[4] Intuitively, a "natural" EFT should only involve quantities that are small, but not too small, relative to the cutoff. An immediate consequence of this is that mass terms, which have coefficients proportional to powers of the cutoff, cannot appear in (2). Thus naturalness is typically formulated in terms of the following condition:[5]

> EFTs must be *natural*, meaning that all possible masses must be forbidden by symmetries.

Note that this does not preclude the existence of massive objects (fields, particles, etc.) in an EFT description of low-energy phenomena. Rather, it constrains such descriptions to those in which mass terms are generated by broken high-energy symmetries. Thus, according to the Standard Model, massive vector bosons (the W and Z bosons) exist at low energies (with respect to the appropriate cutoff) due to electroweak symmetry breaking, even though gauge invariance prohibits massive terms in the electroweak action, and massive fermions exist similarly due to chiral symmetry breaking.[6]

At this point, the following qualifications should be made concerning the above Wilsonian approach to top-down EFTs:

(a) First, in Step (I), the identification of the appropriate heavy and light field variables is not always self-evident. For example, the weakly coupled EFT for quantum chromodynamics (QCD) known as chiral perturbation

---

[4] See, e.g., Neubert (2006, 155).
[5] Polchinski (1993, 9). Another way to motivate this restriction is by noting that mass terms correspond to gaps in the energy spectrum insofar as such terms describe excitations with finite rest energies that cannot be made arbitrarily small. These gaps create problems when taking a smooth low-energy limit (in the sense of a smooth renormalization group evolution of parameters). Thus, for Weinberg (1996, 145), renormalization group theory can only be applied to EFTs that are massless or nearly massless.
[6] While these aspects of the Standard Model suggest it can be viewed as a natural EFT, other aspects famously preclude this view. In particular, terms representing massive scalar particles like the Higgs boson are not protected by any symmetry and thus should not appear in an EFT. That they do, and that the order of the Higgs term is proportional to the electroweak cutoff, generates the "hierarchy problem" for the Standard Model.

theory is written in terms of pion fields as opposed to quark and gluon fields; and Landau's Fermi liquid theory of conductors can be written as an EFT of weakly interacting quasiparticles, as opposed to strongly interacting electrons (Manohar 1997, 321).[7]

(b) Second, in Step (I), the path integral over the heavy fields is typically performed in practice using a saddle-point approximation (Dobado et al. 1997, 4). This involves expanding the action of the high-energy theory about a given configuration of the heavy fields chosen to be a "saddle point" (or "stationary point"; i.e., a global extremum). To second order in this expansion, the integral over the heavy fields takes the form of a Gaussian integral, which has a well-defined solution.

(c) Third, in Step (II), the dimensional method of justifying the finite dependence of an EFT on its high-energy theory is based on using the free theory to determine units, and this assumes the high-energy theory is weakly coupled. Strong interactions may have the effect of changing the scaling behavior of terms.

(d) Finally, the dimensional assignments in Step (II) work when using the EFT to make simple "tree-level" calculations of the values of observed quantities. However, for higher-order "loop" corrections to such calculations, scaling based on dimensional analysis may break down, and one may have to appeal to a particular renormalization scheme in order to justify the explicit reliance of an EFT on a finite number of parameters (see section 4 below).[8]

2.2 Bottom-Up

The procedure outlined above for constructing an EFT requires having in one's possession the action (or Lagrangian density) of a high-energy theory. In some cases of interest, the fundamental high-energy theory is not known, but an EFT is, nonetheless, still constructible. One simply begins with the operator expansion (2) and includes all terms consistent with the naturalness constraint and with the symmetries and interactions assumed to be relevant at the given energy scale. One can then determine how these terms scale when a given cutoff is raised (as opposed to lowered). Examples of such "bottom-up" EFTs include the Fermi theory of low-energy weak interactions (as it was originally constructed); and, in the view of many physicists, the Standard Model itself (see, e.g., Hartmann 2001 for discussion). Another example is effective field theoretic formulations of general relativity in

---

[7] Another example is Non-Relativistic QCD (NRQCD), which is an EFT of quark/gluon bound systems for which the relative velocity is small. The low-energy fields are obtained by splitting the gluon field into four modes and identify three of these as light variables. Rothstein (2003, 61) describes this process of identification as an "art form" as opposed to a systematic procedure.

[8] Tree-level calculations are contributions to the perturbative expansion of a physical quantity (like a scattering cross-section) that do not involve integrating over the internal momenta of virtual processes. Loop calculations, on the other hand, involve possibly divergent integrals over internal momenta and are typically associated with higher-order corrections to tree-level calculations. (The terminology is based on the graphical representation of perturbative calculations by Feynman diagrams.)

which the Hilbert action is identified as the first term of the expansion (2) (Burgess 2004).

2.3 Example: Low-Energy Superfluid Helium-4 Film

An example of a top-down EFT constructed via the method of section 2.1 is the low-energy theory of a superfluid helium-4 (^4He) film. The effective Lagrangian density that describes this system is formally identical to the Lagrangian density for (2+1)-dimensional quantum electrodynamics (QED_3). This is somewhat surprising, given that the underlying "high-energy" theory is nonrelativistic. This example, in which a relativistic EFT might be said to *emerge* from a nonrelativistic high-energy theory, will be instructive in the discussion of the concept of emergence in EFTs in section 6 below.

At low temperatures, the liquid state of ^4He becomes a superfluid characterized by dissipationless flow and quantized vortices. The phase transition between the normal liquid and superfluid states is encoded in an order parameter that takes the form of a macroscopic wave function $\varphi_0 = (\rho_0)^{1/2} e^{i\theta}$ describing the coherent ground state of a Bose condensate with density ρ_0 and coherent phase θ. An appropriate Lagrangian density describes *nonrelativistic* neutral bosons (viz., ^4He atoms) interacting via a spontaneous symmetry breaking potential with coupling constant κ (Zee 2003, 175, 257),

$$\mathcal{L}_{4He} = i\varphi^\dagger \partial_t \varphi - (1/2m)\partial_i \varphi^\dagger \partial_i \varphi + \mu \varphi^\dagger \varphi - \kappa(\varphi^\dagger \varphi)^2, \quad i = 1,2,3. \tag{4}$$

Here m is the mass of a ^4He atom, and the term involving the chemical potential μ enforces particle number conservation. This is a thoroughly nonrelativistic Lagrangian density invariant under a global $U(1)$ symmetry and Galilean transformations. We now consider (4) as representing an underlying "high-energy" theory and seek to construct a low-energy EFT via the top-down approach outlined above.

To investigate the low-energy behavior of (4), one first needs to identify appropriate dynamical variables by means of which a distinction can be made between high- and low-energy degrees of freedom.[9] Since the ground state φ_0 is a function only of the phase, low-energy excitations take the form of phase fluctuations. This suggests rewriting the field variable φ in terms of density and phase variables $\varphi = (\rho)^{1/2} e^{i\theta}$ and identifying the high-energy degrees of freedom with the density ρ. The next task is to integrate the density field out of (4).[10] This can be done by expanding the variables as $\rho = \rho_0 + \delta\rho$, $\theta = \theta_0 + \delta\theta$, where $\delta\rho, \delta\theta$ represent small fluctuations in the density and phase about their stationary ground state values ρ_0, θ_0. Substituting these into (4), one obtains the equivalent of (2) for the

---

[9] The following exposition draws on Wen (2004, 82–83; 259–264) and Zee (2003, 257–258; 314–316).
[10] Formally this involves calculating the functional integral $e^{iS_{eff}[\theta]} = \int \mathcal{D}\rho e^{iS_{4He}[\theta,\rho]}$, where $S_{eff}[\theta]$ is the effective low-energy action, and $S_{4He}[\theta,\rho] = \int d^4x \mathcal{L}_{4He}$ is the action of the high-energy theory. As mentioned at the end of section 2.1, such integrals can be calculated using a saddle-point approximation, which, in this context, is equivalent to the semiclassical expansion method outlined above (Schakel 2008, 75).

effective Lagrangian density: $\mathcal{L}_{4He} = \mathcal{L}_0[\rho_0, \theta_0] + \mathcal{L}'_{4He}[\delta\rho, \delta\theta]$, where the first term reproduces (4) without the interaction term, and the second term describes fluctuation contributions. The high-energy fluctuations $\delta\rho$ can be eliminated by deriving the Euler-Lagrange equations of motion for the density variable and solving for $\delta\rho$. After substitution back into $\mathcal{L}'_{4He}$, and in two dimensions, the result is,

$$\mathcal{L}'_{4He} = (1/4\kappa)(\partial_t \theta)^2 - (\rho_0/2m)(\partial_i \theta)^2 + \cdots, \quad i = 1, 2, \tag{5}$$

with $\delta\theta$ replaced by θ for the sake of notation. Ignoring the higher-order terms, (5) formally describes a scalar field propagating at speed $c^2 = 2\kappa\rho_0/m$. For units in which $c = 1$, it can be rewritten as

$$\mathcal{L}'_{4He} = (1/4\kappa)\eta^{\mu\nu}\partial_\mu \theta \partial_\nu \theta + \cdots, \quad \mu, \nu = 0, 1, 2, \tag{6}$$

where $\eta^{\mu\nu}$ is the (2+1)-dim Minkowski metric. Equation (6) is manifestly Lorentz invariant; in fact, it is formally identical to the Lagrangian density for a massless scalar field propagating in (2+1)-dim Minkowski spacetime. To obtain QED$_3$, consider the first term in (5). Insofar as this represents the kinetic energy density, one can identify a superfluid velocity variable by $v_i \equiv (1/m)\partial_i \theta$. The fact that the macroscopic wave function is unique up to phase then entails that the superflow in a multiply-connected domain is quantized,

$$\oint \vec{v} \cdot d\vec{x} = (1/m) \oint \vec{\partial}\theta \cdot d\vec{x} = (1/m) 2\pi q, \tag{7}$$

around a closed path encircling a "hole," where q is an integer. Such holes may be interpreted as *vortices*—points where the real and imaginary parts of φ_0 vanish.[11] Then (7) entails that the superflow about a vortex is quantized. More important in this context, (7) suggests an analogy with Gauss's Law in which a vortex plays the role of a charge carrier and the superfluid velocity plays the role of the electric field. To further cash out this analogy, note that in two dimensions, the magnetic field is a scalar, whereas the electric field is a 2-vector. This motivates the following identifications:

$$B \equiv \partial_0 \theta = -2\kappa(\rho - \rho_0) \tag{8a}$$

$$E_i \equiv (1/m)\epsilon_{ij}\partial_j \theta = \epsilon_{ij} v_j \tag{8b}$$

in which the magnetic field is identified with the density, and the electric field with the superfluid velocity (here ϵ_{ij} is the skew volume 2-form). Substituting into (6), one obtains the Lagrangian density for sourceless QED$_3$

$$\mathcal{L}'_{QED_3} = (1/4\kappa)\eta^{\mu\sigma}\eta^{\nu\lambda}F_{\sigma\lambda}F_{\mu\nu} + \cdots, \quad \mu, \nu = 0, 1, 2, \tag{9}$$

where $F_{\mu\nu} = \partial_\mu A_\nu - \partial_\nu A_\mu$, with the potential A_μ defined by $E_i = \partial_0 A_i - \partial_i A_0$, $B = \partial_1 A_2 - \partial_2 A_1$. One may further note that (7) entails that the density for "elementary"

---

[11] More precisely, vortices are soliton solutions to the equations of motion of (4) characterized by $\varphi = f(r)e^{i\theta}$, with boundary conditions $f(0) = 0$ and $f(r) \to \psi_0$, as $r \to \infty$. Intuitively, these conditions describe a localized wave with finite energy that does not dissipate over time.

vortices ($q = \pm 1$) is given by $(1/2\pi)\vec{\partial} \times \vec{\partial}\theta$. This can be identified as the 0th component of a vortex current $j_v^\mu = (1/2\pi)\epsilon^{\mu\nu\lambda}\partial_\nu\partial_\lambda\theta$, where $\epsilon^{\mu\nu\lambda}$ is the skew volume 3-form.[12] This vortex current is the dual of the electromagnetic current, insofar as adding a source term $A_\mu j_v^\mu$ to (9) and extremizing with respect to A_μ produces the Maxwell equations with a source.

In summary, we started with the nonrelativistic Lagrangian density (4) for a superfluid ^{4}He film and found that, to lowest order, its EFT takes the form of the relativistic Lagrangian density for (2+1)-dim quantum electrodynamics. This was motivated by the formal similarity between vortex quantization (7) and Gauss's Law. This similarity was exploited in terms of a *duality* transformation under which vortices become the sources of a gauge field formally identical to the Maxwell field. Under a literal interpretation of this dual representation (8), low-energy excitations of a superfluid ^{4}He film take the form of electric and magnetic fields, the former being given by the superfluid velocity, and the latter being given by the superfluid density. Moreover, topological defects (i.e., elementary vortices) take the form of charge-carrying electrons.

3. Renormalizability and Predictability

Historically, the Wilsonian approach to EFTs outlined in section 2.1 had its origin in the development of renormalization group (RG) techniques by Wilson and others in the 1970s (see, e.g., Huggett and Weingard 1995; Cao and Schweber 1993). These techniques were originally developed to study the low-energy behavior of condensed matter systems and were subsequently applied to the problem of renormalization in relativistic quantum field theories; i.e., the appearance of integrals that diverge at high energies when one uses a quantum field theory to calculate the values of observable quantities. This is related to the issue of predictability, insofar as a theory that "blows up" at high energies cannot be used to make high-energy predictions. This section considers the issues of renormalizability and predictability in the context of EFTs. In particular, given that typical EFTs are not renormalizable, how does this affect their ability to make predictions?

In the RG approach to renormalization, the intent is to analyze the behavior of a theory at different energy scales s. One thus uses a scale-dependent momentum cutoff $\Lambda(s)$ as the basis for an initial distinction between high- and low-energy modes, and the heavy modes with respect to an initial energy Λ are then integrated out of the theory. The cutoff is now lowered to $\Lambda(s) = s\Lambda$ and the parameters of the theory are then rescaled to formally restore the cutoff back to Λ. Successive iterations of this procedure generate a flow in the space of parameters of the theory. Scale-dependent parameters can then be classified as relevant (shrinking

---

[12] Note that the form of j_v^μ contracts over skew and symmetric indices; however, it is not identically zero, since for vortices, θ is not a globally defined function.

in the high-energy limit as $s \to \infty$), irrelevant (growing as $s \to \infty$), or marginal (constant under scale transformation).[13] A theory is now said to be renormalizable if it contains no irrelevant parameters. Intuitively, such a theory is cutoff independent, insofar as its parameters become independent of $\Lambda(s)$ in the high-energy limit $s \to \infty$. A nonrenormalizable theory, on the other hand, is one in which there are (scale-dependent) irrelevant parameters. Such parameters cannot be ignored at high energies and thus contribute to ultraviolet divergent integrals.

EFTs can be either renormalizable or nonrenormalizable in the above sense, depending on whether they contain irrelevant terms, although typically the construction outlined in section 2.1 above produces an infinite number of the latter. However, as eluded to in section 2.1, the appearance of an infinite number of irrelevant terms in an EFT need not signal a breakdown in predictability. After Manohar (1997, 322), the effective Lagrangian density associated with (2) can be represented schematically by the sum:

$$\mathcal{L}_{eff} = \mathcal{L}_{\leq D} + \mathcal{L}_{D+1} + \mathcal{L}_{D+2} + \ldots, \tag{10}$$

where $\mathcal{L}_{\leq D}$ contains terms with dimension $\leq D$, $\mathcal{L}_{D+1}$ contains terms with dimension $D+1$, and so on, and, as in section 2.1 above, D is the dimension of spacetime (recall that an operator with dimension δ is deemed irrelevant, relevant, or marginal, depending on whether δ is greater than, less than, or equal to D, respectively). In this sum, each summand contains a *finite* number of terms with coefficients that are powers of the ratio (s/Λ). The first summand consists of a finite number of relevant and/or marginal terms to order zero in (s/Λ) (thus such terms are scale-independent). Each summand thereafter contains a finite number of irrelevant terms to a higher order in (s/Λ) (thus such terms are scale-dependent). A renormalizable Lagrangian density consists of only the first summand, thus when it is used to derive predictions, they will be scale-independent. Nonrenormalizable Lagrangians include irrelevant terms, and predictions derived from them will be scale-dependent. In general, to compute the value of an observable quantity to a given order r in (s/Λ), one should retain terms up to $\mathcal{L}_{D+r}$.

To consider how this analysis of renormalizability relates to predictability, note first that renormalizability, as defined above, is predicated on the property of being scale-independent. A renormalizable theory is independent of energy scale, whereas a nonrenormalizable theory is not. So in order to articulate the relation between renormalizability and predictability, one needs to articulate the relation between scale-independence and predictability. An extreme view might require scale-independence (and hence renormalizability) to be a necessary condition for predictability. The argument might run something like this: if a theory is scale-dependent, then using a cutoff to regulate divergent integrals will be of no help, insofar as (a) the cutoff must be taken to infinity at the end of the day; and (b) doing so will cause scale-dependent terms (which are well-behaved at low-energies

---

[13] In this context, relevant terms are called "super renormalizable," irrelevant terms are called "nonrenormalizable," and marginal terms are called "renormalizable."

with respect to the cutoff) to blow up. One intuition underlying this argument is that the cutoff must, in fact, be taken to infinity at the end of the day; otherwise, we would not end up with the continuum theory we began with. This appears to be the argument underlying Huggett and Weingard's response to their "Problem Number two of understanding renormalization," namely, "why do actual physical theories depend on only a finite number of parameters? A slogan: why is the world renormalisable?" (Huggett and Weingard 1995, 179). Their answer to this problem is the following:

> purely relevant trajectories terminate in the continuum limit—call this "asymptotic safety,". Any irrelevant dependent theories ...do not terminate in this way and are not asymptotically safe. They generate indefinitely long trajectories with ever varying [parameters], either periodically or ever growing. Either way, unphysical singularities are likely. Thus, while asymptotically safe relevant theories are potentially physical, irrelevant theories are not—just the result we hoped for to answer the second question. (Huggett and Weingard 1995, 183)

If one takes "potentially physical" to mean "scale-independent," then Huggett and Weingard's claim that irrelevant (i.e., nonrenormalizable) theories are not potentially physical is correct. However, if one takes "potentially physical" to mean "capable of producing finite predictions," then Huggett and Weingard's claim does not go through: nonrenormalizable theories are capable of producing finite predictions, with the qualification that such predictions are scale-dependent.

Manohar suggests that there is nothing wrong with such a notion of predictability, insofar as there is no reason to expect potentially physical theories to be scale-independent:

> A non-renormalizable theory is just as good as a renormalizable theory for computations, provided one is satisfied with a finite accuracy. ...While exact computations are nice, they are irrelevant. Nobody knows the exact theory up to infinitely high energies. Thus any realistic calculation is done using an effective field theory. (Manohar 1997, 322)

Burgess likewise suggests that the distinction between a renormalizable (viz., scale-independent) theory and a nonrenormalizable (viz., scale-dependent) theory is a matter of degree rather than kind:

> Because...only a finite number of terms in $\mathcal{L}_{eff}$ contributes to any fixed order in $[s/\Lambda]$, and these terms need appear in only a finite number of loops, it follows that only a finite amount of labor is required to obtain a fixed accuracy in observables. Renormalizable theories represent the special case for which

> it suffices to work only to zeroth order in the ratio $[s/\Lambda]$. This can be thought of as the reason why renormalizable theories play such an important role throughout physics. ...Thus, although an effective Lagrangian is not renormalizable in the traditional sense, it nevertheless is predictive in the same way a renormalizable theory is. (Burgess 2007, 349)

The suggestion here is that, to the extent that scale-dependent predictions and scale-independent predictions are both calculated in the same manner (i.e., by applying an appropriate renormalization scheme to divergent integrals), they are of the same kind. Thus, "nonrenormalizable theories are not fundamentally different from renormalizable ones. They simply differ in their sensitivity to more microscopic scales which have been integrated out" (Burgess 1998, 13). This tolerant view of nonrenormalizable theories, and in particular, the predictability of EFTs, has arguably become the norm among physicists. What it implies about the ontology of EFTs, and in particular, the nature of the Wilsonian cutoff Λ, will have to wait until section 5. Section 4 provides a brief review of two explicit ways of deriving predictions using EFTs. My ultimate claim in section 5 will be that the method one chooses to derive predictions from an EFT (i.e., the renormalization scheme one adopts) will influence the possible ways of interpreting it.

4. On Renormalization Schemes and Types of EFTs

As Manohar (1997, 326) observes, knowing the Lagrangian density of a quantum field theory is not enough to calculate the values of observable quantities. To accomplish the latter (using perturbative techniques) requires expanding the Green's function that represents a particular observable quantity in an infinite series in which, typically, divergent integrals appear.[14] This is the problem of renormalization in quantum field theory. Thus, in addition to knowing the Lagrangian density of a quantum field theory, one needs to specify a renormalization scheme. This is a method that specifies (1) a means of regulating divergent integrals and (2) a means of subtracting the associated infinities in a systematic way. There are a number of different methods that accomplish this, two of which are important in the context of interpreting EFTs. The first adopts momentum cutoff regularization and a mass-dependent method of subtraction and is used (at least implicitly) in the Wilsonian approach to constructing EFTs (outlined above in section 2). The second adopts dimensional regularization and a mass-dependent method of subtraction and is associated with what Georgi (1992, 1; 1993, 215) has called "continuum EFTs."

---

[14] A Green's function is a vacuum expectation value of field operators.

4.1 Mass-Dependent Schemes and Wilsonian EFTs

In a Wilsonian EFT, the explicit appearance of the cutoff L that defines the border between the low-energy physics and the high-energy physics suggests employing it as a means to regulate the particular type of divergent integrals that appear in calculations of the values of observable quantities. Given such a divergent integral of the schematic form $\int_0^\infty d^D p \kappa(p)$, where D is the dimension of spacetime and $\kappa(p)$ is a particular function of momentum p, one can insert the cutoff Λ and rewrite the integral as the sum of a finite piece and an infinite piece:

$$\int_0^\Lambda d^D p \kappa(p) + \int_\Lambda^\infty d^D p \kappa(p). \qquad (11)$$

For the types of divergent integrals under consideration, the infinite piece can be absorbed into a redefinition of the parameters of the theory through the introduction of renormalization constants. It turns out that, in this method of regularization, these constants are dependent on the heavy masses that appear in the high-energy theory; hence, the manner in which they are defined is referred to as a *mass-dependent* subtraction scheme.[15]

There are two main advantages of employing this type of renormalization scheme in the context of EFTs. First, it is conceptually consistent with the image of an EFT as a low-energy approximation to a high-energy theory based on a restriction of the latter to a particular energy scale. This scale is explicitly represented by the cutoff Λ, which thus plays a *double role* in designating the appropriate energy scale and in cutting off divergent integrals. The second advantage of using this renormalization scheme is that it guarantees that the Decoupling Theorem holds, given a few other assumptions.

The Decoupling Theorem is due to Appelquist and Carazzone (1975). Hartmann (2001) describes it thusly:

> For two coupled systems with different energy scales m_1 and m_2 ($m_2 > m_1$) and described by a renormalizable theory, there is always a renormalization condition according to which the effects of the physics at scale m_2 can be effectively included in the theory with the smaller scale m_1 by changing the parameters of the corresponding theory. (Hartmann 2001, 283)

This theorem is a formal guarantee of the informal EFT "ideology" of Polchinski (1993, 6), stated above in section 2.1. Hartmann (2001, 284) is careful to note that it requires that there is an underlying high-energy theory that is renormalizable and that different mass scales exist in this theory. Moreover, as Georgi (1992, 3)

---

[15] More precisely, a mass-dependent subtraction scheme is one in which anomalous dimensions and renormalization group β functions explicitly depend on μ/M, where μ is the renormalization scale and M is the heavy mass (Georgi 1993, 221).

indicates, the renormalization condition that the theorem refers to is, in fact, a mass-dependent subtraction scheme.

The above advantages of cutoff regulated, mass-dependent renormalization schemes are balanced by the following disadvantages:

(1) A momentum cutoff regularization method violates Poincaré invariance of the underlying high-energy theory, as well as any gauge invariance it may possess.
(2) Mass-dependent subtraction schemes typically prevent the justification, based on dimensional analysis, that allows one to ignore the potentially infinite number of irrelevant terms in the effective action (2) from being extended from tree-level calculations to higher-order loop corrections. The reason is that in mass-dependent schemes, the simple tree-level dependence of irrelevant terms on orders of $1/\Lambda$ can break down when doing higher-order loop corrections. In particular, in these higher-order corrections, the dependence of irrelevant terms on the cutoff may be of order 1 (in general, such terms have a power law dependence on the cutoff), and thus such terms cannot be ignored (Manohar 1997, 327–328; Pich 1998, 14). Note that this does not prevent loop calculations from proceeding in mass-dependent schemes; rather it makes them more difficult. (Manohar 1997, 329)

4.2 Mass-independent Schemes and Continuum EFTs

To address problem (2), many authors suggest adopting a mass-independent renormalization scheme. In this type of scheme, the dimensional parameter μ (analogous to the momentum cutoff Λ in the cutoff approach) only appears in loop corrections in logarithms, and not powers, thus the relevant integrals are small at scales much smaller than the heavy fields (Manohar 1997, 238; Pich 1998, 15). This allows one to effectively ignore the contributions of irrelevant terms, not only at tree-level as naive dimensional analysis allows but also for higher-order loop corrections as well.

Mass-independent renormalization schemes are typically associated with the method of regulating divergent integrals known as dimensional regularization. This method takes advantage of the mathematical fact that the particular types of divergent integrals that arise in quantum field theoretic calculations, again represented schematically by $\int_0^\infty d^D p \kappa(p)$, will converge for sufficiently small values of D. Formally, one lets $D = 4 - \epsilon$ in the integral (where ϵ is a very small constant), and then analytically continues D to 4. This process picks up poles in D-dimensional momentum space, and these can be absorbed into a redefinition of the parameters of the theory. In this case, such redefinitions are independent of the masses, hence the term *mass-independent subtraction scheme.*

In the context of EFTs, there are two main advantages of employing a mass-independent renormalization scheme based on dimensional regularization. First, dimensional regularization respects Poincaré and gauge invariance. Second, as indicated above, mass-independent renormalization schemes allow one to truncate the effective action (2) to a finite list of terms, not only for tree-level calculations but also for higher-order loop calculations. However, it turns out that this simplification comes at the cost of having terms that explicitly include the heavy fields appear in this finite list (see, e.g., Burgess 2004, 19). This has the following consequences:

(1′) Many authors consider mass-independent schemes to violate the "spirit" of an EFT, to the extent that the latter is based on the notion of a cutoff, below which the physics is explicitly described by only the light fields (Burgess 2004, 19; Burgess 2007, 343; Polchinski 1993, 5).

(2′) Perhaps more importantly, the presence of heavy field terms in an effective action employing a mass-independent renormalization scheme prevents the application of the Decoupling Theorem (Georgi 1993, 225; Manohar 1997, 329). As mentioned above in section 3.1, the latter holds only for mass-dependent schemes.

It turns out that problem (2′) can be addressed by inserting decoupling by hand into an EFT that employs a mass-independent scheme, but this requires a slight reconceptualization of the nature of an EFT. This results in what Georgi (1992, 1; 1993, 215) refers to as a "continuum EFT."

How can an EFT be constructed without initial appeal to a cutoff? Briefly, for top-down constructions, the initial momentum splitting of the fields in a Wilsonian EFT and the integration over the heavy modes is replaced in a continuum EFT by the following steps (after Georgi 1993, 228; see also Burgess 2004, 19–20; Burgess 2007, 344):

(I′) Start with a dimensionally regularized theory with Lagrangian density $\mathcal{L}_H(\chi,\phi) + \mathcal{L}(\phi)$ at a large scale s, where $\mathcal{L}(\phi)$ describes the light fields and $\mathcal{L}_H(\chi,\phi)$ describes everything else (where χ are the heavy fields of mass M). Now evolve the theory to lower scales using the renormalization group: This allows you to go from scale s to scale $s - ds$ without changing the content of the theory.

(II′) When s gets below M, the effective theory is changed to a new one without the heavy fields: $\mathcal{L}(\phi) + \delta\mathcal{L}(\phi)$, where $\delta\mathcal{L}(\phi)$ encodes a "matching correction" that includes any new nonrenormalizable interactions that may be required. The matching correction is made so that the physics of the light fields is the same in the two theories at the boundary $s = M$. To explicitly calculate $\delta\mathcal{L}(\phi)$, one expands it in a complete set of local operators in the same manner that the expansion (2) for Wilsonian EFTs is performed:

$$\mathcal{L}_{\text{eff}} = \mathcal{L}(\phi) + \sum_n \delta\mathcal{L}^n(\phi). \tag{12}$$

Dimensional analysis can now be applied to determine the scaling behavior of the terms in (12), in the same way it is applied in Wilsonian EFTs. Again, in the case of a continuum EFT, this analysis is valid not just for tree-level calculations but also for higher-order loop calculations as well.

To summarize and compare, in the construction of a Wilsonian EFT, the heavy fields are first integrated out of the underlying high-energy theory and the resulting Wilsonian effective action is then expanded in a series of local operator terms. The cutoff Λ in a Wilsonian EFT plays a double role: first, through the definition of the heavy and light fields, in explicitly demarcating the low-energy physics from the high-energy physics; and second, in regulating divergent integrals in the calculation of observable quantities. In the construction of a continuum EFT, the heavy fields are initially left alone in the underlying high-energy theory, which is first evolved down to the appropriate energy scale. The continuum EFT is then constructed by completely removing the heavy fields from the high-energy theory, as opposed to integrating them out; and this removal is compensated for by an appropriate matching calculation (this latter is ultimately responsible for the appearance of heavy modes into the operator expansion (12)). In a continuum EFT, the first role of the Wilsonian cutoff Λ is played by the renormalization scale s that demarcates the low-energy physics from the high-energy physics. The second role of the Wilsonian cutoff is dropped, the procedure of dimensional regularization taking its place.

This last observation suggests the motivation for Georgi's phrase "continuum EFT." In a Wilsonian EFT, the regularization of divergent integrals is performed by restricting the range of momentum variables in integrals over momentum space. The Fourier transform equivalent of this procedure is a restriction of the range of coordinate variables in integrals over coordinate space. Hence regularization in a Wilsonian EFT is analogous to placing a high-energy continuum theory on a discrete lattice, and this is the reason why momentum cutoff regularization violates Poincaré invariance. In contrast, in a "continuum EFT," the regularization of divergent integrals is performed by calculating them in a continuous spacetime of dimension D (the Fourier transform equivalent of D-dimensional momentum space). This is the reason why dimensional regularization does not violate Poincaré invariance.

5. Ontological Implications

Different renormalization schemes ultimately all agree on the values of physical quantities. In particular, both mass-dependent and mass-independent schemes will, at the end of the day, agree on all empirically measured quantities.[16] Thus a Wilsonian EFT for a given physical system is empirically equivalent to that system's continuum EFT. On the other hand, the fact that these types of EFT place different emphasis on the nature of the cutoff suggests they can be interpreted as

---

[16] Burgess (2004, 20) explains this in the following way: "the difference between the cutoff- and dimensionally regularized low-energy theory can itself be parameterized by appropriate local effective couplings within the low-energy theory. Consequently, any regularization-dependent properties will necessarily drop out of final physical results, once the (renormalized) effective couplings are traded for physical observables."

telling us different things about the world. I now consider some of the implications this has for debates over the ontological status of EFTs.

5.1 Decoupling and Quasi-Autonomous Domains

Cao and Schweber emphasize the role of the Decoupling Theorem in understanding the ontological significance of EFTs:

> Thus, with the decoupling theorem and the concept of EFT emerges a hierarchical picture of nature offered by QFT [quantum field theory], one that explains why the description at any one level is so stable and is not disturbed by whatever happens at higher energies, and thus justifies the use of such descriptions. (Cao and Schweber 1993, 64)
>
> In this picture, the [physical world] can be considered as layered into quasi-autonomous domains, each layer having its own ontology and associated 'fundamental law'. (Cao and Schweber 1993, 72)

They further suggest that EFTs entail "an antifoundationalism in epistemology and an antireductionism in methodology" (Cao and Schweber 1993, 69). Huggett and Weingard (1995, 187) interpret this as a view that "holds that nature is described by a genuinely never-ending tower of theories, and that the competing possibilities of unification and new physics should be abandoned."

Hartmann (2001, 298) claims that "Cao and Schweber's talk of quasi-autonomous domains rests on the validity of the decoupling theorem," and then rightly points out that the latter is not necessarily valid in all cases in which EFTs exist. In particular, it requires the existence of an underlying high-energy renormalizable theory with different mass scales. This vitiates Cao and Schweber's antifoundationalism and/or antireductionism, if they are taken to entail that there is no underlying theory.

On the other hand, Castellani (2002) appears to endorse an aspect of the "quasi-autonomous domain" interpretation: "The EFT approach in its extreme version provides a level structure ('tower') of EFTs, each theory connected with the preceding one (going 'up' in the tower) by means of the [renormalization group] equations and the matching conditions at the boundary" (Castellani 2002, 263).

It appears that the participants in this debate are talking past each other, having implicitly adopted different notions of an EFT. Cao and Schweber, in particular, are sometimes ambiguous on what concept of EFT they are employing. On the one hand, in their emphasis on the Decoupling Theorem and momentum cutoff regularization, they appear to adopt a Wilsonian notion of an EFT. For instance, in the context of discussing the ontological significance of momentum cutoff regularization, they observe that "the following can be stated for other regularization schemes ...but not for dimensional regularization, which is more formalistic and irrelevant to the point discussed here" (Cao and Schweber 1993, 92 n. 17). On the other hand, immediately before introducing the EFT-inspired hierarchical picture of nature, they describe an EFT in the following terms: "The EFT can be obtained by deleting all heavy fields from the complete renormalizable theory and suitably redefining the coupling constants, masses, and the scale of the Green's functions, using the renormalization group equations" (Cao and Schweber 1993, 64).

This appears to be a description of the construction of a continuum EFT as outlined in section 3.2. Hartmann (2001) makes it evident in his critique of Cao and Schweber that he implicitly has adopted the Wilsonian notion of an EFT. Finally, Castellani implicitly adopts a continuum notion of an EFT, describing its construction as one based on matching conditions (see, e.g., Castellani 2002, 262).

Note that the notion of EFT one adopts should be important in this debate. The Decoupling Theorem was proven in the context of Wilsonian EFTs of a particular kind; namely, those for which there exists an underlying renormalizable high-energy theory with different mass scales. Thus, if by "EFT" Cao and Schweber mean "Wilsonian EFT," then Hartmann's critique goes through: Wilsonian EFTs do not, by themselves, support an ontology of quasi-autonomous domains. On the other hand, the Decoupling Theorem fails for continuum EFTs, but, arguably this does not prevent them from supporting a well-defined notion of quasi-autonomous domains. This is because in continuum EFTs, decoupling is inserted "by hand" in the form of matching calculations. Thus, if by "EFT" Cao and Schweber mean "continuum EFT," then, arguably, Hartmann's critique does not go through: continuum EFTs are, by themselves, capable of supporting an ontology of quasi-autonomous domains. Hence, provided by "EFT," Castellani means "continuum EFT," her endorsement of such an ontology is justified.[17]

5.2 Realistic Interpretations of the Cutoff

In typical expositions of EFTs, emphasis is placed on a realistic interpretation of the cutoff. Many authors claim that such a realistic interpretation is what separates the contemporary concept of an EFT from older views of renormalization (Hartmann 2001, 282; Castellani 2002, 261; Grinbaum 2008, 37). Under these older views, a cutoff, when it occurred in accounts of QFTs, only played a role as a regulator of integrals and was taken to infinity at the end of the day (if this resulted in a finite theory, then the theory was deemed to be renormalizable). The contemporary concept of an EFT, so the story goes, is based on viewing the cutoff realistically in a different role; namely, as a means of demarcating low-energy physics from high-energy physics. By now it should be obvious from the discussion in section 4 that this standard account is only half the story; it, perhaps unfairly, privileges

---

[17] Grinbaum (2008, 40), notably, makes a distinction between the "strict" form of decoupling associated with the Appelquist-Carazzone theorem, and a "milder" empirical decoupling thesis, which evidently is to be associated with the matching calculations involved in the construction of continuum EFTs.

Wilsonian EFTs and the double role the cutoff plays in their construction, over continuum EFTs.

This is not to say that a realistic interpretation of the cutoff cannot be made in the context of continuum EFTs. Recall that in continuum EFTs, the role that the Wilsonian cutoff Λ plays in demarcating low-energy physics from high-energy physics is played by a scaling variable s (i.e., the renormalization scale that appears in the renormalization group equations). Certainly this scaling variable can be realistically interpreted, perhaps as a basis for an ontology of quasi-autonomous domains; and it might even be referred to as a cutoff, in this context. The important point is that, in a continuum EFT, this scaling variable does not play the second role that the Wilsonian cutoff Λ plays; namely, as a regulator of divergent integrals.

Thus to realistically interpret the cutoff in an EFT could mean one of two things:

(a) The Wilsonian regulator Λ should be realistically interpreted.
(b) The constant that demarcates low-energy physics from high-energy physics (given by Λ in Wilsonian EFTs and by a particular value of s in continuum EFTs) should be realistically interpreted.

As the discussion at the end of section 4.2 suggests, adopting (a) might motivate an ontology in which space is discrete: momentum cutoff regularization is analogous to placing a continuum theory on a discrete lattice. But such an ontology is not forced upon us by a realistic interpretation of the cutoff. Simply put, (b) does not entail (a). One can adopt a realistic interpretation of the cutoff in a continuum EFT and, at the same time, an ontology in which spacetime is continuous.

This conclusion affects part of a recent debate over interpretations of quantum field theory (QFT). Wallace (2006) adopts an interpretation of QFT in which a cutoff is inserted and realistically interpreted and justifies it by appealing to features of EFTs (see, e.g., 43). Fraser (2009) criticizes this "cutoff variant" of QFT in the following way: "If the cutoffs are taken seriously, then they must be interpreted realistically; that is, space is really discrete and of finite extent according to the cutoff variant of QFT" (Fraser 2009, 552). Fraser suggests this makes cutoff QFT unsatisfactory: if the cutoff is not taken seriously, then cutoff QFT reduces to "infinitely renormalized QFT," which is the standard textbook account in which cutoffs, when they appear, are taken to infinity in renormalization schemes. An appeal to Haag's theorem then indicates this textbook account is inconsistent. On the other hand, if the cutoff is taken seriously, then cutoff QFTs avoid Haag's theorem (which, among other things, requires Poincaré invariance); but they entail space is discrete and finite, and "nobody defends the position that QFT provides evidence that space is discrete and the universe is finite" (Fraser 2009, 552).[18]

Once again, being clear on the type of EFT one adopts on which to base the cutoff variant of QFT will make a difference in this debate. If by "cutoff QFT" one means "Wilsonian EFT," then Fraser's critique, arguably, goes through. However, if by "cutoff QFT" one means "continuum EFT," then the above argument does not go

---

[18] See also Huggett and Weingard (1995, 178) for similar intuitions.

through: continuum EFTs support an ontology in which spacetime is continuous.[19] Thus, provided one can demonstrate that continuum EFTs avoid Haag's theorem, a cutoff version of QFT based on continuum EFTs is a viable alternative to the "formal variant" (i.e., axiomatic QFT) that Fraser (2009, 538) advocates.

5.3 EFTs and Approximation

Finally, two further examples of how the distinction between types of EFTs is important for issues of interpretation involve the notions of idealization and approximation. The fact that Wilsonian EFTs support an ontology in which space is discrete suggests to Fraser (2009, 564) that the cutoff variant of QFT is an indispensable idealization: "[It] ...is an idealization in the sense that the possible worlds in which QFT is true are presumed to be worlds in which space is continuous and infinite," and "[t]his idealization is indispensable insofar as it is not possible to remove the cutoffs entirely" (since this would turn the cutoff variant into the infinitely renormalized variant). But, again, continuum EFTs do not idealize space as discrete, hence a version of cutoff QFT based on continuum EFTs is not an idealization of this type.

Relatedly, Castellani (2002, 260, 263) suggests that EFTs are "intrinsically approximate and context-dependent." An EFT, under this view, is an approximation of an underlying high-energy theory and is valid only within a specified energy range. Now, arguably, Cao and Schweber's ontology of quasi-autonomous domains is intended in part to address claims of this type. Under Cao and Schweber's view, an EFT describes a quasi-autonomous domain by means of a complete description of phenomena within a given energy range, independent for the most part of descriptions at higher or lower energies (Cao and Schweber 1993, 64, are careful to explain how high-energy effects do make themselves present in relevant and marginal terms of an effective Lagrangian density). Thus, the discussion in section 5.1 entails that EFTs need not be interpreted as intrinsically approximate, provided one adopts continuum EFTs as the object of one's interpretation.

6. EFTs and Emergence

In the example in section 2.3 above, the EFT of a superfluid ^4He film took the form (to lowest order) of quantum electrodynamics in $(2 + 1)$ dimensions. The duality transformations (8) suggested that, at low energies, the density of a superfluid ^4He

---

[19] This suggests that the categories of empirically equivalent variants of QFT that Fraser (2009, 538) identifies should be expanded. Her "cutoff QFT" category might be split into "cutoff regularized QFT" and "dimensionally regularized QFT."

film behaves like a magnetic field, its velocity behaves like an electric field, and vortices behave like charge-carrying electrons. This example of a relativistic EFT of a condensed matter system, and others like it, have suggested to some physicists that novel phenomena (fields, particles, symmetries, spacetime, etc.) emerge in the low-energy limit of these systems.[20] On the other hand, in the physics literature, references to emergence are typically not associated with EFTs of relativistic QFTs.[21] In the philosophy of physics literature, the converse is true: philosophers of physics have considered notions of emergence related to EFTs of relativistic QFTs, but have paid little attention to emergence in EFTs of condensed matter systems.[22] In this section I take it as a given that the formal nature of an EFT is identical in both contexts and consider the extent to which notions of emergence are applicable, regardless of context.

Consider, first, the view from philosophy of physics: Cao and Schweber, for instance, associate the antireductionism of their quasi-autonomous domains interpretation of EFTs with a notion of emergence: "taking the decoupling theorem and EFT seriously would entail considering the reductionist ...program an illusion, and would lead to its rejection and to a point of view that accepts emergence, hence to a pluralist view of possible theoretical ontologies" (Cao and Schweber 1993, 71). Likewise, Castellani (2002, 263) suggests that the EFT approach "provides a level structure of theories where the way in which a theory emerges from another ...is in principle describable by using RG [Renormalization Group] methods and matching conditions at the boundary." On the other hand, Castellani argues that the EFT approach does not imply antireductionism, insofar as antireductionism is to be associated with the denial of some type of intertheoretic relation: "The EFT schema, by allowing definite connections between theory levels, actually provides an argument against the basic antireductionist claim" (Castellani 2002, 265). Note that while Cao and Schweber take emergence to be descriptive of ontologies (properties, objects, etc.), Castellani suggests emergence be viewed as a relation between theories. In both cases, however, the emphasis is on autonomy. Cao and Schweber's emergent ontologies are restricted to quasi-autonomous domains, each described by a distinct EFT. Castellani's emergent theories stand in "definite connections" with each other, but assumedly not so definite as to warrant the label of reduction.

This section will only be concerned with the extent to which the intertheoretic relation between a top-down EFT and its high-energy theory supports a notion of emergence; thus the approach taken will be to view emergence as a relation between theories. Batterman (2002, 115) has suggested that, under

---

[20] For instance, Zhang (2004, 669) reviews "examples of emergence in condensed matter systems" that take the form of relativistic EFTs, including the QED$_3$ case. These and other examples in the condensed matter literature are discussed in Bain (2008).

[21] See Castellani (2002) for a review of the 1960s?1970s debates between solid-state physicists and particle physicists over the concepts and status of reduction and emergence.

[22] Both philosophers and physicists have considered notions of emergence in condensed matter systems exhibiting spontaneously broken symmetries. But, as section 6.1 below suggests, this context is distinct from the context in which emergence might be associated with EFTs.

a received view in the philosophical literature, such a relation holding between an emergent theory T' and an underlying theory T can mean any or all of the following:

(a) The phenomena of T' cannot be reduced to T.
(b) The phenomena of T' cannot be predicted by T.
(c) The phenomena of T' are causally independent of those of T.
(d) The phenomena of T' cannot be explained by T.

The initial task of this section will be to consider the extent to which the intertheoretic relation between an EFT and its high-energy theory supports these notions of autonomy.

6.1 The EFT Intertheoretic Relation

Sections 2.1 and 4.2 above described the general steps in the construction of a Wilsonian and a continuum EFT, respectively. Although differing in their details, these steps have the following general form: (I) One first identifies and then systematically eliminates high-energy degrees of freedom; and then (II) one expands the resulting effective Lagrangian density (or effective action) in terms of local operators. The intertheoretic relation defined by this procedure has one very important characteristic in the context of a discussion of notions of emergence; namely, its *relata* are distinct theories.

To see this, consider the following consequences of Steps (I) and (II):

(1) First, the low-energy degrees of freedom of the EFT are typically formally distinct from the high-energy degrees of freedom. This suggests they admit distinct ontological interpretations (recall, for instance, some of the examples from section 2: pions versus quarks, quasiparticles versus electrons, and electric and magnetic fields versus ^4He atoms).
(2) Second, the EFT Lagrangian density typically is formally distinct from the high-energy Lagrangian density.

Manohar makes (2) clear in the context of the Fermi EFT of the weak force:

> It is important to keep in mind that the effective theory is a different theory from the full theory. The full theory of the weak interactions is a renormalizable field theory. The effective field theory is a non-renormalizable field theory, and has a different divergence structure from the full theory. (Manohar 1997, 327)

As another example of (2), consider the EFT of a superfluid ^4He film described in section 2.3. Above a critical temperature, the system consists of a nonrelativistic normal liquid. As the temperature is lowered below the critical value, a phase transition occurs, accompanied by a spontaneously broken symmetry, and the system

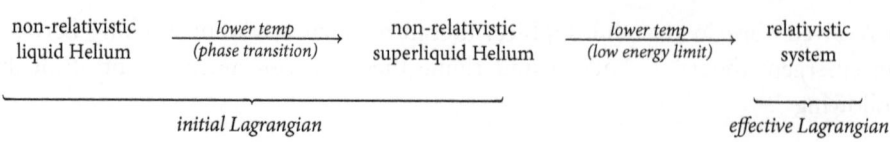

Figure 6.1 The relation between the initial Lagrangian and the effective Lagrangian for superfluid helium.

enters the superfluid phase. If the temperature is lowered further, its constituents can be described in terms of a relativistic EFT. Importantly, both the normal liquid and the superfluid, as well as the phase transition and the spontaneously broken symmetry, are all encoded in a single Lagrangian density (4). All of these states and processes can thus be said to be described by a single theory. On the other hand, the low-energy relativistic system is encoded in the effective Lagrangian density (9), which is sufficiently formally distinct from (4) to warrant viewing it as a different theory (see figure 6.1, after Bain 2008, 313).

Note that the claim that an EFT and its high-energy theory are distinct theories is not intended to be based simply on the fact that there is a formal distinction between their respective Lagrangian densities. It is not the case that, in general, there is a 1-1 correspondence between Lagrangian densities and theories. For instance, simply changing the interaction term in a Lagrangian density does not, arguably, change the theory it is intended to represent (consider the theory of Newtonian particle dynamics applied to different interactions). However, in the case of an EFT and its high-energy theory, the difference between the two Lagrangian densities is substantial enough to warrant the assumption that one is dealing with two distinct theories. In the ^4He case, the contrast is between a nonrelativistic Lagrangian density and a relativistic Lagrangian density: whereas in Manohar's example, the contrast is between a renormalizable Lagrangian density and a nonrenormalizable Lagrangian density. Moreover, as (1) indicates, in both cases, the dynamical variables of the EFT are distinct from those of the high-energy theory.

With the above proviso in mind, in the Lagrangian formalism, a difference in the form of the Lagrangian density entails a difference in the Euler-Lagrange equations of motion for the relevant dynamical variables. One might thus argue that an EFT T' is *derivationally independent* from its associated high-energy theory T, insofar as a specification of the equations of motion of T (together with pertinent initial and/or boundary conditions) will fail to specify solutions to the equations of motion of T'. More generally, the steps involved in the construction of an EFT typically involve approximations and heuristic reasoning. In the case of Wilsonian EFTs, recall that the initial integral over the heavy fields typically involves a saddle-point approximation about the free theory, and even before such an approximation can be constructed, in both the Wilsonian and continuum EFT cases, the task of identifying the relevant high-energy variables must be accomplished. This suggests that, in general, it will be difficult, if not impossible, to reformulate the steps involved in the construction of an EFT (of either the Wilsonian or continuum types) in the form of a derivation.

6.2 Senses of Autonomy

Given that the intertheoretic relation between an EFT and its associated high-energy theory is characterized by derivational independence, what does this suggest about the sense in which the former is autonomous from the latter?

(a) *Reductive Autonomy*. One might first argue that the relation between an EFT and its high-energy theory cannot be characterized by notions of reduction based on derivability. On the standard (Nagelian) account of reduction, for instance, a necessary condition for a theory T' to reduce to another T is that T' be a *definitional extension* of T (see, e.g., Butterfield and Isham 1999, 115). This requires first that T and T' admit formulations as deductively closed sets of sentences in a formal language (i.e., it assumes a *syntactic* conception of theories), and second that an extension T^* of T can be constructed such that the theorems of T' are a subset of the theorems of T^* (i.e., it requires that T' is a *sub-theory* of T^*). Formally, T^* is constructed by adding to T a definition of each of the nonlogical symbols of T'. One might now argue that this cannot be done in the case of a high-energy theory and its EFT. As noted above, in the Lagrangian formalism, differences in the Lagrangian densities representing two theories entail differences in the theories' Euler-Lagrange equations of motion. If one adopts the view that such equations represent the theory's dynamical laws, then the dynamical laws of an EFT and its high-energy theory are different, and a difference in dynamical laws entails a difference in theorems derived from these laws. Thus, an EFT is not a sub-theory of its high-energy theory; hence, one cannot say that an EFT reduces to its high-energy theory on this view of reduction.[23]

Note that the above argument does not depend essentially on a syntactic conception of theories. For instance, under a semantic conception of theories, a typical claim is that a theory reduces to another just when models of the first can be embedded in models of the second. This will not suffice to reduce an EFT to its high-energy theory so long as the embedding is required to preserve dynamical laws (and if it is not, then it is unclear whether the term "reduction" for such an embedding is appropriate[24]).

(b) *Predictive Autonomy*. Predictive autonomy between an EFT and its high-energy theory would seem to be another consequence of derivational independence. Given that the relation between an EFT and its high-energy theory T cannot be described in terms of a derivation in which T is implicated, the phenomena that the EFT describes cannot be derived, and hence predicted, on the basis of T.

---

[23] Butterfield and Isham (1999, 122) observe that the standard definition of supervenience can be characterized in terms of an *infinitistic* definitional extension; thus neither can it be said that an EFT *supervenes* (in this sense) on its associated high-energy theory.

[24] Admittedly this assumes that, whatever else reduction amounts to, it is essentially nomic in nature.

(c) *Causal Autonomy.* Whether derivational independence of an EFT from its high-energy theory entails causal independence will depend on one's concept of causation. To demonstrate the causal independence of an EFT from its high-energy theory, one would have to provide an account of how the phenomena governed by the EFT are not implicated in the causal mechanisms associated with the relevant high-energy phenomena. The example of superfluid ^{4}He films is instructive here. Under one interpretation, this EFT suggests that low-energy "ripples" of a superfluid ^{4}He film behave like relativistic electric and magnetic fields. Insofar as ripples in a substrate are implicated in the causal mechanisms that govern the substrate, this suggests causal links between the phenomena of the EFT and the high-energy theory. On the other hand, if one's view of causation is such that the existence of a causal relation requires the existence of a nomic connection (embodied in a dynamical law, say), then one might argue that to the extent to which an EFT and its high-energy theory are nomically independent (in the sense, perhaps, of possessing distinct dynamical laws), they are causally independent, too.

(d) *Explanatory Autonomy.* Whether the phenomena described by an EFT can be explained in terms of the high-energy theory will obviously depend on the notion of explanation one adopts. Arguably, explanatory autonomy will obtain on any account of explanation that requires the *explanandum* to be the product of a derivation in which the *explanans* is implicated; and to the extent to which an EFT is causally independent of its high-energy theory T, its phenomena cannot be causally explained by T.

6.3 Emergence and Limiting Relations

Evidently there is room to maneuver in addressing the question of whether the intertheoretic relation between an EFT and its high-energy theory can be described in terms of a notion of emergence, at least if such a notion is related to standard accounts of reduction, prediction, causation, and/or explanation. On the other hand, Batterman (2002) has offered a nonstandard account of emergence based on the failure of a limiting relation between two theories. This section considers its applicability in the context of an EFT and its high-energy theory.

Batterman's notion of emergence is associated with the failure of what he refers to as the "physicists' sense" of reduction. Under this notion, a "coarse" theory T_c reduces to a "more refined" theory T_f provided a limit of the schematic form $\lim_{\varepsilon \to 0} T_f = T_c$ can be shown to exist, where e is a relevant parameter.[25] The novelty of emergent properties, according to Batterman, is "a result of the singular nature of the limiting relationship between the finer and coarser theories that are relevant to the phenomenon of interest" (2002, 121). This singular nature, when it exists, is indicative of the existence of a real "physical singularity," according to Batterman.

---

[25] Batterman's (2002, 78) example is the reduction of special relativity to classical mechanics in the limit $v/c \to 0$, where v is the velocity of a given physical system and c is the speed of light.

Thus: "The proposed account of emergent properties has it that genuinely emergent properties, as opposed to 'merely' resultant properties, depend on the existence of physical singularities" (Batterman 2002, 125).

For Batterman, then, there are two necessary conditions for the existence of an emergent property in the context of a fundamental (more refined) theory T and a less fundamental (coarse) theory T':

(a) The physicists' notion of reduction must hold between T and T'; i.e., there must be a limiting relation between T and T'.
(b) The limiting relation must fail in the context with which the emergent property is identified; in particular, there must be a physical singularity associated with the emergent property.

As an example of two theories that satisfy these conditions, Batterman considers thermodynamics (TD) and statistical mechanics (SM). With some qualification, it is possible to define an intertheoretic relation between the TD and SM descriptions of a physical system in terms of the thermodynamic limit $N, V \to \infty$ while $N/V = $ constant, where N and V are the number of particles and volume of the system (Batterman 2002, 123). This limit fails for a thermodynamic system at a critical point at which it undergoes a phase transition. At such a point, the correlation length associated with the system (roughly, the measure of the correlation between spatially separated states) becomes infinite. For Batterman, this is an example of a physical singularity, and he is thus motivated to identify properties associated with phase transitions as emergent properties. (In the ^4He example of section 2.3, such properties would correspond to the highly correlated phenomena associated with superfluidity.)

Importantly, the intertheoretic relation between TD and SM in this context can be modeled by renormalization group (RG) techniques. The thermodynamic limit generates an RG flow in the parameter space of a TD system. This is analogous to how a scale-dependent momentum cutoff L(s) generates an RG flow in the parameter space of an RQFT, as described above in section 3. A TD system at a critical point is then represented by a fixed point in its RG flow. This is a point at which the parameters of the theory remain unchanged under further RG rescaling; i.e., they become scale invariant. As Batterman explains, at a critical point,

> there is a loss of a characteristic length scale. This leads to the hypothesis of scale invariance and the idea that the large scale features of a system are virtually independent of what goes on at a microscopic level. In the thermodynamic case we see that the bulk properties of the thermodynamic systems are independent of the detailed microscopic, molecular constitution of the physical system. (Batterman 2005, 243)

In the discussion in section 3 above, a fixed point in the RG flow associated with an RQFT is the explicit indication that the theory is scale-independent and hence renormalizable. For such a theory, the low-energy properties of its constituents are independent of its detailed high-energy constitution. As in the TD/SM case, this also represents the loss of a characteristic scale, in this case an energy scale. Thus, the features of a renormalizable theory are independent of what goes on at large energies. And just as in the TD/SM case, scale invariance in a renormalizable RQFT is associated with the existence of physical singularities. In this case, a physical singularity is associated with an observable quantity (like a scattering cross section) that is represented by a divergent Green's function.

This analogy between the intertheoretic relation between TD and SM, on the one hand, and the relation between the low-energy and high-energy sectors of a renormalizable RQFT, on the other, suggests that Batterman's notion of emergence might be applicable in the latter case. To make this analogy more explicit, consider the following summaries of the relevant features of these examples:

Example 1: T = statistical mechanics (SM). T' = thermodynamics (TD). The limiting relation is the thermodynamic limit: $N, V \to \infty$ while N/V = constant.

(i) The thermodynamic limit fails at a fixed point in the associated RG flow in the sense that, at a fixed point, there is no link between the bulk TD properties and the microscopic SM properties. This is a manifestation of scale independence.

(ii) A physical singularity associated with the failure of the thermodynamic limit is a diverging correlation length. Emergent properties are properties associated with the system at the fixed point.

Example 2: T = renormalizable continuum RQFT. T' = cutoff-regulated RQFT. The limiting relation is the continuum limit: $\Lambda(s) \to \infty$. More precisely, to further the analogy with Example 1, the continuum limit can be given schematically by $\Lambda(s) \to \infty$, [bare parameters] $\to \infty$, while [renormalized parameters] = [bare parameters]/$\Lambda(s)$ = constant.[26]

(i) The continuum limit fails at a fixed point in the associated RG flow in the sense that, at a fixed point, there is no link between the low-energy cutoff theory and the high-energy continuum theory. This is a manifestation of scale independence.

(ii) A physical singularity associated with the failure of the continuum limit is represented by a diverging Green's function. Emergent properties are properties associated with the system at a fixed point. In principle, these are properties constructed out of relevant operators.

---

[26] See Stone (2000, 204) for the condensed matter context. The bare parameters are the parameters of the theory before rescaling is performed to restore the cutoff back to its initial value after one iteration of the RG transformations. The renormalized parameters are the rescaled parameters.

Fraser has pointed out the following disanalogy between Examples 1 and 2: "whereas the description of a system as containing an infinite number of particles furnished by [statistical mechanics] is taken to be false, the description of space as continuous and infinite that is furnished by QFT with an infinite number of degrees of freedom is taken to be true" (Fraser 2009, 565). Thus, in Example 1, the limiting relation is taken to be an idealization, whereas in Example 2 it is not. It would appear, however, that this disanalogy is not relevant to Batterman's notion of emergence, to the extent that the latter is associated with the necessary conditions (a) and (b) above. Condition (a) requires simply that a limiting relation exist, but it says nothing about the status of this relation; in particular, whether it is taken to be an idealization or not.

This suggests that the properties associated with the values of observable quantities constructed from the Green's functions of a renormalizable RQFT are emergent in Batterman's sense. The question now is: To what extent does Example 2 offer insight into the nature of emergence in the context of EFTs?

Two observations appear to be relevant in this context. First, not all EFTs are associated with renormalizable high-energy theories. For those that are not, Batterman's notion of emergence cannot be supported without further ado. Second, even in the case of an EFT with an associated renormalizable high-energy theory, the EFT will typically be formally distinct from the latter. This is a result of the second step in the construction of an EFT (for both Wilsonian and continuum versions) in which the effective Lagrangian density is constructed via a local operator expansion. In an RG analysis of a renormalizable continuum RQFT, this step is replaced with a parameter-rescaling procedure by means of which the low-energy cutoff Lagrangian density is transformed back into the initial form of the original Lagrangian density. The upshot is that T and T' in Example 2 are formally identical, whereas an EFT and its high-energy theory are not. Simply put, the cutoff-regulated RQFT of Example 2 is not the same mathematical object as an EFT associated with a renormalizable high-energy theory.

This suggests that further work needs to be done if Batterman's notion of emergence is to be applied in the context of an EFT and its high-energy theory.

7. CONCLUSION

Two general conclusions seem appropriate from this review of effective field theory. First, the discussion in sections 4 and 5 suggests that, in order to understand how EFTs can be interpreted, one needs to understand the methods that physicists use in applying them. By focusing attention on different renormalization schemes that practicing physicists actually employ, one can discern two types of empirically equivalent EFTs—Wilsonian EFTs and continuum EFTs. These are nontrivial examples of empirically equivalent theories insofar as, in the context of a given high-energy theory, they make the same low-energy predictions, but they suggest different ontologies. Continuum EFTs support an ontology of quasi-autonomous domains, whereas Wilsonian EFTs do not. Continuum EFTs support an ontology that includes a continuous spacetime, whereas Wilsonian EFTs require space to be discrete and finite. These features of Wilsonian EFTs have contributed to the view that EFTs in general engage in idealization and are inherently approximate. The fact that continuum EFTs do not engage in such idealizations suggests that EFTs do admit interpretations in which they are not considered inherently approximate.

The second conclusion one may draw from this review is that, if one desires to associate (some aspect of) the intertheoretic relation between an EFT and its (possibly hypothetical) high-energy theory with a notion of emergence, then more work has to be done. In the context of standard accounts of emergence, the relevant feature of the EFT intertheoretic relation is that it supports a notion of derivational autonomy (i.e., an EFT cannot be said to be a derivation of its associated high-energy theory). But just how derivational autonomy can be linked with a notion of emergence will depend on such things as how one articulates additional concepts such as reduction, explanation, and/or causation. Section 6 also demonstrated that the EFT intertheoretic relation does not support Batterman's (2002) more formal notion of emergence based on the failure of a limiting relation between two theories. Such a failure of a limiting relation does occur between a renormalizable high-energy RQFT and a cutoff-regulated theory obtained from it by renormalization group techniques, but this is a different context than the one in which an EFT is obtained from a high-energy theory. Again, the relevant property of the EFT intertheoretic relation here is that it is a relation between formally distinct, derivationally autonomous, theories.

References

Appelquist, T., and J. Carazzone (1975). Infrared singularities and massive fields. *Physical Review* D11: 2856–2861.
Bain, J. (2008). Condensed matter physics and the nature of spacetime. In *The Ontology of Spacetime*, Vol. 2, ed. D. Dieks, 301–329. Amsterdam: Elsevier Press.
Batterman, R. (2002). *The devil in the details: Asymptotic reasoning in explanation, reduction, and emergence.* Oxford: Oxford University Press.
———. (2005). Critical phenomenon and breaking drops: Infinite idealization in physics. *Studies in History and Philosophy of Modern Physics* 36: 225–244.
Burgess, C. P. (1998). "An Ode to Effective Lagrangians". Available online at arXiv: hep-ph/9812470v1.
Burgess, C. P. (2004). Quantum gravity in everyday life: General relativity as an effective field theory. *Living Reviews in Relativity.* http://www.livingreviews.org/lrr-2004–5.

———. (2007). An introduction to effective field theory. *Annual Review of Nuclear and Particle Science* 57: 329–367.

Butterfield, J., and C. Isham (1999). On the emergence of time in quantum gravity. In *The Arguments of Time*, ed. J. Butterfield, 111–168. Oxford: Oxford University Press.

Campbell-Smith, A. and N. Mavromatos (1998). "Effective Gauge Theories, the Renormalization Group, and High-Tc Superconductivity," *Acta Physica Polonica B* 29: 3819–3870.

Cao, T., and S. Schweber (1993). The conceptual foundations and the philosophical aspects of renormalization theory. *Synthese* 97: 33–108.

Castellani, E. (2002). Reductionism, emergence, and effective field theories. *Studies in History and Philosophy of Modern Physics* 33: 251–267.

Dobado, A., A. Gomez-Nicola, A. L. Maroto, and J. P. Pelaez (1997). *Effective Lagrangians for the standard model*. Berlin: Springer.

Fraser, D. (2009). Quantum field theory: Underdetermination, inconsistency, and idealization. *Philosophy of Science* 76: 536–567.

Georgi, H. (1992). Thoughts on effective field theory. *Nuclear Physics B* (Proceedings Supplements) 29B, C: 1–10.

———. (1993). Effective field theory. *Annual Review of Nuclear and Particle Science* 43: 209–252.

Grinbaum, A. (2008). On the eve of the LHC: Conceptual questions in high-energy physics. philsci-archive.pitt.edu, deposited on June 27, 2008. http://philsci-archive.pitt.edu/archive/00004088/.

Hartmann, S. (2001). Effective field theories, reductionism and scientific explanation. *Studies in History and Philosophy of Modern Physics* 32: 267–304.

Huggett, N., and R. Weingard (1995). The renormalization group and effective field theories. *Synthese* 102: 171–194.

Manohar, A. (1997). Effective field theories. In *Perturbative and nonperturbative aspects of quantum field theory*, Lecture Notes in Physics, Vol. 479/1997, 311–362. Berlin: Springer. Available online at arXiv: hep-ph/9606222.

Neubert, M. (2006). Effective field theory and heavy quark physics. In *Physics in $D \geq 4$. TASI 2004. Proceedings of the Theoretical Advanced Study Institute in Elementary Particle Physics*, ed. J. Terning, C. Wagner, and D. Zeppenfeld, 149–194. Singapore: World Scientific. Available online at arXiv: hep-ph/0512222v1.

Pich, A. (1998). Effective field theory. arXiv: hep-ph/9806303v1.

Polchinski, J. (1993). Effective field theory and the Fermi surface. In *Proceedings of 1992 Theoretical Advanced Studies Institute in Elementary Particle Physics*, ed. J. Harvey and J. Polchinski. Singapore: World Scientific. arXiv: hep-th/92110046.

Rothstein, I. (2004). TASI Lectures on effective field theories. arXiv: hep-ph/0308266v2.

Schakel, A. (2008). *Boulevard of broken symmetries: Effective field theories of condensed matter*. Singapore: World Scientific.

Stone, M. (2000). *The physics of quantum fields*. Berlin: Springer.

Wallace, D. (2006). In defence of naivete: The conceptual status of Lagrangian quantum field theory. *Synthese* 151: 33–80.

Wen, X.-G. (2004). *Quantum field theory of many-body systems*. Oxford: Oxford University Press.

Weinberg, S. (1996). *The Quantum Theory of Fields, Volume 2: Modern Applications*. Cambridge: Cambridge University Press.

Zee, A. (2003). *Quantum field theory in a nutshell.* Princeton: Princeton University Press.

Zhang, S.-C. (2004) To see a world in a grain of sand. In *Science and ultimate reality: Quantum theory, cosmology and complexity,* ed. J. D. Barrow, P. C. W. Davies, and C. L. Harper, 667–690. Cambridge: Cambridge University Press.

CHAPTER 7

THE TYRANNY OF SCALES

ROBERT BATTERMAN

1. Introduction

In this essay I will focus on a problem in physics and applied mathematics. This is the problem of modeling across scales. Many systems, say a steel girder, manifest radically different, dominant behaviors at different length scales. At the scale of meters, we are interested in its bending properties, its buckling strength, etc. At the scale of nanometers or smaller, it is composed of many atoms, and features of interest include lattice properties, ionic bonding strengths, etc. To design advanced materials (such as certain kinds of steel), materials scientists must attempt to deal with physical phenomena across 10+ orders of magnitude in spatial scales. According to a recent (2006) NSF research report, this "tyranny of scales" renders conventional modeling and simulation methods useless as they are typically tied to particular scales (Oden 2006, p. 29). "Confounding matters further, the principal physics governing events often changes with scale, so the models themselves must change in structure as the ramifications of events pass from one scale to another" (Oden, pp. 29–30). Thus, even though we often have good models for material behaviors at small and large scales, it is often hard to relate these scale-based models to each other. Macroscale models represent the integrated effects of very subtle factors that are practically invisible at the smallest, atomic, scales. For this reason it has been notoriously difficult to model realistic materials with a simple bottom-up-from-the-atoms strategy.

I would especially like to thank Mark Wilson for many helpful discussions. Conversations with Penelope Maddy, Julia Bursten, and Nic Fillion are also much appreciated. This research was supported, in part, by a grant from the Social Sciences and Humanities Research Council of Canada.

The widespread failure of that strategy forced physicists interested in overall macro-behavior of materials toward completely top-down modeling strategies familiar from traditional continuum mechanics.[1]

A response to the problem of the "tyranny of scales" would attempt to exploit our rather rich knowledge of intermediate micro- (or meso-) scale behaviors in a manner that would allow us to bridge between these two dominant methodologies. Macroscopic scale behaviors often fall into large common classes of behaviors such as the class of isotropic elastic solids, characterized by two phenomenological parameters—so-called elastic moduli. Can we employ knowledge of lower scale behaviors to understand this universality—to determine the moduli and to group the systems into classes exhibiting similar behavior? This is related to engineering concerns as well: Can we employ our smaller scale knowledge to better design systems for optimal macroscopic performance characteristics?

The great hope that has motivated a lot of recent research into so-called "homogenization theory" arises from a conviction that a "between-scales" point of view, such as that developed by Kadanoff, Fisher, and Wilson in the renormalization group approach to critical phenomena in fluids and magnets, may very well be the proper methodological strategy with which to begin to overcome the tyranny of scales. A number of philosophers have recently commented on the renormalization group theory, but I believe their focus has overlooked what is truly novel about the methodological perspective that the theory employs.

Philosophical discussions of the applicability of mathematics to physics have not, in my opinion, paid sufficient attention to contemporary work on this problem of modeling across scales. In many instances, philosophers hold on to some sort of ultimate reductionist picture: whatever the fundamental theory is at the smallest, basic scale, it will be sufficient in principle to tell us about the behavior of the systems at all scales. Continuum modeling on this view represents an *idealization*—as Feynman has said, "a smoothed-out imitation of a really much more complicated microscopic world" (Feynman, Leighton, and Sands 1964, p. 12). Furthermore, the suggestion is that such models are in principle eliminable.

There is a puzzle however. Continuum model equations such as the Navier-Stokes equations of hydrodynamics or the equations for elastic solids work despite the fact that they completely (actually, almost completely—this is crucial to the discussion below) ignore small scale or atomistic details of various fluids. The recipe (I call it "Euler's continuum recipe") by which we construct continuum models is safe: if we follow it, we will most always be led to empirically adequate successful equations characterizing the behavior of systems at the macroscopic level. Why? What explains the safety of this recipe? Surely this requires an answer. Surely, the answer must have something to do with the physics of the modeled systems at smaller scales. If such an answer cannot be provided, we will be left with a kind of skepticism: without such an answer, we cannot expect anything like a unified conception of applied mathematics' use of continuum idealizations.[2] If an

---

[1] For related discussions, see Mark Wilson's forthcoming *Physics Avoidance and Other Essays*.
[2] See Maddy (2008) for a forceful expression of this skeptical worry.

answer is forthcoming, then we have to face the reductionist picture mentioned above. Will such an answer—an answer that explains the robustness and safety of employing continuum modeling—support the view that continuum models are mere conveniences, only pragmatically justified, given the powerful simplifications gained by replacing large but finite systems with infinite systems? As noted, many believe that a reductionist/eliminitivist picture is the correct one. I maintain that even if we can explain the safety and robustness of continuum modeling (how this can be done is the focus of this essay), the reductionist picture is mistaken.

It is a mistaken picture of how science works. My focus here is on a philosophical investigation that is true to the actual modeling practices of scientists. (I am not going to be addressing issues of what might be done in principle, if not in practice.) The fact of the matter is that scientists do not model the macroscale behaviors of materials using pure bottom-up techniques.[3] I suggest that much philosophical confusion about reduction, emergence, atomism, and antirealism follows from the absolute choice between bottom-up and top-down modeling that the tyranny of scales apparently forces upon us. As noted, recent work in homogenization theory is beginning to provide much more subtle descriptive and modeling strategies. This new work calls into question the stark dichotomy drawn by the "do it in a completely bottom-up fashion" folks and those who insist that top-down methods are to be preferred.

The next section discusses the proposal that the use of continuum idealizations present no particular justificatory worries at all. Recent philosophical literature has focused on the role of continuum limits in understanding various properties of phase transitions in physical systems such as fluids and magnets. Some authors, particularly Jeremy Butterfield (2011) and John Norton (2011), have expressed the view that there are no particularly pressing issues here: the use of infinite limits is perfectly straightforwardly justified by appeal to pragmatic considerations. I argue that this view misses an important difference in methodology between some uses of infinite limits and those used by renormalization group arguments and homogenization theory.

In section 3, I present an interesting historical example involving nineteenth century attempts to derive the proper equations governing the behavior of elastic solids and fluids. A controversy raged throughout that century concerning the merits of starting from bottom-up atomic description of various bodies in trying to arrive at empirically adequate continuum equations. It turns out that the bottom-up advocates lost the debate. Correct equations apparently could only be achieved by eschewing all talk of atomic or molecular structure, advocating instead a top-down approach supplemented, importantly, with experimentally determined data. In section 4, I formulate the tyranny of scales as the problem, just mentioned, of trying to understand the connection between recipes for modeling at atomic scales (Euler's discrete recipe) and Euler's continuum recipe appropriate for continuum models. Finally, I present a general discussion of work on homogenization that

---

[3] Those who think that the renormalization group provides a bottom-up explanation of the universality of critical phenomena, e.g. Norton (2011), are mistaken, as we shall see below.

provides at least the beginning of an answer to the safety question and to the problem of bridging scales between the atomic and the continuum. This research can be seen as allaying skeptical worries about a unified applied mathematical methodology regarding the use of continuum idealizations of a certain kind.

2. Steel Beams, Scales, Scientific Method

Let us consider the steel girder in a bit more detail. In many engineering applications steel displays linear elasticity. This is to say that it obeys Hooke's Law—its strain is linearly proportional to its stress. One phenomenological parameter related to its stress/strain (i.e., stiffness) properties is Young's modulus appearing in the equations of motion for solids, as well as in equilibrium and variational equations. At scales of 1 meter to 10^{-3} meters, say, the steel girder appears to be almost completely homogeneous: zooming in with a small microscope will reveal nothing that looks much different. In fact, there appears to be a kind of local scale invariance here.[4] So for behaviors that take place within this range of scales, the steel girder is well-modeled or represented by the Navier-Cauchy equations:

$$(\lambda + \mu)\nabla(\nabla \cdot \mathbf{u}) + \rho \nabla^2 \mathbf{u} + \mathbf{f} = 0.[5] \qquad (1)$$

The parameters λ and μ are the "Lamé" parameters and are related to Young's modulus.

Now jump from this large-scale picture of the steel to its smallest *atomic* scale. Here the steel, for typical engineering purposes, is an alloy that contains iron and carbon. At this scale, the steel exhibits highly ordered crystalline lattice structures. It looks nothing like the homogeneous girder at the macroscales that exhibits no crystalline structure. Somehow between the lowest scale of crystals on a lattice and the scale of meters or millimeters, the low-level ordered structures must disappear. But that suggests that properties of the steel at its most basic, atomic level cannot, by themselves, determine what is responsible for the properties of the steel at macroscales. I will discuss this in more detail below.

In fact, the story is remarkably complex. It involves appeal to various geometrical properties that appear at *microscales* intermediate between the atomic and the macro,[6] as well as a number of other factors such as martensitic transformations.[7] The symmetry breaking is effected by a combination of point defects, line defects,

---

[4] "Local" in the sense that the invariance holds for scales of several orders of magnitude but fails to hold if we zoom in even further, using x-ray diffraction techniques, for example.

[6] I call these intermediate scales "microscales" and the structures at these scales "microstructures" following the practice in the literature, but it may be best to think of them as "mesoscopic."

[7] These latter are transformations that take place under cooling when a relatively high symmetry lattice such as one with cubic symmetry loses symmetry to become tetragonal. Some properties of steel girders therefore depend crucially on dynamical changes that take place at scales in between the atomic and the macroscopic (Phillips 2001, p. 547–8).

slip dislocations, and higher dimensional wall defects that characterize interfacial surfaces. All of these contribute to the *homogenization* of the steel we see and manipulate at the macroscale. And, of course, in engineering contexts the macro features (bending properties, for example) are the most important—we do not want our buildings or bridges to collapse.

2.1 Reduction, Limits, Continuum Models

A simpler case than steel involves trying to connect the finite statistical mechanical theory of a fluid at the atomic scale to the thermodynamic continuum theory at macro scales.[8] The relationship between statistical mechanics and thermodynamics has received a lot of attention in the recent philosophical literature. Debates about intertheoretic reduction, its possibility, and its nature have all appealed to examples from thermodynamics and statistical mechanics. Many of these discussions, in the recent literature, have focused on the nature and potential emergence of phase transitions in the so-called thermodynamic limit[9] (Butterfield 2011a; Menon and Callender 2012; Belot 2005; Bangu 2009). What role does the thermodynamic limit play in connecting theories? What role does it play in understanding certain particular features of thermodynamic systems? It will be instructive to consider the role of this limit in a more general context than that typical of the literature. This is the context in which we consider the generic problem of upscaling from atomic to laboratory scales, as in the case of the steel girder discussed above. In doing this, I hope it will become clear that many of the recent philosophical discussions miss crucial features of the methodology of applying limits like the thermodynamic limit.

Before turning to the debates about the use of the thermodynamic limit and the justification of using infinite limits to understand the goings on in finite systems, I think it is worthwhile to step back to consider, briefly, some general issues about theory reduction. As mentioned above, many philosophers and physicists tacitly (and sometimes explicitly) maintain some sort of in principle reductionist point of view. I do not deny that maybe in some as yet to be articulated sense there may be an in principle bottom-up story to be told. However, appeals to this possibility ignore actual practices and furthermore are never even remotely filled out in any detail. Typically the claim is simply: "The fundamental theory (whatever it is, quantum mechanics, quantum field theory, etc.), because it is fundamental (whatever that ultimately means), *must* be able to explain/reduce everything."

Nagel's seminal work (1961) considered the reduction of thermodynamics to statistical mechanics to be a straightforward and paradigm case of intertheoretic

---

[8] Though simpler than the case of understanding how atomic aspects of steel affect its phenomenological properties, this is, itself, a difficult problem for which a Nobel prize was awarded.
[9] This is the limit in which the number of particles N in a system approaches infinity in such a way that the density remains constant—the volume has to go to infinity at the same time as the number of particles.

reduction. On his view, as is well known, one derives the thermodynamic laws from the laws of statistical mechanics employing so-called bridge laws connecting terms/predicates appearing in the reduced theory with those appearing in the reducing theory.[10] In several places I have argued that this Nagelian strategy and its variants fail for many cases of so-called reduction (Batterman 1995, 2002). I have argued that a limiting sense of reduction in which, say, statistical mechanics "reduces to" thermodynamics in an appropriate limit (if it does) provides a more fruitful conception of intertheoretic reduction than the Nagelian strategies where the relation seems to go the other way around: on the Nagelian strategies one has it that thermodynamics reduces to statistical mechanics, in the sense of deductive derivation. There are a number of reasons for thinking the nonNagelian, "limiting," sense of reduction is a superior sense of reduction. For one, there is the difficulty of finding the required definitional connections that the bridge laws are meant to embody.[11] But in addition, the kinds of connections established between theories by taking limits do not appear to be expressible as definitional extensions of one theory to another. In many cases, the limits involved are singular, and even when they are not, the use of mathematical limits invokes mathematics well beyond that expressible in the language of first order logic—a characteristic feature of Nagel's view of reduction and of its neoNagelian refinements.

Despite these arguments a number of authors have recently tried to argue that reduction should be understood in Nagelian terms; that is, as the definitional extension of one theory to another. Jeremy Butterfield and Nazim Bouatta, for example,

> ...take reduction as a relation between theories (of the systems concerned). It is essentially deduction; though the deduction is usually aided by adding appropriate definitions linking two theories' vocabularies. This will be close to endorsing the traditional philosophical account [Nagel's], despite various objections levelled against it. The broad picture is that the claims of some worse or less detailed (often earlier) theory can be deduced within a better or more detailed (often later) theory, once we adjoin to the latter some appropriate definitions of the terms in the former. ...So the picture is, with D standing for the definitions: $T_b \& D \Rightarrow T_t$. In logicians' jargon T_t is a *definitional extension* of T_b. (Butterfield and Bouatta 2011)

In the current context the more *basic, better* theory (statistical mechanics) is T_b and the reduced, *tainted* theory (thermodynamics) is T_t.[12]

---

[10] See Batterman 2001 and 2006 for surveys of this and more sophisticated strategies.

[11] In the present example, it is hard indeed to see how to define or identify a nonstatistical quantity such as temperature or pressure in thermodynamics with a necessarily statistical quantity or set of quantities in the reducing statistical mechanics. (See Sklar 1993, Chapter 9.)

[12] I believe that the use of the evaluative terms "better," "worse," and "tainted" reflects an inherent prejudice against nonreductionist points of view. In particular, as one of the issues is whether a more detailed (atomic)

Butterfield and Bouatta obviously are not moved by the objections to the Nagelian scheme that I briefly mentioned above. I suggest though, as we delve a bit more deeply into the examples of phase transitions and of the steel girder, that we keep in mind the question of whether the continuum account of the bending behavior of the steel can be reduced to the theory of its atomic constituents in the sense that we can derive that continuum behavior from the "better," "more detailed," and "later" atomic theory. Even if we extend the logicians' sense of deduction (as definitional extension) beyond that of first order logic so as to include inferences involving mathematical limits, will such a deduction/reduction be possible?

So the real question, as both of these examples employ continuum limits, concerns why the use of such limits is justified. The debate about the justification of the use of infinite limits and, ultimately, about reduction concerns whether the appeal to limits can in the end be eliminated. It is a pressing debate, because no party thinks that at the most fundamental level, the steel girder is a continuum. And no party thinks that a tea kettle boiling on the stove contains an infinite number of molecules. What justifies our employing such infinite idealizations in describing and explaining the behaviors of those systems?

For Butterfield there is a "Straightforward Justification" for the use of infinite limits in physical modeling.

> This Justification consists of two obvious, very general, broadly instrumentalistic, reasons for using a model that adopts the limit $N = \infty$: mathematical convenience, and empirical adequacy (up to a required accuracy). So it also applies to *many* other models that are almost never cited in philosophical discussions of emergence and reduction. In particular, it applies to the many classical continuum models of fluids and solids, that are obtained by taking a limit of a classical atomistic model as the number of atoms N tends to infinity (in an appropriate way, e.g. keeping the mass density constant). (2011, p. 1080)

He continues by emphasizing two "themes" common to the use of many different infinite models:

> The first theme is abstraction from finitary effects. That is: the mathematical convenience and empirical adequacy of many such models arises, at least in part, by abstracting from such effects. Consider (a) how transient effects die out as time tends to infinity; and (b) how edge/boundary effects are absent in an infinitely large system.
>
> The second theme is that the mathematics of infinity is often much more convenient than the mathematics of the

theory is really better for explanatory, predictive, and modeling concerns, this way of speaking serves to block debate before it can get started.

> large finite. The paradigm example is of course the convenience of the calculus: it is usually much easier to manipulate a differentiable real function than some function on a large discrete subset of $\mathbb{R}$ that approximates it. (2011, p. 1081)

The advantages of these themes are, according to Butterfield, twofold. First, it may be easier to know or determine the limit's value than the actual value primarily because of the removal of boundary and edge effects. Second, in many examples of continuum modeling we have a function defined over the finite collection of atoms or lattice sites that oscillates or fluctuates and so can take on many values. In order to employ the calculus we often need to "have each *value* of the function defined as a limit (namely, of values of another function)" (pp. 1081–82). Butterfield seems to have in mind the standard use of *averaging* over a "representative elementary volume" (REV)[13] and then taking limits $N \to \infty$, volume going to zero, so as to identify a continuum value for a property on the macroscale. In fact, he cites continuum models of solids and fluids as paradigm examples:

> For example, consider the mass density varying along a rod, or within a fluid. For an atomistic model of the rod or fluid, that postulates N atoms per unit volume, the average mass-density might be written as a function of both position x within the rod or fluid, and the side-length L of the volume L^3 centred on x, over which the mass density is computed: $f(N, x, L)$. Now the point is that for fixed N, this function is liable to be intractably sensitive to x and L. But by taking a continuum limit $N \to \infty$, with $L \to 0$ (and atomic masses going to zero appropriately so that quantities like density do not "blow up"), we can define a continuous, maybe even differentiable, mass-density function $\rho(x)$ as a function of position—and then enjoy all the convenience of the calculus.
>
> So much by way of showing in general terms how the use of an infinite limit $N = \infty$ can be justified—but not mysterious! At this point, the general philosophical argument of this paper is complete! (p. 1082)

So for Butterfield most of the discussions concerning the role, and particularly the justification, of the use of the thermodynamic limit in the debates about phase transitions have generated a lot of hot air. The justification, on his view, for employing such limits in our modeling strategies is largely pragmatic—for the sake of convenience. In addition, there is, as he notes, the further concern that the use of such limits be empirically adequate—getting the phenomena right to within appropriate error bounds. Much of his discussion concerns showing that the use of such

---

[13] I have taken this terminology from Hornung (1997, p. 1).

limits can most always be shown to be empirically adequate in this sense (Butterfield 2011). Unfortunately, I think that sometimes things are more subtle than the straightforward justification admits. In fact, there are good reasons to think that the use of the thermodynamic limit in the context of the renormalization group (RG) explanation of critical phenomena—one of the cases he highlights—fails to be justified by his own criteria. It is a different methodology, one that does not allow for the sort of justificatory story just told. The straightforward story as described above cannot be told for the RG methodology for the simple reason that that story fails to be empirically adequate in those contexts.

One can begin to understand this by making a distinction between what might be called "ab initio" and "post facto" computational strategies. Butterfield's remarks about the mass density in a rod (say a steel girder) in one sense appear to endorse the ab initio strategy. Consider a model of the rod at the scale of atoms where the atoms lock together on a crystal lattice. The limit averaging strategy has us increase the size of the lattice until we have, in effect, a perfect crystal of infinite extent. This lets us ignore boundary effects as he notes. The limiting average density that we arrive at using this ab initio (atomic only) strategy will actually be grossly incorrect at higher scales. This is because, at higher micro (meso) scales real iron contains many structures such as dislocations, grain boundaries, and other metastabilities that form within its mass and that energetically allow local portions of the material at these higher scales to settle into stable modes with quite different average densities. See figure 7.1. These average densities will be quite different than the ab initio calculations from the perfect crystal. What special features hidden within the quantum chemistry of iron bond allow those structures to form? We really don't know. But until we gain some knowledge of how those structures emerge, we will not be able accurately to determine computationally the bulk features of steel girders in the way described. Values for Young's modulus and fracture strength that we may try determine on the basis of this ab initio reasoning will be radically at variance with the actual measured values for real steel.

On the other hand, if we possessed a realistic model of steel at all length scales, then we could conceivably define a simple average over a representative volume (at a much higher scale than the atomic). But this post facto calculation would rely upon complete data about the system at all scales. No limits would be involved whatsoever. Perhaps some super genius may someday in principle propose an incredibly detailed model of iron bonding that would allow the calculation of the macro parameters like Young's modulus in a kind of ab initio mode imagined by Butterfield, but such a hypothetical project is certainly not the aim of the RG techniques that are under consideration here.

Such ab initio calculations provide wrong answers because they cannot "see" the energetically allowed local structural configurations that steel manifests at larger scales. On the other hand, if we are investigating materials that (for whatever reason) display nice scaling relationships across some range of scales (as steel does for scales 8–10 orders of magnitude above the atomic), then we will be able to employ RG type techniques to determine the various universality classes (characterized by

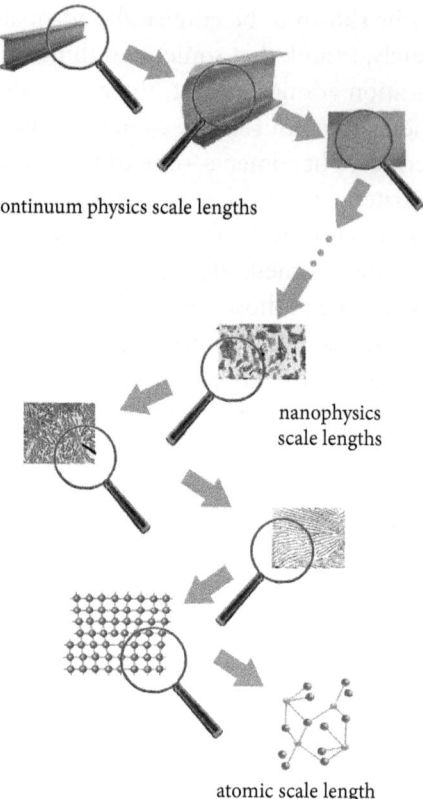

Figure 7.1 Microstructures of steel

the phenomenological parameters—Young's modulus, e.g.) into which they must fall. Thus the RG methodology, unlike the ab initio REV averaging strategy, provides a rationale for evading extreme bottom-up computations so as to gain an understanding of why steel, for example, only requires a few effective parameters to describe its behavior at macroscales.

While there surely are cases in which averaging is appropriate, and the straightforward justification may be plausible, there are other cases, as I have been arguing, in which it is not. In order to further elucidate this point, I will say a bit about what the RG argument aims to do. I will then give a very simple example of why one should, in many instances, expect the story involving averaging over a representative volume element (REV) to fail. In fact, the failure of this story is effectively the motivation behind Wilson's development of the distinct RG methodology. More generally, if our concern is to understand why continuum models such as the Navier-Cauchy equation are safe and robust, the straightforward justification will miss what is most crucial.

I have discussed the RG in several publications (Batterman 2002; 2005; 2011). Butterfield (2011) and Butterfield and Bouatta (2011) present concise descriptions as well. For the purposes here, as noted earlier, I am going to present some of the details with a different emphasis than these other discussions have provided.

In particular, I want to stress the role of the RG as part of a methodology for upscaling from a statistical theory to a hydrodynamic/continuum theory. In so doing, I follow a suggestion of David Nelson (2002, pp. 3–4) who builds on a paper of Ken Wilson (1974). The suggestion is that entire phases of matter (not just critical phenomena) are to be understood as determined by a "fixed point" reflecting the fact that "universal physical laws [are] insensitive to microscopic details" (2002, p. 3). Specifically, the idea is to understand how details of the atomic scale physics get encoded (typically) into a few phenomenological parameters that appear in the continuum equations governing the macroscopic behavior of the materials. In a sense, these phenomenological parameters (like viscosity for a fluid, and Young's modulus for a solid) characterize the appropriate "fixed point" that defines the class of material exhibiting universal behavior despite potentially great differences in microscale physics.

Let us consider a ferromagnet modeled as a set of classical spins σ_i on a lattice—the Ising model. In this model, neighboring spins tend to align in the same direction (either up or down: $\sigma_i = \pm 1$). Further, we might include the effect of an external magnetic field B. Then the Hamiltonian for the Ising model is given by

$$H[\{\sigma_i\}] = -J \sum_{\langle i,j \rangle} \sigma_i \sigma_j + \mu B \sum_i \sigma_i,$$

with the first sum over nearest neighbor pairs of spins, μ is a magnetic moment. A positive value for the coupling constant J reflects the fact that neighboring spins will tend to be aligned, both up or both down.

For ferromagnets we can define an order parameter—a function of the net magnetization for the system—whose derivative exhibits a discontinuity or jump at the so-called critical or Curie temperature, T_c. Above T_c, in zero magnetic field, the spins are not correlated due to thermal fluctuations and so the net magnetization is zero. As the system cools down to the Curie temperature, there is singularity in the magnetization (defined as a function of the free energy). (See figure 7.2.) The

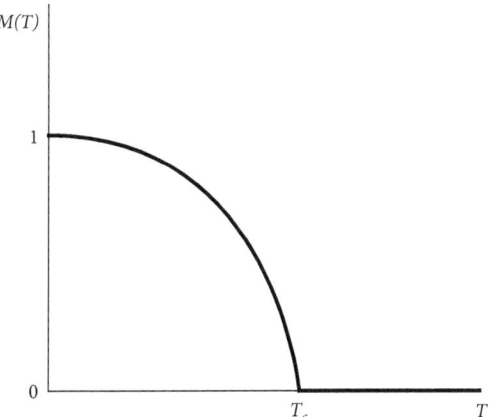

Figure 7.2 Spontaneous magnetization at T_c

magnetization exhibits power law behavior near that singularity characterized by the relation

$$M \propto |t|^\beta,$$

where t is the reduced temperature $t = \frac{T-T_c}{T_c}$. It is a remarkable fact that physically quite distinct systems—magnets modeled by different Hamiltonians, and even fluids (whose order parameter is the difference between vapor and liquid densities in a container)—all exhibit the same power law scaling near their respective critical points: The number β is universal and characterizes the phenomenological behavior of a wide class of systems at and near criticality.[14]

The RG provides an explanation for this universal behavior; and in particular, it allows one theoretically to determine the value for the exponent β. For the 3-dimensional Ising model, that theoretical value is approximately .33. Experimentally determined values for a wide class of fluids and magnets are found in the range .31–.36. So-called "mean field" calculations predict a value of .5 for β (Wilson 1974, p. 120). A major success of the RG was its ability to correct mean field theory and yield results in close agreement with experiment. In a mean field theory, the order parameter M is defined to be the magnetic moment felt at a lattice site due to the average over all the spins on the lattice. This averaging ignores any large-scale fluctuations that might (and, in fact, are) present in systems near their critical points. The RG corrects this by showing how to incorporate fluctuations at all length scales, from the atomic to the macro, that play a role in determining the macroscopic behavior (specifically the power law dependence—$M \propto |t|^\beta$) of the systems near criticality. In fact, near criticality the lattice system will contain "bubbles" (regions of correlated spins—all up or all down) of all sizes from the atomic to the system size. As Kadanoff notes, "[f]rom this picture we conclude that critical phenomena are connected with fluctuations over all length scales between ξ [essentially the system size] and the microscopic distance between particles" (Kadanoff 1976, p. 12).

So away from criticality, below the critical temperature, say, the lattice systems will look pretty much *homogeneous*.[15] For a system with $T \ll T_c$ in figure 7.2 we would have relatively large correlated regions of spins pointing in the same direction. There might be only a few insignificantly small regions where spins are correlated in the opposite direction. This is what is responsible for there being a positive, nonzero, value for M at that temperature. Now suppose we were interested in describing a large system like this away from criticality using the continuum limit as understood by Butterfield above. We would choose a representative elementary volume of radius L around a point x. The volume is small with respect to the system size ξ, but still large enough to contain many spins. Next we would average the quantity $M(N, x, L)$ over that volume and take the limits $N \to \infty$, $L \to 0$ so as to obtain the proper

---

[14] See Kadanoff (2000) and Batterman (2002, 2005, 2011) for details.
[15] Systems above the critical temperature will also appear homogeneous as the spins will be uncorrelated, randomly pointing up and down.

continuum value and so that we would be able to model the actually finite collection of spins using convenient continuum mathematics.

But near the critical temperature (near T_c) the system will look heterogeneous—exhibiting a complicated mixture of two distinct phases as in figure 7.3. Now we face a problem. In fact, it is the problem that effectively undermined the mean field approach to critical phenomena. The averaging method employing a representative elementary volume element misses what is most important. For one thing, we will need to know how to weight the different phases as to their import for the macroscopic behavior of the system. In other words, were we to perform the REV averaging, all of the physics of the fluctuations responsible for the coexisting bubbles of up spins and bubbles of down spins would be ignored.

Here is a simple example to see why this methodology will often fail for heterogeneous systems (Torquato 2002, p. 11). Consider a composite material consisting of equal volumes of two materials, one of which is a good electrical conductor and one of which is not. A couple of possible configurations are shown in figure 7.4.

Suppose that the dark, connected phase is the good conductor. If we were to proceed using the REV recipe, then, because the volume fractions are the same, we would grossly underestimate the bulk conductivity of the material in the left configuration and grossly underestimate its bulk insulating capacities in the right configuration. REV averaging treats only the volume fraction and completely misses microstructural details that are relevant to the bulk (macroscale) behavior of the material. In this simple example, the microstructural feature that is relevant is the topological connectedness of the one phase vs. the other—that is, the details about the boundaries between the two phases. Note that the fact that boundaries play an important role serves to undermine the first "theme" of the Straightforward Justification for the use of limits; namely, that taking the limits enable us to remove edge and boundary effects. To the contrary, these can and do play very important roles in determining the bulk behavior of the materials.

One might object that all one needs to do to save the REV methodology would be to properly weight the contribution of the different phases to the overall average. But this is not something that one can do a priori or through ab initio calculations

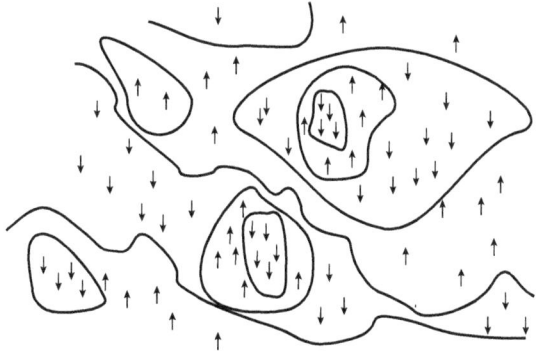

Figure 7.3 Bubbles within bubbles within bubbles ...(after Kadanoff 1976, pp. 11–12)

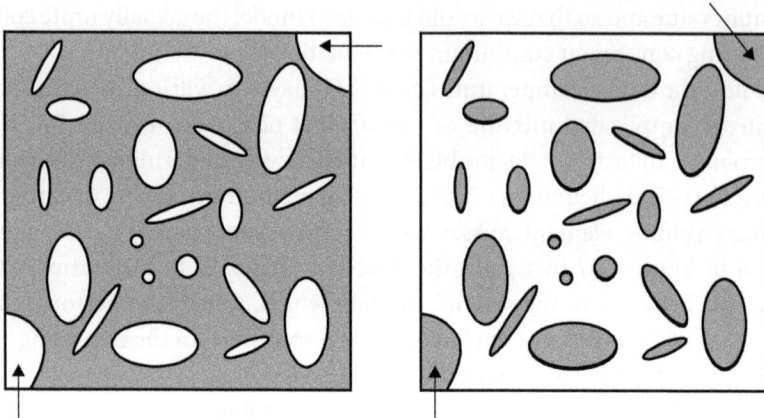

Figure 7.4 50–50 volume mixture

appealing to details and properties of the individual atoms at the atomic scale. Even worse, note the partial blobs at the corners marked by the arrows in figure 7.4. How large are the complete blobs of which they are a part? We do not know because the limited scale of the window (size L of the REV) does not allow us to "see" what is happening at large scales. It is entirely possible (and in the case of critical phenomena actually the case) that these partial blobs will be part of larger connected regions only visible at greater scale lengths. They may be dreaded *invaders from a higher scale*.[16] If such invaders are present, then we have another reason to be wary of limiting REV averaging methods—we will grossly fail to estimate the effective conductivity of the material at macroscales. On the other hand, if we have some nice scaling data about the behavior of material of the sort exploited by the RG, we may well gain enough of a handle on the material's overall behavior to place its conductivity in a firm universality class with other materials that scale in similar ways.

As noted above, in more complicated situations, such as the steel girder with which we began, microstructural features include mesoscale dislocations, defects of various kinds, and martensitic transformations. If we engaged in a purely bottom-up lattice view about steel, paying attention only to the structures for the pure crystal lattice, then we would get completely wrong estimates for its total energy, for its average density, and for its elastic properties. The relevant Hamiltonians require terms that simply do not appear at the smallest scales.[17]

The upshot, then, is that the straightforward justification for the use of infinite limits will miss exactly what is important for understanding what is going on for systems at and near criticality. There, they no longer appear homogeneous across a large range of scales. If we are to try to connect (and thereby extract) correct phenomenological macroscopic values for appropriate parameters (e.g., β) we need to consider

---

[16] Thanks to Mark Wilson for the colorful terminology!
[17] See (Phillips 2001).

structures that exist at scales greater than the fundamental/basic/atomic. Again, what does this say about the prospects for an overall reductionist understanding of the physics of systems viewed at macroscales?

The RG considers such intermediate scales by including in the calculations the effects of fluctuations or equivalently, the fact that bubbles within bubbles of different phases appear near criticality. *We need methods that tell us how to homogenize heterogeneous materials.* In other words, to extract a continuum phenomenology, we need a methodology that enables us to upscale models of materials that are heterogeneous at small scales to those that are homogeneous at macroscales, as is evidenced by the fact that only a very small number of phenomenological parameters are required to characterize their continuum level behaviors. It appears, then, that the straightforward justification of the use of continuum limits needs to be reconsidered or replaced in those contexts where the materials of interest exhibit heterogeneous microstructures.

In section 5 I will say a bit more about the nature and generality of this different methodology. In the next section, I present a historical discussion, one aim of which is to illustrate that this debate about modeling across scales is not, in the least bit, new. Furthermore, the discussion should give pause to those who think continuum models are ultimately unnecessary. This is the story of deriving appropriate continuum equations for the behavior of elastic solids and gave rise to a controversy that lasted for most of the nineteenth century.

3. Bridging across Scales: A Historical Controversy

Why are the Navier-Stokes equations named after Navier and Stokes? The answer is not as simple as "they both, independently, arrived at the same equation." In fact, there are differences between the equation Navier first came up with and that derived by Stokes. The differences relate to the assumptions that each employed in his derivation, but more importantly, these different assumptions actually led to different equations. Furthermore, the difference between the equations was symptomatic of a controversy that lasted for most of the nineteenth century (de Boer 2000, p. 86).

While the Navier-Stokes equation describes the behavior of a viscous fluid, the controversy has its roots in the derivation of equations for the behavior of an elastic solid. I intend to focus on the latter equations and only at the end make some remarks about the fluid equations.

The controversy concerned the number of material constants that were required to describe the behavior of elastic solids. According to Navier's equation, a single constant marked a material as isotropic elastic. According to Stokes and Green, two constants were required. For anisotropic elastic materials (where symmetries cannot be employed) the debate concerned whether the number of necessary constants was 15 or 21. This dispute between, respectively, "rari-constancy" theorists and "multi-constancy" theorists depended upon whether one's approach to the elastic solid equations started from a hypothesis to the effect that solids are composed of interacting molecules or from the hypothesis that solids are continuous.

Navier's derivation began from the hypothesis that the deformed state of an elastic body was to be understood in terms of *forces acting between individual particles or molecules that make up the body*. Under this assumption, he derived equations containing only one material constant ϵ.

Navier's equations for an elastic solid are as follows (de Boer 2000, p. 80):

$$\epsilon \left(\Delta u + 2 \frac{\partial \Theta}{\partial x} \right) + \rho X = \rho \frac{\partial^2 u}{\partial t^2}, \tag{2}$$

$$\epsilon \left(\Delta v + 2 \frac{\partial \Theta}{\partial y} \right) + \rho Y = \rho \frac{\partial^2 v}{\partial t^2}, \tag{3}$$

$$\epsilon \left(\Delta w + 2 \frac{\partial \Theta}{\partial z} \right) + \rho Z = \rho \frac{\partial^2 w}{\partial t^2}. \tag{4}$$

Here ϵ, Navier's material constant, reflects the molecular forces that are supposed to turn on when external forces are applied to the body. x, y, z are the coordinates representing the location of a *material point* in the body.[18] u, v, w are the displacement components in the directions x, y, z; X, Y, Z represent the external accelerations (forces) in the directions x, y, z; $\Delta = \frac{\partial^2}{\partial x^2} + \frac{\partial^2}{\partial y^2} + \frac{\partial^2}{\partial z^2}$ is the Laplace operator; $\Theta = \frac{\partial u}{\partial x} + \frac{\partial v}{\partial y} + \frac{\partial w}{\partial z}$ is the volume strain; and ρ is the material density.

Cauchy also derived an equation for isotropic elastic materials by starting from a molecular hypothesis similar to Navier's. However, his equation contains the correct number of material constants (two). It is instructive to write down Cauchy's equations and to discuss how, essentially, a mistaken, inconsistent derivational move on his part yielded a more accurate set of equations than Navier.

Cauchy's equations for an elastic solid are as follows (de Boer 2000, p. 81) (compare with equation (1)):

$$(R+A)\left(\Delta u + 2 \frac{\partial \Theta}{\partial x} \right) + \rho X = \rho \frac{\partial^2 u}{\partial t^2}, \tag{5}$$

$$(R+A)\left(\Delta v + 2 \frac{\partial \Theta}{\partial y} \right) + \rho Y = \rho \frac{\partial^2 v}{\partial t^2}, \tag{6}$$

$$(R+A)\left(\Delta w + 2 \frac{\partial \Theta}{\partial z} \right) + \rho Z = \rho \frac{\partial^2 w}{\partial t^2}. \tag{7}$$

---

[18] Note that in continuum mechanics, generally, a material point or "material particle" is not an atom or molecule of the system; rather it is an imaginary region that is large enough to contain many atomic subscales (whether or not they really exist) and small enough relative to the scale of field variables characterizing the impressed forces. Of course, as noted, Navier's derivation did make reference to atoms.

R, A are the two material constants. Cauchy noted, explicitly, that when $A = 0$ his equations agree with Navier's when $R = \epsilon$[19] (de Boer 2000, p. 81). How did Cauchy arrive at a different equation than Navier, despite starting, essentially, from the same molecular assumptions about forces? He did so by assuming that, despite the fact that he is operating under the molecular hypothesis, he can, in his derivation replace certain summations by integrations. In effect, he actually employs a *continuum condition* contradictory to his fundamental starting assumption.[20]

George Green, in 1839, published a study that arrived at the correct equations—essentially (5)–(7)—by completely eschewing the molecular hypothesis. He treated the entire body as composed of "two indefinitely extended media, the surface of junction when in equilibrium being a plane of infinite extent."[21] He also assumed that the material was not crystalline and, hence, isotropic. Then using a principle of the conservation of energy/work he derived, using variational principles of Lagrangian mechanics, his multi-constant equation.

Finally, following the discussion of Todhunter and Pearson (1960), we note that Stokes's work supported the multi-constancy theory in that he was able to generalize his equations for the behavior of *viscous fluids* to the case of elastic solids by making no distinction between a viscous fluid and a solid undergoing permanent—plastic—deformation. "He in fact draws no line between a plastic solid and a viscous fluid. The formulae for the equilibrium of an isotropic plastic solid would thus be bi-constant" (Todhunter and Pearson 1960, p. 500). This unification of continuum equations lends further support to the multi-constancy theory.

The historical debate represents just the tip of the iceberg of the complexity surrounding both theoretical and experimental work on the behavior of the supposedly simpler, isotropic, cases of elastic solids. Nevertheless, the multi-constancy theory wins the day for appropriate classes of structures. And, derivations that start from atomic assumptions fail to arrive at the correct theory. It seems that here may very well be a case where a continuum point of view is actually superior: bottom-up derivation from atomistic hypotheses about the nature of elastic solid bodies fails to yield correct equations governing the macroscopic behavior of those bodies. There are good reasons, already well understood by Green and Stokes, for eschewing such reductionist strategies.

This controversy is important for the current project for the following reason. Green and Stokes were moved by the apparent scaling or homogeneity observed in elastic solids and fluids. That is, as one zooms in with reasonable powerful microscopes one sees the steel to be the same at different magnifications; likewise for the fluid. Green and Stokes then extrapolated this scale invariance to hold at even larger magnifications—at even smaller scales. We now know (and likely they suspected) that this extrapolation is not valid beyond certain scale lengths—the atomistic nature of the materials will begin to show itself. Nevertheless, the

---

[19] I have fixed a typographical error in these equations.
[20] See Todhunter and Pearson (1960, pp. 224 and 235–27) for details. Note also how this limiting assumption yields *different and correct* results in comparison to the finite atomistic hypotheses.
[21] Cited in Todhunter and Pearson (1960, p. 495).

continuum modeling was dramatically successful in that it predicted the correct number and the correct character of the phenomenological constants.

De Boer reflects on the reasons for why this controversy lasted so long and was so heated:

> Why was so much time spent on molecular theory considerations, in particular, by the most outstanding mechanics specialists and mathematicians of the epoch? One of the reasons must have been the temptation of gaining the constitutive relation for isotropic and anisotropic elastic continua directly from pure mathematical studies and simple mechanical principles;[22] It was only later realized that Hooke's *generalized law* is an assumption, and that the foundation of the linear relation had to be supported by experiments. (2000, pp. 86–87)

The upshot of this discussion is reflected in de Boer's emphasis that the constitutive equations or special force laws (Hooke's law) are dependent, for their very form, on experimental results. So a simple dismissal of continuum theories as "in principle" eliminable, as reducible, and merely pragmatically justified, is mistaken. Of course, the phenomenological parameters, like Young's modulus (related to Navier's ϵ), must encode details about the actual atomistic structure of elastic solids. But it is naive, indeed, to think that one can, in any straightforward way derive or deduce from atomic facts what are the phenomenological parameters required for continuum model of a given material. It is probably even more naive to think that one will be able to derive or deduce from those atomic facts what are the actual values for those parameters for a given material.

This historical discussion and the intense nineteenth-century debate between the rari- and multi-constancy theorists apparently supports the view that there is some kind of fundamental incompatibility between small scale and continuum modeling practices. That is, it lends support to the stark choice one must apparently make between bottom-up and top-down modeling suggested by the tyranny of scales.

A modern, more nuanced, and better informed view challenges this consequence of the tyranny of scales and will be discussed in section 5. However, such a view will not, in my opinion, bring much comfort to those who believe the use of continuum models or idealizations is only pragmatically justified. A modern statement supporting this point of view can be found in (Phillips 2001):

> [M]any material properties depend upon more than just the identity of the particular atomic constituents that make up the material.…[M]icrostructural features such as point defects, dislocations, and grain boundaries can each alter the measured macroscopic "properties" of a material. (pp. 5–8)

---

[22] This is the temptation promised by an ultimate reductionist point of view.

It is important to reiterate that, contrary to typical philosophical usage, "microstructural features" here is not synonymous with "atomic features"! Defects, dislocations, etc. exist at higher scales.

In the next section I will further develop the stark dichotomy between bottom-up modeling and top-down modeling as a general philosophical problem arising between different recipes for applying mathematics to systems exhibiting different properties across a wide range of scales.

4. Euler's Recipes: Discrete and Continuum

4.1 Discrete

Applied mathematical modeling begins with an attempt to write down an equation governing the system exhibiting the phenomenon of interest. In many situations, this aim is accomplished by starting with a general dynamical principle such as Newton's second law: $\mathbf{F} = m\mathbf{a}$. Unfortunately, this general principle tells us absolutely nothing about the material or body being investigated and, by itself, provides no model of the behavior of the system. Further data are required and these are supplied by so-called "special force laws" or "constitutive equations."

A recipe, due to Leonhard Euler, for finding an appropriate model for a system of particles proceeds as follows (Wilson 1974):

1. Given the class of material (point particles, say), determine the kinds of special forces that act between them. Massive particles obey the constitutive gravitational force: $\mathbf{F}_G = G \frac{m_i m_j}{r_{ij}^2}$. Charged particles additionally will obey the Coulomb force law: $\mathbf{F}_E = k_e \frac{q_i q_j}{r_{ij}^2}$.
2. Choose Cartesian coordinates along which one decomposes the special forces.
3. Sum the forces acting on each particle along the appropriate axis.
4. Set the sum for each particle i equal to $m_i \frac{d^2 x}{dt^2}$ to yield the total force on the particle.

This yields a differential equation that we then employ (= try to solve) to further understand the behavior of our point particle system. Only rarely (for very few particles or for special symmetries) will this equation succumb to analytical evaluation. In many instances, further simplification employing mathematical strategies of variable reduction, averaging, etc. enable us to gain information about the behavior of interest.

4.2 Continuum

As we saw in section 3, Cauchy had a role in the derivation of equations for elastic solids. We note again that he was lucky to have arrived at the correct equations, given that he started with a bottom-up derivation in mind. Nevertheless, Cauchy was an important figure in the development of continuum mechanics: it turns out that at macroscales, forces within a continuum can be represented by a single second-rank tensor, despite all of the details that appear at the atomic level. This is known as the Cauchy stress tensor (Philips 2001, p. 39). The analog of Newton's second law, for continua is the principle of balance of linear momentum. It is a statement that "the time rate of change of the linear momentum is equal to the net force acting on [a] region Ω".[23]

$$\frac{D}{Dt} \int_\Omega \rho v\, dV = \int_{\partial \Omega} t\, dA + \int_\Omega f\, dV. \tag{8}$$

Here $\partial \Omega$ is the boundary of the region Ω, t is the traction vector representing surface forces (squeezings, for instance), and f represents the body forces such as gravity. The left-hand side of equation (8) is the time rate of change of linear momentum. The material time derivative, D/Dt, is required because in addition to explicit time dependence of the field, we need to consider the fact that the material itself can move into a region where the field is different.

As with Euler's discrete recipe, equation (8) requires input from constitutive equations to apply to any real system.

> Whether our interest is in the description of injecting polymers into molds, the evolution of Jupiter's red spot, the development of texture in a crystal, or the formation of vortices in wakes, we must supplement the governing equations of continuum mechanics with some constitutive description. (Phillips 2001, p. 51)

For the case of a steel girder, considered in the regime for use in constructing bridges or buildings we need the input that it obeys something like Hooke's law—that its stress is linearly related to its strain. In modern terminology, we need to provide data about the Cauchy stress tensor. For isotropic linear elastic solids, symmetry considerations come into play and we end up with equation (1)—the Navier-Cauchy equation that characterizes the equilibrium states of such solids:

$$(\lambda + \mu)\nabla(\nabla \cdot \mathbf{u}) + \rho \nabla^2 \mathbf{u} + \mathbf{f} = 0.$$

The "Lamé" parameters (related to Young's modulus) express the empirical details about the material response to stress and strain.

4.3 Controversy

A question of pressing concern is why the continuum recipe should work at all. We have seen in the historical example that it does, and in fact, we have seen that were we

---
[23] See Phillips (2001, pp. 41–42).

simply to employ the discrete (point particle) recipe, we would not arrive at the correct results. In asking why the continuum recipe works on the macroscale, we are asking about the relationship between the dynamical models that track the behavior of individual atoms and molecules and equations like those of Navier, Stokes, Cauchy, and Green that are applicable at the scale of millimeters. Put slightly differently, we would like an account of why it is safe to use the Cauchy momentum equation in the sense that it yields correct equations with the appropriate (few) parameters for broadly different classes of systems—from elastic solids to viscous fluids.

From the point of view of Euler's continuum recipe, one derives the equations for elastic solids, or the Navier-Stokes equations, independently of any views about the molecular or atomic makeup of the medium. (In the nineteenth century the question of whether matter was atomistic had yet to be settled.)

To ask for an account of why it is safe to use the continuum recipe for constructing macroscale models is to ask for an account of the robustness of that methodology. The key physical fact is that the bulk behaviors of solids and fluids are almost completely insensitive to the actual nature of the physics at the smallest scale. The "almost" here is crucial. The atomic details that we do not know (and, hence, do not explicitly refer to) when we employ continuum recipe are encoded in the small number of phenomenological parameters that appear in the resulting equations—Young's modulus, the viscosity, etc. So the answer to the safety question will involve showing how to determine the "fixed points" characterizing broad classes of macroscopic materials—fixed points that are characterized by those phenomenological parameters. Recall the statement by Nelson cited above in section 2.1. In the context of critical phenomena and the determination of the critical exponent β, this upscaling or connection between the Euler's discrete and continuum recipes is accomplished by the renormalization group. In that context, the idea of a critical point and related singularities plays an important role. But Nelson's suggestion is that upscaling of this sort should be possible even for classes of systems without critical points. For example, we would like to understand *why Young's modulus is the appropriate phenomenological parameter for classifying solids as linear elastic, despite rather severe differences in the atomic structure of members of that class.* Finding answers to questions of this latter type is the purview of so-called "homogenization" theory, of which one can profitably think the RG to be a special case.

In the next section, I will spend a bit more time on the RG explanation of the universality of critical behavior, filling in some gaps in the discussion in section 2.1. And, I will try to say something about general methodology of upscaling through the use of homogenization limits.

5. A Modern Resolution

To begin, consider a problem for a corporation that owns a lot of casinos. The CEO of the corporation needs to report to the board of trustees (or whomever) on the expected profits for the corporation. How is she to do it? Assuming (contrary to fact) that casino gaming is fair, she would present to the board a Gaussian or normal probability distribution showing the probabilities of various profits and losses, with standard deviations that would allow for statistical predictions as to expected profits and losses. She may also seek information as to how to manipulate the mean and variance so as to guarantee the likelihood of greater profits for less risk, etc. The Gaussian distribution is a function characterized by two parameters—the mean μ and the variance σ^2. Where will the CEO get the values for the mean and variance? Most likely by *empirically investigating* the actual means and variances displayed over the past year by the various casinos in the corporation. Consider figure 7.5. Should the CEO look to the individual gambles or even to collections of individual gambles of different types in particular casinos? A bottom-up reductionist would say that all of the details about the corporation as a whole are to be found by considering these details. But, in fact, (i) she should not focus too much on spatiotemporal local features of a single casino: suppose someone hits the jackpot on a slot machine. Likely, many people will run to that part of the casino, diminishing profits from the roulette wheels and blackjack tables, and skewing the prediction of the actual mean and variance she is after. Nor (ii) would it be wise to focus too much on groups of casinos say in a particular geographic area (such as Las Vegas) over casinos owned in another area (such as Atlantic City). After all, different tax structures in these different states and municipalities play an important role as well. Such intermediate structures and environmental considerations are crucial—consider again the bubbles within bubbles structures that characterize the heterogeneities at lower scales in the case of the universality of critical phenomena. The CEO needs to look at large groups of collections of casinos where there is evident scaling and self-similarity. Apparent scaling behavior and self-similarity at large scales is an indication of homogeneity. Thus, as with our steel girder, empirical data (at large scales) is required to determine the values of the relevant parameters.

Now why should she think that these two parameters—properties of collections of casinos offering different and varied kinds of games (roulette, poker, blackjack, slots, etc.)—are the correct ones with which to make the presentation to the board? Equivalently, *why she should employ a* Gaussian *probability distribution (it is uniquely defined by the mean and variance) in the first place, as opposed to some other probability distribution?* The answer is effectively provided by an RG argument analogous to that which allows us to determine the functional form of the order parameter M near criticality—that it scales as $|t|^\beta$ near criticality. It is an argument that leads us to expect behavior in accord with the central limit theorem. There are deep similarities between the arguments for why the functional form with exponent β is universal and why Gaussian or central limiting behavior is so ubiquitous. In the former case, the RG demonstrates that various systems all flow to a single fixed point in an abstract space of Hamiltonians or coupling constants. That fixed point determines the universality class that is characterized by the scaling exponent β. Similarly, the Gaussian probability distribution is a fixed point for a wide class of probability distributions under a similar renormalization group transformation. (For details see Batterman

Figure 7.5 Gambles within gambles within gambles ...

2010 and Sinai 1992.) Thus, the answer to why the mean μ and the variance σ^2 are the relevant parameters depends upon an RG, limiting argument. Generalizing, one should expect related argument strategies to tell us why the two elastic "constants" (related to Young's modulus) are the correct parameters with which to characterize the universality class of elastic solids. The appeal to something like central limiting behavior is characteristic of homogenization theory and distinguishes this line of argumentation from that employing REV averaging techniques.

In fact, the difference between averaging and homogenization is related to the difference between the law of large numbers and the central limit theorem: averaging or first order perturbation theory "can often be thought of as a form (or consequence) of the law of large numbers." Homogenization or second order perturbation theory "can often be thought of as a form (or consequence) of the central limit theorem" (Pavliotis and Stuart 2008, pp. 6–7).

Here is a brief discussion that serves to motivate these connections. Consider a sum function of independent and identically distributed random variables, Y_i: $S(n) = \sum_{i=1}^{n} Y_i$. The sample average $\overline{S(n)} = 1/n \sum_{i=1}^{n} Y_i$ converges to the mean or expected value μ. The strong law of large numbers asserts that

$$Pr\left(|\overline{S(n)} - \mu| > \epsilon\right) = 0.$$

As such it tells us about the first moment of the random variable ($\overline{S(n)}$—the average. The central limit theorem by contrast tells us about the second moment of the normalized sum $(\overline{S(n)})$; that is it tells us about the behavior of fluctuations about the average μ. It says that for $n \to \infty$ the probability distribution of $\sqrt{n}(\overline{S(n)} - \mu)$ converges to the normal or Gaussian distribution $\mathcal{N}(0,\sigma^2)$, with mean 0 and variance $\frac{\sigma^2}{n}$ where σ is the standard deviation of the Y_i's.[24]

Thus again we see that in the probabilistic scenario, as in the case of critical phenomena, we must to pay attention to the fact that collections of gambles (bubbles) contribute to the behavior of the system at the macroscale. Once again, we need to pay attention to fluctuations about some average behavior, and not just the average behavior itself.

Furthermore, a similar picture is possible regarding the upscaling of our modeling of the behavior of the steel girder with which we started. Compare the two cases, figure 7.6, noting that here too only a small number of phenomenological parameters are needed to model the continuum/macroscale behavior. (E is Young's modulus and I is the area moment of inertia of a cross-section of the girder.)

The general problem of justifying the use of Euler's continuum recipe to determine the macroscopic equation models involves connecting a statistical/discrete theory in terms of atoms or lattice sites to a hydrodynamic or continuum theory. Much effort has been spent on this problem by applied mathematicians and materials scientists. And, as I mentioned above, the RG argument that effectively determines the continuum behavior of systems near criticality is a relatively simple example of this general homogenization program.

In hydrodynamics, for example Navier-Stokes theory, there appear density functions, $\rho(x)$, that are defined over a continuous variable x. These functions exhibit no atomic structure at all. On the other hand, for a statistical theory, such as the Ising model of a ferromagnet, we have seen that one defines an order parameter (a magnetic density function) $M(x)$ that is the average magnetization in a volume surrounding x that contains many lattice sites or atoms. The radius of the volume, L, is intermediate between the lattice constant (or atomic spacing) and the correlation

---

[24] Proofs of the central limit theorem that involve moment generating functions $M(t)$ for the component random variables Y_i make explicit that there is an asymptotic expansion in a small parameter t, where truncation of the series at first order gives the mean, and truncation of the series at second order gives the fluctuation term. Hence the connection between these limit theorems and first and second order perturbation theory. In fact, two limits are involved: the limit as the small parameter $t \to 0$ and the limit $n \to \infty$.

THE TYRANNY OF SCALES 279

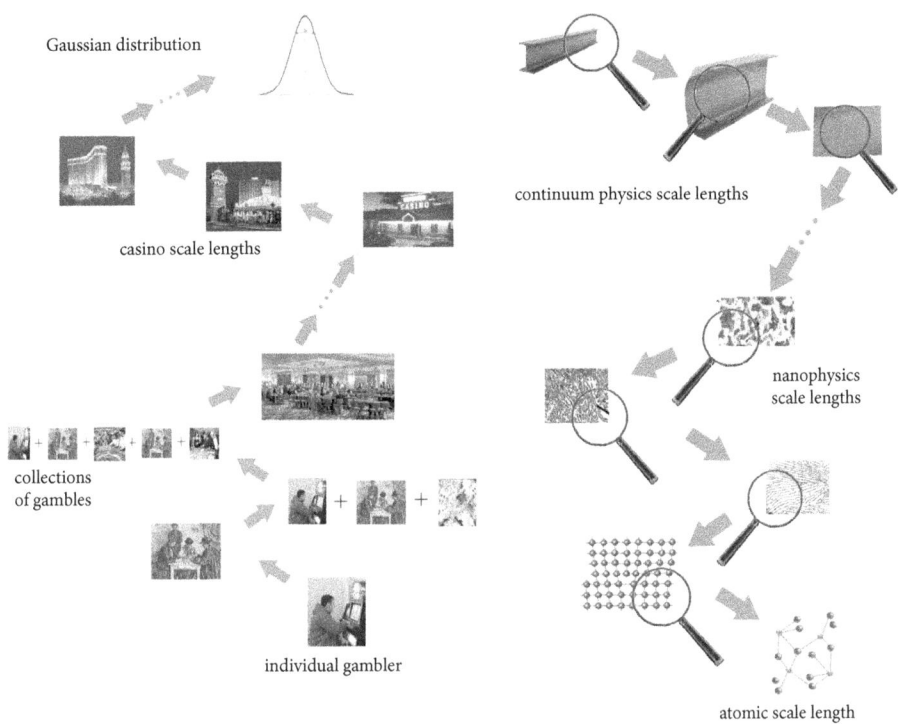

Figure 7.6 Gaussian and steel—few (macro) parameters: $[\mu, \sigma^2]$; $[E, I]$

length ξ: ($a \ll L \ll \xi$). As noted in section 2.1 this makes the order parameter depend upon the length L (Wilson 1974, p. 123).

A crucial difference between the hydrodynamic (thermodynamic) theory and the statistical theory is that the free energy in the former is determined using the single magnetization function $M(\mathbf{x})$. In statistical mechanics, on the other hand, the free energy is "a weighted average over all possible forms of the magnetization $M(\mathbf{x})$." (Wilson 1974, p. 123) This latter set of functions is parameterized by the volume radius L. On the statistical theory due originally to Landau, the free energy defined as a function of $M(\mathbf{x})$ takes the following form:

$$F = \int \left([\nabla M(\mathbf{x})]^2 + RM^2(\mathbf{x}) + UM^4(\mathbf{x}) - B(M(\mathbf{x})) \right) d^3(\mathbf{x}), \tag{9}$$

where R and U are (temperature dependent) constants and B is a (possibly absent) external magnetic field. (Wilson 1974, p. 122) This (mean field) theory predicts the wrong value, $1/2$, for β–the critical exponent. The problem, as diagnosed by Wilson, is that while the Landau theory can accommodate fluctuations for lengths $\lambda < L$ in its definition of M as an average, it cannot accommodate fluctuations of lengths L or greater.

> A sure sign of trouble in the Landau theory would be the dependence of the constants R and U on L. That is, suppose one sets up a procedure for calculating R and U

which involves statistically averaging over fluctuations with wavelengths $\lambda < L$. If one finds R and U depending on L, this is proof that long-wavelength fluctuations are important and Landau's theory must be modified. (p. 123)

The RG account enables one to exploit this L-dependence and eventually derive differential equations (RG) for R and U as functions of L that allow for the calculation of the exponent β in agreement with experiment. The key is to calculate and compare the free energy for different averaging sizes L and $L+\delta L$. One can proceed as follows[25]: Divide $M(x)$ in the volume element into two parts:

$$M(\mathbf{x}) = M_H(\mathbf{x}) + mM_{fl}(\mathbf{x}). \qquad (10)$$

M_H is a hydrodynamic part with wavelengths of order ξ and M_{fl} is a fluctuating part with wavelength between L and $L+\delta L$. The former will be effectively constant over the volume.

By performing a single integral over m—the scale factor in (10)—we get an iterative expression for the free energy for the averaging size $L+\delta L$, $F_{L+\delta L}$, in terms of the free energy for the averaging size L:

$$e^{-F_{L+\delta L}} = \int_{-\infty}^{\infty} e^{-F_L} dm. \qquad (11)$$

In effect, one finds a step by step way to include all the fluctuations—all the physics—that play a role near criticality. One moves from a statistical theory defined over finite N and dependent on L to a hydrodynamic theory of the continuum behavior at criticality. "Including all of the physics" means that the geometric structure of the bubbles within bubbles picture gets preserved and exploited as one upscales from the finite discrete atomistic account to the continuum model at the scale of ξ—the size of the system. That is exactly the structure that is wiped out by the standard REV averaging, and it is for that reason that Landau's mean field theory failed.

5.1 Homogenization

Continuum modeling is concerned with the effective properties of materials that, in many instances, are microstructurally heterogeneous. These microstructures, as noted, are not always to be identified with atomic or lowest scale "fundamental" properties of materials. Simple REV averaging techniques often assume something like that, but in general the effective, phenomenological properties of materials are not simple mixtures of volume fractions of different composite phases or materials. Many times the microstructural features are geometric or topological including (in addition to volume fractions) "surface areas of interfaces, orientations, sizes, shapes, spatial distributions of the phase domains; connectivity of the phases; etc."

---

[25] Details in Wilson (1974, pp. 125–27).

(Torquato 2002, p. 12). In trying to bridge the scales between the atomic domain and that of the macroscale, one needs to connect rapidly varying local functions of the different phases to differential equations characterizing the system at much larger scales. Homogenization theory accomplishes this by taking limits in which the local length (small length scale) of the heterogeneities approaches zero in a way that preserves (and incorporates) the topological and geometric features of the microstructures.

Most simply, and abstractly, homogenization theory considers systems at two scales: ξ, a macroscopic scale characterizing the system size, and a microscopic scale, a, associated with the microscale heterogeneities. There may also be applied external fields that operate at yet a third scale Λ. If the microscale, a, is comparable with either ξ or Λ, then the modeler is stuck trying to solve equations at that smallest scale. However, as is often the case, if $a \ll \Lambda \ll \xi$, then one can introduce a parameter

$$\epsilon = \frac{a}{\xi}$$

that is associated with the fluctuations at the microscale of the heterogeneities—the local properties (Torquato 2002, pp. 305–6). In effect, then one looks at a family of functions u_ϵ and searches for a limit $u = \lim_{\epsilon \to 0} u_\epsilon$ that tells us what the effective properties of the material will be at the macroscale.

Figure 7.7 illustrates this. The left box shows the two scales a and ξ with two phases of the material K_1 and K_2. The homogenization limit enables one to treat the heterogeneous system at scale a as a homogeneous system at scale ξ with an effective material property represented by K_e. For an elastic solid like the steel girder, K_e would be the effective stiffness tensor and is related experimentally to Young's modulus. For a conductor, K_e would be the effective conductivity tensor that is related experimentally to the parameter σ—the specific conductance—appearing

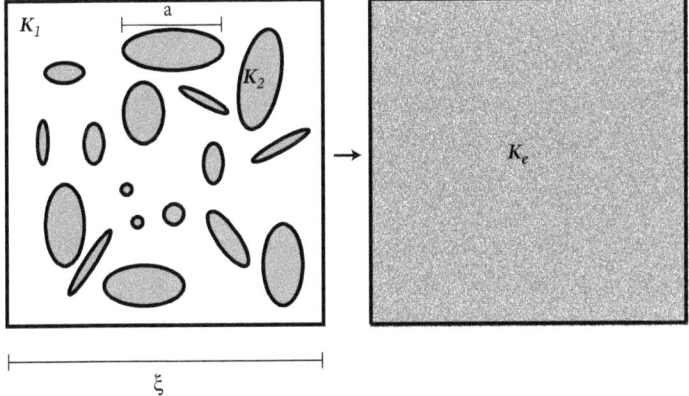

Figure 7.7 Homogenization limit (after Torquato 2002, pp. 2, 305–6)

in Ohm's law:

$$J = \sigma E,$$

where J is the current density at a given location x in the material and E is the electric field at x. At the risk of being overly repetitive, *note that in these and other cases, it is unlikely that the effective material property K_e will be a simple average.*

Let me end this brief discussion of homogenization by highlighting what I take to be a very important concept for the general problem of upscaling. This is the concept of an order parameter and related functions. The notion of an order parameter was introduced in our discussion of continuous phase transitions in thermodynamics, and the statistical mechanical explanations of certain of their features. In effect, the order parameter is a microstructure (mesoscopic scale) dependent function introduced to codify the phenomenologically observed transition between different states of matter. As we have seen, the magnetization M represented in figure 7.2 is introduced to reflect the fact that at the Curie temperature the systems goes from an unordered phase, above T_c to an ordered phase, below T_c. In this context, the divergences and nonanalyticities at the critical point play an essential role in determining the fixed point that characterizes the class of systems exhibiting the same scaling behavior: $M \propto |t|^\beta$. But, again following Nelson's suggestion, entire classes of systems such as the class of linear elastic solids are also characterized by "fixed points" represented by a relatively few phenomenological parameters like Young's modulus.

It is useful to introduce an order-like parameter in this more general context of upscaling where criticality is not really an issue. For example, consider the left image in figure 7.7. In upscaling to get to the right image, one can begin by defining indicator or characteristic functions for the different phases as a function of spatial location (Torquato 2002, pp. 24–5). For instance, if the shaded phase occupies a region U_s in the space, then an indicator function of that phase is given by

$$\chi^s(x) = \begin{cases} 1, & \text{if } x \in U_s \\ 0, & \text{otherwise}. \end{cases}$$

One can also introduce indicator functions for the interfaces or boundaries between the two phases.[26] Much information can then be determined by investigating n-point probability functions expressing the probabilities that n locations $x_1, \ldots, x_n$ are to be found in regions occupied by the shaded phase.[27]

$$S_n^s(x_1, \ldots, x_n) = Pr\{\chi^s(x_1) = 1, \ldots, \chi^s(x_n = 1)\}.$$

In this way many features, other than simple volume fraction, that exist at microscales can be represented and employed in determining the homogenization limit for complex heterogeneous systems. The introduction of such field variables, correlation functions, etc., allow us to characterize the heterogeneous structures

---

[26] These will be generalized distribution functions.
[27] See Torquato (2002) for a detailed development of this approach.

above the atomic scales. In some cases, such as the bubbles within bubbles structure of the different phases at a continuum phase transition, much of this additional apparatus will not be necessary. (Though, of course, it is essential to take into consideration that structure in that particular case.) But for many more involved upscaling problems such as steel, the additional mathematical apparatus will be critical in determining the appropriate effective phenomenological theory at the continuum level. As we have seen these microstructures are critical for an understanding of how the phenomenological parameters at the continuum scale emerge.

The main lesson to take from this all-too-brief discussion is that physics at these micro/meso-scopic scales need to be considered. Bottom-up modeling of systems that exist across a large range of scales is not sufficient to yield observed properties of those systems at higher scales. Neither is complete top-down modeling. After all, we know that the parameters appearing in continuum models must depend upon details at lower scale levels. The interplay between the two strategies—a kind of mutual adjustment in which lower scale physics informs upper scale models and upper scale physics corrects lower scale models—is complex, fascinating, and unavoidable.

6. Conclusion

The solution to the tyranny of scales problem has been presented as one of seeing if it is possible to exploit microstructural scale information (intermediate between atomic scales and macroscopic scales) to bridge between two dominant and apparently incompatible modeling strategies. These are the traditional bottom-up strategies associated with a broadly reductionist account of science and pure top-down strategies that held sway in the nineteenth century and motivated the likes of Mach, Duhem, Maxwell, and others. Despite great progress in understanding the physics of atomic and subatomic particles, the persistence of continuum modeling has led to heated debates in philosophy about emergence, reduction, realism, etc. We have canvassed several different attitudes to the apparent ineliminability of continuum level modeling in physics. On the one hand, there is the view of Butterfield and others, that the use of continuum limits represents nothing more than a preference for the mathematical convenience of the infinite. Another possible view, coming out of the tyranny of scales, suggests a kind of skepticism: we need both atomic scale models and continuum scale models that essentially employ infinite idealizations. However, a unified account of applied mathematics that incorporates both the literally correct atomic models and the essentially idealized continuum models seems to be beyond our reach.[28]

---

[28] See Maddy (2008) for a good discussion of this point of view among other interesting topics about the applicability of mathematics.

I claim that neither of these attitudes is ultimately acceptable. Butterfield et al. are wrong to believe that continuum models are simply mathematical conveniences posing no real philosophical concerns. This position fails to respect some rather deep differences between kinds of continuum modeling. In particular, the strategies employed in the renormalization group and in homogenization theory differ significantly from those employed in standard representative elementary volume (REV) averaging scenarios. The significance of Wilson's renormalization group advance was exactly to point out why such REV methods fail and how that failure can be overcome. The answer, as we have seen, is to pay attention to "between" scale structures as in the case of the bubbles within bubbles picture of what happens at phase transitions. Incorporating such structures—features that cannot be understood as averages over atomic level structures—is exactly the strategy behind upscaling attempts that connect Euler-type discrete modeling recipes to Euler-type continuum recipes. Homogenization lets us give an answer to why the use of the continuum recipe is safe and robust. It provides a satisfactory justification for the use of such continuum models, but not one that is "straightforward" or pragmatically motivated. As such, homogenization provides the beginning of an account of applied mathematics that unifies the radically different scale-dependent modeling strategies.

I have also tried here to focus attention on a rather large subfield of applied mathematics that should be of interest to philosophers working on specific issues of modeling, simulation, numerical methods, and idealizations. In addition, understanding the nature of materials in terms of homogenization strategies can inform certain questions about the nature of physical properties and issues about realism. For instance, we have seen that many materials at macroscales are characterized by a few phenomenological parameters such as the elastic constants. Understanding the nature of materials requires understanding why these constants and not others are appropriate, as well as understanding from where the constants arise. One important lesson is that many of these material defining parameters are not simply dependent upon the nature of the atoms that compose the material. There is a crucial link between structure at intermediate scales and observed properties at the macroscale.

It may do to end with an nice statement (partially cited earlier) from Rob Phillips's excellent book *Crystals, Defects, and Microstructures* (2001) expressing this point of view.

> Despite the power of the idea of a material parameter, it must be greeted with caution. For many features of materials, certain "properties" are not *intrinsic*. For example, both the yield strength and fracture toughness of a material depend upon its internal constitution. That is, the measured material response can depend upon microstructural features such as the grain size, the porosity, etc. Depending upon the extent to which the material has been subjected to prior working and annealing, these properties can vary considerably. Even a seemingly

elementary property such as the density can depend significantly upon that material's life history. The significance of the types of observations given above is the realization than many material properties depend upon more than just the identity of the particular atomic constituents that make up that material.... [M]icrostructural features such as point defects, dislocations, and grain boundaries can each alter the measured macroscopic "properties" of a material. (pp. 5–8)

Philosophers who insist that bottom-up explanations of the macroscopic properties of materials are desirable to the exclusion of top-down modeling considerations are, I think being naive, similar to those who maintain that top-down continuum type modeling strategies are superior. The tyranny of scales appears to force us to choose between these strategies. However, new work on understanding the problem of upscaling or modeling across scales suggests that both types of strategies are required. Our top-down considerations will inform the construction of models at lower scales. And our bottom-up attempts will likewise induce changes and improvements in the construction of higher scale models. Mesoscopic structures cannot be ignored and, in fact, provide the bridges that allow us to model across scales.

References

Sorin Bangu. Understanding thermodynamic singularities: Phase transitions, data, and phenomena. *Philosophy of Science*, 76(4):488–505, 2009.

Robert. W. Batterman. Theories between theories: Asymptotic limiting intertheoretic relations. *Synthese*, 103:171–201, 1995.

Robert W. Batterman. Intertheory Relations in Physics. *The Stanford Encyclopedia of Philosophy*, http://plato.stanford.edu/entries/physics-interrelate/, 2001.

Robert W. Batterman. *The Devil in the Details: Asymptotic Reasoning in Explanation, Reduction, and Emergence*. Oxford Studies in Philosophy of Science. Oxford University Press, New York, 2002.

Robert W. Batterman. Critical phenomena and breaking drops: Infinite idealizations in physics. *Studies in History and Philosophy of Modern Physics*, 36:225–244, 2005a.

Robert W. Batterman. Response to Belot's "Whose devil? Which details?". *Philosophy of Science*, 72(1):154–163, 2005b.

Robert W. Batterman. *Encyclopedia of Philosophy*, chapter Reduction. Macmillan Reference, Detroit, 2nd edition, 2006.

Robert W. Batterman. On the explanatory role of mathematics in empirical science. *The British Journal for the Philosophy of Science*, doi = 10.1093/bjps/axp018:1–25, 2009.

Robert W. Batterman. Reduction and renormalization. In Gerhard Ernst and Andreas Hüttemann, editors, *Time, Chance, and Reduction: Philosophical Aspects of Statistical Mechanics*. Cambridge University Press, Cambridge, 2010.

Robert W. Batterman. Emergence, singularities, and symmetry breaking. *Foundations of Physics*, 41(6):1031–1050, 2011.

Gordon Belot. Whose devil? which details? *Philosophy of Science,* 71(1):128–153, 2005.

Jeremy Butterfield. Less is different: Emergence and reduction reconciled. *Foundations of Physics,* 41(6):1065–1135, 2011a.

Jeremy Butterfield and Nazim Bouatta. Emergence and reduction combined in phase transitions. http://philsci-archive.pitt.edu/id/eprint/8554, 2011b.

Reint de Boer. *Theory of Porous Media: Highlights in Historical Development and Current State.* Springer, Berlin, 2000.

R. Feynman, R. Leighton, and M. Sands. *The Feynman Lectures on Physics,* volume 2. Addison-Wesley, Reading, Massachusetts, 1964.

Ulrich Hornung, editor. *Homogenization and Porous Media,* volume 6 of *Interdisciplinary Applied Mathematics.* Springer, New York, 1997.

Leo P. Kadanoff. Scaling, universality, and operator algebras. In C. Domb and M. S. Green, editors, *Phase Transitions and Critical Phenomena,* volume 5A. Academic Press, San Diego, 1976.

Leo P. Kadanoff. *Statistical Physics: Statics, Dynamics, and Renormalization.* World Scientific, Singapore, 2000.

Penelope Maddy. How applied mathematics became pure. *The Review of Symbolic Logic,* 1(1):16–41, 2008.

Tarun Menon and Craig Callender. Turn and face the strange ... Ch-Ch-Changes: Philosophical questions raised by phase transitions, This Volume 2012.

Ernest Nagel. *The Structure of Science: Problems in the Logic of Scientific Explanation.* Harcourt, Brace, & World, New York, 1961.

David R. Nelson. *Defects and Geometry in Condensed Matter Physics.* Cambridge University Press, Cambridge, 2002.

John Norton. Approximation and idealization: Why the difference matters. http://philsci-archive.pitt.edu/8622/, 2011.

J. T. Oden (Chair). Simulation based engineering science—an NSF Blue Ribbon Report. www.nsf.gov/pubs/reports/sbes_final_report.pdf, 2006.

Grigorios A. Pavliotis and Andrew M. Stuart. *Multiscale Methods: Averaging and Homogenization.* Texts in Applied Mathematics. Springer, New York, 2008.

Rob Phillips. *Crystals, Defects, and Microstructures: Modeling across Scales.* Cambridge University Press, Cambridge, 2001.

Ya G. Sinai. *Probability Theory: An Introductory Course.* Springer-Verlag, Berlin, 1992. Trans.: D. Haughton.

Lawrence Sklar. *Physics and Chance: Philosophical Issues in the Foundations of Statistical Mechanics.* Cambridge University Press, Cambridge, 1993.

Isaac Todhunter and Karl Pearson (Ed.). *A History of the Theory of Elasticity and of the Strength of Materials from Galilei to Lord Kelvin,* volume 1: Galilei to Saint-Venant 1639–1850. Dover, New York, 1960.

Salvatore Torquato. *Random Heterogeneous Materials: Microstructure and Macroscopic Properties.* Springer, New York, 2002.

Kenneth G. Wilson. Critical phenomena in 3.99 dimensions. *Physica,* 73:119–128, 1974.

Mark Wilson. Mechanics, classical. In *Routledge Encyclopedia of Philosophy.* Routledge, London, 1998.

CHAPTER 8

SYMMETRY

SORIN BANGU

1. Introduction

Always a fascination for the human mind, symmetry plays a fundamental role in modern physics. Ancient Greek thinkers introduced the concept (and the word, $συμμετρια$) from which the modern one derives;[1] their interest, however, was mainly aroused by the aesthetic connotations attached to this notion (harmony, good proportion, unity). Later on symmetry considerations became instrumental in the domain of mathematized physical science. The work of Johannes Kepler illustrates the lure of mathematical harmony exemplarily: he famously attempted, in his *Mysterium Cosmographicum* (1596), to devise a theoretical model of the solar system by drawing inspiration from geometrical relations. Kepler was particularly impressed by the austere beauty of the image of the spherical shells inscribed within, and circumscribed around, the five Platonic solids (tetrahedron, cube, octahedron, icosahedron, and dodecahedron). More precisely, he sought to demonstrate a correspondence between the distances of the planets from the sun and the radii of these shells; he was convinced that God's blueprint of the universe reflects an agreement

---

I thank Robert Batterman, Margaret Morrison, Steven French, Katherine Brading, James R. Brown, Matthew Donald, Bruce Schumm, Ranpal Dosanjh, and David Schroeren for constructive criticism of earlier drafts. I am of course responsible for any errors left.

[1] For discussions of symmetry before the Greeks, see Robson (2008, 43–48) on Mesopotamian mathematics, third and early second millennia B.C. Mainzer (2005, especially ch. 1) surveys the role of symmetry in a variety of cultural spaces. Hon and Goldstein (2008) aim to rewrite the conceptual history of the idea of symmetry, taking issue with a series of received views.

between the observed ratios of the maximum and minimum radii of the planets and the geometrical ratios calculated for the nested Platonic solids.

Yet, as the celebrated physicist Freeman Dyson once noted, "This model is a supreme example of misguided mathematical intuition" (1964, 130). The sought correlations did not exist, since there are discrepancies between the predictions and the observed data. A similar belief in the perfection of the circle also hindered Kepler's efforts to find out the correct (elliptical) orbits of the planets. While sometimes leading researchers astray, it is beyond doubt that this type of belief has very often helped them a great deal. It appears that Kepler's confidence that the universe was created following mathematical symmetry principles has in the end been useful in discovering the laws of planetary motion. In fact, it turns out that his second law[2] does establish an important astronomical relation that is grounded on a symmetry (the time-invariance of angular momentum).

An aesthetic notion of symmetry certainly played a role in the early evolution of physics, since cases other than Kepler's can be easily documented. But it is the more precise notion of symmetry as *invariance* that is truly fundamental for the modern period. At an intuitive level we speak of symmetrical geometrical figures, or symmetrical concrete shapes, such as snowflakes; what we mean, however, is that they have invariance properties. For instance, a square has the property that there are ways to manipulate it such that the end result of the manipulation is a square identical to the initial one. Such transformations can be a reflection in one of its diagonals, or a clockwise 90° rotation about its center; by comparison, a circle is invariant under arbitrary amounts of rotation about its center.

Rotations and reflections of a geometrical figure are just a particular type of transformation, performed on a particular kind of entity. As has been observed, this idea can be generalized naturally: in addition to visualizable, concrete entities (such as squares, disks, or snowflakes), abstract entities can be manipulated too—and found to be invariant. For instance, the *relations* holding within a certain configuration of objects can be invariant under permuting the objects or under uniformly shifting the objects' positions in space; also, the *form* of a mathematical expression can be invariant under certain mathematical transformations.

Once this generalization was effected, the next major conceptual advance in the study of invariance was the observation that one can consider the set of all transformations that leave an entity (either abstract or concrete) unchanged, and then define the operation of composition of transformations on this set. A simple example is the set of transformations that leave the square invariant, together with the composition operation. There are eight such transformations: four clockwise 90°, 180°, 270°, and 360° rotations and four reflections (two in the diagonals and two in the lines joining the middle of two opposite sides). This set, call it R, is closed under successive composition of these transformations: performing two successive transformations (either rotations or reflections) that leave the square invariant is equivalent to performing one (either rotation or reflection), with this resultant

---

[2] This is also known as the "equal areas" law: a segment connecting the Sun and a planet on an elliptical orbit sweeps out equal areas in equal time intervals.

transformation being included in R too. Moreover, there is a special member of this set, a clockwise rotation of 360°, which has no compositional effect when following or preceding other transformations from the set. And, for any transformation t in R there is another transformation in the set that can be composed with t to produce the effect of a 360° rotation. Finally, we note that the operation of transformation is also associative. The set R and the operation of composition of transformations form a specific mathematical structure, called a *group*. In this case, the group is called the *symmetry group* of the square.

As we have seen, we can manipulate both concrete and abstract entities. This latter kind of manipulation is relevant to physics: if the invariant object is a certain mathematical function associated with the evolution of a physical system—the Lagrangian (or, more precisely, its time integral, the *action*)—then important consequences follow from its invariance. They have to do with the existence of conservation laws, in a way made precise by Noether's theorem (discussed in section 3). This generalization is in the spirit of Herman Weyl's remark that the deeper significance of symmetry for modern physics comes from the fact that "we no longer seek this harmony in static forms like regular solids, but in dynamic laws" (Weyl 1952, 77).

Group theory is the branch of mathematics that studies the most general properties of structures like R. Unlike analysis or differential geometry, the connection of this mathematical theory to physics was underappreciated until the beginning of the twentieth century.[3] But, as has been pointed out (Brading and Castellani 2003, 4–5; 2007, sect. 5), earlier attempts to link physics and mathematical transformations can be documented. The case they discuss is that of C. G. Jacobi's canonical transformation theory, developed in the context of the "analytical" version of classical mechanics elaborated by Lagrange, d'Alembert, Liouville, Poisson, Hamilton, and others.

The essence of the difficulty confronting the analytical formulation of mechanics was that the canonical Hamiltonian equations of motion could not be integrated directly. For the conservative systems, what was needed was a (canonical) transformation that would turn the Hamiltonian into a function of new variables, the goal being to transform the equations of motion into equivalent ones—and thus the initial problem into an equivalent but simpler one (i.e., one in which the canonical equations, in the new system of coordinates, can be integrated).[4] What grounds this approach is the powerful methodological symmetry principle "Same problem, same solution!" (discussed in van Fraassen (1991, 25) and (1989, ch. 10)). This transformational strategy is the conceptual ancestor of the line of research pursued by the famous Göttingen mathematicians Felix Klein, Hermann Weyl, David Hilbert, and Emmy Noether, who pioneered a new way to conceive of the aims and methods of the physical science—as the study of the invariant properties of theories.

---

[3] Dyson (1964, 129) reports a conversation between O. Veblen and J. Jeans in 1910 about the reformation of the mathematics curriculum at Princeton. Jeans was of the opinion that "we may as well cut out group theory. This is a subject which will never be of any use in physics."

[4] This follows Lanczos (1949, 229–239), which contains the technical details.

The goal of the following survey is to highlight some of the themes, problems, and arguments that justify viewing symmetry and invariance as important topics in physics and in the philosophy of physics. The approach taken here (one of several possible ones[5]) is to focus on the impressive *methodological* and *heuristic* effectiveness of symmetry thinking. Although methodology and heuristics are granted center stage, the discussion will branch off naturally toward a variety of related issues, especially traditional metaphysical and epistemological queries about scientific classification, explanation, prediction, ontology, and unification. More concretely, as we just saw with Jacobi, specific problems in physics are sometimes solved by simplifying them as a result of operating certain invariance transformations. Or, as is the case with Einstein's theories of special and general relativity (STR and GTR henceforth), laws of nature and even whole theories are selected by imposing invariance constraints. Furthermore, by requiring that a certain type of symmetry hold locally (as opposed to globally), one discovers that the most natural attempt to comply with this constraint leads to the introduction of a new field, which happens to be the electromagnetic field (this is the famous gauge argument, to be outlined in section 3).

Mathematical symmetries are also an indispensable classificatory tool in particle physics. Their taxonomic function brings with it two main benefits. The first is that order can be imposed over the huge variety of elementary particles. The second comes from the use of these schemes of classification as guides toward the fundamental ontology: while the known particles neatly fitted the schemes, physicists also perceived "gaps" in these schemes, that is, positions for which no corresponding particle was detected. The existence of these gaps suggests that the physical particles that would fill them might themselves exist, as discussed in section 5.

Moreover, symmetry considerations can operate at a higher level, in the form of overarching methodological or heuristic principles. We have already encountered the seemingly unassailable dictum "Same problem, same solution!" constraining the types of answers proposed to physical questions. Even more general is the famous Principle of Sufficient Reason, which is often invoked as a justification for attempts to discover factors responsible for breaking a symmetry. A related, but more specific, methodological rule is the so-called "Curie Principle," which, by urging that any asymmetry in the effects is reflected in an asymmetry of the cause (or, equivalently, that no asymmetry arises spontaneously), suggests a direction of research when dealing with certain problems in physics.[6]

---

[5] Excellent surveys of the symmetry theme are already available, a few by philosophers (e.g., Brading and Castellani 2003b, 2007; Morrison 2008) and many more by physicists (Coughlan and Dodd 1991, ch. 6; Icke 1995; entries in Francoise, Naber, and Tsun 2006; Zee 2007, etc.), so a challenge for the present survey is to minimize the overlap. The existence of common themes is, however, unavoidable.

[6] The name of the principle comes from the French physicist Pierre Curie, who formulated it while working on the properties of crystals. See Curie (1894). For recent discussions, see Ismael (1997), Earman (2004a), and Brading and Castellani (2007, sect. 2).

In addition to its heuristic-methodological role, symmetry has a more specific epistemological function too. It is quite natural to try to capture the very notion of *objectivity* in terms of invariance, and a number of recent authors have tried to do this in a variety of ways (Nozick 2001; Kosso 2003; Debs and Redhead 2007). When judged in connection to physics, the key link between invariance and objectivity is the idea that, roughly speaking, what is truly, objectively real must look the same independent of the perspective from which it is described. In other words, it should be invariant under changing the frame of reference of the observer. The experimental side of science also offers good examples of the conceptual connection between objectivity and invariance: the result of an experiment qualifies as a piece of objective knowledge insofar as the experiment can be replicated. And this means that the result is robust, or invariant under changing laboratories, lab technicians, countries, political systems, and so on. This epistemological relation between invariance and objectivity is not restricted to physics, but occurs naturally in ethics as well: our moral judgments should not be influenced by (i.e., remain invariant under changes in) the social status, race, ethnicity, nationality, and so on of the persons involved. It is widely accepted that these invariance properties are important constituents of an objective ethical assessment.

Finally, while symmetry is no doubt a pivotal notion in modern science, one should not forget that there are important aspects of recent physics in which it is an *a*symmetry, or the breaking of a symmetry, that must receive special attention. One such example is the asymmetry between past and future, as observed in the behavior of entropy in thermodynamics and statistical mechanics.[7] Another example is the asymmetry between matter and antimatter, an asymmetry effect incorporated in the weak interactions.[8] Equally important, spontaneous symmetry breaking (SSB) plays a central role in the Standard Model for particle physics.

This chapter is structured as follows. Section 2, after introducing the classification of symmetries, focuses on discrete symmetries. Section 3 begins with a presentation of the main result holding for continuous symmetries (Noether's theorem) and develops naturally toward an examination of the so-called "gauge argument." The aim of section 4 is to further the general case for the fertility of the gauge idea but also to offer some details on the more recent concrete uses of symmetry in physics, such as in achieving electroweak unification and in the ongoing hunt for the Higgs boson. Section 5 discusses a concrete situation in which symmetry arguments led to the spectacular predictions of elementary particles in the 1960s. Section 6 enlarges the perspective, taking up the theme of the connection between laws and symmetries, aiming to explore Wigner's idea that certain symmetries can be thought of as "superprinciples" (Wigner 1967, 43) able to explain the laws themselves.

---

[7] See Sklar (1993) for a lucid exposition.
[8] The theory of this type of interaction cannot yet explain a key, observed, asymmetry—namely why there is significantly more matter than antimatter in the universe.

2. Classifying Symmetries

The literature reviewing the role of symmetry in physics, by philosophers and physicists alike (Brading and Castellani 2003b, 2007; Morrison 2008; Coughlan and Dodd 1991, ch. 6, etc.), uses the following classificatory divisions. First, there is the distinction between (i) spacetime symmetries and symmetries that do not involve spacetime, (ii) between continuous and discrete symmetries, and (iii) between local ("gauge") and global ("rigid") symmetries. Another important distinction is the one introduced by Eugene Wigner (1967) between geometrical and dynamical symmetries. Geometrical symmetries are universal, spacetime invariances of laws of nature—all laws of physics have to be Lorentz (Poincaré) invariant; dynamical symmetries, on the other hand, are invariances of the laws governing the specific interactions found in nature (weak, strong, electromagnetic, and gravitational). After I sketch the variety of options physicists have in classifying particular invariances in section 2.1, I discuss the discrete symmetries (charge conjugation, parity, and time-reversal), and then their relations (section 2.2). Continuous symmetries and the local vs. global distinction will be examined in section 3; I will return to Wigner's views in the final section.

When an individual symmetry is considered, it usually falls into more than one category. For instance, isospin symmetry (to be discussed below) is not a spacetime symmetry while being a global symmetry (in the sense, to be explained, that the proton-neutron transformation is effected everywhere at once). By contrast, the grounding symmetry principle of Einstein's GTR, the invariance of the laws of physics under transformations of coordinates depending on arbitrary functions of space and time, is of course a spacetime symmetry, a local symmetry, and, moreover, a dynamical one (this being the very example that prompted Wigner to introduce the geometrical-dynamical distinction (Wigner 1967, 23)). Yet, when referring to spatiotemporal symmetries physicists usually have in mind the symmetries of STR, expressed mathematically as the 10-parameter Poincaré group. Another typical example of a symmetry unrelated to spatiotemporal transformations of coordinates is charge-conjugation symmetry, which holds when systems transform into themselves upon swapping particles for antiparticles (e.g., the electron and the positron). This last example brings up another way to classify symmetries, into continuous and discrete.

2.1 Discrete Symmetries: Parity, Charge Conjugation, Time-Reversal

Intuitively, the distinction of continuous vs. discrete symmetries can be captured in terms of the simple example presented in section 1, the symmetries of a square vs. the symmetries of a circle. They both have rotational symmetry, but while the square is invariant only under discrete amounts of rotation (multiples of 90°), the circle remains invariant under any amount of rotation about an axis passing through its center.

Not all spatiotemporal symmetries are continuous. An interesting case of discrete symmetry that is also spatiotemporal is *parity*, or space inversion. The operation involved in the parity symmetry (the operator is typically denoted by **P**) is a reflection through the origin of the coordinate system. The transformation simply reverses the spatial coordinates of an event from (x, y, z) to (–x, –y, –z).

Importantly, the way the parity transformation has been described above amounts to understanding it in an "active" way: it is the system that undergoes the transformation. We can conceive of the transformation as a "passive" one too; in this case the transformation is applied to the coordinate system (used in the system's description), not to the system itself. Thus, the parity transformation amounts to turning a left-handed coordinate system into a right-handed one. In general, physical systems (such as a single particle or a collection of them) that remain the same after a parity transformation are said to have "even" parity; if they do not, they are assigned "odd" parity. More concretely, the parity transformation operator **P** acts on wave-functions as follows. If r is the spatial coordinates vector, we have, by definition, that $\mathbf{P}|\psi(r,t)\rangle = |\psi(-r,t)\rangle$. Those states that are eigenstates of the parity operator have definite parity, indicated as even by the eigenvalue +1 (i.e., satisfying $\mathbf{P}|\psi(r,t)\rangle = |\psi(r,t)\rangle$), or, if the eigenvalue is −1, as odd ($\mathbf{P}|\psi(r,t)\rangle = -|\psi(r,t)\rangle$).

While parity is not always conserved (e.g., in processes involving weak interaction, such as the beta decay of cobalt-60), physicists recognize an important epistemic payoff associated with its conservation. When the system is subjected to a type of interaction that obeys this symmetry (and all three other types of physical interactions—strong, electromagnetic, and gravitational—do obey it), then changes of parity state are forbidden. Thus, parity conservation imposes a constraint on the possible evolutions of the system in question.[9]

We have already encountered another important discrete symmetry, charge conjugation symmetry; its operator is typically denoted by **C**, and the charge conjugation transformation is $\mathbf{C}|\psi(r)\rangle = \pm|\psi(r)\rangle$. Similar to the parity symmetry, wave-functions of systems can have either even or odd charge conjugation symmetry. The photon wave-function, for instance, is odd. This helps determine the parity of a particle that previously decayed into two photons: it must have even parity (the product of the parities of the two photons), if the interaction responsible for the decay conserves this symmetry. This piece of information is extremely valuable: it forbids other types of processes to happen, in particular the decay of that particle into an odd-charge conjugated state.

---

[9] Coughlan and Dodd (1991, 44–49) provide further technical details. Parity violation (demonstrated experimentally by a team led by Wu, in 1957) has reignited the interest in the discussions on the structure of physical space and the nature of chiral objects, going back to Kant's attempts to account for the difference between the "incongruent counterparts" by appeal to their relation to absolute space ("incongruent counterparts" are objects that are mirror images of each other but are not superposable through rigid motion, e.g., a right and left glove). See Nerlich (1994) for an introduction and Hoefer (2000), Huggett (2003), and Pooley (2003) for recent discussions.

The third important discrete symmetry is time-reversal, whose operator is designated as **T**. Generally speaking, this invariance means that the direction of the flow of time is *irrelevant* in fundamental interactions.[10] To say that a system is invariant under **T** amounts to saying that if the system evolves from an initial state to a final one, then the reversal of the direction of motion of its components is possible and will bring the system from the final state back to the initial one (mathematically speaking, we simply replace the expression for time with its negative version). Particles' collision and their time-reversed twin, the decay, are typical contexts in which this symmetry can be demonstrated.

2.2 The CPT Theorem

The **C** and **P** discrete symmetries can operate together, as a product (or composite) symmetry; one way to combine them is in the form of a **CP** transformation (i.e., charge conjugation and space reversal). If this symmetry is to hold, then, naturally, the laws of physics should be invariant under two operations: interchanging a particle with its antiparticle, and spatial inversion of left and right. But this symmetry is violated, as has been discovered by Christenson, Cronin, Fitch, and Turlay (1964) by studying the decays of neutral kaons. More precisely, what they showed was that weak interactions violate both the charge-conjugation symmetry **C** and the mirror reflection symmetry **P**, and also their combination. While this violation might sound like bad news, it turns out that it is intimately linked with another asymmetry mentioned above: the dominance of matter over antimatter in the known universe. However, once the third operation (time-reversal **T**) is taken into account, the final product **CPT** is an *exact* symmetry—as far as we can tell for now. The claim has testable consequences, and two are usually stressed. First, that particles and antiparticles have the same masses and lifetimes—and, as a corroboration of the **CPT** result, this has been confirmed through many experiments over the years. Second, that some sort of compensation rule is in effect: when one symmetry (or a pair of them) is broken, the other(s) cancel the violation out so that the final composite **CPT** symmetry remains intact (Coughlan and Dodd 1991, 48; see also Greaves 2010, for some philosophical puzzles associated with this symmetry).

I shall stop here with the review of discrete symmetries and in the next section I will turn to continuous symmetries. The reason they must be given special attention is the abovementioned connection between these symmetries and a special class of laws of nature—the conservation laws—as established by a famous theorem proved by Emmy Noether in 1918 (presented in section 3.1). This discussion will progress naturally toward another important topic, the distinction between the global (or

---

[10] One can recall here Feynman's famous proposal to understand antiparticles as particles moving backward in time, or, in other words, that the time-reversal operation applied to a particle state would turn it into the corresponding antiparticle state (Feynman 1985). For details and criticism, see Arntzenius and Greaves (2009). This discussion is related to an earlier debate between Albert and Malament with regard to classical electromagnetism. Albert (2000) argued that classical electromagnetism is not time-reversal invariant, while Malament (2004) defended the standard view, according to which the theory does possess this feature.

"rigid") and local, or gauge, symmetries, followed by the so-called "gauge argument" (section 3.2).

3. Continuous Symmetries, Conservation Laws, and the "Gauge Argument"

3.1 Noether's Theorem

In a nutshell, the theorem of interest here[11] maintains that for every continuous global symmetry of the Lagrangian there is a conservation law (and vice versa, though this second claim is not made in the original theorem). It is clear, even from this rough formulation, that this result is not explicitly available in standard Newtonian classical mechanics. It presupposes the conceptual framework of the so-called "analytical" mechanics, in which the Lagrangian function L (in essence, the difference between the kinetic and potential energy) plays the central role.

An important methodological innovation lies behind the Lagrangian formulation of mechanics: the step-by-step Newtonian description of physical systems is abandoned, and an overall approach is adopted. Within the Newtonian scheme, the aim is to compute what the system (say, a moving particle) will do in the next infinitesimal time interval. Within the Lagrangian scheme, the issue is tackled from a different angle: we take an overall view of the trajectory of the particle and determine it all at once. The actual path is singled out as the one that satisfies certain minimization constraints. While in terms of predictive power the Lagrangian and Newtonian versions of mechanics are taken to be equivalent, physicists tend to think that results such as Noether's make the former preferable to the latter;[12] however, recent analyses of this relation (such as Mark Wilson's, in this volume) recommend more caution in assigning such priorities.

The Lagrangian methodology can be broken down into two steps. After defining the function L, a functional S—called the *action*—is introduced. S is defined for each possible path ("history") connecting the initial and the final spacetime positions of a particle, as the time integral of the Lagrangian. Second, the actual path followed by the particle is selected: among all possible paths, this is the one that minimizes S. The application of this minimization (extremum) constraint—Hamilton's "Principle of the Least Action"—leads to the Euler-Lagrange equations:

$$\partial L/\partial q = d/dt(\partial L/\partial \dot{q})$$

---

[11] Noether's original 1918 paper contains in fact two theorems. The first one, briefly discussed here, concerns the ("global") invariance of the action under a Lie group characterized by a finite number of parameters; the second ("local") concerns an infinite dimensional Lie group. For the full version of the two theorems, see p. 3 of Tavel's (1971) translation of Noether (1918) at http://arxiv.org/PS_cache/physics/pdf/0503/0503066v1.pdf.

[12] In addition to this, the Lagrangian framework permits a more natural passage to quantum mechanics—which was in fact worked out by Feynman in the 1940s, in his path integral formalism.

Physicists recognize important heuristic gains obtained from reconceiving classical mechanics in this way. Unlike the Newtonian approach, the analytical approach requires the construction of a single quantity L, a scalar, which yields the equations of motion. Moreover, due to the introduction of the so-called "generalized" coordinates in the analytical scheme (usually denoted by q), the Euler-Lagrange equations preserve the same form upon switching from Cartesian coordinates to any other general set of coordinates. Finally, the observation that brings us closer to Noether's theorem is that in this formalism the link between symmetries and conservation laws becomes easily noticeable. If we require that the Lagrangian L be independent of a certain coordinate q (which is to require that L is *invariant* under the transformations of this coordinate) then, in mathematical terms, this is tantamount to saying that its corresponding partial derivative is zero:

$$\partial L / \partial q = 0$$

Now, if we look at the Euler-Lagrange identity, it is evident that the time derivative appearing on its right hand

$$d/dt(\partial L / \partial \dot{q})$$

has to be nil too, which amounts to the statement of a conservation law: namely, that the quantity

$$\partial L / \partial \dot{q}$$

does not vary in time. In particular, if we return to usual Cartesian coordinates and plug in

$$L(x, \dot{x}) = 1/2(m\dot{x}^2) - V(x)$$

we get that

$$0 = \partial L / \partial x = d/dt(\partial L / \partial \dot{x})$$

and, since

$$\partial L / \partial \dot{x} = m\dot{x}$$

is the linear momentum p, it follows that this quantity is conserved. In other words, invariance under translations in space implies conservation of linear momentum.

The lesson to draw from these simple considerations is that we actually do not need Noether's theorem to establish some straightforward, but important, implications like the one above. Since the relation between some of the familiar symmetries and the conservation laws was not a surprise at the time she communicated her work, the main reason for which the theorem was praised was its generality (Brading and Brown 2003, 89, 98): roughly put, it proves the most general fact that *if* the action integral S is invariant under a continuous (Lie) group of transformations (characterized by a finite number s of parameters) then, *if* the Euler-Lagrange equations are satisfied, *then* there exist s conserved "currents." Yet, as Brading and

Brown (2003, 92) point out, Noether's original concern had to do with the following (again, more general) question, the so-called "variational problem" (which is in fact the title of her paper): given a smooth infinitesimal transformation of the independent or dependent variables (appearing in the Lagrangian), under what conditions does the action remain invariant? More precisely, her aim was to find the general conditions that the variables must satisfy, if the first-order functional variation of the action ∂S vanishes (assuming that the region of integration in the integral defining the action is arbitrary). The more familiar implications—such as the one holding between the invariance under translations in time and the conservation of energy, or between the invariance under spatial rotations and the conservation of angular momentum—follow immediately, as applications of the general result. Furthermore, the theorem is general in yet another way. Loosely speaking, it turns out that the details of the action are irrelevant: if two different actions remain invariant under the same transformation, the same conservation law corresponds to both (Zee 2007, 119–120).[13]

In Noether's obituary, Einstein placed her theorem in the special category of "spiritual formulas." This praise was meant to convey the point that the result amounts to significantly more than a technically brilliant achievement: it is a profound insight into the order of nature. In addition to this gain in understanding, the practical-heuristic use of the theorem is equally impressive. Examples are easy to find, especially in the high-energy domains. As Zee (2007, 119–120) explains, these are situations in which physicists knew the continuous symmetry governing a certain physical situation, and thus could confidently begin to look for a conserved quantity (as we will see, isospin is a case in point). The reverse case is also possible: sometimes the physicists did not have any clue as to what the action was but were able to identify experimentally certain conserved quantities, so they inferred that some corresponding symmetries must exist. The theorem is then invoked to give them hints about what the action might be. In fact, this second situation is illustrated by the famous example of the electric charge. That this quantity is conserved had been known for a long time, and Noether's theorem indicates that a certain symmetry must correspond to it. Indeed, in 1927, Fritz London, building up on some ideas advanced by Weyl in the early 1920s, showed that the conservation of charge follows from global *phase* invariance (that is, invariance of the physics under an arbitrary shift in the complex phase of the wave-function). Weyl's previous idea was to propose *scale* invariance—i.e., the requirement that the physical laws do not change if the scale of all length measurements is shifted by the same amount. This insight, although ultimately incorrect, was nothing short of revolutionary—it marked the first appearance of the concept of "gauge." The context in which Weyl did this work was his reflections on possible ways to generalize the Riemannian geometry of Einstein's GTR (for more details, see Ryckman 2003).

---

[13] For recent work on the philosophical implications of the Lagrangian formalism for symmetries and other related issues, see Butterfield 2006 and Smith 2008.

3.2 The Gauge Argument

It is worth beginning the following brief exposition of the so-called "gauge" argument (and "gauge" principle) by noting that, in and of itself, the requirement of global phase invariance—i.e., the requirement that the physics does not change upon the multiplication of the wave-function by a constant wave factor—is rather uncontroversial. The more interesting question is what happens when a related but stronger constraint is imposed, namely *local* phase invariance. (Historically, the locality idea was envisaged in analogy with the symmetry grounding GTR, the requirement of invariance under arbitrary curvilinear coordinate transformations.) Thus, the original thought was to study those situations in which the phase factor is not held constant (same everywhere), but is allowed to vary with each spacetime point (hence the global/local distinction).[14]

Following Quigg (1983, 45–47) and Martin (2003, 42–43), more detail can be filled in the above general description. If we start with the Lagrangian L for a free complex scalar field $\psi(x)$, the corresponding action will be invariant ($\partial L = 0$) under global transformations of the form $\psi \to e^{iq\theta}\psi$; $\overline{\psi} \to e^{-iq\theta}\overline{\psi}$, where θ is constant (the corresponding group is the abelian Lie group U(1)). When the equations of motion are satisfied, Noether's theorem gives us the corresponding conserved current (where q can be identified as the electric charge). As noted, the next step is the localization of the transformation; we now take θ to be $\theta(x)$, a function of spacetime coordinates. It is immediate that L is no longer invariant under the corresponding transformation $\psi(x) \to e^{iq\theta(x)}\psi(x)$, $\overline{\psi}(x) \to e^{-iq\theta(x)}\overline{\psi}(x)$, since simple computations show that the derivatives $\partial_\mu \theta(x)$ do not vanish. Therefore, if the local invariance is to be preserved, L has to be modified. Instead of L we consider a new Lagrangian, $L* = L - J^\mu A_\mu$, where $J^\mu = q\overline{\psi}\gamma^\mu\psi$ (the factors γ^μ are the Dirac matrices). This new Lagrangian is invariant, and what contributed to securing the invariance was the introduction of the ("compensatory") field A_μ (the so-called "gauge potential"), which transforms as $A_\mu(x) \to A_\mu(x) - \partial_\mu \theta(x)$. It is in virtue of this behavior that A_μ can be reinterpreted as the (familiar) electromagnetic potential. Thus, the consequences of imposing this "local" symmetry are quite astonishing: roughly speaking, one realizes that a natural route to take is to introduce a new field that has the properties of the electromagnetic field! It is as if there is a "gauge logic" of nature (Martin 2003, 43). This field must have an infinite range, so its quantum must be massless (to obey the time-energy uncertainty relation: massive particles—the quanta of the fields—decay quickly, and can travel only short distances).[15] The surprise is that this is what actually happens in nature: the quantum is the photon. The situation is conceptually intriguing, since it looks like

---

[14] Interesting questions to ask here are (i) to what extent gauge symmetries are actually observable, as well as (ii) whether it makes sense to apply to them the distinction between the active and passive ways of understanding a transformation (introduced in section 2.1). See Kosso (2000) and Brading and Brown (2004) for a discussion.

[15] However, mass and life-time are only indirectly linked. (The proton is heavy and lives more than 10^{28} years.) To be more precise, it is the Compton wavelength that is inversely proportional to the mass of the particle ($\lambda = h/mc$), and thus directly linked to its range.

something (a physically significant object) has been gotten from nothing (mere mathematical re-description).

A rehearsal of the main philosophical problems raised by this argument is in order. Physicists reflecting on the "power of the gauge" usually endorse this feeling of mystery and surprise, especially in their more popular presentations (e.g., Schumm 2004). Most philosophers, however, adopt a more circumspect attitude (see Brown 1999; Teller 2000; Earman 2003a; Healey 2007; etc.), expressing, in some cases, serious reservations about this power. Martin (2003) provides a recent comprehensive analysis of this issue, along the following lines. First of all, it is not clear what are the metaphysical and epistemological grounds upon which to demand local gauge invariance; or, in other words, it is far from clear why one should embrace the Yang-Mills gauge principle—"*every continuous symmetry of nature is a local symmetry*" (Mills 1989, 496; emphasis in original). A series of reasons are traditionally advanced in the textbooks, but a quick review of the philosophical literature (some of it mentioned above) reveals that they are not found entirely satisfying. In their 1954 paper, Yang and Mills argue that the idea of local symmetries is "more consistent with the concept of localized fields" (1954, 191; see also Auyang 1995). More generally, physicists emphasize that locality is required on the basis of STR precluding instantaneous communication between distant spacetime locations. Yet, it is objected, it is not immediately evident whether the local–global distinction grounding the gauge argument perfectly mirrors the locality constraint as imposed by STR.

Another topic that caught philosophers' attention is the uniqueness of the compensatory modification described above. Some deny this feature altogether; Martin, for instance, argues that "the modification is not uniquely dictated by the demand of local gauge invariance. There are infinitely many gauge-invariant terms that might be added to the Lagrangian if gauge invariance were the only input to the argument" (2003, 44). In other words, what is questioned here is the real power of this invariance demand: if taken in isolation, local gauge invariance does not dictate the form of the field; this uniqueness is achieved only by also requiring (i) Lorentz invariance, and (ii) that the final outcome be a renormalizable theory.[16] Thus, Martin suggests, the renormalizability of the theory (roughly, the mathematical technique by which it is ensured that the theory delivers finite values as predictions of the quantities of interest, such as the electric charge) should be granted a more prominent role in the evaluation of the outcome of the gauge argument.

One can even ask why the gauge-invariance constraint is more significant than renormalizability in the economy of the argument. Thus, a different angle of attack against the argument is possible, according to which renormalizability is in fact the pivotal feature of the model; gauge invariance becomes only one among a few valuable features of a quantum theory. Finally, a good deal of the discussion in the philosophical literature is devoted to clarifying to what extent the argument

---

[16] Moreover, it is demanded that we write as simple a theory as possible, hoping that the notorious vagueness of the simplicity constraint can somehow be satisfied.

presents us with a situation in which an interaction field is somehow "generated" out of sheer mathematical formalism.[17] Earman (2003a, 157) summarizes the grounds for a reserved attitude toward the "magic" of the gauge as follows:

> I am in agreement with Martin ([2003]) who finds the "getting something from nothing" character of the gauge argument too good to be true. In particular, a careful look at applications of this argument reveals that a unique theory of the interacting field results only if some meaty restrictions on the form of the final Lagrangian are implicitly in operation; and furthermore, the kind of locality needed for the move from the "global symmetry" (invoking Noether's first theorem) to the "local symmetry" (invoking Noether's second theorem) is not justified by an appeal to the no-action-at-a-distance sense of locality supported by relativity theory. Not only is there no magic to be found in the gauge argument, but the "gauge principle" that prescribes a move from global to local symmetries for interacting fields can be viewed as output rather than input: for example, it can be viewed as the product of a self-consistency requirement (see, for example, Wald 1986) or as a consequence of the requirement of renormalizability (see, for example, Weinberg 1974).

4. Gauge Theories, Unification, Symmetry Breaking, and the Higgs Field

4.1 The Fertility of the Gauge Idea

Electromagnetism can be understood as the first illustration of the application of the (heuristic) effectiveness of the gauge principle.[18] But physicists were able to find

---

[17] Note that it is the "kinetic" component $-1/4 F_{\mu\nu} F^{\mu\nu}$ of the Lagrangian for the full theory (also featuring an "interacting" component), which, as Martin nicely puts it, "imbues the field with its own existence, accounting for the presence of non-zero electromagnetic fields, for the propagation of free photons" (2003, 43). See Quigg (1983, 45–48) for the technical details. But, technicalities aside, one of the main complaints against this standard story has been that this generation talk is misleading, as the gauge field is put in by hand. For discussion, see Brown (1999). A number of further issues arise, having to do with the (in)determinist character of a gauge theory. A source of concern is the identification of those quantities that are actually "physical," as opposed to mere artifacts of description. The discussions in the literature focus on Einstein's "hole argument" (Earman and Norton 1987; Butterfield 1989; Belot 1996, esp. chs. 5, 6, 7; 1998; Saunders 2002; etc.; for a recent introduction, see Norton 2008). Equally pressing is the question about the right ontological interpretation that should be given to those quantities that are not gauge-invariant, the so-called (by Redhead 2003) "surplus structure."

[18] We can say this with the benefit of hindsight; electromagnetism, as a theory, of course pre-dates gauge.

a larger significance of the local gauge invariance idea. As Mills (1989) narrates, C. N. Yang was, in the mid-1950s, among the very few who realized the potential of this insight; he envisaged the thought that all fundamental physical theories could be formulated as gauge theories. Back then, the only conservation law seemingly similar to that of the electric charge was the conservation of a quantity called "isospin" (more on this below). In fact, the scheme Yang had in mind was an analogy with the local gauge invariance idea that worked so well in the electromagnetic case: as we saw, for the strong interaction the role of the electric charge would be played by isospin, and, once the local gauge invariance constraint was in place, new (gauge) fields needed to be introduced, having the "gluing" role assigned to the electromagnetic field in electrodynamics.

The isospin idea has been developed along the following lines.[19] Physicists were intrigued by two indisputable experimental findings. First, neutrons and protons have approximately the same mass.[20] Second, despite the fact that they have different electric charges (neutrons are neutral while protons are positive), they are bound together inside the nucleus by the strong nuclear force. This means, in physicists' jargon, that the strong force is "charge-blind." Hence, their charge could not be of physical significance, as far as the strong force is concerned—that is, once other types of interactions, the electromagnetic one in particular, are "switched off." For this reason, a natural idea was to conceive of their different charges as mere different "labels" applied to them. Heisenberg, who introduced the isospin idea in 1932, proposed to treat protons and neutrons as different states (labeled n and p) of the same particle, called the *nucleon*. In this way, the pair $n - p$ can be understood either as referring to two particles (the orientation of the isospin distinguishing them), or as describing two different states of the same particle, the nucleon. Thus, isospin is introduced by analogy with the electron spin; just as the spin of the electron can have two orientations along the third axis, so too the nucleon can appear in two isospin states (the positive eigenvalue indicates a proton, while the negative one a neutron).

This invariance of strong interactions under neutron-proton permutations simply means that the proton and the neutron are indistinguishable[21]—when looked at from the viewpoint of this kind of interaction. (I mention this aspect now, since, as we will see below, it is important for classificatory reasons.) As with all symmetries, the proton-neutron isospin symmetry is captured mathematically in terms of a group structure; the group involved here is $SU(2)$. The elements of this group "rotate" protons into neutrons (and vice versa); nucleons are thus "mixtures" of these two components. Yang and Mills built up on this idea and required invariance under local redefinitions of these components of the nucleon. In accordance with the "gauge logic" presented above, this invariance led to the introduction of a massless spin-1 gauge particle called ρ, which, because of the possibility that the nucleon

---

[19] This presentation follows Wilkinson (1969).
[20] The current explanation of the difference is in terms of the mass difference between the *up* and *down* quark.
[21] See French (2008) for identity and individuality in quantum theory, and French and Rickles (2003) on some subtleties of the permutation symmetry.

might change its electric charge during interactions, had to exist in three charged states—positive, negative and neutral.

But these particles (gauge bosons) could not be found in nature, so a new theory had to be advanced. As gradually became evident, the strong interactions are governed by a bigger symmetry and, consequently, the structure of interest had to be a bigger group. Roughly speaking, the story unfolds as follows (Coughlan and Dodd 1991, 60). First, a new characteristic of strong interactions has been discovered—the conservation of a new quantity, dubbed by Gell-Mann "strangeness." Second, once this new quantum number was considered in addition to the isospin number, the new symmetry governing these interactions has been found to be SU(3). Thus, it turned out that the original theories about isospin have in fact been invoking the symmetry above its fundamental level. Nevertheless, those theories were of crucial importance in the development of the ideas, and the symmetries, underlying the ultimately successful theory: the gauge theory based on the SU(3) group, quantum chromodynamics (QCD), which postulates quarks (of various "colors") as the basic entities and the ultimate constituents of hadrons (the generic name for particles participating in strong interactions).[22]

Not only strong interactions, but weak interactions too can be treated within the gauge framework.[23] The analogy with isospin is also heuristically helpful in this context, the conserved quantity being called "weak isospin." The key physical invariance grounding the theory is that the weak interaction is blind to distinctions between the neutrino and the electron—it only "sees" a generic lepton. Mathematically, the structure of interest is the group of weak isospin, which, as it happens, is the previously introduced SU(2). So, after overcoming a series of false starts and various experimental difficulties,[24] one can say that at the beginning of the 1970s the strong, electromagnetic and weak interactions were described by relatively well-understood gauge theories. An important theoretical breakthrough, in which symmetry considerations had a crucial role to play, came in 1967–68 when Steven Weinberg and, independently, Abdus Salam, advanced a *unified* model of the electromagnetic

---

[22] To clarify: Gell-Mann's so-called "Eightfold Way" SU(3)-based theory mentioned at the beginning of this paragraph is not QCD as developed later on. The degrees of freedom of the Eightfold Way are not the degrees of freedom of the SU(3)-based QCD—though the group is the same, SU(3). This later theory postulates three different types of strong-force charge (the *red, green,* and *blue* quarks). The former SU(3) space (where only *global* invariance required) is a different entity than the SU(3) space of strong charge, which is under the constraint of *local* ("gauge") invariance. Within the former theory, we only categorize nonfundamental collections of quarks (for more, see section 5). It is the latter theory which is the currently accepted *dynamical* account of the strong nuclear force. Yang and Mills (see the previous paragraph) attempted to make a dynamical theory out of the SU(2) isospin space, but we can now see that this is clearly wrong-headed, since protons and neutrons are not fundamental particles.

[23] This account follows Coughlan and Dodd (1991).

[24] Part of this story provided social-constructivists with a case to uphold their position. As we will see below (next footnote), the experimental demonstration of the so-called "weak neutral currents" would have corroborated the unified model. Analyzing this episode, Andy Pickering (1998, 136) writes: "There I argue that the acceptability of the weak neutral current (and hence of the associated interpretative practices) was determined by the opportunities its existence offered *for future* experimental *and* theoretical practice in particle physics. Quite simply, particle physicists accepted the existence of the neutral current because they could see how to ply their trade more profitably in a world in which the neutral current was real. The key idea here is that of a symbiotic relationship between experimenters and theorists, the two distinct professional groupings within particle physics."

and weak interactions. Their model drew on previous work by Sheldon Glashow and, in a nutshell, was a gauge theory whose gauge group consisted in "pasting" together the two groups governing the component interactions: $U(1) \times SU(2)$. The Glashow-Salam-Weinberg (GSW) model, shown to be renormalizable by 't Hooft and Veltman in 1971, was meant to describe the interactions of leptons through the exchange of weak bosons and photons.[25] While the thought that the two types of interactions might receive a unified treatment was in the air for a while (since two of the weak bosons bear electrical charge), the new, distinctive feature of the model was the incorporation of a "mechanism" meant to ensure that the W and Z bosons acquire mass, while the photon remains massless—and today this is called the "Higgs mechanism" after Peter Higgs, the physicist who proposed it in 1964.[26]

4.2 Symmetry Breaking to the Rescue: Electroweak Unification and the Higgs Mechanism

One way to solve the mass problem for the Standard Model was to borrow an idea from condensed matter physics, the spontaneous breaking of a symmetry (SSB). More precisely, it was the work done on phase transitions in (quantum) statistical mechanics that offered the best analogy. Upon a drop in temperature below a certain value, liquid water freezes; the formation of ice crystals brings the system in a stable state (free energy is minimized), but the rotational symmetry of molecules available in the liquid state is lost (or "broken"). In essence, SSB claims that the massiveness of the particles arise in an analogous fashion: right after the Big Bang, when the value of the available energy dropped under a critical value, the unified (and symmetric) electroweak interaction broke into what we observe today, the electromagnetic and

---

[25] Slightly more technically, the situation is as follows. After the development of the $U(1) \times SU(2)$ GSW unified ("standard") model (SM), one should distinguish between the photon of the electromagnetic interaction (as described in quantum electrodynamics QED) and the quantum of the $U(1)$ symmetry of the Standard Model, which is the so-called B^0 field. A somewhat similar point holds for the quanta of weak interaction; the $SU(2)$ field quanta are the previously known W^+ and W^- weak field quanta, but the GSW model also predicts a third, neutral W^0. Experimental results (such as the impossibility to tell whether a certain interaction is the result of exchanging B^0s or W^0s) led to the idea that those interactions leaving the electrical charge of the particles involved unchanged must take place through the exchange of a *composite* of the two neutral quanta. Remarkably, the model combines precise amounts of B^0 and W^0 and recovers the properties of the (neutral) photon of quantum electrodynamics. The percentage of each in the mixture is known, being given by the *Weinberg mixing angle* θ_w. For the photon (call it A), $A = W^0 \sin\theta_w + B^0 \cos\theta_w$. The "leftovers" of each B^0 and W^0 make up a new electrically neutral field quantum which is responsible for the weak nuclear interaction as we observe it: the Z^0 boson, where $Z^0 = W^0 \cos\theta_w + B^0 \sin\theta_w$. (For more details, see Coughlan and Dodd 1991, 100; as they explain it, the masses of the Ws and of Z_0s depend on θ_w, which is such that $\sin^2\theta_w \approx 0.23$.) Note, however, that the B^0 field quanta is "unphysical." If, for instance, one would somehow manage to produce a B^0, it would decay rapidly, and what one would see in the detector is either that it lives forever (as a photon), or it decays quickly as a Z^0. The experimental discovery (in 1973) that weak interactions can also take place via the exchange of a neutral weak field quanta is the topic of the essay by Pickering mentioned in the previous footnote.

[26] See Higgs (1964). A number of other physicists (R. Brout, F. Englert, G. Guralnik, C. R. Hagen, and T. Kibble) have presented ideas similar to Higgs's.

weak components. Particles' interaction with the Higgs field pervading the whole space "slowed them down," inducing effects similar to inertial effects, and thus mimicking their possession of mass. Importantly, this is a general solution to the mass problem: all particles must interact with this "dragging" field—not only the W and Z bosons, but electrons and protons too, the top quark, and so on—to *appear* as massive (the more intensely they couple to the Higgs field, the more massive they appear). The photon, on the other hand, does not couple to the Higgs field at all, and thus appears as massless. More precisely, the Higgs field is a doublet of fields (Φ_{upper}, Φ_{lower}), and the theory claims that right after the Big Bang nature chose one of these components (the lower one) as being the field pervading the whole universe.[27] As this choice is arbitrary, the symmetry of the Higgs doublet is in fact preserved; it is not really broken, only hidden.

Models borrowed from classical mechanics have also been heuristically instrumental in developing the SSB insight.[28] Imagine a small ball on top of a Mexican hat; this is a symmetric configuration, but not the one of minimum energy. Similar to the freezing example, the ground state is asymmetric; it is the state toward which the system tends to evolve as a result of a small perturbation, the one in which the ball tumbles down and settles into the rim in an arbitrary position (one among the infinite number of positions available).[29] In the quantum context, SSB occurs (roughly speaking) when the Lagrangian is symmetric but the ground state is not. When tried as a solution to the mass problem, the idea of a spontaneous breaking of a *global* symmetry works only up to a point. Mathematical manipulations—in fact, redefinitions of the two component fields—do deliver the sought outcome, that is, the needed mass is "generated"; yet, as it turns out, the main drawback comes from a result known as the Goldstone theorem (see Goldstone 1961, 1962), which claims in essence that the spontaneous breaking of a (global) symmetry results in the appearance of a massless spin-0 boson, the so-called "Goldstone boson." But such an entity does not exist in nature, so this specific way to exploit SSB had to be abandoned. The main insight could be rescued though, by demanding that the Mexican-hat-shaped Lagrangian be invariant under *local* gauge transformations. As expected, the logic of gauge has it that this requirement can be satisfied if a massless gauge boson is introduced. In this new context, however, the old trick works. Upon redefining the two component fields, physicists achieved exactly what they needed: the Goldstone boson disappears, as if absorbed into the massless gauge boson just added[30]—which thus acquires mass (since, as physicists' say, it "eats" the Goldstone boson). For this story to work, the element still in

---

[27] The specific choice is dictated by the so called "Higgs potential."
[28] Liu (2003) discusses SSB in the classical context.
[29] Additional difficulties occur in the matter of SSB because *only* systems with infinite degrees of freedom can undergo such "phase transitions." For philosophical discussion, see Liu (2001), Batterman (2001, 2005), Callender (2001), Ruetsche (2003, 2006), and Bangu (2009).
[30] More precisely, the Goldstone boson becomes the third polarization state of the mass-acquiring boson. The technical details on which this account draws can be found in the textbooks (e.g., Quigg 1983; Aitchison and Hey 1989; Coughlan and Dodd 1991).

need of experimental confirmation is the actual existence of the quanta of the Higgs field (the Higgs boson), one of the emerging redefined fields. As it happens, the present essay is being written during the time when experimentalists at Fermilab's Tevatron and at CERN's Large Hadron Collider are searching for the Higgs boson; as of May 2012, its existence has yet to be either confirmed or ruled out.

While various events associated with this biggest scientific experiment ever performed make the popular news regularly, what fascinates physicists and philosophers of physics alike are a series of extremely intriguing theoretical features of the Higgs field. It is generally acknowledged that this field, were it to exist, would be rather unique. The Higgs mechanism is indeed the *simplest* mechanism having the necessary features (gives mass to the gauge bosons and is incorporable in a gauge theory). It has a nonzero (renormalized) vacuum expectation value (of 246 GeV) and, as Gunion, Stange, and Willenbrock point out, "despite the simplicity of the standard Higgs model, it does not appear to be a candidate for a fundamental theory. The introduction of a fundamental scalar field is *ad hoc*; the other fields in the theory are spin-one gauge fields and spin-half fermion fields. . . . The standard Higgs model accommodates, but does not explain, those features of the electroweak theory for which it is responsible" (1996, 24). The worry here is akin to the more general and familiar philosophical antirealist concern that the postulation of entities and structures guided by pragmatic values might generate a conflict, in this case between simplicity and explanatory power.

Even when told in this oversimplified form, this story invites a diversity of philosophical questions, mainly concerning three issues: (i) the ontological implications of the electroweak unification, (ii) the nature of the Higgs mechanism, and (iii) the right epistemological attitude toward the hidden/spontaneously broken symmetries. In essence, the literature in this area emphasizes the serious difficulties encountered in the process of clarifying and interpreting the conceptual moves made by the theoreticians. In particular, a key epistemological problem stands out: insofar as it is impossible to "switch off" the Higgs field, we cannot really know what happens with the particles' mass in its absence.

Moreover, what are the metaphysical lessons to draw from the electroweak unification? This is unclear given that the pasting together of the two theories and the mixing of the fields does not seem to be much more than a mathematical, formal operation, doing rather little to vindicate the idea that we have discovered true unity in nature, as Georgi (1989), Maudlin (1996), and Morrison (2000) discuss. Furthermore, one can ask whether this case of unification lends support or actually undermines the idea that unification and explanation are connected (Morrison 2000). Another source of concern is the actual meaning of SSB and the (methodological) reasons, if any, for which we should give priority to symmetric, as opposed to asymmetric, laws (Earman 2003b, 2004b). Qualms are also raised with regard to the empirical grounds for believing in the existence of underlying hidden symmetries given that what we observe is asymmetric phenomena (Morrison 2000, 2003). Topics prompted by the analysis of the Higgs mechanism range from the difficulty of getting a grip on its gauge-invariant content (Earman) to the question of whether we should interpret it realistically or not (Morrison).[31]

5. Symmetries, Classification, and Prediction

Ontological[32] issues such as the ones discussed above are not the only area where symmetry considerations are important. They are also essential in classification and prediction; more precisely, they are instrumental in framing a new form of prediction called, by the physicists themselves, "prediction from multiplet structure" (Lipkin 1966). While not unknown to philosophers of science, this idea has not been fully investigated yet. Below I present it very briefly, by drawing on a concrete example, the prediction of the "omega minus" hadron by Gell-Mann and Ne'eman in 1962.

Since Weyl and Wigner's important successes in applying group theory to quantum mechanics, physicists have tried to replicate their methods and take inspiration from their insights. In particular, internal symmetries have become an indispensable tool in the classification of elementary particles and, as we will see, equally important achievements have followed. Wigner is credited with introducing the idea that the 2-dimensional space defined by the proton and neutron has a mathematical correspondent in the 2-dimensional irreducible representation of the group $SU(2)$. This insight is now exploited more generally: in mathematical terms, an elementary particle is conceived to be a physical system whose possible states transform into each other according to some representation of the appropriate symmetry group—in the case first discussed by Wigner, the group is $SU(2)$ and the specific way in which these transformations take place is described mathematically in terms of the *irreducible representations* of $SU(2)$. There is an intuitive way to grasp this relation (Castellani 1998): the elementarity of a particle (system) is mirrored by the irreducibility of the representation of a certain group, where the elements of the group are the transformations that govern the interaction into which the particle enters.[33] Given the symmetry group governing a physical system, the superposed states of the system (in particular, the proton*ness* and the neutron*ness* states) transform into each other according to the irreducible representations of the group. But what is the connection between the physics and the mathematics more precisely? At the physical level, there are the transformations (of superposed states),

---

[31] More recent work on the Higgs mechanism is Lyre (2008).
[32] This section draws on Bangu (2008).
[33] The literature on the (Wignerian) group theoretic approach to the constitution of physical objects has been growing in the last decade, when a variety of approaches have been attempted. See Castellani (1998) and especially the work on ontic structural realism by French (1998), French and Ladyman (2010), and Ladyman (2009), esp. sect. 4 and the bibliography therein.

and these transformations are expressed mathematically as operators acting on the state space. The eigenvalues of these operators supply the invariant numbers (to be used as labels) for identifying, or classifying, the irreducible representations of the group.[34]

Given their similarity, the proton and the neutron were thought to occupy the same place in a scheme of classification. On the face of things, one might wonder why this is important. The answer is that the epistemic context of particle physics is a special one. The taxonomical aspects are far from trivial here because elementary particles, by their very elementary nature, lack the great number of properties displayed by the medium-sized physical objects (color, shape, texture, etc.). Therefore, insofar as these typical classificatory criteria cannot be found in the micro-world, taxonomies in fundamental physics are very hard to come by, so any unambiguous criterion able to contribute to particles' identification and classification is welcome.[35] These criteria are usually supplied by the particles' associated sets of quantum numbers (mass, charge, spin, isospin, strangeness, etc.), which describe their conservation properties under various sets of transformations and thus determine their positions in multiplets. These multiplets, in turn, are mathematically determined as bases of irreducible representations of different groups of transformations (such as $U(1)$, $SU(2)$, or $SU(3)$). In addition to ordering the multitude of particles, the schemes of classification based on symmetries have been put to a different, and somewhat surprising, use: they have been instrumental in making predictions. Perhaps the most famous one is the prediction of a new hadron, called "Omega minus."

Once the physicists were provided with a multiplet scheme of classification, the prediction of the particles filling out the places left unoccupied in the multiplet came "as a matter of course" (Lipkin 1966, 25–26, 53). Physicists even introduced a term to refer to it, Lipkin calling this idea "prediction from the multiplet structure." How does this work? As noted, $SU(2)$ is not the group describing the strong interactions; they are governed by a bigger symmetry, (the flavor) $SU(3)$. Similar to what happens with the $SU(2)$ group, the dimensionalities of $SU(3)$'s irreducible representations $(1, 8, 10, 27, \ldots)$ give the cardinality of the sets of hadron multiplets.

The definitive success of this classificatory strategy came in 1964 with the detection of a new particle that completed the spin-3/2 baryon decuplet. The main idea behind this prediction is rather simple: given the classification scheme for the already known spin-3/2 baryons, the unoccupied, apparently superfluous, entry in the scheme was taken as a guide to the existence of a new particle. It was exactly this "surplus" that suggested the existence of new physical reality (in the form of

---

[34] In particular, physicists associate these labels (e.g. -1/2 and +1/2 in the isospin case) with the values of the invariant properties (isospin) characterizing physical systems (in this case, the doublet neutron-proton). Wigner (1959) derives a formula that encodes the general form of the representations. For a more modern approach, see Joshi (1982, 131).

[35] The diversity and the large number of particles had always bothered the high-energy physicists. Willis Lamb voiced this uneasiness in his Nobel speech, in which he reminded the public of a popular saying in the particle physics community: anyone who discovers a new particle ought be punished by a $10,000 fine (instead of being awarded a Nobel Prize!)

new particles, to fill in gaps in multiplets). Reminding of Mendeleev's prediction of new chemical elements from his table, Murray Gell-Mann and Yuvaal Ne'eman's predictive reasoning can be extracted from the rather detailed account of Ne'eman and Kirsh:[36]

> In 1961 four baryons of spin 3/2 were known. These were the four resonances $\Delta^-, \Delta^0, \Delta^+, \Delta^{++}$, which had been discovered by Fermi in 1952. It was clear that they could not be fitted into an octet, and the eightfold way predicted that they were part of a decuplet or of a family of 27 particles. A decuplet would form a triangle in the S—I_3 [strangeness-isospin] plane, while the 27 particles would be arranged in a large hexagon. (According to the formalism of SU(3), supermultiplets of 1, 8, 10 and 27 particles were allowed.) In the same year (1961) the three resonances Σ (1385) were discovered, with strangeness -1 and probable spin 3/2, which could fit well either into the decuplet or the 27-member family.
>
> At a conference of particle physics held at CERN, Geneva, in 1962, two new resonances were reported, with strangeness -2, and the electric charge -1 and 0 (today known as the Ξ (1530)). They fitted well into the third course of both schemes (and could thus be predicted to have spin 3/2). On the other hand, Gerson and Shoulamit Goldhaber reported a "failure": in collisions of K^+ or K^0 with protons and neutrons, one did not find resonances. Such resonances would indeed be expected if the family had 27 members. The creators of the eightfold way, who attended the conference, felt that this failure clearly pointed out that the solution lay in the decuplet. They saw the pyramid being completed before their very eyes [see figure 8.1].

> Only the apex was missing, and with the aid of the model they had conceived, it was possible to describe exactly what the properties of the missing particle should be! Before the conclusion of the conference Gell-Mann went up to the blackboard and spelled out the anticipated characteristics of the missing particle, which he called "omega minus" (because of its negative charge and because omega is the last letter of the Greek alphabet). He also advised the experimentalists to look for that particle in their accelerators. Yuval Ne'eman

---

[36] From Ne'eman and Kirsh (1996, 202–203). For more details on Gell-Mann and Ne'eman's work, see their (1964). This collection also contains the Brookhaven experimental report "Observation of a Hyperon with Strangeness Minus Three" (*Phys. Rev. Letters* 12 (1964)), which describes the details of the detection of the omega minus.

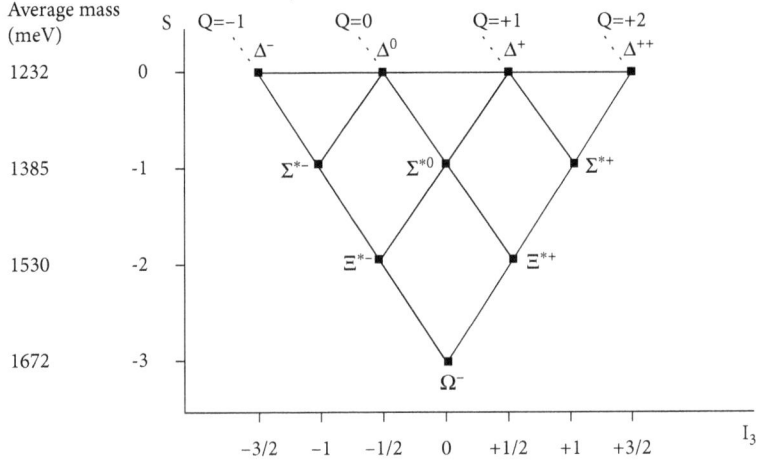

Figure 8.1 A spin-3/2 baryon decuplet

had spoken in a similar vein to the Goldhabers the previous evening and had presented them in a written form with an explanation of the theory and the prediction.

The Gell-Mann and Ne'eman predictive reasoning started off with the observation that each of the upper nine positions in the symmetry scheme[37] has a physical interpretation. In other words, each of these positions gives the physical coordinates of a known spin-3/2 baryon. Now, based on this regularity, it was conjectured that a new physical law might be in sight, claiming that the SU(3) scheme describes the correct symmetry of spin-3/2 baryons. But, were this the case, there should have been 10 baryons, not only nine, since the scheme contains 10 positions. So, a question was raised about the existence of the tenth particle. Gell-Mann noticed that the apex is formally/mathematically similar to the other nine positions, and this is so because it is, like them, an element of the scheme. Therefore, he made the prediction that the apex position has a physical interpretation too: in other words, that the coordinates of this position describe a tenth spin-3/2 baryon as well.[38] However, as is easy to notice, an extra premise, not explicitly stated in the text, is of course needed in order to complete the reasoning. This is the idea that the existence of a baryon having the predicted characteristics is not forbidden by the laws of physics, and thus *can* occur in nature (even if it was not detected so far).[39]

This line of reasoning is supposed to answer the question asked by an experimentalist ready to perform the detections (and by the politicians in charge of

---

[37] More precisely, the "scheme" refers to the "10-dimensional representation of the group SU(3)" pictured above.
[38] Its mass is 1672 MeV, strangeness −3, spin 3/2 and 0 isospin in the z-direction.
[39] The so-called "totalitarian principle" (attributed to Gell-Mann), according to which "what is not forbidden *must* occur" is of notoriety in the particle physics community. It is unclear, however, whether this dictum (reminiscent of the ancient Principle of Plenitude—stating, roughly, that given an infinite time, all genuine possibilities actualize) played an important role in this episode. Its converse—when it seems that an event can happen but it does not, look for a conservation law that precludes it—is also a well-known heuristic tool.

approving the expenses with the accelerators!), namely, "What are the grounds to believe (a) that there is a new entity in nature, and also (b) that this entity has the predicted physical characteristics?" On closer inspection, however, the reasoning involved here is very peculiar. It bears little resemblance to other famous predictions such as Leverrier and Adams's prediction of the planet Neptune in 1846, or Wolfgang Pauli's postulation of the neutrino in 1931. Unlike these two predictions, the omega minus one is essentially formal, because the criteria of similarity are given in mathematical terms (Steiner 1998). The heuristic-ontological role of the mathematical scheme of classification is thus crucial; mathematics does not play merely an eliminable, descriptive, or computational role, since in this case it seems that the predictive argument cannot be reconstructed without invoking the mathematical features of the physical description. This appears to be a rather startling example of the "unreasonable effectiveness of mathematics" (Wigner 1960), though further analysis is certainly needed in order to sort out what is unreasonable about the success of this, and other, applications.[40]

6. Final Remarks: "The Reversal of a Trend"

While the present survey has touched upon a relatively wide range of arguments and issues, a central theme is worth discussing before the end: the emergence of a new relation, without precedent in the history of physics, between symmetries and the laws of nature—an idea advocated by Wigner, who in turn attributes it to Einstein. Wigner highlights it in a number of his philosophical writings, a particularly sharp formulation of it being given in his Nobel lecture of 1963.

Wigner begins by drawing attention to an important distinction, which, he suggests, seems to have been first perceived by Newton: roughly, this is the separation between a certain kind of general statement expressing a regularity holding between events (a law of nature), and more specific statements taking the form of descriptions of current states of affairs (the initial conditions).[41] To Wigner, the distinction is methodologically crucial: its introduction simply delineates the very object of the physical science, as it amounts to "the specification of the explainable" (1967, 39); moreover, this specification "may have been the greatest discovery of physics so far." So, according to him, it is wrong to say that physics explains "nature," if this means that the focus of physical theorizing is on accounting for some specific states of affairs. The aim of physics is rather "to explain the regularities in the behaviour of objects" (1967, 39). He puts this contrast in historical terms, mentioning Kepler, who, as we saw, was also concerned with finding what determines the specific magnitudes of the planetary orbits, in addition to his search for the laws of motion.

---

[40] See Steiner (1998, 2005), French (2000), Wilson (2000), Bangu (2006), Maddy (2007, part IV.2), Batterman (2006, 2010), and Pincock (2010) for a variety of recent perspectives on this issue.
[41] See Ryckman (2008) for discussion.

On the other hand, Newton had restricted his interest to searching only for the explanation of the regularities, the laws of motion (1967, 39–40).

If physicists' interest lies in describing, explaining, and predicting regular behavior, then the separation of laws from initial conditions becomes necessary, as the laws do not also specify their initial conditions. While the former are "precise beyond anything reasonable," "we know virtually nothing" about the latter (1967, 40), as they contain "a strong element of randomness" (1967, 41–42).[42] Thus, in the Wignerian scheme the object of physics is to discover the laws of nature, which describe regular correlations between events. A new event is predicted and explained if such a correlation, or law, is known. But, what about the laws themselves? The question is thus whether there might exist "a *superprinciple* which is in a similar relation to the laws of nature as these are to events" (1967, 43, emphasis added). That is, just as the laws constrain what events might take place in the world, such a superprinciple would constrain, or determine, the laws—in other words, it would have the role of a meta-law which would somehow explain the laws.[43] While Wigner never states this heuristic superprinciple explicitly, it is pretty clear that he understands it as incorporating certain constraints, which he calls "invariance principles," ("symmetry principles" or "invariance transformations").

Wigner explains the role of the superprinciple as follows. Suppose that events A, B, C, . . . entail the occurrence of another event X—this is the general form of a law of nature. The question is whether there are transformations that turn A, B, C into A′, B′, C′, . . . and X into X′ such that if A′, B′, C′, . . . obtain, then X′ obtains too—so, once these transformations are found, new laws of nature are established. Next, he identifies three types of such "invariance transformations." First, there are the Euclidean transformations, where the primed events are identical to the unprimed ones except for a shift in the location in space. In particular, the spatial relations within the configurations of primed and unprimed events are retained. Second, there are time displacements: the primed events are the same as the unprimed ones, but they occur at a different time, and the time intervals in the primed and unprimed configurations are retained. Third, there is the uniform motion transformation: when assessed from the perspective of a uniformly moving coordinate system, the primed events appear to be identical to the unprimed events (Wigner 1967, 43). Having identified the components of the superprinciple (of which the last invariance was crucial in Einstein's formulation of STR), Wigner points out that "the use of the set of invariance principles which is surely most important at present" is as a test for "the validity of possible laws of nature" (1967, 46). More precisely, "a law of nature can be accepted as valid only if the correlations which it postulates are consistent with the accepted invariance principles" (1967, 46)—and, historically, the first illustration of this insight is Einstein's 1905 construction of STR.[44] In fact, Wigner stresses that viewed from this perspective, Einstein's work on STR is

---

[42] Wigner also discusses to what extent the initial conditions are arbitrary; see pp. 40–41.
[43] For symmetries as meta-laws, see Lange (2007).
[44] See Norton (2004) and the bibliography therein for historical details and philosophical discussion of Einstein's methodology prior to, and related to, his 1905 STR paper.

methodologically revolutionary: it marks "the reversal of a trend" (1967, 5). Before it, the principles of invariance were seen merely as interesting features to be noted when examining the laws; after it, a new methodological move became available: "it is now natural for us to derive the laws of nature and to test their validity by means of the laws of invariance, rather than to derive the laws of invariance from what we believe to be the laws of nature" (1967, 5).

Even a cursory glance should support the view that symmetry is an extremely generous foundational topic in physics and philosophy. It prompts questions about the relation between the formal or mathematical structures and the constitution of the world (does the world really exhibit symmetrical structures, or they are just an artifact of our description of it?), and it also challenges us to make sense of the success of symmetry thinking at a methodological-heuristic level. Regardless of the perspective from which this topic is approached, understanding the power of symmetry is certainly one of the most important tasks for both philosophers of science and philosophically-minded physicists.

References

Aitchison, L. J. R., and A. J. G. Hey (1989). *Gauge theories in particle physics.* Bristol: IOP Publishing.
Albert, D. (2000). *Time and chance.* Cambridge, MA: Harvard University Press.
Arntzenius, F., and H. Greaves (2009). Time reversal in classical electromagnetism. *British Journal for the Philosophy of Science* 60(3): 557–584.
Auyang, S. (1995). *How is quantum field theory possible?* Oxford: Oxford University Press.
Bangu, S. (2006). Steiner on the applicability of mathematics and naturalism. *Philosophia Mathematica* 14(1): 26–43.
——. (2008). Reifying mathematics? Prediction and symmetry classification. *Studies in History and Philosophy of Modern Physics* 39(2): 239–258.
——. (2009). Understanding thermodynamic singularities: Phase transitions, data and phenomena. *Philosophy of Science* 76(4): 488–505.
Batterman, R. W. (2001). *The devil in the details: Asymptotic reasoning in explanation, reduction, and emergence.* Oxford: Oxford University Press.
——. (2005). Critical phenomena and breaking drops: Infinite idealizations in physics. *Studies in History and Philosophy of Modern Physics* 36: 225–244.
——. (2006). On the specialness of special functions (The nonrandom effusions of the divine mathematician). *British Journal for the Philosophy of Science* 58: 263–286.
——. (2010). On the explanatory role of mathematics in empirical science. *British Journal for the Philosophy of Science* 61: 1–25.
Belot, G. (1996). Whatever is never and nowhere is not: Space, time, and ontology in classical and quantum gravity. Ph.D. diss., University of Pittsburgh. Available at http://sitemaker.umich.edu/belot/files/dissertation.pdf.
——. (1998). Understanding electromagnetism. *British Journal for the Philosophy of Science* 49: 531–555.

Brading, K., and H. Brown (2003). Symmetries and Noether's theorems. In *Symmetries in physics: Philosophical reflections*, ed. K. Brading and E. Castellani, 89–109. Cambridge: Cambridge University Press.

———. (2004). Are gauge symmetry transformations observable? *British Journal for the Philosophy of Science* 55(4): 645–665.

Brading, K., and E. Castellani, eds. (2003a). *Symmetries in physics: Philosophical reflections*. Cambridge: Cambridge University Press.

———. (2003b). "Introduction." In Brading and Castellani (2003a), 1–18.

———. (2007). Symmetry in classical physics. In *Handbook of the philosophy of physics*, ed. J. Butterfield and J. Earman, 1331–1367. Amsterdam: Elsevier.

Brown, H. (1999). Aspects of objectivity in quantum mechanics. In *From physics to philosophy*, ed. J. Butterfield and C. Pagonis, 45–71. Cambridge: Cambridge University Press.

Butterfield, J. (1989). The hole truth. *British Journal for the Philosophy of Science* 40: 1–28.

———. (2006). On symmetries and conserved quantities in classical mechanics. In *Physical theory and its interpretation*, ed. W. Demopoulos and I. Pitowsky, 43–99. Dordrecht: Springer.

Callender, C. (2001). Taking thermodynamics too seriously. *Studies in History and Philosophy of Modern Physics* 32: 539–553.

Castellani, E. (1998). Galilean particles: An example of constitution of objects. In *Interpreting bodies: Classical and quantum objects in modern physics*, ed. E. Castellani, 181–194. Princeton: Princeton University Press.

Christenson, J., J. Cronin, V. Fitch, and R. Turlay (1964). Evidence for the 2π decay of the K Meson. *Phys. Rev. Letters* 13: 138.

Coughlan, G. D., and J. E. Dodd (1991). *The ideas of particle physics*. 2d ed. Cambridge: Cambridge University Press.

Curie, P. (1894). Sur la symétrie dans les phénomènes physiques : Symétrie d'un champ électrique et d'un champ magnétique. *Journal de Physique* 3(1) : 393–417.

Debs, T., and M. Redhead (2007). *Objectivity, invariance, and convention: Symmetry in physical science*. Cambridge, MA: Harvard University Press.

Dyson, F. (1964). Mathematics in the physical sciences. *Scientific American* 211(3): 129–146.

Earman, J. (2003a). Tracking down gauge: An ode to the constrained Hamiltonian formalism. In *Symmetries in physics: Philosophical reflections*, ed. K. Brading and E. Castellani, 140–162. Cambridge: Cambridge University Press.

———. (2003b). Rough guide to spontaneous symmetry breaking. In *Symmetries in physics: Philosophical reflections*, ed. K. Brading and E. Castellani, 334–345. Cambridge: Cambridge University Press.

———. (2004a). Curie's principle and spontaneous symmetry breaking. *International Studies in the Philosophy of Science* 18: 173–198.

———. (2004b). Laws, symmetry, and symmetry breaking: Invariance, conservation principles, and objectivity. *Philosophy of Science* 71(5): 1227–1241.

Earman, J., and J. Norton (1987). What price spacetime substantivalism? The hole story. *British Journal for the Philosophy of Science* 38: 515–525.

Feynman, R. (1985). *QED*. Princeton: Princeton University Press.

Francoise, J. L., G. L. Naber, and T. S. Tsun, eds. (2006). *Encyclopedia of mathematical physics*. Berlin: Springer.

French, S. (1998). On the withering away of physical objects. In *Interpreting bodies: Classical and quantum objects in modern physics*, ed. E. Castellani, 93–113. Princeton: Princeton University Press.

———. (2000). The reasonable effectiveness of mathematics: Partial structures and the application of group theory to physics. *Synthese* 125: 103–120.

———. (2008). Identity and individuality in quantum theory. In *The Stanford encyclopedia of philosophy* (Fall 2008 edition), ed. E. Zalta. Available at http://plato.stanford.edu/archives/fall2008/entries/qt-idind/.

French, S., and J. Ladyman (2010). In defence of ontic structural realism. In *Scientific structuralism*, ed. P. Bokulich and A. Bokulich, 25–43. Boston: Boston Studies in the Philosophy of Science, Springer.

French, S., and D. Rickles (2003). Understanding permutation symmetry. In *Symmetries in physics: Philosophical reflections*, ed. K. Brading and E. Castellani, 212–238 Cambridge: Cambridge University Press.

Gell-Mann, M., and Y. Ne'eman, eds. (1964). *The eightfold way.* New York: W. A. Benjamin.

Georgi, H. (1989). Grand unified theories. In *The New Physics*, ed. P. Davies, 425–446. New York: Cambridge University Press.

Goldstone, J. (1961). Field theories with superconductor solutions. *Nuovo Cimento* 19: 154–164.

Goldstone, J., et al. (1962). Broken symmetries. *Physical Review* 127: 965–970.

Greaves, H. (2010). Towards a geometrical understanding of the CPT Theorem. *British Journal for the Philosophy of Science* 61: 27–50.

Gunion, J. F., A. Stange, and S. Willenbrock (1996). Weakly-coupled Higgs bosons. In *Electroweak symmetry breaking and the new physics at the TeV scale*, ed. T. L. Barklow, S. Dawson, H. E. Haber, and J. L. Siegrist, 23–124. Singapore: World Scientific Publishing.

Healey, R. (2007). *Gauging what's real: The conceptual foundations of contemporary gauge theories.* Oxford: Oxford University Press.

Higgs, P. W. (1964). Broken symmetries and the masses of gauge bosons. *Phys. Rev. Lett.* 13: 508–509.

Hoefer, C. (2000). Kant's hands and Earman's pions: Chirality arguments for substantival space. *International Studies in the Philosophy of Science* 14: 237–256.

Hon, G., and B. R. Goldstein (2008). *From summetria to symmetry: The making of a revolutionary scientific concept.* Berlin: Springer.

Huggett, N. (2003). Mirror symmetry: What is it for relational space to be orientable? In *Symmetries in physics: Philosophical reflections*, ed. K. Brading and E. Castellani, 281–289. Cambridge: Cambridge University Press.

Icke, V. (1995). *The force of symmetry.* Cambridge: Cambridge University Press.

Ismael, J. (1997). Curie's principle. *Synthese* 110: 167–190.

Joshi, A. W. (1982). *Elements of group theory for physicists.* 3d ed. Hoboken, NJ: John Wiley & Sons.

Kosso, P. (2000). The empirical status of symmetries in physics. *British Journal for the Philosophy of Science* 51(1): 81–98.

———. (2003). Symmetry, objectivity, and design. In *Symmetries in physics: Philosophical reflections*, ed. K. Brading and E. Castellani, 410–421. Cambridge: Cambridge University Press.

Ladyman, J. (2009). Structural realism. In *The Stanford encyclopedia of philosophy* (Summer 2009 edition), ed. E. Zalta. Available at http://plato.stanford.edu/archives/sum2009/entries/structural-realism/.

Lanczos, C. (1949). *The variational principles of mechanics.* Toronto: University of Toronto Press.

Lange, M. (2007). Laws and meta-laws of nature: Conservation laws and symmetries. *Studies in History and Philosophy of Modern Physics* 38: 457–481.

Lipkin, H. (1966). *Lie groups for pedestrians*. Amsterdam: North-Holland Publishing Company.

Liu, C. (2001). Infinite systems in SM explanations: Thermodynamic limit, renormalization (semi-) groups, and irreversibility. *Philosophy of Science* 68: S325–S344.

———. (2003). Spontaneous symmetry breaking and chance in a classical world. *Philosophy of Science* 70: 590–608.

Lyre, H. (2008). Does the Higgs mechanism exist? *International Studies in the Philosophy of Science* 22(2): 119–133.

Maddy, P. (2007). *Second philosophy*. Oxford: Oxford University Press.

Mainzer, K. (2005). *Symmetry and complexity: The spirit and beauty of nonlinear science*. Singapore: World Scientific Publishing.

Malament, D. (2004). On the time reversal invariance of classical electromagnetic theory. *Studies in History and Philosophy of Modern Physics* 35B (2): 295–315.

Martin, C. (2003). On continuous symmetries and the foundations of modern physics. In *Symmetries in physics: Philosophical reflections*, ed. K. Brading and E. Castellani, 29–60. Cambridge: Cambridge University Press.

Maudlin, T. (1996). On the unification of physics. *Journal of Philosophy* 93: 129–144.

Mills, R. (1989). Gauge fields. *American Journal of Physics* 57: 493.

Morrison, M. (2000). *Unifying scientific theories: Physical concepts and mathematical structures*. Cambridge: Cambridge University Press.

———. (2003). Spontaneous symmetry breaking: Theoretical arguments and philosophical problems. In *Symmetries in physics: Philosophical reflections*, ed. K. Brading and E. Castellani, 346–362. Cambridge: Cambridge University Press.

———. (2008). Symmetry. In *The Routledge companion to philosophy of science*, ed. S. Psillos and M. Curd, 468–478. London: Routledge.

Ne'eman, Y., and Y. Kirsh (1996). *The particle hunters*. 2d ed. Cambridge: Cambridge University Press.

Nerlich, G. (1994). *The shape of space*. 2d ed. Cambridge: Cambridge University Press.

Noether, E. (1918). Invariante variationsprobleme. *Konigliche Gesellschaft der Wissenschaften zu Gottingen. Mathematisch-Physikalische Klasse. Nachrichten*, 235–257. Trans. M. A. Tavel (1971). Invariant Variation Problems. In *Transport Theory and Statistical Physics*, I, pp. 186–207.

Norton, J. (2004). Einstein's investigations of Galilean covariant electrodynamics prior to 1905. *Archive for History of Exact Sciences* 59: 45–105.

———. (2008). The hole argument. In *The Stanford encyclopedia of philosophy* (Winter 2008 edition), ed. E. Zalta. http://plato.stanford.edu/archives/win2008/entries/spacetime-holearg/.

Nozick, R. (2001). *Invariances: The structure of the objective world*. Cambridge, MA: Harvard University Press.

Pickering, A. (1998). Against putting the phenomena first: The discovery of the weak neutral current. In *Scientific knowledge: Basic issues in the philosophy of science*, ed. J. Kourany, 135–152. Stamford, CT: Wadsworth Publishing Company. Reprinted from *Studies in History and Philosophy of Science* 15(2): 85–117.

Pincock, C. (2010). Applicability of mathematics. In *Internet encyclopedia of philosophy*. Available at http://www.iep.utm.edu/math-app/.

Pooley, O. (2003). Handedness, parity violation, and the reality of space. In *Symmetries in physics: Philosophical reflections*, ed. K. Brading and E. Castellani, 250–280. Cambridge: Cambridge University Press.

Quigg, C. (1983). *Gauge theories of the strong, weak, and electromagnetic interactions*. Reading, MA: Addison-Wesley.

Redhead, M. (2003). The interpretation of gauge symmetry. In *Symmetries in physics: Philosophical reflections*, ed. K. Brading and E. Castellani, 124–140. Cambridge: Cambridge University Press.

Robson, E. (2008). *Mathematics in Ancient Iraq: A social history*. Princeton: Princeton University Press.

Ruetsche, L. (2006). Johnny's so long at the ferromagnet. *Philosophy of Science* 73: 473–486.

———. (2003). A matter of degree: Putting unitary inequivalence to work. *Philosophy of Science* 70: 1329–1342.

Ryckman, T. A. (2003). The philosophical roots of the gauge principle: Weyl and transcendental phenomenological idealism. In *Symmetries in physics: Philosophical reflections*, ed. K. Brading and E. Castellani, 61–88. Cambridge: Cambridge University Press.

———. (2008). Invariance principles as regulative ideals: From Wigner to Hilbert. *Royal Institute of Philosophy Supplement* 83: 63–80.

Saunders, S. (2002). Indiscernibles, general covariance, and other symmetries. In *Revisiting the foundations of relativistic physics: Festschrift in honour of John Stachel*, ed. A. Ashtekar, D. Howard, J. Renn, S. Sarkar, and A. Shimony, 151–173. Dordrecht: Kluwer.

Schumm, B. (2004). *Deep down things*. Baltimore, MD: Johns Hopkins University Press.

Sklar, L. (1993). *Physics and chance*. Cambridge: Cambridge University Press.

Smith, S. (2008). Symmetries and the explanation of conservation laws in the light of the inverse problem in Lagrangian mechanics. *Studies in the History and Philosophy of Modern Physics* 39: 325–345.

Steiner, M. (1998). *The applicability of mathematics as a philosophical problem*. Cambridge, MA: Harvard University Press.

———. (2005). Mathematics—Application and applicability. In *Oxford handbook of philosophy of mathematics and logic*, ed. S. Shapiro, pp. 625–61. Oxford: Oxford University Press.

Teller, P. (2000). The gauge argument. *Philosophy of Science* 67: S466–S481.

van Fraassen, B. C. (1989). *Laws and symmetry*. Oxford: Oxford University Press.

———. (1991). *Quantum mechanics: An empiricist view*. Oxford: Clarendon Press.

Wald, R. M. (1986). Spin-two fields and general covariance. *Physical Review D* 33: 3613–3625.

Weinberg, S. (1974). Recent progress in gauge theories of the weak, electromagnetic and strong interactions. *Reviews of Modern Physics* 46: 255–277.

Weyl, H. (1952). *Symmetry*. Princeton: Princeton University Press.

Wigner, E. (1959). *Group theory and its application to the quantum mechanics of atomic spectra*. London: Academic Press.

———. (1960). The unreasonable effectiveness of mathematics in the natural sciences. *Communications in Pure and Applied Mathematics* 13(1): 1–14. Reprinted in Wigner (1967), 222–238.

———. (1963). Events, laws of nature, and invariance principles. Nobel Lecture, December 12, 1963. Reprinted in *Science* 145(3636): 995–999 and Wigner (1967), 38–51. References are given to the 1967 version.

———. (1967). *Symmetries and reflections*. Bloomington: Indiana University Press.

Wilkinson, D. H., ed. (1969). *Isospin in nuclear physics.* Amsterdam: North-Holland Pub. Co.

Wilson, M. (2000). The unreasonable uncooperativeness of mathematics in the natural sciences. *The Monist* 83(2): 296–314.

Wu, C. S., Ambler, E., Hayward, R. W., Hoppes, D. D.; Hudson, R. P. (1957). Experimental Test of Parity Conservation in Beta Decay. *Physical Review* 105(4): 1413–1415.

Yang, C. N., and R. Mills (1954). Conservation of isotopic spin and isotopic gauge invariance. *Physical Review* 96(1): 191–195.

Zee, A. (2007). *Fearful symmetry: The search for beauty in modern physics.* Princeton: Princeton University Press.

CHAPTER 9

SYMMETRY AND EQUIVALENCE

GORDON BELOT

1. INTRODUCTION

My topic is the relation between two notions, that of a symmetry of a physical theory and that of the physical equivalence of two solutions or models of such a theory. In various guises, this topic has been widely addressed by philosophers in recent years.[1]

As I intend to use the term here, a symmetry of a theory is a map or transformation that leaves invariant the structure used to encode the laws of the theory. The notion of physical equivalence that I have in mind is as follows: two solutions (models) of a physical theory are *physically equivalent* if and only if, for each possible physical situation, the two are equally well- or ill-suited to represent that situation. The first of these notions is a formal one, in the sense that in specifying the formalism of a theory, one specifies its symmetries. The second is an interpretative notion: two solutions that are physically equivalent according to one interpretation of a given formalism may be inequivalent according to another.

Part of the interest in these notions among philosophers derives from the following line of thought. In an influential discussion, John Earman argued that in

---

Versions of this essay were given (under a different and better title) in Chicago, Ann Arbor, and Montreal—thanks to all those present. For helpful comments and discussion, thanks to Dave Baker, Bob Batterman, Katherine Brading, Caro Brighouse, Kevin Coffey, Hilary Greaves, Juha Pohjanpelto, Peter Railton, Laura Ruetsche, Jamie Tappenden, and Bas van Fraassen. Special thanks to Old Man Batterman for his great patience and editorial sagacity—even if he doesn't know a good title when he sees one.

[1] See, e.g., Baker (2010), Brading and Castellani (2007), Dasgupta (2010), Debs and Redhead (2007), Healey (2009), Ismael and van Fraassen (2003), North (2010), and Roberts (2008).

the special case of spacetime symmetries, facts about symmetries place interesting constraints on good interpretative practice.[2] Consider the case of time translation by b temporal units. In the first instance, this operation is a map that takes each point of spacetime to a point of spacetime b units later than itself. But we can also think of time translation as acting in an obvious way on fields living on spacetime. Earman observes that in this context, a transformation like time translation is a symmetry of a theory if and only if it maps solutions of the theory to solutions of the theory. And he goes on to argue that under any reasonable interpretation, a transformation of this kind ought to be a symmetry of the theory if and only if it is a symmetry of the geometric structure of spacetime.[3] In particular, it follows from Earman's arguments that if one focuses on theories in which spacetime symmetries are the only relevant symmetries, then under any reasonable interpretation, solutions are physically equivalent if and only if they are related by a symmetry—in this way formal facts place interesting constraints on (good) interpretation.

It is natural to wonder how much of this picture carries over if one moves beyond spacetime symmetries. Much recent philosophical writing on symmetries and related topics seems to assume that the answer is: all of it.[4] For the following two doctrines play an important (if often implicit) role in this literature.

D1. The symmetries of a classical theory are those transformations that map solutions of the theory's equation of motion to solutions of the theory's equation of motion.
D2. Two solutions of a classical theory's equation of motion are related by a symmetry if and only if they are physically equivalent, in the sense that they are equally well- or ill-suited to represent any particular physical situation.

Here D1 is a formal condition, D2 a constraint on good interpretative practice. Combined, they would yield a strong and interesting constraint on (good) interpretation.

However, it is not difficult to see that the combination of D1 and D2 is an unhappy one. Consider any classical theory. Let u_1 and u_2 be solutions of the theory's equation of motion. Consider the transformation T on the space of solutions that maps u_1 to u_2, u_2 to u_1, and leaves every other solution where it is. According to the first doctrine above, T is a symmetry of the theory. The second doctrine above then implies that u_1 and u_2 are physically equivalent. So the two doctrines jointly imply that any two solutions of any classical theory are physically equivalent—and hence

---
[2] See Earman (1989, §3.4).
[3] If spacetime geometry fails to be invariant under some such transformation that is a symmetry of the laws, then one is positing unnecessary geometric structure (think of Newton on absolute space). Conversely, if one's spacetime structure is invariant under transformations that are not symmetries of the laws, then one is cheating somehow or other, employing some structure in one's formalism to break a spacetime symmetry without being willing to pay the ontological price (think of someone whose theory has a point mass sproinging back and forth about the origin in the Cartesian plane, but who is unwilling to posit a privileged point of space or any physical structure—matter, field, force, etc.—other than the single moving point-particle).
[4] I won't do anything to substantiate this claim here (let us not descend into recriminations at this time!), beyond saying that, like many a rant, this one is directed in part at earlier versions of its author.

that each classical theory is more or less useless because it is unable to discriminate among the systems for which it provides good models.[5]

There are many strategies for encoding the laws of a theory in a mathematical form—and each such strategy is associated with a formal notion of symmetry. D1 encapsulates the notion of symmetry associated with the most spare and naïve method of encoding the laws of a theory. It is natural to wonder whether some more sophisticated relative of D1 might form a fruitful combination with D2. There has been some recent discussion of this question among philosophers, with some authors adopting an optimistic view, arguing or assuming that some more sophisticated relative to D1 will do the trick, and some authors adopting a pessimistic view, according to which the notion of physical equivalence is not closely linked with a formal notion of symmetry (even when we restrict to well-behaved interpretations).[6]

My project here is to examine the viability of D1 and D2 and to say a bit about the source of their prima facie plausibility. In the next section, I briefly sketch the framework in which I will proceed. Section 3 is devoted to a discussion of D1 and its inadequacies. Sections 4 and 5 give a brief overview of some of the more sophisticated alternatives to D1. Section 6 argues that none of the notions of symmetry considered combine satisfactorily with D2. Section 7 presents my cautiously pessimistic conclusions.

2. Stage-Setting

Consider a classical theory in which a history of a physical system is represented by a function $u: M \to W$; here M is the manifold parameterized by the independent variables of the theory and W is the manifold parameterized by the dependent variables of the theory.[7] Thus M might correspond to time and W to the space

---

[5] Our world contains many systems that can be treated (for all practical purposes) as being isolated. It follows from the conjunction of D1 and D2 that if a theory provides equally good models of two isolated systems, then *any* model of the theory must provide equally good representations of each of these systems—which is to say that the theory is just about useless in application.

[6] Since none of the authors I have in mind address the topic in quite the same terms employed here (or in quite the same terms as one another), the following attributions should perhaps be taken with a grain of salt. On the optimistic side: Baker (2010, §1) conjectures that D2 is true relative to some generalization of the Hamiltonian approach discussed in §5 below; Brading and Castellani (2007, §4.1 and 8.2) claim that D2 is true relative to the notion of a classical symmetry discussed in §4 below; Roberts (2008, fn. 3) holds that D2 is true for the sort of version of D1 discussed in fn. 11 below. On the pessimistic side: Ismael and van Fraassen (2003) take D1 to be the only philosophically salient notion of symmetry—and thus think that in order to get around the sort of problems noted above, one should take models to be physically equivalent if and only if they are related by a symmetry and agree in (roughly speaking) their directly perceivable features; Healey (2009) likewise notes some of the problems surrounding D1, mentions that there are other formal notions of symmetry available, then pursues an approach that builds physical equivalence into its notion of symmetry; Dasgupta (2010) argues that a notion of symmetry appropriate to D2 must involve notions like that of a mental state.

[7] More generally, here and throughout, one could take histories to be represented by sections of a fiber bundle $E \to M$ with typical fiber W. Almost any classical field can be thought of this way. For example, a gauge field, standardly represented by a connection one-form on a principle bundle $P \to M$, can be represented instead

of possible configurations of some system of particles. Or M might correspond to spacetime and W to the space of values available to some field at each point of spacetime. The laws of the theory are given by an equation,

$$\Delta[u] = 0,$$

where Δ is a (linear or nonlinear) differential operator.[8] The job performed by such an equation is to single out from a large space $\mathcal{K}$ of functions that represent "kinematically possible" histories of the system those that represent dynamically possible histories. Typically, $\mathcal{K}$ is specified by specifying the independent and dependent variables of the theory and the degree of regularity (smoothness etc.) of candidate functions, as well as any (asymptotic) boundary conditions that they must satisfy. We call the subspace $\mathcal{S} \subset \mathcal{K}$ consisting of those u that satisfy the equation of the theory the *space of solutions*.

3. A Recipe for Disaster

Here is a recipe for arriving at the characterization of symmetries of classical theories that is embodied in the doctrine D1 discussed above: begin with the standard abstract notion of a symmetry; reflect on the case of spacetime symmetries; extrapolate.

Symmetries, abstractly speaking. There is a basic notion of symmetry, elaborated in various ways in the various branches of mathematics. Consider a *structure*—a set of objects, D, equipped with some relations, R_1, ..., and functions, f_1, ... A *symmetry* of this structure is a one-to-one and onto map $F: D \to D$ that preserves all of the relations and functions between the objects.[9]

Spacetime symmetries. Suppose that we want to check whether a given field theory is invariant under a spacetime transformation such as time-translation. Then all we have to do is to check to make sure that whenever $u(x, t)$ is a solution of the theory's equation of motion, so is $u(x, t - b)$, for any real number b. That is, we define in the obvious way an operator on the set of kinematically possible fields that implements the putative symmetry, then check to see whether it maps solutions to solutions.

Generalization. Suppose that we think of a classical theory as a structure consisting of a large set of functions $\mathcal{K}$ with a distinguished subset $\mathcal{S}$. Then we would expect

---

by a section of a certain affine bundle $J^1(P)/G \to M$ associated with P; see, e.g., Kolář, Michor, and Slovák (1993, §17.4).

[8] So the left-hand side of this equation is the function on M that results from applying Δ to u and the right-hand side is the zero function.

[9] I.e., $R(x_1, \ldots, x_n)$ if and only if $R(F(x_1), \ldots, F(x_n))$; and $f(x_1, \ldots, x_n) = y$ if and only if $f(F(x_1), \ldots, F(x_n)) = F(y)$.

the symmetries of such a theory to be those one-to-one and onto maps from $\mathcal{K}$ to itself that preserve $\mathcal{S}$. Spacetime symmetries fit this pattern exactly! This suggests that we ought to think of classical theories in the way just outlined.

Something has gone wrong. Following this recipe leads to the following

> Fruitless Definition: A *symmetry* of a differential equation Δ is a one-to-one and onto map $T : u \in \mathcal{K} \mapsto T(u) \in \mathcal{K}$ that preserves the space of solutions, in the sense that $u \in \mathcal{S}$ if and only if $T(u) \in \mathcal{S}$.

As an attempt to capture the ordinary notion of a symmetry of an equation (and hence of a theory), this is a disaster. Ordinarily, symmetries of theories are hard to come by. But some remarkable theories have atypically large symmetry groups (relative to the number of degrees of freedom of the systems that they treat). The definition above effaces this sort of distinction between theories. For if we allow arbitrary permutations of the solutions of a theory to count as symmetries, then the size of a theory's group of symmetries depends only on the size of its space of solutions.[10]

All of this is liable to strike fans of the Fruitless Definition as cheap and misleading: after all, since the spaces involved have topological and differential structure, surely one should restrict attention to continuous or smooth transformations of the space of kinematic possibilities? This is fair enough—but it would not make much difference. At best, it would lead us to identify the symmetries of a theory with something like the family of smooth permutations of the space of solutions, rather than with the full family of permutations of that space.[11] But this would still lead us to the conclusion that every theory had an enormous group of symmetries that depended only on relatively coarse features of its space of solutions.[12]

---

[10] Don't mathematicians sometimes offer up characterizations of symmetries along these lines? Yes—but only when speaking loosely and heuristically. Thus in the introduction to Olver's influential textbook on symmetries of differential equations, we are told that: "Roughly speaking, a symmetry group of a system of differential equations is a group which transforms solutions of the system to other solutions" Olver (1993, xviii)—see also, e.g., Bluman and Anco (2002, 2) and Klainerman (2008, 457). But on the next page we are told that once "one has determined the symmetry group of a system of differential equations, a number of applications become available. To start with, one can directly use the defining property of such a group and construct new solutions to the system from known ones." Of course, one cannot do this if one's notion of a symmetry is given by the Fruitless Definition—one needs to be working with one of the more specialized notions that are the focus of Olver's book, some of which are described below. And likewise for the other applications on Olver's list.

[11] Roughly speaking: for non-freaky Δ, one expects $\mathcal{S}$ to sit inside $\mathcal{K}$ as (something like) a submanifold; if ϕ_1 and ϕ_2 are (generic) points in $\mathcal{S}$ then there will be a diffeomorphism $F : \mathcal{S} \to \mathcal{S}$ with $\phi_2 = F(\phi_1)$; and one expects to be able to extend F to a diffeomorphism $\tilde{F} : \mathcal{K} \to \mathcal{K}$. (Slightly more carefully: one expects that any obstructions to extending F in this way are going to be technical in nature, and not detract from the main point.)

[12] Consider, by way of illustration, the Newtonian theory of three gravitating point particles of distinct masses. Here a point in the space of kinematically possible fields, $\mathcal{K}$, essentially assigns each of the particles a worldline in spacetime (without worrying about whether these worldlines jointly satisfy the Newtonian laws of motion). The space of solutions, $\mathcal{S}$, is the 18-dimensional submanifold of $\mathcal{K}$ consisting of points corresponding to particle motions obeying Newton's laws. So one expects that any diffeomorphism from $\mathcal{S}$ itself can be extended to a suitably nice map from $\mathcal{K}$ to itself. But for any solutions $u_1, u_2 \in \mathcal{S}$, we can find a diffeomorphism from $\mathcal{S}$ to itself that maps u_1 to u_2, so we again find that arbitrary pairs of solutions are related by symmetries. This seems unacceptable, since we ordinarily think of this theory as having a relatively small symmetry group (consisting just of spacetime symmetries).

4. Symmetries of Differential Equations

Laws of classical physics are given by differential equations—and there are a number of ways that a differential equation can be encoded in a structure whose symmetries can be investigated. D1 corresponds to the most flat-footed of these. In this section and the next we briefly survey some more sophisticated alternatives: the focus in this section is on approaches that directly encode the equation of motion of the theory in a structure whose symmetries can be identified with the symmetries of the theory; in the next section, we consider approaches that take a detour via a Lagrangian or Hamiltonian formulation of the theory.

There is no settled, definitive notion of a symmetry of a differential equation—rather there are a family of related notions.[13] Three are especially important: the notion of a classical symmetry, the notion of a generalized symmetry, and the notion of a nonlocal symmetry. I will sketch the first of these and then describe how the other two are related to it. A few more details concerning classical symmetries are presented in an appendix below.

Let us focus on the field-theoretic case: the independent variables of the theory parameterize the spacetime manifold M; a configuration of the field is represented by a function $u : M \to W$; and the dynamical laws of the theory are encoded in a kth-order partial differential equation Δ. Roughly speaking, the classical symmetries of such a theory can be characterized as follows. Let E be the manifold $M \times W$ that is parameterized by the independent and dependent variables of the theory taken together. Any diffeomorphism $\bar{d} : E \to E$ will induce a transformation d from the space of kinematically possible fields of the theory to itself (this claim is unpacked in the appendix below). A map d that arises in this way is a *classical symmetry* of the equation if and only if it maps solutions to solutions.[14] So classical symmetries are, roughly speaking, transformations of the space of kinematically possible fields that: (a) map solutions to solutions; and (b) are suitably local, in the sense that they arise from smooth transformations of the dependent and independent variables of the theory. Classical symmetries are also known as *Lie symmetries* or *point symmetries*. The *spacetime symmetries* of a theory are those classical symmetries that arise from transformations of the independent variables of the theory that leave the dependent variables untouched.[15] (For particle theories in fixed spacetime backgrounds, the spacetime symmetries are those translation, rotations, boosts, and dilatations that map solutions to solutions).

---

[13] For an historical overview and references, see Olver (1993, 172ff. and 374ff.). There is of course a trade-off to be made between fecundity and generality. In practice, the notions that mathematicians and physicists find interesting are far more restrictive than the fruitless notion considered above. But there would appear to be no feeling that there is a *correct* notion of a symmetry of a differential equation—plausibly, for every interesting such notion, there is a yet more general one that is still interesting.

[14] Two subtleties are glossed over in the text. (i) In general d may map a kinematically possible field to an object that is a sort of partially-defined multiply-valued field rather than a kinematically possible field (see fn. 60 below). (ii) The story is (yet) more complicated in the important special case where $\dim W = 1$.

[15] Note, in particular, that we make no appeal here to metric tensors and the like—we just look for transformations of the independent variables that leave invariant the equations of motion. (Some subtleties arise at this point for certain types of fields; see, e.g., Kolář, Michor, and Slovák (1993) on gauge natural bundles.)

The classical symmetries of an equation form a family much more restricted than that picked out by D1—and typically this means that solutions related by a classical symmetry share many salient features. In typical cases, the classical symmetries of a theory of n Newtonian particles are just the spacetime symmetries—so solutions are related by a symmetry if and only if they have the particles instantiating the same sequence of relative distance relations.[16] In the case of general relativity, the classical symmetries are generated by spacetime diffeomorphisms and by scale transformations—so that, roughly speaking, two solutions are related by a symmetry if and only if they agree about the pattern of ratios of distances instantiated.[17]

The notion of a classical symmetry is a special case of the basic mathematical notion of symmetry: the classical symmetries of a differential equation are the structure-preserving maps for a certain structure associated with the equation. Here is the story in briefest outline (a few more details are provided in the appendix below).[18] One constructs a space, $J^k(E)$ (the kth *jet bundle* over E) a point of which is specified, intuitively speaking, by specifying a point x of spacetime and the values of the field and its partial derivatives through order k at x. $J^k(E)$ comes equipped with some geometric structure, $\mathcal{C}$ (the *Cartan distribution*), that ensures that the specification of a point of $J^k(E)$ lives up to this intuitive picture. Our differential equation Δ determines a submanifold $\mathcal{E} \subset J^k(E)$—since for each point of spacetime Δ imposes a constraint on the values of the fields and their partial derivatives at that point. So we have a structure consisting of a large space, $J^k(E)$, a geometric structure $\mathcal{C}$ on $J^k(E)$, and a distinguished subspace $\mathcal{E}$ of $J^k(E)$. In accord with our template for defining symmetries of structures, we take a symmetry of this contraption to be a one-to-one and onto map $F : J^k(E) \to J^k(E)$ that respects $\mathcal{C}$ and maps points of $\mathcal{E}$ to points of $\mathcal{E}$. It turns out that the resulting notion coincides with the notion of a classical symmetry as characterized above.

In order to characterize generalized and nonlocal symmetries, it is helpful (as well as more honest—see fn. 60 below) to describe the relevant notions by characterizing their infinitesimal generators.[19]

From this perspective, classical symmetries (investigated by Sophus Lie and others in the late nineteenth century) are, roughly speaking, transformations that map solutions to solutions and whose infinitesimal generators depend only on the independent and dependent variables of the theory.

Emmy Noether introduced a substantially more general notion at the beginning of the twentieth century.[20] Her *generalized symmetries* (also known as *local* or *higher* symmetries or as *Lie–Bäcklund transformations*) are, roughly speaking, transformations that map solutions to solutions and whose infinitesimal generators depend on

---

[16] Hydon (2000, example 4.4).
[17] See Anderson and Torre (1996).
[18] For what follows, see, e.g., Krasil'shchik and Vinogradov (1999, ch. 3).
[19] Unfortunately, discrete symmetries threaten to go missing when one adopts this perspective. For one approach to this problem, see Hydon (2000, ch. 11).
[20] See, e.g., Olver (1993, ch. 5) or Krasil'shchik and Vinogradov (1999, ch. 4).

derivatives of the fields, as well as on the independent and dependent variables of the theory.[21] These arise as the symmetries of a fancier structure that can be used to encode the differential equation of interest: again the equation is represented as a submanifold of an ambient space equipped with a Cartan structure; this time the ambient space is an infinite jet bundle—the specification of a point of which involves specifying a point of spacetime together with the values at that point of a function and all of its partial derivatives.

Every classical symmetry is also a generalized symmetry. Some equations have no nonclassical generalized symmetries (the Einstein field equations of general relativity are an example).[22] But many equations have generalized symmetries that are not classical symmetries. Striking examples include:

The Kepler Problem. This is the problem of determining the motion of Newtonian massive particle moving in the external gravitational field of a fixed massive body.[23] In addition to the "obvious" conserved quantities for this system—energy and angular momentum—there is a "hidden" one, the Lenz–Runge vector.[24] Associated with this hidden conserved quantity is a nonclassical generalized symmetry of the Kepler problem (more on this below).[25]

The KdV Equation. The *Korteweg–de Vries equation* is an equation used to model waves in shallow water. We can think of it as an equation governing a field $u(x,t)$ on a two-dimensional spacetime:

$$\frac{\partial u}{\partial t} + \frac{\partial^3 u}{\partial x^3} + u\frac{\partial u}{\partial x} = 0.$$

The only classical symmetries of this equation are spacetime and scaling symmetries; but it has an infinite-dimensional family of generalized symmetries.[26]

---

[21] To get a feeling for what this means, consider the sort of gauge transformations that normally arise in presentations of Maxwell's theory: if we take the vector potential $A(x)$ as our field, then the theory is invariant under infinitesimal transformations of the form $A \mapsto A + \varepsilon d\Lambda$ where $\Lambda(x)$ is a real-valued function on spacetime. Now suppose that Λ is a map then when fed a kinematically possible A returns a real-valued function $\Lambda[A]$ on spacetime. If for each spacetime point x and each A, the value of $\Lambda[A]$ at x depends only on x and on $A(x)$, then the infinitesimal transformation $A \mapsto A + \varepsilon d\Lambda[A]$ corresponds to a classical symmetry of Maxwell's theory; if the value of $\Lambda[A](x)$ depends also on a finite number of derivatives of A at x, then this map is a generalized symmetry of Maxwell's theory. See the discussion of generalized gauge symmetries in Pohjanpelto (1995) and in Torre (1995). For a thoroughly worked-out example involving only finitely many degrees of freedom, see Cantwell (2002, §14.4.1).

[22] See Anderson and Torre (1996).

[23] Note that the Kepler problem contains all of the dynamics of the honest Newtonian two-body problem. See, e.g., Goldstein, Poole, and Safko (2002, §3.1).

[24] The Lenz–Runge vector is $-\frac{q}{|q|} + \frac{1}{\mu} p \times (q \times p)$, were q is the position of the moving particle, p is its momentum, and μ is a constant that depends on the masses.

[25] For the generalized symmetry of the Kepler problem, see Lévy-Leblond (1971, §5.B). Note that this system also admits a classical symmetry that is in a sense associated with the Lenz–Runge vector; see Prince and Eliezer (1981).

[26] For the classical and generalized symmetries of the Korteweg–de Vries equation, see Olver (1993, 125ff. and 312ff.).

Yet more general are the nonlocal symmetries that have been investigated in recent decades.[27] These are, roughly speaking, transformations that map solution to solutions and whose infinitesimal generators are allowed to depend on nonlocal functionals (such as integrals) of the fields, as well as on their derivatives and on the dependent and independent variables of the equation.[28] They arise as symmetries of a structure that results when the differential equation is encoded in a certain extension of the infinite jet bundle. Many examples of nonlocal symmetries are known—even the Kepler problem admits nonlocal symmetries that are not generalized symmetries.[29]

5. Variational and Hamiltonian Symmetries

The governing equations of many physical theories can be given a Lagrangian or Hamiltonian treatment (although this sometimes requires some craft or flexibility). For such theories, it is natural to consider the symmetries of the structures employed in the Lagrangian or Hamiltonian treatment—which need not coincide with the symmetries of the equation itself.

5.1 Variational Symmetries

There are two main styles of Lagrangian formalism, sometimes called the *dynamical* and the *covariant* approaches, which coincide for theories with finitely many degrees of freedom but differ for field theories.[30] Dynamical approaches are quite similar in spirit to the Hamiltonian approach considered below.[31] Under covariant approaches, which are our concern here, the basic space of interest is a jet bundle over the space of dependent and independent variables of the theory (as in the treatment of classical symmetries sketched above) and the Lagrangian is now to be thought of as an object that when fed a kinematically possible field u gives in return a d-form $L[u]$ on the spacetime M (here $d = \dim M$).[32] Each such Lagrangian is

---

[27] See, e.g, Krasil'shchik and Vinogradov (1999, ch. 6).
[28] For a thoroughly worked-out example in which an infinitesimal symmetry and the corresponding finite symmetry both depend on an integral over space of the dependent variable of the theory, see Cantwell (2002, §16.2.2.1).
[29] See, e.g., Leach, Andriopoulos, and Nucci (2003).
[30] For these labels and for an investigation of the relation between the two approaches, see Castrillón López and Marsden (2008). For introductions to the two approaches, see Abraham and Marsden (1985, §3.5ff. and Example 5.5.9).
[31] Under such approaches, one works with a space of instantaneous states (which will be infinite-dimensional in the field-theoretic case), equips this space with a real-valued function, L (the Lagrangian), and employs a variational principle to find those curves in the space of states that correspond to dynamically possible histories of the system.
[32] For instances of this styles of approach, see, e.g., Castrillón López and Marsden (2008) and Krasil'shchik and Vinogradov (1999, ch. 5).

associated with a differential equation, whose solutions correspond to those fields that satisfy a certain variational principle for the Lagrangian. A (classical or generalized) symmetry of an equation arising from a Lagrangian is called a *variational symmetry* if it leaves the Lagrangian invariant. Even if a theory admits a Lagrangian formulation, some symmetries of the underlying equation may not show up as variational symmetries—as a rule, Galilean boosts and scaling symmetries are often not variational.[33] Modulo niceties about boundary conditions, Noether's theorem assures us that each one-parameter family of variational symmetries is associated in a canonical way with a conservation law.[34]

5.2 Hamiltonian Symmetries

Let us restrict attention to the ideal case in which specifying an initial data set (= a possible instantaneous dynamical state) for the equation of the theory determines a unique solution defined at all times. Then, speaking roughly and heuristically, a Hamiltonian treatment amounts to the following. The *phase space* of the theory is the space, $\mathcal{I}$, of all initial data sets. The *Hamiltonian*, H, of the theory is the real-valued function on the phase space that assigns to each point of the phase space the energy of the corresponding physical state. The phase space can be equipped with a geometric structure, ω, with the following marvelous feature: together H and ω determine a family of curves in $\mathcal{I}$, exactly one passing through each point; each of these curves corresponds to a solution of the theory's equation of motion, in the sense that the two objects pick out the same sequence of instantaneous dynamical states; and for any such solution there is a corresponding curve of this kind.[35] So the structure $(\mathcal{I}, \omega, H)$ in effect encodes the differential equation of the theory. It is natural to investigate the *Hamiltonian symmetries* of the theory: those one-to-one and onto maps from $\mathcal{I}$ to itself that preserve both ω and H.[36] The Hamiltonian

---

[33] On scale transformations, see Olver (1993, 255). A standard remedy is to introduce the notion of a *divergence symmetry*, a transformation that leaves the Lagrangian invariant up to a total divergence; many interesting symmetries are divergence symmetries but not variational symmetries, including boosts of Newtonian systems and the conformal symmetries of the wave equation; see Olver (1993, 278–281). Scaling symmetries are more subtle. Scale transformations are symmetries of general relativity, but are neither variational nor divergence symmetries; see Anderson and Torre (1996, §2.B). Rescaling of space and time is a symmetry of the wave equation that is neither a variational nor a divergence symmetry, although there is a related scale transformation that acts on the dependent variables, as well as the independent variables, which is a divergence symmetry (but not a variational symmetry); see Olver (1993, Examples 2.43, 4.15, and 4.36).

[34] But: certain types of variational (or divergence) symmetries of theories whose initial value problems are ill-posed are associated with so-called trivial conservation laws; see Olver (1993, 342–346) on Noether's second theorem. And: there exist techniques for associating conservation laws with symmetries that do not rely on Noether's theorem; see, e.g., Bluman (2005).

[35] ω is a *symplectic form*—a closed, nondegenerate two-form. ω and H determine a vector field X_H on $\mathcal{I}$: X_H is the vector field that when contracted with ω yields the one-form dH. Integrating this vector field gives the curves mentioned in the text. Note that there is a canonical recipe for constructing $\mathcal{I}$, H, and ω given a Lagrangian treatment of the theory.

[36] Note that the symplectic space $(\mathcal{I}, \omega)$ has a vast family of symmetries. Suppose that we are interested in a Newtonian theory of finitely many particles. Then $\mathcal{I}$ is finite-dimensional, but the family of smooth permutations of $\mathcal{I}$ that preserve ω is infinite-dimensional—it is only when we restrict attention to transformations that also preserve H that we end up with something like what we want. Something similar is of course true in ordinary quantum mechanics: while the family of unitary transformations of a Hilbert space will be very

version of Noether's theorem assures us (fussing about boundary conditions aside) that there is a conserved quantity associated with each one-parameter family of Hamiltonian symmetries of a theory.

6. Symmetry and Physical Equivalence

Why was it (in hindsight, in one sense) a mistake for Newton to postulate absolute space? Because he thereby postulated more spacetime structure than was required for his dynamics—boosts are symmetries of his laws of motion but not of the spatiotemporal structure that he postulated. Reflection on this example and others motivates the principle that sound interpretative practice requires that the spacetime symmetries of one's ontology ought to include the spacetime symmetries of one's preferred theory. Conversely, it seems like something has gone wrong if the spacetime symmetry group of one's equations of motion is more restricted than the spacetime symmetry group of one's ontology—for in this case, one would be shirking one's duty by in effect employing geometrical structure in one's dynamical theory while not being willing to pay the ontological cost.

It is tempting to think that this picture ought to generalize: Why should spacetime symmetries be special? At the level of slogans, the idea is easy to state—our interpretative practice should be guided by the principle that the symmetries of a theory's laws and the symmetries of its ontology should coincide. It is not obvious how to give a precise and general formulation of the idea. But one thing that seems clear is that any such formulation would have as a consequence the second doctrine discussed in the opening section of this chapter:

> D2. Two solutions of a classical theory's equation of motion are related by a symmetry if and only if they are physically equivalent, in the sense that they are equally well- or ill-suited to represent any particular physical situation.[37]

We have seen that disaster follows if this doctrine is combined with the notion that the symmetries of a classical theory are the maps that send solutions to solutions. But having set aside that notion, we can ask whether D2 can be safely combined with one of the more nuanced and discriminating notions of symmetry on offer.

---

large, the family of such transformations that preserve a given Hamiltonian will be quite small—and only the latter is a good candidate for the symmetry group of a theory. The situation is more perplexing in the case of fancier quantum theories. On the one hand, an arbitrary C^*-algebra automorphism is pretty clearly the analogue of an arbitrary symplectic or unitary transformation and so is not a good candidate to be a symmetry of a theory: indeed, in many cases of interest any two states are related by such an automorphism; see Kishimoto, Ozawa, and Sakai (2003). On the other hand, in some contexts it is not possible to identify symmetries of a theory with those C^*-algebra automorphisms that preserve the Hamiltonian because there is no Hamiltonian operator available at the C^*-algebra level; on this point, see Ruetsche (2011, §12.3).

[37] In truth, in order to formulate a plausible doctrine in the neighborhood of D2, one should probably work with infinitesimal symmetries—otherwise it is easy to concoct counterexamples by deleting points from a theory's space of solutions.

Indeed, one can give various plausibility arguments in favor of theses in the neighborhood of D2.

(a) Intuitively speaking, two solutions are related by a symmetry if and only if they are interchangeable by the lights of the theory's formalism. And surely if everything has been set up properly (i.e., just the right ingredients have been built into the formalism of the theory), two solutions that are formally interchangeable should also have identical representational capacities—and so should be physically equivalent in the present sense.

(b) Further, one might think that there is a pretty good reason for thinking that two solutions related by a classical symmetry of a theory must be physically equivalent. For such symmetries can be thought of as smooth transformations of the dependent and independent variables of the theory that preserve the form of the equations of the theory. How could our representational practices reasonably distinguish between two solutions of a theory that are related by a reparameterization of the theory's variables to which the equations of the theory are themselves indifferent?

(c) That this is the right way to go is also suggested by consideration of familiar cases. No one denies that solutions of a theory's equations are physically equivalent if they are related by a spacetime symmetry of the theory (i.e., by a symmetry that transforms the independent variables of the theory without affecting the dependent variables).[38] A similar consensus exists regarding paradigm cases of (global and local) *internal symmetries* (that involve transformations of dependent variables of the theory that do not affect the independent variables). Thus, if one has a theory involving a complex scalar field such that global phase transformations of the form $\phi(x) \mapsto e^{i\theta}\phi(x)$ (θ a constant) are symmetries, then one regards solutions related by such a transformation as physically equivalent. Similarly, one regards as physically equivalent solutions of Maxwell's equations in vector potential form that are related by local gauge transformations of the form $A^\mu(x) \mapsto A^\mu(x) + d\Lambda(x)$ (here Λ is a smooth function on spacetime, so in general the transformation effected on the dependent variables varies from spacetime point to spacetime point). Why should other symmetries be different?

Unfortunately, these mutually reinforcing half-arguments do not add up to much: each of the general notions of symmetry that we have considered leads to unpalatable consequences when plugged into D2.

Consider first the notion of a classical symmetry of a differential equation. As noted above, most equations admit relatively few such symmetries, so that solutions related by a classical symmetry share a great deal in common. But there exist

---

[38] There is disagreement over the question whether there are distinct physical possibilities related by shifts and the like. But that is a different question.

equations whose symmetry groups are so large that they act transitively on the space of solutions (i.e., for any two solutions, u_1 and u_2, there is a classical symmetry of the equation that maps u_1 to u_2). This sort of shocking behavior can be found in some of the most basic theories of classical physics.

(i) A single Newtonian free particle. In this case, the spacetime symmetries of the theory are given by the Galilei group, and any solution can be mapped onto any other by such a symmetry.[39]

(ii) The harmonic oscillator. In one spatial dimension, an arbitrary solution takes the form $q = A \cos t + B \sin t$. The theory admits a one-parameter group of classical symmetries that acts on solutions by changing the value of B, and another such group that changes the value of A—so any solution can be mapped to any other by a symmetry of the theory.[40]

(iii) Linear homogeneous partial differential equations—such as the heat equation, the wave equation, the source-free Maxwell equations, etc. Corresponding to any solution u_0 of such an equation, there is a classical symmetry $T_{u_0} : u \mapsto u + u_0$.[41] And, of course, for any two solutions u_1 and u_2 of a linear equation, there is a solution u_0 such that $u_2 = u_1 + u_0$. So any two solutions of a linear homogeneous partial differential equation are related by a symmetry.

If we maintain that any two solutions related by a classical symmetry are physically equivalent, then each of these theories will be unable to discriminate among the systems it provides good models for—it will consider them to have the same physics. That is presumably the right verdict in the case of the theory of a free Newtonian particle—if we follow the policy of regarding solutions related by a spacetime symmetry as physically equivalent, then no degrees of freedom remain in this theory once such symmetries are factored out. But the other examples are very different: under any ordinary reading, they admit solutions that represent situations in which nothing is happening (the oscillator is permanently immobile at the origin, the field is in a ground state) and others that represent situations in which plenty is going on (the oscillator is continually sproinging around, energy in the form of heat or waves is propagating). An approach to understanding physical theories that leaves us unable to see these distinctions is not something we can live with. So D2 is false if understood as a thesis concerning classical symmetries of differential equations.

---

[39] In fact, the classical symmetries for this theory are not exhausted by the Galilei symmetries—already in two spacetime dimensions, where the Galilei group is three-dimensional, the classical symmetry group for the free particle is eight-dimensional. See, e.g., Duarte, Duarte, and Moreira (1987).

[40] See, e.g., Lutzky (1978, §3). Note that the complete classical symmetry group of the one-dimensional oscillator is eight-dimensional, and so outstrips the group of spacetime symmetries. Note also that all of these results carry over, mutatis mutandis, to the case of a time-dependent oscillator in n spatial dimensions; see Prince and Eliezer (1980).

[41] See, e.g., Hydon (2000, 145). For the classical symmetries of the heat equation and the wave equation, see, e.g., Olver (1993, Examples 2.41 and 2.43). For symmetries of the source-free Maxwell equations, see Anco and Pohjanpelto (2008).

Further, employing generalized or nonlocal symmetries would not help: the problems just noted would remain, and new ones of the same ilk would crop up.[42]

Nor does it help to shift attention to variational or Hamiltonian symmetries. Consider first variational symmetries. The class of (classical or generalized) variational symmetries of an equation is more restrictive than the class of (classical or generalized) symmetries of an equation. And that has some benefits: the variational symmetries of the wave equation do not include addition-of-an-arbitrary-solution (and do not act transitively on the space of solutions).[43] But: the problem with the harmonic oscillator remains—for any two solutions, there is a variational symmetry that maps one to the other.[44] So the class of variational symmetries is still in some respects too generous to underwrite D2. It is also in some respects too restrictive: for example, neither the boost symmetry of classical mechanics nor the scaling symmetry of general relativity is a variational symmetry.[45] But it is hard to deny that two solutions related by such a symmetry are physically equivalent in the relevant sense.[46]

Further, recall that not every equation of motion admits a Lagrangian treatment.[47] True, with some ingenuity one can often find surrogates for such equations that do admit Lagrangian treatment (e.g., by replacing the variables of the original theory by suitable potentials, or by introducing nonphysical fields).[48] But it is hard to see why one should have to take such detours in order to understand the connection between symmetry and physical equivalence.

What about Hamiltonian symmetries, then? This class is again too restrictive in some respects. As in the variational case, scaling symmetries are typically not Hamiltonian symmetries. Galilean boosts cause the same sort of trouble—since they typically leave the potential energy of a system invariant while altering its kinetic energy, they fail to preserve the Hamiltonian. So if one took solutions to be physically equivalent only if related by a Hamiltonian symmetry, then one would have to violate the general principle that solutions of classical theories are physically equivalent if related by a spacetime symmetry.[49]

---

[42] E.g., the group of nonlocal symmetries of the Kepler problem acts transitively on solutions; see Leach, Andriopoulos, and Nucci (2003).

[43] See Olver (1993, Example 4.15).

[44] See Lutzky (1978) and Prince and Eliezer (1980).

[45] What happens if in place of variational symmetries, we consider divergence symmetries? Some problems go away, others reappear. Good news: Galilean boosts and certain types of scaling transformations count as symmetries (see fn. 33 above). Bad news: addition-of-an-arbitrary solution is a divergence symmetry of the wave equation; Olver (1993, Example 4.36). So if we count solutions related by a divergence symmetry as physically equivalent, we have to view every pair of solutions of the wave equation as being physically equivalent.

[46] In the case of general relativity, denying that solutions g and $c \cdot g$ (c a positive constant) related by a scale transformation are physically equivalent is especially difficult for those who deny that there are possible worlds that agree about distance ratios but disagree about matters of absolute distance—for according to such philosophers there is only one world that is adequately represented by g and $c \cdot g$ taken together, and it is not easy to see how one of the two solutions could have a better claim than the other to represent that world.

[47] The problem of identifying those which do so is known as *the inverse problem of the calculus of variations* or as *Helmholtz's problem*. For an introductory survey, see Prince (2000).

[48] See, e.g., Ibragimov and Kolsrud (2004), Olver (1993, Exercises 5.35, 5.36, and 5.46), Rosen (1966), and Sorkin (2002).

[49] Just as one might move from variational symmetries to divergence symmetries (see fn. 33 above), one might consider transformations of a system's phase space that leave invariant the set of Hamiltonian trajectories

At the same time, the class of Hamiltonian symmetries is also too generous in some respects. In the Hamiltonian setting we are at least safe from the threat of the group of symmetries of a theory acting transitively on the space of states (since Hamiltonian symmetries preserve the Hamiltonian and theories ordinarily allow states of differing energy). But we nonetheless still run into cases in which we are unwilling to consider states or solutions related by Hamiltonian symmetries as physically equivalent.

Consider the Kepler problem. This theory has a number of Hamiltonian symmetries, some of which are spacetime symmetries and some of which are not. The spacetime symmetries include time translation (associated with conservation of momentum) and rotation (associated with conservation of angular momentum). As usual, we want to regard solutions related by spacetime symmetries as physically equivalent—in the present case, this means that we regard solutions as physically equivalent if the corresponding elliptical orbits are of the same shape (i.e., have the same eccentricity and have equally long major axes). The further Hamiltonian symmetries of the Kepler problem are associated with the conservation of the Lenz–Runge vector. If two solutions are related by one of these symmetries, then the corresponding ellipses have equally long major axes, but (in general) have different eccentricities and different orientations in space.[50] The upshot is that if we take being related by a Hamiltonian symmetry to imply physical equivalence, then we must take solutions of the (negative energy) Kepler problem to be physically equivalent if and only if they correspond to ellipses with equally long major axes.[51] But we do not normally regard highly eccentric orbits and perfectly circular orbits as being physically equivalent.[52]

We run into the same problem with the harmonic oscillator: the family of Hamiltonian symmetries acts transitively on the surfaces of constant energy in the space of initial data—so two solutions of the theory are related by a Hamiltonian symmetry if and only if they represent the system as having the same total energy.[53]

---

without worrying about whether they also leave the Hamiltonian itself invariant. Suitably interpreted, this should manage to capture Galilean boosts in Newtonian mechanics and the scaling symmetry of the Kepler problem; see Abraham and Marsden (1985, 446 f.) and Prince and Eliezer (1981, §5). Of course, it also includes the various undesirable characters that already count as Hamiltonian symmetries (see below).

[50] See Morehead (2005). This symmetry corresponds to a generalized symmetry of the equations of motion.

[51] For the negative energy Kepler problem, the length of the major axis determines the energy of a solution, so any Hamiltonian symmetry leaves this quantity invariant; see, e.g., Goldstein, Poole, and Safko (2002, §3.7).

[52] (1) One might think that we would do so if scaling transformations were up for grabs, since one can indeed transform a circular orbit into an eccentric orbit by rescaling one coordinate axis while leaving the others invariant. But this sort of rescaling does not preserve the Hamiltonian of the Kepler problem (because it changes the length of the major axis of some solutions). (2) The discussion above glosses over an interesting subtlety. At the infinitesimal level, one does indeed find a large set of symmetries of the Kepler problem. But the space of initial data features a singular set of points (corresponding to situations in which the two particles collide). And the existence of this singular set provides an obstruction to integrating the infinitesimal symmetries into a group action; see Cushman and Bates (1997, 74). However, there exist ways of regularizing the singularities of the Kepler problem—and these constructions lead to a family of finite symmetries that act transitively on surfaces of constant energy. For discussion and references, see Cushman and Bates (1997, ch. 2) or Guillemin and Sternberg (1990, §2.7). (3) Note that the fact that the Hamiltonian symmetry group of the Kepler problem is larger than its spacetime symmetry group plays an important role in the quantum theory of the hydrogen atom; see Jauch and Hill (1940) or Guillemin and Sternberg (1990, §7).

[53] Consider the two-dimensional case. There are four conserved quantities, corresponding to four independent Hamiltonian symmetries. A surface of constant energy is a three-sphere in the phase space and the group

But two solutions of the harmonic oscillator can have the same energy while corresponding to quite different motions—for example, in the two-dimensional case, one might have the particle moving back and forth on a line segment while the other has it executing a circular motion.[54]

Or, again, consider the Korteweg–de Vries equation. This equation governs a real-valued field u in one spatial dimension. Under its usual interpretation, this field describes the dynamics of a shallow body of water with one horizontal dimension, with $u(x,t)$ giving the depth of the water above spatial point x at time t. Consider a particular solution, u_0. How large is the family of solutions physically equivalent to u_0? Well, it surely includes solutions related to u_0 by spacetime symmetries, and perhaps also scaling symmetries. But the family of generalized symmetries of the equation and the family of Hamiltonian symmetries are both infinite-dimensional—so it seems clear that each must contain many symmetries that relate physically inequivalent solutions.[55]

7. Outlook

Where does this leave us? There are various interesting formal notions of symmetry applicable to classical physical theories. But none of the standard ones are suited to underwrite a principle like D2 that makes a direct link between the being related by a symmetry and being physically equivalent.

I leave it as a challenge to the reader to identify a general and interesting formal notion of symmetry that renders D2 true—or, better, to identify a family X of symmetries such that two solutions of a theory that are related by a symmetry are physically equivalent if and only if they are related by a symmetry that belongs to X.

Above we have seen that we ordinarily take spacetime symmetries to belong to X. Further, we have seen that there are classical symmetries, generalized symmetries, nonlocal symmetries, variational symmetries, and Hamiltonian symmetries that are not in X. We have also seen that X contains symmetries that fall outside

---

of Hamiltonian symmetries is the group $U(2)$, which acts transitively on the energy surfaces. For details, see, e.g., Goldstein, Poole, and Safko (2002, §9.8), Cushman and Rod (1982), or Cushman and Bates (1997, ch. 2).

[54] Again, the fact that the symmetry group of the classical system is $U(2)$ rather than just the spacetime symmetry group plays an important role in the quantum theory; see Jauch and Hill (1940).

[55] In the examples just considered, it is pretty clear that one does not want to count every pair of solutions related by a generalized symmetry as being physically equivalent. Does one ever want to count solutions as physically equivalent that are related by a generalized symmetry that is not a classical symmetry? Yes—for instance, when the solutions in question are also related by a respectable classical symmetry. Consider, e.g., the generalized gauge transformations described in fn. 21 above—if two solutions are related by such a symmetry, then they are also related by an ordinary gauge transformation.

Are there pairs of solutions related by a generalized symmetry (but not by any classical symmetry) that one would want to consider physically equivalent? That appears to be a more difficult question. Part of the difficulty lies in the fact that what one has in practice are the infinitesimal generators of generalized symmetries: it is in general a nontrivial task to find the corresponding group actions; see, e.g., Olver (1993, 297ff.). Further, even in cases where the corresponding groups of transformations can be determined, their physical interpretation can be obscure; see, e.g., Olver (1984, 136f.).

of the classes of spacetime symmetries, variational symmetries, and Hamiltonian symmetries.

Perhaps it is possible to find some formal notion of symmetry that combines well with D2. But it is hard to be optimistic—certainly, the ways of encoding the content of laws that are most appealing to mathematicians and physicists appear to lead to notions of symmetry that are coolly indifferent to considerations of representational equivalence. So it appears that the sort of constraint that knowledge of the symmetries of a theory places on the range of reasonable interpretations of that theory may well be more modest than one might have hoped.[56]

Appendix

Here are a few more details about classical symmetries of differential equations. Let us begin by describing $J^k(E)$.[57] For present purposes, it is easiest to work in terms of coordinates.[58]

> A point y in M is specified by specifying a d-tuple of real numbers, $(y_1,\ldots,y_d)$ (the coordinates of y).
>
> A point (y,v) in E is specified by specifying a $(d+m)$-tuple of real numbers, $(y_1,\ldots,y_d;v_1,\ldots,v_m)$ (the coordinates in M of y plus the coordinates in W of $u(y)$, the value of the field u at spacetime point y).
>
> A point (y,v,p) in $J^k(E)$ is specified by specifying a $\left[d+m\binom{d+k}{k}\right]$-tuple of real numbers, $(y_1,\ldots,y_d;v_1,\ldots,v_m;p_1,\ldots,p_n)$, where the p_i are just numerous enough to be regarded as a list of the values of the partial derivatives (from order one through k) of u by the spacetime coordinates.

When we think of $J^k(E)$ in this way, it provides a natural way to encode the content of any k^{th} order partial differential equation whose dependent and independent variables parameterize E. For think what such a differential equation does: for each point x of spacetime, the equation imposes a constraint on the values of the field and its partial derivatives through order k when evaluated at x. But, roughly speaking, that is to say that such a differential equation determines a submanifold of $\mathcal{E} \subset J^k(E)$ and that the content of the equation is exhausted by the structure of $\mathcal{E}$ as a submanifold of $J^k(E)$.

---

[56] Of course, some interesting weaker relative of D2 might be true. E.g., for all that has been said here, being related by a Hamiltonian symmetry that corresponds to a classical symmetry of their equation of motion may be a sufficient condition for two solutions to be physically equivalent.

[57] For what follows, see, e.g., Krasil'shchik and Vinogradov (1999, ch. 3).

[58] But note that everything described below can be done in a respectable global and coordinate-independent fashion. See, e.g., Saunders (2008).

Now we are getting somewhere: we can identify symmetries as one-to-one and onto maps from $J^k(E)$ to itself that preserve $\mathcal{E}$ and all relevant structure of the ambient space $J^k(E)$. But what does this structure amount to?

We know that $J^k(E)$ is a manifold and that we are working in a setting in which everything is at least a bit smooth—so it is natural to require our symmetries to be appropriately smooth diffeomorphisms from $J^k(E)$ to itself.

There is one further kind of structure on $J^k(E)$ that matters in the present context. Recall that we have said that, intuitively, the extra variables that we add in moving from E to $J^k(E)$ are supposed to correspond to the values of the partial derivatives of the field with respect to the spacetime coordinates. We need to put some structure on $J^k(E)$ in order to enforce this intuitive demand.

In order to get a feeling for what is required here, consider the following line of thought. Let $u: M \to W$ be a kinematically possible field. The content of u is encoded in its graph, Γ_u, the subset of E consisting of pairs for the form $(x, u(x))$, for $x \in M$. Of course, the map $x \in M \mapsto (x, u(x)) \in \Gamma_u$ is a diffeomorphism. Further, a sufficiently smooth submanifold $\Gamma \subset E$ corresponds to a kinematically possible u in this way if and only if the projection map $\bar{\pi}: (x, v) \in \Gamma \mapsto x \in M$ is a diffeomorphism from Γ onto M.

Now, each kinematically possible u also determines a diffeomorphism from M to $J^k(E)$. We define the *k-jet* of u to be the map $j[u]: x \in M \to (x, u(x), p) \in J^k(E)$, where $p = (p_1, \ldots, p_n)$ encodes the values at x of the partial derivatives of u through order k. The k-jet of u is a diffeomorphism onto its image J_u. There is of course a projection map $\pi: (x, v, p) \in J^k(E) \mapsto x \in M$, whose restriction to J_u is the inverse of the jet $j[u]$.

It may be tempting to suppose that any sufficiently smooth submanifold $K \subset J^k(E)$ such that $\pi: K \to M$ is a diffeomorphism onto its image arises as the J_u for some kinematically possible u. But this is false. For consider such a K. For each spacetime point $x \in M$, K includes exactly one point (x, v, p). So, questions of boundary conditions aside, K does determine a kinematically possible u—for each spacetime point x just set $u(x) = v$, where (x, v, p) is the unique point in K that π sends to x. But there is no guarantee that $J_u = K$, precisely because we have not yet built into the structure of $J^k(E)$ any connection between the p_i and the partial derivatives of the field—so in general, if (x, v, p) is a point in K and u is the kinematically possible field determined by K, there is no reason to expect that the values of the partial derivatives of u at x are given by p (that is, in general one expects that $j[u](x) \neq (x, v, p)$ so that $J_u \neq K$).

There is an elegant solution to this problem. One imposes on $J^k(E)$ some geometric structure, in the form of the *Cartan distribution,* $\mathcal{C}$, which singles out at each point of $J^k(E)$ a distinguished subspace of dimension $\left[d + m\binom{d+k-1}{d-1}\right]$ in the $\left[d + m\binom{d+k}{k}\right]$-dimensional tangent space at that point. The Cartan distribution has the following beautiful feature: a submanifold K of $J^k(E)$ that is diffeomorphic to M via the natural projection map $\pi : (x, v, p) \mapsto x$ is of the form J_u for some kinematically possible u if and only at any point of K, all tangent vectors pointing along K lie in the privileged subspace picked out at that point by $\mathcal{C}$. That is: the Cartan distribution encodes our intuitive constraint that the p_i should correspond to the values of the partial derivatives of fields.[59]

Putting all of this together: a symmetry (in the present sense) of our equation is a diffeomorphism from $J^k(E)$ to itself that preserves: (i) the submanifold $\mathcal{E} \subset J^k(E)$ that encodes the equation; and (ii) the Cartan distribution $\mathcal{C}$. It now turns out that (except in the special case where $m = 1$) the only such diffeomorphisms arise via diffeomorphisms from E to itself (this claim will be unpacked in the next paragraph). And of course, since each kinematically possible field u can be identified with the submanifold $j[u](M) \subset J^k(E)$, any diffeomorphism from $J^k(E)$ to itself acts in a natural way on the space of kinematically possible fields—and (subtleties aside) any diffeomorphism from $J^k(E)$ to itself that preserves $\mathcal{E}$ maps solutions to solutions. So: the symmetries (in the present sense) of a differential equation are the transformations of the space of kinematically possible solutions that map solutions to solutions and that are suitably local, in the sense of depending only on the independent and dependent variables of the theory.

There are two unfinished bits of business. The first is to unpack the notion of a diffeomorphism from $J^k(E)$ to itself arising via a diffeomorphism from E to itself. Let $\bar{d} : E \to E$ be a diffeomorphism. We seek to define a corresponding diffeomorphism $d : J^k(E) \to J^k(E)$. We proceed by selecting an arbitrary $(x, v, p) \in J^k(E)$ and showing how to find $d(x, v, p)$. Let u be a kinematically possible field such that $u(x) = v$ and $j[u](x) = (x, v, p)$ (i.e., (x, v, p) gives the value of u and its partial derivatives at x). Let Γ_u be the graph of u in E (i.e., the set of points of E of the form $(x, u(x))$, for all $x \in M$). Γ_u is a smooth submanifold of E that projects diffeomorphically to M under the map $\bar{\pi} : (x, v) \mapsto x$. Now, since $\bar{d}$ is a diffeomorphism from E to itself, the set $\bar{d}(\Gamma_u)$ is also a smooth submanifold of E. Sadly, $\bar{d}(\Gamma_u)$ need not project diffeomorphically to M. But let us ignore that detail, and pretend that it does—and hence corresponds to a kinematically possible u^*.[60] Then we can define $d(x, v, p) := j[u^*](x')$, where $x' = \bar{\pi}(\bar{d}(x, v))$. The resulting map $d : J^k(E) \to J^k(E)$ is a well-defined diffeomorphism (in particular, it is independent of the choices we made along the way).

The final piece of unfinished business is to note that one typically works with fields that have well-defined transformations laws under changes of coordinates on spacetime: so a spacetime diffeomorphism $\bar{d} : M \to M$ induces in a natural way a diffeomorphism $\bar{D} : E \to E$.[61] The corresponding transformation $D : \mathcal{K} \to \mathcal{K}$ is

---

[59] The space of vectors picked out by $\mathcal{C}$ at a point of $J^k(E)$ coincides with the vectors annihilated by the family of one-forms on $J^k(E)$ that enforce the differential relations that require the p_i to be the derivatives of the components of u with respect to the x_j.

[60] To do everything honestly, we would need to: (i) introduce a generalized notion of a solution; (ii) work locally; or (iii) shift our focus to the infinitesimal symmetries.

[61] This is one point at which the story can become more complicated for fields that are not given by tensors. See, e.g., Kolář, Michor, and Slovák (1993) on gauge natural bundles.

determined as above. And we find, as expected, that D is a classical symmetry if and only if it maps solutions to solutions.

References

Abraham, R., and J. Marsden (1985). *Foundations of mechanics.* 2d ed. Cambridge, MA: Perseus.
Anco, S., and J. Pohjanpelto (2008). Generalized symmetries of massless free fields on Minkowski space. *Symmetry, Integrability and Geometry: Methods and Applications* 4: 004.
Anderson, I., and C. Torre (1996). Classification of local generalized symmetries for the vacuum Einstein equations. *Communications in Mathematical Physics* 176: 479–539.
Baker, D. (2010). Symmetry and the metaphysics of physics. *Philosophy Compass* 5: 1157–1166.
Bluman, G. (2005). Connections between symmetries and conservation laws. *Symmetry, Integrability and Geometry: Methods and Applications* 1: 011.
Bluman, G., and S. Anco (2002). *Symmetry and integration methods for differential equations.* Berlin: Springer–Verlag.
Brading, K., and E. Castellani (2007). Symmetries and invariances in classical physics. In *Philosophy of physics*, Part B, ed. J. Butterfield and J. Earman, 1331–1367. Amsterdam: Elsevier.
Cantwell, B. (2002). *Introduction to symmetry analysis.* Cambridge: Cambridge University Press.
Castrillón López, M., and J. Marsden (2008). Covariant and dynamical reduction for principal bundle field theories. *Annals of Global Analysis and Geometry* 34: 263–285.
Cushman, R., and L. Bates (1997). *Global aspects of classical integrable systems.* Basel: Birkhäuser.
Cushman, R., and D. Rod (1982). Reduction of the semisimple 1:1 Resonance. *Physica D* 6: 105–112.
Dasgupta, S. (2010). Symmetries in physical reasoning. Unpublished manuscript.
Debs, T., and M. Redhead (2007). *Objectivity, invariance, and convention: symmetry in physical science.* Cambridge, MA: Harvard University Press.
Duarte, L., S. Duarte, and I. Moreira (1987). One-dimensional equations with the maximum number of symmetry generators. *Journal of Physics A* 20: L701–L704.
Earman, J. (1989). *World enough and spacetime: Absolute versus relational theories of space and time.* Cambridge, MA: MIT Press.
Goldstein, H., C. Poole, and J. Safko (2002). *Classical mechanics.* 3d ed. New York: Addison–Wesley.
Guillemin, V., and S. Sternberg (1990). *Variations on a theme by Kepler.* Providence, RI: American Mathematical Society.
Healey, R. (2009). Perfect symmetries. *British Journal for the Philosophy of Science* 60: 697–720.
Hydon, P. (2000). *Symmetry methods and differential equations: A beginner's guide.* Cambridge: Cambridge University Press.

Ibragimov, N., and T. Kolsrud (2004). Lagrangian approach to evolution equations: Symmetries and conservation laws. *Nonlinear Dynamics* 36: 29–40.

Ismael, J., and B. van Fraassen (2003). Symmetry as a guide to superfluous theoretical structure. In *Symmetries in physics: Philosophical reflections*, ed. K. Brading and E. Castellani, 371–392. Cambridge: Cambridge University Press.

Jauch, J., and E. Hill (1940). On the problem of degeneracy in quantum mechanics. *Physical Review* 57: 641–645.

Kishimoto, A., N. Ozawa, and S. Sakai (2003). Homogeneity of the pure state space of a separable C^*-algebra. *Canadian Mathematical Bulletin* 46: 365–372.

Klainerman, S. (2008). Partial differential equations. In *The Princeton companion to mathematics*, ed. T. Gowers, J. Barrow-Green, and I. Leader, 455–483. Princeton: Princeton University Press.

Kolář, I., P. Michor, and J. Slovák (1993). *Natural operations in differential geometry*. Berlin: Springer-Verlag.

Krasil'shchik, I., and A. Vinogradov, eds. (1999). *Symmetries and conservation laws for differential equations of mathematical physics*. Providence, RI: American Mathematical Society.

Leach, P., K. Andriopoulos, and M. Nucci (2003). The Ermanno–Bernoulli constants and the representations of the complete symmetry group of the Kepler problem. *Journal of Mathematical Physics* 44: 4090–4106.

Lévy-Leblond, J.-M. (1971). Conservation laws for gauge-variant Lagrangians in classical mechanics. *American Journal of Physics* 39: 502–506.

Lutzky, M. (1978). Symmetry groups and conserved quantities for the harmonic oscillator. *Journal of Physics A* 11: 249–258.

Morehead, J. (2005). Visualizing the extra symmetry of the Kepler problem. *American Journal of Physics* 73: 234–239.

North, J. (2010). Structure in classical mechanics. Unpublished manuscript.

Olver, P. (1984). Conservation laws in elasticity: II. Linear homogeneous isotropic elastostatics. *Archive for Rational Mechanics and Analysis* 85: 131–160.

———. (1993), *Applications of Lie groups to differential equations*. 2d ed. Berlin: Springer-Verlag.

Pohjanpelto, J. (1995). Symmetries, conservation laws, and Maxwell's equations. In *Advanced electromagnetism: foundations, theory and applications*, ed. T. Barrett and D. Grimes, 560–589. Singapore: World Scientific.

Prince, G. (2000). The inverse problem in the calculus of variations and its ramifications. In *Geometric approaches to differential equations*, ed. P. Vassiliou and I. Lisle, 171–200. Cambridge: Cambridge University Press.

Prince, G., and C. Eliezer (1980). Symmetries of the time-dependent N-dimensional oscillator. *Journal of Physics A* 13: 815–823.

———. (1981). On the Lie symmetries of the classical Kepler problem. *Journal of Physics A* 14: 587–596.

Roberts, J. (2008). A puzzle about laws, symmetries, and measurability. *British Journal for the Philosophy of Science* 59: 143–168.

Rosen, N. (1966). Flat space and variational principle. In *Perspectives in geometry and relativity: Essays in honor of Václav Hlavatý*, ed. B. Hoffmann, 325–327. Bloomington: Indiana University Press.

Ruetsche, L. (2011). *Interpreting quantum theories*. Oxford: Oxford University Press.

Saunders, D. (2008). Jet manifolds and natural bundles. In *Handbook of global analysis*, ed. D. Krupka and D. Saunders, 1035–1068. Amsterdam: Elsevier.

Sorkin, R. (2002). An example relevant to the Kretschmann–Einstein debate. *Modern Physics Letters A* 17: 695–700.

Torre, C. (1995). Natural symmetries and Yang–Mills equations. *Journal of Mathematical Physics* 36: 2113–2130.

CHAPTER 10

INDISTINGUISHABILITY

SIMON SAUNDERS

By the end of the nineteenth century the concept of particle indistinguishability had entered physics in two apparently quite independent ways: in statistical mechanics, where, according to Gibbs, it was needed in order to define an extensive entropy function; and in the theory of black-body radiation, where, according to Planck, it was needed to interpolate between the high frequency (Wien law) limit of thermal radiative equilibrium, and the low frequency (Rayleigh-Jeans) limit. The latter, of course, also required the quantization of energy, and the introduction of Planck's constant: the birth of quantum mechanics.

It was not only quantum mechanics. Planck's work, and later that of Einstein and Debye, foreshadowed the first quantum field theory as written down by Dirac in 1927. Indistinguishability is essential to the interpretation of quantum fields in terms of particles (Fock space representations), and thereby to the entire framework of high-energy particle physics as a theory of local interacting fields.

In this chapter, however, we confine ourselves to particle indistinguishability in low-energy theories, in quantum and classical statistical mechanics describing ordinary matter. We are also interested in indistinguishability as a *symmetry*, to be treated in a uniform way with other symmetries of physical theories, especially with spacetime symmetries. That adds to the need to study permutation symmetry in classical theory—and returns us to Gibbs and the derivation of the entropy function.

The concept of particle indistinguishability thus construed faces some obvious challenges. It remains controversial, now for more than a century, whether classical particles can be treated as indistinguishable; or if they can, whether the puzzles raised by Gibbs are thereby solved or alleviated; and if so, how the differences between quantum and classical statistics are to be explained. The bulk of this chapter is about these questions. In part they are philosophical. As Quine remarked:

> Those results [in quantum statistics] seem to show that there is no difference even in principle between saying of two elementary particles of a given kind that they are in the respective places *a* and *b* and that they are oppositely placed, in *b* and *a*. It would seem then not merely that elementary particles are unlike bodies; it would seem that there are no such denizens of spacetime at all, and that we should speak of places *a* and *b* merely as being in certain states, indeed the same state, rather than as being occupied by two things. (Quine 1990, 35)

He was speaking of indistinguishable particles in quantum mechanics, but if particles in classical theory are treated the same way, the same questions arise.

This chapter is organized in three sections. The first is about the Gibbs paradox and is largely expository. The second is on particle indistinguishability, and the explanation of quantum statistics granted that classical particles just like quantum particles can be treated as permutable. The third is about the more philosophical questions raised by sections 1 and 2, and the question posed by Quine. There is a special difficulty in matters of ontology in quantum mechanics, if only because of the measurement problem.[1] I shall, so far as is possible, be neutral on this. My conclusions apply to most realist solutions of the measurement problem, and even some nonrealist ones.

1. The Gibbs Paradox

1.1 Indistinguishability and the Quantum

Quantum theory began with a puzzle over the statistical equilibrium of radiation with matter. Specifically, Planck was led to a certain combinatorial problem: For each frequency v_s, what is the number of ways of distributing an integral number N_s of "energy elements" over a system of C_s states (or "resonators")?

> The distribution of energy over each type of resonator must now be considered, first, the distribution of the energy E_s over the C_s resonators with frequency v_s. If E_s is regarded as infinitely divisible, an infinite number of different distributions is possible. We, however, consider—and this is the essential point—E_s to be composed of a determinate number of equal finite parts and employ in their determination the

---

[1] See the chapters by David Wallace and Guido Bacciagalluppi, this volume.

natural constant $h = 6.55 \times 10^{-27}$ erg sec. This constant, multiplied by the frequency, ν_s, of the resonator yields the energy element $\Delta\epsilon_s$ in ergs, and dividing E_s by $h\nu_s$, we obtain the number N_s, of energy elements to be distributed over the C_s resonators. (Planck 1900, 239)[2]

Thus was made what is quite possibly the most successful single conjecture in the entire history of physics: the existence of Planck's constant h, postulated in 1900 in the role of energy quantization.

The number of distributions Z_s, or *microstates* as we shall call them, as a function of frequency, was sought by Planck in an effort to apply Boltzmann's statistical method to calculate the energy-density $\overline{E}_s$ of radiative equilibrium as a function of temperature T and of Z_s. To obtain agreement with experiment he found

$$Z_s = \frac{(N_s + C_s - 1)!}{N_s!(C_s - 1)!}. \tag{1}$$

The expression has a ready interpretation: it is *the number of ways of distributing N_s indistinguishable elements over C_s distinguishable cells*—of noting only how many elements are in which cell, not which element is in which cell.[3] Equivalently, the microstates are *distributions invariant under permutations*. When this condition is met, we call the elements *permutable*.[4] Following standard physics terminology, they are *identical* if these elements, independent of their microstates, have exactly the same properties (like charge, mass, and spin).

Planck's "energy elements" at a given frequency were certainly identical; but whether it followed that they should be considered permutable was hotly disputed. Once interpreted as particles ("light quanta"), as Einstein proposed, there was a natural alternative: Why not count microstates as distinct if they differ in which particle is located in which cell, as had Boltzmann in the case of material particles? On that count the number of distinct microstates should be:

$$Z_s = C_s^{N_s}. \tag{2}$$

Considered in probabilistic terms, again as Einstein proposed, if each of the N_s elements is assigned one of the C_s cells at random, independent of each other, the number of such assignments will be given by (2), each of them equiprobable.

But while (2) gave the correct behavior for $\overline{E}_s$ in the high-frequency limit (Wien's law), it departed sharply from the Planck distribution at low frequencies. Eq.(1) was empirically correct, not (2). The implication was that if light was made of particles

---

[2] I have used a different notation from Planck's for consistency with the notation in the sequel.
[3] A microstate as just defined can be specified as a string of N_s symbols "p" and $C_s - 1$ symbols "|" (thus, for example, $N_s = 3$, $C_s = 4$, the string p||pp| corresponds to one particle in the first cell, none in the second, two in the third, and none in the fourth). The number of distinct strings is $(N_s + C_s - 1)!$ divided by $(C_s - 1)!N_s!$, because permutations of the symbol "|" among themselves or the symbol "p" among themselves give the same string. (This derivation of (1) was given by Ehrenfest in 1912.)
[4] I take "indistinguishable" and "permutable" to mean the same. But others take "indistinguishable" to have a broader meaning, so I will give up that word and use "permutable" instead.

labeled by frequency, they were particles that could not be considered as independent of each other at low frequencies.[5]

Eq.(1) is true of bosons; bosons are represented by totally symmetrized states in quantum mechanics and quantum field theory; totally symmetrized states are entangled states. There is no doubt that Einstein, and later Schrödinger, were puzzled by the lack of independence of light-quanta at low frequencies. They were also puzzled by quantum nonlocality and entanglement. It is tempting to view all these puzzles as related.[6] Others concluded that light could not after all be made of particles, or that it is made up of both particles and waves, or it is made up of a special category of entities that are not really objects at all.[7] We shall come back to these questions separately.

For Planck's own views on the matter, they were perhaps closest to Gibbs's.[8] Gibbs had arrived at the concept of particle indistinguishability quite independent of quantum theory. To understand this development, however, considerably more stage-setting is needed, in both classical statistical mechanics and thermodynamics, the business of sections 1.2 to 1.4. (Those familiar with the Gibbs paradox may skip directly to section 1.5.)

1.2 The Gibbs Paradox in Thermodynamics

Consider the entropy of a volume V of gas composed of N_A molecules of kind A and N_B molecules of kind B.[9] It differs from the entropy of a gas at the same temperature and pressure when A and B are identical. The difference is:

$$-kN_A \log N_A - kN_B \log N_B + k(N_A + N_B) \log(N_A + N_B) \tag{3}$$

where k is Boltzmann's constant, $k = 1.38 \times 10^{-16}$ erg K^{-1}. The expression (3) is unchanged no matter how similar A and B are, even when in practice the two gases cannot be distinguished; but it must vanish when A and B are the same. This is the Gibbs paradox in thermodynamics.

It is not clear that the puzzle as stated is really paradoxical, but it certainly bears on the notion of identity—and on whether identity admits of degrees. Thus, Denbigh and Redhead argue:

> The entropy of mixing has the same value ... however alike are the two substances, but suddenly collapses to zero when they are the same. It is the absence of any "warning" of the impending catastrophe, as the substances are made more and more similar, which is the truly paradoxical feature. (Denbigh and Redhead 1989, 284)

---

[5] The *locus classicus* for this story is Jammer (1966), but see also Darrigol (1991).
[6] Or as at bottom the same, as argued most prominently by Howard (1990).
[7] As suggested by Quine. See French and Krause (2006) for a comprehensive survey of debates of this kind.
[8] See Planck (1912, 1921) and, for commentary, Rosenfeld (1959).
[9] This section largely follows van Kampen (1984).

The difficulty is more severe for those who see thermodynamics as founded on operational concepts. Identity, as distinct from similarity under all practical measurements, seems to outstrip any possible experimental determination.

To see how experiment does bear on the matter, recall that the classical thermodynamic entropy is an *extensive* function of the mass (or particle number) and volume. That is to say, for real numbers λ, the thermodynamic entropy S as a function of N and V scales linearly:

$$S(\lambda N, \lambda V, T) = \lambda S(N, V, T), \ \lambda \in \mathbb{R}.$$

By contrast the pressure and temperature are *intensive* variables that do not scale with mass and volume. The thermodynamic entropy function for an ideal gas is:

$$S(P, T, N) = \frac{5}{2} Nk \log T - Nk \log P + cN \tag{4}$$

where c is an arbitrary constant. It is extensive by inspection.

The extensivity of the entropy allows one to define the analogue of a density—entropy per unit mass or unit volume—important to nonequilibrium thermodynamics, but the concept clearly has its limits: for example, it is hardly expected to apply to gravitating systems, and more generally ignores surface effects and other sources of inhomogeneity. It is to be sharply distinguished from *additivity* of the entropy, needed to define a total entropy for a collection of equilibrium systems each separately described—typically, as (at least initially) physically isolated systems. The assumption of additivity is that a total entropy can be defined as their sum:

$$S_{A+B} = S_A(N_A, V_A, T_A) + S_B(N_B, V_B, T_B).$$

It is doubtful that any general statement of the second law would be possible without additivity. Thus, collect together a dozen equilibrium systems, some samples of gas, homogeneous fluids or material bodies, initially isolated, and determine the entropy of each as a function of its temperature, volume, and mass. Energetically isolate them from external influences, but allow them to interact with each other in any way you like (mechanical, thermal, chemical, nuclear), so long as the result is a new collection of equilibrium systems. Then the second law can be expressed as follows: the sum of the entropies of the latter systems is equal to or greater than the sum of the entropies of the former systems.[10]

Now for the connection with the Gibbs paradox. The thermodynamic entropy difference between states 1 and 2 is defined as the integral, over any *reversible* process[11] that links the two states, of dQ/T, that is as the quantity:

$$\Delta S = S_2 - S_1 = \int_1^2 \frac{dQ}{T}$$

---

[10] See, e.g., Lieb and Yngvason (1999) for a statement of the second law at this sort of level of generality.
[11] Meaning a process which at any point in its progress can be reversed, to as good an approximation as is required. Necessary conditions are that temperature gradients are small and effects due to friction and turbulence are small (but it is doubtful these are sufficient).

where dQ is the heat transfer. If the insertion or removal of a partition between A and B is to count as a reversible process, then from additivity and given that negligible work is done on the partition it follows there will be no change in entropy, so no entropy of mixing. This implies the entropy must be extensive. Conversely extensivity, under the same presupposition, and again given additivity, implies there is no entropy of mixing.

Whether the removal of a partition between A and B should count as a reversible process is another matter: surely not if means are available to tell the two gases apart. Thus, if a membrane is opaque to A, transparent to B, under compression work $P_A dV$ must be done against the partial pressure P_A in voiding one part of the cylinder of gas A (and similarly for B), where:

$$P_A = \frac{N_A}{N_A + N_B} P, \; P_B = \frac{N_B}{N_A + N_B} P.$$

The work dW required to separate the two gases isothermally at temperature T is related to the entropy change and the heat transfer by:

$$dQ = dW + TdS.$$

Using the equation of state for the ideal gas to determine $dW = PdV$

$$PV = NkT$$

where $N = N_A + N_B$, the result is the entropy of mixing, Eq.(3). However, there can be no such semi-permeable membrane when the two gases are identical,[12] so in this case the entropy of mixing is zero.

Would it matter to the latter conclusion if the differences between the two gases were sufficiently small (were ignored or remained undiscovered)? But as van Kampen argues, it is hard to see how the chemist will be led into any practical error in ignoring an entropy of mixing, if he cannot take mechanical advantage of it. Most thermodynamic substances, in practice, are composites of two or more substances (typically, different isotopes), but such mixtures are usually treated as homogeneous. In thermodynamics, as a science based on operational concepts, the meaning of the entropy function does not extend beyond the competencies of the experimenter:

> Thus, whether such a process is reversible or not depends on how discriminating the observer is. The expression for the entropy depends on whether or not he is able and willing to distinguish between the molecules A and B. This is a paradox only for those who attach more physical reality to the entropy than is implied by its definition. (Van Kampen 1984, 307)

---

[12] At least in the absence of Maxwell demons: see section 3.1.

A similar resolution of the Gibbs paradox was given by Jaynes (1992). It appears, on this reading, that the entropy is not a real physical property of a thermodynamic system, independent of our knowledge of it. According to Van Kampen, it is attributed to a system on the basis of a system of *conventions*—on whether the removal of a partition is to be counted as a reversible process, and on whether the entropy function for the two samples of gas is counted as extensive. That explains why the entropy of mixing is an all-or-nothing affair.

1.3 The Gibbs Paradox in Statistical Mechanics

Thermodynamics is the one fundamental theory of physics that might lay claim to being based on operational concepts and definitions. The situation is different in statistical mechanics, where the concept of entropy is not limited to equilibrium states, nor bound to the concept of reversibility.

There is an immediate difficulty, however; for the classical derivations of the entropy in statistical mechanics yield a function that is not extensive, even as an idealization. That is, classically, there is always an entropy of mixing, even for samples of the same gas. If the original Gibbs paradox was that there was no entropy of mixing in the limit of identity, the new paradox is that there is.[13]

To see the nature of the problem, it will suffice to consider the ideal gas, using the Boltzmann definition of entropy, so-called.[14] The state of a system of N particles is represented by a set of N points in the 6-dimensional one-particle phase space (or μ-space), or equivalently, by a single point in the total $6N$-dimensional phase space Γ^N. A *fine-graining* of Γ^N is a division of this space into cells of equal volume τ^N (corresponding to a division of μ-space into cells of volume τ, where τ has dimensions of [momentum]^3[length]^3). A *coarse-graining* is a division of Γ^N into regions with a given range of energy. For weakly interacting particles these regions can be parameterized by the one-particle energies ϵ_s, with N_s the number of particles with energy in the range $[\epsilon_s, \epsilon_s + \Delta\epsilon_s]$, and the coarse-graining extended to μ-space as well. These numbers must satisfy:

$$\sum_s N_s = N; \sum_s N_s \epsilon_s = E \tag{5}$$

where E is the total energy. Thus, for any fine-grained description (microstate) of the gas, which specifies how, for each s, N_s particles are distributed over the fine-graining, there is a definite coarse-grained description (macrostate) which only specifies the number in each energy range. Each macrostate corresponds to a definite volume of phase space.

We can now define the Boltzmann entropy of a gas of N particles in a given microstate: *it is proportional to the logarithm of the volume, in Γ^N, of the corresponding*

---

[13] I owe this turn of phrase to Jos Uffink.
[14] Boltzmann defined the entropy in several different ways; see Bach (1990).

macrostate. In this the choice of τ only effects an additive constant, irrelevant to entropy differences.

This entropy is computed as follows. For each s, let there be C_s cells in μ-space of volume τ bounded by the energies $\epsilon_s, \epsilon_s + \Delta\epsilon$, containing N_s particles. Counting microstates as distinct if they differ in which particles are in which cells, we use (2) for the number of microstates, each with the same phase space volume τ^{N_s}, yielding the volume:

$$Z_s \tau^{N_s} = C_s^{N_s} \tau^{N_s}. \tag{6}$$

The product of these quantities (over s) is the N-particle phase-space volume of the macrostate $N_1, N_2, ..., N_s, ..$ for just one way of partitioning the N particles among the various one-particle energies. There are

$$\frac{N!}{N_1!...N_s!...} \tag{7}$$

partitionings in all. The total phase space volume W^B ("B" is for Boltzmann) of the macrostate $N_1, N_2, ..., N_s, ..$ is the product of terms (6) (over s) and (7):

$$W^B = \frac{N!}{N_1!..N_s!....} \prod_s C_s^{N_s} \tau^{N_s} \tag{8}$$

and the entropy is:

$$S^B = k \log W^B = k \log \left[\frac{N!}{N_1!..N_s!....} \prod_s C_s^{N_s} \tau^{N_s} \right].$$

From the Stirling approximation for x large, $\log x! \approx x \log x - x$:

$$S^B \approx kN \log N + k \sum_s N_s \log \frac{C_s}{N_s} + kN \log \tau. \tag{9}$$

By inspection, this entropy function is not extensive. When the spatial volume and particle number are doubled, the second and third expressions on the RHS scale properly, but not the first. This picks up a term $kN \log 2$, corresponding to the 2^N choices as to which of the two sub-volumes contains which particle.

One way to obtain an extensive entropy function is to simply subtract the term $kN \log N$. In the Stirling approximation (up to a constant scaling with N and V) that is equivalent to dividing the volume (8) by $N!$. But with what justification? If, after all, permutations of particles did not yield distinct fine-grained distributions, the factor (7) would not be divided by $N!$; it would be set equal to unity. Call this *the N! problem*. This is itself sometimes called the Gibbs paradox, but is clearly only a fragment of it. It is the main topic of sections 1.5 and 2.1.

1.4 The Equilibrium Entropy

Although not needed in the sequel, for completeness we obtain the equilibrium entropy, thus making the connection with observable quantities.[15]

A system is in equilibrium when the entropy of its coarse-grained distribution is a maximum; that is, when the entropy is stationary under variation of the numbers $N_s \to N_s + \delta N_s$, consistent with (5), that is from (9):

$$0 = \delta \overline{S}^B = \sum_s [\delta N_s \log C_s - \delta N_s \log N_s - \delta N_s] \tag{10}$$

where

$$\sum_s \delta N_s = 0; \quad \sum_s \delta N_s \epsilon_s = 0. \tag{11}$$

If the variations δN_s were entirely independent, each term in the summand (10) would have to vanish. Instead introduce Lagrange multipliers a, β for the respective constraint equations (11). Conclude for each s:

$$\log C_s - \log N_s - \alpha - \beta \epsilon_s = 0.$$

Rearranging:

$$N_s = C_s(e^{-\alpha - \beta \epsilon_s}). \tag{12}$$

Substituting in (9) and using (5) gives the equilibrium entropy $\overline{S}^B$:

$$\overline{S}^B = kN \log N + k \sum_s N_s(\alpha + \beta \epsilon_s) + kN \log \tau$$

$$= kN \log N + kN\alpha + k\beta E + kN \log \tau. \tag{13}$$

The values of α and β are fixed by (5) and (12). Replacing the schematic label s by coordinates on phase space for a monatomic gas $\vec{x}, \vec{p}$, with ϵ_s the kinetic energy $\frac{1}{2m}\vec{p}^2$, the sum over N_s in the first equation of (5) becomes:

$$e^{-\alpha} \int_V \int e^{-\frac{\beta}{2m}\vec{p}^2} d^3x\, d^3p = N.$$

The spatial integral gives the volume V; the momentum integral gives $(2\pi m/\beta)^{3/2}$, so

$$e^{-\alpha} = \frac{N}{V}(2\pi m/\beta)^{-3/2}.$$

From the analogous normalization condition on the total energy (the second constraint (5)), substituting (12) and given that for an ideal monatomic gas $E = \frac{3}{2}NkT$, deduce that $\beta = \frac{1}{kT}$. Substituting in (13), the equilibrium entropy is:

$$\overline{S}^B(N, V, T) = Nk \log V + \frac{3}{2}Nk \log 2\pi mkT + \frac{3}{2}Nk + Nk \log \tau.$$

---

[15] For a textbook derivation using our notation, see, e.g., Hercus (1950).

It is clearly not extensive. Compare Eq.(4), which using the equation of state for the ideal gas takes the form (the Sackur-Tetrode equation):

$$S(N, V, T) = Nk \log \frac{V}{N} + \frac{3}{2} Nk \log 2\pi mkT + cN$$

where c is an arbitrary constant. They differ by the term $Nk \log N$, as already noted.

1.5 The $N!$ Puzzle

The $N!$ puzzle is this: What justifies the subtraction of the term $Nk \log N$ from the entropy? Or equivalently, what justifies the division of the phase space volume Eq.(8) by $N!$? In fact it has a fairly obvious answer (see section 2.1): classical particles, if identical, should be treated as permutable, just like identical quantum particles. But this suggestion has rarely been taken seriously.

Much more widely favored is the view that quantum theory is needed. Classical statistical mechanics is not after all a correct theory; quantum statistical mechanics (Eq.(1)), in the dilute limit $C_s \gg N_s$, gives:

$$Z_s = \frac{(N_s + C_s - 1)!}{N_s!(C_s - 1)!} \approx \frac{C_s^N}{N_s!}$$

yielding the required correction to (6) (setting (7) to unity). Call this the *orthodox* solution to the $N!$ puzzle.

This reasoning, so far as it goes, is perfectly sound, but it does not go very far. It says nothing about why particles in quantum theory but not classical theory are permutable. If rationale is offered, it is that classical particles are localized in space and hence are distinguishable (we shall consider this in more detail in the next section); and along with that, that the quantum state for identical particles is unchanged by permutations.[16] But how the two are connected is rarely explained.

Erwin Schrödinger, in his book *Statistical Thermodynamics*, did give an analysis:

> It was a famous paradox pointed out for the first time by W. Gibbs, that the same increase of entropy must not be taken into account, when the two molecules are of the same gas, although (according to naive gas-theoretical views) diffusion takes place then too, but unnoticeably to us, because all the particles are alike. The modern view [of quantum mechanics] solves this paradox by declaring that in the second case there is no real diffusion, because exchange between like particles is not a real event—if it were, we should have to take account of it statistically. It has always been believed that Gibbs's paradox embodied profound thought. That it was intimately linked up

[16] Statements like this can be found in almost any textbook on statistical mechanics.

with something so important and entirely new [as quantum
mechanics] could hardly be foreseen. (Schrödinger 1946, 61)

Evidently, by "exchange between like particles" Schrödinger meant the sort of thing that happens when gases of classical molecules diffuse—the trajectories of individual molecules are twisted around one another—in contrast to the behavior of quantum particles, which do not have trajectories, and so do not diffuse in this way. But why the exchange of quantum particles "is not a real event" (whereas it is classically) is lost in the even more obscure question of what quantum particles really are.

Schrödinger elsewhere said something more. He wrote of indistinguishable particles as "losing their identity," as "non-individuals," in the way of units of money in the bank (they are "fungible"). That fitted with Planck's original idea of indistinguishable quanta as elements of energy, rather than material things—so, again, quite unlike classical particles.

On this point there seems to have been wide agreement. Schrödinger's claims about the Gibbs paradox came under plenty of criticism, for example, by Otto Stern, but Stern remarked at the end:

> In conclusion, it should be emphasized that in the foregoing remarks classical statistics is considered in principle as a part of classical mechanics which deals with individuals (Boltzmann). The conception of atoms as particles losing their identity cannot be introduced into the classical theory without contradiction. (Stern 1949, 534)

This comment or similar can be found scattered throughout the literature on the foundations of quantum statistics.

There is a second solution to the $N!$ puzzle that goes in the diametrically-opposite direction: it appeals only to classical theory, precisely assuming particle distinguishability. Call this the *classical* solution to the puzzle.

Its origins lie in a treatment by Ehrenfest and Trkal (1920) of the equilibrium conditions for molecules subject to disassociation into a total of N^* atoms. This number is conserved, but the number of molecules N_A, N_B,... formed of these atoms, of various types $A, B,...$ may vary. The dependence of the entropy function on N^* is not needed since this number never changes: it is the dependence on N_A, N_B,... that is relevant to the extensivity of the entropy for molecules of type $A, B,...$, which can be measured. By similar considerations as in section 1.3, the number of ways the N^* atoms can be partitioned among N_A molecules of type A, N_B molecules of type B, ... is the factor $N^*!/N_A!N_B!....$ This multiplies the product of all the phase space volumes for each type of molecule, delivering the required division by $N_A!$ for molecules of type A, by $N_B!$ for molecules of type B, and so on (with the dependence on N^* absorbed into an overall constant).

A similar argument was given by van Kampen (1984), but using Gibbs's methods. The canonical ensemble for a gas of N^* particles has the probability distribution:

$$W(N^*, q, p) = f(N^*) e^{-\beta H(q,p)}.$$

Here (q, p) are coordinates on the $6N^*$-dimensional phase space for the N^* particles, which we suppose are confined to a volume V^*, H is the Hamiltonian, and f is a normalization constant. Let us determine the probability of finding N particles with total energy E in the sub-volume V (so $N' = N^* - N$ are in volume $V' = V^* - V$). If the interaction energy between particles in V' and V is small, the Hamiltonian H_{N^*} of the total system can be approximately written as the sum $H_N + H_{N'}$ of the Hamiltonians for the two subsystems. The probability density $W(N, q, p)$ for N particles as a function of $\langle N, q, p \rangle = \langle \vec{q}_1, \vec{p}_1; \vec{q}_2, \vec{p}_2; ....; \vec{q}_N, \vec{p}_N \rangle$ where $\vec{q}_i \subset V$ is then the marginal on integrating out the remaining N' particles in V', multiplied by the number of ways of selecting N particles from N^* particles. The latter is given by the binomial function:

$$\binom{N^*}{N} = \frac{N^*!}{(N^* - N)! N!}.$$

The result is:

$$W(N, q, p) = f(N^*) \binom{N^*}{N} e^{-\beta H_N(N, q, p)} \int_{V'} e^{-\beta H_{N'}(N', q', p')} dq' dp'.$$

In the limit $N^* \gg N$, the binomial is to a good approximation:

$$\frac{N^*!}{(N^* - N)! N!} \approx \frac{N^{*N}}{N!}.$$

The volume integral yields $V'^{N^* - N}$. For noninteracting particles, for constant density $\rho = V'/N'$ in the large volume limit $V' \gg V$ we obtain:

$$W(N, q, p) \approx f(N^*, V^*) \frac{z^N}{N!} e^{-\beta H_N(q, p)}$$

where z is a function of ρ and β. It has the required division by $N!$

Evidently this solution to the $N!$ puzzle is the same as in Ehrenfest and Trkal's derivation: extensivity of the entropy can only be obtained for an open system, that is, for a proper subsystem of a closed system, never for a closed one—and it follows precisely because the particles are nonpermutable. The tables are thus neatly turned.[17]

Which of the two, the orthodox or the classical, is the "correct" solution to the $N!$ puzzle? It is tempting to say that both are correct, but as answers to different questions: the orthodox solution is about the thermodynamics of real gases, governed by quantum mechanics, and the classical solution is about the consistency of a hypothetical classical system of thermodynamics that in reality does not exist.

---

[17] For another variant of the Ehrenfest-Trkal approach, see Swendsen (2002, 2006). (However, Swendsen does not acknowledge the restriction of the result to open systems. See further Nagle (2004).)

But on either line of reasoning, identical quantum particles are treated as radically unlike identical classical particles (only the former are permutable). This fits with the standard account of the departures of quantum from classical statistics: they are explained by permutability. These are the claims challenged in Part 2.

2. INDISTINGUISHABILITY AS A UNIFORM SYMMETRY

2.1 Gibbs' Solution

There is another answer as to which of the two solutions to the $N!$ puzzle is correct: *neither*. The $N!$ puzzle arises in both classical and quantum theories and is solved in exactly the same way: by passing to the quotient space (of phase space and Hilbert space, respectively). This is not to deny that atoms really are quantum mechanical, or that measurements of the dependence of the entropy on particle number are made in the way that Ehrenfest et al. envisaged; it is to deny that the combinatorics factors thus introduced are, except in special cases, either justified or needed.

Gibbs, in his *Elementary Principles in Statistical Mechanics*, put the matter as follows:

> If two phases differ only in that certain entirely similar particles have changed places with one another, are they to be regarded as identical or different phases? If the particles are regarded as indistinguishable, it seems in accordance with the spirit of the statistical method to regard the phases as identical. (Gibbs 1902, 187)

He proposed that the phase of an N-particle system be unaltered "by the exchange of places between similar particles." Phases (points in phase space) like this he called *generic* (and those that are altered, *specific*). The state space of generic phases is the *reduced phase space* Γ^N/Π_N, the quotient space under the permutation group Π_N of N elements. In this space points of Γ^N related by permutations are identified.

The suggestion is that even classically, the expressions (6) and (7) are wrong. (7) is replaced by unity (as already noted): there is just one way of partitioning N permutable particles among the various states so as to give N_s particles to each state. But (6) is wrong too: it should be replaced by the volume of reduced phase space corresponding to the macrostate (for s), the volume

$$\frac{(C_s\tau)^{N_s}}{N_s!}.$$

For the macrostate $N_1, N_2, ..., N_s, ..$ the total reduced volume, denote W^{red} is:

$$W^{red} = \prod_s \frac{C_s^{N_s} \tau^{N_s}}{N_s!} = \frac{W^B}{N!}. \tag{14}$$

The derivation does not depend on the limiting behavior of Eq.(1), or on the assumption of equiprobability or equality of volume of each fine-grained distribution (and is in fact in contradiction with that assumption, as we shall see).

Given (14), there is no entropy of mixing. Consider a system of particles all with the same energy ϵ_s. The total entropy before mixing is, from additivity:

$$S_A + S_B = k \log \left(\frac{C_A^{N_A} \tau^{N_A}}{N_A!} \frac{C_B^{N_B} \tau^{N_B}}{N_B!} \right). \tag{15}$$

After mixing, if A and B are identical:

$$S_{A+B} = k \log \frac{(C_A + C_B)^{N_A+N_B} \tau^{N_A+N_B}}{(N_A + N_B)!}. \tag{16}$$

If the pressure of the two samples is initially the same (so $C_A/N_A = C_B/N_B$), the quantities (15), (16) should be approximately equal[18]—as can easily be verified in the Stirling approximation. But if A and B are not identical, and permutations of A particles with B particles is not a symmetry, we pass to the quotient spaces under Π_{N_A} and Π_{N_B} separately and take their product, and the denominator in (16) should be $N_A!N_B!$. With that $S_A + S_B$ and S_{A+B} are no longer even approximately the same.

Gibbs concluded his discussion of whether to use generic or specific phases with the words, "The question is one to be decided in accordance with the requirements of practical convenience in the discussion of the problems with which we are engaged" (Gibbs 1902, 188). Practically speaking, if we are interested in defining an extensive classical entropy function (even for closed systems), use of the generic phase (permutability) is clearly desirable. On the other hand, integral and differential calculus is simple on manifolds homeomorphic to $\mathbb{R}^{6N}$, like Γ^N; the reduced phase space Γ^N/Π_N has by contrast a much more complex topology (a point made by Gibbs). If the needed correction, division by $N!$, can be simply made at the end of a calculation, the second consideration will surely trump the first.

2.2 Arguments against Classical Indistinguishability

Are there principled arguments against permutability thus treated uniformly, the same in the classical as in the quantum case? The concept of permutability can certainly be misrepresented. Thus, classically, of course, it makes sense to move atoms about so as to interchange one with another, for particles have definite trajectories;

---

[18] Should they be exactly equal? No, because it is an additional constraint to insist, given that $N_A + N_B$ particles are in volume $V_A + V_B$, that exactly N_A are in V_A and N_B in V_B.

in that sense an "exchange of places" must make for a real physical difference, and in that sense "indistinguishability" cannot apply to classical particles.

But that is not what is meant by "interchange"—Schrödinger was just misleading on this point. It is interchange of points in phase space whose significance is denied, not in configuration space over time. Points in phase-space are in 1:1 correspondence with the dynamically allowed trajectories. A system of N particles whose trajectories in μ-space swirl about one another, leading to an exchange of two or more of them in their places in space at two different times, is described by each of $N!$ points in the $6N$-dimensional phase space Γ^N, each faithfully representing the same swirl of trajectories in μ-space (but assigning different labels to each trajectory). In passing to points of the quotient space Γ^N/Π_N there is therefore no risk of descriptive inadequacy in representing particle interchange in Schrödinger's sense.

Another and more obscure muddle is to suppose that points of phase space can only be identified insofar as they are all traversed by one and the same trajectory. That appears to be the principle underlying van Kampen's argument:

> One could add, as an aside, that the energy surface can be partitioned in $N!$ equivalent parts, which differ from one another only by a permutation of the molecules. The trajectory, however, does not recognize this equivalence because it cannot jump from one point to an equivalent one. There can be no good reason for identifying the Z-star [the region of phase space picked out by given macroscopic conditions] with only one of these equivalent parts. (Van Kampen 1984, 307)

But if the whole reason to consider the phase-space volumes of macrostates in deriving thermodynamic behavior is because (say by ergodicity) they are proportional to the amount of time the system spends in the associated macrostates, then, just because the trajectory cannot jump from one point to an equivalent one, it should be enough to consider only one of the equivalent parts of the Z-star. We should draw precisely the opposite conclusion to van Kampen.

However van Kampen put the matter somewhat differently—in terms, only, of probability:

> Gibbs argued that, since the observer cannot distinguish between different molecules, "it seems in accordance with the spirit of the statistical method" to count all microscopic states that differ only by a permutation as a single one. Actually it is exactly opposite to the basic idea of statistical mechanics, namely that the probability of a macrostate is given by the measure of the Z-star, i.e. *the number of corresponding, macroscopically indistinguishable microstates*. As mentioned ... it is impossible to justify the $N!$ as long as one restricts

oneself to a single closed system. (van Kampen 1984, 309, emphasis added).

Moreover, he speaks of probabilities of macroscopically indistinguishable microstates, whereas the contentious question concerns *microscopically* indistinguishable microstates. The contentious question is whether microstates that differ only by particle permutations, with all physical properties unchanged—which are in this sense indistinguishable—should be identified.

Alexander Bach in his book *Classical Particle Indistinguishability* defended the concept of permutability of states in classical statistical mechanics, understood as the requirement that probability distributions over microstates be invariant under permutations. But what he meant by this is the invariance of functions on Γ^N. As such, as probability measures, they could never provide complete descriptions of the particles (unless all their coordinates coincide)—they could not be concentrated on individual trajectories. He called this the "deterministic setting." In his own words:

> **Indistinguishable Classical Particles Have No Trajectories.**
> The unconventional role of indistinguishable classical particles is best expressed by the fact that in a deterministic setting no indistinguishable particles exist, or—equivalently—that indistinguishable classical particles have no trajectories. Before I give a formal proof I argue as follows. Suppose they have trajectories, then the particles can be identified by them and are, therefore, not indistinguishable. (Bach 1997, 7)

His formal argument was as follows. Consider the coordinates of two particles at a given time. in one dimension, as an extremal of the set of probability measures $M_+^1(\mathbb{R}^2)$ on $\mathbb{R}^2$ (a 2-dimensional configuration space), from which, assuming the two particles are impenetrable, the diagonal $D = \{<x, x> \in \mathbb{R}^2, x \in \mathbb{R}\}$ has been removed. Since indistinguishable, the state of the two particles must be unchanged under permutations (permutability), so it must be in $M_{+,sym}^1(\mathbb{R}^2)$, the space of symmetrized measures. It consists of sums of delta functions of the form:

$$\mu_{x,y} = \frac{1}{2}\left(\delta_{<x,y>} + \delta_{<y,x>}\right), \quad <x,y> \in \mathbb{R}^2 \backslash D$$

But no such state is an extremal of $M_+^1(\mathbb{R}^2)$.

As already remarked, the argument presupposes that the coordinates of the two particles defines a point in $M_+^1(\mathbb{R}^2)$, the unreduced space, rather than in $M_+^1(\mathbb{R}^2/\Pi_2)$, the space of probability measures over the reduced space $\mathbb{R}^2/\Pi_2$. In the latter case, since $M_+^1(\mathbb{R}^2/\Pi_2)$ is isomorphic to $M_{+,sym}^1(\mathbb{R}^2)$, there is no difficulty.

Bach's informal argument above is more instructive. Why not use the trajectory of a particle to identify it, by the way it twists and turns in space? Why not indeed: if that is all there is to being a particle, you have already passed to a trajectory in the quotient space Γ^N/Π_N, for those related by permutations twist and turn in exactly the same way. The concept of particle distinguishability is not about the trajectory or the one-particle state: it is about the label of the trajectory or the one-particle state, or equivalently, the question of which particle has that trajectory, that state.[19]

2.3 Haecceitism

Gibbs's suggestion was called "fundamentally idealistic" by Rosenfeld, "mystical" by van Kampen, "inconsistent" by Bach; they were none of them prepared to see in indistinguishability the rejection of what is on first sight a purely metaphysical doctrine—that after every describable characteristic of a thing has been accounted for, there still remains the question of *which* thing has those characteristics.

The key word is "every"; describe a thing only partly, and the question of which it is of several more precisely described things is obviously physically meaningful. But microstates, we take it, are maximal, complete descriptions. If there is a more complete level of description it is the microstate as given by another theory, or at a deeper level of description by the same theory, and to the latter our considerations apply.

The doctrine, now that we have understood it correctly, has a suitably technical name in philosophy. It is called *haecceitism*. Its origins are medieval if not ancient, and it was in play, one way or another, in a connected line of argument from Newton and Clarke to Leibniz and Kant. That centered on the need, given symmetries, including permutations, not just for symmetry-breaking in the choice of initial conditions,[20] but for a choice among haecceistic differences—in the case of continuous symmetries, among values of absolute positions, absolute directions, and absolute velocities. All parties to this debate agreed on haecceitism. These choices were acts of God, with their consequences visible only to God (Newton, Clarke); or they were humanly visible too, but in ways that could not be put into words—that could only be grasped by "intuition" (Kant); or they involved choices not even available to God, who can only choose on the basis of reason; so there could be no created things such as indistinguishable atoms or points of a featureless space (Leibniz).[21]

So much philosophical baggage raises a worry in its own right. If it is the truth or falsity of haecceitism that is at issue, it seems unlikely that it can be settled by any empirical finding. If that is what the extensivity of the entropy is about, perhaps extensivity has no real physical meaning after all. It is, perhaps, itself metaphysical—or conventional. This was the view advocated by Nick Huggett when he first drew the comparison between Boltzmann's combinatorics and haecceitism.[22]

---

[19] These considerations apply to quantum particles too, when described in terms of the de Broglie-Bohm pilot-wave theory. For the latter, see Bacciagaluppi, this volume.

[20] As in, e.g., a cigar-shaped mass distribution, rather than a sphere. Of course, this is not really a breaking of rotational symmetry, in that each is described by relative angles and distances between masses, invariant under rotations.

[21] For more on this vein, see Saunders (2003). For a compilation of original sources and commentary, see Huggett (1999b).

[22] See Huggett (1999a). It was endorsed shortly after by David Albert in his book *Time and Chance* (Albert 2000, 47–48).

But this point of view is only remotely tenable if haecceitism is similarly irrelevant to empirical questions in quantum statistics. And on the face of it that cannot be correct. Planck was, after all, led by experiment to Eq. (1). Use of the unreduced state space in quantum mechanics rather than the reduced (symmetrized) space surely has direct empirical consequences.

Against this two objections can be made. The first, following Reichenbach (1956), is that the important difference between quantum and classical systems is the absence in quantum theory of a criterion for the re-identification of identical particles over time. They are, for this reason, "non-individuals" (this links with Schrödinger's writings[23]). This, rather than any failure of haecceitism, is what is responsible for the departures from classical statistics.[24] The second, following Post (1963) and French and Redhead (1988), is that haecceitism must be consistent with quantum statistics (including Planck's formula) because particles, even given the symmetrization of the state, may nevertheless possess "transcendental" individuality, and symmetrization of the state can itself be understood as a dynamical constraint on the state, rather than in terms of permutability.

Of these the second need not detain us. Perhaps metaphysical claims can be isolated from any possible impact on physics, but better, surely, is to link them with physics where such links are possible. Or perhaps we were wrong to think that haecceitism is a metaphysical doctrine: it just means nonpermutability, it is to break the permutation symmetry. The converse view is to respect this symmetry.

As for the first, it is simply not true that indistinguishable quantum particles can never be reidentified over time. Such identifications are only exact in the kinematic limit, to be sure, and even then only for a certain class of states; but the ideal gas is commonly treated in just such a kinematic limit, and the restriction in states applies just as much to the reidentification of identical quantum particles that are not indistinguishable—that are not permutable—but that are otherwise entangled.

This point needs some elaboration. Consider first the case of nonpermutable identical particles. The N particle state space is then $\mathcal{H}^N = \mathcal{H} \otimes \mathcal{H} \otimes .. \otimes \mathcal{H}$, the N-fold tensor product of the one-particle state space $\mathcal{H}$. Consider states of the form:

$$|\Phi\rangle = \underbrace{|\phi\rangle_a \otimes |\phi_b\rangle \otimes ... \otimes \overbrace{|\phi_c\rangle}^{k\text{-factors}} \otimes ... \otimes |\phi_d\rangle}_{N\text{-factors}} \quad (17)$$

where the one-particle states are members of some orthonormal basis (we allow for repetitions). The kth-particle is then in the one-particle state $|\phi_c\rangle$. The ordering of the tensor-product breaks the permutation symmetry. If the particles are only weakly interacting, and the state remains a product state, the kth-particle can also be assigned a one-particle state at later times, namely the unitary evolute of $|\phi_c\rangle$. Even if more than one particle has the initial state $|\phi_c\rangle$, still it will be the case that each

---

[23] See Schrödinger (1984, 207–210). The word "individual" has also been used to mean an object answering to a unique description at a single time (as "absolutely discernible" in the terminology of Saunders (2003, 2006b).
[24] As recently endorsed by Pooley (2006 section 8).

particle in that state has a definite orbit under the unitary evolution. It is true that in those circumstances it would seem impossible to tell the two orbits apart, but the same will be true of two classical particles with exactly the same representative points in μ-space.[25]

Now notice the limitation of this way of speaking of particles as one-particle states that are (at least conceptually) identifiable over time: it does not in general apply to superpositions of states of the form (17)—as will naturally arise if the particles are interacting, even starting from (17). In general, given superpositions of product states, there is no single collection of N one-particle states, or orbits of one-particle states, sufficient for the description of the N particles over time. In these circumstances no definite histories, no orbits of one-particle states, can be attributed to identical but distinguishable particles.

Now consider identical permutable quantum particles (indistinguishable quantum particles). The state must now be invariant under permutations, so (for vector states):

$$U_\pi |\Phi\rangle = |\Phi\rangle \tag{18}$$

for every $\pi \in \Pi^N$, where $U : \pi \to U_\pi$ is a unitary representation of the permutation group Π^N. Given (18), $|\Phi\rangle$ must be of the form:

$$|\Phi\rangle = c \sum_{\pi \in \Pi_N} |\phi_{\pi(a)}\rangle \otimes |\phi_{\pi(b)}\rangle \otimes \ldots \otimes |\phi_{\pi(c)}\rangle \otimes \ldots \otimes |\phi_{\pi(d)}\rangle \tag{19}$$

and superpositions thereof. Here c is a normalization constant, $\pi \in \Pi_N$ is a permutation of the N symbols $\{a, b, \ldots, c, ..d\}$ (which, again, may have repetitions), and as before, the one-particle states are drawn from some orthonormal set in $\mathcal{H}$. If noninteracting, and initially in the state (19), the particle in the state $|\phi_c\rangle$ can *still* be reidentified over time—as the particle in the state which is the unitary evolute of $|\phi_c\rangle$.[26] That is to say, for entanglements like this, one-particle states can still be tracked over time. It is true that we can no longer refer to the state as that of the kth particle, in contrast to states of the form (17), but that labeling—unless shorthand for something else, say the lattice position of an atom in a crystal—never had any physical meaning. As for more entangled states—for superpositions of states of the form (19)—there is of course a difficulty; but it is the same difficulty as we encountered for identical but distinguishable particles.

Reichenbach was therefore right to say that quantum theory poses special problems for the reidentification of identical particles over time, and that these

---

[25] One might in classical mechanics add the condition that the particles are impenetrable; but one can also, in quantum mechanics, require that no two particles occupy the same one-particle state (the Pauli exclusion principle). See sections 2.5, 3.3.

[26] As we shall see, there is a complication in the case of fermions (section 3.3), although it does not affect the point about identity over time.

problems derive from entanglement; but not from the "mild"[27] form of entanglement required by symmetrization itself (as involved in states of the form (19)), of the sort that explains quantum statistics. On the other hand, this much is also true: permutability does rule out appeal to the reduced density matrix to distinguish each particle in time (defined, for the kth particle, by taking the partial trace of the state over the Hilbert space of all the particles save the kth). Given (anti)symmetrization, the reduced density matrices will all be the same. But it is hard to see how the reduced density matrix can provide an operational as opposed to a conceptual criterion for the reidentification of one among N identical particles over time.

What would an operational criterion look like? Here is a simple example: a helium atom in the canister of gas by the laboratory door is thereby distinguished from one in the high-vacuum chamber in the corner, a criterion that is preserved over time. This means: the one-particle state localized in the canister is distinguished from the one in the vacuum chamber.

We shall encounter this idea of reference and reidentification by location (or more generally by properties) again, so let us give it a name: call it *individuating reference*, and the properties concerned *individuating properties*. In quantum mechanics the latter can be represented in the usual way by projection operators. Thus if P_{can} is the projector onto the region of space Δ_{can} occupied by the canister, and P_{cham} onto the region Δ_{cham} occupied by the vacuum chamber, and if $|\chi_1\rangle, |\chi_2\rangle$ are localized in Δ_{can} (and similarly $|\psi_1\rangle, |\psi_2\rangle$ in Δ_{cham}), then even in the superposition (where $|c_1|^2 + |c_2|^2 = 1$)

$$|\Phi\rangle = c_1 \frac{1}{\sqrt{2}} (|\chi_1\rangle \otimes |\psi_1\rangle + |\psi_1\rangle \otimes |\chi_1\rangle)$$
$$+ c_2 \frac{1}{\sqrt{2}} (|\chi_2\rangle \otimes |\psi_2\rangle + |\psi_2\rangle \otimes |\chi_2\rangle)$$

one can still say there is a state in which one particle is in region Δ_{can} and one in Δ_{cham} (but we cannot say which); still we have:

$$(P_{can} \otimes P_{cham} + P_{cham} \otimes P_{can})|\Phi\rangle = |\Phi\rangle. \tag{20}$$

If the canister and vacuum chamber are well-sealed, this condition will be preserved over time. Individuating properties can be defined in this way just as well for permutable as for nonpermutable identical particles.

It is time to take stock. We asked whether the notion of permutability can be applied to classical statistical mechanics. We found that it can, in a way that yields the desired properties of the statistical mechanical entropy function, bringing it in line with the classical thermodynamic entropy. We saw that arguments for the unintelligibility of

---

[27] The terminology is due to Penrose (2004, 598). See Ghirardi and Marinatto (2004) and Ghirardi, Marinatto, and Weber (2002) for the claim that entanglement due to (anti)symmetrization is not really entanglement at all.

classical permutability in the literature are invalid or unsound, amounting, at best, to appeal to the philosophical doctrine of haecceitism. We knew from the beginning that state-descriptions in the quantum case should be invariant under permutations, and that this has empirical consequences, so on the most straightforward reading of haecceitism the doctrine is false in that context. Unless it is emasculated from all relevance to physics, haecceitism cannot be true a priori. We wondered if it was required or implied if particles are to be reidentified over time, and found the answer was no to both, in the quantum as in the classical case. We conclude: permutation symmetry holds of identical classical particles just as it does of identical quantum particles, and may be treated in the same way, by passing to the quotient space.

Yet an important lacuna remains, for among the desirable consequences of permutation symmetry in the case of quantum particles are the departures from classical statistics—statistics that are unchanged in the case of classical particles. Why is there this difference?

2.4 The Explanation of Quantum Statistics

Consider again the classical reduced phase-space volume for the macrostate $N_1, N_2, ..., N_s, ..$, as given by Eq.(14):

$$W^{red} = \prod_s \frac{C_s^{N_s} \tau^{N_s}}{N_s!}. \tag{21}$$

In effect, Planck replaced the one-particle phase-space volume element τ, hitherto arbitrary, by h^3, and changed the factor Z_s by which it was multiplied to obtain:

$$W^{BE} = \prod_s \frac{(N_s + C_s - 1)! h^{3N_s}}{N_s!(C_s - 1)!}. \tag{22}$$

Continuing from this point, using the method of sections 1.3 and 1.4 one is led to the equilibrium entropy function and equation of state for the ideal Bose-Einstein gas. The entire difference between this and the classical ideal gas is that for each s, the integer C_s^N is replaced by $(N_s + C_s - 1)!/(C_s - 1)!$ What is the rational for this? It does not come from particle indistinguishability (permutability); that has already been taken into account in (21).

Let us focus on just one value of s, that is, on N_s particles distributed over C_s cells, all of the same energy (and hereinafter drop the subscript s). At the level of the fine-grained description, in terms of how many (indistinguishable) particles are in each (distinguishable) cell, a microstate is specified by a sequence of fine-grained occupation numbers $< n_1, n_2, ... , n_C >$, where $\sum_{j=1}^{C} n_j = N$; there are many such corresponding to the coarse-grained description (N, C) (for a single value of s).

INDISTINGUISHABILITY

Their sum is

$$\sum_{\substack{\text{all sequences } <n_1,..,n_C> \\ \text{s.t. } \sum_{k=1}^{C} n_k = N}} 1 = \frac{(N+C-1)!}{N!(C-1)!} \tag{23}$$

as before. But here is another mathematical identity:[28]

$$\sum_{\substack{\text{all sequences } <n_1,..,n_C> \\ \text{s.t. } \sum_{k=1}^{C} n_k = N}} \frac{1}{n_1!...n_C!} = \frac{C^N}{N!}. \tag{24}$$

In other words, the difference between the two expressions (21) and (22), apart from the replacement of the unit τ by h^3, is that *in quantum mechanics every microstate $<n_1, n_2, ..., n_C>$ has equal weight, whereas in classical mechanics each is weighted by the factor $(n_1!...n_C!)^{-1}$*.

Because of this weighting, a classical fine-grained distribution where the N particles are evenly distributed over the C cells has a much greater weight than one where most of the particles are concentrated in a small handful. In contrast, in quantum mechanics, the weights are always the same. Given that "weight," one way or another, translates into statistics, particles weighted classically thus tend to repel, in comparison to their quantum mechanical counterparts; or put the other way, quantum particles tend to bunch together, in comparison to their classical counterparts.

That is what the weighting does, but why is it there? Consider figure 10.1a, for $N = 2$, $C = 4$. Suppose, for concreteness, we are modeling two classical, non-permutable identical coins, such that the first two cells correspond to one of the coins landing heads (H), and the remainder to that coin landing tails (T) (and similarly for the other coin).[29] The cells along the diagonal correspond not just to both coins landing heads or both landing tails—they are cells in which the two coins have *all* their fine-grained properties the same. For any cell away from the diagonal, there is a corresponding cell that differs only in which coin has which fine-grained property (its reflection in the diagonal). Their combined volume in phase space is therefore twice that of any cell on the diagonal.

The same is true in the reduced phase space, figure 10.1b. For $N = 3$ there are three such diagonals; cells along these have one half the volume of the others. And there is an additional boundary, where all three particles have the same fine-grained properties, each with one sixth their volume. The weights in Eq.(24) follow from the structure of reduced phase space, as faithfully preserving ratios of volumes of microstates in the unreduced space. As explained by Huggett (1999a), two classical identical coins, if permutable, still yield a weight for $\{H, T\}$ twice that of the weight

---

[28] A special case of the multinomial theorem (see, e.g., Rapp 1972, 49–50).
[29] Of course, for macroscopic coins, the assumption of degeneracy of the energy is wildly unrealistic, but let that pass.

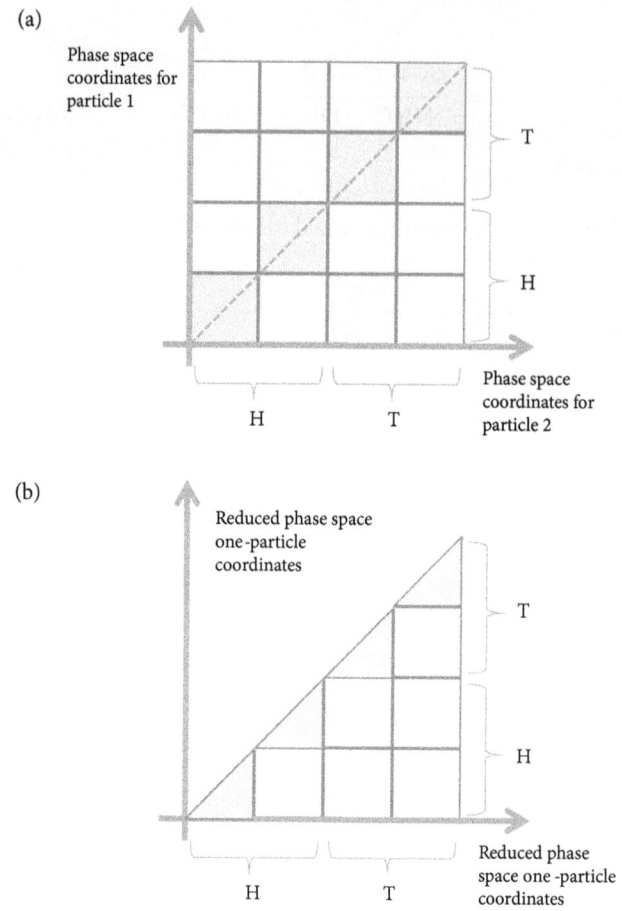

Figure 10.1 Phase space and reduced phase space for two particles

for $\{H,H\}$ or $\{T,T\}$, just as for nonpermutable coins, that is with probabilities one-half, one-quarter, and one-quarter respectively.

Contrast quantum mechanics, where subspaces of Hilbert space replace regions of phase space, and subspace dimension replaces volume measure. Phase space structure, insofar as it can be defined in quantum theory, is derivative and emergent. Since the only measure available is subspace dimension, each of a set of orthogonal directions in each subspace is weighted precisely the same—yielding, for the symmetrized Hilbert space, Eq.(23) instead.[30]

But there are two cases when subspace dimension and volume measure are proportional to one another—or rather, for we take quantum theory as fundamental, for when phase-space structure, complete with volume measure, emerges from

---

[30] One way of putting this is that in the quantum case, the measure on phase space must be discrete, concentrated on points representing each unit cell of "volume" h^3. For early arguments to this effect, see Planck (1912), Poincaré (1911, 1912).

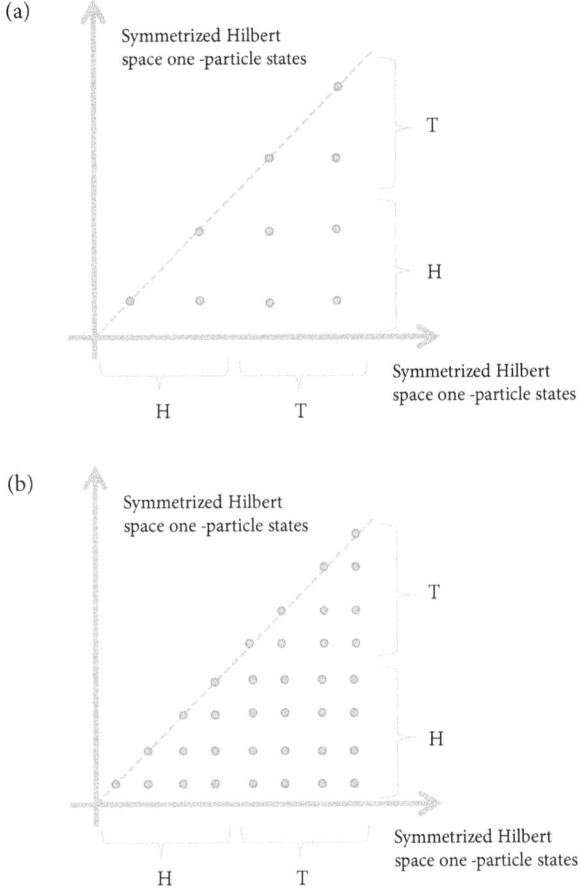

Figure 10.2 Discrete measures for Hilbert space

quantum theory.[31] One is in the limit $C \gg N$, when the contribution from the states along the diagonals is negligible in comparison to the total (figure 10.2b), and the other is when the full Hilbert space for nonpermutable particles is used. That is why permutability makes a difference to statistics in the quantum case but not the classical: for $N \approx C$, as in figure 10.2a, the dimensionality measure departs significantly from the volume measure (in figure 10.2a, as five-eighths to one-half). For $N = 2$, $C = 2$ there are just three orthogonal microstates, each of equal weight. Take two two-state quantum particles (qubits) as quantum coins, and the probabilities $\{H, H\}, \{T, T\}, \{H, T\}$ are all one-third.

Is there a remaining puzzle about quantum statistics—say, the nonindependence of permutable quantum particles, as noted by Einstein? Statistical independence fails, in that the state cannot be specified for $N - 1$ particles, independent of the state of the Nth, but that is true of classical states on reduced phase space

---

[31] See Wallace, this volume.

too (or, indeed, for permutation-invariant states on the unreduced phase space; see Bach 1997). Find a way to impose a discrete measure on a classical permutable system, and one can hope to reproduce quantum statistics as well (Gottesmann 2005). Quantum holism has some role to play in the explanation of quantum statistics, but like entanglement and identity over time, less than meets the eye.

2.5 Fermions

We have made almost no mention so far of fermions. In fact most of our discussion applies to fermions too, but there are some differences.

Why are there fermions at all? The reason is that microstates in quantum theory are actually rays, not vector states $|\phi_c\rangle$, that is, they are 1-dimensional subspaces of Hilbert space. As such they are invariant under multiplication by complex numbers of unit norm. If only the ray need be invariant under permutations, there is an alternative to Eq.(18), namely:

$$U_\pi |\Psi^{FD}\rangle = e^{i\theta} |\Psi^{FD}\rangle \qquad (25)$$

where $\theta \in [0, 2\pi]$. Since any permutation can be decomposed as a product of permutations π_{ij} (that interchange i and j), even or odd in number, and since $\pi_{ij}\pi_{ij} = \mathbb{I}$, it follows that (18) need not be obeyed after all: there is the new possibility that $\theta = 0$ or π for even and odd permutations, respectively. Such states are *antisymmetrized*, that is, of the form:

$$|\Psi^{FD}\rangle = \frac{1}{\sqrt{N!}} \sum_{\pi \in \Pi_N} \text{sgn}(\pi) |\phi_{\pi(a)}\rangle \otimes |\phi_{\pi(b)}\rangle \otimes \ldots \otimes |\phi_{\pi(c)}\rangle \otimes \ldots \otimes |\phi_{\pi(d)}\rangle \qquad (26)$$

where $\text{sgn}(\pi) = 1\ (-1)$ for even (odd) permutations, and superpositions thereof.

An immediate consequence is that, unlike in (19), every one-particle state in (26) must now be orthogonal to every other: repetitions would automatically cancel, leaving no contribution to $|\Psi^{FD}\rangle$. Since superpositions of states (19) with (26) satisfy neither (18) nor (25), permutable particles in quantum mechanics must be of one kind or the other.[32]

The connection between phase space structure and antisymmetrization of the state is made by the Pauli exclusion principle—the principle that no two fermions can share the same complete set of quantum numbers, or equivalently, have the same one-particle state. In view of the effective identification of elementary phase space cells of volume h^3 with rays in Hilbert space, fermions will be constrained so that no two occupy the same elementary volume. In other words, in terms of microstates in phase space, the n_k's are all zeros or ones. In place of Eq.(23), we obtain for the number of microstates for the coarse-grained distribution $\langle C, N \rangle$ (as

---

[32] This is to rule out parastatistics—representations of the permutation group that are not one-dimensional (see, e.g., Greiner and Müller 1994). This would be desirable (since parastatistics have not been observed, except in 2-dimensions, where special considerations apply), but I doubt that it has really been explained.

before, for a single energy level s):

$$\sum_{\substack{\text{fine grainings } n_k \in [0,1] \\ \text{s.t. } \sum_{k=1}^{C} n_k = N}} 1 = \frac{C!}{(C-N-1)!}. \tag{27}$$

Use of (27) in place of (1) yields the entropy and equation of state for the Fermi-Dirac ideal gas. It is, of course, extensive. A classical phase space structure emerges from this theory in the same limit $C \gg N$ (for each s) as for the Bose-Einstein gas, when the classical weights for cells along the diagonals are small in comparison to the total. Away from this limit, whereas for bosons their weight is too small (as suppressed by the factor $(n_1!...n_C!)^{-1}$), for fermions their weight is too large (as not suppressed enough; they should be set equal to zero). Thus fermions tend to repel, in comparison to non-permutable particles.[33]

3. Ontology

The explanation of quantum statistics completes the main argument of this chapter: permutation symmetry falls in place as with any other exact symmetry in physics, and applies just as much to classical systems of equations that display it as to quantum systems.[34] In both cases *only quantities invariant under permutations are physically real*. This is the sense in which "exchange between like particles is not a real event"; it has nothing to do with the swirling of particles around each other, it has only to do with haecceistic redundancies in the mathematical description of such particles, swirling or otherwise. Similar comments apply to other symmetries in physics, where instead of haecceistic differences one usually speaks of coordinate-dependent distinctions.

In both classical and quantum theory state-spaces can be defined in terms only of invariant quantities. In quantum mechanics particle labels need never be introduced at all (the so-called "occupation number formalism")—a formulation recommended by Teller (1995). Why introduce quantities (particle labels) only to deprive them of physical significance? What is their point if they are permutable? We come back to Quine's question and to eliminativism.

There are two sides to this question. One is whether, or how, permutable particles can be adequate as ontology (section 3.1), and link in a reasonable way with philosophical theories of ontology (sections 3.2 and 3.3). The other question is whether some other way of talking might not be preferable, in which permutability as a symmetry does not even arise (section 3.4).

---

[33] The situation is a little more complicated, as antisymmetry in the spin part of the overall state forces symmetry in the spatial part—which can lead to spatial bunching (this is the origin of the homopolar bond in quantum chemistry).
[34] But see Gordon Belot (this volume) for pitfalls in defining such symmetries.

3.1 The Gibbs Paradox, Again

A first pass at the question of whether permutable entities are really objects is to ask how they may give rise to nonpermutable objects. That returns us to the Gibbs paradox in the sense of section 1.2: How similar do objects have to be to count as identical?

On this problem (as opposed to the $N!$ problem) section 2 may seem a disappointment. It focused on indistinguishability as a symmetry, but the existence of a symmetry (or otherwise) seems just as much an all-or-nothing affair as identity. But section 2 did more than that: it offered a microscopic dynamical analysis of the process of mixing of two gasses.

In fact, not even the $N!$ problem is entirely solved, for we would still like to have an extensive entropy function even where particles are obviously nonidentical, say in the statistical behavior of large objects (like stars), and of small but complex objects like fatty molecules in colloids.[35] In these cases we can appeal to the Ehrenfest-Trkal-van Kampen approach, but only given that we can arrive at a description of such objects as distinguishable: How do we do that, exactly?

The two problems are related, and an answer to both lies in the idea of individuating properties, already introduced, and the idea of phase-space structure as emergent, already mentioned. For if particles (or bound states of particles) acquire some dynamically stable properties, there is no reason that they should not play much the same role, in the definition of effective phase-space structure, as do intrinsic ones. Thus two or more nonidentical gases may arise, even though their elementary constituents are identical and permutable, if all the molecules of one gas have some characteristic arrangement, different from those of the other. The two gases will be nonidentical only at an effective, emergent level of description to be sure, and permutation symmetries will still apply at the level of the full phase-space. The effective theory will have only approximate validity, in regimes where those individuating properties are stable in time. Similar comments apply to Hilbert-space structures.[36]

In illustration, consider again figure 10.1b for two classical permutable coins. Suppose that the dynamics is such that one of the coins always rotates about its axis of symmetry in the opposite direction to the other. This fact is recorded in the microstate: each coin not only lands either heads (H) or tails (T), but lands rotating one way (A) or the other (B). It follows that certain regions of the reduced phase space are no longer accessible, among them the cells on the diagonal for which all the properties of the two coins are the same (shaded, figure 10.3a). By inspection, the available phase space has the effective structure of an unreduced phase space for *distinguishable* coins, the A coin and the B coin (figure 10.3b). It is tempting to

---

[35] This problem afflicts the orthodox solution to the Gibbs paradox, too (and was raised as such by Swendsen 2006).

[36] Further, even the familiar intrinsic properties of particles (like charge, spin, and mass) may be state-dependent: string theory and supersymmetric theories provide obvious examples. See Goldstein et al. (2005a,b) for the argument that all particles may be treated as permutable, identical or otherwise.

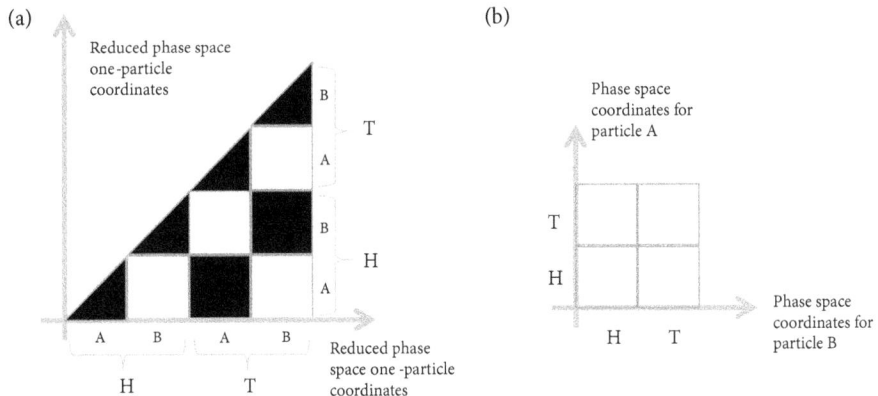

Figure 10.3 Individuating properties as particle labels

add "even if there is no fact of the matter as to which of the coins is the A coin, and which is the B coin," but there is another way of putting it: the coin that is the A coin is the one rotating one way, the B coin is the one rotating the other way.[37]

The elimination of the diagonals makes no difference to particle statistics (since this is classical theory), but analogous reasoning applies to the quantum case, where it does. Two quantum coins, thus dynamically distinguished, will land one head and one tail with probability one half, not one third.

The argument carries over unchanged in the language of Feynman diagrams. Thus, the two scattering processes depicted in figure 10.4 cannot (normally) be dynamically distinguished if the particles are permutable. Correspondingly, there is an interference effect that leads to a difference in the probability distributions for scattering processes involving permutable particles from those for distinguishable particles. But if dynamical distinctions A and B can be made between the two particles, stable over time (in our terms, if A and B are individuating properties), the interference terms will vanish, and the scattering amplitudes will be the same as for distinguishable particles.[38]

The same procedure can be applied to $N = N_A + N_B$ coins, N_A of which rotate one way and N_B the other. The result for large N_A, N_B is an effective phase space representation for two nonidentical gases A and B, each separately permutable, each with an extensive entropy function, with an entropy of mixing as given by (3). And it is clear this representation admits of degrees: it is an effective representation, more or less accurate, more or less adequate to practical purposes.

But by these means we are a long way from arriving at an effective phase space theory of N distinguishable particles. That would require, at a minimum, N distinct individuating properties of the kind we have described—at which point, if used in an effective phase space representation, the original permutation symmetry will

---

[37] For further discussion, see section 3.3. Whether the A coin after one toss is the same as the A coin on another toss (and likewise the B coin) will make a difference to the effective dynamics.

[38] There is, however, more to say about indistinguishability and path integral methods. I do not pretend to do justice to this topic here.

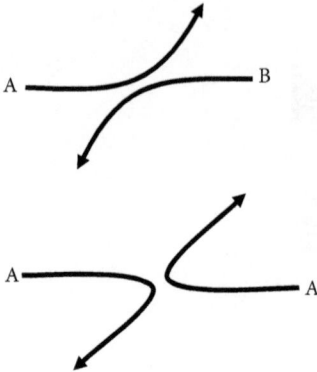

Figure 10.4 Feynman diagrams for particle scattering

have completely disappeared. But it is hardly plausible (for microscopic systems), when N is large, that a representation like this can be dynamically defined. Even where there are such individuating properties, it is hard to see what purposes their introduction would serve—their dynamical definition—unless it is to model explicitly a Maxwell demon.[39] It may be no harm is done by starting ab initio with a system of distinguishable particles. On this point we are in agreement with van Kampen. But it must be added: we should recognize that the use of unreduced phase space, and the structure $\mathbb{R}^{6N}$ underlying it, is in general a mathematical simplification, introducing distinctions in thought that are not instantiated in the dynamics.

That seems to be exactly what Gibbs thought on the matter. He had, recall, an epistemological argument for passing to reduced phase space—that nothing but similarity in qualities could be used to identify particles across members of an ensemble of gasses—but he immediately went on to say:

> And this would be true, if the ensemble of systems had a simultaneous objective existence. *But it hardly applies to the creations of the imagination.* In the cases which we have been considering . . . . it is not only possible to conceive of the motion of an ensemble of similar systems simply as possible cases of the motion of a single system, but it is actually in large measure for the sake of representing more clearly the possible cases of the motion of a single system that we use the conception of an ensemble of systems. The perfect similarity of several particles of a system will not in the least interfere with the identification of a particular particle in one case with a particular particle in another. (Gibbs 1902, 188, emphasis added)

---

[39] The memory records of such a demon in effect provide a system of individuating properties for the N particles.

If pressed, it may be added that a mathematician can always construct a domain of objects in set theory, or in one-to-one correspondence with the real numbers, each number uniquely represented.[40] Likewise for reference to elements of nonrigid structures, which admit nontrivial symmetries—for example, to a particular one of the two roots of -1 in the complex number field, or to a particular orientation on 3-dimensional Euclidean space, the left-handed orientation rather than the right-handed one.[41] But it is another matter entirely as to whether reference like this, in the absence of individuating properties, can carry over to physical objects. The whole of this chapter can be seen as an investigation of whether it can in the case of the concept of particle; our conclusion is negative.

The lesson may well be more general. It may be objects in mathematics are always objects of singular thought, involving, perhaps, an irreducible indexical element. If, as structuralists like Russell and Ramsey argued, the most one can hope for in representation of physical objects is structural isomorphisms with objects of direct acquaintance, these indexical elements can be of no use to physics. It is the opposite conclusion to Kant's.

3.2 Philosophical Logic

A second pass at our question of whether permutable entities can be considered as objects is to ask whether they can be quantified over in standard logical terms. Posed in this way, the question takes us to language and objects as values of bound variables. Arguably, the notion of object has no other home; physical theories are not directly about objects, properties, and identity in the logical sense (namely equality).

But if we are to introduce a formal language, we should be clear on its limits. We are not trying to reproduce the mathematical workings of a physical theory in formal terms. That would be an ambitious, but hardly novel undertaking; it is the one proposed by Hilbert and Russell, that so inspired Carnap and others in the early days of logical empiricism. Our proposal is more modest. The suggestion is that by formalization we gain clarity on the ontology of a physical theory, not rigor or clarity of deduction—or even of explanation. But it is ontology subject to symmetries: in our case, permutability. We earlier saw how invariant descriptions and invariant states (under the permutation group) suffice for statistical mechanics, suffice even for the description of individual trajectories; we should now see how this invariance is to be cashed out in formal, logical terms.

Permutability of objects, as a symmetry, has a simple formal expression: predicates should be invariant (have the same truth value) under permutations of values of variables. Call such a predicate *totally symmetric*.

---

[40] For further discussion, see Muller and Saunders (2008). (Set-theory, of course, yields rigid structures par excellence.)

[41] This was also, of course, a key problem for Kant. For further discussion, and an analysis of the status of mirror symmetry given parity violation in weak-interaction physics, see Saunders (2007).

Restriction to predicates like these certainly seems onerous. Thus take the simple case where there are only two things, whereupon it is enough for a predicate to be totally symmetric that it be symmetric in the usual sense. When we say:

(i) Buckbeak the hippogriff can fly higher than Pegasus the winged horse

the sentence is clearly informative, at least for readers of literature on mythical beasts; but "flies higher" is not a symmetric predicate. How can we convey (i) without this asymmetry?

Like this: by omitting use of proper names. Let us suppose our language has the resources to replace them with Russellean descriptions, say with "Buckbeak-shaped" and "Pegasus-shaped" as predicates (individuating predicates). We can then say in place of (i)

(ii) x is Buckbeak-shaped and y is Pegasus-shaped and x can fly higher than y.

But now (ii) gives over to the equally informative totally symmetric complex predicate:

(iii) x is Buckbeak-shaped and y is Pegasus-shaped and x can fly higher than y, or y is Buckbeak-shaped and x is Pegasus-shaped and y can fly higher than x.

The latter is invariant under permutation of x and y. Prefacing by existential quantifiers, it says what (i) says (modulo uniqueness), leaving open only the question of which of the two objects is the one that is Buckbeak-shaped, rather than Pegasus-shaped, and vice versa. But continuing in this way—adding further definition to the individuating predicate–the question that is left open is increasingly empty. If no further specification is available, one loses nothing in referring to that which is Buckbeak-shaped, that which is Pegasus-shaped (given that there are just the two); or to using "Buckbeak" and "Pegasus" as mass terms, like "butter" or "soil." We then have from (iii):

(iv) There is Buckbeak and there is Pegasus and Buckbeak can fly higher than Pegasus, or there is Buckbeak and there is Pegasus and Buckbeak can fly higher than Pegasus.

With "Pegasus" and "Buckbeak" in object position, (iv) is not permutable; it now says the same thing twice. We have recovered (i).

How does this work when there are several other objects? Consider the treatment of properties as projectors in quantum mechanics. For a one-particle projector P there corresponds the N-fold symmetrized projector:

$$P \otimes (I-P) \otimes ... \otimes (I-P) + (I-P) \otimes P \otimes (I-P)... \otimes (I-P) + ... + (I-P) \otimes ... \otimes (I-P) \otimes P$$

where there are N factors in each term of the summation, of which there are $\binom{N}{1} = N$. For a two-particle projector of the form $P \otimes Q$ (where P and Q are either the same or orthogonal), the symmetrized projector is likewise a sum over products of projectors and their complements (N factors in each), but now there will be $\binom{N}{2} = N(N-1)/2$ summands. And so on. The obvious way to mimic these constructions in the predicate calculus, for the case of N objects, is to define, for each one-place predicate A, the totally symmetric N-ary predicate:

(v) $(Ax_1 \wedge \neg Ax_2 \wedge ... \wedge \neg Ax_N) \vee (\neg Ax_1 \wedge Ax_2 \wedge \neg Ax_3 \wedge ... \wedge \neg Ax_N) \vee .... \vee$
$(\neg Ax_1 \wedge ... \wedge \neg Ax_{N-1} \wedge Ax_N)$.

The truth of (v) (if it is true) will not be affected by permutations of values of the N variables. It says only that exactly one particle, or object, satisfies A, not which particle or object does so. The construction starting with a two-place predicate follows similar lines; and so on for any n-ary predicate for $n \leq N$. Disjuncts of these can be formed as well.

Do these constructions tell us all that we need to know? Indeed they must, given our assumption that the N objects are adequately described in the predicate calculus without use of proper names, for we have:

> Theorem 1 Let $\mathcal{L}$ be a first-order language with equality, without any proper names. Let S be any $\mathcal{L}$−sentence true only in models of cardinality N. Then there is a totally symmetric N-ary predicate $G \in \mathcal{L}$ such that $\exists x_1...\exists x_N\, Gx_1...x_N$ is logically equivalent to S.

(For the proof see Saunders 2006a.) Given that there is some finite number of objects N, anything that can be said of them without using proper names (with no restriction on predicates) can be said of them using a totally symmetric N-ary predicate.[42]

On the strength of this, it follows we can handle uniqueness of reference in the sense of the "that which" construction, as well, "the unique x which is Ax." In Peano's notation it is the object $\iota x Ax$. Following Russell, it is contextually defined by sentences of the form:

(vi) the x that is an A is a B

or $B(\iota x)Ax$, which is cashed out as:

(vii) $\exists x(Ax \wedge \forall y(Ay \rightarrow y = x) \wedge Bx)$.

From Theorem 1 it follows that (vii), supplemented by information on just how many objects there are, is logically equivalent to a sentence that existentially quantifies over a totally symmetric predicate. It says that a thing which is A is a B, that something is an A, and that there are no two distinct things that are both A, without ever saying which of N things is the thing which is A. (v) shows how.

How much of this will apply to quantum particles? All of it. Of course definite descriptions of objects of definite number are rarely needed in talk of atoms, and rarely available. Individuating properties at the macroscopic level normally provide indefinite descriptions of one of an indeterminate number of particles. So it was

---

[42] This construction was overlooked by Dieks and Lubberdink (2011) in their criticisms of the concept of classical indistinguishable particles. They go further, rejecting indistinguishability even in the quantum case (they consider that particles only emerge in quantum mechanics in the limit where Maxwell-Boltzmann statistics hold sway-where individuating predicates in our sense can be defined.

earlier; I was talking of any old helium atom in the canister by the door, any old helium atom in the vacuum chamber, out of an indeterminate number in each case. But sometimes numbers matter: a handful of atoms of plutonium in the wrong part of the human body might be very bad news indeed. Even one might be too many.

Nor need we stop with Russellean descriptions, definite or otherwise. There are plenty of other referential devices in ordinary language that may be significant. It is a virtue of passing from the object level, from objects themselves (the "material mode," to use Carnap's term), to talk of objects (the "formal mode"), that the door is open to linguistic investigations of quite broad scope. Still, in agreement with Carnap and with Quine, our litmus test is compatibility with elementary logic and quantification theory.

To conclude: in the light of Theorem 1, and the use of individuating properties to replace proper names, nothing is lost in passing from nonpermutable objects to permutable ones. There is no loss of expressive content in talking of N permutable things, over and above what is lost in restricting oneself to the predicate calculus and abjuring the use of names. That should dissipate most philosophical worries about permutability.

There remains one possible bugbear, however, namely identity in the logical sense (what we are calling equality). Quantum objects have long been thought problematic on the grounds that they pose insuperable difficulties to any reasonable account of logical equality—for example, in terms of the principle of identity of indiscernibles (see below). To this one can reply, too bad for an account of equality;[43] the equality sign can be taken as primitive, as is usual in formal logic. (That is to say, in any model of $\mathcal{L}$, if a language with equality, the equals sign goes over to equality in the set-theoretic sense.) But here too one might do better.

3.3 Identity Conditions

If physical theories were (among other things) directly about identity in the logical sense, an account of it would be available from them. It is just because physical theories are not like this (although that could change) that I am suggesting the notion of object should be formalized in linguistic terms. It is not spelled out for us directly in any physical theory.

But by an "account of equality" I do not mean a theory of logical equality in full generality. I mean a theory of equality only of physical objects, and specific to a scientific language. It may better be called an account of identity conditions, contextualized to a physical theory.

Given our linguistic methods, there is an obvious candidate: exhaustion of predicates. That is, if $F...s...$ if and only if $F...t...$, for every predicate in $\mathcal{L}$ and for every predicate position of F, then s and t are equal. Call this $\mathcal{L}$-equality, denote "$s =_\mathcal{L} t$." It is clearly a version of Leibniz's famous "principle of identity of

---

[43] See Pniower (2006) for arguments to this effect.

indiscernibles". This is often paraphrased as the principle that objects which share the same properties, or even the same relational properties, are the same, but this parsing is unsatisfactory in an important respect. It suggests that conjuctions of conditions of the form

$$\forall y(Fsy) \leftrightarrow \forall y(Fty) \tag{28}$$

are sufficient to imply that x and y are equal, but more than this is required for exhaustion of predicates. The latter also requires the truth of sentences of the form:

$$\forall y(Fsy \leftrightarrow Fty). \tag{29}$$

These are the key to demonstrating the nonequality of many supposed counterexamples to Leibniz's principle (see Saunders 2003).

$\mathcal{L}$-equality is the only defined notion of equality (in first-order languages) that has been taken seriously by logicians.[44] It satisfies Gödel's axioms for the sign "=," used in his celebrated completeness proof for the predicate calculus with equality, namely the axiom scheme:

Leibniz's law $s = t \to \bigwedge_{F \in \mathcal{L}} (F..s.. \leftrightarrow F..t...)$

together with the scheme $s = s$. Since one has completeness, anything true in $\mathcal{L}$ equipped with the sign "=" remains true in $\mathcal{L}$ equipped with the sign "$s =_\mathcal{L} t$." The difference between $\mathcal{L}$-equality and primitive equality cannot be stated in $\mathcal{L}$.[45]

But the notion that we are interested in is not $\mathcal{L}$-equality, sameness with respect to every predicate in $\mathcal{L}$, but *sameness with respect to invariant predicates constructible in $\mathcal{L}$*, denote "$\mathcal{L}^*$". Call equality defined in this way *physical equality*, denote "$=_{\mathcal{L}^*}$." With that completeness is no longer guaranteed, but our concern is with ontology, not with deduction.

In summary, we have:

physical equality $s =_{\mathcal{L}^*} t \underset{\text{def}}{=} \bigwedge_{F \in \mathcal{L}^*} (Fs \leftrightarrow Ft)$

and, as a necessary condition for physical objects (the *identity of physical indiscernibles*):[46]

IPI $s =_{\mathcal{L}^*} t \to s = t.$

If $s \neq_{\mathcal{L}^*} t$, we shall say s and t are *(physically) discernible*; otherwise *(physically) indiscernible*.

There are certain logical distinctions (first pointed out by Quine) for equality in our defined sense that will prove useful. Call s and t *strongly discernible* if for an open sentence F in one free variable, Fs and not Ft; call s and t *weakly discernible* (respectively *relatively discernible*) if for an open sentence F in two free variables

---

[44] It was first proposed by Hilbert and Bernays (1934); it was subsequently championed by Quine (1960, 1970).
[45] For further discussion, see Quine (1970, 61–64), and, for criticism, Wiggins (2004, 184–188).
[46] For further discussion of this form of the principle of identity of indiscernibles, see Muller and Saunders (2008, 522–23).

Fst but not *Fss* (respectively, but not *Fts*). Objects that are only weakly or relatively discernible are discerned by failure of conditions of the form (29), not (28).

Of these, as already mentioned, weak discernibility is of greater interest from both a logical and physical point of view. Satisfaction of any symmetric but irreflexive relation is enough for weak discernibility: $\neq$ and $\neq_{\mathcal{L}*}$ are prime examples. And many simple invariant physical relations are symmetric and irreflexive: for example, having nonzero relative distance in a Euclidean space (a relation invariant under translations and rotations). Thus, take Max Black's famous example of identical iron spheres s, t, one mile apart, in an otherwise empty Euclidean space. The spheres are weakly discerned by the relation D of being one mile apart, for if Dst is true, it is not the case that $Dxs \leftrightarrow Dxt$ for any x, since Dst but not Dss (or Dtt), so $s \neq_{\mathcal{L}*} t$. And, fairly obviously, if $\mathcal{L}^*$ contains only totally symmetric predicates, physical objects will be at most weakly discernible.

Here as before "s" and "t" are terms, that is variables, functions of variables, or proper names. What difference do the latter make? Names are important to discernibility under $\mathcal{L}$-equality. Thus if it is established that s and t are weakly $\mathcal{L}$-discernible, then, if "s" or "t" are proper names, they are absolutely $\mathcal{L}$-discernible. In the example just given, if Dst and "s" is a proper name, then Dsx is true of t but not s. But the presence of names in $\mathcal{L}$ makes no difference to $\mathcal{L}^*$-discernibility (discernibility by totally symmetric predicates). Thus, even if $Dxy \in \mathcal{L}^*$, on entering a proper name in variable position one does not obtain a one-place predicate in $\mathcal{L}^*$. Permutable objects are only weakly discernible, if discernible at all. We may never say of permutables which of them is a such-and-such; only that there is a such-and-such.

It remains to determine whether permutable particles are discernible at all. In the classical case, assuming particles are impenetrable, they are always some nonzero distance apart, so the answer is positive. Impenetrability also ensures that giving up permutability, and passing to a description of things that are particle states or trajectories, they will be at least weakly discernible. Typically they will be strongly discernible, but as Black's two spheres illustrate (supposing they just sit there), not always.

It is the quantum case that presents the greater challenge; indistinguishable quantum particles have long been thought to violate any interesting formulation of Leibniz's principle of indiscernibles.[47] But in fact the same options arise as in the classical case. One can speak of that which has such-and-such a state, or orbit, and pass to states and orbits of states as things, giving up permutability. One-particle states or their orbits, like classical trajectories, will in general be absolutely discernible, but sometimes only weakly discernible—or (failing impenetrability) not even that. Or retaining permutability, one can speak of particles as being in one or other states, and of N particles as being in an N-particle state, using only totally symmetric predicates, and satisfying some totally irreflexive relation.

On both strategies there is a real difficulty in the case of bosons, at least for elementary bosons. On the first approach, there may be two bosonic one-particle states, each exactly the same; on the second, there seems to be no general symmetric

---

[47] See French and Krause (2006) for this history.

and irreflexive relation that bosons always satisfy. But the situation is different when it comes to fermions. On the first approach, given only the mild entanglement required by antisymmetrization, one is guaranteed that of the N one-particle states, each is orthogonal to every other, so objects as one-particle states are always absolutely discernible; and on the second approach, again following from antisymmetrization, an irreflexive symmetric relation can always be defined (whatever the degree of entanglement). Some further comments on each.

The first strategy is not without its difficulties. To begin with, even restricting to only mildly entangled states, *which* one-particle states are to be the objects replacing particles is ambiguous. The problem is familiar from the case of the singlet state of spin: neglecting spatial degrees of freedom the antisymmetrized state is

$$|\Psi_0\rangle = \frac{1}{\sqrt{2}}\left(|\psi_+^z\rangle \otimes |\psi_-^z\rangle - |\psi_-^z\rangle \otimes |\psi_+^z\rangle\right) \qquad (30)$$

where $|\psi_\pm^z\rangle$ are eigenstates of spin in the z direction. But this state can equally be expanded in terms of eigenstates of spin in the y direction, or of the z direction: Which pair of absolutely discernible one-particle states is present, exactly?

The problem generalizes. Thus, for arbitrary orthogonal one-particle states $|\phi_a\rangle$, $|\phi_b\rangle$, and a two-fermion state of the form:

$$|\Phi\rangle = \frac{1}{\sqrt{2}}(|\phi_a\rangle \otimes |\phi_b\rangle - |\phi_b\rangle \otimes |\phi_a\rangle) \qquad (31)$$

define the states (the first is just a change of notation):

$$|\phi_+^1\rangle = |\phi_a\rangle, \ |\phi_-^1\rangle = |\phi_b\rangle) \qquad (32)$$

$$|\phi_+^2\rangle = \frac{1}{\sqrt{2}}(|\phi_a\rangle + |\phi_b\rangle), \ |\phi_-^2\rangle = \frac{1}{\sqrt{2}}(|\phi_a\rangle - |\phi_b\rangle)$$

$$|\phi_+^3\rangle = \frac{1}{\sqrt{2}}(|\phi_a\rangle + i|\phi_b\rangle), \ |\phi_-^3\rangle = \frac{1}{\sqrt{2}}(i|\phi_a\rangle + |\phi_b\rangle).$$

They yield a representation of the rotation group. One then has, just as for components of spin:

$$|\Phi\rangle = \frac{1}{\sqrt{2}}(|\phi_+^1\rangle| \otimes |\phi_-^1\rangle - |\phi_-^1\rangle| \otimes |\phi_+^1\rangle$$

$$= \frac{1}{\sqrt{2}}(|\phi_+^2\rangle| \otimes |\phi_-^2\rangle - |\phi_-^2\rangle| \otimes |\phi_+^2\rangle = \frac{1}{\sqrt{2}}(|\phi_+^3\rangle| \otimes |\phi_-^3\rangle - |\phi_-^3\rangle| \otimes |\phi_+^3\rangle$$

and an ambiguity in attributing one-particle states to the two particles arises with (31) as with (30). I shall come back to this in section 3.4.

This difficulty can be sidestepped at the level of permutable particles, however. In the case of (30), we may weakly discern the particles by the relation "opposite spin," with respect to any direction in space (Saunders 2003, 2006b; Muller and Saunders 2008). Thus if σ^x, σ^y, σ^z are the Pauli spin matrices, the self-adjoint operator

$$\sigma^x \otimes \sigma^x = \sigma^y \otimes \sigma^y = \sigma^z \otimes \sigma^z \qquad (33)$$

has eigenvalue -1 in the singlet state $|\Psi_0\rangle$, with the clear interpretation that the spins are anticorrelated (with respect to any direction in space). Asserting this relation does not pick out any direction in space, no more than saying Black's spheres are one mile apart picks out any position in space.

For the construction in the generalized sense (32), define projection operators onto the states $|\phi_\pm^k\rangle$

$$P_\pm^k = P_{|\phi_\pm^k\rangle}, \ k = 1, 2, 3$$

and define the self-adjoint operators:

$$(P_+^k - P_-^k) \otimes (P_+^k - P_-^k), \ k = 1, 2, 3.$$

Each has eigenvalue -1 for $|\Phi\rangle$, and likewise picks out no "direction" in space (i.e. the analogue of (33) is satisfied). Moreover, one can define sums of such in the case of finite superpositions of states of the form (31), by means of which fermions can be weakly discerned.

On the strength of this, one can hope to weakly discern bosons that are composites of fermions, like helium atoms. And even in the case of elementary bosons, self-adjoint operators representing irreflexive, symmetric relations required of any pair of bosons have been proposed.[48] The difficulty of reconciling particle indistinguishability in quantum mechanics with the IPI looks well on its way to being solved.

3.4 Eliminativism

We are finally in a position to address the arguments for and against eliminativism—that is, for and against renouncing talk of permutable objects in favor of nonpermutable objects defined in terms of individuating properties, whether points in μ-space, trajectories, one-particle states, or orbits of one-particle states. The gain, usually, is absolute discernibility. On the other hand, we have found that quantification over permutable objects satisfies every conservative guideline we have been able to extract from elementary logic (with the possible exception of identity conditions for elementary bosons). There seems to be nothing wrong with the logic of weak discernibles. And there remains another conservative guideline: we should maintain standard linguistic usage where possible.

That stacks the odds against eliminativism, for talk of particles, and not just of one-particle states, is everywhere in physics. But even putting this to one side, eliminativism would seem to fare poorly, for (anti)symmetrized states are generically entangled, whereupon no set of N one-particle states will suffice for the description of N particles. And as we have seen in section 3.3, where such a set is available it may be non-unique.

---

[48] See Muller and Seevink (2009). Their idea is to use certain commutator relations that could not be satisfied were there only a single particle.

Against this there are two objections. The first is that we anyway know the particle concept is stretched to breaking point in strongly interacting regimes. There the best we can say is that there are quantum fields, and, perhaps, superpositions of states of different particle number. Where the latter can be defined, one can talk of modes of quantum fields instead. In the free-field limit, or as defined by a second-quantization of a particle theory,[49] such modes are in one-to-one correspondence with one-particle states (or, in terms of Fourier expansions of the fields, in correspondence with "generalized" momentum eigenstates). The elimination of particles in favor of fields and modes of fields is thus independently motivated.

The second objection is that we cannot lightly accept indeterminateness in attributing a definite set of N one-particle states to an N-particle system, for it applies equally to particles identified by individuating properties. That is, not even the property of being a bound electron in a helium atom in the canister by the corner, and being one in the vacuum chamber by the door, hold unambiguously. The construction (32) applies just as much to (20).

But *this* difficulty we recognize as a fragment of the measurement problem. Specifically, it is the *preferred basis problem*: Into what states does a macroscopic superposition collapse (if there is any collapse)? Or, if macroscopic superpositions exist: What singles out the basis in which they are written? Whatever settles this question (decoherence, say) will dictate the choice of basis used to express the state in terms of macroscopic individuating properties.

Whether such a choice of basis—or such a solution to the preferred basis problem—can extend to a preferred basis at the microscopic level is moot. It depends, to some extent, on the nature of the solution (decoherence only goes down so far). Of course it is standard practice in quantum theory to express microscopic states in terms of a basis associated with physically interpreted operators (typically generators of one-parameter spacetime symmetry groups, or in terms of the dynamical quantities that are measured). The use of quantum numbers for bound states of electrons in the atom, for energy, orbital angular momentum, and components of angular momentum and spin—in conventional notation, quadruples of numbers $\langle n, l, m_l, m_s \rangle$—is a case in point. When energy degeneracies are completely removed (introducing an orientation in space) one can assign these numbers uniquely. The Pauli exclusion principle then dictates that every electron has a unique set of quantum numbers. Use such quadruples as names and talk of permutable particles can be eliminated.

It is now clearer that the first objection adds support to the second. Quadruples of quantum numbers provide a natural replacement for particles in atoms; modes of quantum fields (and their excitation numbers) provide a natural replacement for particles involved in scattering. And in strongly interacting regimes, even modes of quantum fields give out (or they have only a shadow existence, as with virtual particles). All this is as it should be. Our inquiry was never about fundamental

---

[49] For a discussion of the relation between second quantized and free-field theories (fermionic and bosonic respectively), see Saunders (1991, 1992).

ontology (a question we can leave to a final theory, if there ever is one), but with good-enough ontology, in a definite regime.

In the regime we are concerned with, stable particles of ordinary matter whose number is conserved in time, there is the equivalence between one-particle states and modes of a quantum field already mentioned. Let us settle on a preferred decomposition of the field (or preferred basis) in a given context. But suppose that context involves nontrivial entanglement: Can entanglements of particles be understood as entanglements of modes of fields?

Surely they can—but on pain of introducing many more modes of the field than there were particles, and a variable number to boot. As with one-particle states so modes of the field: in a general entanglement, arbitrarily many such modes are involved, even given a preferred decomposition of the field, whereas the number of particles is determinate. Just where the particle concept is the most stable, in the regime in which particle number is conserved, eliminativism in favor of fields and modes of fields introduces those very features of the particle concept that we found unsatisfactory in strongly interacting regimes. That speaks against eliminativism.

This does not, of course, militate against the reality of quantum fields. We recognize that permutable particles are emergent from quantum fields, just as nonpermutable particles are emergent from permutable ones. Understood in this way, we can explain a remaining fragment of the Gibbs paradox—the fact that particle identity, and with it permutation symmetry, can ever be exact. How is it that intrinsic quantities, like charge and mass, are identically the same? (their values are real numbers, note). The answer is that for a given particles species, the particles are one and all excitations of a single quantum field—whereupon these numerical identities are forced, and permutation symmetry has to obtain. The existence of exact permutation symmetry, in regimes in which particle equations are approximately valid, is therefore explained, and with it particle indistinguishability.

References

Albert, D. (2000). *Time and Chance*. Cambridge, MA: Harvard University Press.
Bach, A. (1990). Boltzmann's probability distribution of 1877. *Archive for the History of the Exact Sciences* 41: 1–40.
———. (1997). *Indistinguishable classical particles*. Berlin: Springer.
Darrigol, O. (1991). Statistics and cominbatorics in early quantum theory, II: Early symptoma of indistinghability and holism. *Historical Studies in the Physical and Biological Sciences* 21: 237–298.
Denbigh, K., and M. Redhead (1989). Gibbs' paradox and non-uniform convergence. *Synthese* 81: 283–312.
Dieks, D., and A. Lubberdink (2011). How classical particles emerge from the quantum world. *Foundations of Physics* 41, 1051–1064. Available online at arXiv:quant-ph/1002.2544v1.
Ehrenfest, P., and V. Trkal (1920). Deduction of the dissociation equilibrium from the theory of quanta and a calculation of the chemical constant based on this. *Proceedings*

of the Amsterdam Academy 23: 162–183. Reprinted in P. Bush, ed., *P. Ehrenfest, Collected scientific papers.* Amsterdam: North-Holland, 1959.

French, S., and D. Krause (2006). *Identity in physics: A historical, philosophical, and formal analysis.* Oxford: Oxford University Press.

French, S., and M. Redhead (1988). Quantum physics and the identity of indiscernibles. *British Journal for the Philosophy of Science* 39: 233–246.

Ghirardi, G., and L. Marinatto (2004). General criterion for the entanglement of two indistinguishable states. *Physical Review* A70: 012109.

Ghirardi, G., L. Marinatto, and Y. Weber (2002). Entanglement and properties of composite quantum systems: A conceptual and mathematical analysis. *Journal of Statistical Physics* 108: 49–112.

Gibbs, J. W. (1902). *Elementary principles in statistical mechanics.* New Haven: Yale University Press.

Goldstein, S., J. Taylor, R. Tumulka, and N. Zanghi (2005a). Are all particles identical? *Journal of Physics* A38: 1567–1576.

———. (2005b). Are all particles real? *Studies in the History and Philosophy of Modern Physics* 36: 103–112.

Gottesman, D. (2005). Quantum statistics with classical particles. Available online at http://xxx.lanl.gov/cond-mat/0511207.

Greiner, W., and B. Müller (1994). *Quantum Mechanics: Symmetries.* 2d ed. New York: Springer.

Hercus, E. (1950). *Elements of thermodynamics and statistical thermodynamics.* Melbourne: Melbourne University Press.

Hilbert, D., and P. Bernays (1934). *Grundlagen der Mathematik,* Vol. 1, Berlin: Springer.

Howard, D. (1990). "Nicht sein kann was nicht kein darf", or the prehistory of EPR, 1909–1935: Einstein's early worries about the quantum mechanics of compound systems. In *Sixty-two years of uncertainty: Historical, philosophical and physical inquiries into the foundations of quantum mechanics,* A. Miller. New York: Plenum.

Huggett, N. (1999a). Atomic metaphysics. *Journal of Philosophy* 96: 5–24.

———. (1999b). *Space from Zeno to Einstein: Classical readings with a contemporary commentary.* Cambridge, MA: Bradford Books.

Jammer, M. (1966). *The conceptual development of quantum mechanics.* McGraw-Hill.

Jaynes, E. (1992). The Gibbs paradox. In *Maximum-entropy and Bayesian methods,* ed. G. Erickson, P. Neudorfer, and C. R. Smith. Dordrecht: Kluwer. 1–21.

van Kampen, N. (1984). The Gibbs paradox. *Essays in theoretical physics in honour of Dirk ter Haar,* ed. W.E. Parry. Oxford: Pergamon Press. 303–312.

Lieb, E., and J. Yngvason (1999). The physics and mathematics of the second law of thermodynamics. *Physics Reports* 310: 1–96. Available online at arXiv:cond-mat/9708200 v2.

Muller, F., and S. Saunders (2008). Distinguishing fermions. *British Journal for the Philosophy of Science* 59: 499–548.

Muller, F., and M. Seevink (2009). Discerning elementary particles. *Philosophy of Science* 76: 179–200. Available online at arxiv.org/PS_cache/arxiv/pdf/0905/0905.3273v1.pdf.

Nagle, J. (2004). Regarding the entropy of distinguishable particles. *Journal of Statistical Physics* 117: 1047–1062.

Penrose, R. (2004). *The Road to Reality.* London: Vintage Press.

Planck, M. (1900). Zur Theorie des Gesetzes der Energieverteilung im Normalspectrum. *Verhandlungen der Deutsche Physicalishe Gesetzen* 2: 202–204. Translated in D. ter Haar, ed., *The old quantum theory.* Oxford: Pergamon Press, 1967.

———. (1912). La loi du rayonnement noir et l'hypothèse des quantités élémentaires d'action. In *La Théorie du Rayonnement et les Quanta: Rapports et Discussions de la Résunion Tenue à Bruxelles, 1911*, ed. P. Langevin and M. de Broglie. Paris: Gauthier-Villars.

———. (1921). *Theorie der Wärmesrahlung*. 4th ed. Leipzig: Barth.

Pniower, J. (2006). Particles, Objects, and Physics. D.Phil. thesis, University of Oxford. Available online at philsci-archive.pitt.edu/3135/1/dphil-bod.pdf.

Poincaré, H. (1911). Sur la theorie des quanta. *Comptes Rendues* 153: 1103–1108.

———. (1912). Sur la theorie des quanta. *Journal de Physique* 2: 1–34.

Pooley, O. (2006). Points, particles, and structural realism. In *The Structural Foundations of Quantum Gravity*, ed. D. Rickles, S. French, and J. Saatsi. Oxford: Oxford University Press. 83–120.

Post, H. (1963). Individuality and physics. *The Listener*, 10 October, 534–537, reprinted in *Vedanta for East and West* 132: 14–22 (1973).

Quine, W. van (1960). *Word and object*. Cambridge, MA: Harvard University Press.

———. (1970). *Philosophy of logic*. Cambridge, MA: Harvard University Press.

———. (1990). *The pursuit of truth*. Cambridge, MA: Harvard University Press.

Rapp, D. (1972). *Statistical mechanics*. New York: Holt, Rinehart and Winston.

Reichenbach, H. (1956). *The direction of time*. Berkeley: University of California Press.

Rosenfeld, L. (1959). Max Planck et la definition statistique de l'entropie. *Max-Planck Festschrift 1958*. Berlin: Deutsche Verlag der Wissenschaften. Trans. as Max Planck and the statistical definition of entropy. In R. Cohen and J. Stachel, eds., *Selected papers of Leon Rosenfeld*. Dordrecht: Reidel, 1979.

Saunders, S. (1991). The Negative Energy Sea. In *Philosophy of Vacuum*, ed. S. Saunders and H. Brown. Oxford: Clarendon Press. 67–107.

———. (1992). Locality, Complex Numbers, and Relativistic Quantum Theory. *Proceedings of the Philosophy of Science Association*, Vol. 1, 365–380.

———. (2003). Physics and Leibniz's principles. In *Symmetries in physics: New reflections*, ed. K. Brading and E. Castellani. Cambridge: Cambridge University Press.

———. (2006a). On the explanation of quantum statistics. *Studies in the History and Philosophy of Modern Physics* 37: 192–211. Available online at arXiv.org/quant-ph/0511136.

———. (2006b). Are quantum particles objects? *Analysis* 66: 52–63. Available online at philsci-archive.pitt.edu/2623/.

———. (2007). Mirroring as an a priori symmetry. *Philosophy of Science* 74: 452–480.

Schrödinger E. (1946). *Statistical thermodynamics*. Cambridge: Cambridge University Press.

———. (1984). What is an elementary particle? *Collected Papers*, Vol. 4, Österreichische Akademie der Wissenschaften. Reprinted in E. Castellani, ed., *Interpreting Bodies: classical and quantum objects in modern physics*. Princeton: Princeton University Press, 1998.

Stern, O. (1949). On the term $N!$ in the entropy. *Reviews of Modern Physics* 21: 534–35.

Swendson, R. (2002). Statistical mechanics of classical systems with distinguishable particles. *Journal of Statistical Physics* 107: 1143–66.

———. (2006). Statistical mechanics of colloids and Boltzmann's definition of the entropy. *American Journal of Physics* 74. 187–190.

Teller, P. (1995). *An interpretative introduction to quantum field theory*. Princeton: Princeton University Press.

Wiggins, D. (2004). *Sameness and substance renewed*. Oxford: Oxford University Press.

CHAPTER 11

UNIFICATION IN PHYSICS

MARGARET MORRISON

1. Introduction and Background

What exactly is unification and what form does it take in physics? Typically when this question is asked we think about high-energy physics and the search for a Theory of Everything (TOE). But what would such a theory look like and what kind of unification would it encompass? Again, the preliminary answer is that it would bring together the four forces of nature and show that they are low-energy manifestations of the same force. But, would such a theory involve deductions from a few simple laws or would it require several free parameters and complex models to apply it in concrete situations? If it is the latter, at what point are we willing to claim the theory presents a "unified" account of the phenomena? Much of the discussion surrounding the operation of the Large Hadron Collider (LHC) and theories like string theory and quantum gravity suggests that the immediate goals of unification in physics involve finding the Higgs particle, determining its nature and properties, and somehow bringing gravity into the framework of the Standard Model. The former goal has been achieved; in July 2012 CERN announced that two experiments using the Large Hadron Collider, ATLAS and CMS, had both amassed strong statistical evidence (around 5 sigma) for a new particle with a mass of roughly 126 GeV which is consistent with Standard Model predictions for the Higgs Boson. However, that the Standard Model provides a unified account of the electromagnetic, weak and strong forces under one theory is far from clear and the recent discovery of the Higgs boson may not necessarily solve that problem.

In order to see why this is the case, we need to go back to our initial question regarding what unification is and how we should characterize a "unified theory." Put differently—how should we understand the drive for unification in physics? Although these questions are of a philosophical nature, they are directly connected with scientific theory and experiment—and, they are the questions that will occupy the bulk of this essay. No account of unification would be complete without an investigation into its associated difficulties, and in the context of high-energy physics that focus will be partly on the role of effective theories as a way of dealing with phenomena at different energy levels. Interestingly the mathematics involved in constructing effective theories points to a different kind of unification in physics, a unification at the level of phenomena that had been poorly understood prior to the use of the renormalization group (RG) techniques. Our discussion will also focus on this type of unification, referred to as "universality," and its relation to the techniques used in the high-energy context.

The first systematic unification in physics was Newtonian mechanics, which brought together terrestrial phenomena (e.g., tides) and celestial phenomena (the moon, planets, etc.) by showing how they mutually influenced each other's motion and were subject to the same force law—universal gravitation. In that sense Newtonian physics represented a grand unification in that it accounted for all the phenomena in the heavens and on earth. More recently, the quest for a theory of everything—a grand unification—involves showing that when energies are high enough, the forces (interactions), while very different in strength, range, and the types of particles on which they act, become one and the same force. The fact that these interactions are known to have many underlying mathematical features in common suggests that they can all be accounted for by a unified field theory. Such a theory describes elementary particles in terms of force fields that further unify all the interactions by treating particles and interactions in a technically and conceptually similar way. It is this theoretical framework that allows for the prediction that measurements made at a certain energy level will supposedly indicate that the four separate forces are low-energy manifestations of a single force. Aspects of this unification are what will hopefully be revealed in the LHC, the biggest and most complicated physics experiment ever seen. The successful experiments responsible for the discovery of the Higgs particle—the confirmatory link in the theory that unifies the weak and electromagnetic forces— will also be important for discovering other aspects of the Standard Model, which includes the strong force and of which the electroweak theory is a part.

In many cases of unification not only is there an ontological reduction where different phenomena, usually forces, are seen as one and the same, but the mathematical framework(s) used to describe the fields associated with these forces facilitates their description in a unified theory. Specific types of symmetries serve an important function in these contexts, not only in the construction of quantum field theories (QFT) but also in the classification of particles: classifications that can lead to new predictions and novel ways of understanding properties like quantum numbers. Hence, in order to address issues about unification and reduction in contemporary physics we must also address the way that symmetries support the development of unified theoretical frameworks.

Despite the association of reduction and unification, there are clear cases where the reductionist ideal has not been met and unification has involved a synthesis where the phenomena have remained largely independent but are nevertheless described using the same theory. The electroweak theory is a case in point. The theory unifies the weak and electromagnetic forces under the SU(2) x U(1) symmetry group via a mixing of the fields, but the carriers of the forces (particles) remain distinct. Contrast this with the unification of electromagnetism and optics where Maxwell's theory showed the identity of light and electromagnetic waves.[1] The other issue relevant in the case of synthetic unity is, of course, the role of free parameters. The electroweak theory contains one free parameter, the Weinberg angle, which represents the mixing of the fields and yields the masses for the W and Z bosons. The Standard Model, by contrast, contains somewhere in the range of 26 such parameters, which have to be put in by hand and whose values are extracted from experimental data. This issue of free parameters is extremely important because the raison d'être for unification is the ability to account for a variety of phenomena using a few general principles. The addition of free parameters not only erodes that capability but undermines inferences about the identification of phenomena (like forces) on the basis of their description under a single theory. In other words, it casts doubt on the idea that nature itself is unified.[2]

The search for a theory of everything that would incorporate gravity presupposes, in some sense, that the Standard Model has unified the weak, strong, and electromagnetic forces. But as I mentioned above, the problem of free parameters and the fact that the theory is an amalgam of three different symmetry groups SU(3) x SU(2) x U(1) rather than a single group speaks against the idea that this is a truly unified theory. Moreover, the search for a TOE presumes that gravity is a force like the others when according to General Relativity it is very unlike the others in that there are no particles that couple to the gravitational field and act as force carriers; the effects of gravitation are ascribed to spacetime curvature instead of a force per se. Some of the most prominent attempts to incorporate gravity into a unified framework with quantum mechanics include string theory or others related to supersymmetry (SUSY) and loop quantum gravity, all of which face theoretical difficulties.[3]

---

[1] The first unification in Maxwell's theory was in terms of a reduction of the electromagnetic and luminiferous aethers.

[2] For an extensive treatment of different types of unification in physics, as well as the way that mathematical structures are used as unifying tools in biology, see Morrison (2000). See also Maudlin (1996) for a discussion of unification in physics.

[3] Perhaps the most cited problem with string theory is that it has a huge number of equally possible solutions, called string vacua, that may be sufficiently diverse to explain almost any phenomena one might observe at lower energies. If so, it would have little or no predictive power for low-energy particle physics experiments. Other criticisms include the fact that it is background dependent, requiring a specific starting point. This is incompatible with general relativity, which is background independent. The problems associated with loop quantum gravity also involve computational difficulties in making predictions directly from the theory and the fact that its description of spacetime at the Planck scale has a continuum limit that is not compatible with general relativity. Obviously, there are many more detailed issues here that I have not mentioned. For more

Those problems aside, the other threat to the unificationist picture of physics comes from the failure of reduction in a different context, specifically in the case of condensed matter physics where many of the phenomena are described as emergent. This picture is exemplified by Anderson's remark that "the ability to reduce everything to simple fundamental laws does not imply the ability to start from those laws and reconstruct the universe.... The behaviour of large and complex aggregates of particles... is not to be understood in terms of a simple extrapolation of the properties of a few particles. Instead at each level of complexity entirely new properties appear" (1972, 393). Examples of emergent phenomena in condensed matter physics include superconductivity and superfluidity. The defining feature of these phenomena is that their behavior or existence, for that matter, cannot be explained, predicted, or reduced to their micro constituents and the laws that govern them. So, while superconductivity involves the pairing of electrons, its essential features (e.g., infinite conductivity) do not depend on microphysical details related to that pairing. This decoupling of physics at different energy levels that is characteristic of emergent phenomena has also been a prominent feature of quantum field theories where effective theories containing appropriate degrees of freedom are used to describe physical phenomena occurring at a chosen length scale, while ignoring substructure and degrees of freedom at shorter distances (or, equivalently, at higher energies). Indeed, much of high-energy physics is now dominated by the use of effective field theories (EFTs), and the decoupling theorem of Appelquist and Carazzone (1975) has often been understood as a basis for interpreting physical reality as constituting a hierarchy of layers that are quasi-autonomous.[4] As we shall see, this in itself need not speak against the *possibility* of reduction and unification. The question is whether the discoveries at the LHC will change physics sufficiently such that effective theories will no longer be a theoretical requirement.

As I mentioned above, this emphasis on emergence and effective theories has produced a new and different kind of unity that is often not considered in the context of unification in physics. What I have in mind is the explanation of what is termed "universal behavior" by the renormalization group methods developed by Kenneth Wilson (1971, 1975) and others. Before RG, there was no account of why systems as different as magnets and superfluids shared the same critical exponents and displayed the same behavior near a second-order phase transition. RG explained this phenomenon by showing that the differences between them are related to irrelevant observables that play no role in the explanation of behavior near critical point. In other words, features of the system that are responsible for the similarities in behavior are largely independent of microphysical structure. Phenomena that

---

discussion, see Dine (2007) on string theory and supersymmetry and Rovelli (2007) on quantum gravity. See Smolin (2001) for a popular account of the latter.

[4] In order to analyze a physical problem, it is necessary to isolate the relevant details or choice of variables that will capture the physics one is interested in. Since this will involve separated energy scales, we can study low-energy dynamics independently of the details of high-energy interactions. The procedure is to identify the parameters that are very large (small) compared with the relevant energy scale of the physical system and put them to infinity (zero). We can then use this as an approximation that can be improved by adding corrections induced by the neglected energy scales as small perturbations.

share the same critical exponents are said to belong to the same universality class. This grouping of phenomena into universality classes exhibits a type of unification that is the antithesis of reductive unity insofar as the microphysical constituents are irrelevant to the universal properties or behavior to be explained. The RG is also an important component in the effective field theory program in particle physics, so interesting questions arise related to the unity of method in these two very distinct domains. I will have more to say about these questions below.

In order to illustrate, extend, and clarify these issues I want to begin by discussing examples of the two different types of unification mentioned above—reductive and synthetic unity. In particular I will address not just the ontological features involved in each case but also the role of mathematics in constructing unified theories. In that sense our discussion will focus on both the epistemic and ontological features of unification and reduction. While reductive unity exemplifies the goals of unification by illustrating the identify of different types of phenomena, its synthetic counterpart presents a rather different picture in that it unifies phenomena under the same theory but falls short of identifying them as one and the same. From there I will go on to discuss the challenges facing the unification picture from effective theories and emergent phenomena. Finally I discuss the way that RG techniques have facilitated an understanding of similarities among very different types of phenomena, indicating a new type of unity in physics that had been largely ignored and previously inexplicable.

2. Reductive Unity: Maxwell's Electrodynamics

The development of Maxwell's electrodynamics is interesting in that the theory was initially formulated using a completely fictitious aether model from which was derived a wave equation that led to the identification of electromagnetic and light waves. This model was given up in later formulations of the theory but its most important feature was the incorporation of a phenomenon known as the displacement current which was responsible for the transmission of electric waves through space, thereby producing the effect of having a closed circuit between two conductors. The aether model explained how the displacement of electricity took place, but it was this notion of electric displacement that was the key to producing a field-theoretic account of electromagnetism. The idea at the foundation of Maxwell's theory was Faraday's account of electromagnetism in terms of lines of force filling space. Prior to that, it was thought that electromagnetic force resided in material bodies and could only be transmitted through some type of mechanical interaction. The notion that the seat of electromagnetic charge was the field rather than matter was both revolutionary and controversial.

In order to disengage the theory from its questionable origins, later versions relied on what Maxwell described as "firmly established empirical facts" together with a few general dynamical principles as characterized by the abstract mathematical structure of Lagrangian mechanics.[5] That structure, unlike the mechanical model, provided no explanatory account of how electromagnetic waves were propagated through space nor any understanding of the nature of electric charge. The new unified theory based on that abstract dynamics entailed no ontological commitment to the existence of forces or structures that could be seen as the source of electromagnetic phenomena. The displacement current was retained as a basic feature (one of the equations), but no mechanical hypothesis was put forward regarding its nature.

The question that immediately arises is how we should view this type of unification given that the initial model was fictitious and the later version had no underlying ontological foundation capable of grounding the apparent reduction. In order to address that question I want to draw attention to the role that mathematical structures play in the unifying process and how those structures were able to facilitate a reductive unity without implying an ontological unity in nature. As it turns out, Maxwell's theory did accomplish the latter but no evidence for that was forthcoming until the production of electromagnetic waves by Hertz in 1888.

Maxwell's use of the Lagrangian approach was due primarily to its generality, which makes it applicable in a variety of contexts; and it was ultimately this feature that made it especially suited to unifying different phenomena/domains. In addition to the importance of these types of mathematical structures for unification, I also want to highlight what I see as the mark of a truly unified theory—the presence of a specific theoretical quantity/parameter that represents the theory's ability to reduce, identify, or synthesize two or more processes within a single theoretical framework. What I have in mind here is the idea that one particular parameter functions as a manifestation of the reduction of different phenomena to one specific kind, or its presence produces a theoretical context wherein different phenomena can be unified. In Maxwell's theory the displacement current plays just such a role. It figures prominently as a fundamental quantity in the field equations, and without it there could be no notion of a quantity of electricity crossing a boundary, no derivation of the electromagnetic wave equation and hence no field theoretic basis for electromagnetism. In other words, displacement is responsible for creating the field theoretic picture that allows Maxwell to identify light and electromagnetic waves as field theoretical processes. As we shall see below, the Weinberg angle in the electroweak theory functions as the "unifying parameter" in that it represents the mixing of the weak and electromagnetic fields. Without such a parameter, we simply have a theory that can accommodate different kinds of phenomena but without any relation or connection between them.

---

[5] Maxwell (1965, 1: 564). The experimental facts concerned the induction of currents by increases or decreases in neighboring currents, the distribution of magnetic intensity according to variations of a magnetic potential and the induction of statistical electricity through dielectrics.

To see exactly why electrodynamics qualifies as a reductive unification and to illustrate the differences with synthetic unity let me give a brief overview of the evolution of the theory from its origins in the aether model to its abstract dynamical formulation. Tracing some of these details is important because in both reductive and synthetic unification there is a reliance on mathematical frameworks as a specific type of unifying tool (Lagrangian mechanics in the Maxwellian case and gauge theory in the electroweak case) yet the outcomes are very different in each case. In other words, the generality in the application of these frameworks to diverse phenomena does not entail anything specific about the type of unity that is produced. The latter is solely a product of the specific way in which the phenomena are brought together. The difference between reductive and synthetic unity is an important feature in establishing ontological claims about unity in nature; hence, the details of how each is achieved are an important part of articulating how, exactly, unification in physics ought to be understood. In other words, we can sometimes construct unified theories but whether there is evidence for unity in nature is a different matter.

2.1 From Fictional Models to a Unified Theory

Maxwell's describes his 1861–62 paper, "On Physical Lines of Force," as an attempt to "examine" electromagnetic phenomena from a mechanical point of view and to determine what tensions in, or motions of, a medium were capable of producing the observed mechanical phenomena (Maxwell 1965, 1: 467). Faraday had described electromagnetic phenomena as lines of force permeating space rather than the result of an interaction among material bodies. At the time Thomson had developed an account of magnetism that involved the rotation of molecular vortices in a fluid aether, an idea that led Maxwell to hypothesize that in a magnetic field the medium (or aether) was in rotation around the lines of force, the rotation being performed by molecular vortices whose axes were parallel to the lines. In order to specify the forces that caused the medium to move and to account for electric currents, Maxwell needed to provide an explanation of the transmission of rotation of the vortices; something he achieved via his aether model. The specific details of this early version of the model are not important here but what is important is how the second of his aether models, developed to account for electrostatics, facilitated the derivation of his theory of light.

In order to explain charge and to derive the law of attraction between charged bodies, Maxwell constructed an elastic solid model in which the aetherial substance formed spherical cells endowed with elasticity. The cells were separated by electric particles whose action on the cells would result in a kind of distortion. Hence, the effect of an electromotive force was to distort the cells by a change in the positions of the electric particles. Because changes in displacement involved a motion of electricity, Maxwell argued that they should be "treated as" currents (1965, 1: 491).

That gave rise to an elastic force that set off a chain reaction. Maxwell saw the distortion of the cells as a displacement of electricity within each molecule, with the total effect over the entire medium producing a "general displacement of electricity in a given direction" (Maxwell 1965, 1: 491). Understood literally, the notion of displacement meant that the elements of the dielectric had changed positions.

Displacement also served as a model for dielectric polarization; electromotive force was responsible for distorting the cells, and its action on the dielectric produced a state of polarization. When the force was removed, the cells would recover their form and the electricity would return to its former position (Maxwell 1965, 1: 492). The amount of displacement depended on the nature of the body and on the electromotive force.

Because the phenomenological law governing displacement expressed the relation between polarization and force, Maxwell was able to use it to calculate the aether's elasticity (the coefficient of rigidity), the crucial step that led him to identify the electromagnetic and luminiferous aethers. It is interesting to note that in Parts I and II of "On Physical Lines" there is no mention of the optical aether. However, once the electromagnetic medium was endowed with elasticity, Maxwell relied on the optical aether in support of his assumption: "The undulatory theory of light requires us to admit this kind of elasticity in the luminiferous medium in order to account for transverse vibrations. We need not then be surprised if the magneto-electric medium possesses the same property" (1965, 1: 489). After a series of mathematical steps, which included correcting the equations of electric currents for the effect produced by elasticity and calculating the value for e, the quantity of free electricity in a unit volume, and E, the dielectric constant, he went on to determine the velocity with which transverse waves were propagated through the electromagnetic aether. The rate of propagation was based on the assumption described above—that the elasticity was due to forces acting between pairs of particles.

Using the formula $V = \sqrt{m/\rho}$, where m is the coefficient of rigidity, ρ is the aethereal mass density, and μ is the coefficient of magnetic induction, we have

$$E^2 = \pi m$$

$$\mu = \pi \rho$$

giving us $\pi m = V^2 \mu$, which yields $E = V\sqrt{\mu}$. Maxwell arrived at a value for V that, much to his astonishment, agreed with the value calculated for the velocity of light (V = 310,740,000,000 mm/sec), which led him to remark that: "The velocity of transverse undulations in our hypothetical medium, calculated from the electromagnetic experiments of Kohlrausch and Weber, agrees so exactly with the velocity of light calculated from the optical experiment of M. Fiseau that we can *scarcely avoid the inference that light consists in the transverse undulations of the same medium which is the cause of electric and magnetic phenomena* (1965, 1: 500)."

Maxwell's success involved linking the equation describing displacement (R = $-4\pi E^2 h$) with the aether's elasticity (modeled on Hooke's law), where displacement produces a restoring force in response to the distortion of the cells of the medium. However, $R = -4\pi E^2 h$ is also an electrical equation representing the flow of charge produced by electromotive force. Consequently, the dielectric constant E is both an elastic coefficient and an electric constant. Interpreting E in this way allowed Maxwell to determine its value and ultimately identify it with the velocity of transverse waves traveling through an elastic aether.

In modern differential form, Maxwell's four equations relate the Electric Field (E) and magnetic field (B) to the charge (ρ) and current (J) densities that specify the fields and give rise to electromagnetic radiation—light.

Gauss's law $\nabla \cdot \mathbf{D} = \rho_f$

Gauss's law for magnetism $\nabla \cdot \mathbf{B} = 0$

Faraday's law of induction $\nabla \times \mathbf{E} = -\dfrac{\partial \mathbf{B}}{\partial t}$

Ampère's circuital law with displacement $\nabla \times \mathbf{H} = \mathbf{J}_f + \dfrac{\partial \mathbf{D}}{\partial t}$

D is the displacement field and H the magnetizing field. The first equation, Gauss's law, describes how an electric field is generated by electric charges where the former tends to point away from positive and toward negative charges. More specifically, it relates the electric flux through any hypothetical closed Gaussian surface to the electric charge within the surface. Gauss's law for magnetism states that there are no "magnetic charges" (magnetic monopoles), analogous to electric charges; or, that the total magnetic flux through any Gaussian surface is zero. Faraday's law describes how a changing magnetic field can induce an electric field. Finally, Ampère's law with Maxwell's correction states that magnetic fields can be generated by electrical current (which was the original "Ampère law") and by changing electric fields (Maxwell's correction). Maxwell's correction to Ampère's law is crucial, since it specifies that both a changing magnetic field gives rise to an electric field and a changing electric field creates a magnetic field. Consequently, self-sustaining electromagnetic waves can propagate through space. In other words, it allows for the possibility of "open circuits."[6]

Given the importance of displacement for producing a field-theoretic account of electromagnetism and its role in calculating the velocity of waves, it is obviously the essential parameter in identifying the optical and electromagnetic aethers. In later versions of the theory, the aether was abandoned, but displacement remained as a fundamental quantity. However, its status changed once it was incorporated into the Lagrangian formulation of the theory in that it was no longer associated with an electric/elastic mechanical explanation.

What Maxwell had in fact shown was that given the specific assumptions employed in developing the mechanical details of his model, the elastic properties of the electromagnetic medium were just those required of the luminiferous

---

[6] For a more extensive discussion of this point, see Morrison (2008).

aether by the wave theory of light. Hence, what was effected was the reduction of electromagnetism and optics to the mechanics of *one aether*, rather than a reduction of optics to electromagnetism simpliciter. In that sense, the first form of Maxwell's theory displayed a reductive unity, but the more interesting question is whether, in the absence of the aether, the identification of electromagnetic and optical waves still constitutes a reduction of two different processes to a single natural kind. The answer to this question is complicated by the difficulties that plagued the model, the most serious being the status of electric displacement itself. Not only did it suffer from ambiguities in interpretation, it was not a natural consequence of the model and there was no experimental data that required its postulation. It was introduced purely to facilitate a field theoretic account of electromagnetic processes. Moreover, the equation relating displacement with charge was not explicitly given, and without any "physical" account of the field it became difficult to see just how charge could occur.

Maxwell claimed that his later account entitled "A Dynamical Theory of the Electromagnetic Field" (1865) (DT) was based on experimental facts and general dynamical principles about matter in motion as characterized by the abstract dynamics of Lagrange. The aim of Lagrange's *Mécanique Analytique* (1788) was to rid mechanics of Newtonian forces and the requirement that we must construct a separate acting force for each particle. The equations of motion for a mechanical system were derived from the principle of virtual velocities and d'Alembert's principle.[7] The method consisted of expressing the elementary dynamical relations

---

[7] Given a system described by n generalized coordinates q_i, their velocities $\dot{q}_i$, along with purely holonomic constraints, d'Alembert's principle yields n equations of motion

$$\frac{d}{dt}\left(\frac{\partial T}{\partial \dot{q}_j}\right) - \frac{\partial T}{\partial q_j} - Q_j = 0$$

where $T = \sum_i \frac{1}{2} m_i v_i^2$ is the kinetic energy and

$$Q_j = \sum_i F_i \cdot \frac{\partial r_i}{\partial q_j}$$

is the generalized force corresponding to q_j. For a conservative system, the forces F_i may be written in terms of a potential function $V(r_1, r_2, \ldots)$, such that

$$F_i = -\nabla_i V.$$

Therefore

$$Q_j = -\sum_i \frac{\partial V}{\partial r_i} \cdot \frac{\partial r_i}{\partial q_j} = -\frac{\partial V}{\partial q_j}$$

The equations of motion become

$$\frac{d}{dt}\left(\frac{\partial (T-V)}{\partial \dot{q}_j}\right) - \frac{\partial (T-V)}{\partial q_j} = 0$$

where we have made use of the fact that V depends only on the generalized coordinates q and not their velocities. This motivates the definition of the Lagrangian

$$L = T - V,$$

from which the **Euler-Lagrange** equations follow:

$$\frac{d}{dt}\left(\frac{\partial L}{\partial \dot{q}_i}\right) - \frac{\partial L}{\partial q_i} = 0$$

The utility of the Lagrangian approach is that, by virtue of d'Alembert's use of generalized coordinates, (holonomic) constraint forces do not appear explicitly.

in terms of the corresponding relations of pure algebraic quantities, which facilitated the deduction of the equations of motion. Consequently, insofar as the formal structure is concerned, analytical mechanics, electromagnetism, and wave mechanics can all be deduced from a variational principle, the result being that each theory has a uniform Lagrangian appearance. Velocities, momenta, and forces related to the coordinates in the equations of motion need not be interpreted literally in the fashion of their Newtonian counterparts. This allows for the field to be represented as a connected mechanical system with currents, integral currents, and generalized coordinates corresponding to the velocities and positions of the conductors. In other words, we can have a quantitative determination of the field without knowing the actual motion, location, and nature of the system itself.

Using this method Maxwell went on to derive the basic wave equations of electromagnetism without any special assumptions about molecular vortices, forces between electrical particles, and without specifying the details of the mechanical structure of the field. The 20 equations consisted of three equations each for magnetic force, electric currents, electromotive force, electric elasticity, electric resistance, total currents; and one equation each for free electricity and continuity. This allowed him to treat the aether (or field) as a mechanical system without any specification of the machinery that gave rise to the characteristics exhibited by the potential-energy function.[8]

It becomes clear, then, that the unifying power of the Lagrangian approach lay in the fact that it ignored the nature of the system and the details of its motion. Because very little information is provided about the physical system, it is easier to bring together diverse phenomena under a common framework. Only their general features are accounted for, yielding a unification that, to some extent, is simply a formal analogy between two different kinds of phenomena. The Lagrangian emphasis on energetic properties of a system, rather than its internal structure, became especially important after the establishment of the principle of conservation of energy. In fact, in the *Treatise on Electricity and Magnetism* the notion of "field energy" became the physical principle on which an otherwise abstract dynamics could rest. Maxwell claimed that all physical concepts in "A Dynamical Theory," except energy, were understood to be merely illustrative, rather than substantial. Displacement constituted one of the basic equations and was defined simply as the motion of electricity, that is, in terms of a quantity of charge crossing a designated area. But, if electricity was being displaced, how did this occur? Due to the lack of a mechanical foundation, the idea that there was a displacement of electricity in the field (a charge), without an associated mechanical source or body, became difficult to motivate theoretically. These issues did not pose significant problems for

---

[8] His attachment to the potentials as primary was also criticized, since virtually all theorists of the day believed that the potentials were simply mathematical conveniences having no physical reality whatsoever. To them, the force fields were the only physical reality in Maxwell's theory but the formulation in DT provided no account of this. Today, of course, we know in the quantum theory that it is the potentials that are primary, and the fields are derived from changes in the potentials.

Maxwell himself, since he associated the force fields with the underlying potentials.[9] Because the value of wave propagation for electromagnetic phenomena is equivalent to that for light, the wave equation represents the reduction of electromagnetism and optics, a process that was facilitated by the displacement current.

What methodological lessons about unification can be gleaned from the Maxwell case? At the very least, it shows that theory unification can be a rather complex process that integrates mathematical techniques and broad-ranging physical principles that govern material systems. In addition to the generality of the Lagrangian formalism, its deductive character displays a crucial feature for successful unification: the ability to derive equations of motion for a physical system with a minimum of information. But, as I noted above, this mathematical framework provided little or no insight into specific physical details, leaving the problem of whether to interpret the unification as indicative of a physical unity in nature. This problem is particularly relevant because of the accompanying difficulties with displacement. Because it provides a necessary condition for formulating the field equations, it forms the foundation for a truly unified theory that integrates or reduces various phenomena as opposed to one that simply incorporates more phenomena than its rivals. However, the theory cast in terms of the Lagrangian formalism lacked real explanatory power due to the absence of specific theoretical details. The field equations could account for both optical and electromagnetic processes as the results of waves traveling through space, but there was no theoretical foundation for understanding of how that took place. And, in the absence of any experimental evidence for electromagnetic waves what Maxwell had shown was only that a unification and reduction of electromagnetism and optics was theoretically possible.

Although the unity achieved in "On Physical Lines" and later versions of the theory involved the reduction of optical and electromagnetic processes, the electric and magnetic fields retained their independence; the theory simply showed the interrelationship of the two—where a varying electric field exists, there is also a varying magnetic field induced at right angles, and vice versa. The two together form the electromagnetic field. In that sense the theory united the two kinds of forces by integrating them in a systematic or synthetic way, but their true unification did not take place until 1905 with the Special Theory of Relativity. Maxwell's equations were crucial in motivating Einstein's paper where he noted in the beginning paragraph that a description of a conductor moving with respect to a magnet must generate a consistent set of fields irrespective of whether the force is calculated in the rest

---

[9] The methods used in "A Dynamical Theory" were extended and more fully developed in the *Treatise on Electricity and Magnetism* (TEM), where the goal was to examine the consequences of the assumption that electric currents were simply moving systems whose motion was communicated to each of the parts by certain forces, the nature and laws of which "we do not even attempt to define, because we can eliminate [them] from the equations of motion by the method given by Lagrange for any connected system" sect. 552). Displacement, magnetic induction and electric and magnetic forces were all defined in the *Treatise* as vector quantities (Maxwell 1873, sect. 11, 12), together with the electrostatic state, which was termed the vector potential. All were fundamental quantities for expression of the energy of the field and were seen as replacing the lines of force.

frame of the magnet or that of the conductor. Maxwell's equations generated an asymmetry that was not present in the phenomena (1952, 37).

Without going into the details of the unification provided by special relativity, it is important to point out that the unity of electricity and magnetism was also indicative of something deeper and more pervasive, specifically, a unification of two domains of physics—mechanics and electrodynamics. This latter unification was a realization of the requirement that the laws of physics must assume the same form in all inertial frames. The further mathematization of the event structure of the theory at the hands of Minkowski showed that the relationship between electricity and magnetism could be represented by the transformation properties of the electromagnetic field tensor. If one begins with a field E due to a static charge distribution, with no magnetic field, and transforms to another frame moving with uniform velocity, the transformation equations show that there exists a magnetic field in the moving frame even though none existed in the inertial frame. Hence, the magnetic field appears as an effect of the transformation from one frame of reference to another. In Maxwell's theory the electric and magnetic fields were two entities combined by an angle of interaction, whereas, in the Minkowski formulation the electromagnetic field is one entity represented by one tensor—their separation is merely a frame dependent phenomena.[10]

Maxwell's electrodynamics was the first unified field theory in physics. It exemplified the same type of reductive unity present in Newtonian mechanics, which unified terrestrial and celestial motion under the same force law—universal gravitation. But few if any subsequent cases of unification have demonstrated this kind of reduction; in fact most unified theories are the result of a synthesis of different phenomena under a single theoretical framework. Mathematics continues to be crucial for achieving unification but as we shall see below, in the electroweak case the goal is to use mathematical tools like symmetry for generating a unified dynamics as opposed to a framework for representing existing theoretical relations among different phenomena as in the case of electrodynamics.

3. Synthetic Unity: The Electroweak Theory

The electroweak theory brings together electromagnetism with the weak force in a single relativistic quantum field theory that involves the product of two gauge symmetry groups. From the perspective of phenomenology these two forces are very different. Electromagnetism has an infinite range; whereas, the weak force, which produces radioactive beta decay, spans distances shorter than approximately

---

[10] For an extended discussion of unification in Special Relativity see Morrison (2000).

10^{-15} cm. Moreover, the photon associated with the electromagnetic field is massless, while the bosons associated with the weak force are massive due to their short range. Despite these differences, they do share some common features: both kinds of interactions affect leptons and hadrons; both appear to be vector interactions brought about by the exchange of particles carrying unit spin and negative parity, and both have their own universal coupling constant that governs the strength of the interactions. The electroweak theory is joined with quantum chromodynamics (QCD)—the theory of the strong interactions—to form the Standard Model.

My focus here will be largely on the electroweak theory for several reasons. First, and perhaps most important for our purposes, by examining the structure of the electroweak theory it is possible to illustrate the nature of unification in a way that is not possible with the larger Standard Model. The electroweak theory involves a combination of the SU(2) group governing isospin/weak interactions and the U(1) group of electromagnetism. The mixing of these fields is represented by the Weinberg angle $sin^2\theta$, which is a free parameter whose value is determined experimentally. The Standard Model structure involves the addition of the SU(3) symmetry group that governs the color charged fermions (quarks) to form the SU(3) x SU(2) x U(1) group. The SU(3) color group corresponds to the local symmetry whose gauging gives rise to quantum chromodynamics—the theory that governs the strong force (QCD). In addition to some of the outstanding theoretical problems with the Standard Model, such as the origin of the masses and mixings of the quarks and leptons, the most significant problem from the "unification" perspective comes in the application of the theory, which involves significantly more free parameters than electroweak—approximately 26 in total.[11]

While the incompatibility of the Standard Model with gravity, and until recently the status of the Higgs boson, are often cited as stumbling blocks for unification, it is the internal structure of the Standard Model itself that undermines its status as a unified theory. Moreover, finding the Higgs particle will only partly rectify the problems. By contrast, the electroweak theory has only one free parameter and involves more than a simple pasting together of the two different force fields under a combined symmetry group. As we shall see, it furnishes an account of the mixing of the fields that involves a synthetic unity that is simply not possible in the current version of the Standard Model. That is not to say that the electroweak theory is without its own difficulties, but only that it clearly qualifies as a unified theory in a way that the Standard Model does not. Below I discuss some of these theoretical issues as they arise in the context of the electroweak theory and their relation to the larger Standard Model, but first let me turn to a more detailed discussion of the specifics of the electroweak theory to illustrate the exact nature of the unification and how it was produced.

---

[11] There are two different types of SU(3) symmetry: the one that acts on the different colors of quarks, which is an exact gauge symmetry mediated by gluons, and the flavor SU(3) symmetry, which rotates different flavors of quarks to each other. The latter is an approximate symmetry of the QCD vacuum and hence is not fundamental. It arises as a consequence of the small mass of the three lightest quarks.

A solution to the incompatibility between electromagnetism and the weak force was achieved by postulating the Higgs mechanism, the newly found element of the Standard Model. This facilitated a unification of the fields but does so in a way that leaves the forces more or less distinct. To see how this unity was achieved let me begin by discussing how gauge symmetry functions as a unifying structure and go on to show how this type of unification presents us with a very different picture than the reductive unity provided by Newtonian mechanics and Maxwellian electrodynamics.[12]

3.1 Symmetry as a Tool for Unification

In physics, a gauge theory is a type of field theory where the Lagrangian is invariant under a continuous group of local transformations known as gauge transformations. These transformations form a Lie group, which is the symmetry group or the gauge group with an associated Lie algebra of group generators. For each group generator there necessarily arises a corresponding vector field called the gauge field, which is included in the Lagrangian to ensure its invariance under the local group transformations. Simply put: in a gauge theory there is a group of transformations of the field variables (gauge transformations) that leaves the basic physics of the quantum field unchanged. This condition, called gauge invariance, gives the theory a certain symmetry, which governs its equations. Hence, the structure of the group of gauge transformations in a particular gauge theory entails general restrictions on how the field described by that theory can interact with other fields and elementary particles. This is the sense in which gauge theories are sometimes said to "generate" particle dynamics—their associated symmetry constraints specify the form of interaction terms. The symmetry associated with electric charge is a local symmetry where physical laws are invariant under a local transformation. This involves an infinite number of separate transformations that are different at every point in space and time. But by introducing new force fields that transform in certain ways and interact with the original particles in the theory, a local invariance can be restored.

To see how local gauge invariance is related to physical dynamics consider the following: if we write the non-relativistic Schrodinger equation

$$[1/2m(-ih\nabla - eA)^2 + e\Phi + V]\psi = ih\, \partial\psi/\partial t$$

(where the canonical momentum operator $p_\mu - eA_\mu$ is replaced by the quantum operator $-ih\nabla - eA$), then after a phase change an additional gradient term proportional to $e\nabla\lambda$ emerges, the result of the operator $-ih\nabla$ acting on the transformation wave function.

---

[12] For a more comprehensive discussion of symmetry and its uses in physics, see Bangu (this volume), as well as the edited collection by Brading and Castellani (2003) and Morrison (1995; 2000).

This additional term spoils the local phase invariance, which can then be restored by introducing the new gauge field A_μ. The gauge transformation:

$$A_\mu \to A_\mu - \partial_\mu \lambda$$

cancels out the new term. This new gauge field is simply the vector potential defining the electromagnetic field. A different choice of phase at each point can be accommodated by interpreting A_μ as the connection relating phases at different points. In other words, the choice of a phase function $\lambda(x)$ will not affect any observable quantity as long as the gauge transformation for A_μ has a form that allows the phase change and the change in potential to cancel each other. What this means is that we cannot distinguish between the effects of a local phase change and the effects of a new vector field.

The combination of the additional gradient term with the vector field A_μ prescribes the form of the interaction between matter and the field because A_μ provides the connections between phase values at nearby points. The phase of a particle's wave function can be identified as a new physical degree of freedom that is dependent on spacetime position. In fact, it is possible to show that from the conservation of electric charge one can, given Noether's theorem, choose a symmetry, and the requirement that it be local forces us to introduce a gauge field, which turns out to be the electromagnetic field. The structure of this field, which is dictated by the requirement of local symmetry, in turn dictates, almost uniquely, the form of the interaction, that is, the precise form of the forces on the charged particle and the way in which the electric-charge current density serves as the source for the gauge field.

In Maxwell's theory the basic field variables are the strengths of the electric and magnetic fields, which may be described in terms of auxiliary variables (e.g., the scalar and vector potentials). The gauge transformations in this theory consist of certain alterations in the values of those potentials that do not result in a change of the electric and magnetic fields. This gauge invariance is preserved in quantum electrodynamics (QED) where the phase transformations are one-parameter transformations and form a one-dimensional Abelian group (meaning that any two transformations commute)—in this case the U(1) group of a U(1) gauge symmetry.

Symmetry groups, however, are more than simply mathematizations of certain kinds of transformations. In the non-Abelian case (non-commutative transformations) the mathematical structure of the symmetry group determines the structure of the gauge field and the form of the interaction. In these more complicated situations, there are several wave functions or fields transforming together, as in the case of SU(2) and SU(3) transformations, which involve unitary matrices acting on multiplets.[13] These symmetries are internal symmetries and typically are associated with families of identical particles. In each case the conserved quantities are simply the quantum numbers that label the members of the multiplets (such as isospin

---

[13] These can also be thought of as phase transformations where the phase is considered a matrix quantity. See Aitchinson and Hey (1989) for a discussion of this topic.

and color), together with operators that induce transitions from one member of a multiplet to another. Hence, the operators correspond, on the one hand, to the conserved dynamical variables (isospin, etc.) and, on the other hand, to the group of transformations of the symmetry group of the multiplets.[14]

The extension of gauge invariance beyond electromagnetism began with the work of Yang and Mills (1954) who generalized it to the conserved quantity isospin (violated in electromagnetic and weak interactions), which allows the proton and neutron to be considered as two states of the same particle. Here a local gauge invariance means that although we can, in one location, label the proton as the "up" state of isospin, and the neutron as the "down" state, the up state need not be the same at another location. But because the SU(2) symmetry group that governs isospin is also the group that governs rotations in a three-dimensional space, the "phase" is replaced by a local variable that specifies the direction of the isospin. However, it was not until the work of Schwinger (1957) that any significant connection was made between the weak and electromagnetic forces. Schwinger's approach was to begin with some basic principles of symmetry and field theory, and go on to develop a framework for fundamental interactions derived from that fixed structure. As we saw above, in QED it was possible to show that from the conservation of electric charge, one could, on the basis of Noether's theorem, assume the existence of a symmetry, and the requirement that it be local forces one to introduce a gauge field, which turns out to be just the electromagnetic field. The symmetry structure of the gauge field dictates, almost uniquely, the form of the interaction; that is, the precise form of the forces on the charged particle and the way in which the electric charge current density serves as the source for the gauge field. The question was how to extend that methodology beyond quantum electrodynamics to embody weak interactions.

3.2 From Mathematics to Physics

Because of the mass differences between the weak force bosons and photons a different kind of symmetry was required if electrodynamics and the weak interaction were to be unified and the weak and electromagnetic couplings related. Due to the mass problem, it was thought that perhaps only partial symmetries—invariance of only part of the Lagrangian under a group of infinitesimal transformations—could relate the massive bosons to the massless photon. In 1961 Glashow developed a model based on the SU(2) x U(1) symmetry group, which required the introduction of an additional neutral boson Z_s, which couples to its own neutral lepton current

---

[14] Isospin actually refers to similar kinds of particles considered as two states of the same particle in particular types of interactions. For example, the strong interactions between two protons and two neutrons are the same, which suggests that for strong interactions they may be thought of as two states of the same particle. So, hadrons with similar masses, but differing in terms of charge, can be combined into groups called multiplets and regarded as different states of the same object. The mathematical treatment of this characteristic is identical with that used for spin (angular momentum). The SU(2) group is the isospin group and is also the symmetry group of spatial rotations that give rise to angular momentum.

J_s. By properly choosing the mass terms to be inserted into the Lagrangian, Glashow was able to show that the singlet neutral boson from U(1) and the neutral member of the SU(2) triplet would mix in such a way as to produce a massive particle B (now identified as Z^0) and a massless particle that was identified with the photon. But, in order to retain Lagrangian invariance gauge theory requires the introduction of only massless particles. As a result the boson masses had to be added to the theory by hand, making the models phenomenologically accurate but destroying the gauge invariance of the Lagrangian, thereby ruling out the possibility of renormalization. Although gauge theory provided a powerful tool for generating an electroweak model, unlike electrodynamics, one could not reconcile the physical demands of the weak force for the existence of massive particles with the structural demands of gauge invariance. Both needed to be accommodated if there was to be a unified theory, yet they were mutually incompatible.

Hopes of achieving a true synthesis of weak and electromagnetic interactions came a few years later with Steven Weinberg's (1967) idea that one could understand the mass problem and the coupling differences of the different interactions by supposing that the symmetries relating the two interactions were exact symmetries of the Lagrangian that were somehow broken by the vacuum. These ideas originated in the early 1960s and were motivated by work done in solid state physics on superconductivity. But, if the electroweak and the electromagnetic theory were truly unified and mediated by the same kind of gauge particles, then how could such a difference in the masses of the bosons and the photons exist? In order for the electroweak theory to work, it had to be possible for the gauge particles to acquire a mass in a way that would preserve gauge invariance.

The answer to these questions was provided by the mechanism of spontaneous symmetry breaking. From work in solid state physics, it was known that when a local symmetry is spontaneously broken the vector particles acquire a mass through a phenomenon that came to be known as the Higgs mechanism (Higgs, 1964a&b). This principle of spontaneous symmetry breaking implies that the actual symmetry of a system can be less than the symmetry of its underlying physical laws; in other words, the Hamiltonian and commutation relations of a quantum theory would possess an exact symmetry while physically the system (in this case the particle physics vacuum) would be nonsymmetrical. In order for the idea to have any merit one must assume that the vacuum is a degenerate state (i.e., not unique) such that for each unsymmetrical vacuum state there are others of the same minimal energy that are related to the first by various symmetry transformations that preserve the invariance of physical laws. The phenomena observed within the framework of this unsymmetrical vacuum state will exhibit the broken symmetry even in the way that the physical laws appear to operate. Although there is no evidence that the vacuum state for the electroweak theory is degenerate, it can be made so by the introduction of the Higgs mechanism, which is an additional field with a definite but arbitrary orientation in the isospin vector space. The orientation breaks the symmetry of the vacuum.

The Higgs field (or its associated particle the Higgs boson) is really a complex SU(2) doublet consisting of four real fields, which are needed to transform the massless gauge fields into massive ones. A massless gauge boson like the photon has two orthogonal spin components transverse to the direction of motion while massive gauge bosons have three including a longitudinal component in the direction of motion. In the electroweak theory the W^{+-} and the Z^0, which are the carriers of the weak force, absorb three of the four Higgs fields, thereby forming their longitudinal spin components and acquiring a mass. The remaining neutral Higgs field is not affected and should therefore be observable as a particle in its own right. The Higgs field breaks the symmetry of the vacuum by having a preferred direction in space, but the symmetry of the Lagrangian remains invariant. So, the electroweak gauge theory predicts the existence of four gauge quanta, a neutral photon-like object, sometimes referred to as the X^0 and associated with the U(1) symmetry, as well as a weak isospin triplet W^{+-} and W^0 associated with the SU(2) symmetry. As a result of the Higgs symmetry breaking mechanisms the particles W^{+-} acquire a mass and the X^0 and W^0 are mixed so that the neutral particles one sees in nature are really two different linear combinations of these two. One of these neutral particles, the Z^0, has a mass while the other, the photon, is massless. Since the masses of the W^{+-} and Z^0 are governed by the structure of the Higgs field they do not affect the basic gauge invariance of the theory. The so-called "weakness" of the weak interaction, which is mediated by the W^{+-} and the Z^0, is understood as a consequence of the masses of these particles.

We can see from the discussion above that the Higgs phenomenon plays two related roles in the theory. It explains the discrepancy between the photon and the intermediate vector boson masses—the photon remains massless because it corresponds to the unbroken symmetry subgroup U(1) associated with the conservation of charge, while the bosons have masses because they correspond to SU(2) symmetries that are broken. Second, the avoidance of an explicit mass term in the Lagrangian allows for gauge invariance and the possibility of renormalizability. With this mechanism in place the weak and electromagnetic interactions could be unified under a larger gauge symmetry group that resulted from the product of the SU(2) group that governed the weak interactions and the U(1) group of electrodynamics.[15]

From this very brief sketch, one can get at least a snapshot of the role played by the formal, structural constraints provided by gauge theory/symmetry in the development of the electroweak theory. I now want to turn to the specific kind of unity that emerged in this context.

The point I want to emphasize regarding the electroweak unification is that the unity achieved was largely structural rather than substantial and as a result does not fit with the ideal of reducing elements of the weak and electromagnetic force to the same basic entity. In the case of electrodynamics, the generality provided by the Lagrangian formalism allowed Maxwell to unify electromagnetism and optics without providing any specific details about how the electromagnetic waves were

---

[15] In order to satisfy the symmetry demands associated with the SU(2) group and in order to have a unified theory (i.e., have the proper coupling strengths for a conserved electric current and two charged W fields), the existence of a new gauge field was required, a field that Weinberg associated with a neutral current interaction that was later discovered in 1973. For a discussion of the difficulties surrounding the neutral current experiments, see Galison (1987) and Pickering (1984).

produced or how they were propagated through space. However, in addition to the structural aspects of the unification, light and electromagnetic waves were thought to be identical; hence the reductive aspect of the unification. The SU(2) × U(1) gauge theory furnishes a similar kind of structure; it specifies the form of the interactions between the weak and electromagnetic forces but provides no causal account as to why the fields must be unified. In this case both the electromagnetic and weak forces remain essentially distinct; the unity that is supposedly achieved results from the unique way in which these forces interact. Hence, with respect to the unifying process the core of the theory is really the representation of the interaction or mixing of the various fields. Because the fields remain distinct, the theory retains two distinct coupling constants, q associated with the U(1) electromagnetic field and g with the SU(2) gauge field. In order to make specific predictions for the masses of the W^{+-} and Z^0 particles, one needs to know the value for the Higgs ground state $|\Phi_0|$. Unfortunately, this cannot be directly calculated, since its value depends explicitly on the parameters of the Higgs potential and at the time the theory was formulated little was known about the properties of the field.[16]

In order to rectify the problem, the coupling constants are combined into a single parameter known as the Weinberg angle θ_w. The angle is defined from the normalized forms of A^{em} and Z^0 which are respectively:

$$A^{em}_\mu = \frac{gA_\mu - qW^3_\mu}{\sqrt{g^2 + q^2}}$$

$$Z^0_\mu = \frac{qA_\mu + gW^3_\mu}{\sqrt{g^2 + q^2}}$$

The mixing of the A_μ gauge field of U(1) and the new neutral gauge field W^3_μ is interpreted as a rotation through θ_w i.e.,

$$\sin\theta_w = \frac{q}{\sqrt{g^2 + q^2}}; \quad \cos\theta_w = \frac{g}{\sqrt{g^2 + q^2}}$$

By relating the weak coupling constant g to the Fermi coupling constant G one obviates the need for the quantity $|\Phi_a|$ (the value of the Higgs ground state). The masses can now be defined in the following way:

$$M^2_{z^0} = \frac{M^2_W}{\cos^2\theta_w}; \quad M^2_W = \frac{g^2}{2G} = \frac{e^2}{2G\sin^2\theta_w} = \frac{(37.4 GeV)^2}{\sin^2\theta_w}$$

In order to obtain a value for θ_w, one needs to know the relative sign and values of g and q; the problem however is that they are not directly measurable. Instead one must measure the interaction rates for the W^{+-} and Z^0 exchange processes and then extract values for g, q, and θ_w. What θ_w does is fix the ratio of U(1) and SU(2) couplings, and in order for the theory to be unified θ_w must be the same for all

---

[16] Indeed, despite its discovery, the properties of the Higgs boson and whether it is a single particle of a family or particles remains largely unknown. Further data from CERN will hopefully reveal these features and what their impact will be on the Standard Model.

processes. Despite this rather restrictive condition, the theory itself does not provide direct values for the Weinberg angle and hence does not furnish a full account of how the fields are mixed (i.e., the degree of mixing is not determined by the theory). More important, the mixing is not the result of constraints imposed directly by gauge theory itself; rather it ultimately depends on the assumption that leptons can be classified as weak isospin doublets governed by the SU(2) symmetry group. The latter requires the introduction of the new neutral gauge field W^3 in order to complete the group generators, that is, a field corresponding to the isospin operator τ. This is the field that combines with the neutral photon-like X^0 to produce the Z^0 necessary for the unity.

We can see then that the use of symmetries to categorize various kinds of particles and their interaction fields is much more than simply a phenomenological classification; in addition it allows for a kind of particle dynamics to emerge. In other words, the symmetry group provides the foundation for the locally gauge-invariant quantum field theory. Hence, given the assumption about isospin, the formal restrictions of the symmetry groups and gauge theory can be deployed in order to produce a formal model showing how these gauge fields could be unified. The crucial feature that facilitates this interaction is the non-Abelian structure of the group rather than something derivable from phenomenology of the physics. Although the Higgs mechanism is a crucial part of the physical dynamics of the theory and necessary for a unified picture to emerge, the framework within which the unification is realized results from the constraints of the isospin SU(2) group and the non-Abelian structure of the field.

To summarize: gauge theory serves as a unifying tool by specifying the form for the strong, weak, and electromagnetic fields. In that sense it functions in a global way to restrict the class of acceptable theories and in a local way to determine specific kinds of interactions, producing not only unified theories but also a unified method. But it is not simply the presence of a unifying *method* or *structure* that is required for theory unification. As we saw in with electrodynamics, the displacement current was the crucial theoretical parameter that allowed Maxwell to formulate a field theoretic account of electromagnetism and to calculate the velocity of wave propagation. The Higgs mechanism facilitates the unification in the electroweak theory by providing the symmetry-breaking mechanism that creates the boson masses; however, it does not explain the mixing of the fields. That mixing was possible through the identification of leptons with the SU(2) isospin symmetry group and represented in the Weinberg angle θ_w. Employing gauge-theoretical constraints, one could then generate the dynamics of an electroweak model from the mathematical framework of gauge theory.

But in what sense does this mixing represent a unification? Because of the neutral-current interactions, the old measure of electric charge given by Coulomb's law (which supposedly gives the total force between electrons) was no longer applicable. Owing to the contribution from the new weak interaction, the electromagnetic potential A_μ^{em} could not be just the gauge field A_μ but had to be a linear combination of the U(1) gauge field and the W_μ^3 field of SU(2). Hence, the mixing was necessary if the electromagnetic potential was to have a physical interpretation in the new theory. So although the two interactions are integrated under a framework that results from a combination of their independent symmetry groups, there is a genuine unity, not merely the conjunction of two theories. A reconceptualization of the electromagnetic potential and a new dynamics emerged from the mixing of the fields. Although this synthesis retains an element of independence for each domain, it also yields a broader theoretical framework within which their integration can be achieved. So, despite the lack of reduction, the electroweak theory nevertheless provides a unified account of the two fields.

4. Problems and Prospects: Electroweak Unification and Beyond

As we noted above, the crucial parameter in the electroweak theory is the Weinberg angle, or as it is sometimes called, the weak mixing angle but its value is not predicted from within the theory and needs to be extracted from parity-violating neutral-current experiments. The electroweak theory has enjoyed overwhelming successes with predictions holding over a range of distances from 10^{-18} m to more than 10^8 m. It has predicted the existence and properties of weak neutral current interactions, the properties of the gauge bosons W^{+-} and Z^0 that mediate neutral and charge current interactions, and required the fourth quark flavor—charm. The recent discovery of the Higgs boson provides the missing link for the electroweak theory but there is still a great deal left unanswered. With a large amount of data still unanalysed, questions remain as to whether the discovery points to a simple Higgs particle or a more complex entity in a larger family of Higgs particles. The standard model predicts that the Higgs boson lasts for only a very short time before it decays into other well known particles. These decay patterns are the data relevant for the discovery. The decay channels, five studied by CMS, yielded a signal with statistical significance at 4.9 above background. The combined fit to the two most sensitive and high resolution channels (photos and leptons) yielded a statistical significance of 5 sigma. What this means is that the probability of the background alone fluctuating up by this amount or more is about one in three million. Further data are required to measure properties like the decay rates in various channels as well as the spin and parity. These will determine whether the observed particle is the Higgs boson as predicted by the standard model, a more complicated version of it or the result of new physics beyond it.

Some of the other problems facing the electroweak theory specifically include the fact that it accommodates but does not predict or explain fermion masses and mixings (elementary fermions are quarks and leptons while composite fermions are baryons that include protons and neutrons). The CKM (Cabibbo-Kobayashi-Maskawa) framework, which represents quark mixing using a 3×3 unitary matrix, describes CP violation but does not explain its origin.[17] The mass of the neutrino, which is implied as a result of the discovery of neutrino flavor mixing, also requires an extension of the current electroweak theory, since specific values are determined by Yukawa couplings of fermions to the Higgs field rather than being set by the theory itself.[18] There are several other problems related to the instability of the Higgs sector to large radiative corrections as well as the lack of any candidates to explain the cold dark matter required for structure formation in the early universe. The Higgs boson, however, is unlikely to provide an explanation for dark matter since the latter must be stable with a very long lifetime and the Higgs decays very rapidly. The favoured explanation is the least massive supersymmetric particle because it cannot decay any further; but, despite the enormous quantity of data from the LHC there is as yet no evidence for the existence of any supersymetric particles. Many of these issues speak to the incompleteness of the Standard Model in general and against the view that it provides a unified description of the strong, weak, and electromagnetic forces. But, some of these issues are also related to the connection between the electroweak theory and the larger context of the Standard Model.

For example, the CP violation mentioned above is one such problem. Quantum chromodynamics (QCD) does not seem to break the CP symmetry even though the electroweak theory does. Although there are natural terms in the QCD Lagrangian that can break the CP symmetry, experiments do not indicate any CP violation in the QCD sector. One of the reasons the CP problem is troublesome is that it leaves unanswered the question of why the universe does not consist of equal parts matter and antimatter. In fact, it is possible to show that one of the conditions required for the current imbalance is CP violation during the first seconds after the Big Bang. Other explanations require the imbalance to be present from the beginning, which is far less plausible. Although the violation of CP symmetry has been verified in the case of the weak force, it only accounts for a small portion of the violation required to explain the matter in the universe. The fact that this discrepancy is not even predicted by the Standard Model suggests a rather serious gap or incompatibility with the electroweak sector.

And, there are other more pressing problems for electroweak theory itself. In addition to the fermion mass problem, there are also the mixing angles that parameterize the discrepancies between neutrino mass eigenstates and those in the quark sector. Although the Higgs boson may be responsible for fermion masses, there is nothing in the electroweak theory that can or will determine the couplings of the Higgs particles to fermions; and in that sense the theory is seriously incomplete.[19]

---

[17] CP is a symmetry that states that the laws of physics should be the same if a particle were interchanged with its antiparticle (C symmetry, or charge conjugation symmetry), and left and right were swapped (P symmetry, or parity symmetry). In addition to its role in weak interactions, it also plays an important role in the attempts of cosmology to explain the dominance of matter over antimatter in the Universe.

[18] The Yukawa interaction describes the coupling between the Higgs field and massless quark and electron fields. Through spontaneous symmetry breaking, the fermions acquire a mass proportional to the vacuum expectation value of the Higgs field.

[19] For an extended discussion of these and other problems facing the electroweak theory, see Quigg (2009). My discussion borrows from his exposition.

Another equally serious concern is the gauge hierarchy problem, which refers to the marked difference between fundamental parameters like masses and couplings that are contained in the Lagrangian and the values that are measured experimentally. Typically the latter are related to the former via renormalization but in many cases there are cancellations between the fundamental quantity and quantum corrections that involve short distance physics. The problem is that very often the details of physics at short distances are largely unknown. More specifically, the gauge hierarchy problem relates to the fact that the weak force is 10^{32} times stronger than gravity. This discrepancy gives rise to the question of why the Higgs boson or the weak scale (at 100 GeV) is so much smaller than the Planck scale (at 10^{19} GeV). The weak scale is given by the vacuum expectation value of the Higgs, $VEV = 246$ GeV, but it is not naturally stable under radiative corrections. The radiative corrections to the Higgs mass, which result from its couplings to gauge bosons, Yukawa couplings to fermions, and its self-couplings, result in a quadratic sensitivity to the ultraviolet cutoff. Hence, if the Standard Model were valid up to the Planck scale, then m_h and therefore the minimum of the Higgs potential would be driven to the Planck scale by the radiative corrections. To avoid this one has to adjust the Higgs bare mass in the Standard Model Lagrangian to one part in 10^{17}. This is called "unnatural fine-tuning" where naturalness is defined in terms of the magnitude of quantum corrections where the bare value and the quantum correction appear to have an unexpected cancellation that gives a result much smaller than either component.

The issue of fine-tuning is important here because, as we saw above, the mass of the Higgs boson is not given by the theory and without fine-tuning the mass would be so large as to undermine the internal consistency of the electroweak theory. Hence, the question becomes whether additions to the Standard Model or any new physics will still require fine-tuning. The answer will depend on what further data from the LHC will reveal about the nature of the Higgs boson and what additional particles might be discovered. Implicit in the reasoning that leads to the fine-tuning is the unsubstantiated assumption that very little physics other than renormalization group scaling exists between the Higgs scale and the grand unification energy which are separated by roughly 11 orders of magnitude (known as the "big dessert" assumption). If this is true, then it would seem that fine-tuning is something we need to live with, at least for the time being.[20] Of course, depending on the specific findings at the Higgs scale the need for fine-tuning may very well be obviated.

Another instance of the hierarchy problem, and one that is a more serious violation of the naturalness requirement, involves the cosmological constant. Observations of an accelerating universe imply the existence of a small but nonzero cosmological constant. But, the essential fact is that the observed vacuum energy

---

[20] Of course, the notion of naturalness here is not something that can be given a precise definition, since it is relative to the gaps in our theoretical knowledge of physics at high energies. Since the quantum correction includes effects from high energy, there is an uncertainty about their extent and validity. At energies beyond that for which our theories are valid, new physics may emerge making the quantum corrections depend entirely on the energy scale. Hence, the notion of naturalness can be thought of as scale relative.

density must be extremely small—a few milli-electronvolts. However, if we take v, the Higgs potential which is roughly 246 GeV, and insert the current lower bound on m_H, the Higgs mass which is 126 GeV, then the Higgs field contribution to the vacuum energy density is roughly 54 orders of magnitude greater than the upper bound inferred from the cosmological constant. If there are other, heavier Higgs fields, the problem is even worse.

It seems clear from our discussion that the often cited problem of trying to adapt the quantum field theoretic framework to general relativity is simply one of several problems facing the Standard Model. Indeed, many of the pressing theoretical difficulties are generated from within the structure of the theory itself. The task of finding the Higgs particle is intimately connected with the possibility of discovering "new physics" beyond the Standard Model that would explain or rectify the origins of the hierarchy problem, among others. Given the list of unanswered questions that arise from the electroweak theory and its connection with the Standard Model, it is reasonably clear that nothing like a unified understanding of the electromagnetic, weak, and strong forces is available from our present theories. So, while the discovery of the Higgs boson has verified an important part of the electroweak theory, it will not necessarily solve the outstanding internal problems facing the theory.

In order for physics beyond the Standard Model to regulate the Higgs mass, and restore naturalness, its energy scale must be around the TeV. Most of the alternative theories that offer solutions to the problem imply that new physics will be discovered at the LHC, the most popular candidate being weak scale supersymmetry. Supersymmetry (SUSY) relates particles of one spin to other "superpartners" that differ by half a unit. In a theory with an unbroken supersymmetry, every type of boson has a corresponding type of fermion with the same mass and internal quantum numbers and vice versa. Because the superpartners of the Standard Model particles have not been observed, if supersymmetry exists it must be broken thereby allowing the superparticles to be heavier than their corresponding Standard Model particles. There are currently many models proposed to explain SUSY breaking, as well as models that incorporate weakly interacting massive particles that serve as candidates for dark matter.[21]

The other bonus supplied by supersymmetry is its ability to unify the different coupling constants at a high-energy scale. Currently, within the framework of the Standard Model, there is no single energy at which they all become equal. However, incorporating supersymmetry changes the rate at which the couplings vary with energy, allowing them to be unified at a single point. If supersymmetry exists close to the TeV scale, it allows for a solution of the hierarchy problem because the superpartners of the Standard Model particles, having different statistics, contribute to the radiative corrections to the Higgs mass with the opposite sign. In the limit of exact supersymmetry, all corrections to m_h cancel.

---

[21] A potential problem for SUSY breaking is whether it can be accomplished in a "natural" way. Because there seems to be no obvious way to break supersymmetry far below the grand unification energy, this problem, in some sense, is simply a reincarnation of the hierarchy problem.

In the quest to unify the four forces into a single fundamental framework—a TOE—Supersymmetry also includes a theory of quantum gravity that would unite general relativity and the Standard Model. Currently, the two predominant approaches to quantum gravity are string theory and loop quantum gravity (LQG). For string theory to be consistent, supersymmetry appears to be required at some level (although it may be a strongly broken symmetry).[22] Loop quantum gravity, in its current formulation, predicts no additional spatial dimensions as in the case of string theory or anything else about particle physics. Nor does LQG require any assumptions about supersymmetry.[23] Experimental evidence at the LHC confirming supersymmetry in the form of supersymmetric particles could provide support for string theory, since supersymmetry is one of its required components. However, the outlook isn't bright. Consistency of the standard model demanded that the Higgs could not be too massive but because the superpartners (particles predicted by supersymmetry) are supposed to be only slightly heavier than the mass of the Higgs, it was assumed that once the Higgs was found the superpartners would also be in evidence. Moreover, they were supposed to be produced in much greater numbers. Because none has been found a possible explanation is that their mass is an order of magnitude heavier than the Higgs, making them currently inaccessible; but that value is inconsistent with the standard model account. Hence, many versions of string theory that predict certain low mass superpartners will need to be significantly revised.

As was the case with Maxwell's electrodynamics at the time of its construction, the electroweak theory and the Standard Model in general are by no means free of theoretical difficulties. The experiments at the Large Hadron Collider in CERN will probe the electroweak symmetry breaking sector to determine whether the properties of the newly discovered particle are consistent with those predicted for the standard model Higgs boson. Although the electroweak theory successfully unifies the weak and electromagnetic fields, the broader theoretical implications create significant problems that serve to undermine its ability to furnish a theoretically coherent account, that is, one that is consistent with other well-established theoretical claims in particle physics and cosmology. Consequently, despite its unifying power its overall epistemic status is not wholly unproblematic.

---

[22] A string is an object with a finite spatial extent that has an intrinsic tension in the same way that a particle has intrinsic mass. The presence of an intrinsic tension means that string theory possesses an inherent mass scale, a fundamental parameter with the dimensions of mass that defines the energy scale at which "stringy" effects (effects associated to the oscillation of the string) become important. The various oscillation modes of the string are effectively localized in its immediate neighborhood and behave like elementary particles with different masses related to the oscillation frequency of the string. Because a string is like a collection of infinitely many point particles, constrained to fit together to form a continuous object, it has infinitely many degrees of freedom. Consequently, its associated quantum theory required the existence of several spatial dimensions (26). The invention of superstring theory—a string with extra degrees of freedom that make it supersymmetric—has reduced that number to 11.

[23] LQG incorporates many of the important aspects of general relativity, but differs from the latter in its quantization of space and time at the Planck scale, as in quantum mechanics. In other words, the space containing all physical phenomena is itself quantized. Lee Smolin, one of the originators of LQG, has proposed that a loop quantum gravity theory incorporating either supersymmetry or extra dimensions, or both, be called "loop quantum gravity II."

5. Effective Field Theories, Renormalization, and a New Type of Unification

As we have seen above, much of what falls under the title "unification" in high-energy physics involves a synthesis under the product of different symmetry groups rather than the kind of reductive unity characteristic of Newtonian mechanics and electrodynamics. More generally, the failure of the unification/reduction strategy in particle physics has given way to the effective field theory (EFT) program where the "theory" incorporates only the particles that are important for the energy levels or distance scales being investigated. Because the theory is valid only below the masses of the heavy particles, it must be superseded by another effective theory on that energy scale or a complete fundamental theory. The predominance of effective theories is sometimes seen as evidence against reduction and the goal of unification but many, including Weinberg, claim EFTs can be interpreted as simply low-energy approximations to a more fundamental theory (e.g., string theory) thereby allowing one to embrace EFTs while remaining loyal to the reductivist/unification goal. The alternative involves the "tower" of EFTs, where there may be no end to the process, just more and more scales as the energies get higher. Moreover, the lack of experimental evidence and difficulties associated with unification that necessitate the use of EFTs may no longer be an issue once the LHC starts producing sufficient data.

Regardless of the future output from the LHC, philosophical questions arise concerning the epistemic and ontological status of unity given the theoretical problems mentioned above and the prevalence of EFTs in many other areas of physics besides high energy. An examination of the evidence from both experiment and theorizing suggests the following characteristics of unity: it is something that can be achieved in certain local contexts, it is characterizable in different ways, but cannot be extended to a "unity of nature" that is systematically defined. None of this speaks against the possibility of grand unification but the question that we, as philosophers, need to address is how to interpret the evidence at hand, particularly the extensive role of EFTs. Several authors have contributed to this debate including Hartmann (2001), who claims that good scientific research can be characterized by a fruitful interaction between fundamental theories, phenomenological models, and effective field theories. All of them have their appropriate functions in the research process, and all of them are indispensable, complementing each other and hanging together in a coherent way. Cao and Schweber (1993) take a more radical approach, claiming that the current situation is evidence for a pluralism in theoretical ontology, antifoundationalism in epistemology, and antireductionism in methodology.

But what exactly are the implications of these claims and are they borne out by the evidence? Consider, for instance, methodological antireductionism and pluralistic ontologies; no one would deny that low and high-energy domains involve not only different kinds of phenomena but also different methodologies in the sense that the reductionism inherent in the search for fundamental theories has been largely unsuccessful in treating many phenomena in the low-energy domain. While recognizing that low and high-energy domains have rather different goals and require different techniques, it is important to note that they also both make use of effective theories and renormalization group (RG) methods. In that sense, there is a unity of method, especially where the latter is concerned. But that in itself is not philosophically interesting unless we can point to reasons why the method should work so well in two rather disparate domains. In other words, is there some other sense of unity in physics that accounts for the success of RG methods?

Before addressing that question, it is important to note that the development of RG methods also revealed a rather different kind of unity that been previously inexplicable, namely, the way that different phenomena such as liquids and magnets exhibit the same type of behavior near critical points regardless of differences in their microstructure. These phenomena are grouped together into universality classes and share the same critical exponents—parameters that characterize phase transitions. These critical exponents at, for example, the liquid-gas transition are independent of the chemical composition of the fluid. The predictions of universal behavior based on RG methods result from the fact that thermodynamic properties of a system near a phase transition depend only on a small number of features, such as dimensionality and symmetry, and are insensitive to the underlying microscopic properties of the system. Although this kind of unity among different kinds of phenomena is quite distinct from the unification of theories in high-energy physics, in some way the goals are similar—explaining why seemingly different phenomena exhibit the same type of behavior.[24] I will say more about this below but first let me turn to the more general methodological issues of unification as they arise with RG.

The first systematic use of the renormalization group in quantum field theory was by Gell-Mann and Low (1954). A consequence of their approach was that quantum electrodynamics could exhibit a simple scaling behavior at small distances. In other words, quantum field theory has a scale invariance that is broken by particle masses, but these masses are negligible at high energies or short distances provided one renormalizes in the appropriate way. In statistical physics Kadanoff (1966) developed the basis for an application of RG to thermodynamic systems near critical point. This picture also led to certain scaling equations for the correlation functions used in the statistical description, a method that was refined and extended by Wilson (1971). With respect to the unification issue two different questions arise. First: What, if anything, is the unifying thread that connects the *different* RG methods and why can we use RG to describe very different kinds of phenomena?

In some sense this question involves two parts: the first concerns the different mathematical techniques with an eye to articulating a common ground that will underwrite the use of RG in both fields. Once the different techniques have been compared the question is whether there is anything about the *method* itself that facilitates its use in different domains. In other words, once we have illustrated the

---

[24] For a general discussion of RG in the context of explanation more generally, see Batterman (2002).

similarities between the quantum field theoretic approach and that used in statistical physics, will that reveal a unity of method in the two domains? That brings us to the second question: Is there anything common to the *phenomena* themselves such that they can all be treated using the RG approach?

It is important to keep in mind here that I am not simply assuming that because we can use the RG approach as a unifying methodology it also unifies phenomena in a way that shows them to be similar. Rather, a proper answer to the second question involves seeing what similarities might be exhibited between statistical and field theoretic phenomena such that they can both be successfully treated using RG techniques.

There is a brief answer to the first question which can then be spelled out in greater detail, but for our purposes here I will outline just the main point. In order to do that I first need to say a couple of things about the basic idea behind the RG approach. Initially one can think of QFT and statistical physics as having similar kinds of peculiarities that give rise to certain types of problems (e.g., many degrees of freedom, fluctuations, and diverse spatial and temporal scales). The RG framework is significant in its ability to link physical behavior across different scales and in cases where fluctuations on many different scales interact. Hence, it becomes crucial for treating asymptotic behavior at very high (or in massless theories very low) energies (even where the coupling constants at the relevant scale are too large for perturbation theory). In field theory when bare couplings and fields are replaced with renormalized ones defined at a characteristic energy scale μ the integrals over virtual momenta will be cut off at energy and momentum scales of order μ. As we change μ we are in effect changing the scope of the degrees of freedom in the calculations. So, to avoid large logarithms take μ to be the order of the energy E that is relevant to the process under investigation. In other words, the problem is broken down into a sequence of sub-problems with each one involving only a few length scales. Each one has a characteristic length and you get rid of the degrees of freedom you do not need.

Reducing the degrees of freedom gives you a sequence of corresponding Hamiltonians, which can be pictured as a trajectory in a space spanned by the system parameters (temperature, external fields, and coupling constants). So the RG gives us a transformation that looks like this:

$$H' = R[H] \qquad (1)$$

where H is the original Hamiltonian with N degrees of freedom. A wide choice of operators R is possible. Not only is there momentum or Fourier space methods, which are usually associated with field theory, but also what is termed real space renormalization used in statistical physics (cases where there is a definite lattice). The initial version, the Gell-Mann/Low formulation, involved the momentum space approach and hinged on the degree of arbitrariness in the renormalization procedure. They essentially reformulated and renormalized perturbation theory in terms of a cutoff-dependent coupling constant $e(\Lambda)$. For example, e, measured in classical experiments is a property of the very long distance behavior of QED (whereas the natural scale is the Compton wavelength of the electron, $\sim 10^{-11}$cm). G-M/L showed that a family of alternative parameters e_λ could be introduced, any one of which could be used in place of e. The parameter e_λ is related to the behavior of QED at an arbitrary momentum scale λ instead of the low momenta for which e is appropriate. In other words, you can change the renormalization point freely in a QFT and the physics will not be affected. Introducing a sliding renormalization scale effectively suppresses the low-energy degrees of freedom.

The real space approach is linked to the Wilson-Kadanoff method. Kadanoff's account of scaling relations involves a lattice of interacting spins (ferromagnetic transition) and transformations from a site lattice with the Hamiltonian $H_a(S)$ to a block lattice with Hamiltonian $H_{2a}(S)$. Each block is considered as a new basic entity. One then calculates the effective interactions between them and in this way constructs a family of corresponding Hamiltonians. If one starts from a lattice model of lattice size a, one would sum over degrees of freedom at size a while maintaining their average on the sub-lattice of size $2a$ fixed. Starting from a Hamiltonian $H_a(S)$ on the initial lattice, one would generate an effective Hamiltonian $H_{2a}(S)$ on the lattice of double spacing. This transformation is repeated as long as the lattice spacing remains small compared to the correlation length. The key idea is that the transition from $H_a(S)$ to $H_{2a}(S)$ can be regarded as a rule for obtaining the parameters of $H_{2a}(S)$ from those of $H_a(S)$. The process can be repeated with the lattice of small blocks being treated as a site lattice for a lattice of larger blocks.

Close to critical point the correlation length (the distance over which the fluctuations of one microscopic variable are associated with another) far exceeds the lattice constant a, which is the difference between neighboring spins. As we move from small to larger block lattices we gradually exclude the small scale degrees of freedom by averaging out through a process of coarse graining. So, for each new block lattice one has to construct effective interactions and find their connection with the interactions of the previous lattice. What Wilson did was show how the coupling constants at different length scales could be computed, how critical components could be estimated and hence how to understand universality, which follows from the fact that the process can be iterated (i.e., universal properties follow from the limiting behavior of such iterative processes).[25] I will have more to say about these processes in answer to question (2) below.

Initially, this looks like one is doing very different things; in the context of critical phenomena one is interested only in long distance not short distance behavior. In the case of QFT, the renormalization scheme is used to provide an ultraviolet cutoff while in critical behavior the very short wave numbers are integrated out. Moreover, why should scale invariance of the sort found in QFT be important in cases of phase transitions? To answer these questions we can think of the similarities in the following way: in the K-W version the grouping together of the variables referring to different degrees of freedom induces a transformation of the statistical

---

[25] Zinn-Justin (1998) discusses some of the connections between the use of RG in statistical physics and quantum field theory.

ensemble describing the thermodynamic system. Or, one can argue in terms of a transformation of the Hamiltonian. Regardless of the notation, what we are interested in is the successive applications of the transformation that allow us to probe the system over large distances. In the field theoretic case, we do not change the "statistical ensemble" but the stochastic variables do undergo a local transformation whereby one can probe the region of large values of the fluctuating variables. Using the RG equations, one can take this to be formally equivalent to an analysis of the system over large distances.

This formal similarity also provides some clues to why RG can be successfully applied to such diverse phenomena. But here I think we need to look more closely at what exactly the RG method does. In statistical physics we distinguish between two phases by defining an order parameter that has a nonzero value in the ordered phase and zero in the disordered phase (high temperature). In a ferromagnetic transition the order parameter is homogenous magnetization. A nonzero value for the order parameter corresponds to symmetry breaking (here, rotational symmetry). In liquid–gas transition the order parameter is defined in terms of difference in density. In the vicinity of a transition, a system has fluctuations for which one can define a correlation length ξ that increases as $T \to Tc$ (provided all other parameters are fixed). If the correlation length diverges as $T \to Tc$, then the fluctuations become completely dominant and we are left without a characteristic length scale because all lengths are equally important. Reducing the number of degrees of freedom with RG amounts to establishing a correspondence between one problem having a given correlation length and another whose length is smaller by a certain factor. So, we get a very concrete model (hence real space renormalization) for reducing degrees of freedom.

In cases of relativistic quantum field theories like QED, the theory works well for the electron because at long distances there is simply not enough energy to observe the behavior of other charged particles; that is, they are present only at distances very small compared to the electron's Compton wavelength. By choosing the appropriate renormalization scale, the logarithms that appear in perturbation theory will be minimized because all the momenta will be of the order of the chosen scale. In other words, one introduces an upper limit Λ on the allowed momentum equivalent to a microscopic length scale $h/2\pi \Lambda c$. We can think of a change in each of these scales as analogous to a phase transition where the different phases depend on the values of the parameters, with the RG allowing us to connect each of these different scales. So, regardless of whether you are integrating out very short wave numbers or using it to provide an ultraviolet cutoff, the effect is the same in that you are getting the right degrees of freedom for the problem at hand.[26] Hence, because the *formal nature* of the problems is similar in these two domains, one can see why the RG method is so successful in dealing with different phenomena. In the momentum space or field theory approach, we can think of the high-momentum variables as corresponding to short-range fluctuations integrated out. And in the

---

[26] My discussion of these issues borrows from Weinberg (1983).

K-W version the reciprocal of a (the lattice constant which is the difference between neighboring spins) acts as a cutoff parameter for large momenta; that is, it eliminates short wave length fluctuations with wavenumbers close to the cutoff parameter.

The notion that the RG equations and EFTs support ontological pluralism, as suggested by Cao and Schweber, is directly connected to the success of the decoupling theorem (Appelquist and Carazzone 1975). In simple terms the theorem states that if one has a renormalizable theory where some fields have much larger masses compared with others, a renormalization procedure can be found enabling the heavy particles to decouple from the low-energy domain. The low-energy physics is then described by an effective theory that deals only with the particles that are important for the energy level being considered. Using the RG equations, one can delete the heavy fields from the composite system and redefine the coupling constants and masses. However, what is significant here is that the decoupling is, to some extent, only partial. In some cases the heavy particles produce renormalization effects but are suppressed by a power of the relevant experimental energy divided by a heavy mass (the fundamental energy). In that sense the cutoffs represented by the heavy particles define the domain in which the EFT is applicable, that is, the process is mass dependent.[27]

But what about unification? As we saw above, properties near critical point are determined primarily by the correlation length for fluctuations in the order parameter (i.e. blocks of spins within a correlation length of each other will be coherently magnetized). The correlation length diverges on approaching critical point but using the RG equations to reduce the degrees of freedom is in effect reducing the correlation length. As the process is iterated the Hamiltonian becomes more and more insensitive to what happens on smaller length scales. These ideas are important for defining the notion of universality mentioned above—the similar behavior in different kinds of systems in the neighborhood of critical point. An instance of this is the wide variety of liquid-vapor systems whose correlation lengths appear to diverge in precisely the same way as ferromagnets. The systems form a "universality class" that is determined primarily by the nature of the order parameter. The behavior of thermodynamic parameters near critical point is also characterized by what are called critical indices. Phase transitions with the same set of critical indices are said to belong to the same universality class. It is important to point out that this is not simply a case of sharing the same exponents in the way that gravitation and electromagnetism both obey an inverse square law, (exponent -2); that does not show a unity between the forces. A correspondence of exponents whose values are fractions like .63 provides evidence that the microstructure is unimportant. In that

---

[27] See Georgi (1993). The point I want to stress here is that we do not need the decoupling theorem to establish exact results to see why reductionism is problematic. Instead we focus on what the theorem *does* show: that the physics at short distances is not only unimportant at longer length scales but that it is immune from changes that take place there in much the same way that atomic physics is irrelevant to understanding turbulence and the Navier Stokes equations at high Reynolds numbers. In other words, it simply does not matter for these types of problems whether matter becomes discrete at Fermis rather than Angstroms, and it is that fact that causes difficulties for the reductionist picture.

sense the unity among these phenomena has nothing to do with similarity at the level of constituent properties as in the case of unification via reduction.

One of the crucial features of Wilson's work was that it showed that in the long wave-length/large space-scale limit the scaling process leads to a fixed point when the system is at a critical point. The properties of this fixed point determine the critical exponents that characterize the fluctuations at the critical point. The same fixed point interactions can describe a number of different types of systems. RG shows that different kinds of transitions have the same critical exponents and can be understood in terms of the same fixed-point interaction that describes all these systems. What the fixed points do is determine the kinds of cooperative behavior that are possible. So, the important point here is not just the elimination of irrelevant degrees of freedom but also the existence of cooperative behavior and its relation to the order parameter (symmetry breaking) that characterizes the different kinds of systems.

What the renormalization group equations show is that phenomena at critical points have an underlying order. Indeed what makes the behavior of critical point phenomena predictable, even in a limited way, is the existence of certain scaling properties that exhibit "universal" behavior. The problem of calculating the critical indices for these different systems was simplified by using the renormalization group because it shows us that the different kinds of transitions such as liquid–gas, magnetic, alloy, and so on that have the same critical exponents experimentally can be understood in terms of the same fixed-point interaction that describes all these systems. In other words, the RG equations provide a mathematical framework that shows *how* and *why* these phenomena are related to each other.

While the notion of unification defined here is in terms of universality, the final question remains to be answered, namely, whether there is some notion of unification based on a connection between the phenomena in QFT and condensed matter physics that is elucidated via the renormalization group techniques. One possibility is to think of gauge theories characteristic of QFT as exhibiting different phases depending on the value of the parameters. Each phase is associated with a symmetry breaking in the same way that phase change in statistical physics is associated with the order parameter. In statistical physics nature presents us with a microscopic length scale. Cooperative phenomena near a critical point create a correlation length and in the limit of the critical point the ratio of these two lengths tends to ∞. In QFT we introduce an upper limit Λ on the allowed momentum defined in terms of a microscopic length scale $h/2\pi \Lambda c$. The real physics is recovered in the limit in which the artificial scale is small compared to the Compton wavelength of the relevant particles. The ratios of the two length scales Λ/m are tuned toward infinity. In that sense all relativistic QFTs describe critical points with associated fluctuations on arbitrarily many length scales (Weinberg, 1983). And, to that extent we can think of them together with those in condensed matter as exhibiting a kind of generic structure; a structure that is made more explicit as a result of the application of RG techniques. What RG does is expose *physical* structural similarities in the phenomena it treats.

As I said above, the unification associated with universal behavior is very different from what is normally understood when we think of unification in physics. But that is exactly the point I want to stress. Unification is a diverse notion that takes many different forms, some of which are linked with reduction while others are not. Indeed some speak against the very notion of reduction by showing that we can have a unity among phenomena that is completely unrelated to their underlying microstructure. Despite these various ways of understanding unification and the theoretical and experimental difficulties associated with theories of everything, unification remains the goal that drives most if not all of high-energy physics. The question of whether, how, and in what form that goal will be realized and how it relates to a unity in nature is an ongoing aspect of both physics research and philosophical inquiry.

REFERENCES

Aitchinson, I. J. R., and Hey, A. J. (1989). *Gauge theories in particle physics.* Bristol: Adam Hilger.
Anderson, P. (1972). More is different: Broken symmetry and the nature of the hierarchical structure of science. *Science* 177 (4047): 393–396.
Appelquist, T., and Carazzone, J. (1975). Infrared Singularities and Massive Fields, *Physical Review* D11, 2856–2861.
Batterman, R. (2002). *Devil in the details: Asymptotic reasoning in explanation.* Oxford: Oxford University Press.
Brading, K., and Castellani, E. (2003). *Symmetries in physics: Philosophical reflections.* Cambridge: Cambridge University Press.
Cao, T. Y., and Schweber, S. (1993). The conceptual foundations and the philosophical aspects of renormalization theory. *Synthese* 97: 33–108.
Dine, M. (2007). *Supersymmetry and string theory: Beyond the standard model.* Cambridge: Cambridge University Press.
Einstein, A. (1952). On the electrodynamics of moving bodies. In *The principle of relativity: A collection of original memoirs on the special and general theory of relativity*, trans. W. Perrett and G. B. Jeffrey, 37–65. New York: Dover. (Originally published 1905).
Galison, P. (1987). *How experiments end.* Chicago: University of Chicago Press.
Gell-Mann, M., and Low, F. E. (1954). Quantum Electrodynamics at Small Distances. *Physical Review* 95(5): 1300-1312.
Georgi, H. (1993). Effective field theory. *Annual Review of Nuclear and Particle Science* 43: 209–252.
Glashow, S. (1961). Partial symmetries of weak interactions. *Nuclear Physics* 22: 579–588.
Hartmann, S. (2001). Effective field theories, reduction and scientific explanation. *Studies in History and Philosophy of Modern Physics* 32B: 267–304.
Higgs, P. (1964a). Broken symmetries, massless particles and gauge fields. *Physics Letters* 12: 132–133.
———. (1964b). Broken symmetries and masses of gauge bosons. *Physical Review Letters* 13: 508–509.

Kadanoff, L. P. (1966). Scaling laws for Ising models near T_c. *Physics* 2: 263.

Lagrange, J. L. (1788). *Mécanique analytique.* Paris.

Maudlin, T. (1996). "On the unification of physics". *Journal of Philosophy* 93(3): 129–144.

Maxwell, J. C. (1873). *Treatise on electricity and magnetism.* 2 vols. Oxford: Clarendon Press. Reprinted 1954 New York: Dover.

———. (1965). *The scientific papers of James Clerk Maxwell.* 2 vols. Edited W. D. Niven. New York: Dover.

Morrison, M. (1995), "The New Aspect: Symmetries as Meta-Laws" in *Laws of Nature: Essays on the Philosophical, Scientific and Historical Dimension,* ed. F. Weinert. Berlin: De Gruyter, 157-90.

———. (2000). *Unifying scientific theories: Physical concepts and mathematical structures.* Cambridge: Cambridge University Press.

———. (2008) "Fictions, Representation and Reality" Mauricio Suarez (ed.) *Fictions in Science: Philosophical Essays on Modelling and Idealization.* London: Routledge, 110-138.

Pickering, A. (1984). *Constructing quarks.* Chicago: University of Chicago Press.

Quigg, C. (2009). Unanswered questions in the electroweak theory. *Annual Review of Nuclear and Particle Science* 59: 506–555.

Rovelli, C. (2007). *Quantum gravity.* Cambridge: Cambridge University Press.

Schwinger, J. (1957). A theory of fundamental interactions. *Annals of Physics* 2: 407–434.

Smolin, L. (2001). *Three roads to quantum gravity.* New York: Basic Books.

Weinberg, S. (1967). A model of leptons. *Physical Review Letters* 19: 1264–1266.

———. (1983). Why the renormalization group is a good thing. In *Asymptotic realms of physics,* Essays in honor of Francis Low, ed. A. Guth, K. Huang, and R. L. Jaffe, 1-19. Cambridge, MA: MIT Press.

Wilson, K. (1971). The renormalization group (RG) and critical phenomena 1. *Physical Review B* 4: 3174.

———. (1975). The renormalization group: Critical phenomena and the Kondo problem. *Reviews of Modern Physics* 47: 773–839.

Yang, C. N., and Mills, R. J. (1954). Conservation of isotropic spin and isotropic gauge invariance. *Physical Review* 96: 191.

Zinn Justin, J. (1998). Renormalization and renormalization group: From the discovery of UV divergences to the concept of effective field theories. In Proceedings of the NATO ASI on *Quantum Field Theory: Perspective and Prospective,* ed. C. de Witt-Morette and J.-B. Zuber, 375–388. Les Houches, France: Kluwer Academic Publishers, NATO ASI Series C 530.

CHAPTER 12

MEASUREMENT AND CLASSICAL REGIME IN QUANTUM MECHANICS

GUIDO BACCIAGALUPPI

In this essay, I shall focus on two of the main problems raising interpretational issues in quantum mechanics, namely the notorious measurement problem (discussed together with the theory of measurement in section 4) and the equally important but not quite as widely discussed problem of the classical regime (discussed together with decoherence in section 3). The two problems are distinct, but they are both intimately related to some of the issues arising from entanglement and density operators, which are thus briefly reviewed in section 2. A few fundamentals are rehearsed in section 1. The essay will aim to be fairly nontechnical in language, but modern in outlook and covering the chosen topics in more depth than most introductory treatments.

The philosophy and foundations of quantum mechanics offer many more examples of live research issues, and much progress has been achieved recently in such traditional approaches as collapse theories, pilot-wave theories and Everett interpretations, and in the (time-honored but recently revived) area of axiomatic reconstructions of the theory. Recent years have seen fascinating advances also in the study of the other great puzzle raised by entanglement, namely quantum mechanical nonlocality. No in-depth coverage of these other topics will be attempted.

1. A Few Fundamentals

1.1 Phenomenology of Measurements

In classical mechanics, measurements are idealized as testing whether a system lies in a certain subset of its phase space. This can be done in principle without disturbing the system, and the result of the test is in principle fully determined by the state of the system. In quantum mechanics, none of these idealizations can be made. Instead: (i) measurements are idealized as testing whether the system lies in a certain (norm-closed) subspace of its Hilbert space;[1] (ii) a measurement in general disturbs a system: more precisely (and in the ideal case), unless the state of the system is either contained in or orthogonal to the tested subspace, the state is projected onto either the tested subspace or its orthogonal complement (this is known as the "collapse" of the quantum state, or the "projection postulate"); (iii) this process is indeterministic, with a probability given by the squared norm of the projection of the state on the given subspace (the "Born rule" or "statistical algorithm" of quantum mechanics).[2]

For instance, take a spin-1/2 system initially in the state

$$|\varphi\rangle = \alpha|+_x\rangle + \beta|-_x\rangle, \qquad (1)$$

where $|+_x\rangle$ and $|-_x\rangle$ are the states of x-spin up and down. If we test for x-spin-up (for the subspace P_{+_x}), the final state will be either $|+_x\rangle$ with probability $|\alpha|^2$, or $|-_x\rangle$ with probability $|\beta|^2$.

Often, one considers testing together a family of mutually orthogonal subspaces.[3] Such a measurement is usually described as measuring a "self-adjoint (linear) operator" (or "observable")

$$A = \sum a_i P_i, \qquad (2)$$

where the (real) numbers a_i are called the eigenvalues of the operator A and are associated with the outcomes of the measurement. The P_i are the projectors onto the given subspaces.[4] These subspaces are called the eigenspaces of A and are the

---

[1] A subspace is a subset that is closed under linear combinations. We shall assume familiarity with the basic concepts of Hilbert spaces.
[2] Terminology varies, and sometimes the terms "collapse postulate" or "projection postulate" include also the Born rule.
[3] Note once and for all that we are not necessarily assuming that these subspaces are one-dimensional. Alternatively, one can think of testing them in succession, in any order. Explicit application of the collapse postulate and the Born rule will show that one will obtain the same results with the same probabilities and the same final state, irrespectively of the order in which the tests are performed.
[4] Linear operators are mappings on the Hilbert space (or a subspace thereof) that map superpositions into the corresponding superpositions. The adjoint of a linear operator A is a linear operator A^* such that $\langle A^*\psi|\varphi\rangle = \langle \psi|A\varphi\rangle$ for all vectors $|\psi\rangle, |\varphi\rangle$ for which the two expressions are well-defined. An operator is self-adjoint iff $A = A^*$. A projection operator P is a self-adjoint operator such that $P^2 = P$. For ease of exposition, we shall mostly confine ourselves to the case of operators with "discrete spectrum" (the sum in (2) is discrete), or even to finite-dimensional Hilbert spaces.

subspaces of all vectors $|\psi_i\rangle$ (the eigenvectors of the operator) such that

$$P_i|\psi\rangle = |\psi\rangle, \qquad (3)$$

or equivalently

$$A|\psi\rangle = a_i|\psi_i\rangle. \qquad (4)$$

This is the origin of the traditional identification of quantum mechanical observables with (self-adjoint) operators.[5]

The collapse postulate then states that upon measurement of A a state $|\psi\rangle$ will collapse onto $P_i|\psi\rangle$ (suitably renormalized), with probability $p_i = \langle\psi|P_i|\psi\rangle$. The quantity

$$\langle A \rangle_\psi := \langle\psi|A|\psi\rangle = \langle\psi|\sum_i a_i P_i|\psi\rangle = \sum_i p_i a_i \qquad (5)$$

is then the average value or expectation value of the operator A in the state $|\psi\rangle$. Note that unless the state is an eigenstate of the operator measured, there is a statistical spread of results, that is, the dispersion of A in the state $|\psi\rangle$,

$$(\Delta A)_\psi := \sqrt{\langle A^2 \rangle_\psi - \langle A \rangle_\psi^2}, \qquad (6)$$

is nonzero.

The association between self-adjoint operators and families of mutually compatible tests may seem purely conventional from the above description. This is not quite so. Self-adjoint operators play a further role in quantum mechanics, namely as (mathematical) generators of the unitary Schrödinger evolution. Now, think of a Stern–Gerlach spin experiment. A Stern–Gerlach magnet produces (approximately) a magnetic field that is inhomogeneous in just one spatial direction. Classically, what such a magnetic field can do is deflect along this direction a particle with nonzero magnetic moment, the amount of the deflection being proportional to the magnetic moment itself. In quantum mechanics, spin operators of the form

$$S = \frac{\hbar}{2} P_+ - \frac{\hbar}{2} P_- \qquad (7)$$

(with P_+ and P_- the projection operators onto the "up" and "down" spin states in some direction) will appear in the Schrödinger evolution that couples the spin of the particle to its position degrees of freedom, and the deflection experienced by the particle will in fact be proportional to the eigenvalue $+\frac{\hbar}{2}$ or $-\frac{\hbar}{2}$. In this sense, the measurement is indeed sensitive to the eigenvalues of the corresponding spin

---

[5] Note that any self-adjoint operator can be decomposed uniquely into a sum (or more generally an integral) of projectors onto a family of mutually orthogonal subspaces. This is the so-called spectral theorem, which in elementary linear algebra is just the diagonalizability of self-adjoint matrices.

operator, and not just to the projections of the state on the mutually orthogonal eigenspaces.[6] This closer relation between a measurement and a single self-adjoint operator will be lost in the case of the generalized measurements discussed in section 4.4.

1.2 Minimal Interpretation and Standard Interpretation

The above phenomenological rules yield a minimal interpretation of the formalism: some laboratory procedures are taken to be state preparations, and others are taken to be tests. Quantum mechanics yields probabilistic relations between states and outcomes of tests (Born rule). And, depending on their outcome, tests are associated with further (preparatory) transformations of the state (collapse postulate). To be sure, the terms "preparation" and "test" (or "measurement") are phenomenological, but in the cases in which we (or the working physicist) would normally apply them, any fundamental approach to quantum mechanics must allow us to recover the usual predictions of the theory, including in particular the fact that future predictions will depend on the previous outcomes in the way specified by the collapse postulate.

A common alternative interpretation of the formalism (often called the "standard" or "orthodox" or "quantum logical" or "Dirac–von Neumann" interpretation: we shall adopt the first of these terms) takes it that a quantum system has certain properties also independently of measurements, namely properties corresponding to tests that the system passes with probability 1. These properties, which are uniquely fixed by the quantum state, can be further identified either with the state itself (or rather the one-dimensional subspace spanned by the vector state)—as is standardly done in the quantum logic literature, most explicitly by Jauch and Piron (1969)—or with an eigenvalue associated with that vector (hence also the name "eigenstate-eigenvalue link," due to Fine (1973), for this interpretational rule).[7] For instance, an electron in a state of spin up in the x-direction will have a property corresponding to the vector $|+_x\rangle$, or, simply, a value $+\frac{\hbar}{2}$ for spin in the x-direction. According to the standard interpretation, a collapse of the quantum state is thus an actual change in the properties of the quantum system.

Assuming that quantum mechanics is meant to apply to any physical system whatsoever, and that there should not be a fundamental difference in the way it

---

[6] Incidentally, note that whether a (classical or quantum) particle moves up or down in a Stern–Gerlach magnetic field will depend also on whether the inhomogeneous magnetic field is stronger at the north pole or at the south pole. Inverting either the gradient or the polarity of the field will invert the direction of deflection of a particle. (Since rotating the apparatus by 180 degrees corresponds to inverting *both* the gradient and the polarity, it has no net effect on the deflection.) Thus the choice of the words "up" and "down" for labeling the results is rather conventional. (The existence of these two different set-ups for measuring spin in the same direction is crucial in discussing contextuality and nonlocality in pilot-wave theory.)

[7] Note that already according to the minimal interpretation, a quantum system described by a vector in Hilbert space has a set of dispositional properties to elicit specific responses with given probabilities in measurement situations (and these are fixed uniquely by the sure-fire disposition to elicit a certain response with probability 1 in a suitable measurement). The standard interpretation further identifies this set of dispositions with an intrinsic property of the system.

is interpreted across different domains, intuitions from the microscopic and the macroscopic domains of application of the theory will pull in different directions. Applying the minimal interpretation to macroscopic systems would mean that such systems will merely appear to have certain properties if measured (the Moon is not there until we look). In this domain, something like the standard interpretation would seem more natural (at least prima facie). On the other hand, applying the standard interpretation to the microscopic domain would mean that measurements appear to induce a discontinuous change in the properties of a microscopic system, in a way that is not necessarily compatible with the Schrödinger equation. This tension is the origin of the measurement problem of quantum mechanics (which we shall eventually discuss in section 4.6).

Obviously, the minimal interpretation is an instrumentalist interpretation, while the standard interpretation involves an ontological commitment to the quantum state. The former could be seen as a stripped-down version of some historically more accurate reading of the "Copenhagen interpretation". Note also that, while Schrödinger clearly had an ontological commitment to the wave function, it is not clear that it could be phrased in the abstract terms of the standard interpretation. He appears to have rather been interested in the 3-dimensional manifestation of his wave functions, in particular in terms of charge density (see also section 3 below). Something like the standard interpretation instead may have been adopted by both Dirac and von Neumann.

2. Density Operators and Reduced States

2.1 Density Operators

Vectors in Hilbert space, as we have seen, define probability measures over the results of measurements of quantum mechanical observables. Indeed, up to phase factors, the association between unit vectors and such probability measures is one-to-one, since it is clear that if two unit vectors differ by other than an overall phase factor, there will be at least one test (the projection onto the subspace spanned by one of them), for which they will define different probabilities.[8]

To get rid of overall phase factors, we can also identify a quantum state defined by the vector $|\psi\rangle$ with the one-dimensional projection operator onto $|\psi\rangle$, denoted by $|\psi\rangle\langle\psi|$, i.e. the linear mapping that takes any vector state $|\varphi\rangle$ to the state $\langle\psi|\varphi\rangle|\psi\rangle$ (the state $|\psi\rangle$ multiplied by the complex number $\langle\psi|\varphi\rangle$). This can be suggestively

---

[8] Note that thinking of Hilbert-space vectors in terms of their associated probability measures also makes readily intelligible why one considers only unit vectors. Indeed, normalization of the vector ensures that the probabilities are correctly normalized (i.e., add up to 1).

written as

$$|\psi\rangle\langle\psi|: \quad |\varphi\rangle \mapsto |\psi\rangle\langle\psi|\varphi\rangle. \tag{8}$$

This identification is particularly useful if one wishes to generalize the notion of a quantum state further. Indeed, it is clear that the probability measures defined by vectors in Hilbert space will not be the most general such probability measures. The set of these measures ought to be a *convex set*, that is, closed under convex sums.

One can write a convex sum of two states corresponding to projection operators, say onto $|\psi_1\rangle$ and $|\psi_2\rangle$ as the operator

$$\rho = p_1|\psi_1\rangle\langle\psi_1| + p_2|\psi_2\rangle\langle\psi_2| \tag{9}$$

that maps any vector $|\varphi\rangle$ to the superposition

$$p_1\langle\psi_1|\varphi\rangle|\psi_1\rangle + p_2\langle\psi_2|\varphi\rangle|\psi_2\rangle, \tag{10}$$

with $p_1 + p_2 = 1$. We can now write the corresponding probability for the system passing a certain test represented by the projection P as

$$p_\rho(P) = \mathrm{Tr}(\rho P). \tag{11}$$

Here $\mathrm{Tr}(\rho P)$ is the symbol for the so-called trace of the operator ρP, defined for any operator A as

$$\mathrm{Tr}(A) = \sum_i \langle\psi_i|A|\psi_i\rangle, \tag{12}$$

with the $|\psi_i\rangle$ forming a basis of the Hilbert space.[9]

As already mentioned in section 1.1, operators of the form $A|\psi\rangle = a_i|\psi_i\rangle$ can be used to classify simultaneous experimental tests for families of mutually orthogonal subspaces. A system will test positively to only one of these tests, and to this test can be associated an eigenvalue of the corresponding operator. Since $\mathrm{Tr}(\rho P_i)$ is the probability for the outcome i in a test of P_i, the expression

$$\mathrm{Tr}(\rho A) = \sum a_i \mathrm{Tr}(\rho P_i) \tag{13}$$

is equal to the expectation value of the self-adjoint operator A.

The operator ρ is known as a *density operator*, because in the expression (11) it plays a role similar to that of a probability density. Note that the one-dimensional projection operators are the *extremal* elements of the convex set of density operators, those that cannot be decomposed further in terms of convex combinations of other density operators.

---

[9] One can check that the definition of the trace is indeed independent of the basis. In finite dimensions and given a matrix representation of A, the trace is simply the sum of the diagonal elements of the corresponding matrix.

Now, it is a deep theorem due to Gleason (1957) that the states defined by density operators are the *most general* probability measures that can be defined over the possible tests that can be (ideally) performed on a quantum system. A probability measure in Gleason's sense, as one would expect, is a positive, normalized mapping that in the finite-dimensional case is additive and in the infinite-dimensional case σ-additive for families of mutually orthogonal projectors.[10]

Quantum mechanical states in the sense of density operators can be alternatively characterized as the most general (linear) expectation value functionals on the self-adjoint operators. This is actually what von Neumann shows in what has come to be known as his no-hidden-variables theorem (von Neumann 1932, pp. 305–324 of the English translation). More precisely, von Neumann takes a state s to be an assignment of an expectation value to each self-adjoint operator A, subject to a continuity requirement (which is vacuous in finite dimensions), a trivial normalization requirement $s(1) = 1$, a positivity requirement and a *linearity* requirement

$$s(aA + bB) = as(A) + bs(B) \tag{14}$$

for any two observables A and B and real numbers a and b. He then proves that the only such expectation functionals on the self-adjoint operators are of the form $\text{Tr}(\rho A)$, with ρ a density operator. That is, the most general states in this sense are indeed the quantum mechanical states.

Von Neumann took this result as showing that there could be no more precise description of ensembles of quantum mechanical systems (in particular no states with zero dispersion for all observables), and thus as ruling out "hidden variables." Note, however, that von Neumann himself explicitly points out that assumption (14) is natural in the context of commuting observables (where we see it is analogous to Gleason's additivity requirement), but is a very nontrivial assumption in the case of noncommuting ones (pp. 308–309). As noted forcibly by Grete Hermann (1935), this vitiates his conclusion about hidden variables.[11]

A very simple geometrical intuition for the convex structure of density operators in the case of spin-1/2 systems can be gained as follows. Imagine mapping each

---

[10] Normalization means $p(1) = 1$, with 1 the identity operator (i.e., the projection onto the whole of the Hilbert space). The theorem holds for quantum systems with Hilbert space of dimension at least 3 (but see the remark at the end of section 4.4 below).

[11] The relevant section 7 in Hermann's essay has been translated into English by M. Seevinck (see http://mpseevinck.ruhosting.nl/seevinck/trans.pdf). The same point was famously made by Bell (1966), who further pointed out the absurdity of requiring linearity of the hypothetical "dispersion-free states" (which would have to assign an eigenvalue to each observable as a definite value). Bell uses the following example: consider the operators σ_x, σ_y and $\sigma_x + \sigma_y$. For a linear, dispersion-free state λ,

$$\langle \sigma_x + \sigma_y \rangle_\lambda = \langle \sigma_x \rangle_\lambda + \langle \sigma_y \rangle_\lambda. \tag{15}$$

But the left-hand side takes the possible values $\pm\sqrt{2}$, while the right-hand side takes the possible values $-2, 0, +2$, so that (15) cannot be satisfied.

state of spin-up in the direction **r** to the corresponding unit vector in three spatial dimensions,

$$\mathbf{r} = \begin{pmatrix} r_x \\ r_y \\ r_z \end{pmatrix}. \tag{16}$$

This mapping between the vector states of a spin-1/2 system and the unit sphere is a bijection (one-to-one and onto). It turns out that it can be extended to an affine isomorphism (i.e., a map that preserves convex combinations). What this means in particular is that for any two vector states $|\psi\rangle$ and $|\varphi\rangle$, which are mapped onto unit vectors **r** and **s** on the sphere, we can map the density operator

$$\rho = \lambda |\psi\rangle\langle\psi| + (1-\lambda)|\varphi\rangle\langle\varphi| \tag{17}$$

to the point $\lambda \mathbf{r} + (1-\lambda)\mathbf{s}$ in the interior of the unit ball in three dimensions.

This representation is known as the Bloch sphere or the Poincaré sphere. We can use it to establish geometrically many propositions about density operators. Here are a few examples. Density operators can be decomposed nonuniquely as convex combinations of vector states, in fact in infinitely many ways, and as combinations of arbitrarily many vector states (even continuously many). On the other hand, for each density operator, there is generally a unique decomposition as a combination of spin-up and spin-down in a single direction (as a combination of antipodal points on the sphere).[12] The only exception is the state that lies at the center of the ball, which is the equal-weight combination of up and down states in any direction ("maximally mixed" state). We also see that the only states that are extremal (also called *pure* states) in the convex set of density operators are indeed the vector states that map to the unit vectors on the sphere.

2.2 Proper and Improper Mixtures

The nonuniqueness in general of convex decompositions of a density operator is one of their most striking features, and a major difference between probability measures in quantum and classical mechanics.

Also in classical mechanics one can introduce states that are convex combinations of the pure states defined by points in the phase space (which correspond to trivial—or "dispersion-free"—probability distributions). These general states are simply probability measures over phase space. But it is always possible to decompose a classical probability measure uniquely as a convex combination of extremal states (a convex set with this property is known as a "simplex"). Indeed, both mathematically and physically, when we deal with a probabilistic state in classical mechanics, we are always dealing with a *statistical* mixture of nonprobabilistic states, that is,

---

[12] Technically, a density operator (in arbitrary dimensions) is a "compact operator," and for such operators a discrete (if not necessarily finite) decomposition as a sum of mutually orthogonal projectors always exists.

probabilities arise through our ignorance of the actual pure state of the system, and any statistical distributions of measurement results are attributable to this same ignorance. There is no possible ambiguity, since the space of classical probability measures is a simplex.

In quantum mechanics, things are different. Even though formally density operators can always be written as "mixtures" (i.e., as convex combinations of pure states), at the very least their nonunique decomposability will introduce an ambiguity in their interpretation. Assuming that in some case a density operator has arisen through our ignorance of the actual pure state of the system, this is not manifest in the form of the density operator. We might know that the spread of results observed in our tests is partly due to our ignorance of what the quantum state actually is, and partly due to the probabilistic nature of the vector states themselves, but knowledge of how to thus "apportion the blame" is knowledge in excess of that encoded in the density operator itself. It corresponds formally not just to the density operator, but to a particular convex decomposition. Unlike the classical case, this decomposition cannot be uniquely retrieved from the state alone.

This feature of quantum mechanical "mixtures" is essential to the question of how they should be understood, especially in the context of our distinction between the minimal and standard interpretations of the theory. There is, however, an even more essential issue for the question of how to understand density operators. Of course, density operators can arise as genuine statistical mixtures of pure quantum states (for instance a state obtained by randomly mixing systems prepared in different pure states). This is generally referred to as a *proper mixture*. So, for instance, if we know that a measurement of spin-x on an electron has been actually carried out, but we are ignorant of the result, then we should apply the collapse postulate, but average over the results (so-called nonselective measurement). In this case we will have a proper mixture of the states $|+_x\rangle$ and $|-_x\rangle$ due to ignorance (we do not know which state we should actually best use for further predictions).[13]

However, there are other cases in which density operators arise that are not thus related to our ignorance, namely as so-called *reduced states*, states of subsystems of a larger system described by an entangled pure state.

Indeed, the phenomenological rules sketched in section 1.1 (collapse postulate and Born rule) turn out to have surprising consequences when applied to the case of entangled states. Take a singlet state of two spin-1/2 systems

$$\frac{1}{\sqrt{2}}\left(|+\tfrac{1}{x}\rangle \otimes |-\tfrac{2}{x}\rangle - |-\tfrac{1}{x}\rangle \otimes |+\tfrac{2}{x}\rangle\right), \tag{18}$$

and test for $P^1_{+x} \otimes P^2_{+x}$. The test will come out negative with probability 1, and the state will be undisturbed, since it lies in a subspace orthogonal to the tested

---

[13] Of course the collapse postulate is a phenomenological rule, so if one does not believe that collapse is fundamental, there is a sense in which proper mixtures cannot be prepared in this way. Nevertheless, any fundamental approach to quantum mechanics, even if it denies the reality of collapse, will have to explain the appearance of the possibility of preparing proper mixtures, just as it will have to explain the appearance of collapse.

one. Now test for $P^1_{+x} \otimes P^2_{-x}$. The result will be $|+\tfrac{1}{x}\rangle \otimes |-\tfrac{2}{x}\rangle$ or $|-\tfrac{1}{x}\rangle \otimes |+\tfrac{2}{x}\rangle$, each with probability 1/2. In this case, we see that the results of the spin measurements performed on the two electrons are perfectly (anti-)correlated. Correlations, albeit weaker, will be observed quite in general if spin is measured along two different directions on the two subsystems (as can be easily checked explicitly). Entanglement thus introduces what appear to be irreducible correlations between results of measurements (even carried out at a distance), and this for a generic pair of tests. This is the origin of nonlocality in quantum mechanics.

On the other hand, performing a measurement (or any other manipulation) on one of a pair of entangled particles does not affect the probability distributions for results of measurements on the other. This is the so-called *no-signaling* theorem. (That is, while conditionalizing on the outcomes of one measurement in general affects the probabilities for the other, conditionalizing on performing the measurement does not.) It is easy to see this in the example: we have perfect anti-correlations between outcomes on the two sides, but averaging over the outcomes on one side yields back the usual 50–50 distribution on the other side. By explicit calculation, one can check the claim in the general case, that is, for measurements along different spin directions on the two sides.

The no-signaling theorem is crucial to our purposes, since it allows us to generalize the description of quantum systems to subsystems of entangled systems. Indeed, although such subsystems cannot be associated with any vector in their Hilbert space, we can assign them a suitable probability measure for each test we may want to carry out on them, because the no-signaling result guarantees that the probability of such a test is well-defined independently of whether any test (or which one) is carried out on the rest of the system. So, we can define a probability measure for a test on a subsystem by simply taking the marginal of the probability measure associated with the entangled state of the total system when the relevant test is paired with an arbitrary test on the rest of the system. But now, because of Gleason's theorem, we know that such a state must be given by a density operator.

Let us see this in a concrete example. Suppose we wish to define the probability for a measurement of spin-x on one of a pair of spin-1/2 systems in some arbitrary entangled state. We can write the state of the pair as

$$\alpha_{++}|+\tfrac{1}{x}\rangle \otimes |+\tfrac{2}{x}\rangle + \alpha_{+-}|+\tfrac{1}{x}\rangle \otimes |-\tfrac{2}{x}\rangle + \alpha_{-+}|-\tfrac{1}{x}\rangle \otimes |+\tfrac{2}{x}\rangle + \alpha_{--}|-\tfrac{1}{x}\rangle \otimes |-\tfrac{2}{x}\rangle. \tag{19}$$

If we were to measure spin-x on both electrons of the pair, the resulting Born-rule probabilities would be

$$p_{++} = |\alpha_{++}|^2 \qquad p_{+-} = |\alpha_{+-}|^2 \qquad p_{-+} = |\alpha_{-+}|^2 \qquad p_{--} = |\alpha_{--}|^2, \tag{20}$$

and averaging over the results for the second electron, we obtain

$$p_+ = |\alpha_{++}|^2 + |\alpha_{+-}|^2 \qquad \text{and} \qquad p_- = |\alpha_{-+}|^2 + |\alpha_{--}|^2. \tag{21}$$

In this way, one can determine the probabilities for arbitrary tests on the first (and similarly on the second) electron, and so associate with it a state in Gleason's sense (a probability measure for any family of mutually orthogonal projections), even though it is not described by a vector in Hilbert space.

A more compact way of thinking of such a state is in terms of a convex combination of the states that one would obtain through the collapse postulate were one to perform a measurement on the other electron. So, for instance, if one were to perform a measurement of spin-x on the first electron, one would obtain the two (normalized) states:

$$\frac{1}{p_+}\left(\alpha_{++}|+_x^1\rangle \otimes |+_x^2\rangle + \alpha_{+-}|+_x^1\rangle \otimes |-_x^2\rangle\right)$$

$$= \frac{1}{p_+}|+_x^1\rangle \otimes \left(\alpha_{++} \otimes |+_x^2\rangle + \alpha_{+-}|-_x^2\rangle\right) \quad (22)$$

and

$$\frac{1}{p_-}\left(\alpha_{-+}|-_x^1\rangle \otimes |+_x^2\rangle + \alpha_{--}|-_x^1\rangle \otimes |-_x^2\rangle\right)$$

$$= \frac{1}{p_-}|-_x^1\rangle \otimes \left(\alpha_{-+} \otimes |+_x^2\rangle + \alpha_{--}|-_x^2\rangle\right). \quad (23)$$

Writing

$$|\psi_+\rangle = \alpha_{++} \otimes |+_x^2\rangle + \alpha_{+-}|-_x^2\rangle \quad (24)$$

and

$$|\psi_-\rangle = \alpha_{-+} \otimes |+_x^2\rangle + \alpha_{--}|-_x^2\rangle, \quad (25)$$

we see that the state of the second electron would collapse to $|\psi_+\rangle$ or $|\psi_-\rangle$ with the probabilities p_+ and p_- (defined by (21)), respectively.

We can now determine the probabilities for any tests on the second electron by taking the weighted average of the probabilities defined by $|\psi_+\rangle$ and $|\psi_-\rangle$, with the weights p_+ and p_-, respectively. We call this the *reduced state* of the second electron, and write it formally as

$$\rho = p_+|\psi_+\rangle\langle\psi_+| + p_-|\psi_-\rangle\langle\psi_-|. \quad (26)$$

This representation makes it explicit that a reduced state is a density operator. Furthermore, the no-signaling theorem shows us explicitly that the representation (26) cannot be unique. If a different measurement were to be carried out on the first electron, then the states (24) and (25) would have to be different, if the total state is entangled, and the corresponding probabilities would generally also be different. As a simple example, take the singlet state (18). Measuring spin in direction **r** on the first electron will collapse the second electron into a state of spin in the same direction **r**, whatever this might be, due to the rotational symmetry of the state.

Thus, the reduced state of an electron from a pair in the singlet state will have the form

$$\rho = \frac{1}{2}|+_r\rangle\langle+_r| + \frac{1}{2}|-_r\rangle\langle-_r|, \tag{27}$$

(in the case of the singlet the probabilities for the different results will always be equal to 1/2) and will be independent of r.[14]

How are we to interpret density operators arising as reduced states of entangled systems? Certainly not as proper mixtures! Indeed, if a composite quantum system is in a pure entangled state, this state cannot be further decomposed as a weighted average of other quantum states, so cannot be interpreted in terms of ignorance. But then, neither can the states of the subsystems be interpreted in terms of ignorance, despite the fact that the subsystems are necessarily described by density operators. Contrapositively, were the subsystems themselves in pure states (and we ignorant of which pure states they were in), then the composite would be in a mixed state, because it would actually be in a product state (but we ignorant of which product state it was in).

A mixed state arising as the reduced state of a subsystem, where the total system is in a pure state, is generally referred to as an *improper mixture*. The reduced state of an electron from an entangled pair is a paradigm example of an improper mixture, so that a decomposition such as (26) should not be taken as indicating that the system is indeed either in the state $|\psi_+\rangle$ or in the state $|\psi_-\rangle$.

At least from the point of view of the minimal interpretation, there is nothing especially problematic about this. Quantum systems have dispositional properties to elicit certain outcomes under certain test circumstances, irrespectively of whether we seek to explain them further. If we do seek to explain these further, the case of subsystems of entangled systems will turn out to be particularly tricky, but from the point of view of the minimal interpretation it is perfectly natural for subsystems of entangled systems to have such dispositional properties. The only aspect of note is that in the case of such subsystems we explain the distributions over outcomes purely in dispositional terms (just as in the case of systems in pure states), while in other cases, we may have reason to analyze the distributions over outcomes partially in terms of ignorance.

Instead, the existence of entanglement and reduced states has rather disquieting consequences for the standard interpretation. Indeed, if the system is neither in the state $|\psi_+\rangle$ nor in the state $|\psi_-\rangle$ (nor in any other state that might appear in a convex decomposition of the density operator of the system), then the system simply lacks the properties that in the standard interpretation are associated with these states. We can still apply the standard interpretation and associate properties of the system with tests that the system will pass with probability 1. In general, however, these properties will no longer correspond to one-dimensional subspaces of the Hilbert space, but only to higher-dimensional ones (the name "eigenstate-eigenvalue link"

---

[14] Geometrically, this is the maximally mixed state at the center of the Bloch sphere.

becomes a bit of a misnomer in this case).[15] In extreme cases, such as with two entangled electrons (where each electron's spin space is itself only two-dimensional), the individual electrons will have *no nontrivial spin properties*: the only test they pass with probability 1 is the trivial one testing the projection onto the whole of the Hilbert space!

2.3 The Bit Commitment Problem

We shall conclude this section with an example illustrating both the notion of density operators and some of the mystery surrounding entangled states. Because a mixed state characterizes all statistical predictions of quantum mechanics for measurements on a system, it is impossible, by means of measurements performed on that system, to distinguish whether a density operator corresponds to a proper mixture or an improper mixture, or which proper mixture (if any) it corresponds to. This can be illustrated with an example from quantum information theory, the so-called bit commitment problem.

The problem is as follows: Alice commits herself to sending Bob a definite bit of information (0 or 1). She then sends it, and Bob receives it. How can he make sure that what she sends is, indeed, what she had committed herself to? (In whatever scheme we devise we must additionally ensure that Bob does not infer the actual bit of information any sooner than when Alice in fact sends it.) Example: Alice and Bob are on the phone, and they decide to bet on something. First Alice tosses a coin. Then Bob chooses heads or tails. Finally, Alice tells him whether it was heads or tails that had come up. The protocol is fair (or safe) if Bob is sure that Alice does not lie and if Alice is sure that Bob did not know the outcome of her toss before he chose heads or tails.

There is an obvious classical solution to this problem (assuming Bob is not an expert lock-picker): Alice writes the result of her toss on a piece of paper (1 for "heads," 0 for "tails"), puts it into a safe, sends the safe to Bob but keeps the key. After Bob has chosen heads or tails, Alice sends the key as well. The question is now whether there is a quantum solution to this problem that is rigorously fair (and could be implemented by sending just a few electrons instead of keys and safes).

Here is an attempt to realize this. (One could also phrase it in terms of polarization states of photons, in which case Alice could send them along a more or less standard optical fiber as used in telecommunications.) Alice takes some random sequence of zeros and ones, say 11000101011110010..., and prepares a collection of electrons as follows. If the result of her coin toss (her "bit commitment") is "heads," she prepares the electrons one after the other as spin-up (for 1) and spin-down (for 0) in the x-direction; if her result is "tails," she does exactly the same, but with spin states in the y-direction. She then sends the electrons, in sequence, to Bob.

---

[15] Technically, these are all those subspaces that contain the (norm-closed) range of the density operator, i.e. the (norm-closure of) the subspace of all vectors that are images of vectors under the linear mapping defined by the density operator.

At this point, Bob has an ensemble of electrons. We assume he knows that Alice has prepared them either in x-spin states or in y-spin states, but since the sequence is random, there are as many up states as down states on average. Since further

$$\frac{1}{2}\big(|+_x\rangle\langle+_x|+|-_x\rangle\langle-_x|\big) = \frac{1}{2}\big(|+_y\rangle\langle+_y|+|-_y\rangle\langle-_y|\big), \tag{28}$$

the ensemble is characterized by the maximally mixed state, irrespective of whether Alice had got heads or tails. As this characterization gives the maximal information Bob can extract from the ensemble by making measurements, he has no way of telling whether Alice has prepared the electrons in x- or y-spin states.

At a later stage, Alice tells Bob which way she had prepared the electrons, together with the random sequence she used. Now Bob can actually check whether Alice is telling the truth. Indeed, if he makes a sequence of measurements on the electrons, in the order they were sent, then, if the direction of his measurements is the same as the one in which they have been prepared, say x, he will reproduce the random sequence told him by Alice; but if the direction of his measurements is the other one, say y, then he will obtain a completely new random sequence which is unrelated to the first (and which Alice could thus not have anticipated). Thus, the fact that no information on top of that provided by the density operator is available, in particular about how a proper mixture has been prepared, provides a "safe" in which the actual information about the result of Alice's toss is inaccessible without a "key."

But the same fact gives Alice also the possibility of cheating. Instead of sending Bob one of the two above proper mixtures, Alice can send him, say, the right-hand electrons from an ensemble of pairs prepared in the singlet state. Since it is impossible for Bob to tell whether the state he receives, namely again the maximally mixed state, is a proper or improper mixture, he sees no difference between this case and the previous case. But in this situation, Alice can wait for Bob to choose heads or tails and *then* perform a sequence of, respectively, spin-x or spin-y measurements, tell Bob she had done that *before* sending him the electrons (as a way of preparing the corresponding proper mixture by way of collapsing the state), and tell him the sequence of results she obtains (exchanging ones and zeros). Since results of spin measurements on pairs of electrons in the singlet state are perfectly anti-correlated, when Bob measures his electrons, he obtains, indeed, the sequence Alice has told him, not suspecting that Alice has just then collapsed the electrons into the states he measures.

Thus, the *indistinguishability* of proper and improper mixtures prevents Bob from finding out that Alice is cheating, while the objective *difference* between proper and improper mixtures (namely, in terms of the state of the composite system) makes all the difference for Alice in enabling her to cheat in the first place!

This situation turns out to be extremely general. If a protocol for bit commitment is based on the idea that a density operator could be one of two different proper mixtures, which information is then disclosed later on, then there always exists a cheating strategy based on the fact that this same density operator could be an improper mixture. This result is called the no-go theorem for safe bit commitment protocols (Lo and Chau 1997; Mayers 1997).

3. CLASSICAL REGIME AND DECOHERENCE

The problem of the classical regime is the question of whether and how the sweeping success of classical physics (in particular on the macroscopic scale) can be explained in quantum mechanical terms. While in the philosophical literature it is the measurement problem that usually takes pride of place, the problem of the classical regime is equally important in assessing the empirical adequacy of quantum theory and its interpretations. In this section we shall look at this problem as it is generally viewed today, through the eyes of decoherence theory. To fix the ideas, however, we start with a couple of early examples of work related to this problem.

Schrödinger (1926) contributed a seminal paper on the classical regime, in which he showed that Gaussian wave packets for the harmonic oscillator maintain their shape and size (narrow in both position and momentum) and follow the trajectories predicted by Newtonian mechanics. He believed this provided the model for the relation between "micromechanics" and "macromechanics." Another early treatment of "classical" trajectories was given by Heisenberg (1927) in his analysis of α-particle tracks as emerging through repeated collapse of the wave function in a bubble chamber. An alternative treatment of α-particle tracks was given by Mott (1929), who showed that the wave function of the combined system of α-particle and gas was concentrated on configurations in which the gas was ionized along straight lines.

These examples (at least in hindsight) represent rather different approaches to understanding the problem of the classical regime, characterized by different (or potentially different) interpretational approaches. Schrödinger had an ontological commitment to the wave function. At the time, he thought of it as representing (or manifesting itself as) a charge density in 3-dimensional space. Thus, in order to recover a classical regime, it is essential in a Schrödinger-like approach to identify quantum states that are both kinematically and dynamically like classical states, that is, for which the classical quantities such as position and momentum are both approximately well-defined and evolve in an approximately classical manner.[16]

As for Heisenberg, it appears that at the time he did not even believe in the existence of wave functions, but only in the transition probabilities between values

---

[16] It is important to add that, at least by 1927, Schrödinger was well aware that this "charge density" was not simply a classical charge density, but a quantity that would (approximately) behave as a classical charge density only in certain respects and/or in the appropriate regime. See in particular his contribution to the 1927 Solvay conference (Schrödinger 1928), and also the discussions in Bacciagaluppi and Valentini (2009, ch. 4, esp. section 4.4) and Bacciagaluppi (2010, esp. section 4).

of (measured) quantum mechanical observables.[17] For such a view, it is essential that the transition probabilities defined by the Born rule reduce approximately to 0 or 1 for results of measurements performed along classical trajectories. Thus, such an approach (if applied consistently throughout, in particular up to the macroscopic scale) arguably aims at an instrumental recovery of the predictions of classical mechanics.

The standard interpretation and the minimal interpretation of quantum mechanics that we have introduced in section 1.2 can be seen as sanitized versions of the approaches by Schrödinger and Heisenberg, respectively. Instead, Mott's treatment is an early example of a decoherence analysis, in which no collapse need be invoked to destroy the interference between the wave components corresponding to the different trajectories. As I see it, a decoherence-based approach is best viewed as interpretationally neutral, but as providing a very powerful tool for any approach to the problem of the classical regime.

A beautiful example of the importance of the problem of the classical regime for foundational issues is given by Einstein's (1953) contribution to the Edinburgh *Festschrift* for Max Born. Einstein describes a macroscopic ball (of 1 mm diameter), bouncing elastically to and fro inside a box along the direction x. The wave function of the ball is given by a standing wave, which fills the entire box, and has a similarly spread-out distribution in momentum. According to Einstein, Born's statistical interpretation provides an adequate description of the situation for an ensemble of systems (at least according to his own reading of Born). However, an individual ball must have a well-defined macroscopic state, and that is not described by the wave function. To the objection that the Schrödinger equation has other solutions, that are sufficiently localized in position and momentum, Einstein replies that these solutions will spread out in time. Einstein considers also two attempts at interpretation of the wave function alternative to Born's. One is de Broglie–Bohm theory, in which a particle will have a well-defined trajectory guided by the wave function.[18] In Einstein's example, however, the velocity of the ball will be equal to zero, so that, in Einstein's view, de Broglie–Bohm theory fails to provide the correct macroscopic description of the ball as bouncing to and fro inside the box. The other one is Schrödinger's idea of the wave function literally describing a wavelike nature of material particles (which in some form Schrödinger had recently returned to). In this case, however, even the macroscopic ball is a wavelike object filling the whole box, and Einstein's verdict is that also Schrödinger's reading of the wave function fails to do justice to the classical regime. Einstein's own conclusion is that a statistical interpretation of the wave function in the sense he attributes to Born is the appropriate interpretation to give to the theory.

---

[17] See in particular Born and Heisenberg's contribution to the 1927 Solvay conference (Born and Heisenberg 1928), and the discussions in Bacciagaluppi (2008), and Bacciagaluppi and Valentini (2009, esp. chs. 3 and 6).
[18] Recall that Bohm (1952) had recently rediscovered and extended the pilot-wave theory by de Broglie (1928).

3.1 Coherent States and Ehrenfest's Theorem

The first question we shall discuss now is the sense in which one could talk of a quantum state as being approximately classical and behaving approximately classically, in the special case of pure quantum states. The obvious candidates are wave functions with a small spread both in position and in momentum (small compared to some macroscopic scale). This was Schrödinger's initial guess as to the appropriate candidates for the description of classical particles in quantum mechanics.

The Heisenberg uncertainty relations give a lower bound for the product of the spreads in position and in momentum, but for sufficiently massive ("macroscopic") systems, this in itself is a very small limitation. For instance, it is compatible with the uncertainty relations that a system has a spread in position of 10^{-13} cm and a spread in momentum of 10^{-13} g cm/s. If the system has a (macroscopic) mass of 1 g, the latter corresponds to a spread in velocity of 10^{-13} cm/s. If we are merely interested in describing our system on such a macroscopic scale, we can reasonably say that the system has both a well-defined position and a well-defined momentum. Note that such a wave function will typically be nonzero everywhere both in position space and in momentum space. "Small spread" means that the "bulk" of the wave function is localized. Indeed, it is well-known that those wave functions that attain the lower bound given by the uncertainty principle are Gaussian wave packets (i.e., they have the shape of Gaussian bell curves when represented either as functions of position or as functions of momentum), and as such have infinite "tails."[19]

It is obvious, on the other hand (as in Einstein's example), that even for very massive systems there are states with macroscopically large spreads. For instance, take ψ_1 and ψ_2 to be two quantum states of a macroscopic system with very small spreads, but with macroscopically different average values of position and momentum, say x and p in one case, x' and p' in the other. Then the state $1/\sqrt{2}(\psi_1 + \psi_2)$ will have spreads of the order of $|x - x'|$ and $|p - p'|$.

An obvious question is thus whether states with small spreads in both position and momentum remain such under the quantum evolution. With regard to this, as already mentioned, Schrödinger (1926) made the following discovery. Gaussian wave functions for a harmonic oscillator (i.e., with the potential proportional to the square of position, e.g., an ideal spring) keep exactly the same shape and move exactly along the classical trajectories. These states, which are both kinematically and dynamically "classical" are called the *coherent states* of the harmonic oscillator.[20] Schrödinger was led by this result to think that all classical behavior could be explained in these terms by quantum mechanics and, indeed, the result can be generalized in various ways. But we shall see that this hope was misplaced.

---

[19] This need not be a problem in itself, say if one interprets the wave function along Schrödinger's lines as manifesting itself in 3-dimensional space as a charge (or mass) density. It may become a problem if the "tails" are themselves highly structured, as will happen in spontaneous collapse theories in the case of measurements or Schrödinger cats, as this allows for an Everettian-style criticism of the idea that such a wave function represents a single copy of a quasi-classical system (i.e., the tail is itself a "tiny" live or dead cat).

[20] They are very important also in quantum optics, because each mode of the electromagnetic field is a harmonic oscillator.

A simple way of generalizing these results, at least in part, is as follows. For short, write $\langle A \rangle$ to mean $\langle \psi(t)|A|\psi(t)\rangle$, that is, the average value of an operator A in the state $|\psi(t)\rangle$ (see equation (5)). Then, with m the mass of the particle, Q and P the position and momentum operators, and $V(Q)$ the operator representing the potential (which is a function of position), one can derive the two parts of Ehrenfest's theorem:

$$\langle P \rangle = m\frac{d}{dt}\langle Q \rangle \tag{29}$$

(the average momentum is mass times the time derivative of the average position), and

$$\frac{d}{dt}\langle P \rangle = -\langle \nabla V(Q) \rangle \tag{30}$$

(the time derivative of the average momentum is equal to the average force). Thus, the average position and momentum almost obey Newton's second law, with the qualification that the classical value of the force at the average position is replaced by the average value of the force. This holds for all quantum states, but if the state has a small spread in position, the average value of the force is approximately equal to the value of the classical force. Thus, a state with a small spread in position will follow an approximately classical trajectory as long as its spread remains small (at least if the external potential in which it moves is uniform enough on the scale over which the state is spread).[21]

Do position spreads remain small? In the case of a Gaussian, any increase in the position spread leads to a decrease in momentum spread and vice versa. Typically, under the unitary evolution, the spread in position increases.[22] In the simple case of no potentials ("free Gaussian"), if the system has macroscopic mass, the spread of the state will remain small for a very long time. For a system with mass 1 g, starting off in a Gaussian state with position spread 10^{-13} cm, it will take 600 years for the spread in position to increase to 10^{-4} cm, and it will take another 6,000 million million years for it to further increase to twice that size. If potentials are present, the spreading can be enhanced or counteracted, for example, if the wave function is in a potential well it may stay trapped there. In the case of the hydrogen atom, the spreading of wave functions was pointed out to Schrödinger by Lorentz in their well-known correspondence of 1926 (published in Przibram 1963). In particular, Lorentz showed that electrons in the hydrogen atom would be spread out over their entire orbits, even for the case of high-energy orbits.

The examples so far are somewhat mixed, and one might think that Schrödinger's intuition might yet prove sound at least for sufficiently macroscopic systems. That is precisely what Schrödinger replied to Einstein upon receipt of his

---

[21] Thus, while we might want to identify kinematically a classical state with one with small spreads in both position and momentum, it is specifically the smallness of the spread in position that determines whether this state will behave classically also in terms of its dynamics (in the sense of Ehrenfest's theorem).

[22] Note that this is not a strict result, but only a *phenomenological arrow of time*, since the Schrödinger equation is time-symmetric.

draft for the Born *Festschrift*, to which Einstein replied that one could repeat the calculation taking not a 1 mm ball but a dust particle, and get a spread-out state within 24 hours![23]

Regardless of the quantitative details, the discussion so far has presupposed that the state of our system always remain a pure state. That is, the time evolution equation of the system (the Schrödinger equation) may include external potential terms, but it includes no interaction terms. If quantum interactions are included, however, the picture changes dramatically. And that is the case we really need to discuss.

3.2 Entanglement with the Environment

If two quantum systems do not interact, the state of each system will evolve (unitarily) within the Hilbert space that describes that system, and the state of the composite system (if initially a product state!) will always retain its product form, $|\psi(t)\rangle|\varphi(t)\rangle$. If the two systems interact, instead, the state of the composite system will evolve (unitarily) within the product Hilbert space, and in general the state of the composite system will have the entangled form

$$\sum_i \alpha_i(t)|\psi_i\rangle|\varphi_i\rangle. \tag{31}$$

We can use this state in the standard way to make predictions for the composite system, as well as for either subsystem (in particular to calculate the spread in position or in momentum of either subsystem). Indeed, a measurement on a subsystem is just a special kind of measurement on the composite system, so the usual formalism applies. Equivalently, as discussed already in section 2.1, we can make predictions for measurements on a subsystem using the reduced state of that subsystem, which is an improper mixture that takes the form, say,

$$\sum_i |\alpha_i|^2 |\psi_i\rangle\langle\psi_i|. \tag{32}$$

In some cases, it may be more convenient to write this as an integral, for instance over Gaussian wave packets centered at different positions (although as mentioned in section 2.1 a decomposition of the form (32) always exists). If the component states are Gaussians with macroscopically different average positions (and/or momenta), the spreads of the state now can be macroscopically large, just as with pure states that are sums of such Gaussian wave packets.

Recall that improper mixtures are not ignorance-interpretable, so that a macroscopically large spread in position or momentum that arises in this way through

---

[23] See Schrödinger to Einstein, no date (but between 18 and 31 January 1953), AHQP microfilm 37, section 005-012 (draft ms.) and 005-013 (carbon copy), and Einstein to Schrödinger, 31 January 1953, AHQP microfilm 37, section 005-014 (both in German).

quantum interactions cannot be discussed away simply by applying an ignorance interpretation to the mixed state. Such a state appears to be genuinely nonclassical.

Thus, we have to ask whether interactions typically lead to mixed states with large spreads, or whether we can find a regime in which these spreads remain small. Now, however, it is clearly the case that quite common interactions do in fact lead to such apparently nonclassical states.

One class of interactions that lead to mixtures of macroscopically different states are measurement interactions, as with Schrödinger's (1935) own example of the cat. Although the scenario is well-known, here is the description of the thought experiment, as given by Schrödinger himself:

> A cat is penned up in a steel chamber, along with the following diabolical device (which must be secured against direct interference by the cat): in a Geiger counter there is a tiny bit of radioactive substance, so small, that perhaps in the course of one hour one of the atoms decays, but also, with equal probability, perhaps none; if it happens, the counter tube discharges and through a relay releases a hammer which shatters a small flask of hydrocyanic acid. If one has left this entire system to itself for an hour, one would say that the cat still lives if meanwhile no atom has decayed. The first atomic decay would have poisoned it. The ψ-function of the entire system would express this by having in it the living and the dead cat (pardon the expression) mixed or smeared out in equal parts.

Such an example clearly provides a link between the problem of the classical regime and the problem of measurement. We shall postpone discussion of the latter, however, since the two problems are distinct. In the case of the measurement problem, we have a special case of failure or apparent failure of classicality at the kinematical level, but the observed behavior of a measuring apparatus (when coupled to the measured system) is actually far from classical (thus we need not worry about recovering classical dynamics). The special twist of the measurement problem is that preparations and measurements are what is needed to apply quantum mechanics in the first place: if it turned out that these could not be analyzed theoretically, the theory would in some sense be undermining itself.[24]

From the point of view of the classical regime, however, something perhaps even more startling happens, namely that very common and spontaneous interactions of a system with its environment lead to the same kind of states with large spreads.

---

[24] See sections 4.1–4.5 for the theoretical discussion of measurements, and section 4.6 for the measurement problem.

To fix the ideas, think at first of a pair of coupled harmonic oscillators and start them off in the nonentangled state

$$|\text{ground}\rangle|\text{first excited}\rangle. \tag{33}$$

Both classically and quantum mechanically, two coupled oscillators will recurringly exchange energy, that is, evolve to and fro between this state and the nonentangled state |first excited⟩|ground⟩. But quantum mechanically, this will happen through intervening stages of the form

$$a(t)|\text{ground}\rangle|\text{first excited}\rangle + b(t)|\text{first excited}\rangle|\text{ground}\rangle, \tag{34}$$

which are entangled; and the single oscillators will be correspondingly in mixtures of their ground and first excited states. As above, these mixed states arise from quantum interactions and the ensuing entanglement. Thus, they do not allow for an ignorance interpretation.

Now imagine a harmonic oscillator coupled to a thermal bath of harmonic oscillators. It will be taking energy from and giving energy to all of them. If initially the oscillator and the bath are unentangled, the recurrence time for disentangling again could be arbitrarily long (or infinite), and in general the state of the oscillator may be a mixture of any of its energy states. Indeed, if the oscillator is assumed to be in thermal equilibrium with its environment, its quantum mechanical description is a mixture of all its energy states. The spread in position and momentum can be calculated in various ways. One rather suggestive way uses the fact that for high temperatures one can rewrite the equilibrium state as a mixture of all possible coherent states of the oscillator, with weights depending on their energy. One can thus picture the oscillator as roughly spread out over the classical trajectories corresponding to the most probable energies (see, e.g., Donald 1998).

This example illustrates very well the following general idea, which I owe to Matthew Donald. While in classical statistical physics we may think of equilibrium states, at least intuitively, as describing our ignorance of the actual microstate of a system, quantum equilibrium states should generally be thought of as improper mixtures: there is no matter of fact about which pure state describes the system, and any macroscopic spreads resulting from the weighted average in the mixture are genuine nonclassical features.

A macroscopic oscillator will clearly not draw in enough energy to be spread out over macroscopic scales, if the environment is, say, at room temperature (a classical oscillator will not start jittering on a macroscopic scale); but as a matter of fact, one can easily think of systems that are much more sensitive to the influence of a thermal environment, and are thus highly problematic from the point of view of justifying an approximate description in terms of classical physics. One example is a molecule of gas in equilibrium in a box. Every such molecule will be spread out over the entire volume of the box (Donald 1998). Thus, deriving classical statistical physics from quantum mechanics is part of the problem of the classical regime (cf. also Wallace 2001).

Another possible example is that of a Brownian particle suspended in a fluid. Our classical intuition is that it is tossed around by the molecules of the fluid, which influence the particle's motion in a very irregular way. If, however, the interaction of the Brownian particle with its environment is treated quantum mechanically, it would seem that its state will be an improper mixture spread over all its classically possible positions.

Radioactive decay always involves entanglement with the environment, and if the emitted radiation causes a carcinogenic mutation that kills a cat, this is only one component in a complicated entangled state (that includes not only the undecayed component, but also components describing decays at different times). The similarity with Schrödinger's cat is not accidental: this is precisely a Schrödinger cat, but arising spontaneously, without the need for the experimenter's "diabolical device."

A little thought will multiply the examples. "Environmental" interactions such as these are clearly ubiquitous. And if this is what they lead to, then it is clear that, at least in its original form, Schrödinger's approach to the problem of the classical regime is doomed to failure.

3.3 Decoherence and the Classical Regime

Luckily, the same interactions that lead to entanglement with the environment also provide at least a crucial ingredient for the resolution of the problem, because they also induce *decoherence* between the various classical components they superpose.[25]

To explain the concept of decoherence, let us first look at a very elementary example, namely the two-slit experiment. One repeatedly sends electrons or other particles through a screen with two narrow slits, the particles impinge upon a second screen, and we ask for the probability distribution of detections on the surface of the screen. In order to calculate this, one cannot just take the probabilities of passage through the slits, multiply with the probabilities of detection at the screen conditional on passage through either slit, and sum over the contributions of the two slits. There is an additional "interference term" in the correct expression for the probability, and this term depends on both of the wave components passing through one or the other slit.

There are, however, situations in which this interference term (for detections at the screen) is not observed, that is, in which the classical probability formula applies. This happens for instance when we perform a detection at the slits, which at least phenomenologically induces a collapse of the wave function. The disappearance of the interference term, however, can happen also spontaneously, when no detection at the screen is performed, for instance if sufficiently many "stray particles" scatter off the electron between the slits and the screen. In this case, the reason why the interference term is not observed is because the electron has become entangled with the stray particles, and the results of any observation on the electron are determined

---

[25] This subsection is mostly based on my entry for the *Stanford Encyclopedia of Philosophy* (Bacciagaluppi 2003).

by its reduced state alone. As in our discussion of reduced states in section 2.2, the probabilities for results of measurements performed only on the electron are calculated as if the wave function had collapsed to one or the other of its two components.

The intuitive picture is one in which the environment monitors the system of interest by continuously "measuring" some quantity characterized by a set of "preferred" states ("eigenstates of the decohering variable"). Interaction potentials are functions of position, so the preferred states will tend to be related to position, or to be in fact joint approximate eigenstates of position and momentum (since information about the time of flight is also recorded in the environment), that is, coherent states. The localization thus achieved can be on a very short length scale, that is, the characteristic length above which coherence is dispersed (coherence length) can be very short. A speck of dust of radius $a = 10^{-5}$ cm floating in air will have interference suppressed between (position) components with a width of 10^{-13} cm. Even more startlingly, the timescales for this process are minute. The above coherence length is reached after a microsecond of exposure to air.

One can thus argue that generically the states privileged by decoherence at the level of components of the quantum state are localized in position or both position and momentum, and therefore kinematically classical. (One should be wary of overgeneralizations, but this is certainly a feature of many concrete examples that have been investigated.)

What about classical dynamical behavior? Interference is a dynamical process that is distinctively quantum, so, intuitively, lack of interference might be associated with classical-like dynamical behavior. To make the intuition more precise, think of the two components of the wave going through the slits. If there is an interference term in the probability for detection at the screen, it must be the case that both components are indeed contributing to the particle manifesting itself on the screen. But if the interference term is suppressed, one can at least formally imagine that each detection at the screen is a manifestation of only one of the two components of the wave function, either the one that went through the upper slit, or the one that went through the lower slit. Thus, there is a sense in which one can recover at least one dynamical aspect of a classical description, a trajectory of sorts: from the source to either slit (with a certain probability), and from the slit to the screen (also with a certain probability). That is, one recovers a "classical" trajectory at least in the sense that formally the probabilities reduce to those of a classical stochastic process.

In the case of continuous models of decoherence with interactions based on the analogy of approximate joint measurements of position and momentum, one can do even better. In this case, the trajectories at the level of the components (the trajectories of the preferred states) will approximate surprisingly well the corresponding classical (Newtonian) trajectories. Intuitively, one can explain this by noting that the preferred states are the states that themselves tend to get least entangled with the environment, so they will tend to follow the Schrödinger equation more or less undisturbed. But in fact, as we have seen from Ehrenfest's theorem, narrow wave packets follow approximately Newtonian trajectories. Thus, the resulting "histories" will be close to Newtonian ones on the relevant scales.[26]

The most intuitive physical examples for this are the observed trajectories of α-particles in a bubble chamber, which are indeed extremely close to Newtonian ones, except for additional tiny "kinks." Indeed, one should expect slight deviations from Newtonian behavior. These are due both to the tendency of the individual components to spread, and to the detection-like nature of the interaction with the environment, which further enhances the collective spreading of the components (a narrowing in position corresponds to a widening in momentum). These deviations appear as noise, that is, particles being kicked slightly off course.[27] Other examples will include trajectories of a harmonic oscillator in equilibrium with a thermal bath (so the decomposition we mentioned above is not just suggestive, but in fact quite accurate), and trajectories of particles in a gas (which are a precondition for then applying classical derivations of thermodynamics from classical statistical mechanics).

Thus we see that decoherence provides us with tantalizingly classical structure, both kinematical and dynamical, at the level of components of the wave function. It is thus natural to assume that it will play a crucial role in any resolution of the problem of the classical regime. Whether it can play such a role and how, however, will depend on the interpretational approach one adopts toward quantum mechanics.

Let us take first the minimal interpretation of the theory, according to which quantum mechanics is about the results of preparations and measurements, and merely provides a probabilistic link between these two. If one adopts this view, the problem of the classical regime is a question about the results of measurements performed on certain "classical," generally macroscopic systems (or possibly certain elements of their environment). Decoherence tells us that it is indeed possible to isolate a classical regime (at least one[28]) for which appropriate measurements will reveal either actual quasi-classical trajectories, or the appearance thereof.

What we mean by this is the following: (a) if the measurements along a quasi-classical trajectory are actually carried out (as in Heisenberg's treatment of α-particle tracks), then the results obtained will "line up" along the quasi-classical trajectories provided by decoherence; but even if (b) the intermediate measurements are not

---

[26] For a review of more rigorous arguments, see, e.g., Zurek (2003, pp. 28–30). Such arguments can be obtained in particular from the Wigner function formalism, as done, e.g., by Zurek and Paz (1994), who apply these results to derive chaotic trajectories in quantum mechanics.

[27] For a very accessible discussion of α-particle tracks roughly along these lines, see Barbour (1999, ch. 20).

[28] The question of uniqueness of a classical or "quasi-classical" regime has been quite hotly debated especially in the "decoherent histories" literature, and it appears that explicit definitions of quasi-classicality always remain too permissive to identify it uniquely. But maybe uniqueness is not strictly necessary (as nowadays often argued in the context of the Everett interpretation). For these issues, see, e.g., Wallace (2008).

Attempts to enforce uniqueness in other ways appear to overshoot the mark. Indeed, various "modal" interpretations based on the biorthogonal decomposition theorem, the polar decomposition theorem, or the spectral decomposition theorem for density operators, select histories uniquely, but end up agreeing with the results of decoherence only in special cases, failing to ensure classicality in general (Donald 1998; Bacciagaluppi 2000).

carried out, and only the final measurement is, one can consistently assign retrospectively the whole trajectory to the system (sometimes merely guessing what the trajectories "must have been"). This distinction is related to what is known as the movability of the Heisenberg "cut" between observer and observed (which we discuss in the next subsection).

One will thus recover the predictions of classical mechanics, but only instrumentally. Indeed, measurements will need to be regarded as primitive even in classical mechanics, and it will be out of measurements that we will reconstruct "objects" that "look" and "behave" classically (the Moon is not there if we do not look).

What of the standard interpretation? In a sense, the problem of the classical regime is more interesting if one adopts this view, because if one manages to derive a classical regime within quantum mechanics in the standard interpretation, then this would recapture also the standard interpretation of classical mechanics (with measurements being derived notions). However, as we have seen, if one rejects a minimal interpretation of the formalism, but has some fuller ontological commitment to the wave function as describing a quantum system itself, then decoherence appears to exacerbate the problematic nature of the classical regime. Indeed, quantum interactions tend to create improper mixtures at the level of the component systems. Therefore, it would appear that they destroy classicality, as in the case of Schrödinger's cat.

As in Einstein's discussion, if one wishes to keep a fuller ontological commitment to the wave function, or to provide a description of individual quasi-classical systems within quantum mechanics, one will have to replace the standard interpretation (or quantum theory itself) with some alternative approach. The same broad frameworks that are usually proposed as relevant to the measurement problem appear to be useful (but note our concluding qualifications in section 5). Today's Everett interpretations are intimately connected with decoherence. Indeed, the revival of Everettian ideas can be traced back to Zeh's work on decoherence from the early 1970s, and was taken up in the philosophy literature arguably starting in the early 1990s with the work of Saunders, and later of Wallace and others. In these modern versions of Everett, either the "many worlds" or the physical correlate of the "many minds" are explicitly identified with the stable structures created by decoherence at the level of components of the universal wave function.[29] Pilot-wave theories along the lines of de Broglie and Bohm also need to address explicitly the problem of the classical regime, since in general the trajectories defined in the theory are highly nonclassical (see, e.g., Holland 1995, ch. 6, and, for a different point of view, Allori and Zanghì 2009). At least in the nonrelativistic particle theory, it would seem that the components preferred by decoherence correspond nicely with the "full" and "empty" waves of the theory. In Einstein's example, the macroscopic ball or dust particle will be decohered by the environment inside the box, and the system will be

---

[29] For a comprehensive collection of recent papers on the Everett interpretation, in particular covering the more modern developments referred to here, see Saunders et al. (2010).

effectively guided by only one of the components running in opposite directions and that form the standing wave when superposed. However, it is less clear whether similar results are available in the case of quantum field theoretic generalizations (see, e.g., Wallace 2008). Finally, spontaneous collapse theories might also be able to take advantage of the structures provided by decoherence (which generally operates on a much faster timescale than spontaneous collapse), but explicit studies combining collapse models and decoherence are notably lacking from the literature.

3.4 Heisenberg's "Cut"

We conclude this section by expanding on the remarks in the last subsection on the Heisenberg "cut." In particular, we wish to make precise in what sense decoherence is relevant to Heisenberg's discussion of the movable cut between observer and observed (or to von Neumann's discussion of measurement chains), and in what sense it is not. This will provide also a good entry into the topic of measurement, treated in the next section.

Especially in the early 1930s, Heisenberg used to emphasize the importance of the movability of the "cut" between the quantum and the classical domains in ensuring the consistency of quantum mechanics (cf. Heisenberg 1930, ch. 4; 1949, pp. 7–21 and 35–46; and especially 1985). Neither Heisenberg nor any of the other founding fathers of quantum mechanics believed in a rigid boundary between a quantum world, to which one could apply quantum mechanics, and a classical world, to which one could apply only classical mechanics and to which the apparatus and the observer belonged. Any parts of the world (including ostensibly "classical" ones) could be treated quantum mechanically if one so wished.[30] Consistency of the theory had to be ensured, according to Heisenberg, in the sense that applying quantum mechanics to a "classical" part of the world should produce the same predictions as if classical mechanics had been used.

At the risk of pre-empting somewhat our discussion of measurements in the next section, let us consider a so-called "measurement chain," say the example discussed by von Neumann (1932) in his chapter on quantum measurement: we measure the temperature of a (quantum) gas using a (classical) thermometer, or we treat the interaction between the gas and the thermometer quantum mechanically, and we observe (classically) the height of the mercury column, or we treat also the interaction between the thermometer and the human retina quantum mechanically, and our brain registers (classically) the image on the retina, or we treat the whole physical process quantum mechanically, and it is only our consciousness that becomes ("classically") aware of the outcome (and collapses the physical

---

[30] While this point is especially clear in Heisenberg's writings, it is clear that it was espoused also by other main exponents of what is known collectively as the Copenhagen interpretation. For instance, Bohr often applies the uncertainty relations to macroscopic pieces of apparatus in his replies to Einstein's critical thought experiments of the period 1927–1935 (Bohr 1949). And Pauli, commenting to Born on Einstein's views, is adamant that under the appropriate experimental conditions also macroscopic objects would display interference effects (Pauli to Born, 31 March 1954, reprinted in Born 1969).

state). Now, there are two senses in which we can establish the consistency of these descriptions.

First, if the successive (quantum or classical) interactions are such as to correlate perfectly the values of the temperature and the values of the quantities that are meant to record the temperature, then it follows straightforwardly that, irrespective of where the collapse postulate and Born rule are applied, one will obtain the same final results with the same probabilities. This is actually the sense in which both Heisenberg and von Neumann are interested in establishing consistency.[31]

Second, we can consider the influence of decoherence. Note that if no decoherence were present, then performing some other measurement on the thermometer (i.e., a measurement incompatible with that of the length of the mercury column), or somewhere further along the measurement chain, would reveal interference terms between the components of the state corresponding to different measured temperatures. The placing of the "cut" would influence the final statistics, just as the timing of the collapse does in the case of the two-slit experiment (collapse behind the slits or at the screen). Conversely, once decoherence has kicked in at the level of the thermometer, there is no further measurement we would be able to perform in practice on the thermometer that could distinguish whether the thermometer is a classical or a quantum system. And similarly for the retina and for the brain of the observer. It is in this stronger sense that decoherence establishes that the location of the cut between the quantum and the classical domain (where the collapse postulate is applied along the measurement chain) is arbitrary.[32]

4. Theory and Problem of Measurement

We now turn to discussing the theory and problem of measurement. We shall start by discussing measurements in some detail, using Stern–Gerlach measurements as our exemplar, and generalizing the phenomenological notion of a measurement (and of a measurable quantity or "observable") using the tools provided by the so-called POV measures. This theoretical discussion will then provide the basis for discussing the measurement problem in section 4.6.

---

[31] Indeed, von Neumann's aim was simply to show that there always exist unitary evolutions that will produce such perfect correlations, in order to establish consistency in this first sense. Heisenberg's discussion, although technically somewhat defective (see the analysis in Bacciagaluppi and Crull 2009), is along similar lines. Note, however, that Heisenberg is particularly interested in the case of the Heisenberg microscope, where the electron interacts with a microscopic ancilla (the photon), and one considers alternative measurements on the ancilla. For Heisenberg's purposes it is thus important that interference is still present and that decoherence does not kick in until later.

[32] The same point is valid if we are talking about the empirical determination of when and where collapse occurs in spontaneous collapse theories. See the nice discussion in Albert (1992, ch. 5).

4.1 Discretized Position Measurements

Quite surprisingly, there is no perfect analogue for the collapse postulate in the case of measurements of continuous quantities, such as position. Naively one would expect a wave function $\psi(x)$ to collapse to a (renormalized) Dirac δ-function centered at some point q, that is, to $\psi(x)\delta(x-q)$, with a Born probability density given by $|\psi(q)|^2$. The problem with this is the mathematical fact that any function that is nonzero at a single point has square integral 0, and is thus identified with the zero vector. Dirac's famous δ-functions are thus not actually quantum states, so, trivially, one cannot collapse a state to a δ-function.

This was recognized already by von Neumann (1932), who used discretization procedures to describe measurements of position. For instance, we can (ideally) test for whether a wave function lies in the subspace of all square-integrable functions that are nonzero in the interval $[x_1, x_2]$. If the test is positive, the original wave function $\psi(x)$ will collapse to $\chi_{[x_1,x_2]}(x)\psi(x)$ (suitably renormalized),[33] with probability

$$\int \chi_{[x_1,x_2]}(x)\psi(x)\,dx. \tag{35}$$

(This suggestion is so obvious that the problematic nature of the collapse postulate for continuous quantities often goes unnoticed.)

Such discretized measurements of position are all we need to analyze explicitly how a spin measurement works, and in fact to generalize it to include more realistic kinds of spin "measurements."

4.2 Ideal Spin Measurements

Let us first describe the case of an ideal measurement of spin. Note that a system with both spin and position degrees of freedom is described using the tensor product of the Hilbert spaces used to describe a "pure" spin-1/2 system (a two-dimensional complex Hilbert space) and a spinless particle (the Hilbert space of Schrödinger's wave functions), just as if one were composing two separate systems.

Take an electron that we assume to be initially in a state

$$|\varphi\rangle \otimes \psi. \tag{36}$$

What this means is that the electron is described as having a spin state, given by the vector $|\varphi\rangle$ in the two-dimensional spin space of the electron, as well as a wave function, ψ.

Now suppose we want to perform a measurement of spin in some given direction, and that with respect to this spin basis, $|\varphi\rangle = \alpha|+\rangle + \beta|-\rangle$, so that (36)

---

[33] The function $\chi_{[x_1,x_2]}(x)$ is the so-called characteristic function of the interval $[x_1, x_2]$, that is, the function that is 1 on the interval and 0 outside.

equals

$$\alpha|+\rangle \otimes \psi + \beta|-\rangle \otimes \psi. \tag{37}$$

If we pass the electron through an ideal Stern–Gerlach magnet, the evolution of the state will be described by the appropriate Schrödinger equation, which is unitary. Therefore, we can consider separately the deflection of the two components and superpose the results. We obtain

$$\alpha|+\rangle \otimes \psi_+ + \beta|-\rangle \otimes \psi_- \tag{38}$$

(where ψ_+ and ψ_- are suitably deflected versions of ψ). We see that the spin degree of freedom of the electron is now entangled with its position degrees of freedom.

We now detect the electron on a screen, that is, perform a position measurement. Indeed, we perform a discretized measurement of position, because we only need to distinguish whether the electron hits the half of the screen associated with the up or down component of the spin (which, as mentioned in footnote 6, depending on the experimental setup might or might not coincide with the upper or lower half of the screen, respectively). Assuming that ψ_+ and ψ_- do not overlap, the standard collapse postulate and Born rule, applied to the detection of the electron on the screen, will yield either

$$|+\rangle \otimes \psi_+ \quad \text{with probability } |\alpha|^2 \tag{39}$$

or

$$|-\rangle \otimes \psi_- \quad \text{with probability } |\beta|^2, \tag{40}$$

and thus we can actually derive the collapse postulate and Born rule for the spin measurement from the collapse postulate and Born rule for the discretized measurement of position.

4.3 "Unsharp" Spin Measurements

Real experiments, however, will not yield exactly the above result. Let us return to the discussion of our Stern–Gerlach example. It is a fact that wave functions, even if at any one time they can be zero outside of a given interval, will (typically) spread instantaneously out to infinity, so that while we could expect the bulk of ψ_+ and ψ_- to be concentrated each on one half of the screen, they will have "tails" spreading out to the "wrong" half of the screen, say

$$\psi_+ = \psi_{++} + \psi_{+-} \quad \text{and} \quad \psi_- = \psi_{--} + \psi_{-+}. \tag{41}$$

Here, ψ_{++} is meant to represent that part of ψ_+ that is distributed over the half-screen associated with the up result, and ψ_{+-} the part that is distributed over the half-screen associated with the down result; and similarly for ψ_{--} and ψ_{-+}. We shall assume for simplicity that

$$\int |\psi_{+-}|^2 = \int |\psi_{-+}|^2 =: \varepsilon, \tag{42}$$

and thus also

$$\int |\psi_{++}|^2 = \int |\psi_{--}|^2 = 1 - \varepsilon. \tag{43}$$

In this case, applying the collapse postulate and Born rule to detecting the electron on the screen yields either

$$\alpha |+\rangle \otimes \psi_{++} + \beta |-\rangle \otimes \psi_{-+} \tag{44}$$

or

$$\alpha |+\rangle \otimes \psi_{+-} + \beta |-\rangle \otimes \psi_{--} \tag{45}$$

(both to be suitably renormalized), with probabilities

$$|\alpha|^2 (1 - \varepsilon) + |\beta|^2 \varepsilon \tag{46}$$

and

$$|\alpha|^2 \varepsilon + |\beta|^2 (1 - \varepsilon), \tag{47}$$

respectively.

We see that the effect of the measurement on the spin state of the electron is no longer simply given by the standard collapse postulate. Indeed, the two possible states of the electron after the measurement are not even product states, so that the spin of the electron is still entangled with its spatial degrees of freedom, and the spin part of the electron is collapsed to an improper mixture: either

$$|\alpha|^2 (1 - \varepsilon) |+\rangle \langle +| + |\beta|^2 \varepsilon |-\rangle \langle -| \tag{48}$$

or

$$|\alpha|^2 \varepsilon |+\rangle \langle +| + |\beta|^2 (1 - \varepsilon) |-\rangle \langle -| \tag{49}$$

with the same probabilities (46) and (47) (also here, we need to suitably renormalize, since the weights in each decomposition need to sum to 1).

Note that these are the states we obtain if we, indeed, know the result of the spin measurement and can select one of these two final states on the basis of the measurement result (thus performing a so-called selective measurement). If we do not know the outcome of the spin measurement, then future predictions for spin measurements on the electron will use a state that is itself a (proper) mixture of the two corresponding density operators. We can obtain this by simply adding the two (unnormalized) states (48) and (49), to yield

$$|\alpha|^2 |+\rangle \langle +| + |\beta|^2 |-\rangle \langle -|. \tag{50}$$

This is now a case of nonselective measurement, in which we obtain a mixed state that is *partially* ignorance-interpretable. But—as in the case of the bit commitment problem of section 2.3—we need to know the past history of the system (how the state has been prepared), in order to know how and how far to interpret this mixed state in terms of ignorance. If the measurement is ideal, then the correct decomposition of the state is in terms of spin-up or spin-down; if the measurement is correctly modeled by the above, the correct decomposition is given in terms of (48) and (49).

In this example (where we have combined an ideal Stern–Gerlach magnet with a more realistic position state), we see that the probabilities in (50) are independent of the shape of the position state (and indeed, of whether it is "ideal" or "realistic"). One easily realizes that even more general transformations on the spin state of the electron can be induced by a detection on the screen, if one considers that the Stern–Gerlach magnetic field itself is not "ideal" (in order to satisfy the Maxwell equations, it cannot be perfectly homogeneous in the directions perpendicular to that of measurement). Or, indeed, if one considers that one could have chosen, at least in principle, any other unitary coupling between the spin and position degrees of freedom of the electron before proceeding to the detection on the screen.

4.4 General Phenomenology of Measurements

The above examples of various kinds of spin measurements serve as perfect illustrations of the general phenomenology and theory of measurements in quantum mechanics. As discussed in section 1.1, measurements are phenomenologically captured by the collapse postulate, which describes *transformations* on the state of the measured system, and the Born rule, which gives the *probabilities* for such transformations. Both rules need to be generalized. We shall sketch this generalization here, but only in the discrete, finite-dimensional case.

Let us first slightly redescribe the collapse postulate and Born's rule for the case of the standard measurements of section 1.1. Take a family of mutually compatible quantum mechanical tests, corresponding to a family of mutually orthogonal subspaces of the Hilbert space. (If they do not span already the whole Hilbert space, we can add to the family the orthogonal complement of their span, corresponding to the system testing negatively to all the tests.) The corresponding projection operators form a so-called (PV, or projection-valued) *resolution of the identity*:

$$\sum_i P_i = 1, \qquad (51)$$

where 1 is the identity operator on the Hilbert space. In this case, we also talk of a PV-observable.[34]

---

[34] Instead of talking of resolutions of the identity, one can also talk of PV "measures," in the sense that (analogously to a probability measure), one can assign to each "event" (subset I of the indices labeling the

In the case of a selective measurement of this PV-observable, with outcome i, the state of the system collapses as:

$$|\psi\rangle \mapsto P_i|\psi\rangle, \tag{52}$$

or more generally, writing ρ for the initial state to cover also the case when it might not be pure:

$$\rho \mapsto P_i \rho P_i \tag{53}$$

(in both cases with suitable renormalization). The probabilities for the collapses (52) and (53) are, respectively, $\langle\psi|P_i|\psi\rangle$ and

$$\mathrm{Tr}(P_i \rho P_i) = \mathrm{Tr}(\rho P_i). \tag{54}$$

(Note that the trace is cyclic, that is, $\mathrm{Tr}(AB) = \mathrm{Tr}(BA)$ for any two operators, and that $P_i^2 = P_i$ for projections.)

In the case of a nonselective measurement, the collapse takes the form

$$\rho \mapsto \sum_i P_i \rho P_i \tag{55}$$

(already normalized, because of (51)).

In the case of the more realistic spin measurements just discussed, instead of the transformation (52), we have a transformation to an (improper) mixture, which we can write as

$$|\psi\rangle\langle\psi| \mapsto \sqrt{1-\varepsilon}P_+|\psi\rangle\langle\psi|\sqrt{1-\varepsilon}P_+ + \sqrt{\varepsilon}P_-|\psi\rangle\langle\psi|\sqrt{\varepsilon}P_- \tag{56}$$

or

$$|\psi\rangle\langle\psi| \mapsto \sqrt{\varepsilon}P_+|\psi\rangle\langle\psi|\sqrt{\varepsilon}P_+ + \sqrt{1-\varepsilon}P_-|\psi\rangle\langle\psi|\sqrt{1-\varepsilon}P_-, \tag{57}$$

depending on the outcome, with probabilities given by the trace of (56) or of (57), respectively.

In the most general case, the transformation (52) or (53) takes the form of a so-called *operation*, or *completely positive map*:

$$\rho \mapsto \sum_j A_{ij} \rho A_{ij}^*, \tag{58}$$

with suitable operators A_{ij} for each outcome i. (If there is only one A_{ij} corresponding to the outcome i, the operation is said to be "pure," because it maps pure states to pure states.)

The corresponding probabilities are given by

$$\mathrm{Tr}\left(\sum_j A_{ij} \rho A_{ij}^*\right) = \mathrm{Tr}\left(\rho \sum_j A_{ij}^* A_{ij}\right) = \mathrm{Tr}(\rho E_i), \tag{59}$$

results) a corresponding projection $\sum_{i \in I} P_i$. One will talk similarly of POV measures when the requirement that the elements of the resolution of the identity be projections is relaxed.

where we have defined the so-called *effect*[35] E_i as

$$E_i := \sum_j A_{ij}^* A_{ij}. \tag{60}$$

In the nonselective case, the transformation (55) becomes

$$\rho \mapsto \sum_i \sum_j A_{ij} \rho A_{ij}^*. \tag{61}$$

And the normalization of the probabilities (59) yields the analogue of (51), namely

$$\sum_i E_i = 1, \tag{62}$$

that is, the E_i form an effect-valued (or POV, or positive-operator-valued) resolution of the identity (or POV-observable).

Interesting special cases are obtained when, as in the case of the spin measurements above, the operations are combinations of the projections from a PV-observable (so-called "unsharp" measurements of the corresponding PV-observable); or when the effects E_i are in fact mutually orthogonal projections, but the corresponding operations are not simple projections, but have a more general form ("disturbing" measurement of the corresponding PV-observable). Other cases of POV-observables can be interpreted as corresponding to sequences of measurements of PV-observables (or of other POV-observables), yet others as corresponding (in certain specific senses) to joint unsharp measurements of incompatible PV-observables. The closer relation between a measurement and a single self-adjoint operator mentioned in section 1.1 is clearly lost in the general case.

These transformations provide the general form of the phenomenological collapse postulate, and the corresponding probabilities the general form of the phenomenological Born rule. The above discussion of spin measurements, however, illustrates also the general theoretical description of such measurements. Indeed, one can show (this is known as the Naimark dilation theorem) that any completely positive map on the states of the measured system can always be obtained by suitable interaction with some other system, followed by a PV-measurement on this other system (i.e., a transformation of the form (53) or (55), where it should again be emphasized that the P_i need not be one-dimensional projections). This other system can be thought of either as a generally microscopic ancillary system or degree of freedom (e.g. the position of the electron in a Stern–Gerlach measurement, or the photon in the Heisenberg microscope), or as an "indicator variable" or "pointer variable" of a generally macroscopic measuring device. We shall see this in detail (for the case of ideal measurements) in section 4.5.

POV-observables provide a very powerful tool for describing the phenomenology of quantum mechanical measurements. And they have become a completely

---

[35] Technically, an effect is a positive operator with spectrum contained in the interval $[0, 1]$.

standard tool in various branches of quantum physics (e.g., quantum information theory).

For instance, it is well-known that using a measurement of a single PV-observable it is impossible to reconstruct completely the quantum state describing an ensemble of systems. (If the PV-observable is spin in some direction, and if the state is pure, say $\alpha|+\rangle + \beta|-\rangle$ in that basis, the measurement statistics will determine only the absolute values of the coefficients α and β, not their relative phases.) But there are single POV-observables (so-called informationally complete observables) that allow such a reconstruction. A simple example is given by the resolution of the identity

$$\frac{1}{3}P^x_+ + \frac{1}{3}P^x_- + \frac{1}{3}P^y_+ + \frac{1}{3}P^y_- + \frac{1}{3}P^z_+ + \frac{1}{3}P^z_-, \tag{63}$$

which intuitively pools together the information provided by measurements of spin in the three directions x, y, and z (and can be seen as one sense of a joint unsharp measurement of the three PV-observables (Cattaneo et al. 1997)). Indeed, such a POV-measurement can be performed simply by throwing a die and measuring spin in the direction x, y, or z depending on whether the die shows up 1, 2, or 3 (mod 3). Note, however, that from the Naimark dilation theorem we also know that there is a single interaction with an ancilla or measuring device that will implement on the electron any set of six operations needed to measure the POV-observable (63).

Note that Gleason's theorem can be formulated also in terms of probability measures over the outcomes of all possible POV experiments, yielding again the quantum mechanical mixed states as the most general states defining probabilities for the outcomes of such experiments (Busch 2003).[36]

We conclude by mentioning one example of continuous POV measurements, the so-called "unsharp" measurements of position, which provide a continuous alternative to von Neumann's discretization procedures. The wave function $\psi(x)$ collapses upon measurement to a wave function

$$\alpha(x-q)\psi(x) \tag{64}$$

(suitably renormalized), where $\alpha(x-q)$ is not a δ-function but a normalized Gaussian centered at q. The probability density for the collapse is given by

$$\int |\alpha(x-q)\psi(x)|^2 dx. \tag{65}$$

This POV-measurement has the intuitive properties of a measurement of position, in that after the collapse the wave function is concentrated around the point q. Readers may recognize this as the family of operations that take place *spontaneously* in the spontaneous collapse theory by Ghirardi, Rimini, and Weber (1986).[37]

---

[36] In this formulation, the theorem holds in all dimensions.
[37] For further details of POV observables, we refer the reader to standard references, e.g. Busch, Grabowski, and Lahti (1995).

4.5 The Standard Model of Measurement

The model of measurement that underlies standard discussions of the measurement problem (although usually phrased mainly in terms of ideal measurements) is directly related to the theoretical description based on the dilation theorem, as follows.

In the ideal case, one takes a basis of eigenvectors of the observable one wishes to measure on the system of interest, $\{|\varphi_i\rangle\}$, and couples it one-to-one to an orthonormal family of states $\{|\psi_j\rangle\}$ of the apparatus, in the sense that for some "ready state" $|\psi_0\rangle$ of the apparatus,

$$|\varphi_i\rangle \otimes |\psi_0\rangle \mapsto |\varphi_i\rangle \otimes |\psi_i\rangle, \tag{66}$$

for all i. This is indeed possible through a single unitary evolution, because it is simply a requirement that orthonormal states be mapped into orthonormal states.

The outcomes of the measurement are assumed to correspond to orthogonal subspaces (not necessarily one-dimensional), or their corresponding projections P_k, each containing one or more of the $|\psi_i\rangle$ (depending on the "resolution" of the measurement). If each outcome corresponds to a single $|\varphi_i\rangle$, the measurement is said to be maximal.[38]

Under this coupling, an arbitrary state of the system will interact with the apparatus in the ready state as

$$\sum_i \alpha_i |\varphi_i\rangle \otimes |\psi_0\rangle \mapsto \sum_i \alpha_i |\varphi_i\rangle \otimes |\psi_i\rangle. \tag{67}$$

If the measurement is maximal, applying the standard collapse postulate to the pointer observable will now yield any one of the states

$$|\varphi_i\rangle \otimes |\psi_i\rangle \tag{68}$$

with probability $|\alpha_i|^2$. If the measurement is non-maximal, more than one $|\psi_i\rangle$ will lie in the subspace associated with the measurement outcome, and the collapse will yield some superposition of the states (68).[39]

More generally, a measurement will involve an arbitrary coupling between the system of interest and the apparatus, so that the final state of the composite will have the form

$$\sum_i \sum_j \alpha_{ij} |\varphi_i\rangle \otimes |\psi_j\rangle, \tag{69}$$

---

[38] Note that the corresponding subspace in the apparatus Hilbert space need not be one-dimensional: in the case of the spin measurements of section 4.2, we had infinite-dimensional projections onto the upper or lower half of the detection screen. Given that the "apparatus" will usually be a macroscopic system, the idea that a reading should correspond to a large subspace of its state space rather than to a single state is quite appealing. A reading ought to correspond rather to a macroscopic state of the apparatus than to a microscopic state, and a macroscopic state could well be represented by an appropriate subspace P_k.

[39] Note that in this case the system is collapsed to an improper mixture of the states $|\varphi_i\rangle$.

or performing the sum over i first,

$$\sum_j \left(\sum_i \alpha_{ij}|\varphi_i\rangle\right) \otimes |\psi_j\rangle. \tag{70}$$

Defining β_j as the norm of $\sum_i \alpha_{ij}|\varphi_i\rangle$, and $|\varphi_j^*\rangle$ as $\frac{1}{\beta_j}\sum_i \alpha_{ij}|\varphi_i\rangle$, (70) can be rewritten as

$$\sum_j \beta_j |\varphi_j^*\rangle \otimes |\psi_j\rangle. \tag{71}$$

Applying the standard collapse postulate to the pointer observable will now yield any one of the states

$$|\varphi_j^*\rangle \otimes |\psi_j\rangle, \tag{72}$$

with the corresponding probability β_j^2 (or some superposition thereof if more than one $|\psi_j\rangle$ lies in the subspace associated with the measurement outcome).

4.6 The Measurement Problem

In this section, we shall build on the theory of measurement we have just sketched, and describe the measurement problem of quantum mechanics. The phrase "measurement problem" denotes a complex of interrelated questions, but we shall take the following to be its core: whether the practical rules of quantum mechanics (collapse postulate and Born rule) are derivable from first principles, by applying the theory (in particular the dynamics of the theory, as given by the deterministic Schrödinger equation) to a measurement situation, i.e. a situation in which we have an appropriate interaction between a system and a measuring apparatus.

As we have seen from generalizing the example of the Stern–Gerlach measurement, a theoretical description of a measurement can indeed be given by coupling the system of interest (the spin of the electron) to some "indicator" variable (the position of the electron on the upper or lower half of the screen). And we have also seen that the collapse postulate and Born rule for the system (in their most general form) can be obtained by applying the collapse and Born rule (in their more restricted form) to the indicator variable itself.

Suppose that from an appropriate application of the Schrödinger equation, and without explicitly invoking the collapse postulate and the Born rule for the indicator variable, one could derive that in the correct fraction of cases the final state after a measurement is given by (68) or (72), rather than by (67) or (71). Then the collapse and the Born rule would be derivable from first principles, irrespective of whether one adopts the minimal or standard interpretation of the theory.

Indeed, under the standard interpretation, states such as (68) or (72)—or even appropriate superpositions of such states—correspond to the apparatus possessing an intrinsic property indicating a definite outcome (a subspace P_k representing an appropriate macrostate of the apparatus). And under the minimal interpretation, these same states mean that the apparatus has a surefire dispositional property to be seen as indicating a definite outcome if somebody looks.

However, these final states do not obtain if the description just given in section 4.5 is correct. Just like in the bit commitment problem of section 2.3, where there is an objective difference between the case in which Alice sends a statistical mixture of electrons in various spin states, and the case in which she sends electrons from entangled pairs (a difference enabling her to cheat), so in the case of the standard model of measurement there is an objective difference between the case of a statistical mixture of states associated with different measurement results and states in which the macroscopic outcome is entangled with the microscopic value of the measured observable. The theoretical description of measurement in terms of a unitary interaction has merely shifted the place of application of the phenomenological rules. Thus, if it is a correct description of the process of measurement, it does not provide a derivation of the collapse postulate and Born rule, whether the interpretation of choice is the minimal interpretation or the standard interpretation.

Before discussing this further, we should pause to consider whether we have been misled by the power of the dilation theorem and been overly rash in adopting this model of measurement. That is, we should see whether the negative result just described is merely an artifact of the model of measurement adopted.

What could count as a derivation of the collapse postulate and Born rule from the Schrödinger equation? At first, it might seem conceptually mistaken even to pose such a question. How can a probabilistic process ever be derived from a deterministic one? According to von Neumann, the differences run even deeper, in that the former is a thermodynamically irreversible process, while the latter is reversible.[40] On the other hand, in the case of classical thermodynamics and statistical mechanics, we are familiar with the claim that a phenomenologically irreversible theory can be reduced (in some appropriate sense) to an underlying reversible one. The obvious first attempt at answering the problem is thus to point out that it is perfectly possible for a deterministic evolution to underpin statistical results, if we consider statistical states (that is, genuinely statistical states, which are proper mixtures) rather than pure states. That is, while the Schrödinger evolution maps pure states into pure states, it is perfectly possible to obtain a final state that is a proper mixture of different readings of the apparatus, if the initial state is not pure but itself a proper mixture.

The intuition here is that the initial state of system and apparatus should be given not by a product of pure states, but more realistically by a state of the form $|\psi_0\rangle \otimes \rho_0$, where ρ_0 is a suitable statistical state of the apparatus. Indeed, any realistic apparatus will arguably be a macroscopic object, and thus its exact microstate will not be specifiable. Instead, the state of the apparatus will be given only in terms

---

[40] Von Neumann's characterization is based on an extensive thermodynamic analysis, which we shall not enter into, but it should be immediately clear that the transformation (61) is not time-reversible.

of certain macroscopic parameters, analogously to the macrostates of statistical mechanics, and thus for instance to be described as lying in some subspace P_0 or as some appropriate proper mixture of microstates.[41]

For simplicity, let us stick to our Stern–Gerlach example, even though the "indicator" variable (the position of the electron) is not itself macroscopic.[42] Imagine that initially we are not able to prepare the wave function ψ for the spatial degrees of freedom of the electron, but only a proper mixture $\rho = \int_\Lambda \beta_\lambda \rho_\lambda d\lambda$, where each ρ_λ corresponds itself to a wave function ψ_λ (a pure state) localized within the spread of the original ψ. Imagine further that ρ decomposes into

$$\rho = \int_{\Lambda_+} \mu_\lambda \rho_\lambda d\lambda + \int_{\Lambda_-} \mu_\lambda \rho_\lambda d\lambda, \tag{73}$$

where the weights μ_λ are related to the coefficients in the decomposition (37) as

$$\int_{\Lambda_+} \mu_\lambda d\lambda = |\alpha|^2 \quad \text{and} \quad \int_{\Lambda_-} \mu_\lambda = |\beta|^2. \tag{74}$$

Here Λ_+ is the set of indices for which $|\psi\rangle \otimes \rho_\lambda$ evolves to some final state entirely contained in the subspace corresponding to $1 \otimes P_+$ (i.e., the projection onto the half of the screen associated with "up"). And correspondingly for Λ_-.

Since the initial mixture was by assumption ignorance-interpretable, also the final state is a proper mixture such that in a fraction $|\alpha|^2$ of cases, the electron has ended up on the half of the screen associated with "up," and in a fraction $|\beta|^2$ of cases, the electron has ended up on half of the screen associated with "down," as desired.

The problem with this obvious strategy is that it does not work in general. Indeed, if the initial state of the electron is not (37) but, say,

$$|\varphi'\rangle = \gamma|+\rangle + \delta|-\rangle, \tag{75}$$

then, even assuming that each ρ_λ still ends up on one or the other half of the screen, it is not clear why the new sets Λ'_+ and Λ'_- into which the set Λ splits should again satisfy (74) with γ and δ substituted for α and β. Indeed, this constraint is impossible to satisfy for all initial spin states $|\varphi\rangle$ if the temporal evolution is unitary.[43] The apparatus would have to conspire to know in advance what spin state it is meant to measure in order to be in the appropriate statistical mixture of microstates that will produce the desired outcomes with the desired frequencies.[44]

---

[41] We shall assume this for the sake of argument, even though we have suggested in section 3.2 that these mixtures might be improper in the first place.

[42] Recall that on the minimal interpretation the indicator variable is merely a variable that if measured will produce the result that the apparatus reads either up or down. The position of the electron fulfills this role perfectly.

[43] Note also that even if this were possible, such a solution to the measurement problem would run into trouble when trying to reproduce the experimental violations of the Bell inequalities, at least unless the microstates of the apparatuses are correlated before the measurements. For a related point see Bacciagaluppi (2012).

[44] Arguably, the only loophole is if one considers models in which the initial correlations between the microstate of the apparatus and the state of the system (and between the microstates of different apparatuses) are

For the case of ideal measurements, the fact that the measurement problem cannot be solved by invoking an initial mixed state of the apparatus was already discussed by von Neumann (1932, section VI.3), who remarks that this idea was often proposed as a solution to the measurement problem.[45] (It is periodically "rediscovered," which only shows that von Neumann's book is often referred to but still not widely read.) Von Neumann used this to support his claim that one needs indeed two different kinds of processes (namely collapse and unitary evolution) to describe the behavior of quantum systems with and without measurements. This "insolubility theorem," as it is now known, has since been widely generalized, in particular to include also measurements of POV observables.[46]

Thus, we are left with our original conclusion, irrespective of the model of measurement we choose. In a nutshell, measurements understood as quantum interactions magnify quantum superpositions to the macroscopic level (because of the linearity of the dynamics), and thus do not lead to the phenomenologically correct behavior (collapse postulate and Born rule). If we apply the Schrödinger equation to describe the measurement process, then we do not obtain states that would seem to include definite measurement results, but superpositions of such states, nor do we obtain any kind of probabilistic distributions over final states.

Under the minimal interpretation, this might not be very satisfactory, but need not be particularly troubling, since the interpretation only seeks to provide an instrumentalist reading of the theory. And in various variants of the Copenhagen interpretation, one can argue that one should in fact expect measurements—or the quantum-classical interface—to display a peculiar behavior. What is essential, on these interpretations, is consistency between different choices of when and where to apply the collapse postulate and the Born rule. And, as we discussed in section 3.3, this consistency is ensured by decoherence, or in a weaker sense by the existence of perfect correlations along a measurement chain.[47]

As a matter of fact, von Neumann considered the measurement problem to be purely the question of whether such consistency (in the weaker sense) could be ensured, and his treatment of the measurement problem consists precisely in showing that unitary evolutions exist that will produce the perfect correlations (i.e., essentially, in showing that the standard model of measurement is well-defined). Collapse could occur when the thermometer records the temperature of the gas, or when the length of the mercury column is recorded in the photons traveling to the

---

explained in retrocausal terms, and thus are no longer conspiratorial. The models of measurement by Schulman (1997) are probably best understood in this way. For a more general discussion of retrocausal models in quantum mechanics, see Price (1996).

[45] Historical puzzle: who is von Neumann referring to? Someone like Schrödinger who suggested matter should be literally described by wave functions? Or something like the early Copenhagen "disturbance" theory of measurement?

[46] See e.g. Fine (1970), Brown (1986), Busch and Shimony (1996), Bassi and Ghirardi (2000) and Bacciagaluppi (2012).

[47] I would suggest, however, that if one considers the in-principle possibility of performing arbitrary measurements unimpeded by decoherence, then problems of consistency arise again in the context of thought experiments of the type of Wigner's friend.

eye, or in our retina, or along the optic nerve, or when ultimately consciousness is involved. If all of these possibilities are equivalent as far as the final predictions are concerned, von Neumann can maintain that collapse is related to consciousness while in practice applying the collapse postulate at a much earlier (and more practical) stage in the description.

From the point of view of the standard interpretation, however, the problem is serious, because in the state (67) or (71) the system and the apparatus are entangled, and the mixed state resulting for the apparatus is not ignorance-interpretable. Thus, the apparatus does not have a reading under the standard interpretation. As we have also seen, in the case of the standard interpretation decoherence does not help; if anything it makes the situation even worse, because it will produce such macroscopic improper mixtures even independently of observer-engineered measurement situations.

Insofar as the standard interpretation is meant as an approach to quantum mechanics that treats it as a fundamental theory, rather than as a phenomenological theory, we see that the standard interpretation fails. In particular, it fails to support a theoretical analysis of the process of measurement that ensures that measurements have definite outcomes, let alone one that enables one to rederive the phenomenological rules for the description of measurements (the collapse postulate and the Born rule). Everett theories, pilot-wave theories and spontaneous collapse theories are again the options of choice if one wishes to provide a solution to the measurement problem rather than a minimalist or (neo)-Copenhagen dissolution,[48] but a detailed discussion of these goes beyond the scope of this essay.

5. Conclusion

We have discussed two of the main interpretational problems of quantum mechanics, both engendered by the nature of quantum mechanical entanglement, and the consequent failure of the ignorance interpretation of reduced states.

The two problems are equally important if one wishes to give a foundationally adequate reading of quantum mechanics. We are not here in the business of discussing what a foundationally adequate reading of quantum mechanics might be. The minimal interpretation, while not being entirely satisfactory, will arguably count as adequate if one has an instrumentalist picture of science. More sophisticated Copenhagen or neo-Copenhagen views may also find it easier to negotiate these two problems.[49] Yet more robust ontological requirements will prompt one to seek a more successful replacement for the standard interpretation of quantum mechanics,

---

[48] As a prominent example of a neo-Copenhagen view, one can take the "quantum Bayesianism" of Fuchs and co-workers (e.g., Fuchs 2010).
[49] Modulo the caveat about Wigner's friend in footnote 47.

with the help of decoherence and usually along the lines of de Broglie–Bohm, collapse or Everett.

A solution to the problem of the classical regime, however, will not automatically be also a solution to the measurement problem (and vice versa). While pieces of apparatus are generally macroscopic systems or arguably at least kinematically classical systems, their dynamical behavior in probing the quantum world is decidedly nonclassical, and solving the dynamical aspects of the measurement problem is thus distinct from deriving approximately Newtonian trajectories. For instance, modern-day Everettians can use the results of decoherence in an extremely effective way, both toward the solution of the problem of the classical regime and toward that of the measurement problem. But the Everettian solution to the measurement problem relies more heavily on a successful derivation of the Born rule (e.g., the decision-theoretic approach proposed by Deutsch (1999) and Wallace (2007)). Should the critics of the Deutsch–Wallace approach prove correct (e.g., Lewis 2010), Everettians might still be lacking a derivation of the Born rule from first principles, and thus a full solution to the measurement problem.

Conversely, a solution to the measurement problem will not automatically be a solution to the problem of the classical regime. For instance, recent developments in de Broglie–Bohm theory have included various proposals for describing at least large portions of the standard model of particle theory (see, e.g., Colin and Struyve 2007; Dürr et al. 2005; Struyve and Westman 2007; and the review in Struyve 2011). Critics, however, argue that the configuration variables in these models (which are guided by the relevant wave functional) are not necessarily decohering variables (Wallace 2008). It may well be, as argued in particular in the "minimalist" model of Struyve and Westman (2007), that there are choices for configuration variables that will ensure that measurement results (suitably construed) will always be well-defined. But should the critics prove correct, the "measured" classical trajectories will be no more real than those of the minimal interpretation, and pilot-wave theorists would still lack a fully satisfactory solution to the problem of the classical regime.

Much progress has been achieved in recent years on the resolution of these two problems, and generally in the philosophy and foundations of quantum mechanics. One should expect to see more in years to come.

References

Albert, D. (1992). *Quantum Mechanics and Experience*. Cambridge, MA: Harvard University Press.

Allori, V., and Zanghì, N. (2009). On the classical limit of quantum mechanics. *Foundations of Physics* 39(1): 20–32.

Bacciagaluppi, G. (2000). Delocalised properties in the modal interpretation of a continuous model of decoherence. *Foundations of Physics* 30: 1431–1444.

———. (2003). The role of decoherence in quantum mechanics. In E. N. Zalta (ed.), *The Stanford Encyclopedia of Philosophy* (Fall 2008 edition, http://plato.stanford.edu/archives/fall2008/entries/qm-decoherence/).

———. (2008). The statistical interpretation according to born and Heisenberg. In C. Joas, C. Lehner, and J. Renn (eds.), *HQ-1: Conference on the History of Quantum Physics* (Vols. I and II), MPIWG preprint series, Vol. 350 (Berlin: MPIWG), (Vol. II) chapter 14, pp. 269–288 (available at http://www.mpiwg-berlin.mpg.de/en/resources/preprints.html).

———. (2010). Collapse theories as beable theories. *Manuscrito* 33(1) (special issue edited by D. Krause and O. Bueno): 19–54.

———. (2012). Insolubility theorems and EPR argument. *European Journal for Philosophy of Science*, forthcoming, DOI: 10.1007/s13194-012-0057-7.

Bacciagaluppi, G., and Crull, E. (2009). Heisenberg (and Schrödinger, and Pauli) on hidden variables. *Studies in History and Philosophy of Modern Physics* 40: 374–382.

Bacciagaluppi, G., and Valentini, A. (2009). *Quantum Theory at the Crossroads: Reconsidering the 1927 Solvay conference*. Cambridge: Cambridge University Press.

Barbour, J. B. (1999). *The end of time*. London: Weidenfeld and Nicolson.

Bassi, A., and Ghirardi, G.C. (2000). A general argument against the universal validity of the superposition principle. *Physics Letters* A 275: 373–381.

Bell, J. S. (1966). On the problem of hidden variables in quantum mechanics. *Reviews of Modern Physics* 38: 447–452. Reprinted in *Speakable and Unspeakable in Quantum Mechanics*. Cambridge: Cambridge University Press, 1–13.

Bohm, D. (1952). A suggested interpretation of the quantum theory in terms of "hidden" variables, I and II. *Physical Review* 85, 166–179 and 180–193.

Bohr, N. (1949). Discussion with Einstein on epistemological problems in atomic physics. In P. A. Schilpp (ed.), *Albert Einstein: Philosopher-scientist*, The Library of Living Philosophers, Vol. VII (La Salle: Open Court), 201–241.

Born, M., ed. (1969). *Albert Einstein, Hedwig und Max Born: Briefwechsel 1916–1955* (Munich: Nymphenburger Verlagshandlung). Translated by I. Born as *The Born–Einstein letters: Correspondence between Albert Einstein and Max and Hedwig Born from 1916–1955* (London: Macmillan, 1971).

Born, M., and Heisenberg, W. (1928). La mécanique des quanta [Quantenmechanik]. In Lorentz (1928), 143–184. Translated (from the original German) in Bacciagaluppi and Valentini (2009), 408–447.

Broglie, L. de (1928). La nouvelle dynamique des quanta. In Lorentz (1928), 105–141. Translated in Bacciagaluppi and Valentini (2009), 374–407.

Brown, H. R. (1986). The insolubility proof of the quantum measurement problem. *Foundations of Physics* 16: 857–870.

Busch, P. (2003). Quantum states and generalized observables: A simple proof of Gleason's theorem. *Physical Review Letters* 91: 120403/1–4.

Busch, P., and Shimony, A. (1996). Insolubility of the quantum measurement problem for unsharp observables. *Studies in History and Philosophy of Modern Physics* 27B: 397–404.

Busch, P., Grabowski, M., and Lahti, P. (1995). *Operational Quantum Physics*. Berlin and Heidelberg: Springer.

Cattaneo, G., Marsico, T., Nisticò, G. and Bacciagaluppi, G. (1997). A concrete procedure for obtaining sharp reconstructions of unsharp observables in finite-dimensional quantum mechanics. *Foundations of Physics* 27: 1323–1343.

Colin, S., and Struyve, W. (2007). A Dirac sea pilot-wave model for quantum field theory. *Journal of Physics* A **40**: 7309–7342.

Deutsch, D. (1999). Quantum theory of probability and decisions. *Proceedings of the Royal Society of London* A **455**: 3129–3137.

Donald, M. (1998). Discontinuity and continuity of definite properties in the modal interpretation. In D. Dieks and P. E. Vermaas (eds.), *The modal interpretation of quantum mechanics* (Dordrecht: Kluwer), 213–222.

Dürr, D., Goldstein, S., Tumulka, R., and Zanghì, N. (2005). Bell-type quantum field theories. *Journal of Physics* A **38**: R1–R43.

Einstein, A. (1953). Elementare Überlegungen zur Interpretation der Grundlagen der Quanten-Mechanik. In *Scientific Papers Presented to Max Born* (Edinburgh: Oliver and Boyd), 33–40. Translated by R. Deltete in *Nature* **356**(2 April 1992): 393–395.

Fine, A. (1970). Insolubility of the quantum measurement problem. *Physical Review* D **2**: 2783–2787.

———. (1973). Probability and the interpretation of quantum mechanics. *British Journal for the Philosophy of Science* **24**: 1–37.

Fuchs, C. A. (2010). QBism, the perimeter of quantum bayesianism. http://arxiv.org/abs/1003.5209.

Ghirardi, G.C., Rimini, A., and Weber, T. (1986). Unified dynamics for microscopic and macroscopic systems. *Physical Review* D **34**: 470–491.

Gleason, A. M. (1957). Measures on the closed subspaces of a Hilbert space. *Journal of Mathematics and Mechanics* **6**(6): 885–893.

Heisenberg, W. (1927). Über den anschaulichen Inhalt der quantentheoretischen Kinematik und Mechanik. *Zeitschrift für Physik* **43**: 172–198.

——— (1930), *Die physikalischen Prinzipien der Quantentheorie.* Leipzig: Hirzel. Translated by C. Eckart and F. C. Hoyt as *The Physical Principles of the Quantum Theory* Chicago: University of Chicago Press, 1930.

——— (1949). *Wandlungen in den Grundlagen der Naturwissenschaft.* Zürich: Hirzel.

——— (1985). 'Ist eine deterministische Ergänzung der Quantenmechanik möglich? In W. Pauli, *Wissenschaftlicher Briefwechsel mit Bohr, Einstein, Heisenberg u.a.*, Band II: *1930–1939*, ed. by K. v. Meyenn, A. Hermann, and V. F. Weisskopf (Berlin and Heidelberg: Springer), 407–418. Translated by E. Crull and G. Bacciagaluppi (with a brief introduction), http://philsci-archive.pitt.edu/8590/.

Hermann, G. (1935). Die naturphilosophischen Grundlagen der Quantenmechanik. *Abhandlungen der Fries'schen Schule* **6**: 75–152. Section 7, Der Zirkel in *Neumanns Beweis*, translated by M. Seevinck (http://mpseevinck.ruhosting.nl/seevinck/trans.pdf).

Holland, P. R. (1995). *The Quantum Theory of Motion: An Account of the de Broglie–Bohm Causal Interpretation of Quantum Mechanics.* Cambridge: Cambridge University Press.

Jauch, J. M., and Piron, C. (1969). On the structure of quantal proposition systems. *Helvetica Physica Acta* **42**: 842–848.

Lewis, P. J. (2010). 'Probability in Everettian quantum mechanics. *Manuscrito* **33**(1) (special issue edited by D. Krause and O. Bueno): 285–306.

Lo, H.-K., and Chau, H. F. (1997). Is quantum bit commitment really possible? *Physical Review Letters* **78**: 3410–3413.

Lorentz, H. A., ed. (1928). *Electrons et photons: Rapports et Discussions du Cinquième Conseil de Physique Solvay.* Paris: Gauthier-Villars.

Mayers, D. (1997) Unconditionally secure quantum bit commitment is impossible. *Physical Review Letters* **78**: 3414–3417.

Mott, N. F. (1929). The wave mechanics of α-ray tracks. *Proceedings of the Royal Society of London* **A 126** (1930, No. 800 of 2 December 1929): 79–84.

Neumann, J. von (1932), *Mathematische Grundlagen der Quantenmechanik*. Berlin: Springer. Translated by R. T. Beyer as *Mathematical Foundations of Quantum Mechanics* (Princeton: Princeton University Press, 1955).

Price, H. (1996). *Time's Arrow and Archimedes' Point*. New York: Oxford University Press.

Przibram, K., ed. (1963). *Briefe zur Wellenmechanik*. Vienna: Springer. Translated by M. J. Klein as *Letters on Wave Mechanics* (New York: Philosophical Library, 1967).

Saunders, S., Barrett, J., Kent, A., and Wallace, D., eds. (2010). *Many Worlds? Everett, Quantum Theory, and Reality*. Oxford: Oxford University Press.

Schrödinger, E. (1926). Der stetige Übergang von der Mikro- zur Makromechanik. *Naturwissenschaften* **14**: 664–666.

———. (1928). Mécanique des ondes [Wellenmechanik]. In Lorentz (1928), 185–213. Translated (from the original German) in Bacciagaluppi and Valentini (2009), 448–476.

———. (1935). Die gegenwärtige Situation in der Quantenmechanik. *Naturwissenschaften* **23**: 807–812, 823–828, 844–849. Translated in J. A. Wheeler and W. H. Zurek (eds.), *Quantum Theory and Measurement* (Princeton: Princeton University Press, 1983), 152–167.

Schulman, L. S. (1997). *Time's Arrows and Quantum Measurement*. Cambridge: Cambridge University Press.

Struyve, W. (2011). Pilot-wave approaches to quantum field theory. *Journal of Physics: Conference Series* **306**: 012047/1–10.

Struyve, W., and Westman, H. (2007). A minimalist pilot-wave model for quantum electrodynamics. *Proceedings of the Royal Society of London* **A 463**: 3115–3129.

Wallace, D. (2001). Implications of quantum theory in the foundations of statistical mechanics. http://philsci-archive.pitt.edu/410/.

———. (2007). Quantum probability from subjective likelihood: Improving on Deutsch's proof of the probability rule. *Studies in History and Philosophy of Modern Physics* **38**: 311–332.

———. (2008). Philosophy of quantum mechanics. In D. Rickles (ed.), *The Ashgate Companion to Contemporary Philosophy of Physics* (Aldershot: Ashgate), 16–98. Preliminary version available as "The quantum measurement problem: State of play (December 2007)," http://philsci-archive.pitt.edu/3420/.

Zurek, W. H. (2003). Decoherence, einselection, and the quantum origins of the classical. *Reviews of Modern Physics* **75**: 715–775.

Zurek, W. H., and Paz, J.-P. (1994). Decoherence, chaos, and the second law. *Physical Review Letters* **72**: 2508–2511.

CHAPTER 13

THE EVERETT INTERPRETATION

DAVID WALLACE

1. INTRODUCTION

The Everett interpretation of quantum mechanics—better known as the Many-Worlds Theory—has had a rather uneven reception. Mainstream philosophers have scarcely heard of it, save as science fiction. In the philosophy of physics it is well known but has historically been fairly widely rejected.[1] Among physicists (at least, among those concerned with the interpretation of quantum mechanics in the first place), it is taken very seriously indeed, arguably tied for first place in popularity with more traditional operationalist views of quantum mechanics.[2]

For this reason, my task in this chapter is twofold. Primarily I wish to provide a clear introduction to the Everett interpretation in its contemporary form; in addition, though, I aim to give some insight into just *why* it is so popular among physicists. For that reason, I begin in section 2 by briefly reprising the measurement problem in a way that (I hope) gives some insight into just why Everett's idea, if workable, is so attractive. In section 3 I introduce that idea and state "the Everett interpretation"—which, I argue in that section, is really just quantum mechanics itself understood in a conventionally realist fashion. In sections 4–10 I explore the

Thanks to Harvey Brown, Eleanor Knox and Simon Saunders for helpful comments on this chapter.

[1] Arguably this has changed, but only in the last decade or so, and more so in the UK than elsewhere. (Students occasionally ask me how the Everett interpretation is perceived outside Oxford; my flippant answer is that there is a significant divide between philosophers who do and do not take it seriously, and the divide is called the Atlantic Ocean.)
[2] This is largely anecdotal; see, however, Tegmark (1998).

consequences of the Everett interpretation via considerations of its two traditional difficulties: the "preferred basis problem" (sections 4–6) and the "probability problem" (sections 8–10). I conclude (sections 11–12) with a brief introduction to other issues in the Everett interpretation and with some further reading.

Little about the Everett interpretation is uncontroversial, but I deal with the controversy rather unevenly. The concepts of *decoherence theory*, as I note in sections 5–6, have significantly changed the debate about the preferred-basis problem, but these insights have only entered philosophy of physics relatively recently, and relatively little in the way of criticism of a decoherence-based approach to Everettian quantum mechanics has appeared as yet (recent exceptions are Hawthorne (2010), Maudlin (2010), and Kent (2010)). By contrast (perhaps because the salient issues are closer to mainstream topics in metaphysics and philosophy of science) the probability problem has been vigorously discussed in the last decade. As such, my discussion of the former fairly uncritically lays out what I see as the correct approach to the definition of the Everett interpretation and to the preferred basis problem. Readers will, I suspect, be better served by forming their own criticisms, and seeking them elsewhere, than by any imperfect attempt of mine to pre-empt criticisms. My discussion of the latter is (somewhat) less opinionated and attempts to give an introduction to the shape of the debate on probability.

I use little technical machinery, but I assume that the reader has at least encountered quantum theory and the measurement problem, at about the level of Albert (1992) or Penrose, (1989, ch. 6).

2. The Measurement Problem

There are philosophical puzzles, perhaps, in how physical theories other than quantum mechanics represent the world, but it is generally agreed that there is no paradox. States of any such theory—be it Newtonian particle mechanics or classical electrodynamics or general relativity—are mathematical objects of some kind: perhaps functions from one space to another, perhaps N-tuples of points in a three-dimensional space, perhaps single points in a high-dimensional, highly-structured space. And, insofar as the theory is correct in a given situation, these states represent the physical world, in the sense that different mathematically defined states correspond to different ways the world can be.[3] There is space for debate as to the nature of this representation—is it directly a relationship between mathematics and the world, or should it be understood as proceeding via some linguistic description of the mathematics?[4]— but these details cause no problems for the straightforward

---

[3] This simplifies slightly: it is frequently convenient—notably in cases involving symmetry—to define the space of states so that the mathematics-to-physics relation is many-to-one, and it is somewhat controversial in some such cases whether it *is* many-to-one (see, e.g., Saunders (2003) and references therein.) Such concerns are orthogonal to the quantum measurement problem, though.

[4] That is: what is the correct view of scientific theories—semantic or syntactic (cf. Ladyman and Ross (2007, 111–118) and references therein).

(naive, if you like) view that a theory in physics is a description, or a representation, of the world.

Quantum mechanics, it is widely held, cannot be understood this way. To be sure, it has a clean mathematical formalism—most commonly presented as the evolution of a vector in a highly structured, high-dimensional complex vector space. To be sure, *some* of the states in that space seem at least structurally suited to represent ordinary macroscopic systems: physicists, at least, seem relaxed about regarding so-called "wave packet" states of macroscopic systems as representing situations where those systems have conventional, classically describable characteristics. But central to quantum mechanics is the superposition principle, and it tells us (to borrow a famous example) that if χ is a state representing my cat as alive, and ϕ is a state representing my cat as dead, then the "superposition state"

$$\psi = \alpha\chi + \beta\phi \qquad (1)$$

is also a legitimate state of the system (where α and β are complex numbers satisfying $|\alpha|^2 + |\beta|^2 = 1$)—and what can *it* represent? A cat that is alive and dead at the same time? An undead cat, in an indefinite state of aliveness? These don't seem coherent ways for the world to be; they certainly don't seem to be ways we observe the world to be.

Nor does the practice of physics seem to treat such states as representing the state of the physical world. Confronted with a calculation that says that the final state of a system after some process has occurred is some superposition like ψ, a theoretician instead declares that the state of the system after the process cannot be known with certainty, but that it has probability $|\alpha|^2$ of being in the macroscopic physical state corresponding to χ, and probability $|\beta|^2$ of being in the macroscopic physical state corresponding to ϕ. (If he is more cautious, he may claim only that it has probabilities $|\alpha|^2$, $|\beta|^2$ of being *observed* to be in those macroscopic physical states.) That is, the theoretician treats the mathematical state of the system less like the states of classical mechanics, more like those of classical *statistical* mechanics, which represent not the way the world *is* but a probability distribution over possible ways it *might be*.

But quantum mechanics cannot truly be understood that way either. The most straightforward way to understand why is via quantum interference—the α and β coefficients in ψ can be real or imaginary or complex, positive or negative or neither, and can reinforce and cancel out. Ordinary probability doesn't do that. Put more physically: if some particle is fired at a screen containing two slots, and if conditional on it going through slot 1 it's detected by detector A half the time and detector B half the time, and if conditional on it going through slot 2 it's likewise detected by each detector half the time, then we shouldn't need to know how likely it is to go through slot 1 to predict that it will have a 50% chance of being detected by A and a 50% chance of being detected by B. But a particle in an appropriately weighted superposition of going through each slot can be 100% likely to be detected at A, or 0% likely to be, or anything in between.

So it seems that our standard approach to understanding the content of a scientific theory fails in the quantum case. That in turn suggests a dilemma: either that standard approach is wrong or incomplete, and we need to understand quantum mechanics in a quite different way; or that approach is just fine, but quantum mechanics itself is wrong or incomplete, and needs to be modified or augmented. Call these strategies "change the philosophy" and "change the physics," respectively.

Famous examples of the change-the-philosophy strategy are the original Copenhagen interpretation, as espoused by Niels Bohr, and its various more-or-less operationalist descendants. Many physicists are attracted to this strategy: they recognize the virtues of leaving quantum mechanics—a profoundly successful scientific theory—unmodified at the mathematical level. Few philosophers share the attraction: mostly they see the philosophical difficulties of the strategy as prohibitive. In particular, attempts to promote terms like "observer" or "measurement" to some privileged position in the formulation of a scientific theory are widely held to have proved untenable.

Famous examples of the change-the-physics strategy are de Broglie and Bohm's pilot-wave hidden variable theory, and Ghirardi, Rimini, and Weber's dynamical-collapse theory. Many philosophers are attracted to this strategy: they recognize the virtue of holding on to our standard picture of scientific theories as representations of an objective reality. Few physicists share the attraction: mostly they see the scientific difficulties of the strategy as prohibitive. In particular, the task of constructing alternative theories that can reproduce the empirical successes not just of nonrelativistic particle mechanics but of Lorentz-covariant quantum field theory has proved extremely challenging.[5]

But for all that both strategies seem to have profound difficulties, it seems nonetheless that one or the other is unavoidable. For we have seen (haven't we?) that if neither the physics of quantum mechanics nor the standard philosophical approach to a scientific theory is to be modified, we do not end up with a theory that makes any sense, far less one that makes correct empirical predictions.

3. Everett's Insight

It was Hugh Everett's great insight to recognize that the apparent dilemma is false—that, *contra* the arguments of section 2, we can after all interpret the bare quantum

---

[5] In the case of dynamical-collapse theories, Tumulka (2006) has produced a relativistically covariant theory for *non-interacting* particles, but to my knowledge there is no dynamical-collapse theory empirically equivalent to any relativistic theory with interactions. There has been rather more progress in the case of hidden variable theories (perhaps unsurprisingly, as these supplement but do not modify the already-known unitary dynamics); for three different recent approaches, see Dürr et al. (2004, 2005) (hidden variables are particle positions), Struyve and Westman (2006) (hidden variables are bosonic field strengths), and Colin (2003) and Colin and Struyve (2007) (hidden variables are local fermion numbers). As far as I know, no such approach has yet been demonstrated to be empirically equivalent to the Standard Model to the satisfaction of the wider physics community.

formalism in a straightforwardly realist way, without either changing our general conception of science or modifying quantum mechanics.

How is this possible? Haven't we just seen that the linearity of quantum mechanics commits us to macroscopic objects being in superpositions, in indefinite states? Actually, no. We have indeed seen that states like ψ—a superposition of states representing macroscopically different objects—are generic in unitary quantum mechanics, but it is actually a non sequitur to go from this to the claim that macroscopic objects are in indefinite states.

An analogy may help here. In electromagnetism, a certain configuration of the field—say, $F_1(x, t)$ (here F is the electromagnetic 2-form) might represent a pulse of ultraviolet light zipping between Earth and the Moon. Another configuration, say $F_2(x, t)$, might represent a different pulse of ultraviolet light zipping between Venus and Mars. What then of the state of affairs represented by

$$F(x, t) = 0.5 F_1(x, t) + 0.5 F_2(x, t)? \qquad (2)$$

What weird sort of thing is this? Must it not represent a pulse of ultraviolet light that is *in a superposition* of traveling between Earth and Moon, and of traveling between Mars and Venus? How can a single pulse of ultraviolet light be in two places at once? Doesn't the existence of superpositions of macroscopically distinct light pulses mean that any attempt to give a realist interpretation of classical electromagnetism is doomed?

Of course, this is nonsense. There is a perfectly prosaic description of F: it does not describe a single ultraviolet pulse in a weird superposition, it just describes two pulses, in different places. And *this*, in a nutshell, is what the Everett interpretation claims about macroscopic quantum superpositions: they are just states of the world in which more than one macroscopically definite thing is happening at once. Macroscopic superpositions do not describe indefiniteness, they describe multiplicity.

The standard terminology of quantum mechanics can be unhelpful here. It is often tempting to say of a given macroscopic system—like a cat, say—that its possible states are all the states in some "cat Hilbert space," $\mathcal{H}^{cat}$. Some states in $\mathcal{H}^{cat}$ are "macroscopically definite" (states where the cat is alive or dead, say); most are "macroscopically indefinite." From this perspective, it is a very small step to the incoherence of unitary quantum mechanics: quantum mechanics predicts that cats often end up in macroscopically indefinite states; even if it makes sense to imagine a cat in a macroscopically indefinite state, we have certainly never seen one in such a state; so quantum mechanics (taken literally) makes claims about the world that are contradicted by observation.

From an Everettian perspective this is a badly misguided way of thinking about quantum mechanics. This $\mathcal{H}^{cat}$ is presumably (at least in the nonrelativistic approximation) some sort of tensor product of the Hilbert spaces of the electrons and atomic nuclei that make up the cat. Some states in this box certainly look like they can represent live cats, or dead cats. Others look like smallish dogs. Others look like the Mona Lisa. There is an awful lot that can be made out of the atomic constituents of a cat, and all such things can be represented by states in $\mathcal{H}^{cat}$, and so calling it a "cat Hilbert space" is very misleading.

But if so, it is equally misleading to describe a macroscopically indefinite state of $\mathcal{H}^{cat}$ as representing (say) "a cat in a superposed state of being alive and being dead." It is far more accurate to say that such a state is a superposition of a live cat and a dead cat.

One might still be tempted to object: very well, but we don't observe the universe as being in superpositions of containing live cats and containing dead cats, any more than we observe cats as being in superpositions of alive and dead. But it is not at all clear that we *don't* observe the universe in such superpositions. After all, cats are the sort of perfectly ordinary objects that we seem to see around us all the time—a theory that claims that *they* are normally in macroscopically indefinite states seems to make a nonsense of our everyday lives. But the *universe* is a very big place, as physics has continually reminded us, and we inhabit only a very small part of it, and it will not do to claim that it is just "obvious" that it is not in a superposition.

This becomes clearer when we consider what actually happens, dynamically, to $\mathcal{H}^{cat}$, to its surroundings, and to those observing it, when it is prepared in a superposition of a live-cat and a dead-cat state. In outline, the answer is that the system's surroundings will rapidly become entangled with it, so that we do not just have a superposition of live and dead cat, but a superposition of extended quasi-classical regions—"worlds," if you like—some of which contain live cats and some of which contain dead cats. If the correct way to understand such superpositions is as some sort of multiplicity, then our failure to observe that multiplicity is explained quite simply by the fact that we live in one of the "worlds" and the other ones don't interact with ours strongly enough for us to detect them.

This, in short, is the Everett interpretation. It consists of two very different parts: a contingent physical postulate, that the state of the Universe is faithfully represented by a unitarily evolving quantum state; and an a priori claim about that quantum state, that if it is interpreted realistically it must be understood as describing a multiplicity of approximately classical, approximately noninteracting regions that look very much like the "classical world."

And this is *all* that the Everett interpretation consists of. There are no additional physical *postulates* introduced to describe the division into "worlds," there is just unitary quantum mechanics. For this reason, it makes sense to talk about *the* Everett interpretation, as it does not to talk about *the* hidden-variables interpretation or *the* dynamical-collapse interpretation. The "Everett interpretation of quantum mechanics" is just quantum mechanics *itself*, "interpreted" the same way we have always interpreted scientific theories in the past: as modeling the world. Someone might be right or wrong *about* the Everett interpretation—they might be right or wrong about whether it succeeds in explaining the experimental results of quantum mechanics, or in describing our world of macroscopically definite objects, or even in making sense—but there cannot be multiple logically possible Everett interpretations any more than there are multiple logically possible interpretations of molecular biology or classical electrodynamics.[6]

This in turn makes the study of the Everett interpretation a rather tightly constrained activity (a rare and welcome sight in philosophy!). For it is not possible to solve problems with the Everett interpretation by changing the interpretative rules or changing the physics: if there are problems with solving the measurement problem Everett-style, they can be addressed only by hard study—mathematical and conceptual—of the quantum theory we have.

Two main problems of this kind have been identified:

1. the *preferred basis problem* (which might better be called the *problem of branching*)—what actually justifies our interpretation of quantum superpositions in terms of multiplicity?
2. The *probability problem*—how is the Everett interpretation, which treats the Schrödinger equation as deterministic, to be reconciled with the probabilistic nature of quantum theory?

My main task in the remainder of this chapter is to flesh out these problems and the contemporary Everettian response to each.

4. The Preferred Basis Problem

If the preferred basis problem is a question ("how can quantum superpositions be understood as multiplicities?"), then there is a traditional answer, more or less explicit in much criticism of the Everett interpretation (Barrett (1999), Kent (1990), Butterfield (1996)): they cannot. That is: it is no good just *stating* that a state like (1) describes multiple worlds: the formalism must be explicitly *modified* to incorporate them. Adrian Kent put it very clearly in an influential criticism of Everett-type interpretations:

> one can perhaps intuitively view the corresponding components [of the wave function] as describing a pair of independent worlds. But this intuitive interpretation goes beyond what the axioms justify: the axioms say nothing about the existence of multiple physical worlds corresponding to wave function components. (Kent, 1990)

This position dominated discussion of the Everett interpretation in the 1980s and early 1990s: even advocates like Deutsch (1985) accepted the criticism and rose to the challenge of providing such a modification.

---

[6] Perhaps in *some* sense there are multiple interpretations of classical electromagnetism: perhaps realists could agree that the electromagnetic field is physically real but might disagree about its nature. Some might think that it was a property of spacetime points; others might regard it as an entity in its own right. I am deeply skeptical as to whether this really expresses a distinction, but in any case, I take it *this* is not the problem that we have in mind when we talk about the measurement problem.

Modificatory strategies can be divided into two categories. *Many-exact-worlds theories* augment the quantum formalism by adding an ensemble of "worlds" to the state vector. The "worlds" are each represented by an element in some particular choice of "world basis" $|\psi_i(t)\rangle$ at each time t: the proportion of worlds in state $|\psi_i(t)\rangle$ at time t is $|\langle\Psi(t)|\psi_i(t)\rangle|^2$, where $|\Psi(t)\rangle$ is the (unitarily evolving) universal state. Our own world is just one element of this ensemble. Examples of many-exact-worlds theories are given by the early Deutsch (1985, 1986), who tried to use the tensor-product structure of Hilbert space to define the world basis,[7] and Barbour (1994, 1999) who chooses the position basis.

In *many-minds theories*, by contrast, the multiplicity is to be understood as illusory. A state like (1) really is indefinite, and when an observer looks at the cat and thus enters an entangled state like

$$\alpha|\text{Live cat}\rangle \otimes |\text{Observer sees live cat}\rangle + \beta|\text{Dead cat}\rangle \otimes |\text{Observer sees dead cat}\rangle$$

(3)

then the observer too has an indefinite state. However: to each physical observer is associated not one mental state, but an ensemble of them: each mental state has a definite experience, and the proportion of mental states where the observer sees the cat alive is $|\alpha|^2$. Effectively, this means that in place of a global "world-defining basis" (as in the many-exact-worlds theories) we have a "consciousness basis" for each observer.[8] When an observer's state is an element of the consciousness basis, all the minds associated with that observer have the same experience and so we might as well say that the observer is having that experience. But in all realistic situations the observer will be in some superposition of consciousness-basis states, and the ensemble of minds associated with that observer will be having a wide variety of distinct experiences. Examples of many-minds theories are Albert and Loewer (1988), Lockwood (1989, 1996), Page (1996), and Donald (1990, 1992, 2002). It can be helpful to see the many-exact-worlds and many-minds approaches as embodying two horns of a dilemma: either the many worlds really exist at a fundamental level (in which case they had better be included in the formalism), or they do not (in which case they need to be explained away as somehow illusory).

Both approaches have largely fallen from favor. Partly, this is on internal, philosophical grounds. Many-minds theories, at least, are explicitly committed to a rather unfashionable anti-functionalism—probably even some kind of dualism—about the philosophy of mind, with the relation between mental and physical states being postulated to fit the interests of quantum mechanics rather than being deduced at the level of neuroscience or psychology. If it is just a *fundamental law* that consciousness is associated with some given basis, clearly there is no hope of a functional *explanation* of how consciousness emerges from basic physics (and hence much,

---

[7] A move criticized on technical grounds by Foster and Brown (1988).
[8] Given that an "observer" is represented in the quantum theory by some Hilbert space many of whose states are not conscious at all, and that conversely almost any sufficiently large agglomeration of matter can be formed into a human being, it would be more accurate to say that we have a consciousness basis for all *systems*, but one with many elements that correspond to no conscious experience at all.

perhaps all, of modern AI, cognitive science, and neuroscience is a waste of time[9]). And on closer inspection, many-exact-worlds theories seem to be committed to something as strong or stronger: if "worlds" are to be the kind of thing we see around us, the kind of thing that ordinary macroscopic objects inhabit, then the relation between those ordinary macroscopic objects and the world will likewise have to be postulated rather than derived.

But more important, both approaches undermine the basic motivation for the Everett interpretation. For suppose that a wholly satisfactory many-exact-worlds or many-minds theory were to be developed, specifying an exact "preferred basis" of worlds or minds. Nothing would then stop us from taking that theory, discarding all but one of the worlds/minds[10] and obtaining an equally empirically effective theory without any of the ontological excess that makes Everett-type interpretations so unappealing. Put another way: an Everett-type theory developed along the lines that I have sketched would really just be a hidden-variables theory with the additional assumption that a continuum of many noninteracting sets of hidden variables exists, each defining a different classical world. (This point is made with some clarity by Bell (1981b) in his classic attack on the Everett interpretation.)

At time of writing, almost no advocate of "the many-worlds Interpretation" actually advocates anything like the many-exact-worlds approach[11] (Deutsch, for instance, clearly abandoned it some years ago) and many-minds strategies that elevate consciousness to a preferred role continue to find favor mostly in the small group of philosophers of physics strongly committed for independent reasons to a nonfunctionalist philosophy of mind. Advocates of the Everett interpretation among physicists (almost exclusively) and philosophers (for the most part) have returned to Everett's original conception of the Everett interpretation as a pure interpretation: something that emerges simply from a realist attitude to the unitarily evolving quantum state.

How is this possible? The crucial step occurred in physics: it was the development of *decoherence theory*.

5. THE ROLE OF DECOHERENCE

A detailed review of decoherence theory lies beyond the scope of this chapter, but in essence, decoherence theory explores the dynamics of systems that are coupled to some environment with a high number of degrees of freedom. In the most common models of decoherence, the "system" is something like a massive particle and the "environment" is an external environment like a gas or a heat bath, but it is equally

---

[9] In fact many adherents of many-minds theories (e.g., Lockwood and Donald) embrace this conclusion, having been led to reject functionalism on independent grounds.
[10] It would actually be a case of discarding all but one *set* of minds—one for each observer.
[11] Barbour (1999) might be an exception; so might Allori et al. (2009), though it is unclear if Allori et al. are actually advocating the interpretation rather than using it to illustrate broader metaphysical themes.

valid to take the "system" to be the macroscopic degrees of freedom of some large system and to take the "environment" to be the residual degrees of freedom of that same system. For instance, the large system might be a solid body, in which case the "system" degrees of freedom would be its centre-of-mass position and its orientation and its "environment" degrees of freedom would be all the residual degrees of freedom of its constituents; or it might be a fluid, in which case the "system" degrees of freedom might be the fluid density and velocity averaged over regions a few microns across.

Whatever the system–environment split, "decoherence" refers to the tendency of states of the system to become entangled with states of the environment. Typically no system state is entirely immune to such entanglement, but certain states—normally the wave-packet states, which have fairly definite positions and momentums—get entangled fairly slowly. Superpositions of such states, on the other hand, get entangled with the environment extremely quickly, for straightforward physical reasons: if, say, some stray photon in the environment is on a path that will take it through point q, then its future evolution will be very different according to whether or not there is a wave-packet localized at q. So if the system is in a superposition of being localized at q and being localized somewhere else, pretty soon system-plus-environment will be in a superposition of (system localized at q, photon scattered) and (system localized somewhere else, photon not scattered). Intuitively, we can think of this as the system being constantly measured by the environment, though this "measurement" is just one more unitary quantum-mechanical process.

Mathematically, this looks something like the following. If $|q,p\rangle$ represents a wave-packet state of our macroscopic system with position q and momentum p, then an arbitrary nonentangled state of the system will have state

$$\int dq dp\, \alpha(q,p)|q,p\rangle, \tag{4}$$

so that if the environment state is initially $|env_0\rangle$, the combined system-plus-environment state is

$$\left(\int dq dp\, \alpha(q,p)|q,p\rangle\right) \otimes |env_0\rangle. \tag{5}$$

But very rapidly (very rapidly, that is, compared to the typical timescales on which the system evolves), this state evolves into something like

$$\int dq dp\, \alpha(q,p)|q,p\rangle \otimes |env(q,p)\rangle, \tag{6}$$

where $\langle env(q,p)|env(q',p')\rangle \simeq 0$ unless $q \simeq q'$ and $p \simeq p'$. In this way, the environment records the state of the system, and it does so quickly, repeatedly, and effectively irreversibly (more accurately, it is reversible only in the sense that other macroscopic-scale processes, like the melting of ice, are reversible).

Why does this matter? Because as long as the environment is constantly recording the state of the system in the wave-packet basis, interference experiments cannot be performed on the system: any attempt to create a superposition of wave-packet states will rapidly be undone by decoherence. The overall quantum system (that is, the system-plus-environment) remains in a superposition, but this has no dynamical significance (and, in particular, cannot be empirically detected) without carrying out in-practice-impossible experiments on an indefinitely large region of the universe in the system's vicinity.

And *this* matters, in turn, because it is interference phenomena that allow the different structures represented by a quantum state in a superposition to interact with one another, so as to influence each other and even to cancel out with one another. If interference is suppressed with respect to a given basis, then evolving entangled superpositions of elements of that basis can be regarded as instantiating multiple independently evolving, independently existing structures. As such, if macroscopic superpositions are decohered—as they inevitably will be—then such superpositions really can be taken to represent multiple, dynamically isolated, macroscopic states of affairs.

For this reason, by the mid-1990s decoherence was widely held in the physics community to have solved the preferred basis problem, by providing a definition of Everett's worlds. (It was just as widely held to have solved the measurement problem entirely, independent of the Everett interpretation; since decoherence does not actually remove macroscopic superpositions, though, it was never clear how decoherence alone was supposed to help.) Philosophers of physics were rather more skeptical (Simon Saunders was a notable exception; cf. Saunders Saunders (1993), 1995), essentially because decoherence seems to fall foul of Kent's criticism: however suggestive it might be, it does not seem to succeed in defining an "explicit, precise rule" (Kent 1990) for what the worlds actually are. For decoherence is by its nature an approximate process: the wave-packet states that it picks out are approximately defined; the division between system and environment cannot be taken as fundamental; interference processes may be suppressed far below the limit of experimental detection but they never quite vanish. The previous dilemma remains (it seems): either worlds are part of our fundamental ontology (in which case decoherence, being merely a dynamical process within unitary quantum mechanics, and an approximate one at that, seems incapable of defining them), or they do not really exist (in which case decoherence theory seems beside the point).

Outside the philosophy of physics, though (notably in the philosophy of mind, and in the philosophy of the special sciences more broadly), it has long been recognized that this dilemma is mistaken, and that something need not be *fundamental* to be *real*. In the last decade, this insight was carried over to the philosophy of physics.

6. Higher-Order Ontology and the Role of Structure

On even cursory examination, we find that science is replete with perfectly respectable entities that are nowhere to be found in the underlying microphysics.

Douglas Hofstader and Daniel Dennett make this point very clearly:

> Our world is filled with things that are neither mysterious and ghostly nor simply constructed out of the building blocks of physics. Do you believe in voices? How about haircuts? Are there such things? What are they? What, in the language of the physicist, is a hole—not an exotic black hole, but just a hole in a piece of cheese, for instance? Is it a physical thing? What is a symphony? Where in space and time does "The Star-Spangled Banner" exist? Is it nothing but some ink trails in the Library of Congress? Destroy that paper and the anthem would still exist. Latin still *exists* but it is no longer a living language. The language of the cavepeople of France no longer exists at all. The game of bridge is less than a hundred years old. What sort of a thing is it? It is not animal, vegetable, or mineral.
>
> These things are not physical objects with mass, or a chemical composition, but they are not purely abstract objects either—objects like the number pi, which is immutable and cannot be located in space and time. These things have birthplaces and histories. They can change, and things can happen to them. They can move about—much the way a species, a disease, or an epidemic can. We must not suppose that science teaches us that every *thing* anyone would want to take seriously is identifiable as a collection of particles moving about in space and time. (Hofstadter and Dennett 1981, 6–7)

The generic philosophy-of-science term for entities such as these is *emergent*: they are not directly definable in the language of microphysics (try defining a haircut within the Standard Model!) but that does not mean that they are somehow independent of that underlying microphysics.

To look in more detail at a particularly vivid example, consider tigers, which are (I take it!) unquestionably real, objective physical objects, even though the Standard Model contains quarks, electrons, and the like, but no tigers. Instead, tigers should be understood as patterns, or structures, *within* the states of that microphysical theory.

To see how this works in practice, consider how we could go about studying, say, tiger hunting patterns. In principle—and only in principle — the most reliable way to make predictions about these would be in terms of atoms and electrons, applying molecular dynamics directly to the swirl of molecules that make up, say, the Kanha National Park (one of the sadly diminishing places where Bengal tigers can be found). In practice, however (even ignoring the measurement problem itself!), this is clearly insane: no remotely imaginable computer would be able to solve the 10^{35} or so simultaneous dynamical equations that would be needed to predict what the tigers would do.

Actually, the problem is even worse than this. For in a sense, we *do* have a computer capable of telling us how the positions and momentums of all the molecules in the Kanha National Park change over time. It is called the Kanha National Park. (And it runs in real time!) Even if, *per impossibile*, we managed to build a computer simulation of the Park accurate down to the last electron, it would tell us no more than what the Park itself tells us. It would provide no explanation of any of its complexity. (It would, of course, be a superb vindication of our extant microphysics.)

If we want to understand the complex phenomena of the Park, and not just reproduce them, a more effective strategy can be found by studying the structures observable at the multi-trillion-molecule level of description of this "swirl of molecules." At this level, we will observe robust—though not 100% reliable—regularities, which will give us an alternative description of the tiger in a language of cell membranes, organelles, and internal fluids. The principles by which these interact will be deducible from the underlying microphysics (in principle at least; in practice there are usually many gaps in our understanding), and will involve various assumptions and approximations; hence very occasionally they will be found to fail. Nonetheless, this slight riskiness in our description is overwhelmingly worthwhile given the enormous gain in usefulness of this new description: the language of cell biology is both explanatorily far more powerful, and practically far more useful, than the language of physics for describing tiger behavior.

Nonetheless it is still ludicrously hard work to study tigers in this way. To reach a really practical level of description, we again look for patterns and regularities, this time in the behavior of the cells that make up individual tigers (and other living creatures that interact with them). In doing so we will reach yet another language, that of zoology and evolutionary adaptationism, which describes the system in terms of tigers, deer, grass, camouflage, and so on. This language is, of course, the norm in studying tiger hunting patterns, and another (in practice very modest) increase in the riskiness of our description is happily accepted in exchange for another phenomenal rise in explanatory power and practical utility.

The moral of the story is: there are structural facts about many microphysical systems which, although perfectly real and objective (try telling a deer that a nearby tiger is not objectively real) simply cannot be seen if we persist in analyzing those systems in purely microphysical terms. Zoology is of course grounded in cell biology, and cell biology in molecular physics, but the entities of zoology cannot be discarded in favor of the austere ontology of molecular physics alone. Rather, those entities are structures instantiated within the molecular physics, and the task of almost all science is to study structures of this kind.

Of *which* kind? (After all, "structure" and "pattern" are very broad terms: almost any arrangement of atoms might be regarded as some sort of pattern.) The tiger example suggests the following answer, which I have previously (Wallace, 2003a, 93) called "Dennett's criterion" in recognition of the very similar view proposed by Daniel Dennett (1991):

> **Dennett's criterion:** A macro-object is a pattern, and the existence of a pattern as a real thing depends on the usefulness—in particular, the explanatory power and predictive reliability—of theories which admit that pattern in their ontology.

Nor is this account restricted to the relation between physics and the rest of science: rather, it is ubiquitous within physics itself. Statistical mechanics provides perhaps the most important example of this: the temperature of bulk matter is an emergent property, salient because of its explanatory role in the behavior of that matter. (It is a common error in textbooks to suppose that statistical-mechanical methods are used only because in practice we cannot calculate what each atom is doing separately: even if we could do so, we would be missing important, objective properties of the system in question if we abstained from statistical-mechanical talk.) But it is somewhat unusual because (unlike the case of the tiger) the principles underlying statistical-mechanical claims are (relatively!) straightforwardly derivable from the underlying physics.

For an example from physics that is closer to the cases already discussed, consider the case of quasi-particles in solid-state physics. As is well known, vibrations in a (quantum-mechanical) crystal, although they can in principle be described entirely in terms of the individual crystal atoms and their quantum entanglement with one another, are in practice overwhelmingly simpler to describe in terms of "phonons"—collective excitations of the crystal that behave like "real" particles in most respects. And furthermore, this sort of thing is completely ubiquitous in solid-state physics, with different sorts of excitation described in terms of different sorts of "quasi-particle"—crystal vibrations are described in terms of phonons; waves in the magnetization direction of a ferromagnet are described in terms of magnons, collective waves in a plasma are described in terms of plasmons, and so on.[12]

Are quasi-particles real? They can be created and annihilated; they can be scattered off one another; they can be detected (by, for instance, scattering them off "real" particles like neutrons); sometimes we can even measure their time of flight; they play a crucial part in solid-state explanations. We have no more evidence than this that "real" particles exist, and indeed no more grip than this on what makes a particle "real," and so it seems absurd to deny that quasi-particles exist—and yet, they consist only of a certain pattern within the constituents of the solid-state system in question.

When *exactly* are quasi-particles present? The question has no precise answer. It is essential in a quasi-particle formulation of a solid-state problem that the quasi-particles decay only slowly relative to other relevant timescales (such as their time of flight) and when this criterion (and similar ones) is met then quasi-particles are

---

[12] For an elementary introduction, see, e.g., Kittel (1996); for a more systematic treatment see, e.g., Tsvelik (2003) or (old but classic) Abrikosov, Gorkov, and Dzyalohinski (1963).

definitely present. When the decay rate is much too high, the quasi-particles decay too rapidly to behave in any "particulate" way, and the description becomes useless explanatorily; hence, we conclude that no quasi-particles are present. It is clearly a mistake to ask *exactly* when the decay time is short enough (2.54 × the interaction time?) for quasi-particles not to be present, but the somewhat blurred boundary between states where quasi-particles exist and states when they don't should not undermine the status of quasi-particles as real, any more than the absence of a precise boundary to a mountain undermines the existence of mountains.

What has all this got to do with decoherence and Everett? Just this: that the branches which appear in decoherence are precisely the kind of entities that special sciences in general tell us to take seriously. They are emergent, robust *structures* in the quantum state, and as such, we have (it seems) as much reason to take them ontologically seriously as we do any other such structure in science—such as those structures that we identify as chairs and tables, cats and dogs and tigers. So—on pain of rejecting the coherence of the special sciences as a whole—we should accept that unitary quantum mechanics is *already* a many-worlds theory: not a many-exact-worlds theory in which the worlds are *part of* the basic mathematical structure, but an emergent-worlds theory in which the worlds are instantiated as higher-level structures within that basic structure.

In this sense, advocacy of the Everett interpretation has come full circle: the rise and fall of many-exact-worlds and many-minds theories has returned us to Everett's original insight that unitary quantum mechanics should be understood as, not modified to become, a many-worlds theory.

7. Aspects of the Probability Problem

Concerns about probability, and attempts to resolve concerns about probability, have been part of the Everett interpretation since its inception, and the bulk of philosophical work on the interpretation continues to focus on this issue, so that I can do no more here than provide an introduction. I will do so by briefly considering three questions that might be (and indeed have been) raised by critics:

1. How can probability even make sense in the Everett interpretation, given that it is deterministic and that all possible outcomes occur?
2. What justifies the actual form of the quantum probability rule in the Everett interpretation?
3. How can the Everett interpretation make sense of the scientific process by which quantum mechanics was experimentally tested?

Before doing so, however, I make two more general observations. First, if there is a problem of probability in the Everett interpretation, then it is an essentially *philosophical* problem. There is no mystery about how probabilistic theories are mathematically represented in theoretical physics: they are represented by a space of states, a set of histories in that space of states (that is, paths through, or ordered sequences of elements drawn from, that space), and a probability measure over those histories (that is, a rule assigning a probability to each subset of histories, consistent with the probability calculus). Given decoherence, quantum mechanics provides all three (at the emergent level where branches can be defined) just fine, using the standard modulus-squared amplitude rule to define the probability of each branch; indeed, historically much of the motivation of the decoherence program was to ensure that the probability calculus was indeed satisfied by the modulus-squared amplitudes of the branches. So a physicist who objected to the rather philosophical tenor of the debates on probability in the Everett interpretation would be missing the point: insofar as he is unconcerned with *philosophical* aspects of probability, he should have no qualms about Everettian probability at all.

Second, it has frequently been the case that what appear to be philosophical problems with probability *in Everettian quantum mechanics* in fact turn out to be philosophical problems with probability *simpliciter*. Probability poses some very knotty philosophical issues, which often we forget just because we are so used to the concept in practice; sometimes it takes an unfamiliar context to remind us of how problematic it can be.

Note that it is of no use for a critic to respond that all the same we have a good practical grasp of probability in the non-Everettian context but that that grasp does not extend to Everettian quantum physics. For that is exactly the point at issue: the great majority, if not all, of the objective probabilities we encounter in science and daily life ultimately have a quantum-mechanical origin, so if the Everett interpretation is correct, then most of our practical experience of probability is with Everett-type probability.

8. Probability, Uncertainty, and Possibility

How can there be probabilities in the Everett interpretation? (asks the critic): there is nothing for them to be probabilities of! Defenders will reply that the probabilities are probabilities of branches (understood via decoherence), but the objection is that somehow it is illegitimate to assign probabilities to the branches, either because probabilities require uncertainty and it makes no sense to be uncertain of which outcome will occur in a theory like Everett's, or because somehow probabilities quantify alternative possibilities and there are no alternative possibilities in the Everett interpretation.

The conciliatory approach here would be to argue that these concepts do after all find a home in Everettian quantum mechanics; that is, to argue that people in an Everettian universe should indeed regard different branches as different alternative possibilities, and be uncertain as to which one will actually occur. To my knowledge this was first argued for by Saunders (1998), via an ingenious thought experiment related to traditional intuition pumps in the philosophy of personal identity; Saunders' goal was to make it intuitive that someone in an Everettian universe should indeed be uncertain about their future, even if they knew the relevant facts about the future branches (though see Greaves (2004) for an attempted rebuttal). Subsequent work (much of it building on Saunders') has tried to go beyond intuitive plausibility and give a positive account of what would ground uncertainty in the Everett interpretation.

I am aware of three broad strategies of this kind. First, and most directly, Lev Vaidman points out (Vaidman 2002) that someone who carried out a quantum measurement *but did not observe the result* would be in a state of genuine (albeit indexical) uncertainty. (There would be multiple copies of the experimenter, some in branches with one result and some in branches with another, but each would be in subjectively identical states.) It is unclear whether this notion of uncertainty (which does not appear to apply to pre-measurement situations) is sufficient to assuage concerns.

An alternative approach via indexical uncertainty—this time also applying to the pre-measurement situation—is to think about branches as four-dimensional rather than three-dimensional entities (thus entailing that branches overlap in some sense[13] prior to whatever quantum event causes them to diverge. Uncertainty is then to be understood as uncertainty as to which four-dimensional branch an observer is part of. For exploration and defense of this position, see Saunders and Wallace (2008a, 2008b), Saunders (2010), and Wilson (2010a, 2010b); for criticism, see Lewis (2007b) and Tappenden (2008).

The third strategy is closely related to the second, but takes its cue from semantics rather than from metaphysics: namely, consider the way in which words like "uncertainty" would function in an Everettian universe (possibly given some theory of semantic content along the "charity" lines advocated by Lewis (1974), Davidson (1973), and others) and argue that they would in fact function in such a way as to make claims like "one or other outcome of the measurement will occur, but not both" actually turn out correct. I explore this idea in Wallace (2005, 2006) and in chapter 7 of Wallace (2012); see also Ismael (2003) for a position that combines aspects of the second and third strategies. Whether such semantical considerations are metaphysically (let alone physically) relevant depends on one's view of metaphysics; Albert (2010), for instance, argues that they are irrelevant.

A conciliatory approach of a rather different kind is to concede that probability has no place in an Everettian world and to show how one can do without it; typically, this is done by arguing that human activity in general, and science in particular, would proceed as if quantum-mechanical mod-squared amplitude was probability, even if "really" it was not. Deutsch (1999) and Greaves (2004) advocate positions

---

[13] In exactly *what* sense is controversial, and the debate arguably overlaps(!) with others in mainstream metaphysics; see Saunders (2010) and Wilson (2010a, 2010b) for further discussion.

of this kind; both regard "probability" as something to be understood decision-theoretically, via an agent's actions. If it can be argued that (rational) agents in an Everettian world would act as if each branch has a certain probability, then (Deutsch and Greaves argue) this is sufficient.

Of course, there is also a decidedly nonconciliatory response available: just to deny the claim that genuine probability requires either alternative probabilities or any form of uncertainty. One seldom hears actual *arguments* for these requirements; typically they are just stated as if they were obvious. And perhaps they are *intuitively* obvious, but it is not clear that this has any particular bearing on anything. Someone who adopts the (hopelessly unmotivated) epistemological strategy of regarding intuitive obviousness as a guide to truth in theoretical physics will presumably have given up on the Everett interpretation long ago in any case.

This response is actually fairly close to Deutsch's and Greaves's position: if it can be argued that mod-squared amplitude functions exactly like probability but lacks certain standardly required philosophical features that probability has, it is open to us just to deny that those philosophical features are required, and to adopt the position that insofar as mod-squared amplitude functions *exactly like* probability, that's all that's required to establish that it *is* probability. This is my own view on the problem,[14] developed *in extenso* in Wallace (2012).

9. The Quantitative Problem

Grant, if only for the sake of argument, that it is somehow legitimate to attach probabilities to branches. There is a further question: Why should those probabilities be required to equal those given by quantum mechanics?

One version of this objection—going right back to Graham (1973)—is that the quantum probabilities *cannot* be the right probabilities, because the right probabilities must give each branch equal probability. There is generally no positive *argument* given for this claim, beyond some gesture to the effect that the versions of me on the different branches are all "equally me"; still, it has a strong intuitive plausibility.

It can, however, be swiftly dismissed. It is possible to argue that the rule is actually inconsistent when branching events at multiple times are considered,[15] but more crucially, decoherence just does not license any notion of branch count. It makes sense, in the presence of decoherence, to say that the quantum state (or some part of it) branches into a part in which measurement outcome X occurs and a part in which it does not occur, but it makes no sense at all to say *how many branches* comprise the part in which X occurs. Study of a given branch at a finer level of detail will inevitably show it to consist of many sub-branches; eventually this will cease to

---

[14] It represents a departure from my position in Wallace (2006).
[15] See Wallace (2012) for details; I learned the argument from David Deutsch in conversation.

be the case as decoherence ceases to be applicable and interference between branches becomes nonnegligible; but there is no well-defined point at which this occurs and different levels of tolerance—as well as small changes in other details of how we define "branch"—lead to wildly differing answers as to how many branches there are. Put plainly, "branch number," insofar as it is defined at all in a given decoherence formalism, is an artifact of the details of that formalism. (And it is not by any means defined in all such formalisms; many use a continuum framework in which the concept makes no sense even inside the formalism. For more details on this and on the general question of branch counting, see chapter 3 of Wallace (2012).)

So much for branch counting. The question remains: What positive justification can be given for identifying mod-squared amplitude with probability? One might answer, as did Simon Saunders in the 1990s (Saunders 1995, 1997, 1998), by rejecting the idea that any "positive justification" is needed: after all, in general we do not argue that the probabilities in a physical theory are what they are (nor indeed, in general, that the other physical magnitudes in a theory have the interpretation they have); we just postulate it. It is not immediately clear why this response is any less justified in the Everett interpretation than in non-Everettian physics; indeed, arguably it works rather better as a postulate, since it is at least clear what categorical, previously understood magnitude is to be identified with probability. By contrast, in *classical physics* the only real candidate seems to be long-run relative frequencies or some related concept, and even establishing that those have the *formal* properties required of probability has proven fraught. The most promising candidate so far is Lewis's "best systems analysis" (Lewis 1986, 55, 128–131), which constructs probabilities indirectly from relative frequencies and related categorical data; even if that analysis succeeded fully, though, it would deliver no more than quantum physics (together with decoherence) has already delivered, namely a set of quantities with the right formal properties to be identified with probability but no further justification for making such an identification.[16]

Papineau (1996, 2010) puts essentially the same point in a more pessimistic way. He identifies two criteria that a theory of probability must satisfy: a "decision-theoretic link" (why do we use probability as a guide to action?) and an "inferential link" (why do we learn about probabilities from observed relative frequencies?) and concedes that Everettian quantum mechanics has no good explanation of why either is satisfied—but, he continues, neither does any other physical theory, nor any other extant philosophical theory of probability. The Everett interpretation (Papineau argues) therefore has no *special* problem of probability.

In fact, in recent years the possibility has arisen that probability may actually be in *better* shape in Everettian quantum mechanics than in non-Everettian physics. Arguments originally given by David Deutsch (1999) and developed in

---

[16] For reasons of space I omit detailed discussion of the parallel tradition in Everettian quantum mechanics of identifying probability via long-run relative frequency (notably by Everett himself (1957) and by Farhi, Goldstone, and Gutmann (1989). I discuss this program in detail in chapter 4 of Wallace (2012); my conclusion is that it works about as well, or as badly, as equivalent classical attempts, though there is no direct Everettian analogue to the best-systems approach.

Wallace (2003b, 2007) suggest that it may be possible to derive the quantum probability rule from general principles of decision theory, together with the mathematical structure of quantum mechanics shorn of its probabilistic interpretation. A fully formalized version of this argument can be found in Wallace (2010) and in chapters 5 and 6 of Wallace (2012).

In philosophical terms, what such arguments attempt to do is to show that rational agents, cognizant of the facts about quantum mechanics and conditional on believing those facts to be true, are required to treat mod-squared amplitude operationally as probability. Specifically, they are required to use observed relative frequencies as a guide to working out what the unknown mod-squared amplitudes are (Papineau's inferential link), and to use known mod-squared amplitudes as a guide to action (his decision-theoretic link).[17]

Space does not permit detailed discussion of this approach to probability, but at essence it relies on the symmetries of quantum mechanics. There is a long tradition of deriving probability from considerations of symmetry, but in the classical case these approaches ultimately struggle with the fact that *something* must break the symmetry, simply to explain why one outcome occurs rather than another. This is, of course, not an issue for Everettian quantum mechanics! From this perspective, the role of decision theory is less central in the arguments than it might appear: its main function is to justify the applicability of probabilistic concepts to Everettian branches at all. (And conversely, if one were concerned *purely* with the question of what the probabilities of each branch were, and prepared to grant that branches *did* have probabilities and that they satisfy normal synchronic and diachronic properties, it is possible to prove the quantum probability rule without any mention of decision theory; cf. Wallace (2012, ch. 4).)

If this last approach to probability works (and fairly obviously, I believe it does), it marks a rather remarkable shift in the debate; probability, far from being something that makes the Everett interpretation unintelligible, becomes something that can be understood in Everettian quantum mechanics in a way which does not seem available otherwise. (See Saunders (2010) for further development of this theme.) I feel obliged to note that it is highly controversial whether the approach does indeed work; for recent criticism, see Albert (2010), Price (2010), and other articles in Saunders et al. (2010).

10. Epistemic Puzzles

The rise of decision-theoretic approaches to Everettian probability (whether to make sense of probability or to derive the probability rule) has led to a new worry about

---

[17] A more precise way of stating both is that the program attempts to show that agents are rationally required to act as if mod-squared amplitude played the objective-probability role in David Lewis's *Principal Principle*; cf. Lewis (1980).

probability in the Everett interpretation. Suppose for the sake of argument that it really can be shown, or legitimately postulated, that someone who accepts the Everett interpretation as correct should behave, at least for all practical purposes, as if mod-squared amplitude were probability. What has that to do with the question of why we should believe the Everett interpretation in the first place? Put another way, how would it license us to interpret the usual evidence for quantum mechanics as evidence for *Everettian* quantum mechanics?

This suggests a division of the probability problem into *practical* and *epistemic* problems (Greaves 2007a), where the former concerns how rational agents should act *given that Everettian quantum theory is correct*, and the latter concerns how evidence bears on the truth of quantum theory in the first place, given that it is to be interpreted à la Everett. Arguably, Deutsch's decision-theoretic program (and my development of it) speaks only to the practical problem; indeed, arguably most of the tradition in thinking about Everettian probability speaks only to the practical problem.

The last decade has seen the development of a small, but complex, literature on this subject. In essence, there are two strategies that have been developed for solving the epistemic problem. The first is highly philosophical: if it can be established that mod-squared amplitude *is probability*, then (it is claimed) no more is required of the Everett interpretation than of any other physical theory as regards showing why probability plugs into our epistemology in the way it does. Strategies of this form rely on a mixture of solutions to the practical problem (cf. section 9), arguments that branching leads to genuine uncertainty about the future and/or genuine probabilities (cf. section 8), and appeal to the no-double-standards principle I mentioned in section 7. The strategy is tacit in Saunders (1998); I defended an explicit version in Wallace (2006) (and, in less developed form, in Wallace (2002)); Wilson (2010b) defends a similar thesis.

The other strategy is significantly more technical and formal: namely, construct a formal decision-theoretic framework to model the epistemic situation of agents who are unsure whether the Everett interpretation is correct, and show that in that situation (perhaps contingent on a solution to the practical problem), agents regard "ordinary" evidence as confirmatory of quantum mechanics in a standard way, even when quantum mechanics is understood according to the Everett interpretation. This strategy was pioneered by Greaves (2004) and brought to a mature state in Greaves (2007a) and Greaves and Myrvold (2010). The latter two papers, on slightly different starting assumptions (including in both cases the Bayesian approach to statistical inference) take it as given that conditional on the Everett interpretation being true, mod-squared amplitude functions as probability in decision-making contexts, and derive that agents will update their personal probability in quantum mechanics via standard update procedures, whether or not quantum probabilities are to be understood in Everettian terms. As such, these arguments take as input a solution to the practical problem (whether postulated or derived via Deutsch's and/or my arguments) and give as output a solution to the epistemic problem. It is also possible (cf. Wallace, 2012, ch. 6) to combine the two strategies into one theorem, which makes standard decision-theoretic assumptions and derives solutions to the epistemic and practical problems in a unified fashion.

11. OTHER TOPICS

While the bulk of contemporary work on the Everett interpretation has been concerned with the preferred-basis and probability problems (and, more generally, has been concerned with whether the interpretation is viable, rather than with its philosophical implications if viable), there are a goodly number of other areas of interest within the Everett interpretation (or, as I would prefer to put it: within quantum mechanics, once it is understood that it should be interpreted Everett-style), and I briefly mention some of these here.

- If Everettian quantum mechanics is only *emergently* a theory of branching universes, what is its *fundamental* ontology, insofar as that question has meaning? That is: what kind of physical entity is represented by the quantum state? Of course, this question can be asked of any approach to quantum theory that takes the state as representing something physically real, but it takes on a particular urgency in the Everett interpretation given that the theory is supposed to be *pure* quantum mechanics, shorn of any additional mathematical structure. For various approaches to the problem, see Deutsch and Hayden (2000), Deutsch (2002), Wallace and Timpson (2007, 2010), Maudlin (2010) (who argues that there is *no* coherent understanding of the Everett interpretation's ontology), Hawthorne (2010) (who is at least sympathetic to Maudlin), Allori et al. (2009), and (in the general context of the ontology of the quantum state) Albert (1996) and Lewis (2004b).
  (I should add one cautionary note: it is very common in the literature to phrase the question as, what is the ontology of the wave-function? But recall that the wave-function is only one of a great many ways to represent the quantum state, and one which is much more natural in nonrelativistic physics than in quantum field theory.)
- It is generally (and in my view correctly) held that the experimental violation of Bell's inequalities[18] shows not just that hidden variable theories must involve superluminal dynamics, but that *any* empirically adequate theory must involve superluminal dynamics.[19] But the Everett interpretation is generally (and again correctly, in my view) viewed as an exception, essentially because it violates a tacit premise of Bell's derivation, that only

---

[18] See, the discussions in e.g., Bell (1981a) or Maudlin (2002).
[19] That the dynamics are thereby required to violate Lorentz covariance does not uncontroversially follow; cf. Myrvold (2002), Wallace and Timpson (2010), and Tumulka (2006).

one outcome actually occurs.[20] There has, however, been rather little exploration of this issue; Bacciagaluppi (2002) is a notable exception.[21]

- There is an ongoing (and somewhat sensationalist) discussion in the literature about so-called "quantum suicide": the idea that an agent in an Everettian universe should expect with certainty to survive any process which third-party observers regard him as having nonzero probability of surviving. The idea has been around in the physics community for a long time (see, e.g., Tegmark (1998); it was first introduced to philosophers by David Lewis, in his only paper on the Everett interpretation (Lewis 2004a) and has been discussed further by Lewis (2000) and Papineau (2003).
- Everett was originally motivated in part by a desire for an interpretation of quantum mechanics that was suitable for cosmology in that it did not assume an external observer. The Everett interpretation has been widely influential in quantum cosmology ever since: for an introduction, see Hartle (2010). It is not *universally* acknowledged that quantum cosmology does require the Everett interpretation, though; for dissenting views (from widely differing perspectives), see Fuchs and Peres (2000), Smolin (1997: 240–266), and Rovelli (2004: 209–222).
- The de Broglie-Bohm "pilot wave" theory (aka Bohmian mechanics) has sometimes been criticized for being "Everett in denial": that is, being the Everett interpretation with some additional epiphenomenal structure. For examples of this criticism, see Deutsch (1996) and Brown and Wallace (2004); for responses, see Lewis (2007a) and Valentini (2010) (see also Brown's (2010) response to Valentini). Allori et al. (2008) can also be read as a response, insofar as it advocates a position on the ontology of a physical theory far removed from that of section 6 and from which the Everett-in-denial objection cannot be made.

12. Further Reading

Saunders et al. (2010) is an up-to-date and edited collection of articles for and against the Everett interpretation, including contributions from a large fraction of the physicists and philosophers involved in the contemporary debate; Saunders's introduction to the book provides an overview of the Everett interpretation complementary to this chapter. Barrett (1999) is a comprehensive guide to discussions of the Everett interpretation in (mostly) the philosophy of physics literature, up to

---

[20] For a more detailed analysis—which gives a slightly different account of why the Everett interpretation is an exception to Bell's result—see Timpson and Brown (2002).
[21] Storrs McCall also explores these issues in developing his approach to quantum mechanics (see, e.g., McCall (2000)); that approach is related to, but not identical to, the Everett interpretation (and, insofar as it relies on an explicit and precise concept of branching without offering a dynamical explication of when branching occurs, arguably fails to solve the measurement problem.)

the late 1990s. DeWitt and Graham (1973) is a classic collection of original papers. Wallace (2012) is my own book-length defense of the Everett interpretation; Wallace (2008) is a review of the measurement problem more generally, focused on the role of decoherence theory. Greaves (2007b) reviews work in the probability problem.

Afterword

I have left undiscussed the often-unspoken, often-felt objection to the Everett interpretation: that it is simply unbelievable. This is because there is little to discuss: that a *scientific theory* is wildly unintuitive is no argument at all against it, as twentieth-century physics proved time and again. David Lewis is memorably reported to have said that he did not know how to refute an incredulous stare; had he been less charitable, he might have said explicitly that an incredulous stare is not an argument, and that if someone says that they are incapable of believing a given theory—philosophical or scientific—they are but reporting on their psychology.

References

Abrikosov, A. A., L. P. Gorkov, and I. E. Dzyalohinski (1963). *Methods of quantum field theory in statistical physics*. New York: Dover. Revised English edition; translated and edited by R. A. Silverman.
Albert, D. Z. (1992). *Quantum mechanics and experience*. Cambridge, MA: Harvard University Press.
———. (1996). Elementary quantum metaphysics. In J. T. Cushing, A. Fine, and S. Goldstein (Eds.), *Bohmian mechanics and quantum theory: An appraisal*, 277–284. Dordrecht: Kluwer Academic Publishers.
———. (2010). Probability in the Everett picture. In S. Saunders, J. Barrett, A. Kent, and D. Wallace (Eds.), *Many worlds? Everett, quantum theory and reality*. Oxford: Oxford University Press.
Albert, D. Z., and B. Loewer (1988). Interpreting the many worlds interpretation. *Synthese* 77: 195–213.
Allori, V., S. Goldstein, R. Tumulka, and N. Zanghi (2008). On the common structure of Bohmian mechanics and the Ghirardi–Rimini–Weber theory. *British Journal for the Philosophy of Science* 59: 353–389.
———. (2009). Many-worlds and Schrödinger's first quantum theory. Forthcoming in *British Journal for the Philosophy of Science*; available online at http://arxiv.org/abs/0903.2211.
Bacciagaluppi, G. (2002). Remarks on space-time and locality in Everett's interpretation. In J. Butterfield and T. Placek (Eds.), *Non-locality and modality*, Vol. 64 of Nato Science Series II. Mathematics, Physics and Chemistry. 105–122. Dordrecht: Kluwer.

Barbour, J. B. (1994). The timelessness of quantum gravity: II. The appearance of dynamics in static configurations. *Classical and Quantum Gravity* 11: 2875–2897.

———. (1999). *The end of time*. London: Weidenfeld and Nicholson.

Barrett, J. A. (1999). *The quantum mechanics of minds and worlds*. Oxford: Oxford University Press.

Bell, J. S. (1981a). Bertlmann's socks and the nature of reality. *Journal de Physique* 42, C2: 41–61. Reprinted in Bell (1987), 139–158.

———. (1981b). Quantum mechanics for cosmologists. In C. J. Isham, R. Penrose, and D. Sciama (Eds.), *Quantum gravity 2: A second Oxford Symposium*. Oxford: Clarendon Press. Reprinted in Bell (1987), 117–138; page references are to that version.

———. (1987). *Speakable and unspeakable in quantum mechanics*. Cambridge: Cambridge University Press.

Brown, H. (2010). Reply to Valentini: "de Broglie-Bohm theory: Many worlds in denial?". In *Many worlds? Everett, quantum theory, and reality*. Oxford: Oxford University Press. 510–517.

Brown, H. R., and D. Wallace (2004). Solving the measurement problem: de Broglie-Bohm loses out to Everett. Forthcoming in *Foundations of Physics*; available online at http://arxiv.org/abs/quant-ph/0403094.

Butterfield, J. N. (1996). Whither the minds? *British Journal for the Philosophy of Science* 47: 200–221.

Colin, S. (2003). Beables for quantum electrodynamics. Available online at http://arxiv.org/abs/quant-ph/0310056.

Colin, S., and W. Struyve (2007). A Dirac sea pilot-wave model for quantum field theory. *Journal of Physics A* 40: 7309–7342.

Davidson, D. (1973). Radical interpretation. *Dialectica* 27: 313–328.

Dennett, D. C. (1991). Real patterns. *Journal of Philosophy* 87, 27–51. Reprinted in D. Dennett, *Brainchildren* (London: Penguin, 1998), 95–120.

Deutsch, D. (1985). Quantum theory as a universal physical theory. *International Journal of Theoretical Physics* 24(1): 1–41.

———. (1986). Interview. In P. Davies and J. Brown (Eds.), *The ghost in the atom*, 83–105. Cambridge: Cambridge University Press.

———. (1996). Comment on Lockwood. *British Journal for the Philosophy of Science* 47: 222–228.

———. (1999). Quantum theory of probability and decisions. *Proceedings of the Royal Society of London* A455: 3129–3137. Available online at http://arxiv.org/abs/quant-ph/9906015.

———. (2002). The structure of the multiverse. *Proceedings of the Royal Society of London* A458, 2911–2923.

Deutsch, D., and P. Hayden (2000). Information flow in entangled quantum systems. *Proceedings of the Royal Society of London* A456: 1759–1774. Available online at http://arxiv.org/abs/quant-ph/9906007.

DeWitt, B., and N. Graham, eds. (1973). *The many-worlds interpretation of quantum mechanics*. Princeton: Princeton University Press.

Donald, M. J. (1990). Quantum theory and the brain. *Proceedings of the Royal Society of London* A427: 43–93.

———. (1992). A priori probability and localized observers. *Foundations of Physics* 22: 1111–1172.

———. (2002). Neural unpredictability, the interpretation of quantum theory, and the mind–body problem. Available online at http://arxiv.org/abs/quant-ph/0208033.

Dürr, D., S. Goldstein, R. Tumulka, and N. Zanghi (2004). Bohmian mechanics and quantum field theory. *Physical Review Letters* 93: 090402.

———. (2005). Bell-type quantum field theories. *Journal of Physics* A38: R1.

Everett, H. I. (1957). Relative state formulation of quantum mechanics. *Review of Modern Physics* 29: 454–462. Reprinted in DeWitt and Graham (1973).

Farhi, E., J. Goldstone, and S. Gutmann (1989). How probability arises in quantum-mechanics. *Annals of Physics* 192: 368–382.

Foster, S., and H. R. Brown (1988). On a recent attempt to define the interpretation basis in the many worlds interpretation of quantum mechanics. *International Journal of Theoretical Physics* 27: 1507–1531.

Fuchs, C., and A. Peres (2000). Quantum theory needs no "interpretation." *Physics Today* 53(3): 70–71.

Graham, N. (1973). The measurement of relative frequency. In B. DeWitt and N. Graham (Eds.), *The many-worlds interpretation of quantum mechanics*. Princeton: Princeton University Press: 229–252.

Greaves, H. (2004). Understanding Deutsch's probability in a deterministic multiverse. *Studies in the History and Philosophy of Modern Physics* 35: 423–456.

———. (2007a). On the Everettian epistemic problem. *Studies in the History and Philosophy of Modern Physics* 38: 120–152.

———. (2007b). Probability in the Everett interpretation. *Philosophy Compass* 38: 120–152.

Greaves, H., and W. Myrvold (2010). Everett and evidence. In S. Saunders, J. Barrett, A. Kent, and D. Wallace (Eds.), *Many worlds? Everett, quantum theory and reality*. Oxford: Oxford University Press. 229–252.

Hartle, J. (2010). Quasiclassical realms. In S. Saunders, J. Barrett, A. Kent, and D. Wallace (Eds.), *Many worlds? Everett, quantum theory, and reality*. Oxford: Oxford University Press: 73–98.

Hawthorne, J. (2010). A metaphysician looks at the Everett interpretation, in Saunders et al. (2010), 144–153.

Hofstadter, D. R., and D. C. Dennett, eds. (1981). *The mind's I: Fantasies and reflections on self and soul*. London: Penguin.

Ismael, J. (2003). How to combine chance and determinism: Thinking about the future in an Everett universe. *Philosophy of Science* 70: 776–790.

Kent, A. (1990). Against many-worlds interpretations. *International Journal of Theoretical Physics* A5: 1745–1762. Expanded and updated version at http://www.arxiv.org/abs/gr-qc/9703089.

———. (2010). One world versus many: The inadequacy of Everettian accounts of evolution, probability, and scientific confirmation. In Saunders, Barrett, Kent, and Wallace (2012), 307–354.

Kittel, C. (1996). *Introduction to solid state physics*. 7th ed. New York: Wiley.

Ladyman, J., and D. Ross (2007). *Every thing must go: Metaphysics naturalized*. Oxford: Oxford University Press.

Lewis, D. (1974). Radical interpretation. *Synthese* 23: 331–344. Reprinted in David Lewis, *Philosophical Papers*, Vol. I. Oxford: Oxford University Press, 1983.

———. (1980). A subjectivist's guide to objective chance. In R. C. Jeffrey (Ed.), *Studies in inductive logic and probability*. Vol. 2. Berkeley: University of California Press.

Reprinted, with postscripts, in David Lewis, *Philosophical Papers*, Vol. 2. Oxford: Oxford University Press, 1986; page numbers refer to this version.
———. (1986). *Philosophical Papers*. Vol. 2. Oxford: Oxford University Press.
———. (2004a). How many lives has Schrödinger's cat? *Australasian Journal of Philosophy* 81: 3–22.
Lewis, P. (2007a). How Bohm's theory solves the measurement problem. *Philosophy of Science* 74: 749–760.
Lewis, P. J. (2000). What is it like to be Schrödinger's cat? *Analysis* 60: 22–29.
———. (2004b). Life in configuration space. *British Journal for the Philosophy of Science* 55: 713–729.
———. (2007b). Uncertainty and probability for branching selves. *Studies in the History and Philosophy of Modern Physics* 38: 1–14.
Lockwood, M. (1989). *Mind, brain and the quantum: The compound 'I'*. Oxford: Blackwell Publishers.
———. (1996). "Many minds" interpretations of quantum mechanics. *British Journal for the Philosophy of Science* 47: 159–188.
Maudlin, T. (2002). *Quantum non-locality and relativity: Metaphysical intimations of modern physics*. 2d ed. Oxford: Blackwell.
———. (2010). Can the world be only wavefunction? In Saunders, Barrett, Kent, and Wallace (2010), 121–143.
McCall, S. (2000). QM and STR: The combining of quantum mechanics and relativity theory. *Philosophy of Science* 67: S535–S548.
Myrvold, W. (2002). On peaceful coexistence: Is the collapse postulate incompatible with relativity? *Studies in the History and Philosophy of Modern Physics* 33: 435–466.
Page, D. N. (1996). Sensible quantum mechanics: Are probabilities only in the mind? *International Journal of Modern Physics* D5: 583–596.
Papineau, D. (1996). Many minds are no worse than one. *British Journal for the Philosophy of Science* 47: 233–241.
———. (2003). Why you don't want to get into the box with Schrödinger's cat. *Analysis* 63: 51–58.
———. (2010). A fair deal for Everettians. In Saunders, Barrett, Kent, and Wallace (2010), 205–226.
Penrose, R. (1989). *The emperor's new mind: Concerning computers, brains and the laws of physics*. Oxford: Oxford University Press.
Price, H. (2010). Decisions, decisions, decisions: can Savage salvage Everettian probability? In Saunders, Barrett, Kent, and Wallace (2010), 369–390.
Rovelli, C. (2004). *Quantum gravity*. Cambridge: Cambridge University Press.
Saunders, S. (1993). Decoherence, relative states, and evolutionary adaptation. *Foundations of Physics* 23: 1553–1585.
———. (1995). Time, decoherence and quantum mechanics. *Synthese* 102: 235–266.
———. (1997). Naturalizing metaphysics. *The Monist* 80(1): 44–69.
———. (1998). Time, quantum mechanics, and probability. *Synthese* 114: 373–404.
———. (2003). Physics and Leibniz's principles. In K. Brading and E. Castellani (Eds.), *Symmetries in physics: Philosophical reflections*, 289–308. Cambridge: Cambridge University Press.
———. (2010). Chance in the Everett interpretation. In Saunders, Barrett, Kent, and Wallace (2010), 181–205.
Saunders, S., and D. Wallace (2008a). Branching and uncertainty. *British Journal for the Philosophy of Science* 59: 293–305.

———. (2008b). Saunders and Wallace reply. *British Journal for the Philosophy of Science* 59: 315–317.

Saunders, S., J. Barrett, A. Kent, and D. Wallace, eds. (2010). *Many worlds? Everett, quantum theory, and reality*. Oxford: Oxford University Press.

Smolin, L. (1997). *The life of the cosmos*. New York: Oxford University Press.

Struyve, W., and H. Westman (2006). A new pilot-wave model for quantum field theory. *AIP Conference Proceedings* 844: 321.

Tappenden, P. (2008). Saunders and Wallace on Everett and Lewis. *British Journal for the Philosophy of Science* 59: 307–314.

Tegmark, M. (1998). The interpretation of quantum mechanics: Many worlds or many words? *Fortschrift Fur Physik* 46: 855–862. Available online at http://arxiv.org/abs/quant-ph/9709032.

Timpson, C., and Brown, H. R. (2002). Entanglement and relativity. In R. Lupacchini and V. Fano (Eds.), *Understanding physical knowledge*, Preprint no. 24, Departimento di Filosofia, Università di Bologna, CLUEB, 2002, 147–166. Available online at http://arxiv.org/abs/quant-ph/0212140.

Tsvelik, A. M. (2003). *Quantum field theory in condensed matter physics*. 2d ed. Cambridge: Cambridge University Press.

Tumulka, R. (2006). Collapse and relativity. In A. Bassi, T. Weber, and N. Zanghi (Eds.), *Quantum mechanics: Are there quantum jumps? and on the present status of quantum mechanics*, 340. American Institute of Physics Conference Proceedings. Available online at http://arxiv.org/abs/quant-ph/0602208.

Vaidman, L. (2002). The many-worlds interpretation of quantum Mechanics. In *the Stanford Encyclopedia of Philosophy* (Summer 2002 edition), ed. Edward N. Zalta. Available online at http://plato.stanford.edu/archives/sum2002/entries/qm-manyworlds.

Valentini, A. (2010). De Broglie-Bohm pilot wave theory: Many worlds in denial? In Saunders, Barrett, Kent, and Wallace (2010), 476–509.

Wallace, D. (2002). Quantum probability and decision theory, revisited. Available online at http://arxiv.org/abs/quant-ph/0211104. (This is a long (70-page) and not-formally-published paper, now entirely superseded by published work and included only for historical purposes.).

———. (2003a). Everett and structure. *Studies in the History and Philosophy of Modern Physics* 34: 87–105.

———. (2003b). Everettian rationality: Defending Deutsch's approach to probability in the Everett interpretation. *Studies in the History and Philosophy of Modern Physics* 34: 415–439.

———. (2005). Language use in a branching universe. Unpublished manuscript; Available online from http://philsci-archive.pitt.edu.

———. (2006). Epistemology quantized: Circumstances in which we should come to believe in the Everett interpretation. *British Journal for the Philosophy of Science* 57: 655–689.

———. (2007). Quantum probability from subjective likelihood: Improving on Deutsch's proof of the probability rule. *Studies in the History and Philosophy of Modern Physics* 38: 311–332.

———. (2008). The interpretation of quantum mechanics. In D. Rickles (Ed.), *The Ashgate companion to contemporary philosophy of physics*, 197–261. Burlington, VT: Ashgate.

———. (2010). How to prove the Born rule. In Saunders, Barrett, Kent, and Wallace (2010), 227–263. Available online at http://arxiv.org/abs/0906.2718.

———. (2012). *The emergent multiverse: Quantum theory according to the Everett interpretation*. Oxford: Oxford University Press: 227–263.

Wallace, D., and C. Timpson (2007). Non-locality and gauge freedom in Deutsch and Hayden's formulation of quantum mechanics. *Foundations of Physics* 37(6): 951–955.

———. (2010). Quantum mechanics on spacetime I: Spacetime state realism. *British Journal for the Philosophy of Science* vol. 61: 697–727. Available online at http://philsci-archive.pitt.edu.

Wilson, A. (2010a). Macroscopic ontology in Everettian quantum mechanics. *Philosophical Quarterly* 60: 1–20.

Wilson, A. (2010b). Modality Naturalized: The Metaphysics of the Everett Interpretation. Ph.D. diss., University of Oxford.

CHAPTER 14

UNITARY EQUIVALENCE AND PHYSICAL EQUIVALENCE

LAURA RUETSCHE

1. INTRODUCTION

By tradition, to quantize a theory of classical mechanics, one constructs a *Hilbert space representation* of a set of magnitudes obeying *canonical commutation relations* (CCRs) characteristic of the theory in question. Similarly, to construct a quantum theory of spin systems, one finds a Hilbert space representation of characteristic *canonical anticommutation relations* (CARs) for spin systems. The representations in question take the form of symmetric Hilbert space operators satisfying the relevant canonical relations. Following standard practice of eliding the distinction between operators and the physical magnitudes (aka observables) they represent, I will call such operators "canonical observables." By tradition, other observables recognized by the theory can be obtained as polynomials of, and limits of sequences of polynomials of, the representation-bearing canonical observables. By tradition, a quantum state is an expectation value assignment to this collection of magnitudes, which is normed, linear, and countably additive. Thus, a Hilbert space representation of the canonical relations circumscribing a quantum theory supplies a *kinematics* for that theory, that is, an account of the states it recognizes as physically possible and the magnitudes it recognizes as physically significant. In standard Hilbert space quantum mechanics, the Schrödinger equation equips such a theory with a dynamics.

Even very simple quantum theories, so realized, harbor provocative difficulties. The quantum theory of two spin $\frac{1}{2}$ systems features entangled states that can be understood to predict distant correlations, alarmingly suggestive of spooky action at a distance. This is the difficulty of quantum nonlocality. The quantum theory of a cat and a radioactive atom, both modeled as bivalent systems, when equipped with what for all the world seems an appropriate model of purely unitary measurement, deposits the cat in a state eerily superposed between life and death. This difficulty is the quantum measurement problem. Foundational discussions of quantum nonlocality and the measurement problem are legion. They are also largely immune to a foundational anxiety that is the mission of this contribution to induce. The anxiety is whether the definite descriptions earlier in the paragraph are appropriate: whether, that is, the quantum theory of two spin $\frac{1}{2}$ systems is *unique*; more generally, whether, there might be *multiple, physically inequivalent* ways to concoct a quantum theory from representations of the canonical relations circumscribing that theory.

The substantial literature on quantum nonlocality and the measurement problem is largely innocent of the uniqueness anxiety because it largely concerns quantum theories that fall within the scope of two results taken to silence that anxiety. These are the *Stone-von Neumann* and *Jordan-Wigner* theorems. According to the former, every Hilbert space representation of the CCRs for a particular classical Hamiltonian theory of finitely many particles is *unitarily equivalent* to every other.[1] For each finite n, the Jordan-Wigner theorem guarantees that representations of the CARs for n spin systems are unique up to unitary equivalence.[2] These results are taken to silence the uniqueness anxiety because unitary equivalence is widely supposed to explicate physical equivalence for (quantum theories obtained via) Hilbert space representations. Given the supposition, the results imply that superficially variant representations of the canonical relations circumscribing a quantum theory are merely variant means of expressing the same quantum kinematics. That is, they are merely variants provided the quantum theory in question concerns a suitably "finite" system.

But not all quantum theories do concern suitably finite systems. Quantum field theory (QFT), and the thermodynamic limit of quantum statistical mechanics (QSM), are quantum theories whose degrees of freedom are infinite in number. A typical QFT comes about as the quantization of a classical field theory, which assigns a field amplitude to *every point of spacetime*. To take QSM to the thermodynamic limit is to allow the number of microconstituents of the system analyzed, and the volume they occupy, go to infinity while the density remains finite. Concerning infinite systems, QFT and the thermodynamic limit of QSM—quantum theories which I will lump together under the heading "QM_∞"—fall outside the scope of the Jordan-Wigner and Stone-von Neumann uniqueness theorems. Indeed, the canonical relations circumscribing a theory of QM_∞ can admit continuously many unitarily inequivalent Hilbert space representations. According to very same criterion of physical equivalence, that warranted the reading of those theorems as results about physical equivalence, theories of QM_∞ can admit infinitely many presumptively physically inequivalent Hilbert space representations.

---

[1] Subject to provisos beautifully explained by Summers 1999.
[2] For an introduction, see Emch 1972, 269–275.

This contribution addresses quantum theories that fall outside the scope of the comforting uniqueness results, with a view toward assessing whether unitary equivalence is an appropriate criterion of physical equivalence for these theories. More emphasis is put on means of assessment and what motivates them than on endorsing and defending a particular answer. Indeed, I will suggest that appropriate criteria of physical equivalence for quantum theories are sensitive to factors that are not obviously criteria of identity for those theories, factors such as the uses to which those theories are being put and the scientific climates in which they find themselves.

I proceed as follows. The next section sketches the Stone-von Neumann and Jordan-Wigner uniqueness results and explicates the assumptions underlying their conventional reading as results about the physical equivalence of quantum theories. Section 3 develops two accessible examples of unitarily inequivalent representations. With these examples in hand, section 5 revisits the case that unitary equivalence is criterial for physical equivalence. Articulating a new-fangled alternative criterion, section 5 also identifies presuppositions favoring the traditional criterion over the new-fangled one. These include presuppositions about which relations between physical observables serve to define other physical observables, as well as presuppositions about what states are physical. The concluding section 6 discusses strategies for securing these presuppositions, observes certain tensions within these strategies, and comments on what to make of this state of play. Throughout, the exposition is informal, with references to more thorough and rigorous discussions supplied. In an effort to keep the discussion self-contained and accessible, a technical interlude (section 4) offers an incomplete and impressionistic introduction to the rudiments of operator algebra theory.

2. The "Uniqueness" Results

2.1 Preliminaries

This contribution derives its dramatic tension from the fact that entrenched criteria of physical equivalence for quantum theories render a verdict of 'inequivalent!' for what otherwise seem to be realizations of the same basic quantum theoretical structure. So our first task will be to announce a criterion of individuation for quantum theories. With that criterion in hand, we can turn to the question of what it takes for superficially different realizations of the theory, so individuated, to be physically equivalent.

What makes a quantum theory the theory it is? There is a consensus among the community of people who work with such theories. Fulling reports that "most theoretical physicists, following Schwinger, regard the *action principle* as fundamental.

Theories are defined by Lagrangians" (1989, 126). By fixing the symplectic structure of a classical theory, the Lagrangian fixes the commutation relations its quantizer seeks to represent.[3] But not all interesting quantum theories descend from real or imagined classical Lagrangians. Haag observes,

> The idea that one must first invent a classical model and then apply to it a recipe called "quantization" has been of great heuristic value. In the past two decades, the method of passing from a classical Lagrangian to a corresponding quantum theory has shifted more and more away from the canonical formalism to Feynman's path integral. This provides an alternative (equivalent?) recipe. There is, however, no fundamental reason why a quantum theory should not stand on its own legs, why the theory could not be completely formulated without regard to an underlying deterministic principle. (1992, 6)

Thus, Fulling proposes a liberalization of Schwinger's conventional wisdom:

> There is, however, another point of view, more consistent with the spirit of axiomatic field theory. *Any* commutation or anticommutation relations consistent with the dynamics can define a possible model.... In this approach a formal theory consists of equations of motion plus commutation rules (or some more general algebraic relations). They need not determine each other, but it is a nontrivial requirement that they be mutually consistent. (Fulling 1989, 126)

Agreeing that quantum theories are to be defined by their characteristic commutation relations and equations of motion, Fulling does not require that these emanate from the same source, or that that source be a Lagrangian.

To keep the discussion tractable, we will focus on the question of physical equivalence for quantum theories specified up to their *kinematics*—that is, their accounts of what states are physically significant, and (thinking of a state as a map from physical magnitudes to their expectation values) what physical magnitudes lie in the scope of those states. For theories specified up to their kinematics, Fulling's refinement of the conventional wisdom identifies canonical commutation or anticommutation relations as a principle of individuation. Where $\mathfrak{R}$ is a set of such relations, let Q_R be the quantum theory circumscribed by those relations. We want to know: When are different kinematical schemes for a theory Q_R physically equivalent?

---

[3] For an introduction, see Wald 1994, ch. 2.

2.2 Quantizing

It will help us address these questions to review some basic strategies for constructing quantum theories.

The state of a classical Hamiltonian system is given by its position and momentum. Take the simplest case of a single system moving on the real line. The position and momentum variables, real numbers q and p, act as coordinates for the phase space M of possible states of the system. M is just the real plane $\mathbb{R}^2$. For more complicated systems—n particles in Euclidean three-space, say—the phase space of possible states is larger ($\mathbb{R}^{2n}$), but constructed along the same principles. A classical Hamiltonian theory with phase space M represents physical magnitudes (aka observables) by functions from M to $\mathbb{R}$. The position and momentum observables for the simplest system are examples: they map points in phase space to their q and p coordinate values, respectively. All other observables pertaining to the system can be expressed as functions of these observables. The Hamiltonian observable H, which usually coincides with the sum of the system's kinetic and potential energies, is of particular significance. Fed into Hamilton's equations, H helps determine dynamically possible trajectories—the system's position and momentum as functions $q(t), p(t)$ of a time variable t—through M. For more complicated systems, the classical observable set and the dynamical prescription are more complicated, but the principles are the same.

The canonical *Hamiltonian quantization recipe* exploits the fact that the collection of classical observables just described exhibits an *algebraic structure*. As smooth functions on phase space, classical observables form a set that is also a vector space over the real numbers. An *algebra* is just a linear vector space endowed with a (not necessarily associative) multiplicative structure (see Kadison and Ringrose 1997 for an introduction). The vector space of classical observables becomes a *Lie algebra* upon being endowed with a multiplicative structure supplied by the *Poisson bracket*. The Poisson bracket $\{f, g\}$ of classical observables $f : M \to \mathbb{R}$ and $g : M \to \mathbb{R}$ is

$$\{f, g\} := \sum_i \left(\frac{\partial f}{\partial q_i} \frac{\partial g}{\partial p_i} - \frac{\partial f}{\partial p_i} \frac{\partial g}{\partial q_i} \right) \tag{1}$$

For the canonical observables p_i and q_j

$$\{p_i, p_j\} = \{q_i, q_j\} = 0, \; \{p_i, q_j\} = -\delta_{ij} \tag{2}$$

The Poisson bracket also affords a particular expeditious expression of Hamilton's equations

$$\frac{dq_i}{dt} = \{q_i, H\}, \; \frac{dp_i}{dt} = \{p_i, H\} \tag{3}$$

A description of the time-evolution of a general observable $f : M \to \mathbb{R}$ follows: $\frac{df}{dt} = \{f, H\}$.

Given its centrality both to the structure of classical observables and to their dynamics, it is tempting to think that the Poisson bracket structure has a great deal to do with making a classical Hamiltonian theory the theory it is.[4]

According to the canonical Hamiltonian quantization recipe, what it takes to quantize such a theory is to find a characteristically quantum mechanical analog of that theory's Poisson bracket structure. In particular, one identifies canonical quantum observables with symmetric operators $\hat{q}_i, \hat{p}_i$ acting on a separable Hilbert space $\mathcal{H}$ and obeying commutation relations corresponding to the classical Poisson brackets of the classical theory. When the classical theory has phase space $\mathbb{R}^{2n}$ and canonical observables q_i and p_i, these CCRs are (where $[\hat{A}, \hat{B}] := \hat{A}\hat{B} - \hat{B}\hat{A}$, $\hat{I}$ is the identity operator, and Planck's constant $\hbar$ is set to one)

$$[\hat{p}_i, \hat{p}_j] = [\hat{q}_i, \hat{q}_j] = 0, [\hat{p}_i, \hat{q}_j] = -i\delta_{ij}\hat{I} \tag{4}$$

For a classical theory with phase space $M = \mathbb{R}^{2n}$, the Hamiltonian quantization recipe is typically realized by the *Schrödinger representation*, set in the Hilbert space $L^2(\mathbb{R}^n)$ of square integrable complex-valued functions of $\mathbb{R}^n$. For $n = 1$, the Schrödinger representation defines $\hat{q}\psi(x) = x\psi(x)$ and $\hat{p}\psi(x) = -i\frac{d\psi(x)}{dx}$.

A story similarly centered on algebraic structures can be told about quantum theories of spin systems. To build the quantum theory of a single spin system, find symmetric operators $\{\hat{\sigma}(x), \hat{\sigma}(y), \hat{\sigma}(z)\}$ acting on a Hilbert space $\mathcal{H}$ to satisfy the *Pauli Relations*, which include

$$[\hat{\sigma}(x), \hat{\sigma}(y)] = i\hat{\sigma}(z), \ [\hat{\sigma}(y), \hat{\sigma}(z)] = i\hat{\sigma}(x), \ [\hat{\sigma}(z), \hat{\sigma}(x)] = i\hat{\sigma}(y) \tag{5}$$

Call the elements $\hat{\sigma}(j)$ satisfying (5) the *Pauli spin observables*. The generalization to n spin systems is straightforward. To build the quantum theory for n spin systems, find for each spin system k a Pauli spin $\hat{\sigma}^k = (\hat{\sigma}^k(x), \hat{\sigma}^k(y), \hat{\sigma}^k(z))$ satisfying the Pauli Relations, expanded to include the requirement that spin observables for different systems commute. A set of operators satisfies the Pauli relations if and only if they satisfy the CARs (see Emch 1972, 271–272); thus (5) can be taken to impose the algebraic structure of the CARs.

2.3 Unitary Equivalence as Physical Equivalence

A Hilbert space representation of the canonical relations $\mathfrak{R}$ circumscribing a quantum theory Q_R takes the form of operators $\hat{C}_i$ acting on a Hilbert space $\mathcal{H}$ to satisfy $\mathfrak{R}$. Having obtained such a representation, the aspiring quantum mechanic cannot rest. She needs to build products and linear combinations of her canonical observables if she is to have a viable theory of physics. Where $\hat{p}$ and $\hat{q}$ are her canonical observables, she will use $\hat{p}^2/2m$ to describe the kinetic energy of a particle of mass m. Where V is the magnitude of some configuration-dependent potential to which that particle is subject, she will use $V\hat{q}$ to describe its potential energy and use a

---

[4] See Butterfield 2007 and Belot 2007 for this idea elaborated.

Hamiltonian function, which is a sum of these, to describe its energy. In addition to such polynomials of the canonical observables, the aspiring quantum mechanic also needs observables that are defined as limits of sequences of other observables—for instance, the unitary Schrödinger evolution operators $\hat{U}(t) = e^{-i\hat{H}t}$ are limits of the Taylor series of polynomials of $\hat{H}$. Described in other terms, the aspiring quantum mechanic needs to build an observable algebra using the canonical observables $\hat{C}_i$ as generators. This means building the algebra from polynomials, and limits of sequences of polynomials, of those observables.

Hereinafter I will use "ordinary QM" to mean the tradition in which the observable algebra generated by representation-bearing canonical observables $\{C_i\}$ acting on $\mathcal{H}$ will coincide with $\mathfrak{B}(\mathcal{H})$, the full set of bounded operators on $\mathcal{H}$. In more technical terms (glossed immediately below), the aspiring quantum mechanic has begun with an *irreducible representation* of $\mathfrak{R}$, and used the *weak operator topology* to determine which sequences of polynomials of the canonical observables converge.

> $\{\hat{C}_i\}$ acting on $\mathcal{H}$ is an *irreducible representation* if and only if the only subspaces of $\mathcal{H}$ invariant under the action of $\{\hat{C}_i\}$ are the 0 subspace and $\mathcal{H}$ itself.

Unless otherwise announced, all representations discussed here will be irreducible.[5]

> A sequence $\hat{A}_i$ of operators on $\mathcal{H}$ converges to an operator $\hat{A}$ in $\mathcal{H}$'s *weak operator topology* if and only if for all $|\psi\rangle, |\phi\rangle \in \mathcal{H}$, $|\langle\psi|(\hat{A} - \hat{A}_i)|\phi\rangle|$ goes to 0 as i goes to ∞.

When I want to emphasize that the algebra $\mathfrak{B}(\mathcal{H})$ is generated by the set $\{\hat{C}_i\}$ of canonical observables representing $\mathfrak{R}$, I will use "$\mathfrak{B}(\mathcal{H})_{\hat{C}_i}$" to designate it.

In ordinary QM, quantum states are normalized, linear, positive, and countably additive maps from $\mathfrak{B}(\mathcal{H})$ to the complex numbers $\mathbb{C}$. Gleason's theorem tells us that states so conceived coincide with the set $\mathbb{T}^+(\mathcal{H})$ of density operators on $\mathcal{H}$, provided the dimension of $\mathcal{H}$ exceeds 2. Each $\hat{W} \in \mathbb{T}^+(\mathcal{H})$ determines a state via the trace prescription, which assigns each self-adjoint $\hat{A} \in \mathfrak{B}(\mathcal{H})$ the expectation value $Tr(\hat{W}\hat{A})$.

Call $(\mathfrak{B}(\mathcal{H})_{\hat{C}_i}, \mathbb{T}^+(\mathcal{H}))$ a *kinematic pair* for a theory of ordinary QM. $(\mathfrak{B}(\mathcal{H})_{\hat{C}_i}, \mathbb{T}^+(\mathcal{H}))$ is an instance of a general scheme $(\mathfrak{Q}, \mathcal{S})$ for kinematic pairs. The first entry gives the algebra of physical magnitudes recognized by a theory, with the theory's observables corresponding to that algebra's self-adjoint elements. The second entry gives the theory's physical states. The next section motivates a criterion of physical equivalence applicable to generic kinematic pairs, then argues that

---

[5] This is a restriction for the sake of simplicity, and a drastic one. Almost generically, QM$_\infty$ systems are described by states associated with *reducible*—that is, not irreducible—representations. See Earman and Ruetsche 2005 for a discussion.

for kinematic pairs of the form $(\mathfrak{B}(\mathcal{H})_{\hat{C}_i}, \mathbb{T}^+(\mathcal{H}))$ favored by ordinary QM, that criterion reduces to unitary equivalence.

2.4 Analyzing Physical Equivalence

The basic type of physical possibility recognized by a quantum theory takes the form of an expectation value assignment to the family of magnitudes recognized by that theory. There is, of course, a further interpretive question of how to understand the nontrivial probabilities implicit in such an expectation value assignment, and in particular of whether multiple, distinct "value states" correspond to each given quantum state. Different interpretations of ordinary QM urge different understandings of the quantum state, and thereby eventuate in different pictures of the set of physical possibilities associated with a quantum theory. But these disagreements occur, as it were, "downstream" from the identification of a kinematic pair on behalf of the theory. A question we can articulate and address without embroiling ourselves in these tendentious questions of interpretation is: When are candidate realizations of a quantum theory Q_R, realizations specified up to kinematic pairs, physically equivalent?

If the content of a physical theory consists in the set of physical possibilities it recognizes, then physical theories have the same content just in case they admit the same set of physical possibilities. On this picture, a necessary criterion for the physical equivalence of $(\mathfrak{Q}, \mathcal{S})$ and $(\mathfrak{Q}', \mathcal{S}')$ is a one-to-one correspondence between the physical possibilities admitted by the first pair and the physical possibilities admitted by the second. Claiming Glymour as an inspiration, Clifton and Halvorson (2001) analyze this demand into two conditions. The first is that this one-to-one correspondence 'preserve expectation values,' in the sense specified by a criterion I will call *PEV*:

> PEV. There are bijections $i_{obs} : \mathfrak{Q} \to \mathfrak{Q}'$ and $i_{state} : \mathcal{S} \to \mathcal{S}'$ such that
>
> $$i_{state}(\omega)(i_{obs}(A)) = \omega(A) \qquad (6)$$
>
> for all $A \in \mathfrak{Q}$ and all $\omega \in \mathcal{S}$.

Whenever two kinematic pairs satisfy PEV, to each state ω of one pair there corresponds a state $i_{state}(\omega)$ in the other such that $i_{state}(\omega)$'s assignment of expectation values to the observable $i_{obs}(A)$ exactly duplicates ω's assignment of expectation values to observable A. Part of how quantum theories characterize possibilities is as maps from physical magnitudes to their expectation values. Kinematic pairs satisfying PEV characterize exactly the same set of possibilities thus conceived. So let us join Clifton and Halvorson in supposing that two theories specified up to kinematic pairs are physically equivalent only if they satisfy PEV by admitting expectation-value-preserving bijections of the sort it demands.[6]

---

[6] What if my theory differs from yours only in a trivial scale transformation? That is, we don't satisfy PEV, but there are bijections $i_{obs} : \mathfrak{Q} \to \mathfrak{Q}'$ and $i_{state} : \mathcal{S} \to \mathcal{S}'$ such that (say) $i_{state}(\omega)(i_{obs}(A)) = 2 \times \omega(A)$. Wouldn't

PEV pays no obvious heed to the "algebraic structure" of the observable sets $\mathfrak{Q}$ and $\mathfrak{Q}'$. Nor does PEV engage the fact that those sets are in some sense descended from realizations of the canonical relations $\mathfrak{R}$ circumscribing the quantum theory Q_R. We might want a criterion of physical equivalence for kinematic pairs that is sensitive to such matters. After all, part of what makes a physical theory *the theory it is* is the functional relationships it posits between the physical magnitudes it recognizes. We have been recognizing this implicitly by identifying quantum theories by appeal to their constitutive CARs/CCRs, which express such relationships. Functional relationships between observables are moreover implicated in a theory's laws: $\hat{H} = \hat{p}^2/2m$ makes the energy of a free system of mass m a function of its momentum; the Schrödinger equation uses the energy of an isolated system to build a family $\hat{U}(t) = e^{-i\hat{H}t}$ of operators describing that system's time evolution, which implies (roughly speaking) that the operator $\hat{H}$ is a limit of a sequence of functions of the evolution operators $\hat{U}(t)$. These considerations suggest that criteria of physical equivalence for quantum theories specified up to kinematic pairs should include a demand to the effect that the algebraic structures of their observable families be suitably isomorphic.

Reflections like the following might tempt one to hope that when one demands the preservation of expectation values by imposing PEV, one gets a suitable isomorphism of algebraic structure for free:

> Where prime and unprimed commodities denote elements identified by the bijections satisfying PEV, suppose that i_{obs} between observable algebras failed to preserve additive algebraic structure. Then there exists observables X, Y such that $(X+Y)' \neq (X'+Y')$. Let us also suppose that the observables $(X+Y)'$ and $X'+Y'$ are different only if there is some state ω' that separates them in the sense that $\omega'[(X+Y)'] \neq \omega'(X'+Y')$. If PEV is satisfied, this separation condition implies $\omega(X+Y) \neq \omega(X+Y)$. But that is impossible. So PEV cannot be satisfied by a bijection i_{obs} that fails to preserve additive algebraic structure.

Similar arguments invoking the separation condition establish that PEV is satisfied only by bijections between observable sets that are linear and preserve their identity elements (see Roberts and Roepstorff 1969, Prop. 3.1).[7] The complete hope is that

---

it be mad to take this failure to satisfy PEV to disqualify our theories from physical equivalence?! I am not sure it would be. Notice that at least one of the theories entertains only states that fail to be normalized. And notice as well that we can restore unitary equivalence by attributing the theorists the same observable algebra but different conventions for coordinating self-adjoint elements of that algebra and measurement procedures. (Thanks to Dave Baker and Bryan Roberts, who independently raised this point.)

[7] Here is one reason linearity is important. Since only the self-adjoint elements of $\mathfrak{Q}$ and $\mathfrak{Q}'$ correspond to observables, PEV should take i_{obs} to be a bijection between the self-adjoint parts $\mathfrak{Q}_{sa}$ and $\mathfrak{Q}'_{sa}$ of those algebras, rather than a bijection between the algebras in their entirety. But for any element Q of $\mathfrak{Q}$, there are $A, B \subset \mathfrak{Q}_{sa}$ such that $Q = A + iB$. So if an i_{obs} restricted to the self-adjoint parts of the observable algebras acts linearly—a supposition the result just cited secures— it induces a bijection between the entire algebras, which bijection we will continue to call i_{obs}.

PEV on its own ensures a suitably "physical" isomorphism between the observable algebras of kinematic pairs that satisfy it. Examples dashing this hope will be provided in section 5. For now, let us take their existence to lend urgency to a (at present vague) demand that physically equivalent kinematic pairs enjoy suitably isomorphic observable algebras, and try to make that demand more precise.

The algebras at issue are each supposed to be generated by elements satisfying the relations $\mathfrak{R}$ circumscribing the quantum theory, relationships such as

$$AB - BA = kI$$

Where C_i satisfying $\mathfrak{R}$ generate the observable algebra $\mathfrak{Q}$ and C'_i satisfying $\mathfrak{R}$ generate the observable algebra $\mathfrak{Q}'$, I contend that a bijection $i_{obs} : \mathfrak{Q} \to \mathfrak{Q}'$ "preserves relevant algebraic structure" only if it enables the primed and unprimed theorists agree about what makes the canonical magnitudes canonical—only, that is, if i_{obs} maps observables realizing $\mathfrak{R}$ in the unprimed theory to observables realizing $\mathfrak{R}$ in the primed theory. Continuing the convention that primed and unprimed commodities are those identified by i_{obs}, this agreement requires that if the unprimed theorist's canonical relations are realized by $AB - BA = kI$, then those observables' primed counterparts also realize the canonical relations: $(AB - BA)' = kI' = A'B' - B'A'$ (where the first and last term are different ways of taking primed counterparts of observables involved in the canonical relations). The linearity of i_{obs} gives us $(AB - BA)' = (AB)' - (BA)'$. We are very close to drawing another conclusion about any i_{obs} that satisfies PEV. Given other features already established for i_{obs}, i_{obs} will qualify as an isomorphism, if only i_{obs} could be shown to be multiplicative.

> Where $\mathfrak{A}$ and $\mathfrak{A}'$ are algebras, a map $\alpha : \mathfrak{A} \to \mathfrak{A}'$ is a *morphism* if and only if α is linear, multiplicative ($\alpha(XY) = \alpha(X)\alpha(Y)$ for all $X, Y \in \mathfrak{A}$), and takes the identity to the identity. α is an *isomorphism* if it is one-to-one.[8]

Alas, we cannot use the separation condition tactic to argue that if i_{obs} satisfies PEV, then $(XY)' = X'Y'$. The catch is that, unlike sums of self-adjoint elements of our observable algebras, products of self-adjoint elements need not themselves be self-adjoint—and so need not themselves be observables, and so need not engage the gears of the separation condition. So shift attention to symmetrized products, elements of the form $XY + YX$, which are self-adjoint. Then a separation condition argument establishes that i_{obs} satisfies PEV only if

$$(XY + YX)' = X'Y' + Y'X' \quad [*]$$

which makes i_{obs} a Jordan homomorphism (Roberts and Roepstorff 1969, Prop. 6.1). But a Jordan homomorphism need not be an isomorphism. In particular, it need not

---

[8] I am suppressing a further condition arising from the fact that a quantum algebras has an adjoint operation *. It is that $\alpha(X^*) = (\alpha(X))^*$. We will want i_{obs} to satisfy this condition because we will want it to identify observables, self-adjoint elements, with observables.

be multiplicative. [*] will be satisfied just in case $(XY)' = X'Y' + Z_{XY'}$ and $(YX)' = Y'X' - Z_{XY'}$, even if $Z_{XY'}$ is a *nonzero* element of the primed algebra. However, our requirement that i_{obs} take canonical observables to canonical observables, along with the separation condition, implied that $(AB - BA)' = (AB)' - (BA)' = A'B' - B'A'$. That cannot be true unless $Z_{AB'} = 0$—unless, that is, i_{obs} acts multiplicatively on canonical observables. Because canonical observables generate the algebra, it follows that i_{obs} act multiplicatively on the algebra, period. And that completes the case that, supposing our observable set is rich enough to separate states, an i_{obs} satisfying PEV preserves relevant physical structure only if it is an isomorphism between observables algebras $\mathfrak{Q}$ and $\mathfrak{Q}'$.

But i_{obs} must accomplish one more task if it is to secure physical equivalence. i_{obs} is an isomorphism between the observable algebras of kinematic pairs. These kinematic pairs cannot be physically equivalent if they do not agree about *what the theory's fundamental canonical magnitudes are*. Thus, they are not physically equivalent unless the isomorphism i_{obs} identifies the canonical magnitudes generating one algebra with the canonical magnitudes generating the other.

Let us consolidate the foregoing reflections on what, beyond satisfying PEV, i_{obs} must accomplish to establish physical equivalence. Where $\hat{C}_i$ satisfying $\mathfrak{R}$ generate the observable algebra $\mathfrak{Q}$ and $\hat{C}'_i$ satisfying $\mathfrak{R}$ generate the observable algebra $\mathfrak{Q}'$, a map i_{obs} preserves relevant physical structure just in case it <u>P</u>reserves <u>A</u>lgebraic <u>S</u>tructure by satisfying

PAS. i_{obs} is an isomorphism between $\mathfrak{Q}$ and $\mathfrak{Q}'$ such that

$$i_{obs}(\hat{C}_i) = \hat{C}'_i \qquad (7)$$

for all i.

Our analysis of physical equivalence for quantum theories specified up to generic kinematic pairs is thus:

> Kinematic pairs $(\mathfrak{Q}, \mathcal{S})$ and $(\mathfrak{Q}', \mathcal{S}')$ for a quantum theory Q_R circumscribed by the relations $\mathfrak{R}$ are physically equivalent if and only if there exist bijections $i_{obs} : \mathfrak{Q} \to \mathfrak{Q}'$ and $i_{state} : \mathcal{S} \to \mathcal{S}'$ satisfying both PEV and PAS.

This is essentially the analysis offered by Clifton and Halvorson, although the justification offered here uses words different from the words used in their justification.[9]

Now we are getting somewhere. It turns out that kinematic pairs of ordinary QM's form, kinematic pairs $(\mathfrak{B}(\mathcal{H}), \mathfrak{T}^+(\mathcal{H}))$ and $(\mathfrak{B}(\mathcal{H}'), \mathfrak{T}^+(\mathcal{H}'))$, satisfy both

---

[9] Clifton and Halvorson (2001) follow Glymour's analysis of physical equivalence, which supposes physical theories to be interpreted as axiomatic systems. For Glymour, such theories are physically equivalent only if intertranslatable in such a way that axioms get translated as axioms and theorems get translated as theorems. Roughly speaking, Clifton and Halvorson assimilate the generators of an observable algebra to axioms and its other elements to theorems, thereby motivating (7) as the axiom-to-axiom demand and PEV as the theorem-to-theorem demand. I take my reconstruction to agree in spirit with theirs.

PEV and PAS if and only if their collections $\{C_i\}$ and $\{C'_i\}$ of canonical operators are unitarily equivalent in the following sense:

> A Hilbert space $\mathcal{H}$, and a collection of operators $\{\hat{O}'_i\}$ is *unitarily equivalent* to $(\mathcal{H}', \{\hat{O}'_i\})$ if and only if there exists a one-to-one, linear, invertible, norm-preserving transformation ("unitary map") $U : \mathcal{H} \to \mathcal{H}'$ such that $U^{-1}\hat{O}'_i U = \hat{O}_i$ for all i.

Here is a sketch of an argument that $(\mathfrak{B}(\mathcal{H}), \mathfrak{T}^+(\mathcal{H}))$ and $(\mathfrak{B}(\mathcal{H}'), \mathfrak{T}^+(\mathcal{H}'))$ satisfy both PEV and PAS if and only if they arise from unitarily equivalent representations of the canonical relations. To see that unitary equivalence is sufficient, notice that the U effecting the unitary equivalence of primed and the unprimed representations of $\mathfrak{R}$ furnishes both a bijection $i_{state}(\hat{W}) = U\hat{W}U^{-1}$ from the unprimed kinematic pair's state space to the primed pair's state space and a bijection $i_{obs}(\hat{A}) = U\hat{A}U^{-1}$ from the unprimed kinematic pair's observable algebra to the primed pair's algebra. The property of unitary maps that $UU^{-1} = U^{-1}U = I$ along with the trace prescription guarantee that these bijections together satisfy PEV. i_{obs} is moreover an isomorphism between the observable algebras, because of truths such as $U(A+B)U^{-1} = UAU^{-1} + UBU^{-1}$. Finally, the isomorphism i_{obs}, which is induced by a map identifying canonical elements of the unprimed algebra with canonical elements of the primed algebra, satisfies (7), and therefore satisfies (PAS). Notice for future reference that because a unitary map between $\mathcal{H}$ and $\mathcal{H}'$ preserves inner products, a sequence $A_i \in \mathfrak{B}(\mathcal{H})$ converges weakly to an element $A \in \mathfrak{B}(\mathcal{H})$ if and only if the sequence $UA_iU^{-1} \in \mathfrak{B}(\mathcal{H})$ converges weakly to an element $UAU^{-1} \in \mathfrak{B}(\mathcal{H})$.

To see that the unitary equivalence of the representations $\{C_i\}$ and $\{C'_i\}$ underlying the kinematic pairs $(\mathfrak{B}(\mathcal{H}), \mathfrak{T}^+(\mathcal{H}))$ and $(\mathfrak{B}(\mathcal{H}'), \mathfrak{T}^+(\mathcal{H}'))$ is necessary for those pairs to satisfy both PEV and PAS, note that $\mathfrak{B}(\mathcal{H})$ and $\mathfrak{B}(\mathcal{H}')$ are Type I von Neumann algebras. Because all isomorphisms between Type I von Neumann algebras are implemented unitarily, an isomorphism i_{obs} satisfies PAS only if it is induced by a unitary map and takes C_i to C'_i. But such an isomorphism is available only if the representations are unitarily equivalent.

To summarize: particularized to kinematic pairs of ordinary QM's sort, the analysis of physical equivalence in terms of PEV and PAS implies that such pairs are physically equivalent if and only if the representations of $\mathfrak{R}$ generating them are unitarily equivalent.

In 1931, von Neumann demonstrated what had been conjectured the previous year by Stone: the *unitary equivalence* (up to multiplicity) of any pair of Hilbert space representations of the CCRs arising from a classical theory with phase space $\mathbb{R}^{2n}$. The Jordan-Wigner theorem likewise establishes the uniqueness, up to unitary equivalence, of Hilbert space representations of the CARs for n degrees of freedom (n finite). Given our analysis of physical equivalence, it follows that once we have settled on the CCRs or CARs circumscribing a quantum theory in the scope of these results, every ordinary QM-ish kinematic pair we can construct on behalf of that theory is physically equivalent to every other. Disagree howsoever we might about the further interpretation of that theory, we can at least agree about its core identity, an identity provided by a kinematic pair that is essentially unique.

3. UNITARY INEQUIVALENCE: SOME EXAMPLES

This section musters several examples of quantum theories falling outside the scope of the Stone-von Neumann and Jordan-Wigner theorems. The canonical relations circumscribing these theories admit unitarily inequivalent representations. Our working analysis of physical equivalence implies that ordinary QM-ish kinematic pairs based on such representations are also *physically inequivalent*. We will raise some worries about whether 'physically inequivalent!' is the right verdict to reach about the representations sketched here.

3.1 The Infinite Spin Chain

An exceedingly simple quantum theory whose CARs admit unitarily inequivalent representations is the theory of infinitely many spin $\frac{1}{2}$ systems in a linear array.[10] As a warmup for our encounter with this theory, consider the quantum theory of a finite number n of spin $\frac{1}{2}$ systems, arranged in a one-dimensional lattice. To construct such a theory, we need only equip each location k in the lattice with a Pauli spin $\hat{\sigma}^k = (\hat{\sigma}^k(x), \hat{\sigma}^k(y), \hat{\sigma}^k(z))$ in such a way that the collection of these Pauli spins satisfies the Pauli relations.

One way to do this employs a vector space $\mathcal{H}$ spanned by a basis whose elements correspond to sequences s_k, where each entry in the sequence takes one of the values ± 1, and k ranges from 1 to n. (Notice that there are finitely many distinct such sequences, because there are only finitely many ways to map a set of finite cardinality into the set $\{+1, -1\}$.) We introduce operators $\hat{\sigma}^j(z), j = 1$ to n in such a way that sequences s_k whose j^{th} entry is ± 1 serve as $\hat{\sigma}^j(z)$ eigenvectors associated with the eigenvalue ± 1. Along with operators $\hat{\sigma}^j(y), \hat{\sigma}^j(x)$, constructed by analogy to their single electron counterparts, these operators provide a representation of the Pauli relations, and thus the CARs, for n.

Of course, we could have constructed a representation of Pauli relations by many other, superficially competing, means. But because we are considering only finitely many spin systems, the Jordan-Wigner theorem guarantees that any representation of the CARs for n is unitarily equivalent to any other.

---

[10] Here I follow Sewell 2002, §2.3, to which I refer the reader for details.

Let us belabor a consequence of that. Imagine that Werner and Erwin each build a representation of the Pauli relations for chain of n spin system. Let $\sigma^k(i)^W$ be the operator on $\mathcal{H}_W$ by which Werner represents the ith component of spin for the kth particle; let $\sigma^k(i)^E$ be the operator on $\mathcal{H}_E$ by which Erwin represents the ith component of spin for the kth particle.[11] If Werner's representation and Erwin's are unitarily equivalent, then there exists a unitary map $U : \mathcal{H}_E \to \mathcal{H}_W$ such that

$$U\sigma^k(i)^E U^{-1} = \sigma^k(i)^W \quad \text{for all } i \in \{x, y, z\}, k \in \{1, 2, \ldots, n\} \tag{8}$$

Because unitary maps are linear and norm preserving, this unitary map not only identifies each Pauli spin operator in Erwin's representation with a Pauli spin operator in Werner's representation, it also extends *in a way that respects the identifications between Pauli spins* to a bijection between the full sets of bounded operators on each theorist's Hilbert space.

The *polarization* of a system will be of particular interest. A system's polarization is described by a vector whose magnitude ($\in [0, 1]$) gives the strength and whose orientation gives the direction of the system's net magnetization. On a single electron, it is represented by an observable $\hat{m}$ whose three components correspond to three orthogonal components of spin. Thus in the $+1$ eigenstate $|+\rangle$ of $\hat{\sigma}(z)$ (understood as the z-component of spin), the polarization has an expectation value of $+1$ along the z axis. For a finite chain of spins, the polarization observable has components $\hat{m}(i)$, $i \in \{x, y, z\}$, that are just the average, over links in the chain, of the corresponding component of spin: $\hat{m}(i) = \frac{1}{n} \sum_{k=1}^{n} \hat{\sigma}^k(i)$. Let $[s_k]_j \in \{\pm 1\}$ denote the jth entry of the sequence s_k. In the basis sequence s_k, the z-component of polarization $\hat{m}(z)$ takes an expectation value of magnitude $\frac{1}{n} \sum_{j=1}^{n} [s_k]_j$. This quantity attains extreme values (of ± 1) for sequences every term of which is the same.

From their representations of the Pauli relations, Erwin and Werner both construct a kinematic pair of the ordinary QM form $(\mathfrak{B}(\mathcal{H}), \mathbb{T}^+(\mathcal{H}))$. The Jordan-Wigner theorem implies that any other representation of the Pauli relations will be unitarily equivalent to Werner's. Werner's and Erwin's kinematic pairs thus satisfy both PEV and PAS. Suppose $\hat{\rho}$ is a state in Werner's state set assigning $\hat{m}(z)$ the expectation value $+1$. Then Erwin's state set must include a state $\hat{\rho}'$ (the image of $\hat{\rho}$ under the isomorphism induced by unitary map implementing the equivalence of the representations) and an observable $\hat{m}(z)'$ (the image of $\hat{m}(z)$ under that map) such that the expectation value of $\hat{m}(z)'$ in the state $\hat{\rho}'$ is $+1$.

Now let us leave the scope of the Jordan-Wigner theorem to consider a doubly infinite chain of spins, its sites labeled by the positive and negative integers $\mathbb{Z} = \{\ldots, -2, -1, 0, 1, 2, \ldots\}$. We are after a representation of the CARs that associates with each site k a Pauli spin satisfying the Pauli relations. But we cannot adapt the strategy adopted for the finite spin chain to do so. Such an adaptation would build the representing Hilbert space from a basis consisting of all possible maps from $\mathbb{Z}$ to

---

[11] For the duration of this explication, I am dropping hats over operators to minimize notational clutter.

{±1}. Because the set of such maps is nondenumerable, the Hilbert space envisioned would be nonseparable, breaking the staunch tradition of using separable Hilbert spaces for physics.[12]

One way to build a *separable* Hilbert space representation is to start with the "base" sequence $[s_k]_j = +1$ for $j \in \mathbb{Z}$, and add all sequences that differ from the base one at only finitely many local sites. The collection of such sequences forms a basis; its members are sequences in which only finitely many +1s appear. A Hilbert space spanned by the basis hosts a representation of the Pauli relations modeled on that of the finite spin chain. In particular, it features operators $\hat{\sigma}^j(z)^+, j \in \mathbb{Z}$. Sequences s_k whose j^{th} entry is ±1 are $\hat{\sigma}^j(z)^+$ eigenvectors associated with the eigenvalue ±1. Call this *the $\mathcal{H}^+$ representation* for short. Much is lost in the abbreviation, for it matters to the algebraic structure of this representation which elements of $\mathfrak{B}(\mathcal{H}^+)$ play the role of which Pauli spins.

A total polarization observable $\hat{m}^+$ can be defined in terms of the $\mathcal{H}^+$ representation if it can be understood to be an element of the observable algebra $\mathfrak{B}(\mathcal{H}^+)$ generated by that representation. To be an element of $\mathfrak{B}(\mathcal{H}^+)$, the polarization observable must be a polynomial of Pauli spins, or the limit (in the appropriate sense) of a sequence of such polynomials. Consider the sequence of partial sums $\hat{m}(z)_N := \frac{1}{2N+1} \sum_{k=-N}^{N} \hat{\sigma}^k(z)^+$, which define the z component of net polarization of finite stretches of the chain. For each N, $\hat{m}(z)_N$ is a polynomial of Pauli spins and so a member of $\mathfrak{B}(\mathcal{H}^+)$. The z component of net polarization for the entire infinite chain would be given by the $N \to \infty$ limit of the sequences of partial sums $\hat{m}(z)_N$, if that limit exists. In $\mathcal{H}^+$'s weak topology, it does.

Recall that a sequence $\hat{A}_i$ of operators on $\mathcal{H}$ converges to an operator $\hat{A}$ in $\mathcal{H}$'s weak operator topology if and only if for all $|\psi\rangle, |\phi\rangle \in \mathcal{H}$, $|\langle\psi|(\hat{A} - \hat{A}_i)|\phi\rangle|$ goes to 0 as i goes to ∞. Considering the sequence $\hat{m}(z)_N$, remark that $\langle s_k|\hat{m}(z)_N|s'_k\rangle = \frac{1}{2N+1} \sum_{j=-N}^{N} \left(\frac{[s_k]_j + [s'_k]_j}{2}\right)$ for basis sequences s_k and s'_k. Because -1 occurs only finitely many times in each basis sequence, $\langle s_k|\hat{m}(z)_N|s'_k\rangle$ will converge to 1 as $N \to \infty$, no matter what s_k and s'_k are. This shows that the sequence $\hat{m}(z)_N$ converges weakly. Its limit is an operator that has every vector in $\mathcal{H}^+$ as an eigenvalue 1 eigenvector (because every element of $\mathcal{H}^+$'s basis is such a vector, and that basis spans the space). In other words, in an ordinary QM-ish observable algebra generated by taking the weak closure of the $\mathcal{H}^+$ representation, the z-component of the global polarization just is the identity operator I^+ on $\mathcal{H}^+$.

---

[12] Simon expresses a commitment to the tradition as he launches into an exposition of those aspects of functional analysis he considers most central to physics: "Throughout, all our Hilbert spaces will be separable unless otherwise indicated. Many of the results extend to non-separable spaces, but we cannot be bothered with such obscurities" (1972, 18). Although there are some ways in which the mathematics of separable Hilbert spaces are "nicer" (for instance, some operator topologies are first-countable), I am not aware of a canonical explanation of the tradition.

The other components of the total polarization can also be defined as weak limits of polynomials of Pauli spins. Each component of the total polarization is an element of the algebra $\mathfrak{B}(\mathcal{H}^+)$, and so an observable in ordinary QM's sense.

It is noteworthy that there are representations of the Pauli relations, modeled on the $\mathcal{H}^+$ representation, for which the convergence constituting the global polarization as a bona fide observable fails to obtain. Consider, for instance, a representation whose "base state" s_k is a sequence for which $\lim\limits_{N \to \infty} |\langle s_k | \hat{m}(z)_N | s_k \rangle|$ does not converge (e.g., the sequence $[s_k]_j = +1$ for $2^n < |j| \leq 2^{n+1}$ and n odd; $[s_k]_j = -1$ otherwise). The $\mathcal{H}^+$ representation hosts a total polarization observable only because $\mathcal{H}^+$'s weak topology facilitates the definition of such an observable. For other representations, this need not be so. Whether there is a global polarization observable hinges on our choice of representation. Supposing kinematic pairs numbering polarization observables among their bona fide magnitudes differ physically from kinematic pairs that do not, this is a hint that when the spin chains become infinite, the choice of representation could have physical significance.

The Jordan-Wigner theorem applies to representations of the CARs for finitely many spin systems. Not applying to the infinite spin chain, the theorem is silent about whether the $\mathcal{H}^+$ representation is unique up to unitary equivalence. Indeed, it is not. Consider, for contrast, a representation set in a Hilbert space whose basis elements correspond to the sequence $[s_k]_j = -1$ for $j \in \mathbb{Z}$, along with all sequences differing from this one in only finitely many places. Operators $\hat{\sigma}^k(i)^-$ satisfying the Pauli relations are introduced in such a way that $[s_k]_j$, the jth entry in the basis sequence s_k, gives the expectation value of $\hat{\sigma}^j(z)$. Call this *the $\mathcal{H}^-$ representation*. By parity of reasoning, the z-component of the total polarization $\hat{m}(z)^-$ is an observable generated by this representation. Enjoying every vector in $\mathcal{H}^-$ as an eigenvalue -1 eigenvector, $\hat{m}(z)^-$ coincides with $-I^-$.

To establish that the $\mathcal{H}^-$ and $\mathcal{H}^+$ representations are not unitarily equivalent, suppose, for contradiction, that they were. Then, we know from the last section, the unitary map between these representations would define bijections i_{obs} and i_{state} between the observable algebras and state sets of the kinematic pairs $(\mathfrak{B}(\mathcal{H}^+), \mathbb{T}^+(\mathcal{H}^+))$ and $(\mathfrak{B}(\mathcal{H}^-), \mathbb{T}^+(\mathcal{H}^-))$, bijections satisfying PEV and PAS. Any bijection i_{obs} that preserves algebraic structure (as demanded by PAS) must map each $\hat{\sigma}^k(z)^+$ in the $\mathcal{H}^+$ representation to $\hat{\sigma}^k(z)^-$ in the $\mathcal{H}^-$. Preserving this correspondence through the sequence of polynomials of Pauli spins whose limit defines the z-component of global polarization, a unitarily implemented i_{obs} obedient to PAS maps $\hat{m}(z)^+$ on the $\mathcal{H}^+$ representation to $\hat{m}(z)^-$ on the $\mathcal{H}^-$ representation. But that is to say that i_{obs} maps the identity operator on the first representation to -1 times the identity operator on the second. It follows that no bijection i_{state} between states of the $\mathcal{H}^+$ and $\mathcal{H}^-$ representations can consort with i_{obs} to preserve expectation values as demanded by PEV. Such an i_{state} would have to identify a state ρ on the $\mathcal{H}^+$ representation with a state $i_{state}(\rho)$ on the $\mathcal{H}^-$ representation in such a way that $i_{state}(\rho)$ assigns $i_{obs}(\hat{m}(z)^+)$ the same value ρ assigns $\hat{m}(z)^+$. But—keeping in mind that all states are linear and normed—ρ assigns $\hat{m}(z)^+$ the value 1, because

$\hat{m}(z)^+$ is the identity operator, whereas any state on the $\mathcal{H}^-$ assigns $i_{obs}(\hat{m}(z)^+)$ the value -1, because $i_{obs}(\hat{m}(z)^+)$ is -1 times the identity operator. Thus, no bijection i_{obs} between the observable sets that preserves algebraic structure can hope also to preserve expectation values. The bijections that would have to exist, if the representations were unitarily equivalent, cannot exist. We conclude that the representations fail to be unitarily equivalent.

This argument that the $\mathcal{H}^+$ and $\mathcal{H}^-$ representations are not unitarily equivalent makes striking an apparent physical difference between the kinematic pairs based on those representations. In an ordinary quantum theory built up from the $\mathcal{H}^+$ representation, states whose polarizations differ from $+1$ in the z direction do not occur. In an ordinary quantum theory built up from the $\mathcal{H}^-$ representation, only such states occur. Thus, the values of global polarization allowed distinguish physically between theories based on the inequivalent representations. Indeed, the rival quantum theories built on those representations can be subject to a critical test in the form of a measurement of the z-component of global polarization. And this physical difference would be expected to persist in finer-grained interpretations of the kinematic pairs corresponding to the representations, supposing those interpretations take systems in eigenstates of an observable to actually possess the corresponding eigenvalue of the observable.

Section 3.3 will raise the question of whether we can with good conscience regard the difference just elucidated to be a genuine physical difference between rival quantum theories.

3.2 The Bead on a Circle

To leave the scope of the Stone-von Neumann theorem, we need only consider the apparently simple system consisting of a single particle constrained to move on the unit circle S_1. The canonical variables of a classical Hamiltonian treatment of this system are its position, given by an angular variable $\phi \in [0, 2\pi]$, and its angular momentum $\ell \in \mathbb{R}$. Thus its configuration space is the circle S_1 and its phase space is the cylinder $S_1 \times \mathbb{R}$.

Before quantizing,[13] a change of variables from the standard cylindrical coordinates (ϕ, ℓ) is in order. We are aiming in our quantization to reproduce the Poisson bracket structure of the classical theory (at least insofar as it applies to canonical observables) in corresponding commutation relations between Hilbert space operators. The variable ϕ is not a classical observable, because classical observables are continuous functions on phase space, but ϕ is not. The discontinuity occurs as ϕ approaches 2π, and occurs because the configuration space of the system is a circle. So we will instead characterize the classical algebraic structure in terms of the

---

[13] A project in which my impressionistic exposition tries to follow the careful treatments of Isham (1983) and Gotay (2000). Thanks are owed to Gordon Belot for help with this. Blame for persisting misunderstandings is not.

variables

$$x = \cos\phi, \quad y = \sin\phi, \quad z = \ell \tag{9}$$

which are continuous on the cylinder.[14]

Among these variables, the standard Poisson brackets are given by

$$\{x,y\} = 0, \quad \{z,x\} = y, \quad \{z,y\} = -x \tag{10}$$

Following the Poisson bracket goes to commutator rule, we build a quantum theory of the particle on the circle, by finding a Hilbert space representation of the Circular Canonical Commutation Relations (CCCRs) corresponding to the Poisson brackets (10):

$$[x,y] = 0, \quad [z,x] = iy, \quad [z,y] = -ix \tag{11}$$

These CCCRs (11) have a standard representation in terms of Hilbert space operators acting on $L^2(S_1)$, the space of functions $\psi(\phi) : S_1 \to \mathbb{C}$ that are square integrable with respect to the measure $\frac{d\phi}{2\pi}$.

$$\hat{x} = \cos\phi, \quad \hat{y} = \sin\phi, \quad \hat{z} = i\frac{d}{d\phi} \tag{12}$$

Happily, the spectrum of $\hat{x}$ and $\hat{y}$ is $[-1,1]$, which is exactly as it should be for a system whose configuration space is the unit circle S_1. Also happily, $\hat{z}$'s spectrum is $2\pi n$, $n \in \{0, 1, 2, ...\}$. That is, it is the angular momentum spectrum for a particle on a circle suggested by our boundary conditions and the de Broglie relations.

The Stone-von Neumann theorem concerns representations of CCRs expressing the quantization of a classical theory with phase space $\mathbb{R}^{2n}$. But the CCCRs are not the CCRs, and the phase space for a bead on the circle is not $\mathbb{R}^{2n}$ for any n, but the cylinder $S_1 \times \mathbb{R}$. Thus the Stone-von Neumann theorem does not imply that the standard representation (12) of the CCCRs is unique up to unitary equivalence. And it is not. Instead, there exists a family representations of the CCCRs labeled by $\theta \in [0, 1]$ (see Isham 1983, 1270–1272 for details). These representations are related to the standard one by:

$$\hat{x}_\theta = \hat{x}, \quad \hat{y}_\theta = \hat{y}, \quad \hat{z}_\theta = i\frac{d}{d\phi} + \theta \tag{13}$$

The $\theta = 0$ member of this family is just the standard representation, which we will call Q_0. For $\theta \neq 0$, the representation Q_θ agrees with the Q_0 representation about the spectra of the configuration observables x and y. But it disagrees about the angular momentum spectrum. In the standard representation Q_0, the angular momentum spectrum is $\{2\pi n\}$ for integer n; in Q_θ, the angular momentum spectrum is $\{2\pi(n - \theta)\}$. Representations Q_θ and $Q_{\theta'}$ are unitarily equivalent only if $\theta = \theta'$. Thus we have a veritable host of unitarily inequivalent representations of the CCCRs.

---

[14] Another approach is to let the configuration variable be unitary instead of self-adjoint. See Lèvy-Leblond 1976.

This indicates a signature of the physical difference between ordinary quantum theories based on unitarily inequivalent representations: their momentum spectra.[15] States possible according to an ordinary quantum theory built up from the Q_0 representation include $\hat{z}_0$ eigenstates in which the bead's angular momentum vanishes. States possible according to an ordinary quantum theory built up from the $Q_{\theta \neq 0}$ representation include no $\hat{z}_\theta$ eigenstates in which the bead's angular momentum vanishes.[16]

The family Q_θ of unitarily inequivalent representations of the CCCRs are not idle mathematical curiosities. They are related to the θ angles of Yang-Mills theory. With minor adjustments, they can be applied to the Bohm-Aharonov effect, wherein an electron translated around an (infinitely extended) solenoid experiences a phase rotation determined by the flux through the solenoid.[17] Excluding the electron from the region of space occupied by the solenoid, we attribute it a configuration space that is $\mathbb{R}^3$ with a cylinder removed. Like the circle, this topologically non-$\mathbb{R}^n$ configuration space frames a Hamiltonian theory that admits a family of inequivalent quantizations. Different members of this family correspond to different fluxes through the solenoid and hence to different phase shifts for the transported electron (see Landsman 1990 for details).

Thinking big, observe that mechanics set in a spatially compact universe (such as, presumably, our own) will have a topologically non-$\mathbb{R}^{2n}$ phase space, placing its quantization outside the scope of the Stone-von Neumann theorem.

3.3 Other Unitarily Inequivalent Representations

The foregoing examples hardly exhaust the field of unitarily inequivalent representations. Other examples drawn from QFT have elicited the attention of philosophers of physics. These include: Fock space representations quantizing the free Klein-Gordon field associated with 'incommensurable' particle notions (Clifton and Halvorson 2001; Arageorgis, Earman, and Ruetsche 2003); the standard Minkowski vacuum representation and infrared coherent representations (Baker 2009); representations associated with different states of broken symmetry (Earman 2004; Liu and Emch 2005); representations generated from one another by time evolution in nonstationary spacetimes (Arageorgis, Earman, and Ruetsche 2002). Examples from the thermodynamic limit of QSM include representations associated with equilibrium states at different temperatures, or associated with different phases at the same temperature (Ruetsche 2003; Emch 2007). Even the simplest quantum mechanical

---

[15] See Dürr et al. for an argument that Bohmian mechanics "provides a sharp mathematical justification" (2006, 791) of the expectation that classical theories with topologically exotic configuration spaces have unitarily inequivalent quantizations.

[16] Appropriately weighted superpositions of $\hat{z}_\theta$ eigenstates might assign expectation value 0 to angular momentum, but can be empirically distinguished from $\hat{z}_\theta$ eigenstates by repeated non-disturbing measurements of angular momentum.

[17] For why this might be interesting, see Belot 1998 or Healey 2007. Dürr et al. 2006 catalogs other topologically non-$\mathbb{R}^{2n}$ phase spaces with physical applications.

system—a single particle confined to the real line—admits unitarily inequivalent representations in the form of the 'position' and 'momentum' representations (Halvorson 2001). Running roughshod over conventional expectations (see Teller 1979), the former makes available exact position eigenstates; the latter does the same for momentum.

Even on their own, the examples developed above should inspire us to interrogate the assumptions underlying our reception of unitary equivalence as criterial for physical equivalence. These include assumptions about the proper configuration of kinematic pairs for quantum theories, as well as the proper analysis of physical equivalence for kinematic pairs generically conceived. When kinematic pairs of ordinary QM's form $(\mathcal{B}(\mathcal{H}), \mathbb{T}^+(\mathcal{H}))$ are subject to criteria of physical equivalence explicated by PEV and PAS, it follows that kinematic pairs based on representations of canonical relations $\mathfrak{R}$ are physically equivalent just in case those representations are unitarily equivalent.

With respect to the bead on the circle, embracing the criterion limits the angular momentum spectrum for the system to the one given by a single member of the family Q_θ of representations. Such a limitation hamstrings the quantum theory's capacity to model the Bohm-Aharonov effect, as well as other phenomena admitting descriptions in terms of "double-valued" wave functions, not to mention the applications catalogued in the penultimate paragraph of section 3.2.

With respect to the infinite spin chain, embracing the criterion requires us to deny that physically possible conditions of the chain include *both* states in which its global polarization takes the value $+1$ in the z-direction and states in which its global polarization takes the value -1 in the z-direction. Such a denial is in tension with the behavior of ferromagnetic substances, which exhibit both kinds of spontaneous magnetization.[18] This exhibition is moreover an example of a critical phenomenon engaging in what is known as universal behavior. The behavior of a wide variety of ferromagnetic substances is described by a phase diagram of a ferromagnet (see figure 14.1) characterized by the same critical exponents. The theory of universal phenomena is described elsewhere in the volume. The present point is that to use a phase diagram such as figure 14.1 to describe a system, thereby bringing it within the ambit of this theory, is to allow it *distinct possible states of spontaneous magnetization*. Taking unitary equivalence to be criterial for physical equivalence cancels this allowance.

Resistance to the criterion also derives from a very different sort of attitude toward the infinite spin chain.[19] It is a commonplace among philosophers of space and time that solutions to the fundamental equations defining a spacetime theory,

---

[18] Another suppressed complication: the ferromagnetic-paramagnetic phase transition does not occur in the 1-d model provided by the infinite spin chain, but does in models of higher dimension. See Emch and Liu 2002 for further discussion.

[19] It will not escape the reader's notice that the morals drawn in this paragraph undermine the considerations of the last paragraph. Since I am here only trying to motivate discontent with unitary equivalence as a criterion of physical equivalence, that is fine with me: whether you are drawn by the consideration of the last paragraph or the present paragraph, you have a reason to be unhappy with the criterion. Since I myself am drawn by both sorts of considerations, I face a real puzzle, which deserves more attention than it gets here.

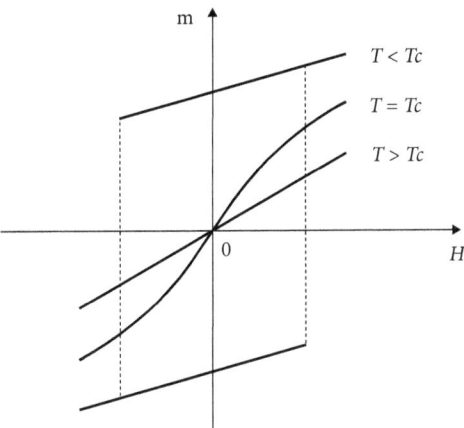

Figure 14.1 Phase diagram for a ferromagnet

solutions connected by a spacetime symmetry of those equations, are physically equivalent.[20] And it is easy to imagine embedding the infinite spin chain into a spacetime theory in such a way that (for instance) the base state of the $\mathcal{H}^+$ representation is connected to the base state of $\mathcal{H}^-$ representation by the space time symmetry of flipping over the z-axis. This symmetry moreover extends to the Hilbert spaces in their entireties, identifying each vector in $\mathcal{H}^+$ with its 'flipflop' in $\mathcal{H}^-$. Implying that each pair of symmetry-connected vectors represents the same physical situation, the metaphysical commonplace implies that the collections of states implemented by density operators on $\mathcal{H}^+$ and on $\mathcal{H}^-$ represent the *very same collection of physical situations*. Accepting unitary equivalence as a criterion of physical equivalence means rejecting the metaphysical commonplace.[21]

This suggests that widespread metaphysical intuitions, as well as the use that practicing physicists make of unitarily inequivalent representations, stand in tension with the established accounts of what quantum theories are and when they are physically equivalent. After a technical interlude, section 5 explores reactions to this distressing circumstance.

4. Technical Interlude

For the sake of formulating and evaluating various responses to the prevalence in QM_∞ of unitary inequivalent representations of the canonical relations identifying a quantum theory, it is time for us to take on board some formal notions. As previously announced, my exposition of these notions will be unrigorous.[22]

---

[20] This sounds simpler than it is; see Belot (this volume).
[21] This problem is addressed systematically in Baker (forthcoming).
[22] Kadison and Ringrose 1997, ch. 4, gives a more formal introduction.

There is an abstract algebraic structure that all concrete Hilbert space representations of the canonical relations $\mathfrak{R}$ circumscribing a quantum theory Q_R have in common. This is the structure of the C^* algebra $\mathfrak{A}_R$ generated by $\mathfrak{R}$. One way to construct $\mathfrak{A}_R$ is to start with a concrete Hilbert space representation of $\mathfrak{R}$, form polynomials of the canonical operators affording that representation, then close in the *uniform* (aka the *norm*) operator topology of the representing Hilbert space.[23] The C^* algebra $\mathfrak{A}_R$ results. (A concrete C^* algebra is a uniformly closed subset of the bounded operators on some Hilbert space.) If starting from another representation of $\mathfrak{R}$, we followed the same recipe to obtain a C^* algebra $\mathfrak{A}'_R$, we would find that $\mathfrak{A}_R$ and $\mathfrak{A}'_R$ are isomorphic: there exists a one-to-one map between the algebras preserving algebraic structure. As representation-independent, this is the algebraic structure shared by all of $\mathfrak{R}$'s Hilbert space representations.

Ordinary QM-ish observable algebras, algebras of the form $\mathfrak{B}(\mathcal{H})$, are weakly closed, and indeed can be regarded (as section 2 presents them) as generated on behalf of a quantum theory Q_R by following the * recipe of the preceding paragraph, but substituting *weak* for uniform closure in the last step. (Because weakly closed sets are also uniformly closed, every $\mathfrak{B}(\mathcal{H})$ is also a C^* algebra.) The criterion of convergence supplied by the uniform topology is harder to satisfy than the criterion of convergence supplied by the weak topology. So there are sequences of Hilbert space operators that converge weakly, but not uniformly. For instance, the sequence of partial sums of n pairwise orthogonal projection operators on a separable infinite dimensional Hilbert space converges weakly to the identity operator as n goes to infinity. That same sequence of partial sums fails to converge uniformly. Given a concrete irreducible representation of canonical relations $\mathfrak{R}$ on some Hilbert space $\mathcal{H}$, its uniform closure, which is isomorphic to the canonical C^* algebra $\mathfrak{A}_R$, can turn out to be a proper subalgebra of $\mathfrak{B}(\mathcal{H})$, the observable algebra ordinary QM generates by weak closure from the concrete representation. (Recall the example of the infinite spin chain. The convergence of the limit that defines the global polarization observable is representation-dependent. That observable belongs to $\mathfrak{B}(\mathcal{H}^+)$, but has no counterpart in $\mathfrak{A}_{CAR}$, the C^* algebra generated as the uniform closure of a representation of the Pauli relations.) One difference between unitarily inequivalent representations of $\mathfrak{R}$ arises from observables that make it into the weak, but not the uniform, closures of concrete representations of $\mathfrak{R}$.

The canonical C^* algebra $\mathfrak{A}_R$ can be considered in abstraction from any concrete Hilbert space, as the algebraic structure all representations of $\mathfrak{R}$ share. But an abstract C^* algebra $\mathfrak{A}$ will always admit a concrete Hilbert space *representation*, a morphism π from $\mathfrak{A}$ *into* $\mathfrak{B}(\mathcal{H})$. Two representations $(\pi, \mathcal{H})$ and $(\pi', \mathcal{H}')$ of the same algebra $\mathfrak{A}$ are *unitarily equivalent* just in case there exists a unitary map $U: \mathcal{H} \to \mathcal{H}'$ such that for each $A \in \mathfrak{A}$, $U\pi(A)U^{-1} = \pi'(A)$. Even if representations π and π' are not unitarily inequivalent, there is still a isomorphism between $\pi(\mathfrak{A}) \subsetneq$

---

[23] 'Closing in the uniform topology' means adding to the algebra the limit points of all uniformly convergent sequences of elements that have made their way into the algebra by other means.

$\mathfrak{B}(\mathcal{H})$ and $\pi'(\mathfrak{A}) \subsetneq \mathfrak{B}(\mathcal{H}')$—just not one implemented unitarily, and so not one that extends to an isomorphism between $\mathfrak{B}(\mathcal{H})$ and $\mathfrak{B}(\mathcal{H}')$ in their entirety.

The foregoing apparatus frames a C^* algebraic approach to quantum theories, which includes the Hilbert space approach of ordinary QM as a special case. The C^* algebraic approach associates the observables of a quantum theory with the self-adjoint elements of a C^* algebra $\mathfrak{A}$ appropriate to that theory. (In the special case of ordinary QM, that algebra takes the form of $\mathfrak{B}(\mathcal{H})$ for some Hilbert space $\mathcal{H}$.) The C^* algebraic approach identifies quantum states on the observable algebra $\mathfrak{A}$ with linear functionals $\omega : \mathfrak{A} \to \mathbb{C}$ that are normed ($\omega(I) = 1$)) and positive ($\omega(A^*A) \geq 0$ for all $A \in \mathfrak{A}$). $\omega(A)$ may be understood as the expectation value of (self-adjoint) $A \in \mathfrak{A}$. These states are uniformly continuous: if A_i is a sequence of elements of $\mathfrak{A}$ that converges uniformly to A and ω is a state on $\mathfrak{A}$, then $\omega(A_i)$ converges in the good-old fashioned sense to $\omega(A)$. Ordinary QM adds an extra codicil to its conception of states: admissible states on $\mathfrak{B}(\mathcal{H})$ need also be *countably additive*—a virtue that lacks a natural representation-independent account. Countably additive states on $\mathfrak{B}(\mathcal{H})$ are ultraweakly continuous: if A_i is a sequence of elements of $\mathfrak{B}(\mathcal{H})$ that converges in $\mathcal{H}$'s ultraweak operator topology[24] to A and ω is a countably additive state, then the expectation values ω assigns A_i converge to the expectation value ω assigns A.

Like the states of ordinary QM, the set of states on a C^* algebra $\mathfrak{A}$ is convex. Its extremal elements—that is, states ω which cannot be expressed as nontrivial convex combinations of other states—are pure states; all other states are mixed.

It is straightforward that a countably additive ordinary QM state implemented by a density operator $\hat{W}$ acting on a Hilbert space $\mathcal{H}$ carrying a representation $\pi : \mathfrak{A} \to \mathfrak{B}(\mathcal{H})$ of an algebra $\mathfrak{A}$ defines a state ω on $\mathfrak{A}$. Simply set $\omega(A) = Tr(\hat{W}\pi(A))$ for all $A \in \mathfrak{A}$. It is gratifying that we can travel in the other direction. Let ω be a state on a C^* algebra $\mathfrak{A}$. Then there exists a Hilbert space $\mathcal{H}_\omega$, a faithful[25] representation $\pi_\omega : \mathfrak{A} \to \mathfrak{B}(\mathcal{H}_\omega)$ of the algebra, and a cyclic[26] vector $|\Psi_\omega\rangle \in \mathcal{H}_\omega$ such that, for all $A \in \mathfrak{A}$, the expectation value the algebraic state ω assigns A is duplicated by the expectation value the Hilbert space state vector $|\Psi_\omega\rangle$ assigns the Hilbert space observable $\pi(A)$. The triple $(\mathcal{H}_\omega, \pi_\omega, |\Psi_\omega\rangle)$ is unique up to unitary equivalence, and known as the state's *GNS representation*.

Clearly, other states ϕ on $\mathfrak{A}$ can be implemented as density operator states on ω's GNS representation. For example any $\hat{W}$ acting on $\mathcal{H}_\omega$ defines a state ϕ on $\mathfrak{A}$ via $\phi(A) := Tr(\hat{W}\pi_\omega(A))$ for all $A \in \mathfrak{A}$. The set of states thus definable as density operators on ω's GNS representation comprise what is known as ω's *folium*.

If $\mathfrak{A}$ admits unitarily inequivalent representations, not every state on $\mathfrak{A}$ lies in ω's folium. Recalling the infinite spin chain, consider, for example, the state on the algebra $\mathfrak{A}_{CAR}$ implemented by the base state of the $\mathcal{H}^+$ representation. This is the state that assigns the z-component of every Pauli spin, as well as the z-component of

---

[24] A near-relative of weak convergence: A_i converges ultraweakly to A just in case for each density operator W, $|Tr(WA_i) - Tr(WA)|$ converges to 0.
[25] π is faithful iff $\pi(A) = 0$ implies $A = 0$ for all $A \in \mathfrak{A}$.
[26] $|\Psi\rangle$ is cyclic for $\pi_\omega(\mathfrak{A})$ means $\{\pi_\omega(\mathfrak{A})|\Psi\rangle\}$ is *dense* in $\mathcal{H}$.

every finite subchain polarization observable, the eigenvalue $+1$. Call this state ω_+. We can regard the $\mathcal{H}^+$ representation as ω_+'s GNS representation. We have argued that in $\mathcal{H}^+$'s weak operator topology, the finite subchain operators $m(z)_N^+$ converge in the $N \to \infty$ limit to the identity operator I^+. This implies that they converge ultraweakly to I^+ as well. A state on $\mathfrak{A}_{CAR}$ that cannot be implemented by a density operator on ω_+'s GNS representation is the state, call it ω_-, implemented by the base state of the $\mathcal{H}^-$ representation. ω_- is the state in which the spin at every site points in the negative z direction. So for each k, $\omega_-(\sigma^k(z)^+) = -1$. Hence for each N, $\omega_-(m(z)_N^+) = -1$. But $\omega_-(I^+)$ had better be $+1$, because ω_- is a state and I^+ is the identity operator. So $m(z)_N^+$ converges ultraweakly to I^+, but $\omega_-(m(z)_N^+)$ does not converge to $\omega_-(I^+)$. But that means ω_- fails to be ultraweakly continuous in $\mathcal{H}^+$. If ω_- were implemented by a density operator on $\mathcal{H}^+$, it would be ultraweakly continuous. ω_- is not implementable by a density operator on ω_+'s GNS representation. Every state in ω_+'s folium is thus implementable. So ω_- lies outside ω_+'s folium.

When the GNS representations π_ω and π_ϕ of two algebraic states are unitarily equivalent, the folia of those algebraic states coincide. If two pure algebraic states ϕ and ω have unitarily inequivalent GNS representations, their folia are *disjoint*: no algebraic state expressible as density matrix on ϕ's GNS representation is so-expressible on ω's and vice versa. (With mixed algebraic states, the situation is more delicate. See Kadison and Ringrose 1997, ch. 4 for details.) *Disjoint states* are states whose folia are disjoint. The states ω_+ and ω_- on $\mathfrak{A}_{CAR}$ are disjoint.

Section 2.3 explicated physical equivalence for generic kinematic pairs by means of the pair of conditions PEV and PAS, and promised an argument that kinematic pairs satisfying the expectation-value preserving condition (PEV) will not in general thereby satisfy the algebraic-structure preserving condition PAS. We are now in a position to give that argument, which underscores the incapacity of PEV on its own to capture a robust sense of physical equivalence. Let $\mathfrak{A}_R$ be the canonical C^* algebra for a quantum theory Q_R, and let ω and ϕ be *disjoint* pure states on it. Q_R's grip on which physical possibilities are which begins with the canonical relations $\mathfrak{R}$ and the expectation values assigned observables satisfying those relations. States assigning different expectation values to canonical observables correspond to different physical possibilities. In order to be disjoint, ω and ϕ have to be *different* states, that is, there has to be some canonical observable regarding whose expectation value ω and ϕ disagree. Now consider two ordinary QM-ish kinematic pairs for Q_R: $(\mathfrak{B}(\mathcal{H}_\omega), \mathbb{T}^+(\mathcal{H}_\omega))$ and $(\mathfrak{B}(\mathcal{H}_\phi), \mathbb{T}^+(\mathcal{H}_\phi))$. Every separable infinite dimensional Hilbert space is isomorphic to every other, and in particular there is a unitary map $U: \mathcal{H}_\omega \to \mathcal{H}_\phi$. This U defines a pair of bijections i_{obs} and i_{state} between observable algebra and state sets of the kinematic pairs: for all $A \in \mathfrak{B}(\mathcal{H}_\omega)$, $i_{obs}(A) = UAU^{-1}$, for all $W \in \mathbb{T}^+(\mathcal{H}_\omega)$, $i_{state}(W) = UWU^{-1}$. It is trivial that these bijections together satisfy PEV. But it is mad to take states identified by the bijections to represent the same physical possibility. No state in the first kinematic pair can reproduce the expectation value assignment to canonical observables of any state in the second kinematic pair. (That is just what it is for the representations to be disjoint.) i_{state} therefore identifies states that act differently on $\mathfrak{A}_R$, states corresponding to different physical possibilities. But i_{state} was supposed to identify states corresponding to the same physical possibility. Something has gone wrong.

A natural diagnosis is that the bijections defined by the unitary map above have gone astray by failing to respect Q_R's algebraic structure. In particular, where C_i are the elements satisfying the constitutive relationships $\mathfrak{R}$ and generating $\mathfrak{Q}_R$, i_{obs} violates the (7) clause of the condition PAS, which requires:

$$i_{obs}(\pi_\omega(C_i)) = \pi_\phi(C_i) \quad \text{for all } C_i \in \mathfrak{A}_R$$

(We know this because if i_{obs} satisfied this condition, the representations would be unitarily equivalent, which by hypothesis they are not.) Violating (7), i_{obs} satisfies PEV in a way that loses track of which observables are canonical. That is why states identified by i_{state} differ in the values they assign canonical observables.

5. WHY UNITARY EQUIVALENCE

5.1 Competing Criteria of Equivalence

The disclosures of the technical interlude inspire an interpretive strategy Arageorgis (1995) has dubbed "Algebraic Imperialism." The Algebraic Imperialist conceives of a theory of QM_∞ not in terms of a particular concrete Hilbert space representation of the canonical relations $\mathfrak{R}$ circumscribing that theory, but in terms of the abstract algebraic structure every such representation shares. That is, the Imperialist supposes the theory's physical magnitudes to be given by the self-adjoint part of the C^* algebra $\mathfrak{A}_R$, and takes possible states to be the set $\mathcal{S}_{\mathfrak{A}_R}$ of states in the algebraic sense on $\mathfrak{A}_R$.

The Algebraic Imperialist equips a theory Q_R with a kinematic pair of the form $(\mathfrak{A}_R, \mathcal{S}_{\mathfrak{A}_R})$. For a quantum theory that, like a typical theory of QM_∞, admits unitarily inequivalent representations, the Imperialist thereby rejects ordinary QM's conception of a kinematic pair as a double $(\mathfrak{B}(\mathcal{H}), \mathbb{T}^+(\mathcal{H}))$ for some concrete Hilbert space $\mathcal{H}$. Generated as the weak closure of a concrete Hilbert space representation of $\mathfrak{R}$, the ordinary QM observable algebra $\mathfrak{B}(\mathcal{H})$ generally contains elements without counterpart in the Imperialist's observable algebra $\mathfrak{A}_R$. And the Imperialist's state set $\mathfrak{A}_R$ generally includes states not implementable by members of the ordinary QM state set $\mathbb{T}^+(\mathcal{H})$. The Imperialist takes more states to be physically possible than can be ratcheted into an ordinary QM account. She also takes fewer observables to be physically significant.

Section 2.3 anointed unitary equivalence a criterion of physical equivalence for kinematic pairs of the form $(\mathfrak{B}(\mathcal{H})_{\hat{C}_i}, \mathbb{T}^+(\mathcal{H}))$ on the grounds that (i) any two such pairs satisfied the demands PEV (preservation of expectation values) and PAS (preservation of algebraic structure) just in case their generating representations were unitarily equivalent, and (ii) the demands PEV and PAS appropriately explicated physical equivalence for theories specified up to kinematic pairs. For the Imperialist, (i) is true, but irrelevant: she assigns theories of QM$_\infty$ kinematic pairs different in kind from those of ordinary QM. She should, however, take (ii) to be both true and relevant. If anything, more sensitive to the physical import of algebraic structure than the advocate of ordinary QM, the Imperialist should welcome the criterion (PAS). Moreover the Imperialist has no special reason to resist the preservation of expectation values as a criterion for physical equivalence. So she should accept section 2.3's gloss of physical equivalence in terms of the pair of conditions (PEV and PAS), but deny that it be applied to kinematic pairs of ordinary QM's sort. Rather, the Imperialist would apply those conditions to kinematic pairs $(\mathfrak{A}_R, S_{\mathfrak{A}_R})$ of the sort she recognizes.

When the criteria PAS and PEV for physical equivalence are applied to kinematic pairs of the sort favored by Imperialists, unitary equivalence emerges as sufficient, but not necessary, for physical equivalence. Let $(\mathfrak{A}, S_\mathfrak{A})$, $(\mathfrak{A}', S_{\mathfrak{A}'})$ be kinematic pairs for a theory Q_R circumscribed by the canonical relations $\mathfrak{R}$, and suppose the algebras $\mathfrak{A}$, $\mathfrak{A}'$ have canonical generators R_i, R'_i, respectively. Assume there is an isomorphism $\alpha: \mathfrak{A} \to \mathfrak{A}'$ such that $\alpha(R_i) = R'_i$ for all i. Because it is an isomorphism on the algebras' generators, α extends to their products, linear combinations and uniform limits (Kadison and Ringrose 1997, Thm 4.1.8). Between $\mathfrak{A}$ and $\mathfrak{A}'$, α provides a bijection $i_{obs}(A) = \alpha(A)$ that satisfies PAS. Conversely, if there is no such isomorphism, there is no bijection satisfying PAS. α also generates a bijection between states: $i_{state}: \omega \to \omega \circ \alpha^{-1}$ where $\omega \circ \alpha^{-1}$ assigns each $A \in \mathfrak{A}$ the value ω assigns $\alpha^{-1}(A)$. Together, i_{state} and i_{obs} satisfy PEV. So kinematic pairs $(\mathfrak{A}, S_\mathfrak{A})$, $(\mathfrak{A}', S_{\mathfrak{A}'})$ satisfy PEV and PAS if and only if there is an isomorphism $\alpha: \mathfrak{A} \to \mathfrak{A}'$ such that $\alpha(R_i) = R'_i$ for all i.

Unitary equivalence emerged as a criterion for physical equivalence for kinematic pairs *cast in ordinary QM form* $(\mathfrak{B}(\mathcal{H}), \mathbb{T}^+(\mathcal{H}))$. It emerged from the analysis of physical equivalence for general kinematic pairs in terms of PAS and PEV. We have just seen that applying that same general analysis to kinematic pairs cast in Algebraic Imperialist form $(\mathfrak{A}, S_\mathfrak{A})$ eventuates in algebraic isomorphism as a criterion of physical equivalence.

Our overarching question is: Is unitary equivalence a suitable criterion of physical equivalence for kinematic pairs realizing quantum theories that fall outside the scope of the Stone-von Neumann and Jordan-Wigner theorems? We now have a provisional answer. It depends. It depends on how we ought to construe the kinematic pairs for such theories—in ordinary QM's form, in Algebraic Imperialist form, or some other way. But which form is the right one? The next, and concluding, section presents some approaches to this question.

6. PRINCIPLES?

The infinite spin chain evoked an anxiety: its possible states should include the states on $\mathfrak{A}_{CAR}$ we have called ω_+ and ω_-, but any theory of ordinary QM reckoning ω_+ to be a possible state of the infinite spin chain is physically inequivalent to any theory of ordinary QM reckoning ω_- to be a possible state. The last section charts a strategy for alleviating this anxiety: ditch ordinary QM's account of quantum kinematics and embrace instead the Imperialist's account. Limiting observables pertaining to the infinite spin chain to self-adjoint elements of $\mathfrak{A}_{CAR}$, we constitute a quantum theory according to which both ω_+ and ω_- are possible states of the chain.

But at a cost. Among the observables we have given up is the global polarization observable, the observable that distinguishes most conspicuously and directly between the states ω_+ and ω_-, the observable that labels an axis of the phase diagram distilling a ferromagnet's universal behavior. The Imperialist strategy for welcoming both ω_+ and ω_- as physical undermines the physics some people would like to do with those states.

Of course, other people—those with a particular, but widespread set of metaphysical scruples about symmetry—would reject any observable that marked a physical difference between ω_+ and ω_-, on the grounds that, as symmetry-connected, ω_+ and ω_- represent the same physical situation. Those committed to such scruples also have reason to resist the ordinary QM picture of quantum theories—ω_+ and ω_- cannot represent the same physical situation if one's a possible state only if the other is not. The Imperialist's picture presents them with no such immediate absurdity. It does, however, require the symmetricians to do some work to avert inanity. Symmetries like the flip-flop symmetry connecting ω_+ and ω_- are implemented by automorphisms of $\mathfrak{A}_{CAR}$. (As the name suggests, an automorphism $\alpha : \mathfrak{A} \to \mathfrak{A}$ of an algebra $\mathfrak{A}$ is just an isomorphism that maps $\mathfrak{A}$ to itself.) It turns out that for every pair of pure states on $\mathfrak{A}_{CAR}$, there is some automorphism that connects them. So if every automorphism implements a symmetry, the symmetricians' commitment to identify symmetry-connected states collapses the state space of the theory to a single point. A physical theory with such a state space is inane. To avert inanity, symmetricians need to distinguish automorphisms that implement symmetries from automorphisms that do not.

Section 5 determined that the explication of physical equivalence for theories of QM_∞ depends on the sorts of kinematic pairs it is appropriate to attribute such theories. We have just seen two principled attitudes toward ω_+ and ω_- which concur in rejecting kinematic pairs of ordinary QM's sort as inappropriate. One attitude reckons ω_+ and ω_- to be physically distinct possible states of the infinite spin chain, on the principle that this arrangement supports explanatory aspirations, exercises in unification, and programs of theory development. The other attitude reckons ω_+ and ω_- to be physically identical states, on the principle that they are connected by a symmetry. United in their departure from ordinary QM, these principles soon diverge. Those in the grip of the explanation-honoring principle cannot abide the Imperialist's account of quantum kinematics, for that account obliterates what they take to be the physically fruitful distinction between ω_+ and ω_-. Those in the grip of the symmetry-honoring principle can live with the Imperialist's account, provided they can supplement their metaphysical principle with another one, a principle able to identify those automorphisms of $\mathfrak{A}_{CAR}$ which implement symmetries. This illustrates a typical predicament: the question of what sort of kinematic pair it is appropriate to attribute theories of QM_∞—the question on whose answer turns the explication of physical equivalence for such theories—is a question to which there are many responses, guided by many principles, some of which are in mutual (or even internal) tension.

We can try to simplify the question about kinematic pairs by breaking it into component questions: What observable algebras should we attribute theories of QM_∞, and what state sets? Take the former question first. Suppose all hands agree that observables realizing the canonical relations $\mathfrak{R}$ circumscribing a quantum theory Q_R are physically significant. Because expressing the relations $\mathfrak{R}$ requires forming linear combinations and products of canonical observables, suppose that all hands agree as well that that the set of physical observables is closed under those algebraic operations. Then all we need to induce agreement among all hands about Q_R's observable algebra is to get all hands to agree about what criterion of convergence is appropriate to use in adding limit points to the algebra generated by the canonical observables. If they agree to uniform convergence, they agree as well that the C^* algebra $\mathfrak{A}_R$ is the observable algebra for Q_R. If they agree to convergence in the weak operator topology of a Hilbert space $\mathcal{H}$ bearing a representation of $\mathfrak{R}$, they agree to $\mathfrak{B}(\mathcal{H})$ as the observable algebra.

Segal (1959) struck on an argument favoring uniform convergence. The conceptual crux of Segal's argument is the claim that if a sequence of observables A_i converges to another observable A, the world had better not be able to get itself into a state ω such that $\omega(A_i)$ fails to converge to $\omega(A)$. Segal used a broadly operationalist outlook to secure this claim. Protocols for measuring A will include ones where we measure many members of the sequence A_i and extrapolate to the limit of the sequence of outcomes obtained. This protocol would break down if $\omega(A_i)$ failed to converge to $\omega(A)$. Lest our experimental protocols lose their grip on the observables they are meant to measure, Segal contends, we should regard as observables only those quantities defined by limiting relationships $A_i \to A$ such that $\omega(A_i)$ converges to $\omega(A)$, *no matter what ω is*. To deal weak convergence the *coup de grâce*, Segal remarks that for any representation $(\pi, \mathcal{H})$ of a canonical algebra $\mathfrak{A}$ admitting unitarily inequivalent representations, there will be sequences $A_i \in \mathfrak{A}$ and states ω on $\mathfrak{A}$ such that $\pi(A_i)$ converges in $\mathcal{H}$'s weak topology, but $\omega(A_i)$ fails to converge. (States ω on $\mathfrak{A}$ not implementable by a density operator state on the representation $(\pi, \mathcal{H})$ will have this feature.) If we added the weak limit of the sequence $\pi(A_i)$ to our catalog of observables, we would recognize an observable whose value we could not gauge even by the most careful sequence of measurements performed on systems in the state ω. To a good operationalist, a quantity whose value eludes experimental assessment is no real quantity. To build an observable algebra containing only genuine quantities, Segal urges, avoid weak limits. By his lights, uniform limits are fine. Because all states on $\mathfrak{A}$ are uniformly continuous, if A_i converges uniformly to A, *no matter what ω is*, $\omega(A_i)$ converges to $\omega(A)$ as well.[27]

---

[27] Ruetsche 2011 uses considerations of Segal's sort to sketch reasons, innocent of operationalism, for coordinating a quantum theory's state space and its observable algebra.

This is an admirably principled argument for attributing Q_R the observable algebra $\mathfrak{A}_R$ generated by uniform closure, rather than any larger algebra $\mathfrak{B}(\mathcal{H})$ generated by weak closure. But as Segal develops it, the argument quite obviously makes covert appeal to a particular answer to a question it was the aim of our simplifying strategy to set aside: What states are physical? If Q_R's physical states were required to reside in the folium of the representation whose weak operator topology was used to obtain $\mathfrak{B}(\mathcal{H})$, then for every weakly convergent sequence of observables $A_i \to A$ and every physical state ω, $\omega(A_i)$ would converge to $\omega(A)$, and the limiting observables would survive Segal's test for significance.[28] Segal's criterion for the significance of an observable cannot operate in isolation from an account of which states are physical. Operating in concert with different accounts of physical states, Segal's criterion makes different judgments about which observables are physical. Operating in isolation from an account of physical states, Segal's criterion makes no judgment at all.

We are after a principled way of settling the question of what sort of kinematic pair $(\mathfrak{Q}, \mathcal{S})$ to attribute a theory Q_R. Even supposing we accept Segal's principle that the continuity properties of physical states march in lockstep with the continuity properties of algebras of observables, we need a principled way to fix either the state set $\mathcal{S}$ or the observable algebra $\mathfrak{Q}$ of Q_R to precipitate a kinematic pair from that principle. The good news is that there are a variety of principles we might invoke to configure either $\mathcal{S}$ or $\mathfrak{Q}$. The bad news is that different principles eventuate in different kinematic pairs, which in turn are suited to different philosophical and physical projects. Adopting a single and uniform principle inevitably stymies some of those projects. Here are some examples of principles, and the projects they sustain and frustrate.

- The Hadamard condition is a principle for identifying physically significant states on the canonical algebra $\mathfrak{A}$ for a quantum field theory on curved spacetime (see Wald 1994, ch. 4 for an account). States that fulfill the Hadamard condition are states for which a point-splitting prescription for assigning the stress energy tensor an expectation value succeeds. Such success is essential to the pursuit of semi-classical quantum gravity, an approach that marries quantum theory to the general theory of relativity by replacing the stress energy tensor T_{ab} in Einstein's field equations with its expectation value $\langle T_{ab} \rangle_\phi$. For a QFT Q_R set in a closed spacetime, the Hadamard condition fosters significant progress on the question of what kinematic pair to assign Q_R. In such a spacetime setting, every Hadamard state is unitarily equivalent to every other. Thus, states satisfying the Hadamard condition are confined to those expressible as density operators on a particular concrete representation $(\pi, \mathcal{H})$ of $\mathfrak{A}$. With states so confined, Segal's coordination

---

[28] The continuity claim follows from the fact that every density operator state on $\mathfrak{B}(\mathcal{H})$ is ultra-weakly continuous.

principle delivers $\mathfrak{B}(\mathcal{H})$—perhaps with the stress-energy observable added 'by hand'—as the observable algebra appropriate to Q_R.

In open spacetimes the story is not so neat: unitarily inequivalent Hadamard states abound. Perhaps more distressingly, when it comes to sustaining the aspirational pursuit of quantum gravity, the Hadamard condition is a mixed bag. One triumph of semi-classical quantum gravity is its role in predicting the "phenomenon" of black hole evaporation. In black hole evaporation, the region of spacetime exterior to a black hole occupies a quantum state that is a thermal state at a temperature related to the surface gravity of the black hole, which eventually radiates its substance away. This Hawking radiation and the associated model of black hole thermodynamics serve fledgling projects in quantum gravity as something like a datum. To be considered plausible, theories of quantum gravity should be able to accommodate and predict black hole evaporation. The catch is that the QFT state that in some respects best models the Hawking radiation exterior to a black hole during its evaporation is a state that violates the Hadamard condition (see Candelas 1980). Dismissing that state from physical relevance frustrates the pursuit of quantum gravity by undermining one of our best models of a "phenomenon" quantum gravity is meant to save.

- Let $\mathfrak{A}$ be the canonical algebra for a QFT Q_R set in a spacetime M. Axiomatic approaches to QFT include an axiom demanding that M's spacetime symmetries be implemented unitarily (see Dimock 1980). We can cast this axiom as a principle that in order for a state ω to be physical, its GNS representation must be one on which M's symmetries are unitarily implementable. Again, given the right sort of M, this principle for identifying physical states helps identify a kinematic pair. Coupled with other axioms for Minkowski spacetime, the principle singles out a privileged vacuum state ω_0 on $\mathfrak{A}$. Taking the physical states of the theory to be those in ω_0's folium and applying Segal's coordination principle, we attribute Q_R the kinematic pair $(\mathfrak{B}(\mathcal{H}_\omega), \mathbb{T}^+(\mathcal{H}_\omega))$.

In more general spacetime settings, the story is not so neat. But even confining attention to Minkowski spacetime, the kinematic pair underwritten by the principle that symmetries be unitarily implementable stymies certain projects in physics. For instance, so-called "infra-red states," invoked in models of soft-photon scattering (see Strocchi 1985, 87), fall outside the pale of physical possibility limned by this kinematic pair. But these states are integral to our modeling and understanding a significant class of experimental particle physics phenomena. Rejecting infrared states on principle limits the explanatory reach of Q_R.

This list can be continued: for instance, insisting that dynamics be well defined mitigates in favor of abstract algebraic approaches when dynamics cannot be implemented unitarily on a fixed Hilbert space (Arageorgis et al. 2002), but in favor of concrete Hilbert space representations when the existence of a dynamics is representation-dependent (see Emch and Knops 1970). But even in its short form, the list suggests a moral that complicates the search for overarching and univocal criteria of physical equivalence appropriate to theories of QM_∞. Those criteria will depend on the sorts of kinematic pair suited to theories of QM_∞. Their capacity for wide application is a virtue of theories of QM_∞. Any recipe that, given $\mathfrak{R}$, generated a kinematic pair for Q_R *without looking for guidance to the details of Q_R's applications*, threatens to erode that virtue.

REFERENCES

Arageorgis, Aristidis (1995). Fields, particles, and curvature: Foundations and philosophical aspects of quantum field theory in curved spacetime. Ph.D. diss., University of Pittsburgh.

Arageorgis., Aristidis, John Earman, and Laura Ruetsche (2002). Weyling the time away: The non-unitary implementability of quantum field dynamics on curved spacetime and the algebraic approach to QFT. *Studies in the History and Philosophy of Modern Physics* 33: 151–184.

—— (2003). Fulling non-uniqueness, Rindler quanta, and the Unruh effect: A primer on some aspects of quantum field theory. *Philosophy of Science* 70: 164–202.

Baker, David J. (2009). Against field interpretations of quantum field theory. *British Journal for the Philosophy of Science* 60: 585–609.

—— (forthcoming). Spontaneous symmetry breaking in QFT. *Philosophy of Science.*

Belot, Gordon (1998). Understanding electromagnetism. *British Journal for the Philosophy of Science* 49: 531–555.

—— (2007). The representation of time and change in classical mechanics. In Butterfield and Earman (2007a), 133–227.

Butterfield, Jeremy, (2007). On symplectic reduction in classical mechanics. In Butterfield and Earman (2007a), 1–131.

Butterfield, Jeremy, and Earman, John, eds. (2007a). *Handbook of philosophy of science: Philosophy of physics,* Vol. 1. Amsterdam: Elsevier.

—— (2007b). *Handbook of philosophy of science: philosophy of Physics,* Vol. 2. Amsterdam: Elsevier.

Candelas, P. (1980). Vacuum polarization in Schwarzschild spacetime. *Physical Review D* 21: 2185–2202.

Clifton, Robert and Hans Halvorson (2001). Are Rindler quanta real? inequivalent particle concepts in quantum field theory. *British Journal for the Philosophy of Science* 52: 417–470.

Dimock, J. (1980). Algebras of local observables on a manifold. *Communications in Mathematical Physics* 77: 219–228.

Dürr, Detlef et al. (2006). Topological factors derived from Bohmian mechanics. *Journal Annales Henri Poincaré* 7: 791–807.

Earman, John (2004). laws, symmetry, and symmetry breaking: Invariance, conservation principles, and objectivity. *Philosophy of Science* 71: 1227–1241.

Earman, John and Laura Ruetsche (2005). Relativistic invariance and modal interpretations. *Philosophy of Science* 72: 557–583.

Emch, Gérard (1972). *Algebraic methods in statistical mechanics and quantum field theory.* (New York: Wiley).

Emch, Gérard (2007). Quantum statistical physics. In Butterfield and Earman (2007b), 1075–1182.

Emch, Gérard and H. J .F. Knops (1970). Pure thermodynamic phases as extremal KMS states. *Journal of Mathematical Physics* 11, 3008–3018.

Emch, Gérard and Liu, Chuang (2002). The logic of Thermostatistical Physics. (Berlin: springer).

Fulling, Stephen A. (1989). *Aspects of quantum field theory in curved space-time.* Cambridge: Cambridge University Press.

Glymour, Clark (1970). Theoretical realism and theoretical equivalence. In *Boston Studies in Philosophy of Science*, Vol. 7, ed R. Buck and R. Cohen, 275–288. Dordrecht: Reidel.

Gotay, M. J. (2000). Obstructions to quantization. in *Mechanics: From theory to computation*, 171–216. New York: Springer.

Haag, Rudolf (1992). *Local quantum physics.* New York: Springer-Verlag.

Halvorson, Hans (2001). "On the nature of continuous physical quantities in classical and quantum mechanics. *Journal of Philosophical Logic* 30: 27–50.

Healey, Richard (2007). *Gauging what's real: The conceptual foundations of contemporary gauge theories* Oxford: Oxford University Press.

Isham, C. J. (1983). Topological and global aspects of quantum theory. In ed B. S. DeWitt and R. Stora. *Les Houches, Session XL: Relativité, groupes et topologie II* Amsterdam: Elsevier.

Kadison, R. V. and J. R. Ringrose (1997). *Fundamentals of the theory of operator algebras*, Vol. 1. New York: Academic Press.

Landsman, N. P. (1990). C^*-algebraic quantizaton and the origin of topological quantum effects. *Letters in Mathematical Physics* 20: 11–18.

Lèvy-Leblond, Jean-Marc (1976). Who is afraid of nonhermitian operators? A quantum description of angle and phase. *Annals of Physics* 101: 319–341.

Liu, Chuang and Emch, Gérard (2005). Explaining quantum spontaneous symmetry breaking. *Studies in History and Philosophy of Modern Physics* 36: 137–163.

Roberts, J. E. and G. Roepstorff (1969). Some basic concepts of algebraic quantum theory. *Communications in Mathematical Physics* 11: 321–338.

Ruetsche, Laura (2003). A matter of degree: Putting unitary inequivalence to work. *Philosophy of Science* 70 [Proceedings]: 1329–1342.

—— (2011). Why be normal? *Studies in History and Philosophy of Modern Physics*, forthcoming.

Segal, Irving E. (1959). The mathematical meaning of operationalism in quantum mechanics. In L. Henkin, P. Suppes, and A. Tarski. *Studies in Logic and the Foundations of Mathematics*, ed., 341–352. Amsterdam: North-Holland.

Sewell, G. (2002). *Quantum mechanics and its emergent Metaphysics.* Princeton: Princeton University Press.

Simon, B. (1972). Topics in functional analysis. In, *Mathematics of contemporary physics*, ed. R.F. Streater. London: Academic Press.

Strocchi, F. (1985). *Symmetry breaking* Berlin: Springer.

Summers, Stephen J. (1999). On the Stone-von Neumann uniqueness theorem and its ramifications. In *John von Neumann and the foundations of quantum physics.* 135–152. Budapest, 1999. *Vienna Circle Institute Yearbook*, 8. Dordrecht: Kluwer.

Teller, Paul (1979). Quantum mechanics and the nature of continuous physical quantities. *Journal of Philosophy* 7: 345–361.

Wald, Robert M. (1994), *Quantum field theory in curved spacetime and black hole thermodynamics*. Chicago: University of Chicago Press

CHAPTER 15

SUBSTANTIVALIST AND RELATIONALIST APPROACHES TO SPACETIME

OLIVER POOLEY

1. Introduction

A significant component of the philosophical interpretation of physics involves investigation of what fundamental kinds of things there are in the world if reality is as physics describes it to be. One candidate entity has proven perennially controversial: *spacetime*.[1] The argument about whether spacetime is an entity in its own right goes by the name of the *substantivalist–relationalist debate*. Substantivalists maintain that a complete catalog of the fundamental objects in the universe lists, in addition to the elementary constituents of material entities, the basic parts of spacetime. Relationalists maintain that spacetime does not enjoy a basic, nonderivative existence. According to the relationalist, claims apparently about spacetime itself are ultimately to be understood as claims about material entities and the possible patterns of spatiotemporal relations that they can instantiate.

Versions of material on which parts of this chapter are based were presented in Vancouver, Leeds, Oxford, and Florence. I am grateful to members of those audiences for questions and criticism. Particular thanks are due to Carl Hoefer, Edward Anderson, Harvey Brown, David Wallace, Ulrich Meyer, Julian Barbour, Cian Dorr, Syman Stevens, and Julia Bursten for comments on drafts of the paper that led to corrections and improvements. Any errors remaining are, of course, mine. Work for this chapter was partially supported by a Philip Leverhulme Prize and MICINN project FI2008-06418-C03-03, which I gratefully acknowledge. Finally, special thanks are due to the editor for his forbearance.

[1] Strictly speaking, the controversy has concerned two candidate entities. Prior to Minkowski's reformulation of Einstein's special theory of relativity in four-dimensional form, the debate was about the existence of space. Since then, the debate has been about the existence of spacetime. For the sake of brevity, I will often only mention spacetime, leaving the "and/or space" implicit.

In his *Principia*, Newton famously distinguishes absolute from relative space and states that the former "of its own nature without reference to anything external, always remains homogeneous and immovable" (Newton 1999, 408). Newton's description of, and arguments for, absolute space are commonly (and rightly) taken to be a statement and defense of substantivalism. In section 2, I consider Newton's reasons for postulating absolute space before examining, in section 3, one of the strongest arguments against its existence, the so-called *kinematic shift argument*. These arguments highlight the close connection between the spatiotemporal symmetries of a dynamical theory and the spacetime ontology that the theory is naturally interpreted as committed to. It turns out that the Galilean covariance of Newtonian mechanics tells against both substantival space and the most obvious relationalist alternative.

With hindsight, the natural substantivalist response to this predicament is to jettison space for spacetime. Section 4 reviews orthodox spacetime substantivalism. The most defensible substantivalist interpretation of Newtonian physics has much in common with a very natural interpretation of relativistic physics. From this perspective, our current best theory of space and time vindicates Newton rather than his relationalist critics, contrary to what early philosophical interpreters claimed (e.g., Reichenbach, 1924).

Section 5 is a skeptical review of two antisubstantivalist themes that motivate some contemporary relationalists. Its conclusion is that the aspiring relationalist's best hope is Ockham's razor, so the focus shifts onto the details of relationalists' proffered alternatives to substantivalism. The move to a four-dimensional perspective expands the range of possibilities available to the classical relationalist. In section 6 I distinguish three strategies that the relationalist can pursue in the face of the challenge posed by Galilean covariance and consider how the corresponding varieties of relationalism fare when one moves from classical to relativistic physics. It turns out that a number of well-known relationalist views find a natural home in this framework.

A review of the substantivalist–relationalist debate cannot get away without mention of Earman and Norton's (in)famous adaptation of Einstein's Hole Argument. In the final section, I highlight what many see as the most promising substantivalist response and relate it to so-called structural realist approaches to spacetime.

2. Newton's Bucket

Newton's best-known discussion of absolute space comes in a scholium to the definitions at the start of the *Principia*. According to the once-standard reading, Newton's purpose in the Scholium is to argue for the existence of substantival space via the existence of absolute motion, which he supposedly takes to be established by his famous bucket experiment and two-globes thought experiment (see, e.g., Sklar, 1974, 182–184). While there is now widespread agreement that this account badly misrepresents Newton's arguments, there is less consensus over how they in fact should be understood.[2] Thanks to Koyré (1965) and Stein (1967), it is now recognized that, in order to understand Newton's Scholium, one has to appreciate that Newton was in large part reacting to Descartes's claims about the nature of motion. I briefly review the relevant Cartesian background before giving an account of Newton's arguments that is essentially in agreement with that of Rynasiewicz (1995).

Descartes was one of the first natural philosophers to put the *principle of inertia*—the claim that bodies unaffected by net external forces remain at rest or move uniformly in a straight line—at the center of his physics (Descartes 1644, II 37, 39). At the same time, and apparently without recognizing the problem, he espoused an account of motion that is hopelessly incompatible with it. Descartes distinguishes motion in an everyday sense of the term from motion in a strict, philosophical sense. Motion in the ordinary sense is said to be *change of place* (ibid., II: 24) and Descartes gives a relational definition of a body's place in terms of that body's position relative to external reference bodies (ibid., II: 10, 13). Which bodies are to be treated as reference bodies is an arbitrary matter. Descartes's ordinary notion of motion is therefore a *relative* one: a given body may be said to be moving uniformly, nonuniformly, or not moving at all, depending on which other bodies are taken to be at rest (ibid., II: 13). In contrast, Descartes's definition of motion in the strict sense, while still relational, was supposed to secure a unique proper motion for each body (ibid., II: 31). True motion is defined as "*the transfer of one piece of matter, or of one body, from the vicinity of the other bodies which are in immediate contact with it, and which are regarded as being at rest, to the vicinity of other bodies*" (ibid., II; 25, emphasis in the original).[3]

Newton gave a single definition of motion, as change of place, but he also recognized two kinds of motion, depending on whether the places in question were the parts of a relative space (defined in terms of distances relative to material reference bodies) or the parts of substantival space. Newton's relative motion, therefore, corresponds closely to Descartes's motion in the ordinary sense. It is the motion we most directly observe and, Newton agreed, it is what we mean by "motion" in everyday contexts. But, he insisted, when it comes to doing physics, we need to abstract from such observations and consider a body's true motion, which, he argued, has to be defined in terms of an independently existing absolute space.

---

[2] For a varied sample of competing interpretations, see Laymon (1978), Rynasiewicz (1995, 2004), and DiSalle (2002, 2006), who goes so far as to claim that Newton was not a substantivalist.

[3] In 1633, on hearing of the Church's condemnation of Galileo for claiming that the Earth moved, Descartes suppressed an early statement of his physics, which did not contain his later relational claims about the nature of motion. It is frequently (and plausibly) conjectured that Descartes's official views on motion were devised to avoid Church censure. However, the precise manner in which Descartes's definitions secure the Earth's lack of true motion suggest that he was genuinely committed to a relational conception of motion. What does the work in securing the Earth's rest is not that, in Descartes's cosmology, there is no relative motion with respect to immediately contiguous bodies (Descartes explicitly says there is such motion; ibid., III: 28); it is that Cartesian true motion is motion with respect to those contiguous bodies that are *regarded as at rest*.

Newton's arguments appeal to alleged "properties, causes, and effects" of true motion. His aim is to show that various species of relative motion, including Cartesian proper motion (though this is not targeted by name), fail to have the requisite characteristics. If one assumes, as Newton tacitly did, that true motion can only be some kind of privileged relative motion or else is motion with respect to an independently existing entity, Newton's preferred option wins by default. That each body has a unique, true motion and that such motion has the purported properties, causes and effects, are unargued assumptions.

Newton's claims about the properties of true motion arguably beg the question against the Cartesian. The arguments from causes and effects are more interesting, both because they connect with physics and because their premises were accepted by the Cartesians. The particular effect of true motion that Newton cites is almost an immediate corollary of the principle of inertia: bodies that are undergoing genuine circular motion "endeavour to recede from the axis" because, at each instant, their natural, inertial motion is along the tangent at that point; they only follow their curved path because of the application of a centripetal force.

Descartes fully endorsed these claims about circular motion. Indeed, they formed a central component of his model of planetary motions and the cosmos (e.g., Descartes, 1644, III: 58–62). His definition of true motion, however, fails to fit these phenomena, as Newton's bucket experiment was designed to illustrate. Newton asks us to consider a water-filled bucket suspended by a wound cord. Once released, as the cord unwinds, the bucket starts to rotate. Initially, the water is at rest and its surface is flat. As friction gradually transfers the bucket's motion to the water, the water's surface becomes ever more concave until its rate of rotation reaches a maximum and it is comoving with the bucket. The concavity of the water's surface reveals its endeavour to recede from the axis of rotation. According to Descartes's definition of true motion, however, the water is at rest both before the bucket is released and at the end of the experiment, when the water and bucket are once again at relative rest, even though the water now manifests an effect of true rotation. It might also seem that Descartes should count the water as truly moving just after the bucket has been released, because it is transferred from the vicinity of bodies in immediate contact with it (viz., the sides of the bucket), even though, at this stage, the water's surface is flat. The effects of true motion are not correlated with true motion as defined by Descartes in the way they are supposed to be. Newton concludes that the effect revealing true rotation "does not depend on the change of position of the water with respect to [immediately] surrounding bodies, and thus true circular motion cannot be [defined in terms of] such changes of position" (Newton 1999, 413).[4]

---

[4] The paragraph describing the bucket experiment completes Newton's arguments for his account of true motion in terms of absolute space but it is not the end the Scholium. After a brief paragraph that explicitly concludes: "Hence relative quantities are not the quantities themselves, whose names they bear, but are only sensible measures of them," there follows a long, final paragraph describing a thought experiment involving two globes attached by a cord in a universe in which no other observable objects exist. The purpose of this thought experiment is not to further argue for absolute space by, e.g., describing a situation in which there is absolute motion (revealed by a tension in the cord) and yet no relative motion whatsoever. Instead, Newton's purpose

The *Principia*'s Scholium on space, time, and motion is no longer our only source for Newton's views on these topics. In the 1960s, a pre-*Principia* manuscript, known after its first line as *De Gravitatione*, was published for the first time (Newton, 2004). In *De Grav*, Newton quite explicitly targets Descartes, and one argument is particularly telling.[5] Newton points out that, according to Descartes's definition of motion, no body has a determinate velocity, and there is no definite trajectory that it follows. From moment to moment a body's motion is defined with respect to those bodies in immediate contact with it, which (for any body in motion) change from moment to moment. There is nothing in this picture that allows us to identify at some time the exact places through which a body has traveled and so *a fortiori* nothing that can tell us whether these places constitute a straight line which the body has traversed at a uniform rate. Descartes's account of true motion, therefore, cannot secure a fact of the matter about whether a body is moving uniformly, as the principle of inertia requires. Newton concludes: "So it is necessary that the definition of places, and hence of local motion, be referred to some motionless being such as . . . space in so far as it is seen to be truly distinct from bodies" (Newton, 2004, 20–21).

Talk of a "being" that is "truly distinct from bodies" indicates that Newton's alternative to Cartesian motion involves a variety of substantivalism. The waters are muddied, however, by Newton's explicit denial in *De Grav* that space is a substance. Newton's position does qualify as a version of substantivalism as defined above: according to Newton, space is a genuine entity of a fundamental kind. Newton's denial that space is a substance comes in a passage where he also denies both that it is merely a property ("accident") and that it is "nothing at all." In fact, of the three categories—substance, accident, or nothing—Newton states that space is closest in nature to substance. His two reasons for denying that space is a substance relate only to how this category was understood in the then-dominant Scholastic tradition. In particular, space was disqualified from being a substance because, on Newton's view, it does not act and because, in a certain rather technical sense, Newton did not regard it as a self-subsistent entity.[6]

In postulating space as an entity with its own manner of existence, Newton was directly following a number of the early modern atomists, such as Patrizi (1943, 227, 240–241), Gassendi (see, e.g., Grant, 1981, 209) and Charleton (1654, 66). These authors all treated space as more substantial than traditional Aristotelian substances. And there are striking structural parallels between some of the arguments in *De Grav* and those in Charleton's book, which we know from one of Newton's early notebooks that Newton had studied. The conclusion must be that in postulating

---

is to demonstrate how true motion can (partially) be empirically determined, despite the imperceptibility of the space with respect to which it is defined: the tension in the cord is a measure of the rate of rotation and, by measuring how this tension changes as different forces are applied to opposite faces of the globes, one can also determine the axis and sense of the rotation.

[5] As has been emphasized by Stein (1967, 269–271); the argument is also singled out by Barbour (1989, 616–617).

[6] Since Newton held that everything that exists exists somewhere, the existence of any other being entails the existence of space (see Stein 2002, 300, n. 32, for further discussion).

substantival space Newton adopted, albeit for truly original reasons, a metaphysical package already very much on the table.

3. THE PUZZLE OF GALILEAN INVARIANCE

For Newton's first two laws of motion to make sense, there needs to be a fact of the matter about whether a body's motion is uniform and, if it is not, a quantitative measure of how it is changing. Newton recognized that Descartes's definitions of motion failed to secure this and signed up to a metaphysics that underwrites the required quantities. There is a sense, though, in which Newton's absolute space underwrites too much. The problem arises because of the symmetries of Newtonian mechanics, in particular its *Galilean invariance.*

3.1 Spacetime and Dynamical Symmetries

The relevant notions of symmetry can be introduced in terms of coordinate transformations. Given our topic, a little care is needed because the substantivalist and the relationalist do not share a conception of a coordinate system. Roughly speaking, a spacetime coordinate system is a map from spacetime into $\mathbb{R}^4$ but, of course, only the substantivalist thinks of spacetime as an genuine entity.[7] The relationalist thinks of a coordinate system as assigning quadruples of real numbers $(t, x, y, z) = (t, \vec{x})$ to material events rather than to spacetime points. And, whereas the substantivalist will view every quadruple of a coordinate system as assigned to something, the relationalist will view some sets of coordinate values (those that the substantivalist thinks of as assigned to unoccupied regions) as simply not assigned to anything at all.

Despite this difference, both substantivalists and relationalists will view certain coordinate systems as *kinematically privileged* in the sense of being optimally adapted to the particular spatiotemporal quantities that they each recognize. In the context of classical mechanics, the natural relationalist alternative to Newton's substantivalism is *Leibnizian relationalism.*[8] According to this view, a possible history of the universe is given by a sequence of *relative particle configurations*: the primitive spatiotemporal facts about the universe are composed solely of facts about the instantaneous relative distances between particles (assumed to obey the constraints of Euclidean geometry) and facts about the time intervals between the successive

---

[7] Recall, also, that for the *Newtonian* substantivalist the basic entity is still space. Until spacetime substantivalism is explicitly introduced in section 4, reference to spacetime points should be understood as reference to ordered pairs of pointlike substantival places with instants of time, and reference to point-like material events as reference to instantaneous states of persisting point particles.

[8] This terminology is established (see. e.g., Maudlin, 1993, 187). No suggestion that the historical Leibniz was a Leibnizian relationalist is intended.

instantaneous material configurations. The ways in which a coordinate system can be adapted to these quantities is straightforward. The time coordinate, t, is chosen so that, for any material events e and e', the difference, $t(e) - t(e')$, corresponds to the temporal interval between e and e', and is positive or negative according to whether e occurs later or earlier than e'. Finally, spatial coordinates are chosen so that, for all particles i,j and for all times, $\sqrt{(x_i - x_j)^2 + (y_i - y_j)^2 + (z_i - z_j)^2} = r_{ij}$, where r_{ij} is the instantaneous inter-particle distance between i and j. Assuming that these distances evolve smoothly over time, one also requires that each particle's spatial coordinates are smooth functions of the time coordinate.

Coordinate systems that encode the Leibnizian relationalist quantities in this way are sometimes known as *rigid Euclidean coordinate systems* (Friedman, 1983, 82). If a particular coordinate system $(t, \vec{x})$ satisfies these constraints then so will any $(t', \vec{x}')$ related to it by a member of the *Leibniz group*[9] of transformations:

$$\vec{x}' = \mathbf{R}(t)\vec{x} + \vec{a}(t),$$
$$t' = t + d. \tag{Leib}$$

$\mathbf{R}(t)$ is an orthogonal matrix that implements a time-dependent rotation. The components of $\vec{a}(t)$ are smooth functions of time that implement an arbitrary time-dependent spatial translation and d is an arbitrary constant that changes the choice of temporal origin.

The manner in which a coordinate system can be adapted to the Newtonian's spatiotemporal quantities is very similar. *Mutatis mutandis*, the substantivalist imposes the same constraints as the relationalist, although now the spatial and temporal distance relations to which the coordinate values are to be adapted hold between the points of spacetime rather than (only) between material events.[10] There are also the substantivalists' trademark "same place over time" facts to encode. Here we simply require that the spatial coordinates of each point of space remain constant. The transformations that relate coordinate systems adapted to the full set of Newtonian spatiotemporal quantities form a proper subgroup of the Leibniz group (it might appropriately be labeled the *Newton group*[11]), since the only rotations and spatial translations that preserve the extra constraint are time-independent:

$$\vec{x}' = \mathbf{R}\vec{x} + \vec{c},$$
$$t' = t + d. \tag{New}$$

Identified in this way, the Leibniz and Newton groups are examples of *spacetime symmetry groups*: they are groups of transformations that *preserve spatiotemporal structure* (as encoded in coordinate systems). A conceptually distinct route to identifying special classes of coordinate transformations goes via the dynamical laws of

---

[9] There is no established label for these transformations. I follow Bain (2004, 350, fn 6); see also Earman (1989, 30–31). Ehlers (1973a, 74) calls this group of transformations the *kinematical group*.
[10] Subject to the qualifications in footnote 7.
[11] Cf. Earman (1989, 34). Ehlers (1973a, 74) follows Weyl in referring to this group of transformations as the *elementary group*.

a particular theory. A *dynamical symmetry group* is a group that *preserves the form of the equations* that express the dynamical laws. Since the Leibnizian relationalist holds that every rigid Euclidean coordinate system is optimally adapted to all the real spatiotemporal quantities, they might expect such coordinate systems to be dynamically equivalent. In other words, it is natural for someone who thinks that the Leibniz group is a spacetime symmetry group to expect it to be a dynamical symmetry group as well. This might indeed be the case if, for example, the dynamical laws dealt directly with the relative distances between bodies. But Newton's laws do not. Instead they presuppose that individual bodies have determinate motions independently of their relations to other bodies. If Newton's laws take their canonical form in a given coordinate system K (which, we may imagine, is adapted to Newtonian space and time), then they will not take the same form in a coordinate system, K', related to K by an arbitrary member of the Leibniz group. The equations that hold relative to K' will involve additional terms corresponding to source-free "pseudo forces" that, the Newtonian maintains, are artefacts of K''s acceleration with respect to substantival space.

3.2 The Kinematic Shift Argument

The mismatch between dynamical symmetries and (what Leibnizian relationalists regard as) spacetime symmetries is a problem for the relationalist. But a similar problem afflicts the Newtonian substantivalist. While the Newton group is a dynamical symmetry group of classical mechanics, it is not the full symmetry group. The equations that express Newton's three laws of motion and particular Newtonian force laws (such as the law of universal gravitation) are invariant under a wider range of coordinate transformations, namely those constituting the *Galilei group*:

$$\vec{x}' = R\vec{x} + \vec{v}t + \vec{c},$$
$$t' = t + d. \qquad\qquad\text{(Gal)}$$

As in (New), the rotation matrix is time-independent, but now uniform time-dependent translations of the spatial coordinates ("boosts") are allowed.

Let's stipulate that two coordinate systems are adapted to the same *frame of reference* if and only if they are related by an element of the Newton group.[12] Two coordinate systems related by a nontrivial Galilean boost are then adapted to different frames of reference. However, if Newton's laws hold with respect to either frame, they hold with respect to both of them. In particular, both frames might be *inertial frames* in that, with respect to them, force-free bodies move uniformly in straight lines.

---

[12] In spacetime terms, the notion of a frame of reference implicit in this stipulation corresponds to the following: a *fibration* of spacetime that specifies a standard of rest; a *foliation* of spacetime that specifies a standard of distant simultaneity; a *temporal metric* on the quotient of spacetime by the foliation; and a *spatial metric* on the quotient of spacetime by the fibration.

This gives rise to the following epistemological embarrassment for the Newtonian substantivalist. Imagine a possible world W' just like the actual world except that, at every moment, the absolute velocity of each material object in W' differs from its actual value by a fixed amount (say, by two meters per second in a direction due North). W' is an example of a world that is *kinematically shifted* relative to the actual universe. Two kinematically shifted worlds are observationally indistinguishable because, by construction, the histories of relative distances between material objects in each world are exactly the same. The worlds differ only over how the material universe as a whole is moving with respect to space. Since substantival space is not directly detectible, this is not an observable difference. Further, the Galilean invariance of Newtonian mechanics means that any two kinematically shifted worlds either both satisfy Newton's laws, or neither does.

The upshot is that the Newtonian substantivalist is committed to the physical reality of certain quantities, *absolute velocities*, that are in principle undetectable given the symmetries of the dynamical laws. Given how W' was specified in the previous paragraph, we know it is not the actual world. But consider a world, W'', just like the actual world in terms of the relative distances between bodies at each moment but in which, at 12 a.m. on 1 January 2000, the absolute velocity of the Eiffel Tower is exactly $527 ms^{-1}$ due North. For all we know, W'' is the actual world.[13] To paraphrase Maudlin (1993, 192), there may be no *a priori* reason why all physically real properties should be experimentally discoverable but one should at least be uneasy about empirically inaccessible physical facts; *ceteris paribus*, one should prefer a theory that does without them.[14]

The conclusion of this section is that the Galilean invariance of Newtonian physics poses something of an interpretative dilemma. On the one hand, to make sense of the successful dynamical laws it seems that we have to acknowledge more spacetime structure than the Leibnizian relationalist is prepared to countenance. On the other hand, Newton's manner of securing a sufficiently rich structure introduces more than is strictly required and therefore underwrites empirically undetectable yet

---

[13] Supposing, for the sake of argument, that the actual universe is Newtonian.

[14] In his correspondence with Clarke (Alexander 1956), Leibniz is sometimes read as offering kinematic-shift arguments somewhat different to the one just sketched. The idea is that kinematically shifted possible worlds would violate the Principle of Sufficient Reason (PSR) and the Principle of the Identity of Indiscernibles (PII). Since these principles are *a priori* true, according to Leibniz, there can be no such plurality of possibilities. A "Leibnizian" argument from the PSR would ask us to consider what reasons God could have had for creating the actual universe rather than one of its kinematically shifted cousins. An argument from the PII would claim that, since kinematically shifted worlds are observationally indistinguishable, they directly violate the PII.

Neither argument is convincing (nor is either faithful to Leibniz; see Pooley, unpublished). The sense of indiscernibility relevant to kinematic shifts is not that which has been the focus of contemporary discussion of the PII. This takes two entities to be indiscernible just if they share all their (qualitative) properties. In general, two kinematically shifted worlds do differ qualitatively; given how the qualitative/nonqualitative distinction is standardly understood, a body's absolute speed is a qualitative property, and differences in absolute velocity are (typically) qualitative differences. Such qualitative differences are empirically inaccessible but, theoretically, they could ground a reason for an all-seeing God's preference for one possibility over another. A PSR dilemma for God is created if we consider kinematically shifted worlds that differ, not in terms of the magnitude of their objects' absolute velocities, but only over their directions. These *are* worlds that are qualitatively indistinguishable. Discussion of how the substantivalist should treat these is postponed until section 7.

allegedly genuine quantities. Absent an alternative way to make sense of Newtonian physics, one might learn to live with Newton's metaphysics. However, the irritant of absolute velocities motivates a search for an alternative. In the next section, I consider a substantivalist way out of the dilemma. Relationalist strategies are explored in section 6.

4. Spacetime Substantivalism

4.1 Neo-Newtonian Spacetime

The substantivalist can do away with unwanted absolute velocities by adopting the essentially four-dimensional perspective afforded by spacetime. In this framework, there is an elegant way of characterizing a spatiotemporal structure that might seem to be neither too weak nor too strong for Newtonian physics.[15] The dynamical quantities of classical mechanics presuppose the simultaneity structure, instantaneous Euclidean geometry and temporal metric common to the Leibnizian relationalist and the Newtonian substantivalist. They additionally require some extra *transtemporal* structure. Geometrically, what is needed is a standard of *straightness* for paths in spacetime (provided in differential geometry by an *affine connection*). The possible trajectories of ideal force-free bodies correspond to those straight lines in spacetime that do not lie within surfaces of simultaneity. These straight lines fall into families of nonintersecting lines that fill spacetime. Each family of lines are the trajectories of the points of the "space" of some inertial frame. The resulting spacetime is known as *Galilean* or *neo-Newtonian spacetime* (see, e.g., Sklar 1974, 202–206; Earman, 1989, 33).

Now, it is one thing to give a nonredundant characterization of the spacetime structure that Newtonian mechanics assumes. It is another to provide a satisfactory account of its metaphysical foundations (that is, of what, in reality, underwrites this structure). The obvious option is to take the unitary notion of spacetime (rather than space and time separately) ontologically seriously. One regards spacetime as something that exists in its own right and which literally has the geometric structure that the affine connection, among other things, encodes. In terms of such ontology, one can provide a metaphysical account of the distinction between absolute and relative motion in a way that respects the physical equivalence of inertial frames.

It will facilitate comparison with relativistic theories to introduce a formulation of Newton mechanics that makes explicit reference to this geometrical structure. In abstract terms, physical theories often have the following general form. A space $\mathcal{K}$ of *kinematically possible models* (KPMs) is first specified. The job of the theory's

---

[15] In fact, it is too strong; see section 6.1 below.

equations (which relate the quantities in terms of which the KPMs are characterized) is then to single out the subspace S of K containing the *dynamically possible models* (DPMs).[16] The KPMs can be thought of as representing the range of metaphysical possibilities consistent with the theory's basic ontological assumptions. The DPMs represent a narrower set of physical possibilities.

In a coordinate-dependent formulation of Newtonian theory like that so far considered, the KPMs might be sets of inextendible smooth curves in $\mathbb{R}^4$ which are nowhere tangent to surfaces of constant t (where $(t, \vec{x}) \in \mathbb{R}^4$). The models assign to the curves various parameters $(m, \ldots)$. Under the intended interpretation, the curves represent possible trajectories of material particles, described with respect to a canonical coordinate system, and the parameters represent various dynamically relevant particle properties, such as mass. The space of DPMs consists of those sets of curves that satisfy the standard form of Newton's equations.

In the *local spacetime formulation* of the theory, one takes the KPMs to be n-tuples of the form $\langle M, t_{ab}, h^{ab}, \nabla_a, \Phi_1, \Phi_2, \ldots \rangle$.[17] M is a four-dimensional differentiable manifold and t_{ab}, h^{ab} and ∇_a are geometric-object fields on M that encode, respectively, the temporal structure (both the simultaneity surfaces and the temporal metric), the Euclidean geometry of instantaneous space and the inertial structure (that is, which paths in spacetime are straight). Together they represent neo-Newtonian spacetime. The "matter fields" Φ_i, which represent the material content of the model, can be curves (maps from the real line into M, representing particle trajectories) or fields (maps from M into some space of possible values, either encoding force potentials or continuous matter distributions).

The theory's DPMs are picked out by a set of equations that relate these various objects. Some of these will involve both the spacetime structure and the matter fields. For example, Newton's second law becomes:

$$F^a = m\xi^n \nabla_n \xi^a. \tag{N2}$$

For those not familiar with the tensor notation, the essential points to note about this equation are the following. F^a and ξ^a are spacetime *four-vectors*. F^a stands for the four-force on the particle. As in more traditional formulations of Newtonian theories, it will be specified by one or more additional equations. ξ^a is the four-velocity of the particle; it is the tangent vector to the particle's spacetime trajectory if that trajectory has been parameterized by absolute time.[18] Note, in particular, the explicit appearance of ∇_a in the equation. $\xi^n \nabla_n \xi^a$ is the four-acceleration of

---

[16] These notions are standard, although terminology varies; see, e.g., Anderson (1967, 74) and Friedman (1983, 48).

[17] A *locus classicus* in the philosophical literature for a discussion of Newtonian theory formulated in this style is Friedman (1983, ch. III). I adopt the following widespread notational conventions: Roman indices from the start of the alphabet do not denote components—they are "abstract indices" merely indicating the type of geometric-object field; Greek indices denote that the components of the objects relative to some spacetime coordinate system are being considered; repeated indices indicate a sum over those indices (the Einstein summation convention).

[18] That is, the trajectory is parametrized so that $t_a \xi^a = 1$, where t_a is a one-form related to the temporal metric via $t_{ab} = t_a t_b$.

the particle; it characterizes how the particle's trajectory deviates from the adjacent tangential straight line in spacetime (that is, from the relevant inertial trajectory).

4.2 Symmetries Revisited

In section 3 the Galilei group and the Leibniz group were introduced as sets of coordinate transformations, and the dynamical symmetries of Newtonian mechanics were characterized in terms of the form invariance of its equations. We now see that the implementation of this approach is not completely straightforward: various formulations of Newtonian mechanics involve different sets of equations, and these can have different invariance properties. In particular, when Newton's laws are recast so as to make explicit reference to the geometrical structure of neo-Newtonian spacetime, the resulting equations are either *generally covariant* (they hold with respect to a set coordinate systems related by smooth but otherwise arbitrary coordinate transformations) or they are *coordinate-independent* (they directly equate certain geometrical objects, rather than the values of the objects' components in some coordinate system).

There is an alternative way to characterize the symmetries of a spacetime theory. Rather than focusing on the theory's equations, one considers their solutions. Suppose, now, that a group G of maps from spacetime to itself has a natural action on the space of KPMs. G is a *symmetry* of the theory if and only if it fixes the solution space $\mathcal{S}$.[19] Note the need to relativize this characterization of symmetry to groups of maps from spacetime to itself. If we were allowed to consider any action of any group on the space of KPMs, the requirement that $\mathcal{S}$ be fixed would be too easily satisfied. In particular, for any two points s_1, s_2 in $\mathcal{S}$, we could find a group action on the space of KPMs such that $\mathcal{S}$ is fixed and s_1 is mapped to s_2. That is, every point in $\mathcal{S}$ would be mapped to every other by some symmetry transformation or other.[20]

If the KPMs and DPMs of Newtonian mechanics are defined according to the first formulation above (as certain classes of curves in $\mathbb{R}^4$), the symmetry group of the theory turns out, as one might have expected, to be the Galilei group. This is not so if the local spacetime formulation of the theory is adopted. The space of KPMs then carries an action of the diffeomorphism group Diff(M): for any $\langle M, t_{ab}, h^{ab}, \nabla_a, \Phi_i \rangle \in \mathcal{K}$ and for any $d \in \text{Diff}(M)$, $\langle M, d^* t_{ab}, d^* h^{ab}, d^* \nabla_a, d^* \Phi_i \rangle \in \mathcal{K}$.[21] It follows from the tensorial nature of the equations that pick out the

---

[19] In general, an action of group G on a space $\mathcal{K}$ is a function $\phi : (g, k) \in G \times \mathcal{K} \mapsto g \cdot k \in \mathcal{K}$ such that, for all $g, h \in G$ and $k \in \mathcal{K}$, $g \cdot (h \cdot k) = (gh) \cdot k$ and $e \cdot k = k$, where e is the identity element of G. To avoid triviality, we should also require that the action is *faithful*, that is, that for any g, if $g \cdot k = k$ for all k, $g = e$. G is a symmetry group if and only if $g \cdot s \in \mathcal{S}$ for all $g \in G$ and $s \in \mathcal{S}$. The symmetry group of a theory characterized in this way is referred to as the theory's *covariance group* by Anderson (1967, 75).

[20] See Gordon Belot's chapter in this volume for further discussion.

[21] The diffeomorphism group Diff(M) is the group of all differentiable one-one mappings from M onto itself. The definition of the map d^*, which acts on geometrical objects on M and is induced by the manifold mapping $d : p \in M \mapsto dp \in M$, will depend on the type of field. For a scalar field, ϕ,

solution subspace that, if $\langle M, t_{ab}, h^{ab}, \nabla_a, \Phi_i\rangle$ satisfies the equations then so does $\langle M, d^*t_{ab}, d^*h^{ab}, d^*\nabla_a, d^*\Phi_i\rangle$ (Earman and Norton, 1987, 520). In other words, the full group Diff(M) fixes S and therefore counts as a symmetry group of this formulation of the theory.

At this point it might look as if the formulation-dependence that afflicts a definition of dynamical symmetries in terms of the invariance of equations has simply been reproduced at the level of models. We can, however, reintroduce the distinction between spacetime and dynamical symmetries in model-theoretic terms. Our characterization of the models of a local spacetime formulation of Newtonian mechanics involved distinguishing those geometric-object fields on M that represent spacetime structure from those that represent the material content of spacetime. Let's write $\langle M, A_1, \ldots, A_n, P_1, \ldots, P_m\rangle$ for a generic spacetime model, where the A_is stand for the fields that represent spacetime structure and the P_js stand for the fields that represent the matter content. Recall that the coordinate-dependent definition of spacetime symmetries given in section 3 required that the encoding of spacetime structure by the coordinate system was preserved. Analogously, we can identify a theory's *spacetime symmetry group* as the set of elements of Diff(M) that are automorphisms of the spacetime structure; that is, we require that $d^*A_i = A_i$ for each field A_i.[22] In the case of our Newtonian theory, the proper subgroup of Diff(M) that leaves each of t_{ab}, h^{ab} and ∇_a invariant is the Galilei group.

An alternative method of singling out a subgroup of Diff(M) focuses on the matter fields rather than the spacetime structure fields. Typically the matter content of a solution will have no nontrivial automorphisms and, if it does, there will be other solutions whose matter content does not share these symmetries. We can, however, ask whether a diffeomorphism acting solely on the matter fields maps solutions to solutions. In other words, we pick out a subgroup of Diff(M) via the requirement that (for a given choice of A_i) for all $\langle M, A_i, P_j\rangle \in S$, $\langle M, A_i, d^*P_j\rangle$ should also be in S.[23] Note that, in this definition, d acts only on the matter fields and not on the spacetime structure fields. The subgroup of diffeomorphisms with this property is sometimes identified as a theory's *dynamical symmetry group* (Earman, 1989, 45–46). Again, in the spacetime formulation of Newtonian theory set in neo-Newtonian spacetime, if this group turns out to be the Galilei group

---

$d^*\phi(dp) := \phi(p)$. The action of d^* on scalar fields can then be used to define its action on tensor fields. For example, for the vector field V, we require that $d^*V(d^*\phi)|_{dp} = V(\phi)|_p$ for all points p and scalar fields ϕ.

[22] More carefully, the requirement that $d^*A_i = A_i$ for each A_i picks out a specific subgroup of Diff(M) *relative to* a particular choice of A_i. Suppose that $\langle M, A_i, P_j\rangle$ and $\langle M, A'_i, P'_j\rangle$ are both models of our theory and that there is a diffeomorphism $\phi : M \to M$ such that $A'_i = \phi^*A_i$ for each A_i, A'_i (all models of a theory set in Galilean spacetime will have this property). Although it will not be true, in general, that $\{d \in \text{Diff}(M) : d^*A_i = A_i\} = \{d' \in \text{Diff}(M) : d'^*A'_i = A'_i\}$, the groups will be *isomorphic* to the same (abstract) group; cf. Earman (1989, 45).

[23] As noted in footnote 22, different choices of A_i will, strictly speaking, yield distinct subgroups of Diff(M) but (for well-behaved theories) these will simply correspond to different representations of the same abstract group.

one has a perfect match between the spacetime symmetry group and the dynamical symmetry group.[24]

The model-theoretic perspective supports the idea that the problems faced by both the Leibnizian relationalist and the Newtonian substantivalist involve mismatches between these two symmetry groups. Whenever the spacetime symmetry group is a proper subset of the dynamical symmetry group, the theory will admit nonisomorphic models whose material submodels are nevertheless isomorphic. (Such models will be related by dynamical symmetries that are not also spacetime symmetries.) This will give rise to supposedly meaningful yet physically undiscoverable quantities: models will differ over some quantities in virtue of different relations of matter to spacetime structure and yet (assuming the material content of the models encompasses all that is observable) such differences will be undetectable. This is exactly the situation of the Newtonian substantivalist, who postulates a richer spacetime structure than that of neo-Newtonian spacetime. Models related by Galilean boosts of their material content differ over undetectable absolute velocities.

The mismatch faced by the Leibnizian relationalist is of the opposite kind: the Galilei group is a proper subset of the Leibniz group. Strictly speaking, this is not a case where the dynamical symmetry group of some theory is smaller than the spacetime symmetry group. For that we would need a spacetime formulation of a theory that (i) was set in so-called Leibnizian spacetime and (ii) had the Galilei group as its dynamical symmetry group (e.g., in virtue of an isomorphism between the set of its matter submodels and those of standard Newtonian theory). But the way in which the equations of the spacetime formulation of standard Newtonian theory single out its matter submodels uses the full structure of neo-Newtonian spacetime. The mismatch between the two groups is precisely what stands in the way of constructing such a relational theory.

Before considering the spacetime substantivalist interpretation of relativistic physics, a brief comment on the relation between the two formulations of Newtonian mechanics that I have been discussing: I first characterized the privileged coordinate systems of the coordinate-dependent form of Newtonian physics as those adapted to the spatiotemporal quantities recognized by the Newtonian. From this perspective, the spacetime formulation of the theory simply makes explicit structure that, while implicit, is no less present in the coordinate-dependent, Galilean-covariant formulation of the theory. Suppose one starts with (the coordinate expressions of) the equations of the spacetime formulation of a Newtonian theory. These equations will be generally covariant. However, one can use the symmetries of the theory's spacetime structure to pick out a special class of coordinate systems in which the values of the components of the fields representing the spacetime structure take on constant or vanishing values. In these coordinate systems the otherwise generally covariant equations apparently simplify. In other words,

---

[24] The dynamical symmetry group of Newtonian theory set in neo-Newtonian spacetime in fact turns out to be a larger group if the theory incorporates gravitation in a field-theoretic way. See section 6.1.1 below.

the Galilean covariant equations *just are* the generally covariant equations, written with respect to coordinate systems that "hide" the objects that represent spacetime structure.[25]

4.3 Relativistic Spacetimes

In the previous section, I maintained that the geometrical structure of neo-Newtonian spacetime featured, implicitly or otherwise, in the various formulations of classical mechanics. A similar claim holds true for special relativity. This is most obvious in generally covariant or coordinate-free formulations of the equations of any specially relativistic theory, where the Minkowski metric structure, encoded by the tensor field η_{ab}, figures explicitly. But it is equally true of the "standard" formulations of the equations that hold true only relative to privileged inertial coordinate systems related by Lorentz transformations. One can think of these coordinate systems as the spacetime analogues of Cartesian coordinates on Euclidean space: the coordinate intervals encode the spacetime distances via the condition $(t_p - t_q)^2 - |\vec{x}_p - \vec{x}_q|^2 = \Delta s^2$. Minkowski geometry is thus implicit in the standard formulation of the laws.

Because of the manner in which spacetime geometry features in the formulation of the laws, substantivalists hold that it explains certain features of the phenomena covered by those laws.[26] Consider, for example, the "twin paradox" scenario. Of two initially synchronized clocks, one remains on Earth while the other performs a round trip at near the speed of light. On its return the traveling clock has ticked away less time than the stay-at-home clock. The geometrical facts behind this phenomenon are straightforward: the inertial trajectory of the stay-at-home clock is simply a longer timelike path than the trajectory of the traveling clock. Note that the substantivalist does not simply assert that the number of a clock's ticks is proportional to the spacetime distance along its trajectory. Clocks are complicated systems the parts of which obey various (relativistic) laws. One should look to these laws for a complete understanding of why the "ticks" of such a system will indeed correspond to equal temporal intervals of the system's trajectory. But since, for the substantivalist, these laws make (implicit or explicit) reference to independently real geometric structure, an explanation that appeals to the laws will, in part, be an explanation in terms of such geometric structure, postulated as a fundamental feature of reality.

---

[25] This might seem like a banal observation but I take it to be significant because it conflicts with prevalent claims about the meaning of preferred coordinates in non-generally covariant theories made by, e.g., Rovelli (2004, 87–88) and Westman and Sonego (2009, 1952–1953). Their conception of the significance of such coordinates implies that there is a difference in kind between the observables of noncovariant and generally covariant theories. On the view outlined above, there is no such difference.

[26] Nerlich (1979, 2010) is staunch advocate of the explanatory role of the geometry of spacetime, realistically construed. He classifies the role of physical geometry in such explanations as noncausal, but, on certain plausible understandings of causation (e.g., Lewis, 2000), it does count as causal (see also Mellor, 1980).

Soon after formulating his special theory of relativity, Einstein began a decade-long quest for a theory of gravitation that was compatible with the new notions of space and time. The triumphant culmination of this effort was his publication of the field equations of his general theory of relativity (GR) in 1915.[27] These equations, known as the Einstein Field Equations (EFEs), relate certain aspects of the *curvature* of spacetime, as encoded in the *Einstein tensor*, G_{ab}, to the matter content of spacetime, as encoded in the *energy momentum tensor*, T_{ab}:[28]

$$G_{ab} = 8\pi T_{ab}. \tag{EFE}$$

One goal above all others guided Einstein's search: the generalization of the "special" principle of relativity to a principle that upheld the equivalence of frames of reference in arbitrary states of relative motion. In this, Einstein was motivated by the belief that the role of primitive inertial structure in explaining phenomena in both Newtonian physics and in special relativity was "epistemologically suspect": real effects, he believed, should be traceable to observable, material causes.

In the early stages of his search, Einstein had a crucial insight, which he thought would play a key role in the implementation of a generalized relativity principle. According to his *Principle of Equivalence*, an inertial frame in which there is a uniform gravitational field is physically equivalent to a uniformly accelerating frame in which there is no gravitational field. It is not hard to see why one might take this principle to extend the principle of relativity from uniform motion to uniform acceleration.[29]

When Einstein first published the field equations of GR, he believed that their general covariance ensured that they implemented a general principle of relativity. Since smooth but otherwise arbitrary coordinate transformations include transformations between coordinate systems adapted to frames in arbitrary relative motion, it might seem that there can be no privileged frames of reference in a generally covariant theory. Almost immediately, Kretschmann (1917) pointed out that this cannot be correct, arguing that it should be possible to recast any physical theory in generally covariant form.

Compared to its predecessors, GR is without doubt a very special theory. But one will do more justice to its conceptual novelty, not less, by seeking as much common ground with previous theories as possible. We have already met generally covariant formulations of Newtonian and specially relativistic physics. As these examples attest, Kretschmann's instincts were sound and it is now well understood that the general covariance of GR does not implement a general principle of relativity. In

---

[27] The genesis of Einstein's general theory has been subject to extensive historical and philosophical scrutiny. For an excellent introduction to the topic, see Janssen (2008).

[28] $G_{ab} := R_{ab} - \frac{1}{2} R g_{ab}$, where R_{ab} (the *Ricci tensor*) and R (the *Riemann curvature scalar*) are both measures of curvature. g_{ab} is the *metric tensor* and encodes all facts about the spatiotemporal distances between spacetime points. R_{ab} and R are officially defined in terms of the Riemann tensor, itself defined in terms of the connection ∇_a. However, since we are considering the unique torsion-free, metric-compatible connection, we can view these quantities as defined in terms of the metric and, indeed, they can be given natural geometric interpretations directly in terms of spacetime distances. T_{ab} encodes the net energy, stress, and momentum densities associated with the material fields in spacetime.

[29] For a critical discussion of Einstein's various formulations of the principle, see Norton (1985).

fact, GR arguably includes privileged frames of reference in much the same way as pre-relativistic theories.[30]

It is illuminating to consider a concrete example. Compare the local spacetime formulation of specially relativistic electromagnetism to its generally relativistic version. Models of the former are of the form $\langle M, \eta_{ab}, F_{ab}, J^a \rangle$, where η_{ab} encodes the Minkowski spacetime distances between the points of M, the covariant tensor field F_{ab} represents the electromagnetic field and the vector field J^a represents the charge current density. The dynamically possible models can be picked out via the coordinate-free form of Maxwell's equations:

$$\nabla_a F^{ab} = -4\pi J^b$$
$$\nabla_{[a} F_{bc]} = 0 \qquad \text{(Maxwell)}$$

The only difference between this theory and the generally relativistic theory is that, in the latter, the Minkowski metric η_{ab} is replaced by a variably curved Lorentzian metric field, g_{ab}. Maxwell's equations remain as constraints on the DPMs of the GR version, and the same relativistic version of Newton's second law (which is formally identical to the nonrelativistic version: equation (N2) of section 4.1) holds in both theories.[31] But, whereas the specially relativistic metric was stipulated to be flat, to be physically possible according to GR, combinations of g_{ab}, F_{ab} and J^a must now obey the EFEs.

In the case of SR, the standard, Lorentz-invariant form of Maxwell's equations is recovered by choosing inertial coordinate systems adapted to the spacetime distances.[32] The only reason that the same cannot be said of GR is that the pattern of spacetime distances catalogued by g_{ab} does not allow for global coordinate systems that encode them. One can, however, always chose a coordinate system that is optimally adapted the spacetime distances in the infinitesimal neighborhood of any point $p \in M$. In any coordinate system in which, at p, $g_{\mu\nu} = \text{diag}(1, -1, -1, -1)$ and $g_{\mu\nu,\gamma} = 0$, the laws governing matter fields will take their standard Lorentz-invariant form at p.

It is in terms of this *strong equivalence principle* that the phenomena that figure in Einstein's original principle are nowadays understood.[33] From a modern perspective, apparent "gravitational fields" have precisely the same status as the potentials that give rise to the centrifugal and coriolis forces in Newtonian physics: they correspond to pseudo forces that are artifacts of the failure of the relevant coordinate systems to be fully adapted to the true geometry of spacetime. The "force" that holds us on the surface of the Earth and the "force" pinning the astronaut to the floor of the accelerating rocket ship are literally of one and the same kind. In

---

[30] I should note that some still hold out against this orthodoxy (e.g., Dieks, 2006).
[31] In this case, the four-force on a particle with charge q and four-velocity ξ^a is given by $qF^a{}_b \xi^b$ and the equation is simply the coordinate-free version of the Lorentz force law.
[32] For example, the components of F_{ab} relative to an inertial coordinate system are $F_{0i} = -E_i$, $F_{ij} = \epsilon_{ijk} B_k$, where E_i and B_i are the components of the electric and magnetic three-vector fields in that frame.
[33] Ehlers (1973b, 18); see Brown (2005, 169–172) for a recent discussion.

GR, gravitational phenomena are not understood as resulting from the action of forces.[34]

5. Reasons to be a Relationalist?

We have seen that substantivalism is recommended by a rather straightforward realist interpretation of our best physics. This physics presupposes geometrical structure that it is natural to interpret as primitive and as physically instantiated in an entity ontologically independent of matter. Although I have only considered nonquantum physics explicitly, the claim is equally true of both nonrelativistic and relativistic quantum theories (Weinstein, 2001). One might therefore wonder: why even try to be a relationalist? For some, the Hole Argument, discussed below in section 7, is the major reason to seek an alternative to substantivalism. Here I wish to review some other antisubstantivalist themes that have motivated relationalists. The conclusion will be that the only strong consideration in favor of relationalism is Ockham's razor: if a plausible relational interpretation of empirically adequate physics can be devised, then the standard reasons for postulating the substantivalist's additional ontology are undermined.

5.1 A Failure of Rationality?

In Section 6.2 I review Julian Barbour's approach to dynamics. According to Barbour, there is something "irrational" about standard Newtonian mechanics. Barbour sets up the problem in terms of the data required for the Newtonian *initial value problem*: given the equations of a Newtonian theory, what quantities must be specified at an instant in order to fix a solution? Following Poincaré, Barbour emphasizes that the natural relational data—instantaneous relative distances and their first derivatives—are almost but not quite enough. In addition, three further parameters, which specify the magnitude and direction of the angular momentum of the entire system, are

---

[34] Some authors favor talk of "tidal forces" or state that there is a real "gravitational field" just where the Riemann tensor is nonzero (e.g., Synge, 1960, ix). As far as I can see, this is simply a misleading way of talking about spacetime curvature and (typically) nothing of conceptual substance is intended by it. For a discussion of some of the pros and cons of identifying various geometrical structures with the "gravitational field," see Lehmkuhl (2008, 91–98). Lehmkuhl regards the metric g_{ab} as the best candidate. My own view is that consideration of the Newtonian limit (e.g., Misner et al., 1973, 445–446) favors a candidate not on his list, viz., *deviation of the metric from flatness*: h_{ab}, where $g_{ab} = \eta_{ab} + h_{ab}$. That this split is not precisely defined and does not correspond to anything fundamental in classical GR underscores the point that, in GR, talk of the "gravitational field" is at best unhelpful and at worst confused. The distinction between background geometry and the graviton modes of the quantum field propagating against that geometry is fundamental to perturbative string theory, but this is a feature that one might hope will not survive in a more fundamental "background-independent" formulation.

needed.[35] Barbour claims that there is something odd about this fact (see, e.g., Barbour, 2011, §2.2).

Fixing the Euclidean relative distances between N particles and their rates of change requires the specification of $6N - 12$ numbers; $6N$ numbers are required to specify their positions and velocities in absolute space. The Newtonian initial value problem, however, requires $6N - 9$ numbers. Six of the $6N$ numbers that specify the particles' absolute positions and velocities can be thought of as specifying the orientation of the system as a whole and the position of its center of mass. States that differ solely in terms of these quantities only differ *nonqualitatively*, in terms of *which particular points of space* are related to the material system in (qualitative) ways common to both situations; the structural pattern of relations between space and matter is shared by both states. Even if such differences are to be regarded as real differences (something to be questioned in section 7), Newtonian physics should not be expected to take them into account (*pace* Barbour, 1999, 83). It does not, after all, single out particular points of space by name. The remaining three parameters can be thought of as specifying the absolute velocity of the center of mass. As we have seen, while the Newtonian substantivalist regards this as a genuinely qualitative matter, the spacetime substantivalist does away with these quantities by embracing neo-Newtonian spacetime. Arguably, $6N - 9$ numbers is precisely what the spacetime substantivalist should expect to be needed as Newtonian initial data.

What, then, is "irrational" about Newtonian mechanics? In Barbour's view it fails to be maximally predictive: *relative to the Leibnizian data*, there is an apparent breakdown of determinism, which is only restored by specifying the global angular momentum. At one level, this simply amounts to a prejudice in favor of the relational quantities over the Newtonian ones. For Barbour, this preference is based on the fact that it is the relational quantities that are "directly observable" (see, e.g., Barbour, 2010, 1280) but (i) direct observability is an extreme criterion for determining ontological commitment and (ii) *instantaneous* relative distances are not directly observable.[36] One can still agree with Barbour's less ambitious point. Without accepting that there is something inherently objectionable about standard Newtonian theory, one might nevertheless prefer a theory that does as well (is at least as explanatory etc.) but with fewer resources. Barbour's observations about initial data highlight that, if an adequate relational theory of the envisaged type can be found, it will be more predictive, because its initial data form a proper subset of those of the Newtonian theory.

---

[35] Instantaneous relative distances and their first derivatives are the natural Leibnizian relational data. As reviewed in section 6.2, Barbour's preferred framework for understanding classical mechanics also dispenses with a primitive temporal metric and an absolute length scale. With respect to these more frugal initial data, five, not three, additional numbers are needed. See Barbour (2011, §2.2).

[36] Nor are distance ratios, Barbour's preferred relational quantities. For an illuminating discussion of how instantaneous quantities are detected only indirectly, in measurements that necessarily take finite time, see Stein (1991, 157).

5.2 The Spacetime Explanation of Inertia

In section 6.3 I consider the *dynamical approach* to special relativity, defended in Brown (2005) and Brown and Pooley (2006). I noted above that substantivalists view the postulated spacetime geometry as explanatory. Brown is suspicious of this doctrine. In assessing it, the putative explanatory roles of affine structure and of metric structure should be treated separately. Here I only consider the former; the latter is discussed in section 6.3.

The idea that affine structure plays a quasi-causal role in explaining the motions of bodies figures significantly in Einstein's criticism of Newtonian mechanics and SR and in his subsequent understanding of GR. Consider Einstein's example from his early review paper on GR, of two fluid bodies separated by a great distance and in relative rotation about the line joining their centres. The two are of the same size, shape, and nature except that one body is spherical whereas as the other is oblate. In Newtonian mechanics the explanation of this difference is that the oblate body, but not the spherical body, is rotating with respect to the inertial frames: the absolute rotation of the oblate body causes its oblateness. Einstein labels the Newtonian spacetime structure with respect to which such rotation is defined as a "merely factitious cause" of the difference; he held that a genuine explanation should instead cite another "observable fact of experience" (Einstein, 1916, 112–113). He initially maintained that this requirement was met in GR because he believed that the theory satisfied what he later called *Mach's Principle* (Einstein, 1918, 241–242): if the metric field g_{ab} were fully determined by the distribution of matter throughout the universe, then the difference between the inertial behavior of the two bodies would be traceable to differences in their relations to distant (observable) masses.

As Einstein soon recognized, GR does not satisfy Mach's Principle so defined.[37] Inertial structure, as encoded in g_{ab}, is *influenced but not determined by* the matter content of spacetime.[38] Einstein ceased to regard the field describing inertial structure as having a secondary status relative to ponderable matter. By 1921, the objection to Newtonian absolute space was no longer that it was invisible. Instead the fact that it *acted without being acted upon* was held up as problematic; a "defect" not shared by the spacetimes of GR (Einstein, 1922, 61–62).[39] At around the same time, Weyl advocated a similar conception of the role of inertial structure (he called it the "guiding field"), which he regarded as "physically real" in both pre-relativistic physics and in GR (Weyl, 1922, §27).

---

[37] De Sitter first pointed out to Einstein that, in addition to specification of T_{ab}, one needs to specify boundary conditions at infinity in order to determine g_{ab}. This prompted Einstein to search for spatially compact solutions to the EFEs and to introduce the cosmological constant to allow for a static, spatially closed universe. This in turn led de Sitter to the discovery of the *de Sitter universe*: a spatially compact vacuum solution to the modified EFEs. See Janssen (2008, §5) for a summary of this episode and for further references.

[38] It is also worth stressing that the stress-energy properties of matter, as encoded in T_{ab}, cannot even be defined independently of g_{ab}; see Lehmkuhl (2011).

[39] The idea that something should be capable of acting if and only if it can also be affected by those things that it can influence is known as the *action–reaction principle* (see Anandan and Brown, 1995, for a discussion).

What does Brown object to in this picture? Consider the "conspiracy of inertia": the relatively simple case of the force-free motions of a collection of Newtonian particles (Brown, 2005, 15, 141). Despite the fact that, *ex hypothesi*, the particles are not influencing one another, they move in a highly coordinated way: there are spacetime coordinate systems with respect to which all the trajectories are straight lines.[40] As we have seen, the coordinate-free geometrical description of this state of affairs regards the trajectories as coinciding with the straight lines of an affine connection on spacetime, the flatness of which allows for the global inertial coordinate systems in terms of which the phenomenon was stated. For the substantivalist, inertial structure is an element of reality that exists independently of the particles and their motions. On this view, aspects of the geometry play a role in explaining the phenomenon because, as stated in the substantivalist version of Newton's first law, the particles' trajectories are constrained as a matter of physical necessity to be aligned with features of this realistically construed geometry. For Brown, this is merely verbal pseudo-explanation. His preferred point of view reverses the arrow of explanation: the geometry is just a *codification* of the phenomenon, which (in pre-relativistic physics) must be taken as primitive.

If the only role of inertial structure was to explain pure inertial motion, Brown's complaint against the substantivalist would have some intuitive force; a flat affine connection could be thought of as a rather direct codification of the regularities manifest in the phenomena via the "coordinative definition" of spacetime geodesics as the trajectories of force-free bodies.[41] Explaining inertial motion, though, is not the real purpose of inertial structure. Inertial structure figures centrally in the explanation of *noninertial* motion. In contrast to the force-free case, the sequence of relative distances between interacting particles manifest over time in a Newtonian universe displays no obvious regularity. It is a rather remarkable fact that, by postulating a highly symmetric geometrical structure in terms of which the motions of individual particles are to be understood, one can provide an elegant explanation, in terms of simple force laws, of the complicated and irregular history of relational quantities. Anyone who is not amazed by *this* conspiracy has not understood it. Since the postulated deep structure is not manifest in the surface phenomena it seems genuinely explanatory and not a mere codification. It is from an application

---

[40] That is, there are coordinate systems with respect to which the particles' spatial coordinates are linear functions of their time coordinates. In Brown's view, "anyone who is not amazed by this conspiracy has not understood it" (Brown, 2005, 15).

[41] The idea that spacetime geodesics are defined as the trajectories of force-free bodies is defended by DiSalle (1995, 327), whom Brown quotes approvingly. Elsewhere Brown, ostensibly to make a point against the substantivalist explanation of inertia, stresses that the principle that the trajectories of force-free bodies are geodesics in fact has limited validity in GR (Brown 2005, 141, see also 161–168). What this observation in fact undermines is a *relationalist* approach to spacetime geometry that tries to define geodesics in terms of "basic physical laws" (DiSalle 1995, 325). More recently, DiSalle makes clear that he differs from the logical positivists in not regarding the coordination of geodesics with free-fall trajectories as a matter of arbitrary stipulation. Instead it is said to be "a kind of discovery, at once physical and mathematical, that . . . the only objectively distinguishable state of motion corresponds to the only geometrically distinctive path in a generally covariant geometry" (DiSalle, 2006, 131–132). Nothing in the substantivalist's metaphysics is inconsistent with this position; it is less clear what other metaphysical views are compatible with it. DiSalle does not share the substantivalist's and relationalist's preoccupation with ontological questions but nor does he offer reasons to see such questions as illegitimate.

of standard inference-to-the-best-explanation reasoning to this type of scenario that substantivalism gets its real support.[42]

Brown suggests that, around 1927, Einstein ceased to assign a quasi-casual role to spacetime in determining the inertial trajectories of bodies (Brown 2005, 161). The alleged reason is that, at this time, Einstein came to recognize that the principle of inertia does not need to be postulated as a basic law in GR; it is instead a theorem.[43] In Brown's view this fact undermines the idea that spacetime structure "in and of itself" acts "directly" on force-free bodies because it shows that, in GR, when such bodies undergo geodesic motion, "such motion is ultimately due to the way the Einstein field $g_{\mu\nu}$ couples to matter, as determined by the field equations" (Brown, 2005, 162–163). Brown's picture is, then, that the relationship of the motions of force-free bodies to inertial structure in GR is radically different to this relationship in Newtonian physics and SR. In the latter theories, inertial structure is a mere codification of basic, mysterious inertial behavior. In the former, it receives a *dynamical* explanation via the coupling of matter fields to the metric, as described by the EFEs.

The substantivalist should not be especially troubled by Brown's claim, for it concedes that, in our best (classical) theory of spacetime, the metric structure of spacetime is a primitive element of reality that plays a role in determining the inertial behavior of bodies. Even so, there are reasons to doubt that the contrast between GR and pre-relativistic theories can really bear the weight Brown demands of it. The derivation of geodesic motion from the EFEs basically involves two steps. First, one notes that the EFEs imply the vanishing of the covariant divergence of the stress energy tensor: $\nabla_a T^{ab} = 0$. Second, one makes various assumptions about the nature of the stress-energy tensor to be associated with a force-free particle that, together with the vanishing of the divergence of stress-energy, can be shown to entail that the particle's trajectory is a geodesic. Now, the second step of this derivation is as applicable in SR as in GR.[44] What difference there is between the theories must therefore concern the status of the conservation principle, $\nabla_a T^{ab} = 0$. This equation, of course, also holds in SR. Further, while in SR it cannot be derived from the *gravitational* field equations (there are none), it is a consequence of the *matter* field equations (as it is in GR also).[45]

In sum, geodesic motion is arguably as much a theorem in SR as it is in GR. An alternative perspective to Brown's is that the "dynamical coupling" of matter fields to inertial structure is in essential respects the same in GR and pre-relativistic theories. It is true that the geodesic theorem also demonstrates that it has limited validity in GR but not in SR (recall footnote 41). But the reason why, for example, rotating bodies do not deviate from geodesics in SR is not that the relationship

---

[42] Note that Einstein had in mind descriptions of interacting systems in different states of acceleration, and not simple inertial motion, when he claimed that "something real has to be conceived as the cause for the preference of an inertial system over a noninertial system" (Einstein, 1924, 16).

[43] See Malament (2010) for a critical discussion of this result.

[44] A closely parallel derivation is also possible in the geometrized form of Newtonian gravity; see Weatherall (2011a,b). This might be taken to further undermine the claim that only in GR is inertia explained.

[45] See, e.g., Trautman (1962, 180–181).

between realistically construed inertial structure and matter is radically different in this theory to that in GR. The reason is simply that, in SR, there is no curvature to which rotating bodies might couple.[46]

We have yet to meet a decisive reason to look for an alternative to substantivalism. Huggett, whose "regularity account" of relational spacetime I consider in section 6.3, states that his reasons for advocating relationalism are the usual ones: "worries about 'Leibniz shifts' and considerations of ontological parsimony that militate against the introduction of bizarre non-material substances" (Huggett, 2006, 41). The kinematic shift has already been dealt with; other Leibniz shifts are discussed in section 7. It is not clear why Huggett regards spacetime as "bizarre." In a pre-relativistic context, one feature of spacetime that might seem odd is its failure, emphasized by Einstein, to obey the action–reaction principle. As we have seen, however, if this is a failing, it is not one shared by the spacetime of GR. That leaves considerations of ontological parsimony: *other things being equal,* a relationalist interpretation of physics might seem preferable to substantivalism because it makes do with fewer metaphysical commitments. The question is: Are other things equal? It is time to examine some concrete relationalist proposals.

6. THREE VARIETIES OF RELATIONALISM

Call the objects to whose existence a theory is committed the *ontology* of the theory. Call the range of (primitive) distinctions that a theory is able to express via its (primitive) predicates and terms—roughly, the set of (primitive) properties and relations to which the theory is committed—the theory's (primitive) *ideology*.[47] In these terms, the problem faced by the Leibnizian relationalist is that classical mechanics employs an ideology that appears to presuppose a substantivalist ontology. Inertial structure is naturally understood in terms of relations that hold of spacetime points; it cannot be understood (straightforwardly) in terms of properties and relations that are instantiated only by material objects. On the other hand, *bona fide* relationalist ideology appears to be too impoverished a basis for an empirically successful alternative to Newtonian theory.

This means that there are two obvious strategies open to relationalists. On the first, they can attempt to expand their ideology so that it underwrites the same

---

[46] Brown's thesis that inertia receives a dynamical explanation only in GR has recently been defended by Sus (2011). Sus emphasizes that in GR the metric is a genuinely dynamical entity and that one can derive $\nabla_a T^{ab} = 0$ from the very equations that govern the metric's behavior. In contrast, SR, as standardly conceived, involves fixed inertial structure whose properties are postulated by *fiat*. However, this difference between the theories is compatible with the theories agreeing on the fundamental reasons why force-free bodies are related to inertial structure in just the way they are.

[47] The terminology is Quine's, who characterizes a theory's ontology as "the objects over which the bound variables of the theory should be construed as ranging in order that the statements affirmed in the theory be true" (Quine 1951, 11).

physical distinctions as substantivalist inertial structure but in a way compatible with a relationalist ontology. To complete this program, the relationalist then needs to reconstrue standard Newtonian theory in terms of these new relationalist quantities.[48] Variants of this strategy are the topic of section 6.1. On the second strategy, the relationalist seeks an alternative theory to Newtonian mechanics that employs only traditional relationalist quantities but is, although empirically distinct from Newtonian theory, nonetheless empirically adequate. Since the inertial frames are empirically determinable (in our neighborhood of this universe), such a theory still needs to account for them, at least as a feature of solutions that could serve as models of the actual world. Unlike theories that result from the first strategy, however, it will not construe them as encoding primitive spatiotemporal properties and relations. The most promising version of this approach is reviewed in section 6.2.

These courses of action correspond closely to the first two of three options identified by Nick Huggett (1999). He sees Newton's globes thought experiment as illustrating that no theory has the following three characteristics. (i) Its spatiotemporal ideology is restricted to Leibnizian relations; (ii) its dynamically allowed histories of such relations are exactly those predicted by Newtonian theory and; (iii) inertial effects supervene on the specified spatiotemporal relations between bodies. Strategies (1) and (2) correspond to relinquishing (i) and (ii) respectively. But, Huggett observes, the relationalist might try to retain (i) and (ii) by dropping requirement (iii) (Huggett 1999, 22–23). This amounts to a third, non-obvious strategy: do not change the theory and do not add to the ideology and yet somehow be a relationalist. This type of "have-it-all" relationalism is the topic of section 6.3.

6.1 Enriched Relationalism

6.1.1 Classical Mechanics

Part of the substantivalist's response to the kinematic shift argument involved replacing persisting space with spacetime. If relationalists likewise adopt a four-dimensional perspective, a number of options richer than Leibnizian relationalism become available. The most straightforward is *Newtonian relationalism* (Maudlin, 1993, 187). Whereas the Leibnizian relationalist posits spatial relations that hold only between simultaneous material events, the Newtonian relationalist simply posits that all material point events stand in spatial distance relations.[49]

---

[48] The need for this second step is emphasized by Earman (1989, 128), though not in precisely these terms.
[49] Related versions of relationalism, according to which absolute velocity (or even absolute position) is interpreted as a primitive, *monadic* property of particles, have been discussed by Horwich (1978, 403) and Friedman (1983, 235) (see also Teller, 1987). In addition to being less natural than the form of Newtonian relationalism identified by Maudlin, they are vulnerable (like Newtonian relationalism) to the kinematic shift argument. The absolute position version is also vulnerable to the static shift argument mentioned in section 7.

If we impose the natural constraints, the embedding of a Newtonian relational history into neo-Newtonian spacetime is fixed up to Galilean transformations.[50] Whether this by itself entitles the relationalist to exploit the full resources of Newtonian dynamics (as Maudlin thinks; 1993, 192–193) need not be resolved, for the Newtonian relationalist can also interpret the dynamical laws directly in terms of relationalist quantities. For example, the absolute velocity of particle i at time t is just the limit, as δt goes to zero, of the directed distance between the instantaneous stage of i at t and its instantaneous stage at $t + \delta t$ divided by δt. The Newtonian relationalist therefore has available a relational understanding of the very quantities that feature in the standard form of Newtonian laws expressed with respect to an inertial frame.

Unfortunately, Newtonian relationalism, like the Newtonian substantivalism from which its ideology is plundered, is vulnerable to the kinematic shift argument.[51] Since absolute velocities are unobservable, the Newtonian relationalist is committed to physically real distinctions that are in principle empirically inaccessible. The Newtonian relationalist must accept that, while only one inertial frame discloses the true Newtonian relational distances, that frame is forever beyond our grasp.

The spacetime substantivalist solves the kinematic shift problem by replacing Newtonian with neo-Newtonian spacetime. There is an obvious relational analogue of this move: since neo-Newtonian spacetime's inertial structure is equivalent to a relation of collinearity between triples of spacetime points, the relationalist can add to their ideology a three-place relation of collinearity between material events. The *neo-Newtonian relationalist* claims that, for three nonsimultaneous events e_1, e_2, e_3, the relation $col(e_1, e_2, e_3)$ holds just if, from the substantivalist perspective, e_1, e_2 and e_3 lie on a single inertial trajectory.[52]

The move solves the kinematic shift problem, but only at the cost of leaving the relationalist's ideology too impoverished to fix the embedding of a relational history into neo-Newtonian spacetime. An example of Maudlin's nicely illustrates the point:

> [C]onsider two particles in a neo-Newtonian spacetime that are uniformly rotating about their common center of mass. Until the first rotation is complete, no triple of occupied event locations are collinear. Even after any number of rotations,

---

[50] The obvious constraints are that the embedding respects the temporal separation between material events and that there is a single congruence of inertial geodesics such that, for any two material events e_1, e_2 located on geodesics from the congruence v_1, v_2, the Newtonian relational distance between e_1 and e_2 equals the (constant) spatial distance between simultaneous points of v_1 and v_2.

[51] This is the principal inadequacy of Newtonian relationalism that Maudlin identifies (1993, 193). Friedman (1983, 235) makes the same criticism of the postulation of a primitive property of "absolute velocity."

[52] Maudlin restricts the extension of *col* to nonsimultaneous events, but there is no reason why mutually simultaneous events should not be included, with $col(e_1, e_2, e_3)$ holding just if the sum of the distances between two of the pairs of events equals the distance between the third pair.

> the collinearity relations among occupied points will be consistent with any periodic rotation, uniform or nonuniform. (Maudlin 1993, 194)

One cannot, therefore, interpret the spacetime coordinates in which the dynamical laws take their standard form as just those coordinates adapted to the neo-Newtonian relationalist's ideology; if the relationalist ontology is sufficiently sparse, this ideology underdetermines the inertial frames.[53]

This kinematical underdetermination is not necessarily fatal to neo-Newtonian relationalism. It means that the neo-Newtonian relationalist cannot simply lay claim to standard Galilean-invariant dynamics. On the strategy we are considering, however, the relationalist succeeds so long as they can identify, in a relationally respectable manner, a set of relational DPMs that correspond to the full set of Newtonian DPMs. Can the neo-Newtonian relationalist find dynamical laws expressed directly in terms of neo-Newtonian relations that achieve this?

As far as I know, no one has seriously attempted to construct such laws. Even so, one knows that any such laws will exhibit a particularly unattractive feature: they will not be expressible as differential equations that admit an initial value formulation.[54] In standard Newtonian theory the specification of the instantaneous positions and velocities of the particles with respect to some inertial frame suffices, via the laws, to determine the particles' relative positions and motions at all times. Strictly speaking, what needs to be specified at an instant transcends the specification of the intrinsic state of that instant: in specifying velocities one is specifying quantities that are ultimately grounded in the pattern of the instantiation of collinearity relations between nonsimultaneous spacetime points. But the relevant points lie in the *infinitesimal neighborhood* of each instant and so can be used, via the usual limiting procedure, to define derivative quantities that are possessed at that instant. What Maudlin's example illustrates is that there can be *finite* stretches of time in a neo-Newtonian relational world such that no nonsimultaneous triples of material events occurring during that time instantiate the collinearity relation. Indeed, this is a generic feature of neo-Newtonian relational worlds of point particles. The collinearity relation, therefore, cannot be used to define derivative quantities that can supplement the instantaneous data definable in terms of Leibnizian relations.

The problem faced by Newtonian relationalism suggested neo-Newtonian relationalism; the trouble faced by neo-Newtonian relationalism suggests another,

---

[53] A similar example involving Minkowski spacetime is discussed by Mundy (1986), Catton and Solomon (1988), and Earman (1989, 168–169). The relations of spacelike separation, lightlike separation and timelike separation determine the structure of Minkowski spacetime up to an overall scale factor. However, these relations instantiated between material events need not fix their embedding into Minkowski spacetime up to Poincaré transformations. The examples discussed by Mundy et al. involve a small finite number of events, but the problem generalizes to certain configurations of continuum many. For example, consider two particles which move so that any two events from distinct trajectories are always spacelike (the events on each trajectory are all mutually timelike). We know that, as $t \to \pm\infty$, the particles must be accelerating in opposite directions, but not much more.

[54] This objection to neo-Newtonian relationalism, reported by Huggett (1999, 26), is again due to Maudlin.

rather desperate, relationalist maneuver. If the problem is the lack of appropriate instantaneous quantities, why not simply co-opt as primitive certain derivative instantaneous quantities available to the substantivalist? This, in essence, is how Sklar proposes the relationalist treat absolute accelerations; as intrinsic, primitive, time-varying properties possessed by particles at every instant of their trajectories (Sklar 1974, 230). Huggett usefully dubs them *Sklarations* (Huggett, 1999, 27).

It is not obvious how the relationalist is supposed to use this additional ideology. Skow notes that simply supplementing Leibnizian relational initial data with Sklarations fails to fix a Newtonian history. Consider, for example, the following pair of two-particle solutions to Newton's theory of gravitation (Skow 2007, 783–784). In one solution the two bodies follow circular orbits about their common center of mass. In the second solution the particles travel in on parabolic paths from spatial infinity to slingshot past each other before heading back out to infinity. Suppose that the distance of closest approach of the particles in the second solution matches the constant separation of the particles in the first. Then, at the moment of closest approach, the second solution matches the first in terms of its Leibnizian initial data and absolute accelerations: the separation between the particles is the same, its rate of change is zero (it is the moment of closest approach) and, because accelerations due to gravity depend only on the masses of particles and the relative separation between them, the Sklarations are the same too.

In fact, this problem is generic. Precisely because, according to any Newtonian theory satisfying Newton's third law of motion, forces and hence absolute accelerations are functions only of the relative distances, they are effectively already included in the Leibnizian initial data. Thus every set of Leibnizian initial data "supplemented" with Sklarations will radically underdetermine the future evolution of any system of interacting Newtonian particles. As we saw in section 5.1, this evolution depends on the overall angular momentum and the Leibnizian initial data, with or without Sklarations, does not tell us what this is.

Skow's assumption about the appropriate initial data for a theory employing Sklarations could be questioned. Why should the Sklar relationalist not include, say, the first time derivatives of Sklarations?[55] What the relationist really needs to provide are some relatively natural equations involving Sklarations that fix their theory's DPMs, for these will determine what the appropriate initial data are.

Sklar himself did not flesh out his proposal. Both Friedman (1983, 234) and Huggett (1999, 27) suggest that the Sklar relationist can simply utilize Newton's second law as expressed in arbitrary rigid Euclidean coordinate systems, that is, coordinate systems adapted to the Leibnizian relational ideology. However, it is not at all straightforward how Sklarations are supposed to feature in such a formulation of Newton's second law.[56] More significantly, Friedman's and Huggett's attempt to

---

[55] Since these are functions of the $\dot{r}_{ij}$s, just as Sklarations are functions of the r_{ij}s, they would not actually be of any help either. The situation changes if higher derivatives are allowed.

[56] Friedman's expression of the law is $F^i/m = \ddot{x}^i + a^i + 2\dot{a}_{ji}\dot{x}^j + \ddot{a}_{ki}x^k$ (Friedman, 1983, 226, eqn 8; I have slightly altered the notation). F^i is the ith component of Newtonian (three-)force on the particle we are

reinterpret the standard equations in terms of Sklarations does not even get off the ground unless Sklarations are additionally constrained to be embeddable in spacetime as four-accelerations. But why should the instantiation of an allegedly primitive monadic property be constrained in this way as a matter of metaphysical necessity? Regarded as a kinematical constraint, the requirement is very fishy from a relationalist perspective. The alternative is to view the constraint only as a restriction on the physically possible, that is, as an additional "law of motion" governing the evolution of Sklarations (and constraining admissible initial data).[57] Either way, a strong suspicion must remain that, in this guise, Sklarations do not constitute a genuine alternative to *accelerations* and the attendant substantivalist commitments they require.

The relationalist needs ideology weaker than Newtonian relations but richer than the neo-Newtonian's collinearity relation. In particular, ideology that is sufficiently richly instantiated in the neighborhood of any instant is needed in order to avoid the initial value problem faced by the neo-Newtonian relationalist. Sklarations might have been expected to provide what was needed but, because the quantities are treated as primitive, their necessary connections with the relational states of the world at earlier and later times is severed. Putting these connections back in by hand looks like substantivalism by another name.

This suggests that to avoid the pitfalls of Sklarations the relationalist should look for ideology that is instantiated by some *n*-tuples of *nonsimultaneous* events. And to avoid the pitfalls of neo-Newtonian relationalism, this ideology should be instantiated by *n*-tuples of nonsimultaneous events in the infinitesimal neighborhood of any instant. Further, any kinematic constraints on the possible instantiation of this ideology should be comprehensible independently of its interpretation, once appropriately embedded, in neo-Newtonian spacetime.

It is certainly possible to specify relational ideology that meets these requirements[58] but at this point we should take a step back and recall the structure of the original dilemma posed by Galilean invariance. A dynamical symmetry group that was larger than the spacetime symmetry group leads to in-principle unobservable quantities; a spacetime symmetry group larger than the dynamical symmetry group requires a nonstandard story about the privileged status of some dynamically preferred frames of reference (for they form a proper subset of those maximally adapted

---

considering and $x^i(t)$ is the *i*th component of its position vector with respect to some rigid Euclidean coordinate system. a^i is the *i*th component of the absolute acceleration of the origin *of the coordinate system* (that is, the Sklaration that a hypothetical particle would have were it comoving with the coordinate origin). $\dot{a}_{ji}$ is the rotation of the coordinate system about its origin with respect to an inertial frame. Thus only the first of the three additional terms on the right-hand side of the equation is directly interpretable in terms of a Sklaration, and then only if we pick a coordinate system that happens to have a particle comoving with its origin. Crucially, we need to be told how to interpret the rotation pseudo-vector $\dot{a}_{ji}$ in terms of Sklarations.

[57] Mundy (1983, 224) even interprets the Euclidean constraints on instantaneous distances similarly, so that his relationalist does not need a primitive notion of geometric possibility over and above that of physical possibility.

[58] I explore some of the options in Pooley (in preparation). As with Sklarations, the required "kinematical" constraints on the instantiation of such relations suggest that the proposals are really substantivalism in disguise.

to the spacetime quantities). I have given the impression that, in the context of classical mechanics, the structures of neo-Newtonian spacetime get things just right, but in fact this is not the case: the dynamical symmetry group of Newtonian physics is in fact larger than the Galilei group.

In inertial frame coordinates, the field-theoretic form of Newton's law of gravitation is expressed by the following equations:

$$\vec{F} = -m\vec{\nabla}\phi \qquad \text{(Grav Force)}$$

$$\nabla^2 \phi = 4\pi G \rho \qquad \text{(Poisson)}$$

where ϕ is the gravitational potential, G is Newton's constant and ρ is the mass density. In the coordinate-free notation of Section 4, these equations become:

$$F^a = -m h^{ab} \nabla_b \phi$$

$$h^{ab} \nabla_a \nabla_b \phi = 4\pi G \rho.$$

and the theory has models of the form $\langle M, t_{ab}, h^{ab}, \nabla_a, \rho, \phi \rangle$. The coordinate-dependent equations are invariant under Galilean transformations, which are also dynamical symmetries in the model-theoretic sense, but these transformations do not exhaust the symmetries. Consider the *Maxwell group*[59] of coordinate transformations:

$$\begin{aligned} \vec{x}' &= R\vec{x} + \vec{a}(t), \\ t' &= t + d. \end{aligned} \qquad \text{(Max)}$$

Like those of the Leibniz group, they involve an arbitrary, time-dependent translation term, $\vec{a}(t)$. Like those of the Newton and Galilei groups, the rotation matrix R is not time-dependent. They therefore correspond to a spacetime structure that embodies an absolute standard of rotation but no general standard of acceleration.[60]

These transformations will also preserve the coordinate-dependent form of the equations of Newtonian gravitation, and map solutions to solutions, so long as the gravitational potential field is also transformed appropriately:

$$\phi \mapsto \phi' = \phi - \vec{x}.\ddot{\vec{a}} + f(t),$$

where $f(t)$ is an arbitrary time-dependent function that is constant on surfaces of simultaneity. It follows that, if d is a diffeomorphism corresponding to such a transformation and $\langle M, t_{ab}, h^{ab}, \nabla_a, \rho, \phi \rangle$ is a model of the spacetime formulation of Newtonian gravity we are considering, then so is $\langle M, t_{ab}, h^{ab}, \nabla_a, d^*\rho, \phi' \rangle$. This means that the neo-Newtonian substantivalist is in precisely the same kind of predicament as the substantivalist who advocated absolute space: their metaphysics grounds physical quantities (in this case absolute accelerations, rather than absolute

---

[59] My terminology again follows Earman (1989, 31) and Bain (2004, 351). Ehlers (1973a, 78–79) discusses the group but leaves it unnamed.

[60] There is no canonical differential-geometric way of capturing this structure. Earman (1989, 32) resorts to an equivalence class of connections whose congruences of geodesics are nonrotating with respect to one another. Saunders (Forthcoming, §7) offers an elegant characterization of a similar but strictly weaker structure.

velocities) that it is impossible in principle to detect. Here is another way to see the problem. Since, for the type of diffeomorphism under consideration, $d^* t_{ab} = t_{ab}$ and $d^* h^{ab} = h^{ab}$, if $\mathcal{M}_1 = \langle M, t_{ab}, h^{ab}, \nabla_a, \rho, \phi \rangle$ is a model of the theory, then so is $\mathcal{M}_2 = \langle M, t_{ab}, h^{ab}, \nabla'_a, \rho, \phi' \rangle$, where $\nabla'_a = (d^{-1})^* \nabla_a$. That is, the laws and a given matter distribution ρ fix the temporal and spatial metric structures, but they leave it underdetermined whether the combination of inertial structure and gravitational force is that given by (∇_a, ϕ) or by (∇'_a, ϕ'). And if we take the postulated inertial structures ontologically seriously, these differences correspond to qualitative differences. For example, in one model a given particle might be force-free and moving inertially; in the other it might be accelerated under a gravitational force.[61]

A natural thought at this point is that $\mathcal{M}_1$ and $\mathcal{M}_2$ are merely mathematically distinct representations of the same physical possibility and that ϕ and ∇_a are *gauge-dependent* quantities. But one cannot simply declare this so by fiat. One should also provide a characterization of a gauge-invariant reality in terms of which the gauge dependent quantities can be understood. It turns out that the substantivalist can, indeed, do this. The solution is *Newton–Cartan theory*, a formulation of Newtonian gravitation first developed by Cartan and Friedrichs.[62] In this theory, just as in GR, gravitational phenomena are not the effects of forces. The flat inertial connection ∇_a is replaced by dynamical inertial structure ∇_a^{NC} (in part) governed by the following generalization of the coordinate-free form of Poisson's equation:

$$R_{ab}^{NC} = 4\pi G \rho t_{ab}, \qquad \text{(Poisson}_{NC}\text{)}$$

which relates the Ricci tensor R_{ab}^{NC} defined by ∇_a^{NC} to the mass density. Our two models of Newtonian gravity set in neo-Newtonian spacetime, $\mathcal{M}_1$ and $\mathcal{M}_2$, correspond to a unique model of Newton-Cartan theory (up to isomorphism). Any given (∇_a, ϕ) pair that solves nongeometrized Newtonian gravity determines a unique dynamical connection but the converse is not true: a given Newton–Cartan connection can always be decomposed into a flat connection and a gravitational potential, but this decomposition is nonunique in a way that corresponds exactly to the underdetermination of gravitational theory set in neo-Newtonian spacetime.[63]

While the problem of the symmetries of Newtonian gravity and its substantivalist solution are relatively well-known, the fact that an enriched-ideology relationalist strategy can also be fruitfully pursued is far less appreciated. When canvassing enriched relationalist options earlier in the section, the operative assumption was that Newtonian dynamics was Galilean invariant. Now that the larger Maxwell group has been recognized as a symmetry group, a reevaluation is needed. The equations of any N-body Newtonian system whose force laws obey Newton's third law can be

---

[61] This problem for Newtonian gravitation set in neo-Newtonian spacetime is well-known. For a related discussion in the philosophical literature, see, e.g., Friedman (1983, 95–97).

[62] This theory is presented as the solution to the problem faced by the Galilean substantivalist by Friedman (1983, 97–104; 120–124). See also Malament (1995), who presents it as a solution to a closely related problem raised by Norton (1993).

[63] Malament (2012, ch. 4) reviews these results and Newton–Cartan theory more generally; see also Bain (2004).

re-expressed as:

$$\ddot{\vec{r}}_{ij} = \frac{1}{m_i} \sum_{k \neq i}^{N} \vec{F}_{ki} - \frac{1}{m_j} \sum_{k \neq j}^{N} \vec{F}_{kj} \qquad (1)$$

where $\vec{r}_{ij} := (\vec{x}_i - \vec{x}_j)$ is the directed relative distance between particles i and j and $\vec{F}_{ij}$ is the force exerted by particle i on particle j (Hood, 1970; see Earman, 1989, 81, for discussion). For the time derivatives of $\vec{r}_{ij}$ to be well defined, the full inertial structure of neo-Newtonian spacetime is not required. All that is needed is a standard of rotation, that is, exactly the spatiotemporal structure invariant under the Maxwell group. Since we are assuming Newton's third law is also satisfied, the only spatial dependence of $\vec{F}_{ij}$ will be on $\vec{r}_{ij}$. It follows that Equation (1) is invariant under the Maxwell group. The only ideology, in addition to Leibnizian-relational quantities, needed to ground a standard of rotation is the transtemporal comparison of the directions of the directed distances between material bodies.[64] For example, the *Maxwellian relationalist* can postulate a primitive four-place relation A on material events such that, when (e_1, e_2) and (e_3, e_4) are pairs of simultaneous events, $A(e_1, e_2, e_3, e_3)$ takes a value between 0 and 2π, to be interpreted as the angle between $\overrightarrow{e_1 e_2}$ and $\overrightarrow{e_3 e_4}$.

What this shows is that the full set of Newtonian solutions *for a finite system of interacting particles* can be given a *bona fide* relationalist interpretation (with or without Newtonian gravitation). With Earman (1989, 81), we should now ask whether the basic idea can generalize to field theory. Maxwellian relationalism for field configurations can easily be implemented using Barbour's best-matching machinery, discussed in section 6.2, so field theory *per se* is not an obstacle.[65] Barbour's machinery, however, is only applicable to spatially finite systems (or systems with appropriate spatial boundary conditions). In such "island universe" scenarios, the Maxwellian invariance of dynamics does not trouble the neo-Newtonian substantivalist, for a preferred inertial connection can be identified via the condition that the total three-momentum of the whole system is constant. Underdetermination only genuinely arises for the neo-Newtonian substantivalist in Newtonian cosmology when one considers, for example, infinite homogeneous matter distributions. As far as I know, no relationalist theory for such situations has been devised. The Maxwellian relationalist seems to be in the unfortunate position of having a

---

[64] As far as I am aware, Simon Saunders was the first to stress that transtemporal comparison of directions are obviously compatible with relationalist ontology. Saunders (Forthcoming) is a recent discussion of related topics. I am grateful to him for discussion. Earman (1989, 78–81) comes close to attributing the basic idea to James Clerk Maxwell, who, when discussing absolute rotation in Maxwell (1877, §104), wrote: "in comparing one configuration of the system with another, we are able to draw a line in the final configuration parallel to a line in the original configuration." Earman's assessment is that "Maxwell's set of parallel directions is, of course, inertial structure, and in modern terms what he seems to be proposing is that neo-Newtonian spacetime is the appropriate arena for the scientific description of motion" (Earman, 1989, 80). However, it is clear that Maxwell here only assumes a standard of parallelism for *spacelike* lines which, as we have seen, does not require the full structure of neo-Newtonian spacetime. Perhaps Earman did not realize how apt his label *Maxwellian spacetime* is.

[65] One "best matches" instantaneous configurations only with respect to rigid translations and not, as Barbour does, by translations, rotations, and dilations. Barbour-type particle theories that do not implement rotations as gauge symmetries have been discussed recently by Anderson (2012, section 2.4).

solution applicable to those cases that are not genuine problems and no solution for the truly troubling cases.

In the context of field theory, there is one relatively easy way out for the relationalist.[66] Recall that the troubles faced by the neo-Newtonian relationalist arose because, in a world of point particles, the three-place collinearity relation typically will not be instantiated by material events in the infinitesimal neighborhood of a given material point. If the relationalist embraces a plenum, this problem goes away. In the context of Newtonian gravity, the relationalist can combine a material plenum with the insight of Newton–Cartan theory and postulate a primitive three-place collinearity relation on material events that holds of triples of material events in a physically possible world just if, in the corresponding substantivalist model, they lie on a geodesic of the substantivalist's dynamical affine connection. Such a Newton–Cartan relationalist still has work to do. The characterization of the position just given made crucial reference to substantivalist models. Can the standard mathematical formalism of Newton–Cartan theory be independently understood in terms of such relational ideology? What are the material fields and why must they constitute a plenum? Similar questions recur in the context of relativistic physics, where fields are no longer optional extras. It is to relativity we now turn.

6.1.2 Relativity

In the context of classical mechanics, the relationalist who pursues the enriched ideology strategy is forced to be creative. Simply co-opting substantivalist ideology (by restricting the domain of possible instantiation to the material events) fails, primarily because of the relative sparseness of the relationalist's ontology in comparison to the substantivalist's plenum of spacetime points. In the context of SR, however, the flat-footed move works. Restricting the substantivalist's ideology—Minkowski spacetime distances—to material events fixes the embedding of a relational history into Minkowski spacetime (up to isomorphism). Further, Minkowski distances conceived of as relationalist ideology can be used to frame dynamical principles directly in relationalist terms.[67]

Central to relativistic mechanics (even if not to relativistic physics in general) is the idea that unaccelerated motion is default behavior and that accelerations are due to forces. In order to lay claim to this picture, the *Minkowski relationalist* needs accounts of both accelerations and forces. In classical mechanics, forces were unproblematic for the relationalist (because they are functions of Leibnizian relational quantities); it was acceleration that proved troublesome. In relativity, the difficulties are reversed.

Consider the standard, coordinate-dependent forms of relativistic laws. The privileged class of coordinate systems relative to which these equations hold are simply those adapted to the Minkowski distance relations between material events (cf. section 4.3). Dynamically significant absolute acceleration, therefore, is simply

---

[66] I am grateful to David Wallace for highlighting this possibility.
[67] The main discussions of a position of this sort are Earman (1989, 128–130) and Maudlin (1993, 196–199).

acceleration relative to the coordinate systems adapted to the relationalist's spatiotemporal distances. In fact, the Minkowski relationalist can do better and give an intrinsic characterization of acceleration. Recall that, in Minkowski spacetime, the inertial trajectories are not structure over and above the spatiotemporal distances; the straight line in spacetime between two temporally separated events is the path of *maximal* temporal distance. This means that a particle will be unaccelerated just if, for any temporally ordered points p, q, r of its trajectory, $I(p,r) = I(p,q) + I(q,r)$ and, conversely, if $I(p,r) > I(p,q) + I(q,r)$, we know that the particle is accelerated between the points p and r (Earman 1989, 129). The four-acceleration of a trajectory at a point just is the intrinsic curvature of the trajectory at that point and so, as for curves in Euclidean space, one can define the acceleration of the particle (both its magnitude and its direction) in terms of such distances.[68]

The Minkowski relationalist treatment of forces is less straightforward. The coordinate-free statement of the second law, $F^a = m\xi^n \nabla_n \xi^a$, is formally the same in classical mechanics and SR and, in both cases, the four-force, F^a, must be a spacelike vector. This formal identity, however, hides a crucial difference. In the neo-Newtonian case, spacelike vectors lie in (that is, are tangent to) surfaces of simultaneity. As a result neo-Newtonian four-forces can be defined in terms of Leibnizian spatial distances, which are intrinsic to such surfaces. In the Minkowski case, if one considers an arbitrary spacelike hyperplane and the accelerations of a number of interacting particles at the points where their trajectories intersect this plane, then, in general, none of these accelerations will be tangent to the hyperplane. It is no accident that in relativistic theories F^a is standardly given as a local function of fields.[69]

Some think that the need to invoke fields is a problem for the relationalist. On one (natural) interpretation, fields are simply assignments of properties or states to the points of spacetime (Field, 1985, 40). Such a view does indeed presuppose substantivalism, but there is an alternative available, and it is one that arguably fits more naturally the language employed by physicists. On this other view, the field itself is reified as a vast, spatiotemporally extended object in its own right.[70] Adopting this second conception of fields does not, by itself, amount to relationalism; many substantivalists will agree that at least some fields are extended objects in spacetime (rather than properties of spacetime).[71] Taking a "relational" view of a field also does not by itself commit one to the view that such a field could exist

---

[68] A treatment of acceleration along these lines can be found in Minkowski's original presentation (Minkowski, 1909, 85–86).

[69] Relativistic theories in which the four-force on a particle at a point is determined directly by the properties of other particles at other spacetime locations are not impossible; Feynman and Wheeler's version of electromagnetism is such a "pure particle theory" (Wheeler and Feynman, 1949). These theories, however, have various unwelcome features, and their empirical adequacy remains an open question; see Earman (1989, 155–158) for discussion.

[70] See Malament (1982, 532, fn 11), who is responding to Field's argument. Other clear expressions of this view can be found in Belot (1999, 45) and Rovelli (2001, 104).

[71] As the rest of this section illustrates, still less does the move trivialize the substantivalist–relationalist debate (*pace* Field 1985, 41), although it does excuse the relationalist from replacing field theories with action-at-a-distance theories.

in the absence of spacetime, or, without spacetime, have the very properties one's theory characterizes it as having (*cf.* footnote 38). The devil will be in the details.

Consider the simple case of a field, ϕ, with just one degree of freedom per spacetime point. The relationalist wishes to view ϕ as an extended, physical entity rather than as an assignment of properties to spacetime. Since spacetime itself is supposed not to exist, this extended object cannot be characterized in terms of the spatiotemporal locations of the various field intensities. Instead, the relationalist should view the field as characterized by the infinite number of facts about the Minkowski distances between its pointlike parts; together these fully characterize the pattern of field intensities.[72] These distances cannot (in practice) be specified directly. But there is nothing relationally improper about describing the field relative to a Lorentzian coordinate system so long as such a chart is thought of as a map directly from the field itself into $\mathbb{R}^4$ that encodes the Minkowski distances.

Consider, now, the substantivalist's presentation of a theory of such a field. The KPMs will be of the form $\langle M, \eta_{ab}, \phi \rangle$ and the DPMs will be picked out via an equation relating ϕ and η_{ab}. Suppose the relationalist's only way to identify dynamically possible field configurations was to use this machinery. Would they then be in the embarrassing position of relying on a substantivalist "fairy tale" without a proper explanation of why it works (Earman 1989, 172)? It does not seem so. That ϕ is the only field in the model reified by the relationalist does not mean that η_{ab} is a fiction. The substantivalist suggests that one understands fields as assigning various properties and relations. In this case, the relationalist agrees. They just disagree about the subject of predication: for the substantivalist it is spacetime itself, for the relationalist it is the one substantival field of the model, ϕ. Equations relating ϕ to the other fields then have a straightforward relationalist reading as claims about the allowed (geometrical) properties of ϕ itself.

So far I have only considered scalar fields. More complex fields can pose additional problems for the relationalist. Standard vector and tensor fields, for example, are not obviously conceptually independent of the structure of the manifold on which they are defined. Their degrees of freedom at a point are normally understood as taking values in the *tangent space* at that point (or in more complex spaces constructed in terms of it), which might appear to presuppose the differentiable structure of the manifold on which the fields are defined. In fact, even characterizing scalar fields normally involves this manifold structure, for one is normally interested in *smooth* fields. In this case, however, it is clear how one can do away with reference to an independent manifold. What one requires (roughly speaking) is that the field's values vary smoothly as a function of the distances between its parts: fields themselves can have the structure of a differentiable manifold in virtue

---

[72] Note that this constitutes an answer to Earman's challenge that the relationalist must provide a "direct characterization" of the reality underlying the substantivalist's description of fields (Earman, 1989, 171).

of these Minkowski distances. Vector and tensor fields, conceived of as substantival entities in their own right, will likewise have a manifold structure, but there is something suspiciously circular about taking the spaces in terms of which a field's degrees of freedom are defined to be themselves defined in terms of that field's own spatiotemporal extension. An alternative is to try to understand the degrees of freedom of some material fields in terms of their interactions with other fields whose relational credentials are not in doubt.[73]

The upshot is that the combination of Minkowski relationalism and a relational interpretation of fields is at least a going concern as an interpretation of SR. The final task for this section is to consider whether the picture can be adapted to GR. The strong similarities between SR and GR stressed in section 4.3 might lead the relationalist to be optimistic. In fact, the move from flat to curved geometric structure, and the manner in which it features in GR, presents a formidable obstacle. Recall that the Minkowski relationalist does not reify the metric field η_{ab}. Instead this field is regarded as cataloging primitive spatiotemporal distances that hold between the parts of *bona fide* material fields. At the level of kinematics, the generalization of this to GR is straightforward. Minkowski distances are simply replaced by those of a curved semi-Riemannian geometry. A crucial consequence of this move is that the distances instantiated between material events need no longer fix (independently of the dynamical laws) all the facts about the geometry of spacetime. In particular, consider an "island universe" involving a matter distribution of finite spatial extent. The spatiotemporal distances instantiated in the history of the material world will not fix the geometry of the empty spacetime regions beyond it.

This is not a problem of principle. After all, the relationalist will claim that there is literally nothing beyond the boundary of the material universe to instantiate one geometry rather than another. There is also no difficulty, in principle, with this type of relationalist regarding geometry as dynamical and as influenced by matter. For example, the laws of a relational theory could lay down how the network of spatiotemporal relations instantiated in some temporally thick slice through the material world determine (together with other dynamically relevant properties of matter) the pattern of spatiotemporal relations instantiated in earlier and later regions of the material universe. The particular difficulty GR poses for the envisaged relationalist involves the combination of these two factors. In GR the geometrical properties of the supposed nonentity beyond the material universe do make a dynamical difference. For example, the entire history of spatiotemporal distances instantiated in our island universe up to some time will not record whether a "gravitational wave" (that is, a propagating ripple in the fabric of spacetime itself) is approaching from outside the system and will thus underdetermine the system's future evolution (Earman, 1989, 130; Maudlin, 1993, 199).

In response, the relationalist could rule out by fiat models with empty regions of spacetime. To do so, however, is not only to give up on the goal of empirical

---

[73] See, for example, Malament (2004, §3), where the tensorial properties of the electromagnetic field F_{ab} are derived from assumptions about its action on charges.

equivalence with standard theory; it is to impose a restriction that is arbitrary by the lights of the relationalist's own theoretical apparatus. The relationalist does not have problems with empty regions per se. What they have problems with is those regions having a determinate geometry that need not supervene on the properties of and relations between matter-filled regions and with the geometry of empty regions playing the dynamical role that GR assigns it. The better "relationalist" move is to treat the metric tensor as a "material" field in its own right, but then, since all parties affirm the existence of a substantival entity whose properties are characterized by g_{ab}, it is not clear what substantive issue remains.[74]

6.2 Barbour's Machian Relationalism

There are two straightforward relationalist responses to the mismatch between relationalist spacetime symmetries and the dynamical symmetries of Newtonian mechanics. The previous section covered one of these: enrich relationalist ideology in order to bring spacetime symmetries into line. This section investigates the other: change the dynamics in order to bring the dynamical symmetries into line. The most thorough and successful development of this strategy is that of Julian Barbour and collaborators. The label *Machian relationalism* is appropriate for three reasons. First, it accords with Barbour's own terminology. He sees the requirement that a theory be maximally predictive with respect to relational initial data (in the sense discussed in Section 5.1) as a precise version of *Mach's Principle* and he takes his approach to dynamics to reveal that GR is in fact a Machian theory. Second, in the context of pre-relativistic particle dynamics, the spacetime quantities Barbour takes as fundamental are even sparser than those of the Leibnizian relationalist: the Euclidean nature of the instantaneous relative distances between particles is accepted as primitive, but the temporal intervals between successive instantaneous configurations are not. In his critique of Newton, Mach (1901, 222–226) claimed that the question of whether a motion is in itself uniform is senseless, on the grounds that a motion can (allegedly) only be judged uniform relative to some other motion or material process.[75] Finally, Barbour's particle theories provide a concrete implementation of Mach's idea that the inertial properties of a body might be understood in terms of that body's relations to the rest of the bodies in the universe, rather than with respect to substantival spacetime structure (Mach 1901, 231-235).

Up to this point I have presented the DPMs of a theory as singled out in terms of differential equations that must be everywhere satisfied within a model by its

---

[74] The relationalist can also question whether one should regard regions of zero field strength as regions where the material field literally does not exist. This might be the natural interpretation of fields that represent "dust" in models of GR, but it is at least controversial for, e.g., the electromagnetic field. The stipulation is yet more problematic when one moves to quantum field theory. I am grateful to David Wallace for pressing this point.

[75] For related reasons, Earman defines *Machian spacetime* to be spacetime with simultaneity structure and Euclidean metrical structure on its simultaneity surfaces but with no temporal metric (Earman 1989, 27–30).

constituent fields and particle trajectories. In some formulations of dynamics, the DPMs are singled out in terms of their relations to other KPMs. Machian relational theories are most illuminatingly developed in this type of framework. Consider, in particular, the Lagrangian formulation of Newtonian mechanics. Central to this framework is a system's *configuration space*, Q, the points of which represent possible instantaneous states of the system. According to the substantivalist, such a state for an N-particle system corresponds to a set of positions for each particle relative to some inertial frame. Q is then $3N$-dimensional. As the system evolves, the point in Q representing the system's instantaneous state traces out a continuous curve. In Lagrangian mechanics, KPMs (that is, metaphysically possible histories) are (monotonically rising) curves in the product space formed from Q and a one-dimensional space, T, representing time. The DPMs are those curves that extremize a particular functional of such histories (the *action*).

This framework can be adapted to Leibnizian and Machian relationalism in a straightforward way. First, since the relationalist's possible instantaneous states correspond to sets of inter-particle distances (rather than positions defined with respect to spacetime structure), the relationalist replaces Q with the *relative configuration space* Q_{RCS}. For N particles, Q_{RCS} is $(3N-6)$-dimensional. In fact, the relationalist might be tempted to go further. Formulating dynamics in terms of Q_{RCS} involves treating transtemporal comparisons of length as primitive. Distinct curves in Q_{RCS} can correspond to exactly similar sequences of Euclidean configurations if some of the corresponding configurations represented in the two curves differ in overall size. This is not true in *shape space* (Q_{SS}), a configuration space of one less dimension that treats only the ratios of distances within the same configuration as physically meaningful.

Second, standard theory distinguishes histories that correspond to a single path in configuration space being traced out at different rates with respect to the primitive temporal metric. The Machian relationalist, in contrast, will view each path in configuration space as corresponding to exactly one possible history. They therefore dispense with T, the space encoding primitive temporal separations, in favor of a "timeless" formulation of dynamics in terms of configuration space alone. One way to do this is to equip the space with a metric. The DPMs are then picked out via a *geodesic principle*: physically possible histories correspond to paths in configuration space of extremal length relative to the metric.

The implementation of this second step can be achieved via a reinterpretation of *Jacobi's Principle*, part of the standard toolkit of Newtonian dynamics.[76] The metric structure of three-dimensional physical space can be used to define a metric on Q known as the *kinetic metric*: $ds_{kin}^2 = \sum_i \frac{m_i}{2} d\vec{x}_i . d\vec{x}_i$. Its geodesics correspond to histories of particles moving inertially. Dynamics is incorporated by multiplying

---

[76] It should be stressed that Barbour initially postulated a Jacobi-like action on purely Machian grounds and only learned of the connections with standard dynamics several years afterward.

ds_{kin}^2 by a conformal factor $F_E = (E - V(\vec{x}_1, \ldots, \vec{x}_N))$. The geodesic principle is then:

$$\delta I = 0, \quad I = \int \sqrt{F_E} \, ds_{\text{kin}} = 2 \int d\lambda \sqrt{F_E T_{\text{kin}}}, \tag{2}$$

$$\text{where} \quad T_{\text{kin}} := \frac{1}{2} \sum_i^N m_i \frac{d\vec{x}_i}{d\lambda} \cdot \frac{d\vec{x}_i}{d\lambda}. \tag{3}$$

Its solutions correspond to Newtonian histories of a system of N particles with a total energy E interacting according to the potential V. T_{kin} looks like the standard Newtonian kinetic energy but note that λ represents an arbitrary parameterization of paths in Q: the path length I is invariant under reparameterizations: $\lambda \mapsto \lambda' = f(\lambda)$, where $df/d\lambda > 0$ but f is otherwise arbitrary. The equations of motion corresponding to (2) are:

$$\frac{d}{d\lambda}\left(\sqrt{\frac{F_E}{T_{\text{kin}}}} m_i \frac{d\vec{x}_i}{d\lambda}\right) = \sqrt{\frac{T_{\text{kin}}}{F_E}} \frac{\partial F_E}{\partial \vec{x}_i}. \tag{4}$$

These simplify dramatically, reducing to the standard form of Newton's second law, if the freedom in the choice of λ is exploited to set $F_E = T_{\text{kin}}$, that is, $E = T_{\text{kin}} + V$. The substantivalist sees imposing this requirement as a way to determine the rate at which the system traces out its path in Q relative to a primitive temporal metric. The Machian sees the equation as defining an emergent temporal metric in terms of the temporal parameter that simplifies the dynamics of the system as a whole (Barbour 1994, 2008).[77]

Jacobi's principle involves a metric on Q. To construct a relational theory, we need a metric on Q_{RCS}. One can be obtained by replacing T_{kin} (a function of velocities defined with respect to inertial structure) with a function of the *relative velocities* $\dot{r}_{ij}$. Theories of this kind were independently discovered on a number of occasions during the twentieth century. They predict mass anisotropy effects (how easy it is to accelerate a body becomes direction dependent) that are ruled out by experiment.[78] It is also not clear how they might generalize to field theory, where analogues of the transtemporal particle identities used in the definition of the $\dot{r}_{ij}$s are absent. Barbour and Bertotti (1982) found a way to surmount both problems.

The (ambitious) relationalist thinks of instantaneous configurations as completely characterized by the ratios of inter-particle separations. A three-dimensional coordinate system encodes such data just if $|\vec{x}_i - \vec{x}_j|/|\vec{x}_m - \vec{x}_n| = r_{ij}/r_{mn}$ for all particles i, j, m, n. If one coordinate system satisfies this constraint, so will any other

---

[77] Note that choosing a simplifying parameter for Equation (4) is quite unlike choosing a time coordinate that is adapted to the spacetime substantivalist's temporal metric. The latter also simplifies the (generally covariant) equations, but these equations explicitly refer to an independent standard of duration. According to the Machian interpretation of Jacobi's Principle, fundamental dynamics is formulated without reference to such an external time. See Pooley (2004, 78–79).

[78] Noteworthy examples are Hofmann (1904), Reissner (1914), Schrödinger (1925), Barbour (1974a), Barbour and Bertotti (1977), and Assis (1989). For further discussion, see Earman (1989, 92–96) and Barbour and Pfister (1995, 107–178). It turns out that Barbour's later particle theories (discussed immediately below) can themselves be formulated in a natural way directly in terms of the right choice of relative coordinates. For details, see Anderson (2012, Chapters 2 and 3), where a wider class of such theories is considered.

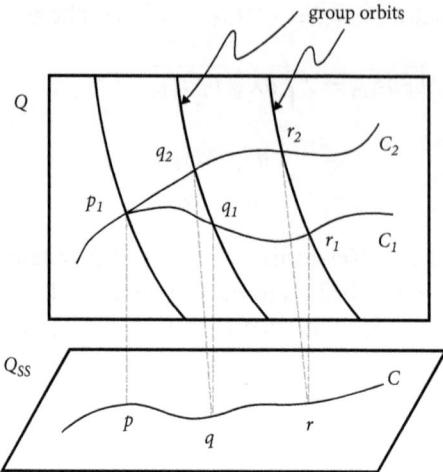

Figure 15.1 Best matching. The curves C_1 and C_2 in Q correspond to the same sequence of relative configurations. q_2 is the point on the orbit containing q_1 that minimizes the distance along the curve from p_1. r_2 is similarly related to q_2. C_2 is the best-matched curve; the length along it gives the length along C, the corresponding curve in Q_{SS}.

related to it by a rigid rotation, translation, or a dilation (an overall change of scale). The relationalist therefore regards the points of Q, not as specifications of positions in some inertial frame, but as natural representations of relational configurations. The representation involves some redundancy: points of Q connected by an element of the *similarity group* (the group of rigid translations, rotations, and dilations) correspond to the same relative configuration. Q is partitioned by the group into sets of such points (the *group orbits*). Consider, now, two paths in Q that correspond to the same sequence of relative configurations. A metric on Q will, in general, assign them different lengths. However, starting from any given point p in Q, one can use the action of the similarity group on Q to define a unique length, by shifting the points of any curve through p along the corresponding group orbits so as to extremize the length assigned to the curve. This is the process Barbour and Bertotti called *best matching*.[79] It is depicted in Figure 15.1.

If best matching is to define a metric on Q_{SS} (the quotient of Q by the similarity group), the metric on Q must have the right properties. In particular, suppose p_1 and p_2 are points on the same group orbit that are widely separated in Q. Consider two paths through p_1 and p_2, respectively, that correspond to the same sequence of relative configurations. Suppose one now best matches these paths, keeping the points p_1 and p_2 fixed. Best matching only leads to a well-defined metric on Q_{SS} if the same result is obtained in each case. The metric $ds^2 = F_E ds^2_{\text{kin}}$ satisfies this requirement provided F_E meets certain conditions.

---

[79] For an informal discussion of the central idea, see Barbour (1999, ch. 7). For the extension to dilations, see Barbour (2003). For more formal and general treatments, see Anderson (2006) and Gryb (2009).

If one first considers best matching just with respect to the Euclidean group (translations and rotations), V must be a function only of the relative distances, r_{ij}. This requirement is satisfied by familiar Newtonian potentials. The corresponding best-matched theories, which take DPMs to be geodesics of the metric induced on Q_{RCS}, have as solutions sequences of relative configurations that correspond to the standard Newtonian solutions with zero overall angular momentum (relative to the center-of-mass frame). The fact that a subset of standard Newtonian solutions is recoverable by this method highlights the fact that the theories provide a relational interpretation of inertial structure: best matching establishes a nonprimitive "equilocality relation," corresponding to the space of the inertial frame in which the system's total linear and angular momenta vanish. Note, also, that the recovery of only a proper subset of the solutions of standard dynamics is arguably a strength of the best-matching theory (assuming solutions capable of modeling the actual world fall within this set). This is because the theory predicts and explains a feature of the world (the vanishing of its overall angular momentum) that is a contingent fact on the orthodox Newtonian view (Pooley and Brown, 2002; Pooley, 2004).[80]

Best matching with respect to dilations imposes a more severe requirement: F_E must be a homogeneous function of the $\vec{x}_i$s of degree -2, in order to compensate for the scaling behavior of ds^2_{kin}. Standard Newtonian potentials do not have this property, but they can nevertheless be incorporated as effective potentials in scale-invariant theories if a weak, epoch-dependent universal force is also included (Barbour, 2003, 1556–7).

Barbour's framework for nonrelativistic particle dynamics, therefore, constitutes a genuinely relationalist (and potentially fruitful) alternative to Newtonian physics as standardly conceived. What one is really interested in, though, is how the program transfers to relativistic physics. The best matching idea can be applied in the context of SR (Barbour and Bertotti 1982, 302–303), but I move straight to a consideration of GR, where the results are truly surprising. The first step is to consider how one might generalize the framework to configurations manifesting a variably curved Riemannian geometry. One confronts the issue, raised in section 6.1.2, of how to deal with the possibility that the geometry of empty space might be both nontrivial and nonreducible to relations between material bodies. Barbour bites the bullet. In the context of GR, the Machian "relationalist" takes the geometry of substantival (instantaneous) space as primitive.

Assume that instantaneous space has the determinate topology of some closed 3-manifold without boundary, Σ. The obvious analogue of Q is then Riem(Σ), the space of Riemannian 3-metrics on Σ. An analogue of Q_{RCS} is *superspace*: the space of *3-geometries*. Two points (Σ, h_{ab}) and (Σ, h'_{ab}) of Riem(Σ) correspond to the same 3-geometry just if they are isometric, that is, just if, for some diffeomorphism

---

[80] Similarly, one can argue that the Machian relationalist is able to explain formal features of the potential, such as its dependence only on the r_{ij}s, that are again nonessential aspects of standard Newtonian theory (Barbour 2011).

d of Σ, $h'_{ab} = d^* h_{ab}$. Superspace is therefore $\text{Riem}(\Sigma)/\text{Diff}(\Sigma)$, the quotient of $\text{Riem}(\Sigma)$ by the group of diffeomorphisms of Σ.

Proceeding as before, one seeks an action principle on superspace defined, via best matching, in terms of a metric on $\text{Riem}(\Sigma)$. In this case, best matching is implemented by diffeomorphisms of Σ. Seeking as direct a parallel as possible with Jacobi's Principle (2) leads to a $\text{Riem}(\Sigma)$ geodesic principle of the form:

$$\delta I = 0, \quad I = \int d\lambda \sqrt{\int d^3 x \sqrt{h} W \int d^3 x \sqrt{h} T}. \tag{5}$$

The first integral inside the square root is the analogue of the conformal factor F_E in (2);[81] the second is the analogue of the (parameterized) kinetic energy. In this case, $T = G_A^{abcd} \dot{h}_{ab} \dot{h}_{cd}$, where $\dot{h}_{ab} = dh_{ab}/d\lambda$ are the metric velocities (with respect to the arbitrary path parameter λ) and the general form of the *supermetric* G_A^{abcd} is $h^{ac} h^{bd} + A h^{ab} h^{cd}$, where A is an arbitrary constant. Best matching with respect to 3-diffeomorphisms is achieved by replacing T with $T_{BM} = G_A^{abcd}(\dot{h}_{ab} - \mathcal{L}_\xi h_{ab})(\dot{h}_{cd} - \mathcal{L}_\xi h_{cd})$ and extremizing with respect to variations in $\dot{\xi}^a$.[82]

Theories of this kind make good sense but they do not provide direct analogues of GR.[83] For these, one needs to consider a $\text{Riem}(\Sigma)$ action principle that involves a subtle but radical difference. It has the form:

$$\delta I = 0 \quad I = \int d\lambda \int d^3 x \sqrt{h W T}, \tag{6}$$

with W and T defined as before. The difference between Principles (5) and (6) is that, in the former, integration over 3-space occurs within a global square root, but in the latter the square root is taken at each point of space and occurs within the spatial integration.[84] Whereas the reparameterization invariance of (5) gives rise to a single constraint, the position of the square root in (6) leads to an infinity of constraints, one associated with each point of space. These must be propagated by the equations of motion if the theory is to be consistent. This happens only if $A = -1$ in G_A^{abcd} and $W = \Lambda + \alpha R$, where R is the scalar curvature tensor of h_{ab}, and α is 0 or ± 1. The choice of $\alpha = 1$ and the imposition of best matching with respect to 3-diffeomorphisms transforms (6) into the action principle for GR found by Baierlein, Sharp, and Wheeler (1962). (Or, strictly speaking, a time-reparameterization analogue of the BSW action, because the BSW action involves Lie derivatives defined with respect to the shift-vector field, rather than with respect to the velocity of 3-vector field. Thanks to Edward Anderson here.) This is dynamically equivalent to

---

[81] W can depend on h_{ab} and its spatial derivatives up to some finite order; the presence of $h := \sqrt{\det h_{ab}}$ is simply to ensure that the integration is invariantly defined.

[82] If ξ^a is an infinitesimal 3-vector field, $(\dot{h}_{ab} - \mathcal{L}_\xi h_{ab})$ is the result of acting on h_{ab} by an infinitesimal diffeomorphism generated by ξ^a. $\mathcal{L}_\xi h_{ab}$, the Lie derivative of h_{ab} with respect to ξ^a, is given by $\mathcal{L}_\xi h_{ab} = \nabla_a \xi_b + \nabla_b \xi_a$, where ∇_a is the derivative operator associated with the unique torsion-free connection compatible with h_{ab}. For the reason why T_{BM} is defined in terms of the Lie derivatives with respect the velocity of a 3-vector field, see Barbour et al. (2002, 3219) and Barbour (2003, §4).

[83] Gryb (2010, 16–18) contains a brief discussion of these theories.

[84] The importance of the distinction between (5) and (6), and the fact that GR could be cast in the form of (6), was first pointed out to Barbour by Karel Kuchař (Barbour and Bertotti 1982, 305).

the standard spacetime action restricted to globally hyperbolic spacetimes. In other words, *without any spacetime presuppositions* and starting with a family of "timeless" action principles for the evolution of 3-geometries of the general form (6), the requirement of mathematical consistency alone (almost) uniquely singles out an action principle corresponding to GR.[85]

The BSW action principle for GR, which formally singles out curves in Riem(Σ), is degenerate: a point and a direction in Riem(Σ) fail to pick out a unique solution. By itself, this is not a problem for the Machian relationalist. An analogous property holds of the best-matched action principles for particle dynamics: given a point and direction in Q, a continuous infinity of curves solve the equations. The reason this is not a drawback in the particle case is that each of these curves corresponds to the same sequence of relative configurations: they project down to a single curve in the quotient of Q by the relevant group. The same is not true for the BSW action. After projecting down from Riem(Σ), one still has a continuum of curves for each point and direction in superspace. Since these curves correspond to *non-isometric* sequences of 3-geometries, and since such 3-geometries are the Machian's fundamental ontology, the Machian is compelled to regard these curves as corresponding to physically distinct histories. The theory is therefore radically indeterministic. The indeterminism is only removed if one can find a way to regard all curves with the same initial data as representations of a single physical history. As we shall see in section 7, the spacetime substantivalist, who regards spatio*temporal* geometry as primitive, can do this, because the different sequences of 3-geometries correspond to different foliations of a single 4-dimensional spacetime. The Machian, however, who regards spacetime geometry as secondary to, and defined in terms of, the dynamical evolution of spatial geometry, has no such option (Pooley, 2001, 16–18).

Fortunately for the Machian, this otherwise devastating underdetermination can be resolved in strictly 3-dimensional terms. In the particle case, the ambitious relationalist eschewed transtemporal scale comparisons and regarded only the shapes of configurations as fundamental. Analogous moves are possible in the context of GR. In particular, in a *conformal 3-geometry* only angles and the ratios of (infinitesimal) distances are regarded as physically fundamental. In terms of Riem(Σ), one regards any two 3-metrics related by a (spatially varying) scale transformation as physically equivalent: $h_{ab} \sim \phi h_{ab}$, $\phi > 0$. *Conformal superspace* is the quotient of Riem(Σ) by such scale transformations (in addition to 3-diffeomorphisms). It can be viewed as analogous to Q_{SS}.

Solutions to the BSW action that share initial data in superspace (that is, sequences of 3-geometries corresponding to different foliations of the same spacetime) do not project down to a unique curve in conformal superspace. However, the

---

[85] This is one of the main results of Barbour et al. (2002), who also claim to recover the equivalence principle and Maxwellian electromagnetism from the constraints that consistency alone places on how matter fields can be added to the theory. The results are extended to Yang Mills theory in Anderson and Barbour (2002). It should be stressed that these results do not amount to a derivation of GR and the equivalence principle from Machian first principles alone. In addition to the choice of (6) over (5), the form of (6) embodies a number of simplicity assumptions, the relaxing of which permits a range of other Machian theories; see Anderson (2007).

equations of GR can be recast so as to determine a unique such curve (Barbour and Ó Murchadha, 2010). Its points correspond to the foliation of the corresponding spacetime by spacelike hypersurfaces of *constant mean extrinsic curvature*. The fact that these geometrically privileged hypersurfaces simplify and make tractable the initial value problem in GR has been known since the work of James York in the 1970s. What the Machian perspective provides is an (alternative) understanding of the relevant equations in terms of a generalization of best matching to (volume preserving) conformal transformations (Anderson et al., 2003, 2005).

The Machian perspective on GR is not relationalist in the sense of this chapter, but it does offer a mathematically and conceptually elegant (and radical) alternative to the standard spacetime perspective. The key issue is not an *ontological* one about the reality of instantaneous spatial (that is, spacetime) points (the theory is naturally understood as committed to them); it concerns the relative priority of spatial versus spatiotemporal *ideology*. Despite Barbour's claims, the local conformal degrees of freedom of CMC spacelike hypersurfaces are not obviously philosophically superior to the standard spacetime quantities: they are not (more) directly observable (recall footnote 36), nor are primitive temporal intervals, or primitive comparisons of distant lengths, somehow inherently suspect. (In fact, an argument could even be made that observability considerations favor spacetime over instantaneous quantities.) Even the parsimony argument in favor of the Machian theory is less clear-cut in GR than in Newtonian mechanics. In GR it is no longer the case that the kinematic structures of the Machian theory are simply a proper subset of those accepted in the spacetime theory.[86] The true test of the Machian program will be its physical fruitfulness, in particular whether, as its advocates hope, it leads to progress in the search for a theory that successfully reconciles quantum mechanics and general relativity.

6.3 Have-it-all Relationalism

The relationalist strategies examined in sections 6.1 and 6.2 involve a certain honesty. They accept that restricted dynamical symmetries betoken spacetime structure with symmetries that are at least as restricted and seek to square this with relationalism, either by showing how such structure can be both primitive and anchored in a relationalist ontology or by seeking new dynamics with expanded symmetries. The approach reviewed in this section is a case of trying to have one's cake and eat it. It seeks a way to reconcile restricted dynamical symmetries with more permissive spacetime symmetries. On this approach, therefore, some of the spacetime structure implicit in the dynamics is judged to have only an effective status, ultimately grounded in a less structured relationalist ontology. Huggett's "regularity approach"

---

[86] There may be good reason to see conformal geometrodynamics as superior to alternative 3 + 1 approaches to GR: its basic quantities are dimensionless and the true degrees of freedom are transparent (Barbour 2011, 24, 39). This, though, does not speak directly to the preferability of a 3 + 1 over a spacetime perspective.

is an explicit proposal about how to do this for Newtonian mechanics. The dynamical approach to special relativity, defended by Brown (2005) and Brown and Pooley (2006), can be understood along similar lines.

6.3.1 The Regularity Approach to Relational Spacetime

Huggett draws inspiration from some remarks of van Fraassen's on the meaning of Newton's laws. Having posed the problem of how the relationalist can account for the privileged status of the inertial frames, van Fraassen seeks to dissolve it by asserting that inertial frames do not have a privileged status at all (van Fraassen 1970, 116). In claiming this, he is not asserting that Newton's laws fail to differentiate between frames of reference. He acknowledges, of course, that they do. His claim, rather, is that there need be nothing more to the inertial frames' being privileged than their being exactly those frames with respect to which certain statements about mass, motion and force hold true.

The difference between this point of view and the standard substantivalist position might seem elusive, but it concerns which facts are to be taken as basic. For the substantivalist, the basic facts include facts about such things as the relative temporal distances between pairs of events, about what counts as a straight spacetime trajectory, and so on. While the dynamical laws are to be understood in terms of such facts, and while the success of those laws is acknowledged as our only evidence for there being such facts, the facts are not to be conceived of as in any way dependent on the dynamical laws. According to the relationalist view now under consideration, they are so dependent. Beyond the facts that the Leibnizian relationalist acknowledges as primitive, the most basic spatiotemporal fact for van Fraassen is the existence of privileged coordinate systems with respect to which the dynamics of matter takes on a particularly simple form. Nonrelational quantities of motion are to be thought of as defined in terms of the very laws in which they feature.

Dynamical laws, and the equations that express them, figure prominently in the characterization of this position. Exactly what it amounts to, therefore, will depend on how laws themselves are to be conceived. Suppose, for example, that laws of nature are held to involve some kind of primitive natural necessity. The position then becomes the claim that the relative distances between all particles in the universe are constrained as a matter of nomological necessity to evolve over time so that they satisfy certain simple equations with respect to a privileged class of coordinate systems. Such a view, while consistent, has little to recommend it over the substantivalist's acceptance at face value of the quantities featuring in the dynamical laws. The relationalist is effectively claiming that relative distances between bodies are constrained to evolve as if each body had an independent quantity of motion that was governed by certain simple laws. This looks like a case where Earman's charge that the relationalist position is "hardly distinguishable from instrumentalism" is justified (Earman 1989, 128). Whether or not that spells trouble for the relationalist, their debate with the substantivalist has been replaced by a more generic dispute and has lost its distinctive character.[87]

Things look more interesting if one adopts a Humean approach to laws. The most promising Humean view is the Mill–Ramsey–Lewis "Best Systems" account, according to which the laws of nature are statements that appear as theorems of those axiom systems true of the totality of Humean facts that best combine the competing virtues of simplicity and strength (see, e.g., Lewis 1973, 72–73; Earman 1986, ch. 5). Without some constraints on admissible vocabulary, the simplicity requirement is not straightforward, because a theory's simplicity appears to be language-dependent. Lewis's later preferred constraint invokes a primitive distinction among properties: the formulations of candidate laws with respect to which simplicity is to be judged are in languages whose predicates denote perfectly natural properties and relations (Lewis, 1983). Huggett's idea, effectively, is that this requirement can be liberalized without becoming vacuous. In particular, it is very plausible that, (i) if one assumes the ontology and ideology of Leibnizian relationalism and (ii) if one allows, as candidate Humean laws, systems formulated in terms of supervenient properties, as well as perfectly natural properties, Newton's laws will constitute by far and away the simplest and strongest systematization of a typical Leibnizian relational history compatible with those laws (Huggett 2006, 48–50).

Unbeknown to Huggett, a parallel liberalization of Lewis's Best Systems prescription had already been outlined by Sider, as a possible response to Kripke's "rotating disks" argument against perdurance (Sider 2001, 230–234). Sider's goal was to ground a distinction between rotating and nonrotating homogeneous matter in the primitive ontology and ideology of the perdurantist (that is, someone who analyzes material persistence in terms of numerically distinct temporal parts of the persisting object located at the different times at which the object exists). The trick is to suppose that Best Systems laws might be formulated in terms of "physical continuants," that is, aggregates of genidentity-interrelated material events where, crucially, the non-Humean genidentity relation is not a primitive relation but supervenes on the total history of Humean properties together with the laws in which it features:

> Consider various ways of grouping stages together into physical continuants. Relative to any such way, there are candidate laws of dynamics. The correct grouping into physical continuants is that grouping that results in the best candidate set of laws of dynamics; the correct laws are the members of this candidate set. (Sider 2001, 230)

---

[87] One might also worry that if the laws are about coordinate systems it will be hard for the relationalist to avoid what Field (1985) calls *heavy duty platonism*. (Thanks to Jeremy Goodman for highlighting this.) I take the role of coordinate systems in the specification of the Humean alternative discussed next to be less problematic.

The comparison of Huggett's and Sider's proposals prompts the following worry. If Huggett's reduction of inertial structure relies on primitive transtemporal particle identity and Sider's reduction of material genidentity relies on primitive inertial structure, one or the other of the reductions must be untenable. In response, the liberal Humean might embrace both moves at once: if one strips facts about transtemporal particle identity from a typical Leibnizian relational history compatible with Newton's laws, it remains very plausible that those laws will form part of any Best System theory of such a world, if one is permitted to express the theory in terms of supervenient genidentities with respect to supervenient privileged coordinate systems. But combing both proposals into a single package highlights a related difficulty. Once the strict requirement that primitive vocabulary should express primitive, perfectly natural properties and relations is relaxed, what governs which quantities are part of the supervenience base and which quantities are supervenient? Why stop at a reduction of genidentity and inertial structure? Why not seek to offer a reductive account of mass and charge too? Why not even seek a reductive account of the temporal metric and instantaneous spatial distances? Once the reduction via the dynamical laws of some apparently natural properties to the others is on the table, we need some principles to determine which properties are ripe for reduction and which are to be part of the basic ideology.[88]

Huggett himself recognizes the issue. He notes that he has included masses and charges but not forces in his supervenience base because "a quantity can only be said to be a force if it plays the right kind of role in the laws and so cannot be metaphysically prior to the laws" (Huggett, 2006, 47). This is a surprising thing for a Humean to say. As Huggett concedes, one might (as many non-Humeans do) say the same about mass and charge. Later, when worrying that his supervenient quantities proposal is "too easy," he cites "serious objections, with a long history, against the supposition of a non-material, physical substance" (ibid., 70) as reason to pursue a reduction that at least allows one to do without spacetime. But, as we have seen, (i) what *prima facie* strong objections there are to substantivalism can be met and (ii), in the context of Newtonian theory, there are relationalist alternatives to Huggett's program that do not suffer from this particular problem for regularity relationalism. Huggett does offer a criterion for determining a point beyond which reduction should not be pursued: the laws should be such that they determine the supervenient quantities in all nomically possible worlds (ibid., §4). However, this

---

[88] If one pursues the program too far, the supervenience base will eventually become too impoverished to subvene Newtonian laws. Suppose, for example, that the only spatiotemporal information one retains is that which is common to all coordinatizations of the particle trajectories obtainable from an initial inertial coordinate system by smooth but otherwise arbitrary coordinate transformations that preserve the timelike directedness of the trajectories (that is, that they are nowhere tangent to surfaces of constant time coordinate). Such topological data includes information about whether any two trajectories ever intersect, and information about how the trajectories are "knotted", but little else. Many Newtonian worlds involving complex histories of relative distances and interactions will be topologically equivalent to histories where all particles maintain constant distance from one another. If one includes only such topological information in the supervenience base, worlds like this will not be worlds where Newton's laws are laws of nature. One might also wonder whether any degree of topological complexity (that is, any degree of complex entwining of the trajectories) will promote Newton's laws to Best System status. Might not simpler, equally strong alternatives always be available?

does not address the possibility that, with respect to the same set of laws, two distinct sets of putative subvening properties might share this property. In such a case, how does one discover which set contains the "real" fundamental properties?

In his 2006 paper, Huggett only considers Newtonian worlds. We should consider how the program looks from the perspective of relativistic physics. Special relativity does not provide a very hospitable arena for the view. The relative attractiveness of regularity relationalism in the context of Newtonian physics is due to a couple of factors. First, the fact that the ideology of the Leibnizian relationalist forms a natural subset of Newtonian ideology means that it is relatively natural to seek a reductive account of the additional (inertial) structure in terms of Leibnizian relations. Second, as reviewed in section 6.1.1, the full neo-Newtonian ideology, when restricted to a (point particle) relationalist ontology, is not sufficient for a relationalist account of standard physics, which undermines one obvious relationalist alternative. Neither factor remains true in the context of SR. In particular, Leibnizian relations are quite unmotivated as a supervenience base; it is far more natural to take the spacetime interval as basic and to understand the spatial distance relations associated with any particular family of simultaneity surfaces in terms of it. Moreover, if the relationalist is happy to countenance spatiotemporal relations between material events as primitive, there is no longer a need for a reduction of some spatiotemporal quantities in terms of others for, as reviewed in section 6.1.2, the Minkowski interval restricted to material events looks like a viable basis for a relational interpretation of standard specially relativistic physics.

Things are more interesting when one moves to GR. One aspect of Huggett's proposal that I have not so far highlighted is that Huggett sees it as a way to allow the geometry of empty space to supervene on the geometrical relations instantiated by material bodies.[89] Consider, for example, a history of instantaneous spatial relations between bodies that are initially Euclidean but that depart from Euclidicity after some moment, perhaps then to return to Euclidicity after some finite further time. Suppose that this history is a solution of a (generalized) Newtonian theory set in a three-dimensional space, G, of a fixed geometry that is everywhere Euclidean except for some finite, geometrically simple, non-Euclidean region. The particles start out in the Euclidean region and eventual stray into the non-Euclidean region. Huggett's idea is that the relationalist can view both the geometry of the total space, and the particles' particular embedding in it, as supervenient on the history of relations via (his liberalized version of) the Best Systems approach to laws. The idea is that the following hypotheses jointly constitute the simplest and strongest systematization of the relational history: (i) the history of instantaneous relations is constrained to be embeddable at all times into G and (ii) the relational history follows, at any moment, from the instantaneous relations and the embedding into G at that moment, together with a certain set of Newtonian laws. In particular, the simplicity requirement fixes G over other more complicated geometries into

---

[89] This aspect of the proposal is scrutinized by Belot (2011, ch. 3).

which the particular relational history can also be embedded, e.g., geometries with additional non-Euclidean regions unsurveyed by the material particles.[90]

Now recall the problem that the variable geometry of empty regions of spacetime can cause for a relationist who would simply restrict spatiotemporal distance relations to material events: a particular partial history of pseudo-Riemannian relations instantiated within an island configuration of material events, together with the laws of GR, might not fix the future history because of the possible influences of the geometry of empty spacetime beyond the material configuration (section 6.1.2). In the natural extension of Huggett's scheme, one takes the entire history of instantiated spatiotemporal relations between material events as the supervenience base. One and only one future evolution of the material world compatible with the considered initial segment and laws of GR is, of course, included in this. The interesting question for the liberalizing Humean is whether, *if facts about the geometry of empty spacetime are allowed to supervene, together with the laws, on the material relational history*, the laws of GR constitute the Best System laws of such a world.

6.3.2 The Dynamical Approach to Relativity

I finish this section by highlighting some of the similarities between the dynamical approach to special relativity, defended by Brown (2005) and by Brown and Pooley (2006), and Huggett's proposal for Newtonian physics. The dynamical approach seeks to offer a reductive account of the Minkowski spacetime interval in terms of the dynamical symmetries of the laws governing matter. It therefore qualifies as a type of relationalism, although this is not something that Brown himself emphasizes.

One of the guiding intuitions behind the dynamical approach concerns explanatory priority. Consider, for example, the relativistic phenomenon of length contraction. Do rods behave as they do in virtue of the spatiotemporal environment in which they are immersed, or are facts about the geometrical structure of spacetime reducible to (*inter alia*) the behavior of rods? And if one opts for the latter point of view, what explanation is to be given of why measuring rods in motion are contracted relative to similarly constituted rods at rest?

Brown reads Bell (1976) as seeking to demonstrate that "a moving rod contracts, and a moving clock dilates, *because of how it is made up and not because of the nature of its spatio-temporal environment*" (Brown 2005, 8, emphasis in the original). And, Brown thinks, Bell was surely right. This, though, is to present a false dichotomy. The substantivalist should claim that a moving rod's contraction reflects both how it is made up and the nature of its spatiotemporal environment.[91] Recall the discussion of the explanatory role of substantival geometry in section 4.3.

---

[90] Simplicity will not determine a unique geometry for G in all cases, but Huggett makes a persuasive case that the underdetermination is benign and that the regularity relationalist should be content to live with the possibility that there may be no determinate fact of the matter about the geometry of physical space (Huggett 2006, 55–56).

[91] This is not to say that every explanatory question one might ask about the phenomenon of length contraction requires an appeal to dynamical laws; in some contexts it is enough to cite the relevant geometrical facts in order to provide an explanation. This is a point explicitly emphasized in Brown and Pooley (2006, 78–79, 82), where paradigm explanatory uses of Minkowski diagrams (e.g., to highlight that observers in relative motion consider different cross-sections of a rod's world tube when judging its length) are said to constitute

The substantivalist should agree that a complex material rod does not conform to the axioms of some geometry simply because that is the substantival geometry in which it is immersed; the rod would not do what it does were the laws governing its microphysical parts different in key respects. But equally, according to the substantivalist, the coordinate-dependent equations that are appealed to in, for example, Bell's toy-model derivation of length contraction make implicit reference, via the choice of coordinate system, to primitive spatiotemporal geometry.

What features of the laws governing the constituents of a rod are responsible for the rod's characteristic relativistic behavior such as its length contraction? In an important sense, the details of the dynamics are irrelevant. If subject to appropriately nondestructive accelerations, rods made of steel, wood, and glass will contract by the same amount, and for the same reason, namely, *the Lorentz covariance of the laws governing their constituents*.[92] In recent discussion of the dynamical approach (e.g., Janssen 2009, Frisch 2011), this point is widely agreed upon. As Frisch emphasizes, what genuine disagreement there is centers on the status of the dynamical symmetries to which such explanations appeal.

For Balashov and Janssen, these are ultimately to be explained in terms of the geometry of spacetime. To the question: "Does the Minkowskian nature of spacetime explain why the forces holding a rod together are Lorentz invariant or the other way around?" they reply: "Our intuition is that the geometrical structure of space (-time) is the *explanans* here and the invariance of the forces the *explanandum*" (Balashov and Janssen, 2003, 340) and Janssen likes to talk of the symmetries of Minkowski geometry as the *common origin* of the symmetries of the various laws governing matter. For geometry to play this role, its instantiation in the physical world had better not depend on facts about the dynamical laws. This is true on the substantivalist view reviewed in section 4.3 but, note, that it is also true on the Minkowski relationalist view discussed in section 6.1.2, which likewise takes both the ideology of the spacetime interval and its satisfying the constraints of Minkowski geometry as primitive.

How does this alleged explanation of dynamical symmetries in terms of spacetime symmetries go? Clearly it will not be any kind of causal explanation. Moreover, as the examples of Galilean (or Maxwellian) invariant Newtonian physics set in Newtonian (or Galilean) spacetime illustrate,[93] the explanation must be compatible with the logical possibility of theories in which there is a mismatch between

---

"perfectly acceptable explanations (perhaps the only acceptable explanations) of the explananda in question." Our emphasis of this fact seems to have been overlooked by some authors (Skow 2006, Frisch 2011).

[92] As it was put in Brown and Pooley (2006, 82): "it is sufficient for these bodies to undergo Lorentz contraction that the laws (whatever they are) that govern the behavior of their microphysical constituents are Lorentz covariant. It is *the fact that the laws are Lorentz covariant* …that explains why the bodies Lorentz contract. To appeal to any further details of the laws that govern the cohesion of these bodies would be a mistake." Janssen's (2009) carefully argued case that phenomena recognized to be kinematical (in his sense) should not be explained in terms of the details of their dynamics is therefore one that we had antecedently conceded. The explanation of the phenomena in terms of symmetries nonetheless deserves the label "dynamical" (though not, as acknowledged in Brown and Pooley (2006, 83), "constructive") because the explanantia are (in the first instance) the *dynamical* symmetries of the laws governing the material systems manifesting the phenomena.

[93] Other examples are provided by Lorentz invariant dynamics set in Newtonian spacetime; see, e.g., Earman (1989, 50–55).

dynamical symmetries and the symmetries of independently postulated spacetime structure (Brown and Pooley 2006, 83–84).

In these cases, the mismatches are all in one direction; the spacetime symmetries are a proper subset of the dynamical symmetries. It might be thought that the substantivalist can readily explain this.[94] On their view, dynamical laws ultimately involve coordinate-independent claims describing how dynamically varying matter is constrained by and adapted to spacetime structure. If the properties of spacetime structure are described explicitly, these laws should be expressible by equations that hold good in any coordinate system. But if the spacetime structure has symmetries that allow for a privileged set of adapted coordinate systems, one expects these equations will (apparently) simplify, as some aspects of the spacetime structure will now be encoded in the coordinate system. Recall that, in coordinate terms, dynamical symmetries are transformations between coordinate systems in which the equations expressing the laws take the same form. If the equations in question are the special, simplified equations, then, on the substantivalist's understanding of these equations, (i) they should hold in all coordinate systems appropriately adapted to spacetime structure, and (ii) they need not hold in others. But, in terms of coordinates, spacetime symmetries just are the transformations between adapted coordinate systems. Hence, the dynamical symmetries should include the spacetime symmetries. And, very crudely, the possibility that dynamical symmetries outstrip spacetime symmetries arises because the dynamical laws governing matter might exploit only some of the spacetime structure, so that the coordinate systems in which dynamics simplifies need be adapted to only some of the structure postulated.[95]

Given the substantivalist's understanding of the coordinate-dependent forms of dynamical equations, therefore, it follows that the symmetries of these equations cannot be more restricted than the symmetries of the full set of postulated spacetime structures. In at least this sense, the substantivalist can explain dynamical symmetries in terms of spacetime symmetries. According to the dynamical approach, however, this gets things exactly the wrong way round. Facts about dynamical symmetries come first and are the ground of true claims about the geometry of spacetime: "the Minkowskian metric is no more than a codification of the behavior of rods and clocks, or equivalently, it is no more than the Kleinian geometry associated with the symmetry group of the quantum physics of the non-gravitational interactions in the theory of matter" (Brown 2005, 9).

If spacetime geometry is to be grounded in the symmetries of the dynamical laws governing matter, it had better be the case that the very idea of such a law and its symmetries does not presuppose spacetime geometry. That it need not do so is particularly clear if a Humean conception of laws is adopted. This will also bring out the parallels with Huggett's proposal. Recall that Huggett's regularity

---

[94] I am grateful to Hilary Greaves for discussion of this point. The story given here can also be told, *mutatis mutandis*, by a relationalist who posits primitive spatiotemporal relations held to satisfy primitive geometrical constraints.

[95] See Earman (1989, 45–47) for a related discussion of the connection between spacetime symmetries and dynamical symmetries.

relationalist postulates primitive Leibnizian relations but no ideology corresponding to inertial structure. The latter is grounded in the existence of a proper subset of the coordinate systems adapted to the Leibnizian relations with respect to which the description of the entire relational history is the solution of particularly simple equations (Newton's laws expressed with respect to inertial frame coordinates). The dynamical approach involves a similar but much more radical move: the metrical relations themselves are to be grounded in exactly the same way.

The idea is best illustrated with a simple example. The advocate of the dynamical approach need not be understood as eschewing all primitive spatiotemporal notions (*pace* Norton 2008). In particular, one might take as basic the "topological" extendedness of the material world in four dimensions. Imagine such a world whose only material dynamical entity has pointlike parts whose degrees of freedom can be modeled by the real numbers. One obtains a coordinate description of such an entity by associating, in a way that respects its local topology, each of its pointlike parts with distinct elements of $\mathbb{R}^4$, and associating with each of these a real number representing the dynamical state of the corresponding part. In other words, we directly map the parts of the material field postulated to be the sole entity in the world into $\mathbb{R}^4$ and choose a way to represent its dynamical state so as to obtain a scalar field on $\mathbb{R}^4$. Different choices of coordinate system will yield different mathematical descriptions. Suppose, now, that for some special choice of coordinate system the description one obtains is the solution of a very simple equation. Moreover, suppose that (i) the descriptions one obtains relative to coordinate systems related to this first coordinate system by Lorentz transformations yield (distinct) descriptions that are solutions of the very same equation but that (ii) descriptions with respect to other coordinate systems, if they can be represented as solutions of equations at all, are solutions of more complicated equations.

If all this were the case, the simplest equation might be considered one of the Humean laws of this world.[96] The Lorentz group's being their dynamical symmetry group is constituted by its being the group that maps between the coordinate systems with respect to which descriptions of the material world satisfy the simple equation. And finally, the spatiotemporal geometry of the world is defined in terms of the invariants of the symmetry group so identified. In particular, for the spatiotemporal interval between two parts of the material world p, q to be I *just is* for $(t_p - t_q)^2 - |\vec{x}_p - \vec{x}_q|^2 = \pm I(p, q)^2$ with respect to the privileged coordinate systems. Spacetime geometry is reduced to a notion of dynamical symmetry that does not presuppose it. The example considered is, of course, very simple, and a number of issues will arise when fleshing out an analogous story for more realistic physics. Some of the choices to be made are highlighted by Norton (2008), who denies the feasibility of exactly this kind of project. Two charges he makes are worth dwelling on.

---

[96] This law could be expressed in a coordinate-independent manner if one introduces an auxiliary device, the Minkowski metric, which would then be "no more than a codification of the Kleinian geometry associated with the symmetry group" of the laws.

First, he considers the case where the world contains several matter fields, each described by a distinct theory. He grants that each of these might be Lorentz invariant. His challenge to the advocate of the dynamical approach (dubbed the "constructivist") is to justify the assumption that the sets of coordinate systems with respect to which these cases of Lorentz invariance are manifest coincide. The simple answer is that the *spatiotemporally coincident parts of distinct matter fields should be assigned the same element of* $\mathbb{R}^4$. The issue is how this relation of coincidence between matter fields is to be understood. For the substantivalist it involves colocation at the same spacetime point. The Minkowski relationalist, who takes interval facts as primitive, can analyze it in terms of these (though not, of course, straightforwardly in terms of the vanishing of the interval, for this will not exclude noncoincident, lightlike related events). What options are open to the constructivist? The most natural is to take spatiotemporal coincidence as primitive (as many relationalists have done; e.g., Rovelli (1997, 194)). After all, the project was to reduce chronogeometric facts to symmetries, not to recover the entire spatiotemporal nature of the world from no spatiotemporal assumptions whatsoever. The constructivist's project might need a primitive notion of "being contiguous," but Norton is wrong to think that it follows from this that constructivists are illicitly committed to the independent existence of spacetime.[97]

The other of Norton's objections that I wish to highlight involves what the constructivist must say about the geometry of empty regions of spacetime and of regions containing homogeneous matter. Suppose some way K of coordinatizing the material world satisfies the type of condition described above. Now suppose that the world contains an empty region of spacetime. Translated into our terms, Norton's point is that any K' that agrees with K on its assignment of coordinates to material events will yield the same description. K' can differ from K in any way one likes over the coordinates it assigns to the empty region. Does this leave the geometry of the empty region indeterminate? Put this way, that there really is no problem here should be obvious: for the constructivist there is *literally nothing* in an empty region and so nothing whose geometrical properties might be indeterminate. The constructivist does not believe in the existence of an independently existing spacetime!

The case of homogeneous matter is more problematic. Now one is supposing there are entities—the material pointlike parts of the homogeneous region—whose spatiotemporal relatedness one would like to be able to enquire after. Suppose that the constructivist has attributed some primitive topological properties to matter. Even

---

[97] More radical options could also be pursued. Starting with the idea that there are no primitive facts about the contiguity or otherwise of distinct material events, one might nonetheless map them into a single copy of $\mathbb{R}^n$. The coincidence of events (which events are to be mapped to the same element in $\mathbb{R}^n$) is then to be thought of as determined in the same manner as the spacetime interval, that is, determined by those coordinatizations that yield total descriptions of all events that satisfy some simple set of equations. Perhaps one could even view the value of n (that is, the dimensionality of spacetime itself) as determined in this way too. As with generalizations of Huggett's proposal (see footnote 88), the more one views as grounded via some kind of Best System prescription, the more unconstrained the problem becomes; it ceases to be plausible that the complexity of the postulated supervenience base will be sufficient to underwrite the target quantities and the laws they obey.

so, we can respect these properties and smoothly alter K to K' within the region to obtain exactly the same description. The constructivist is forced to conclude that for any two material events in the region there is no fact of the matter concerning the interval between them. How bad is this? Note that a number of other geometrical properties will be determinate (because invariant under all coordinate transformations that leave the description of matter unaltered). For example, the spacetime volume of the homogeneous region might be determinate even though the spatiotemporal relatedness of the points within it is not.[98] This is surely a peculiarity of the constructivist's position. But, like Huggett's regularity relationalist in the face of analogous problems (Huggett, 2006, 55-56), they might argue that it is not such a painful bullet to have to bite.

7. Substantivalism in Light of the Hole Argument

For much of the last 25 years, arguments about spacetime substantivalism have been dominated by discussion of the Hole Argument. This is not the place for a thorough review of the sizeable literature that the argument has spawned.[99] Here I wish only to highlight one form of substantivalism that evades the Hole Argument and to emphasize an important disanalogy between the Hole Argument and the arguments against Newtonian and Galilean substantivalism that were considered in earlier sections.

Originally due to Einstein, who used it prior to 1915 to explain away his inability (at that point in time) to find satisfactory generally covariant field equations, the Hole Argument was rehabilitated by John Stachel (1989) before being put to work against spacetime substantivalism by Earman and Norton (1987). Let $\mathcal{M}_1 = \langle M, g_{ab}, T_{ab} \rangle$ be a model of a generally relativistic theory.[100] It follows from the *diffeomorphism invariance* of GR that, for an *arbitrary* diffeomorphism d, $\mathcal{M}_2 = \langle M, d^*g_{ab}, d^*T_{ab} \rangle$ will also satisfy the theory's equations. The natural (though not ineluctable) conclusion is that $\mathcal{M}_1$ and $\mathcal{M}_2$ *jointly* represent spacetimes (call them W_1 and W_2) that are physically possible according to the theory.

In $\mathcal{M}_1$, each $p \in M$ is assigned certain properties encoded by $\langle g_{ab}(p), T_{ab}(p) \rangle$; in $\mathcal{M}_2$, p is assigned the (in general) distinct properties encoded by $\langle d^*g_{ab}(p), d^*T_{ab}(p) \rangle$. But, according to the substantivalist, M represent physical spacetime. This means that (on one natural understanding of how the points of

---

[98] Compare how, on some treatments of vagueness, disjunctions can be determinately true (Fred is either bald or not bald) even though neither disjunct is determinately true.
[99] A good introduction is provided by Norton (2011).
[100] Recall (section 4) that the pseudo-Riemannian metric tensor g_{ab} encodes all of the geometrical properties of spacetime, itself represented by the four-dimensional manifold M. Strictly speaking, the stress–energy tensor T_{ab} does not directly represent the fundamental matter content of the model. This will be represented by other fields, in terms of which T_{ab} is defined.

M represent physical spacetime points) $\mathcal{M}_1$ and $\mathcal{M}_2$ represent one and the same spacetime point as having different properties. This gives us the next ingredient in the argument: the claim that the substantivalist is committed to regarding W_1 and W_2 as *distinct* possible worlds.[101]

The problem is that, if this interpretation of spacetime models is permitted, GR is radically indeterministic. Let d be a *hole diffeomorphism*, a map that it is only nontrivial within a restricted region of M (the so-called hole). Suppose that, relative to the metric of $\mathcal{M}_1$, d is nontrivial only to the future of some spacelike surface, Σ. $\mathcal{M}_1$ and $\mathcal{M}_2$ will then be identical structures up to and including this surface but differ to its future. On the proposed interpretation of $\mathcal{M}_1$ and $\mathcal{M}_2$, they represent spacetimes that are identical up to the spacelike surface represented by Σ but that differ to its future. It follows that the equations of GR, together with a complete specification of the history of the world up to some spacelike surface, fail to fix the future. Earman and Norton do not see this as a problem for substantivalism because they think indeterminism is objectionable per se. Their claim, rather, is that determinism should fail only for reasons of physics and not as the result of a metaphysical commitment and in a theory-independent way (Earman and Norton, 1987, 524).

Note that $\mathcal{M}_1$ and $\mathcal{M}_2$ are *isomorphic* structures. The possibilities they represent, therefore, involve exactly the same patterns of qualitative features. If W_1 and W_2 are distinct possibilities, they differ only over which spacetime points instantiate which of the particular features common to both worlds. In the terminology of modal metaphysics, the difference between the possibilities is merely *haecceitistic* (Kaplan, 1975). Many of the pro-substantivalist responses to the argument make crucial use of this aspect of the setup.

For example, a substantivalist might agree that accepting GR involves a commitment to such haecceitistic distinctions and accept that the theory is indeterministic. However, they might deny that this indeterminism is in any sense troublesome precisely because it is an indeterminism only about which objects instantiate which properties and not about which patterns of properties are instantiated. A closely related response accepts that GR is committed to haecceitistic distinctions but denies that it follows that GR is indeterministic because the correct definition of determinism, it is claimed, is only sensitive to qualitative differences.[102]

The most popular response, however, has been to advocate some variety of *sophisticated substantivalism*, that is, a version of substantivalism that denies the existence of physically possible spacetimes that differ merely haecceitistically. The simplest way to secure this is to endorse *antihaecceitism*, that is, the general denial of merely haecceitistic distinctions between possible worlds.[103]

---

[101] This amounts to a denial of *Leibniz Equivalence*. Earman and Norton take such a denial to be the acid test of substantivalism (Earman and Norton, 1987, 521).

[102] For further discussion of the definition of determinism appropriate to GR, and of the merits of these options, see Butterfield (1989b), Rynasiewicz (1994), Belot (1995), Leeds (1995), Brighouse (1997), and Melia (1999).

[103] This is my preferred option (see Pooley 2006, 99–103). Despite the important differences between them, I take Maudlin (1989), Butterfield (1989a), Maidens (1992), Stachel (1993, 2002), Brighouse (1994), Rynasiewicz

Two arguments discussed earlier in the chapter also involved the claim that, because of the dynamical symmetries of the relevant physical theory, the (relevant stripe of) substantivalist was committed to distinct physically possible worlds, the nonidentity of which was alleged to be problematic. The important difference between these cases and those of the Hole Argument is that the former involve qualitative differences between the relevant worlds. In the case of the kinematic shift, the worlds differ over the absolute velocities assigned to bodies. In the case of Maxwellian invariant dynamics set in Galilean spacetime, they differ over the absolute accelerations assigned to bodies.[104] The fact that these differences are qualitative has two important consequences.

First, that the possibilities differ qualitatively creates an epistemological problem (given that one cannot observationally distinguish between the relevant quantities) that is not present in the case of merely haecceitistic differences.[105] Even if diffeomorphic models of GR are to be interpreted as representing distinct possibilities, there is no substantive fact, about which I could be ignorant despite knowing all the observable facts, concerning which model really represents the actual world. Each model is equally apt, and which model represents the actual world will be a matter of (arbitrary) representational convention. In contrast, models of Galilean invariant physics set in Newtonian spacetime that differ by boosts of their material content are not equally suited to represent any given possibility. Even once representational conventions are fixed, the Newtonian substantivalist does not know whether the model that attributes a velocity of $10 ms^{-1}$ to the Eiffel Tower, the one that attributes $20 ms^{-1}$, or yet some other model, corresponds to the actual world.

Second, the antihaecceitist way out of the Hole dilemma is of no use in the context of the kinematic shift argument. The argument is evaded if any two models related by Galilean boosts can be shown to be different representations of the same state of affairs. Since the models represent qualitatively distinct possibilities according to the Newtonian substantivalist, merely embracing antihaecceitism does not collapse the distinction between them. A substantivalist position that can view Galilean boosted models as distinct representations of one and the same state of affairs requires substantive work, viz., the replacement of Newton's substantival space with neo-Newtonian spacetime. (A similar observation holds concerning the passage from neo-Newtonian to Newton-Cartan spacetime.) This is in contrast to the so-called *static shift* argument against Newton's absolute space, which exploits

---

(1994), Hoefer (1996), and Saunders (2003a) all to deny that the relevant haecceitistic differences correspond to distinct physical possibilities. For several of these authors (though notably not for Maudlin), the commitment follows from a commitment to some kind of antihaecceitism, at least concerning spacetime points, whether on general philosophical grounds (as in Hoefer's case), or as a perceived lesson of the diffeomorphism invariance of the physics (as in Stachel's case).

[104] Note one parallel between the Hole Argument and the argument against Galilean spacetime that exploits the Maxwell group. The fact that the Maxwell group involves a parameter that is an arbitrary function of time means that the Galilean substantivalist interpretation of the models of a Maxwellian invariant theory involves regarding the theory as indeterministic (cf Stein 1977, Saunders, 2003a). The fact that the indeterminism involves qualitative differences (according to the Galilean substantivalist) arguably makes the argument more effective against Galilean substantivalism than the Hole Argument is against GR.

[105] This point is discussed by Horwich (1978), Field (1985), and Maudlin (1993).

the Euclidean symmetries of Newtonian mechanics and compares only models related by *time-independent* rotations or translations.[106] In this case antihaecceitism does collapse the number of relevant physical possibilities to one.

It is enough to note that antihaecceitism is a live view within metaphysics in order to see that substantivalism need not fall to the Hole Argument. More controversial is how well-motivated the position is from the perspective of the interpretation of physics. Here a couple of remarks are in order.

First, as Belot and Earman (1999, 2001) have stressed, several physicists grappling with the conceptual and technical problems of unifying quantum mechanics and general relativity do claim to draw substantive morals from the Hole Argument. What is not clear, however, is whether the genuinely substantive interpretational questions that have come to the fore as a result of work on the quantization of GR have anything to do with the kind of diffeomorphism invariance that lies at the heart of the Hole Argument. One key issue concerns the nature of the "observables" (that is, the genuine physical magnitudes) of diffeomorphism-invariant theories. Another concerns differences between GR and pre-generally relativistic theories. In particular, are the true physical magnitudes of GR essentially different in kind to those of pre-GR theories (when the latter are properly understood)? While Earman (2006a,b) believes that the right answers to these questions will be inconsistent with anything like a substantivalist interpretation of GR, even of the sophisticated variety, it is not obvious that some of the views about the nature of "observables" advocated by the physicists Earman cites, such as those of Rovelli (2002), are incompatible with sophisticated substantivalism.

Second, some of the work on "structural realist" interpretations of spacetime, at least where these do not involve an eliminativism about spacetime points, can be understood as varieties of antihaecceitist substantivalism.[107] It is possible that the development of one of these will provide an additional motivation for sophisticated substantivalism.[108]

---

[106] An argument like this was made by Leibniz in his correspondence with Clarke (Alexander, 1956). That Leibniz makes a precisely parallel argument, exploiting permutation invariance, against the existence of atoms, should give those sympathetic to the static shift argument pause for thought. Consistency should lead one either to embrace or reject both conclusions.

[107] Self-declared structuralist approaches to spacetime that are best described as varieties of substantivalism (in the sense that they include spacetime points among the ground-floor ontology) include those of Stachel (2002, 2006), Saunders (2003a), Esfeld and Lam (2008), and Muller (2011). For an overview of a wider range of structuralist approaches, see Greaves (2011), who gives reasons to be skeptical that a coherent position that does not collapse into sophisticated substantivalism (or relationalism) has yet to be clearly identified. Bain (2006) and Rickles (2008) are two more advocates of spacetime structuralism, not cited by Greaves.

[108] I am attracted to the view that sees individualistic facts as grounded in general facts (Pooley, unpublished). However, as Dasgupta (whose terminology I adopt) has recently stressed (Dasgupta, 2011, 131–134), this requires that one's understanding of general facts does not presuppose individualistic facts. Since the standard understanding of general facts arguably does take individualistic facts for granted, the spacetime structuralist/sophisticated substantivalist must show that they are not illicitly making the same presupposition. (Dasgupta's own view is that something quite radical is needed (2011, 147–152).) The recent literature on "weak discernibility" (see, e.g., Saunders, 2003b) has made much of the fact that numerical diversity facts can supervene on facts statable without the identity predicate even when traditional forms of the Principle of the Identity of Indiscernibles are violated. Note, however, that merely showing that one set of facts supervene on another set of facts is not sufficient to show that the former are grounded in the latter (or even that it is possible to think of them as so grounded).

The upshot of this section is that the substantivalist understanding of spacetime physics, as set out in section 4, is not undermined by the Hole Argument. What, then, should one conclude about the relative merits of substantivalism versus relationalism? In section 5 I considered and rejected two other strands of antisubstantivalist argument that have motivated recent relationalists. That leaves substantivalism as a going concern. What about relationalism? Of the three general strategies outlined, the most promising is the Machian, 3-space approach of Barbour and collaborators. But, recall, this turned out not to be a form of relationalism in the traditional, ontological sense. It does represent an approach that is metaphysically very different from spacetime orthodoxy, but the dividing issue is not the existence of spacetime points but the relative priority of 3-dimensional versus 4-dimensional concepts.

The other two relationalist approaches fare less well. Recognizing that the Maxwell group is a symmetry group of Newtonian physics allows for an intriguing and relatively overlooked form of enriched relationalism, but it does not generalize to relativistic physics. In the context of SR, the restriction of Minkowski distances to a material ontology already provides for a viable, if unexciting, form of relationalism. In the context of GR, however, the same move does not work: in general, the dynamically significant chronometric facts outstrip the chronometric facts about matter, as is most vividly illustrated by the abundance of interesting vacuum solutions.

The relationalist approach reviewed in section 6.3 has not been pursued in the context of GR. Instead, a popular move for relationalists is to treat the metric field as just another material field (see, e.g., Rovelli, 1997, 193–195). This, it turns out, is also the view endorsed by Brown (2005, ch. 9). So, while the "dynamical approach" to relativity provides a reductive account of the metric—that is, a form of have-it-all relationalism—in the context of SR (section 6.3.2) the same is not true, for Brown at least, in GR. Brown stresses that the metric field only gains its usual "chronometrical significance" (that is, only corresponds to the practical geometry manifest by the behavior of material rods and clocks) in virtue of the particular way it dynamically couples to matter, but, as I hope to have made clear, no sensible substantivalist should demur.

What, then, is at stake between the metric-reifying relationalist and the traditional substantivalist? Both parties accept the existence of a substantival entity, whose structural properties are characterized mathematically by a pseudo-Riemannian metric field and whose connection to the behavior of material rods and clocks depends on, *inter alia*, the truth of the strong equivalence principle. It is hard to resist the suspicion that this corner of the debate is becoming merely terminological. At least this much that can be said for the choice of substantivalist language: it underlines an important continuity between the "absolute" spacetime structures that feature in pre-generally relativistic physics and the entity that all sides of the current dispute admit is a fundamental element of reality. To the extent that one should seek to understand the content and success of previous theories in terms of our current best theory, this arguably vindicates the substantivalist interpretation of Newtonian and specially relativistic physics.[109]

References

Alexander, H. G., ed. (1956). *The Leibniz–Clarke correspondence.* Manchester: Manchester University Press.
Anandan, Jeeva, and Brown, Harvey R. (1995). On the reality of spacetime geometry and the wavefunction. *Foundations of Physics* 25: 349–360.
Anderson, Edward (2006). Leibniz–Mach foundations for GR and fundamental physics. In *General relativity research trends,* New York: ed. A. Reimer, Volume 249 of *Horizons in World Physics,* 59–122. New York: Nova Science. http://arxiv.org/abs/gr-qc/0405022v2.
———. (2007). On the recovery of geometrodynamics from two different sets of first principles. *Studies in History and Philosophy of Modern Physics* 38: 15–57.
———. (2012). The Problem of Time and Quantum Cosmology in the Relational Particle Mechanics Arena. http://arxiv.org/abs/1111.1472v2.
Anderson, Edward and Barbour, Julian B. (2002), "Interacting vector fields in relativity without relativity", *Classical and Quantum Gravity* 19: 3249–3262.
Anderson, Edward, Barbour, Julian B., Foster, Brendan, and Ó Murchadha, Niall (2003). Scale-invariant gravity: Geometrodynamics. *Classical and Quantum Gravity,* 20, 1571–1604.
Anderson, Edward, Barbour, Julian B., Foster, Brendan Z., Kelleher, B., and Ó Murchadha, Niall (2005). The physical gravitational degrees of freedom. *Classical and Quantum Gravity* 22: 1795–1802.
Anderson, James L. (1967). *Principles of relativity physics.* New York: Academic Press.
Assis, A. K. T. (1989). On Mach's principle. *Foundations of Physics Letters* 2: 301–318.
Baierlein, Ralph F., Sharp, David H., and Wheeler, John A. (1962). Three-dimensional geometry as carrier of information about time. *Physical Review* 126: 1864–1865.
Bain, Jonathan (2004). Theories of Newtonian gravity and empirical indistinguishability. *Studies in History and Philosophy of Modern Physics* 35: 345–376.
———. (2006). Spacetime structuralism. In *The ontology of spacetime,* ed. D. Dieks, Vol. 1 of *Philosophy and foundations of physics,* 37–65. Amsterdam: Elsevier.
Balashov, Yuri, and Janssen, Michel (2003). Presentism and relativity. *British Journal for the Philosophy of Science* 54: 327–346.
Barbour, Julian B. (1974a). Relative-distance Machian theories. *Nature* 249: 328–329. Misprints corrected in Barbour (1974b).
———. (1974b). *Nature* 250: 606.
———. (1989). *Absolute or relative motion?* Volume 1: *The discovery of dynamics.* Cambridge: Cambridge University Press.
———. (1994). The timelessness of quantum gravity: I. The evidence from the classical theory. *Classical and Quantum Gravity* 11: 2853–2873.
———. (1999). *The end of time: The next revolution in our understanding of the universe.* London: Weidenfeld & Nicholson.
———. (2003). Scale-invariant gravity: Particle dynamics. *Classical and Quantum Gravity* 20: 1543–1570.

---

[109] The skepticism concerning the substantiveness of the debate expressed in this paragraph is therefore not that of Rynasiewicz (1996). For a convincing response to many Rynasiewicz's claims, see Hoefer (1998).

———. (2008). The nature of time. http://arxiv.org/abs/0903.3489v1.
———. (2010). The definition of Mach's principle. *Foundations of Physics* **40**: 1263–1284.
———. (2011). Shape dynamics: An introduction. http://arxiv.org/abs/1105.0183v1.
Barbour, Julian B., and Bertotti, Bruno (1977). Gravity and inertia in a Machian framework. *Nuovo Cimento* **38B**: 1–27.
———. (1982). Mach's principle and the structure of dynamical theories. *Proceedings of the Royal Society, London* **A 382**: 295–306.
Barbour, Julian B., and Ó Murchadha, Niall (2010). Conformal superspace: the configuration space of general relativity. Arxiv preprint: http://arxiv.org/abs/1009.3559v1.
Barbour, Julian B., and Pfister, H. ed. (1995). *Mach's principle: From Newton's bucket to quantum gravity*, Vol. 6 of *Einstein Studies*. Boston: Birkhäuser.
Barbour, Julian B., Foster, Brendan Z., and Ó Murchadha, Niall (2002). Relativity without relativity. *Classical and Quantum Gravity* **19**: 3217–3248.
Bell, John S. (1976). How to teach special relativity. *Progress in Scientific Culture* **1**. Reprinted in Bell (1987).
———. (1987). *Speakable and unspeakable in quantum mechanics*. Cambridge: Cambridge University Press.
Belot, Gordon (1995). New work for counterpart theorists: Determinism. *British Journal for the Philosophy of Science* **46**: 185–195.
———. (1999). Rehabilitating relationism. *International Studies in the Philosophy of Science* **13**: 35–52.
———. (2011). *Geometric possibility*. Oxford: Oxford University Press.
Belot, Gordon and Earman, John (1999). From metaphysics to physics. In *From physics to philosophy*, ed. J. Butterfield and C. Pagonis, 166–186. Cambridge: Cambridge University Press.
———. (2001). Pre-Socratic quantum gravity. In *Physics meets philosophy at the Planck scale*, ed. C. Callender and N. Huggett, 213–55. Cambridge: Cambridge University Press.
Brighouse, Carolyn (1994). Spacetime and holes. In *Proceedings of the 1994 biennial meeting of the Philosophy of Science Association*, ed. D. Hull, M. Forbes, and R. Burian, Vol. 1, 117–225, East Lansing, MI. Philosophy of Science Association.
———. (1997). Determinism and modality. *British Journal for the Philosophy of Science* **48**: 465–481.
Brown, Harvey R. (2005). *Physical relativity: Spacetime structure from a dynamical perspective*. Oxford: Oxford University Press.
Brown, Harvey R., and Pooley, Oliver (2006). Minkowski spacetime: A glorious non-entity. In *The ontology of spacetime*, ed. D. Dieks, Vol. 1 of *Philosophy and Foundations of Physics*, 67–89. Amsterdam: Elsevier.
Butterfield, Jeremy (1989a). Albert Einstein meets David Lewis. In *Proceedings of the 1988 biennial meeting of the Philosophy of Science Association*, ed. A. Fine and J. Leplin, Vol. 2. 65–81. East Lansing, MI: Philosophy of Science Association.
———. (1989b). The hole truth. *British Journal for the Philosophy of Science* **40**: 1–28.
Catton, Philip, and Solomon, Graham (1988). Uniqueness of embeddings and spacetime relationalism. *Philosophy of Science* **55**: 280–291.
Charleton, Walter (1654). *Physiologia Epicuro-Gassendo-Charltoniana: Or a Fabrick of Science Natural, Upon the Hypothesis of Atoms, Founded by Epicurus, Repaired by Petrus Gassendus, and Augmented by Walter Charleton*. Thomas Newcomb, for Thomas Heath, London.
Dasgupta, Shamik (2011). The bare necessities. *Philosophical Perspectives* **25**: 115–160.

Descartes, René (1644). Principia Philosophiae. *The philosophical writings of Descartes*, vol. 1, trans. and ed. J. Cottingham, R. Stoothoff, and D. Murdoch. Cambridge: Cambridge University Press, 1985.

Dieks, Dennis (2006). Another look at general covariance and the equivalence of reference frames. *Studies in History and Philosophy of Modern Physics* **37**: 174–191.

DiSalle, Robert (1995). Spacetime theory as physical geometry. *Erkenntnis* **42**: 317–337.

———. (2002). Newton's philosophical analysis of space and time. In *The Cambridge companion to Newton*, ed. I. B. Cohen and G. E. Smith, 33–56. Cambridge: Cambridge University Press.

———. (2006). *Understanding spacetime*. Cambridge: Cambridge University Press.

Earman, John (1986). *A primer on determinism*. Dordrecht: D. Riedel.

———. (1989). *World enough and spacetime: Absolute versus relational theories of space and time*. Cambridge MA: MIT Press.

———. (2006a). The implications of general covariance for the ontology and ideology of spacetime. In *The ontology of spacetime*, ed. D. Dieks, 3–24. Elsevier.

———. (2006b). Two challenges to the requirement of substantive general covariance. *Synthese* **148**: 443–468.

Earman, John, and Norton, John (1987). What price spacetime substantivalism? The hole story. *British Journal for the Philosophy of Science* **38**: 515–525.

Ehlers, Jürgen (1973a). The nature and structure of spacetime. In *The physicist's conception of nature*, ed. J. Mehra, 71–91. Dordrecht: Reidel.

———. (1973b). Survey of general relativity theory. In *Relativity, astrophysics and cosmology: Proceedings of the summer school held, 14–26 August, 1972 at the Banff Centre, Banff Alberta*, ed. W. Israel, Vol. 38, 1–125. Dordrecht, Holland: Kluwer.

Einstein, Albert (1916). The foundation of the general theory of relativity. *Annalen der Physik* **49**: Reprinted in (Einstein, Lorentz, Weyl and Minkowski, 1952, 109–164).

———. (1918). Prinzipielles zur allgemeinen Relativitätstheorie. *Annalen der Physik* **360**: 241–244.

———. (1922). *The meaning of relativity*. Princeton: Princeton University Press. Four lectures delivered at Princeton University, May 1921; trans. by E. P. Adams.

———. (1924). Über den Äther. *Schweizerische naturforschende Gesellschaft, Verhanflungen* **105**: 85–93. Trans. S. W. Saunders in (Saunders and Brown, 1991, 13–20); page references are to this translation.

Einstein, Albert, Lorentz, H. A., Weyl, H., and Minkowski, H. (1952). *The principle of relativity*. New York: Dover.

Esfeld, Michael, and Lam, Vincent (2008). Moderate structural realism about spacetime. *Synthese* **160**: 27–46.

Field, Hartry (1985). Can we dispense with spacetime? In *Proceedings of the 1984 biennial meeting of the Philosophy of Science Association*, ed. Asquith and Kitcher, Vol. 2, 33–90.

Friedman, Michael (1983). *Foundations of spacetime theories: Relativistic physics and philosophy of science*. Princeton: Princeton University Press.

Frisch, Mathias (2011). Principle or constructive relativity. *Studies in History and Philosophy of Modern Physics* **42**: 176–183.

Garber, Daniel (1992). *Descartes' metaphysical physics*. Chicago and London: University of Chicago Press.

Grant, Edward (1981). *Much ado about nothing: Theories of space and vacuum from the Middle Ages to the scientific revolution*. Cambridge: Cambridge University Press.

Greaves, Hilary (2011). In search of (spacetime) structuralism. *Philosophical Perspectives* **25**: 189–204.

Gryb, Sean (2009). Implementing Mach's principle using gauge theory. *Physical Review D* **80**: 024018.

———. (2010). A definition of background independence. *Classical and Quantum Gravity* **27**: 215018.

Hoefer, Carl (1996). The metaphysics of spacetime substantivalism. *Journal of Philosophy* **93**: 5–27.

———. (1998). Absolute versus relational spacetime: For better or worse, the debate goes on. *British Journal for the Philosophy of Science* **49**: 451–467.

Hofmann, Wenzel (1995 [1904]). Motion and inertia. In Julian B. Barbour and H. Pfister (eds), *Mach's Principle: From Newton's bucket to quantum gravity*, vol. 6 of *Einstein Studies*, 128–133. Boston: Birkhäuser. Trans. J. B. Barbour from *Kritische Beleuchtung der beiden Grundbegriffe der Mechanik: Bewegung und Trägheit und daraus gezogene Folgerungen betreffs der Achsendrehung der Erde des Foucault'schen Pendelversuchs*, Vienna and Leipzig: M. Kuppitsch Wwe.

Hood, C. G. (1970). A reformulation of Newtonian dynamics. *American Journal of Physics* **38**: 438–442.

Horwich, Paul (1978). On the existence of time, space and spacetime. *Noûs*, 12, 397–419.

Huggett, Nick (1999). Why manifold substantivalism is probably not a consequence of classical mechanics. *International Studies in the Philosophy of Science* **13**: 17–34.

———. (2006). The regularity account of relational spacetime. *Mind* **115**: 41–73.

Janssen, Michel (2008). "No success like failure ...": Einstein's quest for general relativity, 1907–1920. To be published in Janssen and Lehner (forthcoming).

———. (2009). Drawing the line between kinematics and dynamics in special relativity. *Studies in History and Philosophy of Modern Physics* **40**: 26–52.

Janssen, Michel, and Lehner, Christoph eds. (forthcoming). *The Cambridge companion to Einstein*. Cambridge: Cambridge University Press.

Kaplan, David (1975). How to Russell a Frege–Church. *Journal of Philosophy* **72**: 716–729.

Koyré, Alexandre (1965). *Newtonian studies*. Chicago: University of Chicago Press.

Kretschmann, E. (1917). Über den physikalischen Sinn der Relativitätspostulate. *Annalen der Physik* **53**: 575–614.

Laymon, Ronald (1978). Newton's bucket experiment. *Journal of the History of Philosophy* **16**: 399–413.

Leeds, S. (1995). Holes and determinism: Another look. *Philosophy of Science* **62**: 425–437.

Lehmkuhl, Dennis (2008). Is spacetime a gravitational field? In *The ontology of spacetime II*, ed. D. Dieks, Vol. 4 of *Philosophy and Foundations of Physics*, 83–110. Amsterdam: Elsevier.

———. (2011). Mass-energy-momentum: Only there because of spacetime? *British Journal for the Philosophy of Science* **62**: 453–488.

Lewis, David K. (1973). *Counterfactuals*. Cambridge, MA: Harvard University Press.

———. (1983). New work for a theory of universals. *Australasian Journal of Philosophy* **61**: 343–377. Reprinted in Lewis (1999).

———. (1999). *Papers in metaphysics and epistemology*. Cambridge: Cambridge University Press.

———. (2000). Causation as influence. *Journal of Philosophy* **97**: 182–197.

Mach, Ernst (1919 [1901]). *The science of mechanics*. 4th ed. Trans. T. J. McCormach. LaSalle, IL: Open Court.

Maidens, Anna (1992). Review of Earman, John S. [1989]: World enough and spacetime: Absolute versus relational theories of space and time. *British Journal for the Philosophy of Science* **43**: 129–136.

Malament, David B. (1982). Review of "Science without numbers: A defense of nominalism" by Hartry H. Field. *Journal of Philosophy* **79**: 523–534.

———. (1995). Is Newtonian cosmology really inconsistent? *Philosophy of Science* **62**: 489–510.

———. (2004). On the time reversal invariance of classical electromagnetic theory. *Studies in History and Philosophy of Modern Physics* **35**: 295–315.

———. (2010). A remark about the "geodesic principle" in general relativity. Available at: http://www.socsci.uci.edu/ dmalamen/bio/papers/GeodesicLaw.pdf.

———. (2012). *Topics in the foundations of general relativity and Newtonian gravitation theory*. Chicago: University of Chicago Press. http://www.socsci.uci.edu/ dmalamen/bio/GR.pdf.

Maudlin, Tim (1989). The essence of spacetime. In *Proceedings of the 1988 biennial meeting of the Philosophy of Science Association*, ed. A. Fine and J. Leplin, Vol. 2. 82–91. East Lansing, MI: Philosophy of Science Association.

———. (1993). Buckets of water and waves of space: Why spacetime is probably a substance. *Philosophy of Science* **60**: 183–203.

Maxwell, James Clerk (1952 [1877]). *Matter and motion*. New York: Dover.

Melia, Joseph (1999). Holes, haecceitism and two conceptions of determinism. *British Journal for the Philosophy of Science* **50**: 639–664.

———. (2003). *Modality*. Chesham: Acumen.

Mellor, D. H. (1980). On things and causes in spacetime. *British Journal for the Philosophy of Science* **31**(3): 282–288.

Minkowski, Herman (1909). Raum und Zeit. *Physikalische Zeitschrift* **10**: 104–111. Trans. W. Perrett and G. B. Jeffrey in (Einstein *et al.*, 1952, 75–91); page references are to this translation.

Misner, Charles W., Thorne, Kip S., and Wheeler, John Archibald (1973). *Gravitation*. San Francisco: W. H. Freeman and Company.

Muller, Fred A (2011). How to defeat Wüthrich's abysmal embarrassment argument against spacetime structuralism. *Philosophy of Science* **78**: 1046–1057.

Mundy, Brent (1983). Relational theories of Euclidean space and Minkowski spacetime. *Philosophy of Science* **50**: 205–226.

———. (1986). Embedding and uniqueness in relational theories of space. *Synthese* **67**: 383–390.

Nerlich, Graham (1979). What can geometry explain? *The British Journal for the Philosophy of Science*, **30**, 69–83.

———. (2010). Why spacetime is not a hidden cause: A realist story. In *Space, time, and spacetime*, ed. V. Petkov, Vol. 167 of *Fundamental Theories of Physics*, 181–91. Heidelberg: Springer.

Newton, Isaac (1684 [2004]). De Gravitatione. In *Philosophical writings*, ed. A. Janiak, 12–39. Cambridge: Cambridge University Press.

———. (1726 [1999]). *Mathematical principles of natural philosophy*. Trans. I. Bernard Cohen and Anne Whitman. Berkeley: University of California Press.

Norton, John D. (1985). What was Einstein's principle of equivalence? *Studies in History and Philosophy of Science Part A* **16**(3): 203–246.

———. (1993). A paradox in Newtonian gravitation theory. In *Proceedings of the 1992 biennial meeting of the Philosophy of Science Association*, ed. M. Forbes, D. Hull, and K. Okruhlik, Vol. 2, 412–420. East Lansing, MI: Philosophy of Science Association.

———. (2008). Why constructive relativity fails. *British Journal for the Philosophy of Science* **59**: 821–834.

———. (2011). The hole argument. In *The Stanford encyclopedia of philosophy* (Fall 2011 edn), ed. E. N. Zalta. http://plato.stanford.edu/archives/fall2011/entries/spacetime-holearg/.

Patrizi, Francesco (1943). On physical space. *Journal of the History of Ideas* 4: 224–245. Trans. Benjamin Brickman.

Pooley, Oliver (2001). Relationism rehabilitated? II: Relativity. http://philsci-archive.pitt.edu/id/eprint/221.

———. (2004). Comments on Sklar's "relationalist metric of time". *Chronos* 6: 77–86. http://philsci-archive.pitt.edu/id/eprint/2915.

———. (2006). Points, particles and structural realism. In *The structural foundations of quantum gravity*, ed. D. Rickles, S. French, and J. Saatsi, 83–120. Oxford: Oxford University Press. Preprint: http://philsci-archive.pitt.edu/2939/.

———. (unpublished). Substantivalism and haecceitism. Unpublished manuscript.

———. (in preparation). *The reality of spacetime*. Oxford: Oxford University Press.

Pooley, Oliver, and Brown, Harvey R. (2002). Relationalism rehabilitated? I: Classical mechanics. *British Journal for the Philosophy of Science* 53: 183–204. Preprint: http://philsci-archive.pitt.edu/id/eprint/220.

Quine, Willard Van (1951). Ontology and ideology. *Philosophical Studies* 2: 11–15.

Reichenbach, Hans (1924). The theory of motion according to Newton, Leibniz, and Huygens. In *Modern philosophy of science*, ed. M. Reichenbach. London: Routledge and Kegan Paul (1959), 46–66.

Reissner, H. (1914). Über die Relativität der Beschleunigungen in der Mechanik [on the relativity of accelerations in mechanics]. *Phys. Z.* 15: 371–375. Trans. J. B. Barbour in Barbour and Pfister (1995), 134–142.

Rickles, Dean (2008). Symmetry, Structure and Spacetime. Amsterdam: Elsevier.

Rovelli, Carlo (1997). Halfway through the woods: Contemporary research on space and time. In *The cosmos of science*. ed. J. Earman and J. Norton, 180–223. Pittsburgh: University of Pittsburgh Press.

———. (2001). Quantum spacetime: What do we know? In *Physics meets philosophy at the Planck scale*, ed. C. Callender and N. Huggett, 101–122. Cambridge: Cambridge University Press. http://arxiv.org/abs/gr-qc/9903045.

———. (2002). Partial observables. *Physical Review D* 65: 124013.

———. (2004). *Quantum gravity*. Cambridge: Cambridge University Press.

Rynasiewicz, Robert A. (1994). The lessons of the hole argument. *British Journal for the Philosophy of Science* 45: 407–436.

———. (1995). Absolute vs. relational theories of space and time: A review of John Earman's "World Enough and Spacetime". *Philosophy and Phenomenological Research* 55: 675–687.

———. (1996). Absolute versus relational spacetime: An outmoded debate? *Journal of Philosophy* 93: 279–306.

———. (2004). Newton's views on space, time, and motion. In *The Stanford encyclopedia of philosophy*, ed. E. N. Zalta.

Saunders, Simon W. (2003a). Indiscernibles, general covariance, and other symmetries. In *Revisiting the foundations of relativistic physics: Festschrift in honour of John Stachel*, ed. A. Ashtekar, D. Howard, J. Renn, S. Sarkar, and A. Shimony. Dordrecht: Kluwer. http://philsci-archive.pitt.edu/id/eprint/459.

———. (2003b). Physics and Leibniz's principles. In *Symmetries in physics: Philosophical reflections*, ed. K. Brading and E. Castellani, 289–307. Cambridge: Cambridge University Press. http://philsci-archive.pitt.edu/2012/.

———. (Forthcoming). Rethinking *Principia*. *Philosophy of Science*.

Saunders, Simon W., and Brown, Harvey R., eds. (1991). *The philosophy of the vacuum*. Oxford: Oxford University Press.

Schrödinger, Erwin (1925). Die Erfüllbarkeit der Relativitätsforderung in der klassischen Mechanik [The possibility of fulfillment of the relativity requirement in classical mechanics]. *Ann. Phys.* **77**: 325–336. Trans. J.B. Barbour in Barbour and Pfister (1995, 147–156).

Sider, Ted (2001). *Four dimensionalism: An ontology of persistence and time.* Oxford and New York: Oxford University Press.

Sklar, Lawrence (1974). *Space, time and spacetime.* Berkeley: University of California Press.

Skow, Bradford (2006). Physical relativity: Spacetime structure from a dynamical perspective. *Notre Dame Philosophical Reviews.* http://ndpr.nd.edu/news/25025-physical-relativity-space-time-structure-from-a-dynamical-perspective/

———. (2007). Sklar's maneuver. *British Journal for the Philosophy of Science* **58**: 777–786.

Stachel, John (1989). Einstein's search for general covariance, 1912–1915. In *Einstein and the history of general relativity*, ed. D. Howard and J. Stachel. Boston: 63–100. Birkhäuser.

———. (1993). The meaning of general covariance. In *Philosophical problems of the internal and external worlds: Essays on the philosophy of Adolph Grünbaum*, ed. J. Earman, A. Janis, and G. Massey, 129–60. Pittsburgh: University of Pittsburgh Press.

———. (2002). "The relations between things" versus "the things between relations": The deeper meaning of the hole argument. In *Reading natural philosophy. Essays in the history and philosophy of science and mathematics*, ed. D. B. Malament, 231–266. Chicago: Open Court.

———. (2006). Structure, individuality and quantum gravity. In *Structural foundations of quantum gravity*, ed. D. Rickles, S. French, and J. Saatsi, 53–82. Oxford: Oxford University Press.

Stein, Howard (1967). Newtonian spacetime. *Texas Quarterly,* **10**, 174–200. Reprinted in Robert Palter, ed., 1970, *The Annus Mirabilis of Sir Isaac Newton 1666–1966* (Cambridge, MA: MIT Press), 258–284.

———. (1977). Some pre-history of general relativity. In *Foundations of spacetime theories*, ed. J. Earman, C. Glymour, and J. Stachel, Vol. 8 of *Minnesota studies in the philosophy of science.* Minneapolis: 3–49. University of Minnesota Press.

———. (1991). One relativity theory and openness of the future. *Philosophy of Science* **58**: 147–167.

———. (2002). Newton's metaphysics. In *The Cambridge companion to Newton*, eds. I. Bernard Cohen and George E. Smith, 256–307. Cambridge: Cambridge University Press.

Sus, Adán (2011). On the explanation of inertia. Unpublished.

Synge, John Lighton (1960). *Relativity: The general theory.* Amsterdam: North-Holland Publication Co.

Teller, Paul (1987). Spacetime as a physical quantity. In *Kelvin's Baltimore lectures and modern theoretical physics*, ed. P. Achinstein and R. Kagon, 425–448. Cambridge, MA: MIT Press.

Trautman, Andrzej (1962). Conservation laws in general relativity. In *Gravitation: An introduction to current research*, ed. L. Witten, 169–198. New York: Wiley.

van Fraassen, Bas C. (1970). *An introduction to the philosophy of space and time.* New York: Columbia University Press.

Weatherall, James Owen (2011a). The motion of a body in Newtonian theories. *Journal of Mathematical Physics* 032502/52:

———. (2011b). On the status of the geodesic principle in Newtonian and relativistic physics. http://philsci-archive.pitt.edu/8662/.

Weinstein, Steven (2001). Absolute quantum mechanics. *British Journal for the Philosophy of Science* **52**(1): 67–73.

Westman, Hans, and Sonego, Sebastiano (2009). Coordinates, observables and symmetry in relativity. *Annals of Physics*, 1585–1611.

Weyl, Hermann (1922). *Space–time–matter*. 4th ed. Trans. H. L. Brose. London: Methuen and Co. Ltd.

Wheeler, John Archibald, and Feynman, Richard Phillips (1949). Classical electrodynamics in terms of direct interparticle action. *Reviews of Modern Physics* **21**: 425–433.

CHAPTER 16

GLOBAL SPACETIME STRUCTURE

JOHN BYRON MANCHAK

1. Introduction

The study of global spacetime structure is a study of the more foundational aspects of general relativity. One steps away from the details of the theory and instead examines the qualitative features of spacetime (e.g., its topology and causal structure).

We divide the following into three main sections. In the first, we outline the basic structure of relativistic spacetime and record a number of facts. In the second, we consider a distinction between local and global spacetime properties and provide important examples of each. In the third, we examine two clusters of global properties and question which of them should be regarded as physically reasonable. The properties concern "singularities" and "time travel" and are therefore of some philosophical interest.

2. Relativistic Spacetime

We take a (relativistic) *spacetime* to be a pair (M, g_{ab}). Here M is a smooth, connected, n-dimensional ($n \geq 2$) manifold without boundary. The metric g_{ab}

I am grateful to Bob Batterman, Erik Curiel, David Malament, and Jim Weatherall for comments on a previous draft.

is a smooth, nondegenerate, pseudo-Riemannian metric of Lorentz signature $(+,-,...,-)$ on M.[1]

2.1 Manifold and Metric

Let (M, g_{ab}) be a spacetime. The manifold M captures the topology of the universe. Each point in the n-dimensional manifold M represents a possible event in spacetime. Our experience tells us that any event can be characterized by n numbers (one temporal and $n-1$ spatial coordinates). Naturally, then, the local structure of M is identical to $\mathbb{R}^n$. But globally, M need not have the same structure. Indeed, M can have a variety of possible topologies.

In addition to $\mathbb{R}^n$, the sphere S^n is certainly familiar to us. We can construct a number of other manifolds by taking Cartesian products of $\mathbb{R}^n$ and S^n. For example, the 2-cylinder is just $\mathbb{R}^1 \times S^1$ while the 2-torus is $S^1 \times S^1$ (see figure 16.1). Any manifold with a closed proper subset of points removed also counts as a manifold. For example, $S^n - \{p\}$ is a manifold where p is any point in S^n.

We say a manifold M is *Hausdorff* if, given any distinct points $p, p' \in M$, one can find open sets O and O' such that $p \in O$, $p' \in O'$, and $O \cap O' = \emptyset$. Physically, Hausdorff manifolds ensure that spacetime events are distinct. In what follows, we assume that manifolds are Hausdorff.[2]

We say a manifold is *compact* if every sequence of its points has an accumulation point. So, for example, S^n and $S^n \times S^m$ are compact while $\mathbb{R}^n$ and $\mathbb{R}^n \times S^m$ are not. It can be shown that every noncompact manifold admits a Lorentzian metric. But there are some compact manifolds that do not. One example is the manifold S^4. Thus, assuming spacetime is four dimensional, we may deduce that the shape of our universe is not a sphere. One can also show that, in four dimensions, if a compact manifold does admit a Lorentzian metric (e.g., $S^1 \times S^3$), it is not simply connected. (A manifold is *simply connected* if any closed curve through any point can be continuously deformed into any other closed curve at the same point.)

We say two manifolds M and M' are *diffeomorphic* if there is a bijection $\varphi : M \to M'$ such that φ and φ^{-1} are smooth. Diffeomorphic manifolds have identical manifold structure and can differ only in their underlying elements.

The Lorentzian metric g_{ab} captures the geometry of the universe. Each point $p \in M$ has an associated tangent space M_p. The metric g_{ab} assigns a length to each vector in M_p. We say a vector ξ^a is *timelike* if $g_{ab}\xi^a\xi^b > 0$, *null* if $g_{ab}\xi^a\xi^b = 0$, and *spacelike* if $g_{ab}\xi^a\xi^b < 0$. Clearly, the null vectors create a double cone structure; timelike vectors are inside the cone while spacelike vectors are outside (see figure 16.1). In general, the metric structure can vary over M as long as it does so smoothly. But it certainly need not vary and indeed most of the examples considered below will have a metric structure that remains constant (i.e., a *flat* metric).

---

[1] In what follows, the reader is encouraged to consult Hawking and Ellis (1973), Geroch and Horowitz (1979), Wald (1984), Joshi (1993), and Malament (2012).
[2] See Earman (2008) for a discussion of this condition.

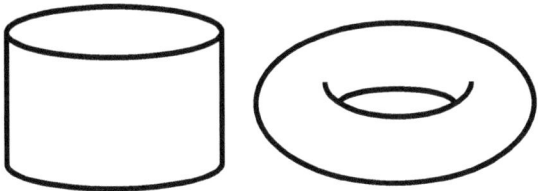

Figure 16.1 The cylinder $\mathbb{R}^1 \times S^1$ and torus $S^1 \times S^1$.

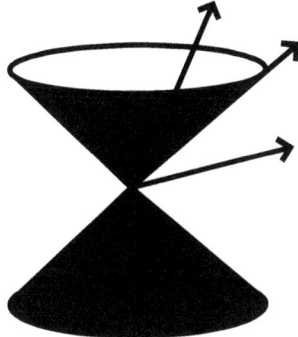

Figure 16.2 Timelike, null, and spacelike vectors fall (respectively) inside, on, and outside the double cone structure.

For some interval $I \subseteq \mathbb{R}$, a smooth curve $\gamma : I \to M$ is *timelike* if its tangent vector ξ^a at each point in $\gamma[I]$ is timelike. Similarly, a curve is *null* (respectively, *spacelike*) if its tangent vector at each point is null (respectively, spacelike). A curve is *causal* if its tangent vector at each point is either null or timelike. Physically, the worldlines of massive particles are images of timelike curves while the worldlines of photons are images of null curves. We say a curve $\gamma : I \to M$ is not *maximal* if there is another curve $\gamma' : I' \to M$ such that I is a proper subset of I' and $\gamma(s) = \gamma'(s)$ for all $s \in I$.

We say a spacetime (M, g_{ab}) is *temporally orientable* if there exists a continuous timelike vector field on M. In a temporally orientable spacetime, a future direction can be chosen for each double cone structure in way that involves no discontinuities. A spacetime that is not temporally orientable can be easily constructed by taking the underlying manifold to be the Möbius strip. In what follows, we will assume that spacetimes are temporally orientable and that a future direction has been chosen.[3]

Naturally, a timelike curve is *future-directed* (respectively, *past-directed*) if all its tangent vectors point in the future (respectively, past) direction. A causal curve is *future-directed* (respectively, *past-directed*) if all its tangent vectors either point in the future (respectively, past) direction or vanish.

Two spacetimes (M, g_{ab}) and (M', g'_{ab}) are *isometric* if there is a diffeomorphism $\varphi : M \to M'$ such that $\varphi_*(g_{ab}) = g'_{ab}$. Here, φ_* is a map that uses φ to "move" arbitrary

---

[3] See Earman (2002) for a discussion of this condition.

tensors from M to M'. Physically, isometric spacetimes have identical properties. We say a spacetime (M', g'_{ab}) is a (proper) *extension* of (M, g_{ab}) if there is a proper subset N of M' such that (M, g_{ab}) and $(N, g'_{ab|N})$ are isometric. We say a spacetime is *maximal* if it has no proper extension. One can show that every spacetime that is not maximal has a maximal extension.

Finally, two spacetimes (M, g_{ab}) and (M', g'_{ab}) are *locally isometric* if, for each point $p \in M$, there is an open neighborhood O of p and an open subset O' of M' such that $(O, g_{ab|O})$ and $(O', g'_{ab|O'})$ are isometric, and, correspondingly, with the roles of (M, g_{ab}) and (M', g'_{ab}) interchanged. Although locally isometric spacetimes can have different global properties, their local properties are identical. Consider, for example, the spacetimes (M, g_{ab}) and (M', g'_{ab}) where $M = S^1 \times S^1$, $p \in M$, $M' = M - \{p\}$, and g_{ab} and g'_{ab} are flat. The two are not isometric but are locally isometric. Therefore, they share the same local properties but have differing global structures (e.g., the first is compact while the second is not). One can show that for every spacetime (M, g_{ab}), there is a spacetime (M', g'_{ab}) such that the two are not isometric but are locally isometric.

2.2 Influence and Dependence

Here, we lay the foundation for the more detailed discussion of causal structure in later sections. Consider the spacetime (M, g_{ab}). We define the two-place relations $\ll$ and $<$ on the points in M as follows: we write $p \ll q$ (respectively, $p < q$) if there exists a future-directed timelike (respectively, causal) curve from p to q. For any point $p \in M$, we define the *timelike future* (domain of influence) of p, as the set $I^+(p) \equiv \{q : p \ll q\}$. Similarly, the *causal future* (domain of influence) of p is the set $J^+(p) \equiv \{q : p < q\}$.

The causal (respectively, timelike) future of p represents the region of spacetime that can be possibly influenced by particles (respectively, massive particles) at p. The timelike and causal pasts of p, denoted $I^-(p)$ and $J^-(p)$, are defined analogously. Finally, given any set $S \subset M$, we define $I^+[S]$ to be the set $\cup \{I^+(p) : p \in S\}$. The sets $I^-[S]$ and $J^+[S]$, and $J^-[S]$ are defined analogously. We shall now list a number of properties of timelike and causal pasts and futures.

For all $p \in M$, the sets $I^+(p)$ and $I^-(p)$ are open. Therefore, so are $I^+[S]$ and $I^-[S]$ for all $S \subseteq M$. However, the sets $J^+(p)$, $J^-(p)$, $J^+[S]$ and $J^-[S]$ are not, in general, either open or closed. Consider Minkowski spacetime[4] and remove one point from the manifold. Clearly, some causal pasts and futures will be neither open nor closed.

By definition, $I^+(p) \subseteq J^+(p)$ and $I^-(p) \subseteq J^-(p)$. And it is clear that if $p \in I^+(q)$, then $q \in I^-(p)$ and also that if $p \in I^-(q)$, then $q \in I^+(p)$. Analogous results hold for causal pasts and futures. We can also show that if either (i) $p \in I^+(q)$ and $q \in J^+(r)$

---

[4] Minkowski spacetime (M, g_{ab}) is such that $M = \mathbb{R}^n$, g_{ab} is flat, and there exist no incomplete geodesics (defined below). See Hawking and Ellis (1973).

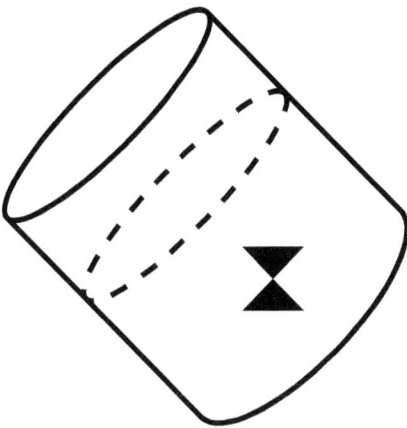

Figure 16.3 Cylindrical Minkowski spacetime containing a closed causal curve (e.g., the dotted line) but no closed timelike curves.

or (ii) $p \in J^+(q)$ and $q \in I^+(r)$, then $p \in I^+(r)$. Analogous results hold for the timelike and causal pasts. From this it follows that $\overline{I^+(p)} = \overline{J^+(p)}$, $\overline{I^-(p)} = \overline{J^-(p)}$, $\dot{I}^+(p) = \dot{J}^+(p)$, and $\dot{I}^-(p) = \dot{J}^-(p)$.[5]

Because future-directed casual curves can have vanishing tangent vectors, it follows that for all p, we have $p \in J^+(p)$ and $p \in J^-(p)$. Of course, a similar result does not hold generally for timelike futures and pasts. But there do exist some spacetimes such that, for some $p \in M$, $p \in I^+(p)$ (and therefore $p \in I^-(p)$). Gödel spacetime is one famous example (Gödel 1949).

We say the *chronology violating region* of a spacetime (M, g_{ab}) is the (necessarily open) set $\{p \in M : p \in I^+(p)\}$. We say a timelike curve $\gamma : I \to M$ is *closed* if there are distinct points $s, s' \in I$ such that $\gamma(s) = \gamma(s')$. Clearly, a spacetime contains a closed timelike curve if and only if it has a nonempty chronology violating region. One can show that, for all spacetimes (M, g_{ab}), if M is compact, the chronology violating region is not empty (Geroch 1967). The converse is false. Take any compact spacetime and remove one point from the underlying manifold. The resulting spacetime will contain closed timelike curves and also fail to be compact.

We define a causal curve $\gamma : I \to M$ to be *closed* if there are distinct points $s, s' \in I$ such that $\gamma(s) = \gamma(s')$ and γ has no vanishing tangent vectors. It is immediate that closed timelike curves are necessarily closed causal curves. But one can find spacetimes that contain the latter but not the former. Consider, for example, Minkowski spacetime (M, g_{ab}), which has been "rolled up" along one axis in such a way that some null curves but no timelike curves are permitted to loop around M (see figure 16.3). Other conditions relating to "almost" closed causal curves will be considered in the next section.

---

[5] In what follows, for any set S, the sets $\overline{S}$, $\dot{S}$, and int(S) denote the closure, boundary, and interior of S, respectively.

Finally, we say the spacetimes (M, g_{ab}) and (M, g'_{ab}) are *conformally related* if there is a smooth, strictly positive function $\Omega : M \to \mathbb{R}$ such that $g'_{ab} = \Omega^2 g_{ab}$ (the function Ω is called a *conformal factor*). Clearly, if (M, g_{ab}) and (M, g'_{ab}) are conformally related, then for all points $p, q \in M$, $p \in I^+(q)$ in (M, g_{ab}) if and only if $p \in I^+(q)$ in (M, g'_{ab}). Analogous results hold for timelike pasts and causal futures and pasts. Thus, the causal structures of conformally related spacetimes are identical.

A point $p \in M$ is a *future endpoint* of a future-directed causal curve $\gamma : I \to M$ if, for every neighborhood O of p, there exists a point $s' \in I$ such that $\gamma(s) \in O$ for all $s > s'$. A *past endpoint* is defined analogously. For any set $S \subseteq M$, we define the *future domain of dependence of S*, denoted $D^+(S)$, to be the set of points $p \in M$ such that every causal curve with future endpoint p and no past endpoint intersects S. The *past domain of dependence of S*, denoted $D^-(S)$, is defined analogously. The entire *domain of dependence of S*, denoted $D(S)$, is just the set $D^-(S) \cup D^+(S)$. If "nothing can travel faster than light," there is a sense in which the physical situation at every point in $D(S)$ depends entirely upon the physical situation on S.

Clearly, we have $S \subseteq D^+(S) \subseteq J^+[S]$ and $S \subseteq D^-(S) \subseteq J^-[S]$. Given any point $p \in D^+(S)$, and any point $q \in I^+[S] \cap I^-(p)$, we know that $q \in D^+(S)$. An analogous result holds for $D^-(S)$. One can verify that, in general, $D(S)$ is neither open nor closed. Consider Minkowski spacetime (M, g_{ab}). If $S = \{p\}$ for any point $p \in M$, we have $D(S) = S$, which is not open. If $S = I^+(p) \cap I^-(q)$ for any points $p \in M$ and $q \in I^+(p)$, we have $D(S) = S$, which is not closed.

A set $S \subset M$ is a *spacelike surface* if S is a submanifold of dimension $n - 1$ such that every curve in S is spacelike. We say a set $S \subset M$ is *achronal* if $I^+[S] \cap S = \emptyset$. One can show that for an arbitrary set S, $\dot{I}^+[S]$ is achronal. In what follows, let S be a closed, achronal set. We have $D^+(S) \cap I^-[S] = D^-(S) \cap I^+[S] = \emptyset$. We also have $\text{int}(D^+(S)) = I^-[D^+(S)] \cap I^+[S]$ and the analogous result for $D^-(S)$. Finally, we have $\text{int}(D(S)) = I^-[D^+(S)] \cap I^+[D^-(S)] = I^+[D^-(S)] \cap I^-[D^+(S)]$.

We say a closed, achronal set S is a *Cauchy surface* if $D(S) = M$. Physically, conditions on a Cauchy surface S (necessarily a submanifold of M of dimension $n - 1$) determine conditions throughout spacetime (Choquet-Bruhat and Geroch 1969). Clearly, if S is a Cauchy surface, any causal curve without past or future endpoint must intersect S, $I^+[S]$, and $I^-[S]$. One can verify that Minkowski spacetime admits a Cauchy surface.

We define the *future Cauchy horizon* of S, denoted $H^+(S)$, as the set $\overline{D^+(S)} - I^-[D^+(S)]$. The *past Cauchy horizon* of S is defined analogously. One can verify that $H^+(S)$ and $H^-(S)$ are closed and achronal. The *Cauchy horizon* of S, denoted $H(S)$, is the set $H^+(S) \cup H^-(S)$. We have $H(S) = \dot{D}(S)$ and therefore $H(S)$ is closed. Also, a nonempty, closed, achronal set S is a Cauchy surface if and only if $H(S) = \emptyset$.

The *edge* of a closed, achronal set $S \subset M$ is the set of points $p \in S$ such that every open neighborhood O of p contains a point $q \in I^+(p)$, a point $r \in I^-(p)$, and a timelike curve from r to q which does not intersect S. A closed, achronal set $S \subset M$ is a *slice* if it is without edge. It follows that every Cauchy surface is a slice. The converse is false. Consider Minkowski spacetime with one point removed from

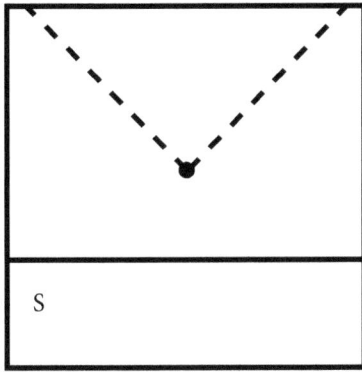

Figure 16.4 Minkowski spacetime with one point removed contains a slice S but no Cauchy surface. The region above the dotted line is not part of $D(S)$.

the manifold. It certainly admits a slice but no Cauchy surface (see figure 16.4). Of course, not every spacetime admits a slice. For a counterexample, consider any spacetime that has a chronology violating region identical to its manifold.

3. Spacetime Properties

We say a property P on a spacetime is *local* if, given any two locally isometric spacetimes (M, g_{ab}) and (M', g'_{ab}), (M, g_{ab}) has P if and only if (M', g'_{ab}) has P. A property is *global* if it is not local. Below, we will introduce and classify a number of spacetime properties of interest.

3.1 Local Properties

The most important local spacetime property is that of being a "solution" to Einstein's equation. There are a number of ways one can understand this property and we shall investigate each of them in what follows.

Let (M, g_{ab}) be a spacetime. Associated with the metric g_{ab} is a unique (torsion-free) derivative operator ∇_a such that $\nabla_a g_{bc} = 0$. Given a smooth curve $\gamma : I \to M$ with tangent field ξ^a, we say a vector η^a, defined at every point in the range of γ, is *parallelly transported* along γ if $\xi^b \nabla_b \eta^a = 0$ (see figure 16.1). We say a smooth curve $\gamma : I \to \mathbb{R}$ is a *geodesic* (i.e., nonaccelerating) if its tangent field ξ^a is such that $\xi^b \nabla_b \xi^a = 0$. Given any point $p \in M$, there is some neighborhood O of p such that any two points $q, r \in O$ can be connected by a unique geodesic contained entirely in O. Such a neighborhood is said to be *convex normal*.

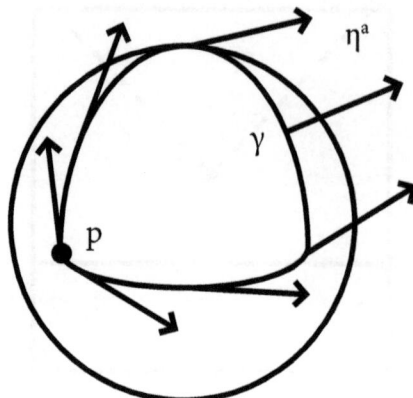

Figure 16.5 The vector η^a is parallelly transported along a closed curve γ. Note that the vector returns to the point p orientated differently.

The derivative operator ∇_a can be used to define the *Riemann curvature tensor*. It is the unique tensor $R^a{}_{bcd}$ such that for all ξ^a, $R^a{}_{bcd}\xi^b = -2\nabla_{[c}\nabla_{d]}\xi^a$.[6] A metric g_{ab} on M is flat if and only if its associated Riemann curvature tensor $R^a{}_{bcd}$ vanishes everywhere on M. The tensors $R^a{}_{bcd}$ and R_{abcd} have a number of useful symmetries: $R^a{}_{b(cd)} = 0$, $R^a{}_{[bcd]} = 0$, $\nabla_{[n}R^a{}_{|b|cd]} = 0$, $R_{ab(cd)} = 0$, $R_{a[bcd]} = 0$, $R_{(ab)cd} = 0$, and $R_{abcd} = R_{cdab}$.

We define the *Ricci tensor* R_{ab} to be $R^c{}_{abc}$ and the *scalar curvature* R to be $R^a{}_a$. The *Einstein tensor* G_{ab} is then defined as $R_{ab} - \frac{1}{2}Rg_{ab}$. It plays a central role in what follows. One can verify that $\nabla^a G_{ab} = 0$.

We suppose that the entire matter content of the universe can be characterized by smooth tensor fields on M. For example, a source-free electromagnetic field is characterized by an anti-symmetric tensor F_{ab} on M, which satisfies Maxwell's equations: $\nabla_{[a}F_{bc]} = 0$, $\nabla^a F_{ab} = 0$. Other forms of matter, such as perfect fluids and Klein-Gordon fields, are characterized by other smooth tensor fields on M.

Associated with each matter field is a smooth, symmetric *energy-momentum tensor* T_{ab} on M. For example, the energy-momentum tensor T_{ab} associated with an electromagnetic field F_{ab} is $F_{an}F^n{}_b + \frac{1}{4}g_{ab}(F^{nm}F_{nm})$. Note that T_{ab} is a function not only of the matter field itself but also of the metric. Other matter fields, such as those mentioned above, will have their own energy-momentum tensors T_{ab}.

Fix a point $p \in M$. The quantity $T_{ab}\xi^a\xi^b$ at p represents the *energy density* of matter as given by an observer with tangent ξ^a at p. The quantity $T^a{}_b\xi^b - T_{nb}\xi^n\xi^b\xi^a$ at p represents the *spatial momentum density* as given by the same observer at p. We require that any energy-momentum tensor satisfy the *conservation*

---

[6] In what follows, square brackets denote anti-symmetrization. Parentheses denote symmetrization. See Malament (2012).

condition: $\nabla^a T_{ab} = 0$. Physically, this ensures that energy-momentum is locally conserved.

Finally, we come to *Einstein's equation*: $G_{ab} = 8\pi T_{ab}$.[7] It relates the curvature of spacetime with the matter content of the universe. In four dimensions, Einstein's equation can be expressed as $R_{ab} = 8\pi(T_{ab} - \frac{1}{2}Tg_{ab})$ where $T = T^a_a$.

Of course, any spacetime (M, g_{ab}) can be thought of as a *trivial solution* to Einstein's equation if T_{ab} is simply defined to be $\frac{1}{8\pi}G_{ab}$. Note that T_{ab} automatically satisfies the conservation condition, since $\nabla^a G_{ab} = 0$. But, in general, the energy-momentum tensor defined in this way will not be associated with any known matter field. However, if the T_{ab} so defined is also the energy-momentum tensor associated with a known matter field (or the sum of two or more energy momentum tensors associated with known matter fields) the spacetime is an *exact solution*. We say an exact solution is also a *vacuum solution* if $T_{ab} = 0$. And, in four dimensions, one can use the alternate version of Einstein's equation to show that $T_{ab} = 0$ if and only if $R_{ab} = 0$.

Between trivial and exact solutions, there are the *constraint solutions*. These are spacetimes whose associated energy-momentum tensors (defined via Einstein's equation) satisfy one or more conditions of interest. Here, we outline three. We say T_{ab} satisfies the *weak energy condition* if, for any future-directed unit timelike vector ξ^a at any point in M, the energy density $T_{ab}\xi^a\xi^b$ is not negative.

We say T_{ab} satisfies the *strong energy condition* if, for any future-directed unit timelike vector ξ^a at any point in M, the quantity $(T_{ab} - \frac{1}{2}Tg_{ab})\xi^a\xi^b$ is not negative. The strong energy condition can be interpreted as the requirement that a certain effective energy density is not negative. Note that, in four dimensions, the strong energy condition is satisfied if and only if the (timelike) *convergence condition*, $R_{ab}\xi^a\xi^b \geq 0$, is also satisfied. This latter condition can be understood to assert that gravitation is attractive in nature.

Finally, we say T_{ab} satisfies the *dominant energy condition* if, for any future-directed unit timelike vector ξ^a at any point in M, the vector $T^a_b\xi^b$ is causal and future-directed. This last condition can be interpreted as the requirement that matter cannot travel faster than light. Indeed, if T_{ab} vanishes on some closed, achronal set $S \subset M$ and satisfies the dominant energy and conservation conditions, then T_{ab} vanishes on all of $D(S)$ (Hawking and Ellis 1973). Clearly, the dominant energy condition implies (but is not implied by) the weak energy condition.

One can show that being a trivial, exact, or vacuum solution of Einstein's equation is a local spacetime property. In addition, being a constraint solution is also a local spacetime property if the constraint under consideration is one of the three energy conditions considered here.

---

[7] Here, we drop the cosmological constant term $-\Lambda g_{ab}$ sometimes added to the left side of the equation for some $\Lambda \in \mathbb{R}$. For more on this term, see Earman (2001).

3.2 Global Properties

A large number of important global properties concern either "causal structure" or "singularities." Here we investigate them.

There is a hierarchy of conditions relating to the causal structure of spacetime.[8] Each condition corresponds to a global spacetime property (the property of satisfying the condition). We say a spacetime satisfies the *chronology* condition if it contains no closed timelike curves (equivalently, $p \notin I^+(p)$ for all $p \in M$). A spacetime satisfies the *causality* condition if there are no closed causal curves (equivalently, $J^+(p) \cap J^-(p) = \{p\}$ for all $p \in M$). As mentioned previously, causality implies chronology but the implication does not run in the other direction (see figure 16.2). The next few conditions serve to rule out "almost" closed causal curves.

We say a spacetime (M, g_{ab}) satisfies the *future distinguishability* condition if there do not exist distinct points $p, q \in M$ such that $I^+(p) = I^+(q)$. The *past distinguishability* condition is defined analogously. One can show that a spacetime (M, g_{ab}) satisfies the future (respectively, past) distinguishability condition if and only if, for all points $p \in M$ and every open set O containing p, there is an open set $V \subset O$ also containing p such that no future (respectively, past) directed causal curve that starts at p and leaves V ever returns to V. We say a spacetime satisfies the *distinguishability* condition if it satisfies both the past and future distinguishability conditions.

Future or past distinguishability implies causality. But the converse is not true. Of course, distinguishability implies past (or future) distinguishability. But one can certainly find spacetimes that satisfy future (respectively, past) distinguishability but not past (respectively, future) distinguishability (Hawking and Ellis 1973).

Consider two distinguishing spacetimes (M, g_{ab}) and (M', g'_{ab}) and a bijection $\varphi : M \to M'$ such that for all $p, q \in M$, $p \in I^+(q)$ if and only if $\varphi(p) \in I^+(\varphi(q))$. One can show (Malament 1977) that φ is a diffeomorphism and $\varphi_*(g_{ab}) = \Omega^2 g'_{ab}$ for some conformal factor $\Omega : M' \to \mathbb{R}$. Thus, if the causal structure of spacetime is sufficiently well-behaved, that structure alone determines the shape of the universe, as well as the metric structure up to a conformal factor.

We say a spacetime satisfies the *strong causality* condition if, for all points $p \in M$ and every open set O containing p, there is an open set $V \subset O$ also containing p such that no causal curve intersects V more than once. If a spacetime (M, g_{ab}) satisfies strong causality, then, for every compact set $K \subset M$, a causal curve $\gamma : I \to K$ must have future and past endpoints in K. Thus, in a strongly causal spacetime, an inextendible causal curve cannot be "imprisoned" in a compact set. Clearly, strong causality implies distinguishability. One can show that the implication does not run in the other direction (Hawking and Ellis 1973).

A spacetime (M, g_{ab}) satisfies the *stable causality* condition if there is a timelike vector field ξ^a on M such that the spacetime $(M, g_{ab} + \xi_a \xi_b)$ satisfies the chronology

---

[8] Although we only consider a small handful here, there are an infinite number of conditions in the causal hierarchy (Carter 1971).

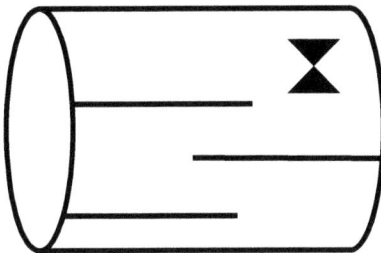

Figure 16.6 Cylindrical Minkowski spacetime with three horizontal lines removed as shown. The spacetime is strongly causal but not stably causal.

condition. Physically, even if the light cones are "opened" by a small amount at each point, the spacetime remains free of closed timelike curves. We say a spacetime (M, g_{ab}) admits a *global time function* if there is a smooth function $t: M \to \mathbb{R}$ such that, for any distinct points $p, q \in M$, if $p \in J^+(q)$, then $t(p) > t(q)$. The function assigns a "time" to every point in M such that it increases along every (nontrivial) future-directed causal curve. An important result is that a spacetime admits a global time function if and only if it satisfies stable causality (Hawking 1969). One can also show that stable causality implies strong causality but the converse is false (see figure 16.6).

The remaining causality conditions not only require that there be no almost closed causal curves but, in addition, that there be limitations on the kinds of "gaps" in spacetime (Hawking and Sachs 1974).

We say a spacetime (M, g_{ab}) satisfies the *causal continuity* condition if it satisfies distinguishability and, for all $p, q \in M$, $I^+(p) \subseteq I^+(q)$ if and only if $I^-(q) \subseteq I^-(p)$. Physically, causal continuity ensures that points that are close to one another do not have wildly different timelike futures and pasts. One can show that causal continuity implies stable causality. The converse is not true. A counterexample can be constructed by taking Minkowski spacetime and excising from the manifold a compact set with nonempty interior. The resulting spacetime satisfies stable causality but not causally continuity.

A spacetime (M, g_{ab}) satisfies the *causal simplicity* condition if it satisfies distinguishability and, in addition, for all $p \in M$, the sets $J^+(p)$ and $J^-(p)$ are closed. One can show that causal simplicity implies causal continuity. The converse is false, since Minkowski spacetime with a point removed from the manifold satisfies causal continuity but not causal simplicity.

Finally, we say a spacetime (M, g_{ab}) satisfies *global hyperbolicity* if it satisfies strong causality and, in addition, for all $p, q \in M$, the set $J^+(p) \cap J^-(q)$ is compact. A fundamental result is that a spacetime satisfies global hyperbolicity if and only if it admits a Cauchy surface (Geroch 1970b). In addition, one can show that the manifold of any spacetime that satisfies global hyperbolicity must have the topology of $\mathbb{R} \times \Sigma$ for any Cauchy surface Σ. Global hyperbolicity implies causal simplicity but the converse is not true. Anti-de Sitter spacetime is one counterexample (Hawking and Ellis 1973).

In sum, we have the following implications (none of which run in the other direction): global hyperbolicity ⇒ causal simplicity ⇒ causal continuity ⇒ stable causality ⇒ strong causality ⇒ distinguishability ⇒ future (or past) distinguishability ⇒ causality ⇒ chronology.

There are a number of senses in which a spacetime may be said to contain a "singularity."[9] Here, we restrict attention to the most important one: geodesic incompleteness. We say a geodesic $\gamma : I \to M$ is *incomplete* if it is maximal and such that $I \neq \mathbb{R}$. We say a future-directed maximal timelike or null geodesic $\gamma : I \to M$ is *future incomplete* (respectively, *past incomplete*) if there is a $r \in \mathbb{R}$ such that $r > s$ for all $s \in I$. A past incomplete geodesic is defined analogously.

Naturally, a spacetime is *timelike geodesically incomplete* if it contains a timelike incomplete geodesic. In a timelike geodesically incomplete spacetime, it is possible for a nonaccelerating massive particle to experience only a finite amount of time. We can define *spacelike* and *null geodesic incompleteness* analogously. Finally, we say that a spacetime is *geodesically incomplete* if it is either timelike, spacelike, or null geodesically incomplete.

If a spacetime has an extension, it is geodesically incomplete. The converse is false. Consider Minkowski spacetime (M, g_{ab}) and let M' be the manifold $M - \{p\}$ for any $p \in M$. Let $\Omega : M' \to \mathbb{R}$ be a conformal factor that approaches zero as the missing point p is approached. The resulting spacetime $(M', \Omega g_{ab|M'})$ is maximal but contains timelike, spacelike, and null incomplete geodesics. Other maximal spacetimes exist which are geodesically incomplete and have a flat metric.[10] In other words, one can have singularities without any spacetime curvature at all. Since there are certainly flat spacetimes that are geodesically complete (e.g., Minkowski spacetime), it follows that geodesic incompleteness is a global property. We mention in passing that the property of being maximal is also global.

Finally, one can show that timelike, spacelike, and null incompleteness are independent conditions in the sense that there are spacetimes that are incomplete in any one of the three types and complete in the other two (Geroch 1968). Additionally, one can show that compact spacetimes are not necessarily geodesically complete (Misner 1963). These two results suggest that geodesic incompleteness fails to mesh completely with our notion of a "hole" in spacetime.

4. Which Properties Are Reasonable?

So far, we have provided examples of a number of spacetime properties. In this section, we ask: Which properties are "physically reasonable"?

---

[9] See Ellis and Schmidt (1977), Geroch, Liang, and Wald (1982), Clarke (1993), and Curiel (1999) for details.
[10] Here is one example. Remove a point from $\mathbb{R}^2$ and take the universal covering space. Let the resulting spacetime manifold have a flat metric.

It is usually taken for granted that "the normal physical laws we determine in our spacetime vicinity are applicable at all other spacetime points" (Ellis 1975, 246). This assumption allows us to stipulate that the local property of being a solution to Einstein's equation is a physically reasonable one. And often this means that we take the energy conditions as necessarily satisfied. However, some have argued that even the energy conditions can be violated in some physically reasonable spacetimes (Vollick 1997).

One global property that is usually taken to be physically reasonable is that spacetime be maximal. Metaphysical considerations seem to drive the assumption. One asks (Geroch 1970a, 262), "Why, after all, would Nature stop building our universe... when She could just as well have carried on?" Of course, such reasoning can be questioned (Earman 1995, Norton (2011)).

What about the global properties concerning singularities and causal structure? Which of them are to be considered physically reasonable?

4.1 Singularities

Much of the work in global structure has concerned singularities. The task has been to show, using fairly conservative assumptions, that all physically reasonable spacetimes must be (null or timelike) geodesically incomplete. The project has produced a number of theorems of this type. Here, we examine an influential one due to Hawking and Penrose (1970).

Three preliminary conditions are crucial and each has been taken to be satisfied by all (or almost all) physically reasonable spacetimes. We shall temporarily adopt these background assumptions in what follows. The first is chronology (no closed timelike curves). The second is the convergence condition ($R_{ab}\xi^a\xi^a \geq 0$ for all unit timelike vectors ξ^a). Recall that the convergence condition is satisfied in four dimensions if and only if the strong energy condition is. In this section, we will restrict attention to four-dimensional spacetimes. The third is the *generic condition*—that each causal geodesic with tangent ξ^a contains a point at which $\xi_{[a}R_{b]cd[e}\xi_{f]}\xi^c\xi^d \neq 0$. Physically, the generic condition requires that somewhere along each causal curve a certain effective curvature is encountered. Although highly symmetric spacetimes may not satisfy the generic condition (e.g., Minkowski spacetime), it is thought to be satisfied by all sufficiently "generic" ones. Now, consider the following statement.

(S) Any spacetime that satisfies chronology, the convergence condition, the generic condition, and _____, must be timelike or null geodesically incomplete.

We seek to fill in the blank with physically reasonable "boundary" conditions that make (S) true. Hawking and Penrose (1970) considered three of them (see also Earman 1999).

First, if the boundary condition is the requirement that there exist a compact slice, (S) is true. So, a "spatially closed" universe is singular if it is physically reasonable. One can show that the existence of a compact slice is a necessary condition for predicting future spacetime events (Manchak 2008). Thus, we have the somewhat counterintuitive result that prediction is possible in a physically reasonable spacetime only if singularities are present.[11]

Second, (S) is true if the boundary condition is the requirement that there exist a trapped surface. A *trapped surface* is a two-dimensional compact spacelike surface T such that both sets of "ingoing" and "outgoing" future-directed null geodesics orthogonal to T have negative expansion at T.[12] Physically, whenever a sufficiently large amount of matter is contained in a small enough region of spacetime, a trapped surface forms (Schoen and Yau 1983).

Third, (S) is true if the boundary condition is the requirement that there is a point $p \in M$ such that the expansion along every future (or past) directed null geodesic through p is somewhere negative. Physically, a spacetime that satisfies this condition contains a contracting region in the causal future (or past) of a point. It is thought that the observable portion of our own universe contains such a region (Ellis 2007).

Additional examples of boundary conditions that make (S) true could be multiplied (Senovilla 1998). And instead of boundary conditions, one can also find causal conditions that make (S) true. We mention one here. It turns out that (S) is true if the causal condition is the requirement that stable causality is *not* satisfied (Minguzzi 2009). Thus, physically reasonable spacetimes (which are assumed to be causally well behaved in the sense that they satisfy chronology) are singular if they are not *too* causally well behaved. One naturally wonders if it is possible for physically reasonable spacetimes to avoid singularities if the chronology condition is dropped. But this seems unlikely (Tipler 1977; Kriele 1990).

A large number of physically reasonable spacetimes (including our own) seem to satisfy at least one of the above mentioned boundary conditions and hence contain singularities. And the worry has been that these singularities can be observed directly—that they are "naked" in some sense. So, one would like to show that all (or almost all) physically reasonable spacetimes do not contain naked singularities. This is the "cosmic censorship" hypothesis. There are a number of ways to formulate the hypothesis (Joshi 1993; Penrose 1999). Here, we consider one.

We do not wish to count a "big bang" singularity as naked and therefore restrict attention to future (rather than past) incomplete timelike or null geodesics. We say a spacetime (M, g_{ab}) is *nakedly singular* if there is a point $p \in M$ and a future incomplete timelike or null geodesic $\gamma : I \to M$ such that the range of γ is contained in $I^-(p)$ (see figure 16.7).

---

[11] For a related discussion, see Hogarth (1997).
[12] The (scalar) *expansion* of a congruence of null geodesics is a bit complicated to define (see Wald 1984). But one can get some idea of the quantity by noting that the expansion of a congruence of timelike geodesics with unit tangent field ξ^a is $\nabla_a \xi^a$.

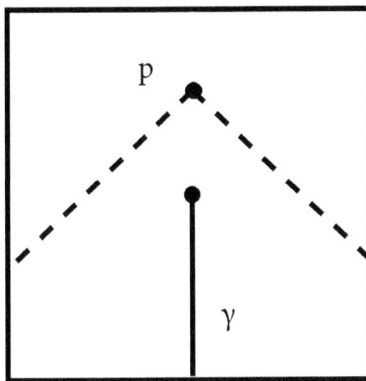

Figure 16.7 Minkowski spacetime with one point removed is nakedly singular. The future incomplete geodesic γ, contained in the timelike past of p, approaches the missing point.

One can show that a nakedly singular spacetime does not admit a Cauchy surface (Geroch and Horowitz 1979). Thus, if all physically reasonable spacetimes are globally hyperbolic, then the cosmic censorship hypothesis is true. And Penrose (1969, 1979) has suggested that one might be able to show the antecedent of this conditional. The idea would be to show that spacetimes that fail to be globally hyperbolic are unstable under certain types of perturbations. However, such a claim is difficult to express precisely (Geroch 1971). And although some evidence does seem to indicate that instabilities are present in nonglobally hyperbolic spacetimes (Chandrasekhar and Hartle 1982), still other evidence suggests otherwise (Morris, Thorne, and Yurtsever 1988).

There is also an epistemological predicament at issue. An observer never can have the evidential resources to rule out the possibility that his or her spacetime is not globally hyperbolic—even under any assumptions concerning local spacetime structure (Manchak 2011b). And how could we ever know that all physically reasonable spacetimes are globally hyperbolic if we cannot even be confident that our own spacetime is?

4.2 Time Travel

If the cosmic censorship hypothesis is false, there are physically reasonable spacetimes that do not satisfy global hyperbolicity. Might there be some physically reasonable spacetimes that do not even satisfy chronology? We investigate the question here.

One way to rule out a number of chronology-violating spacetimes concerns self-consistency constraints on matter fields of various types. Here, we examine source free Klein-Gordon fields. Let (M, g_{ab}) be a spacetime. We say an open set $U \subset M$ is *causally regular* if, for every function $\varphi : U \to \mathbb{R}$ which satisfies $\nabla^a \nabla_a \varphi = 0$, there is a function $\varphi' : M \to \mathbb{R}$ such that $\nabla^a \nabla_a \varphi' = 0$ and $\varphi'_{|U} = \varphi$. We say (M, g_{ab}) is *causally benign* if, for every $p \in M$ and every open set U containing p, there is an open set $U' \subset U$ containing p which is causally regular.

It has been argued that a spacetime that is not causally benign is not physically reasonable. We certainly know that every globally hyperbolic spacetime is causally benign. But although some chronology violating spacetimes are not causally benign, a number of others are (Yurtsever 1990; Friedman 2004).

Given the existence of causally benign yet chronology violating spacetimes, another area of research seems fruitful to pursue. One wonders if chronology violating region can, in some sense, be "created" by rearranging the distribution and flow of matter (Stein 1970). In other words, can a physically reasonable spacetime contain a "time machine" of sorts? Here, we examine one way of formalizing the question given by Earman, Smeenk, and Wüthrich (2009).[13]

First, in order to count as a time machine, a spacetime (M, g_{ab}) must contain a spacelike slice $S \subset M$ representing a "time" before the time machine is switched on. Second, the spacetime must also have a chronology violating region V after the machine is turned on. So we require $V \subset J^+[S]$. Finally, in order to capture the idea that a time machine must "create" a chronology violating region, every physically reasonable maximal extension of $int(D(S))$ must contain a chronology violating region V'.[14] Consider the following statement.

(T) There is a spacetime (M, g_{ab}) with a spacelike slice $S \subset M$ and a chronology violating region $V \subset J^+[S]$ such that every maximal extension of $int(D(S))$ which satisfies _____ contains some chronology violating region V'.

We seek to fill in the blank with physically reasonable "potency" conditions that make (T) true. And we know from counterexamples constructed by Krasnikov (2002) that (T) will be false unless there *is* a potency condition and this condition limits spacetime "holes" in some sense.

But Hawking (1992) has suggested that limiting holes may not be enough. Indeed, he conjectured that all physically reasonable spacetimes are "protected" from chronology violations and provided some evidence for the claim. We say $H^+(S)$ is *compactly generated* if all past directed null geodesics through $H^+(S)$ enter and remain in some compact set. Any spacetime with a slice S such that $H^+(S)$ is nonempty and compactly generated does not satisfy strong causality. And Hawking showed there is no spacetime that satisfies the weak energy condition which has a noncompact slice S such that $H^+(S)$ is nonempty and compactly generated.

But some have argued that insisting on a compactly generated Cauchy horizon rules out some physically reasonable spacetimes (Ori 1993; Krasnikov 1999). And, of course, a slice S need not be noncompact to be physically reasonable. Thus, Hawking's chronology protection conjecture remains an open question.

---

[13] See also Earman and Wüthrich (2010) and Smeenk and Wüthrich (2011).
[14] Here we abuse the notation somewhat. Properly, we require that every physically reasonable maximal extension of $(int(D(S)), g_{ab}|_{int(D(S))})$ must contain a chronology violating region V'.

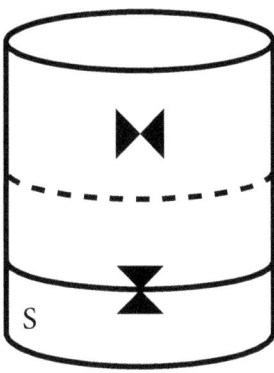

Figure 16.8 Misner spacetime. Every maximal, hole-free extension of *int*(*D*(*S*)) (the region below the dotted line) contains some chronology violating region.

Are there any potency conditions that make (T) true? We say a spacetime (M, g_{ab}) is *hole-free* if, for any spacelike surface S in M there is no isometric embedding $\theta : D(S) \to M'$ into another spacetime (M', g'_{ab}) such that $\theta(D(S)) \neq D(\theta(S))$. Physically, hole-freeness ensures that, for any spacelike surface S, the domain of dependence $D(S)$ is "as large as it can be." And one can show that any spacetime with one point removed from the underlying manifold fails to be hole-free. It has been argued that all physically reasonable spacetimes are hole-free (Clarke 1976; Geroch 1977). And it turns out that (T) is true if the potency condition is hole-freeness (Manchak 2009b). The two-dimensional spacetime of Misner (1967) can be used to prove the result (see figure 16.8).

However, hole-freeness may not be a physically reasonable potency condition after all. Indeed, some maximal, globally hyperbolic models, including Minkowski spacetime, are not hole-free (Manchak 2009a; Krasnikov 2009). But, another more reasonable "no holes" potency condition can be used to make (T) true: the demand that, for all $p \in M$, $J^+(p)$ and $J^-(p)$ are closed (Manchak 2011a). Call this condition *causal closedness* and recall that causal closedness is used, along with distinguishability, to define causal simplicity.

Not only is causal closedness satisfied by all globally hyperbolic models, including Minkowski spacetime, but it is also satisfied by many chronology violating spacetimes as well (e.g., Gödel spacetime, Misner spacetime). In this sense, then, it is a more appropriate condition than hole-freeness. But is causal closedness satisfied by all physically reasonable spacetimes? The question is open. So too is the question of which other potency conditions make (T) true.

5. Conclusion

Here, we have outlined the basic structure of relativistic spacetime. As we have seen, general relativity allows for a wide variety of global spacetime properties—some of them quite unusual. And one wonders which of these properties are physically reasonable.

Early work focused on singularities. Initially, a number of results established that all physically reasonable spacetimes are geodesically incomplete. Next, the relationship between these singularities and determinism was investigated: Can a physically reasonable (and therefore geodesically incomplete) spacetime fail to be globally hyperbolic? The question remains open.

Recently, focus has shifted somewhat toward acausality: Can physically reasonable spacetimes contain closed timelike curves? If so, can these closed timelike curves be "created" in some sense by rearranging the distribution and flow of matter? Again, these questions remain open.

References

Carter, B. (1971). Causal structure in spacetime. *General Relativity and Gravitation* 1: 349–391.
Chandrasekhar, S., and J. Hartle (1982). On crossing the Cauchy horizon of a Reissner–Nordström black–hole. *Proceedings of the Royal Society (London) A*. 348: 301–315.
Choquet–Bruhat, Y., and R. Geroch (1969). Global aspects of the Cauchy problem in general relativity. *Communications in Mathematical Physics* 14: 329–335.
Clarke, C. (1976). Spacetime singularities. *Communications in Mathematical Physics*. 49: 17–23.
——— (1993). *The analysis of spacetime singularities*. Cambridge: Cambridge University Press.
Curiel, E. (1999). The analysis of singular spacetimes. *Philosophy of Science*. 66: S119–S145.
Earman, J. (1995). *Bangs, crunches, whimpers, and shrieks: Singularities and acausalities in relativistic spacetimes*. Oxford: Oxford University Press.
——— (1999). The Penrose–Hawking singularity theorems: History and implications. In *The expanding worlds of general relativity*, Einstein Studies, ed: H. Goerher, J. Renn, and T. Sauer, Vol. 7, Boston: Birkhäuser. 235–267.
——— (2001). Lambda: The constant that refuses to die, *Archives for History of Exact Sciences*, 51: 189–220.
——— (2002). What time reversal invariance is and why it matters, *International Journal for the Philosophy of Science*, 16: 245–264.
——— (2008). Pruning some branches from 'Branching Spacetimes', in *The ontology of spacetime II*, ed. D. Dieks, 187–205, Amsterdam: Elsevier.
Earman, J., and C. Wüthrich (2010). Time Machines. In *Stanford Encyclopedia of Philosophy*, ed. E. Zalta. http: //plato.stanford.edu/entries/time–machine/.
Earman, J., C. Smeenk, and C. Wüthrich, (2009). Do the laws of physics forbid the operation of time machines? *Synthese* 169: 91–124.
Ellis, G. (1975). Cosmology and verifiability. *Quarterly Journal of the Royal Astronomical Society*, 16: 245–264.
——— (2007). Issues in the philosophy of cosmology. In *Handbook of the philosophy of physics*, ed. J. Butterfield and J. Earman, 1183–1286. Oxford: Elsevier.

Ellis, G., and B. Schmidt (1977). Singular spacetimes. *General Relativity and Gravitation* 8: 915–953.

Friedman, J. (2004). The Cauchy problem on spacetimes that are not globally hyperbolic. In *The Einstein equations and the large scale behavior of gravitational fields*, ed. P. Chrusciel and H. Friedrich, 331–346. Boston: Birkhäuser,

Geroch, R. (1967). Topology in general relativity. *Journal of Mathematical Physics* 8: 782–786.

——— (1968). What is a singularity in general relativity? *Annals of Physics* 48: 526–540.

——— (1970a). Singularities. In *Relativity*, ed. M. Carmeli, S. Fickler, and L. Witten, 259–291. New York: Plenum Press.

——— (1970b). Domain of dependence., *Journal of Mathematical Physics* 11: 437–449.

——— (1971). General relativity in the large. *General Relativity and Gravitation*. 2: 61–74.

——— (1977). Prediction in general relativity. In *Foundations of spacetime theories*, Minnesota Studies in the Philosophy of Science Vol. 8, ed. J. Earman, C. Glymour, and J. Stachel, 81–93. Minneapolis: University of Minnesota Press.

Geroch, R., and G. Horowitz (1979). Global structure of spacetimes. In *General Relativity: An Einstein Centenary Survey*, ed. S. W. Hawking and W. Israel, 212–293. Cambridge: Cambridge University Press.

Geroch, R., C. Liang, and R. Wald (1982). Singular boundaries of spacetime. *Journal of Mathematical Physics*. 23: 432–435.

Gödel, K. (1949). An example of a new type of cosmological solutions of Einstein's field equations of gravitation. *Reviews of Modern Physics* 21: 447–450.

Hawking, S. (1969). The existence of cosmic time functions. *Proceedings of the Royal Society A* 308: 433–435.

——— (1992). The chronology protection conjecture. *Physical Review D* 46: 603–611.

Hawking, S., and G. Ellis (1973). *The large scale structure of spacetime*. Cambridge: Cambridge University Press.

Hawking, S., and R. Penrose, (1970). The singularities of gravitational collapse and cosmology. *Proceedings of the Royal Society of London A* 314: 529–548.

Hawking, S., and B. Sachs (1974). Causally continuous spacetimes. *Communications in Mathematical Physics* 35: 287–296.

Hogarth, M. (1997). A remark concerning prediction and spacetime singularities. *Studies in History and Philosophy of Modern Physics* 28: 63–71.

Joshi, P. (1993). *Global aspects in gravitation and cosmology*. Oxford: Oxford University Press.

Krasnikov, S. (1999). Time machines with non-compactly generated cauchy horizons and "Handy Singularities". In *Proceedings of the Eighth Marcel Grossmann Meeting on General Relativity*, ed. T. Piran and R. Ruffini, 593–595. Singapore: World Scientific.

——— (2002). No time machines in classical general relativity. *Classical and Quantum Gravity* 19: 4109–4129.

——— (2009). Even the minkowski space is holed. *Physical Review D* 79: 124041.

Kriele, M. (1990). Causality violations and singularities. *General Relativity and Gravitation*, 22: 619–623.

Malament, D. (1977). The class of continuous timelike curves determines the topology of spacetime. *Journal of Mathematical Physics* 18: 1399–1404.

——— (2012). *Topics in the foundations of general relativity and Newtonian gravitation theory*. Chicago: University of Chicago Press.

Manchak, J. (2008). Is prediction possible in general relativity? *Foundations of Physics*, 38: 317–321.

——— (2009a). Is spacetime hole-free? *General Relativity and Gravitation* 41: 1639–1643.
——— (2009b). On the existence of "time machines" in general relativity. *Philosophy of Science* 76: 1020–1026.
——— (2011a). No no-go: A remark on time machines. *Studies in History and Philosophy of Modern Physics.* 42: 74–76.
——— (2011b). What is a physically reasonable spacetime? *Philosophy of Science* 78: 410–420.
Minguzzi, E. (2009). Chronological spacetimes without lightlike lines are stably causal. *Communications in Mathematical Physics* 288: 801–819.
Misner, C. (1963). The flatter regions of Newman, Unti, and Tamburino's generalized Schwarzschild space. *Journal of Mathematical Physics* 4: 924–937.
——— (1967). Taub–NUT space as a counterexample to almost anything. In *Relativity theory and astrophysics I: relativity and cosmology*, ed. J. Ehlers, 160–169. Providence: American Mathematical Society.
Morris, M., K. Thorne, and U. Yurtsever, (1988). Wormholes, time machines, and the weak energy condition. *Physical Review Letters* 61: 1446–1449.
Norton, J. (2011), "Observationally indistinguishable Spacetimes: A challenge for any inductivist," in G. Morgan, Philosophy of Science Matters. Oxford: Oxford University Press, p. 164–176.
Ori, A. (1993). Must time-machine construction violate the weak energy condition? *Physical Review Letters* 71: 2517–2520.
Penrose, R. (1969). Gravitational collapse: The role of general relativity. *Revisita del Nuovo Cimento*, Serie I, 1: 252–276.
——— (1979). Singularities and time-asymmetry. In *General Relativity: An Einstein Centenary Survey*, ed. S. Hawking and W. Israel, 581–638. Cambridge: Cambridge University Press.
——— (1999). The question of cosmic censorship. *Journal of Astrophysics and Astronomy* 20: 233–248.
Senovilla, J. (1998). Singularity theorems and their consequences. *General Relativity and Gravitation* 30: 701–848.
Schoen, R., and S. Yau (1983). The existence of a black hole due to condensation of matter. *Communications in Mathematical Physics* 90: 575–579.
Smeenk, C., and C. Wüthrich (2011). Time travel and time machines. In *The Oxford Handbook of Time*, ed. C. Callender, 577–630. Oxford: Oxford University Press.
Stein, H. (1970). On the paradoxical time-structures of Gödel. *Philosophy of Science* 37: 589–601.
Tipler, F. (1977). Singularities and causality violations. *Annals of Physics* 108: 1–36.
Vollick, D. (1997). How to produce exotic matter using classical fields. *Physical Review D* 56: 4720–4723.
Wald, R. (1984). *General relativity*. Chicago: University of Chicago Press.
Yurtsever, U. (1990). Test fields on compact spacetimes. *Journal of Mathematical Physics* 31: 3064–3078.

CHAPTER 17

PHILOSOPHY OF COSMOLOGY

CHRIS SMEENK

1. Introduction

Cosmology has made enormous progress in the last several decades. It is no longer a neglected subfield of physics, as it was as recently as 1960; it is instead an active area of fundamental research that can boast of a Standard Model well-supported by observations. Prior to 1965 research in cosmology had a strikingly philosophical tone, with debates focusing explicitly on scientific method and the aims and scope of cosmology (see, e.g., Munitz 1962; North 1965; Kragh 1996). One might suspect that with the maturation of the field these questions have been settled, leaving little room for philosophers to contribute. Although the nature of the field has changed dramatically with an increase of observational knowledge and theoretical sophistication, there are still ongoing foundational debates regarding cosmology's proper aims and methods. Cosmology confronts a number of questions dear to the hearts of philosophers of science: the limits of scientific explanation, the nature of physical laws, and different types of underdetermination, for example. There is an opportunity for philosophers to make fruitful contributions to debates in cosmology and to consider the ramifications of new ideas in cosmology for other areas of philosophy and foundations of physics.

Due to the uniqueness of the universe and its inaccessibility, cosmology has often been characterized as "unscientific" or inherently more speculative than other parts of physics. How can one formulate a scientific theory of the "universe as a whole"?

This essay is dedicated to the memory of Ernan McMullin, whose contributions shaped this and many other areas of philosophy of science, and who generously encouraged my work. For discussion and comments on earlier versions of this paper, I am grateful to Jeremy Butterfield, ErikCuriel, John Earman, David Malament, John Manchak, and Henrik Zinkernagel. Thanks to Nic Fillion for producing the figures. I am particularly indebted to Bob Batterman for his guidance and patience.

Even those who reject skepticism regarding cosmology often assert instead that cosmology can only make progress by employing a distinctive methodology. These discussions, in my view, have by and large failed to identify the source and the extent of the evidential challenges faced by cosmologists. There are no convincing, general no-go arguments showing the impossibility of secure knowledge in cosmology; there are instead specific problems that arise in attempting to gain observational and theoretical access to the universe. In some cases, cosmologists have achieved knowledge as secure as that in other areas of physics—arguably, for example, in the account of big bang nucleosynthesis.

Cosmologists do, however, face a number of distinctive challenges. The finitude of the speed of light, a basic feature of relativistic cosmology, insures that global properties of the universe cannot be established directly by observations (section 5). This is a straightforward limit on observational access to the universe, but there are other obstacles of a different kind. Cosmology relies on extrapolating local physical laws to hold universally. These extrapolations make it possible to infer, from observations of standard candles such as Type Ia supernovae,[1] the startling conclusion that the universe includes a vast amount of dark matter and dark energy. Yet the inference relies on extrapolating general relativity (GR), and the observations may reveal the need for a new gravitational theory rather than new types of matter. It is difficult to adjudicate this debate due to the lack of independent access to the phenomena (section 3). The early universe (section 6) is interesting because it is one of the few testing grounds for quantum gravity. Without a clear understanding of the initial state derived from such a theory, however, it is difficult to use observations to infer the dynamics governing the earliest stages of the universe's evolution. Finally, it is not clear how to take the selection effect of our presence as observers into account in assessing evidence for cosmological theories (section 7).

These challenges derive from distinctive features of cosmology. One such feature is the interplay between global aspects of the universe and local dynamical laws. The Standard Model of cosmology is based on extrapolating local laws to the universe as a whole. Yet, there may be global-to-local constraints. The uniqueness of the universe implies that the normal ways of thinking about laws of physics and the contrast between laws and initial conditions do not apply straightforwardly (section 4). In other areas of physics, the initial or boundary conditions themselves are typically used to explain other things rather than being the target of explanation. Many lines of research in contemporary cosmology aim to explain why the initial state of the Standard Model obtained, but the nature of this explanatory project is not entirely clear. And due to the uniqueness of the universe and the possibility of anthropic selection effects it is not clear what underwrites the assignment of probabilities.

What follows is not a survey of a thoroughly explored field in philosophy of physics. There are a variety of topics in this area that philosophers could fruitfully study, but as of yet the potential for philosophical work has not been fully realized.[2]

---

[1] A "standard candle" is an object whose intrinsic luminosity can be determined; the observed apparent magnitude then provides an accurate measurement of the distance to the object.

[2] This is not to say that there is no literature on the topic, and much of it will be cited below. For more systematic reviews of the literature by someone whose contributions have shaped the field, and which I draw on in the following, see Ellis (1999, 2007).

The leading contributions have come primarily from cosmologists who have turned to philosophical considerations arising from their work. The literature has a number of detailed discussions of specific issues, but there are few attempts at a more systematic approach. As a result, this essay is an idiosyncratic tour of various topics and arguments rather than a survey of a well-charted intellectual landscape. It is also a limited tour and leaves out a variety of important issues—most significantly, the impact of quantum mechanics on issues ranging from the origin of density perturbations in the early universe to the possible connections between Everettian and cosmological multiverses. But I hope that despite these limitations, this survey may nonetheless encourage other philosophers to actualize the potential for contributions to foundational debates within cosmology.

2. Overview of the Standard Model

Since the early 1970s cosmology has been based on what Weinberg (1972) dubbed the "Standard Model." This model describes the universe's spacetime geometry, material constituents, and their dynamical evolution. The Standard Model is based on extending local physics—including general relativity, quantum physics, and statistical physics—to cosmological scales and to the universe as a whole. A satisfactory cosmological model should be sufficiently rich to allow one to fix basic observational relations, and to account for various striking features of the universe, such as the existence of structures like stars and galaxies, as consequences of the underlying dynamics.

The Standard Model describes the universe as starting from an extremely high-temperature early state (the "big bang") and then expanding, cooling, and developing structures such as stars and galaxies. At the largest scales the universe's spacetime geometry is represented by the expanding universe models of general relativity. The early universe is assumed to begin with matter and radiation in local thermal equilibrium, with the stress-energy dominated by photons. As the universe expands, different types of particles "freeze out" of equilibrium, leaving an observable signature of earlier stages of evolution. Large-scale structures in the universe, such as galaxies and clusters of galaxies, arise later via gravitational clumping from initial "seeds." Here I will give a brief sketch of the Standard Model to provide the necessary background for the ensuing discussion.[3]

2.1 Expanding Universe Models

Einstein (1917) introduced a strikingly new conception of cosmology, as the study of exact solutions of general relativity that describe the spacetime geometry of

---

[3] Of the several textbooks that cover this territory, see in particular Peebles (1993); Dodelson (2003); Weinberg (2008); see Longair (2006) for a masterful historical survey of the development of cosmology and astrophysics.

the universe. One would expect gravity to be the dominant force in shaping the universe's structure at large scales, and it is natural to look for solutions of Einstein's field equations (EFE) compatible with astronomical observations. Einstein's own motivation for taking the first step in relativistic cosmology was to vindicate Mach's principle.[4] He also sought a solution that describes a static universe, that is, one whose spatial geometry is unchanging. He forced his theory to accommodate a static model by modifying his original field equations, with the addition of the infamous cosmological constant Λ. As a result Einstein missed one of the most profound implications of his new theory: general relativity quite naturally implies that the universe evolves dynamically with time. Four of Einstein's contemporaries discovered a class of simple evolving models, the Friedman-Lemaître-Robertson-Walker (FLRW) models, that have proven remarkably useful in representing the spacetime geometry of our universe.

These models follow from symmetry assumptions that dramatically simplify the task of solving EFE. They require that the spacetime geometry is both *homogeneous* and *isotropic*; this is also called imposing the "cosmological principle." Roughly speaking, homogeneity requires that at a given moment of cosmic time every spatial point "looks the same," and isotropy holds if there are no geometrically preferred spatial directions. These requirements imply that the models are topologically $\Sigma \times \mathfrak{R}$, visualizable as a "stack" of three-dimensional spatial surfaces $\Sigma(t)$ labeled by values of the cosmic time t. The worldlines of "fundamental observers," taken to be at rest with respect to matter, are orthogonal to these surfaces, and the cosmic time corresponds to the proper time measured by the fundamental observers. The spatial geometry of Σ is such that there is an isometry carrying any point $p \in \Sigma$ to any other point lying in the same surface (homogeneity), and at any point p the three spatial directions are isometric (isotropy).[5]

The cosmological principle tightly constrains the properties of the surfaces $\Sigma(t)$. These are three-dimensional spaces (Riemannian manifolds) of constant curvature, and all of the surfaces in a given solution have the same topology. If the surfaces are simply connected, there are only three possibilities for Σ: (1) spherical space, for the case of positive curvature; (2) Euclidean space, for zero curvature; and (3) hyperbolic space, for negative curvature.[6] Textbook treatments often neglect to mention, however, that replacing *global* isotropy and homogeneity with *local* analogs opens the door to a number of other possibilities. For example,

---

[4] At the time, Einstein formulated Mach's principle as the requirement that inertia derives from interactions with other bodies rather than from a fixed background spacetime. His model eliminated the need for anti-Machian boundary conditions by eliminating boundaries: it describes a universe with spatial sections of finite volume, without edges. See Smeenk (2012) for further discussion.

[5] An isometry is a transformation that preserves the spacetime geometry; more precisely, a diffeomorphism ϕ that leaves the spacetime metric invariant, i.e., $(\phi^* g)_{ab} = g_{ab}$.

[6] A topological space is *simply connected* if, roughly speaking, every closed loop can be smoothly contracted to a point. For example, the surface of a bagel is multiply connected, as there are two different types of loops that cannot be continuous deformed to a point. There is another possibility for a globally isotropic space with constant positive curvature that is multiply connected, namely projective space (with the same metric as spherical space but a different topology). These three possibilities are unique up to isometry. See, e.g., Wolf (2011), for a detailed discussion.

there are models in which the surfaces Σ have finite volume but are multiply connected, consisting of, roughly speaking, cells pasted together.[7] Although isotropy and homogeneity hold locally at each point, above some length scale there would be geometrically preferred directions reflecting how the cells are connected. In these models it is in principle possible to see "around the universe" and observe multiple images of a single object, but there is at present no strong observational evidence of such effects.

Imposing global isotropy and homogeneity reduces EFE—a set of 10 nonlinear, coupled partial differential equations—to a pair of differential equations governing the scale factor $R(t)$ and $\rho(t)$, the energy density of matter. The scale factor measures the changing spatial distance between fundamental observers. The dynamics are then captured by the Friedmann equation:[8]

$$\left(\frac{\dot{R}}{R}\right)^2 = \frac{8\pi G \rho}{3} - \frac{k}{R^2} + \frac{\Lambda}{3}, \tag{1}$$

and (a special case of) the Raychaudhuri equation:

$$3\frac{\ddot{R}}{R} = -4\pi G (\rho + 3p) + \Lambda. \tag{2}$$

$\dot{R}$ means differentiation with respect to the cosmic time t, G is Newton's gravitational constant, and Λ is the cosmological constant. The curvature of surfaces $\Sigma(t)$ of constant cosmic time is given by $\frac{k}{R^2(t)}$, where $k = \{-1, 0, 1\}$ for negative, flat, and positive curvature (respectively). The assumed symmetries force the matter to be described as a perfect fluid with energy density ρ and pressure p.[9] The energy density and pressure are given by the equation of state for different kinds of perfect fluids; for example, for "pressureless dust" $p = 0$, whereas for radiation $p = \rho/3$. Given a specification of the matter content, there exist unique solutions for the scale factor $R(t)$ and the energy density $\rho(t)$ for each type of matter included in the model.

Several features of the dynamics of these models are clear from inspection of these equations. Suppose we take "ordinary" matter to always have positive total stress-energy density, in the sense of requiring that $\rho + 3p > 0$. Then, from (2), it is clear that the effect of such ordinary matter is to decelerate cosmic expansion, $\ddot{R} < 0$—reflecting the familiar fact that gravity is a force of attraction. But this is only so for ordinary matter. A positive cosmological constant (or matter with negative

---

[7] See Ellis (1971) for a pioneering study of this kind of model, and Lachieze-Rey and Luminet (1995) for a more recent review.

[8] EFE are: $G_{ab} + \Lambda g_{ab} = 8\pi T_{ab}$, where G_{ab} is the Einstein tensor, T_{ab} is the stress-energy tensor, g_{ab} is the metric, and Λ is the cosmological constant. Equation (1) follows from the "time-time" component of EFE, and equation (2) is the difference between it and the "space-space" component. (All other components vanish due to the symmetries.) The Raychaudhuri equation is a fundamental equation that describes the evolution of a cluster of nearby worldlines, e.g., for the particles making up a small ball of dust, in response to curvature. It takes on the simple form given here due to the symmetries we have assumed: in the FLRW models the small ball of dust can change only its volume as a function of time, but in general there can be a volume-preserving distortion (shear) and torsion (rotation) of the ball as well.

[9] The stress energy tensor for a perfect fluid is given by $T_{ab} = (\rho + p)\xi_a \xi_b + (p)g_{ab}$, where ξ_a is the tangent vector to the trajectories of the fluid elements.

stress-energy) leads, conversely, to accelerating expansion, $\ddot{R} > 0$. Einstein satisfied his preference for a static model by choosing a value of Λ that precisely balances the effect of ordinary matter, such that $\ddot{R} = 0$. But his solution is unstable, in that a slight concentration (deficit) of ordinary matter triggers run-away contraction (expansion). It is difficult to avoid dynamically evolving cosmological models in general relativity.

Restricting consideration to ordinary matter and setting $\Lambda = 0$, the solutions fall into three types depending on the relative magnitude of two terms on the right-hand side of Eq. (1), representing the effects of energy density and curvature. For the case of flat spatial geometry $k = 0$, the energy density takes exactly the value needed to counteract the initial velocity of expansion such that $\dot{R} \to 0$ as $t \to \infty$. This solution separates the two other classes: if the energy density is greater than critical, there is sufficient gravitational attraction to reverse the initial expansion, and the spatial slices Σ have spherical geometry ($k = +1$); if the energy density is less than critical, the sign of $\dot{R}$ never changes, expansion never stops, and the spatial slices have hyperbolic geometry ($k = -1$).[10] This simple picture does not hold if $\Lambda \neq 0$, as the behavior then depends on the relative magnitude of the cosmological constant term and ordinary matter.

The equations above lead to simple solutions for $R(t)$ for models including a single type of matter: for electromagnetic radiation, $R(t) \propto t^{1/2}$; for pressureless dust $R(t) \propto t^{2/3}$; and for a cosmological constant, $R(t) \propto e^t$.[11] Obviously, more realistic models include several types of matter. The energy density for different types of matter dilutes with expansion at different rates: pressureless dust—$\rho(t) \propto R^{-3}$; radiation—$\rho(t) \propto R^{-4}$; and a cosmological constant remains constant. As a result of these different dilution rates, a complicated model can be treated in terms of a sequence of simple models describing the effects of the dominant type of matter on cosmic evolution. At $t \approx 1$ second, the Standard Model describes the universe as filled with matter and radiation, where the latter initially has much higher energy density. Because the energy density of radiation dilutes more rapidly than that of matter, the initial radiation-dominated phase is followed by a matter-dominated phase that extends until the present. Current observations indicate the presence of "dark energy" (discussed in more detail below) with properties like a Λ term. Supposing these are correct, in the future the universe will eventually transition to a dark-energy-dominated phase of exponential expansion, given that the energy density of a Λ term does not dilute at all with expansion.

FLRW models with ordinary matter have a singularity at a finite time in the past. Extrapolating back in time, given that the universe is currently expanding, Eq. (2) implies that the expansion began at some finite time in the past. The current

---

[10] These are both classes of solutions, where members of the class have spatial sections with curvature of the same sign but different values of the spatial curvature at a given cosmic time.

[11] One can treat the cosmological constant as a distinctive type of matter, in effect moving it from the left to the right side of EFE and treating it as a component of the stress-energy tensor. It can be viewed instead as properly included on the left-hand side as part of the spacetime geometry. This issue of interpretation does not, however, make a difference with regard to the behavior of the solution.

rate of expansion is given by the Hubble parameter, $H = \frac{\dot{R}}{R}$. Simply extrapolating this expansion rate backward, $R(t) \to 0$ at the Hubble time H^{-1}; from Eq. (2) the expansion rate must increase at earlier times, so $R(t) \to 0$ at a time less than the Hubble time before now. As this "big bang" is approached, the energy density and curvature increase without bound. This reflects the instability of evolution governed by EFE: as $R(t)$ decreases, the energy density and pressure both increase, and they both appear with the same sign on the right-hand side of Eq. (2). It was initially hoped that the singularity could be avoided in more realistic models that are not perfectly homogeneous and isotropic, but Penrose, Hawking, and Geroch showed in the 1960s that singularities hold quite generically in models suitable for cosmology. It is essential for this line of argument that the model includes ordinary matter and no cosmological constant; since the Λ term appears in Eq. (2) with the opposite sign, one can avoid the initial singularity by including a cosmological constant (or matter with a negative stress-energy).

One of the most remarkable discoveries in twentieth-century astronomy was Hubble's (1929) observation that the red-shifts of spectral lines in galaxies increase linearly with their distance.[12] Hubble took this to show that the universe is expanding uniformly, and this effect can be given a straightforward qualitative explanation in the FLRW models. The FLRW models predict a change in frequency of light from distant objects that depends directly on $R(t)$.[13] There is an approximately linear relationship between red-shift and distance at small scales for all the FLRW models, and departures from linearity at larger scales can be used to measure spatial curvature.

At the length scales of galaxies and clusters of galaxies, the universe is anything but homogeneous and isotropic, and the use of the FLRW models involves a (usually implicit) claim that above some length scale the average matter distribution is sufficiently uniform. By hypothesis the models do not describe the formation and evolution of inhomogeneities that give rise to galaxies and other structures. Prior to 1965, the use of the models was typically justified on the grounds of mathematical utility or an argument in favor of the cosmological principle, with no expectation that the models were in more than qualitative agreement with observations— especially when extrapolated to early times. The situation changed dramatically with the discovery that the FLRW models provide an extremely accurate description of the early universe, as revealed by the uniformity of the cosmic background

---

[12] Hubble's distance estimates have since been rejected, leading to a drastic decrease in the estimate of the current rate of expansion (the Hubble parameter, H_0). However, the linear redshift-distance relation has withstood scrutiny as the sample size has increased from 24 bright galaxies (in Hubble 1929) to hundreds of galaxies at distances 100 times greater than Hubble's, and as astrophysicists have developed other observational methods for testing the relation (see Peebles 1993, 82–93 for an overview).

[13] The problem is underspecified without some stipulation regarding the worldlines traversed by the observers emitting and receiving the signal. Assuming that both observers are fundamental observers, a photon with frequency ω emitted at a cosmic time t_1 will be measured to have a frequency $\omega' = \frac{R(t_1)}{R(t_2)}\omega$ at a later time t_2. (For an expanding universe, this leads to a red-shift of the light emitted.) Given a particular solution one can calculate the exact relationship between spectral shift and distance.

radiation (CBR, described below). The need to explain why the universe is so strikingly symmetric was a driving force for research in early universe cosmology (see section 6 below).

2.2 Thermal History

Alvy Singer's mother in *Annie Hall* is right: Brooklyn is not expanding. But this is not because the cosmic expansion is not real or has no physical effects. Rather, in the case of gravitationally bound systems such as the Earth or the solar system the effects of cosmic expansion are far, far too small to detect.[14] In many domains the cosmological expansion can be ignored. The dynamical effects of expansion are, however, the central theme in the Standard Model's account of the thermal history of the early universe.

Consider a given volume of the universe at an early time, filled with matter and radiation assumed to be initially in local thermal equilibrium.[15] The dynamical effects of the evolution of $R(t)$ are locally the same as slowly varying the volume of this region, imagining that the matter and radiation are enclosed in a box that expands (or contracts) adiabatically. For some stages of evolution the contents of the box interact on a sufficiently short timescale that equilibrium is maintained through the change of volume, which then approximates a quasi-static process. When the interaction timescale becomes greater than the expansion timescale, however, the volume changes too fast for the interaction to maintain equilibrium. This leads to a departure from equilibrium; particle species "freeze out" and decouple, and entropy increases. Without a series of departures from equilibrium, cosmology would be a boring subject—the system would remain in equilibrium with a state determined solely by the temperature, without a trace of things past. Departures from equilibrium are of central importance in understanding the universe's thermal history.[16]

---

[14] Quantitatively estimating the dynamical effects of the expansion on local systems is remarkably difficult. One approach is, schematically, to imbed a solution for a local system (such as a Schwarzschild solution) into an FLRW spacetime, taking care to impose appropriate junction conditions on the boundary. One can then calculate an upper bound on the effect of the cosmological expansion; the effect will presumably be smaller in a more realistic model, which includes a hierarchy of imbedded solutions representing structures at larger length scales such as the galaxy and the Local Group of galaxies. Because of the nonlinearity of EFE it is surprisingly subtle to make the idea of a "quasi-isolated" system immersed in a background cosmological model precise, and to differentiate effects due to the expansion from those due to changes within the local system (such as growing inhomogeneity). See Carrera and Giulini (2010) for a recent systematic treatment of these issues.

[15] This assumption of local thermal equilibrium as an "initial state" at a given time presumes that the interaction timescales are much less than the expansion timescale at earlier times.

[16] The departures from equilibrium are described using the Boltzmann equation. The Boltzmann equation formulated in an FLRW spacetime includes an expansion term. As long as the collision term (for some collection of interacting particles) dominates over the expansion term then the interactions are sufficient to maintain equilibrium, but as the universe cools, the collision term becomes subdominant to the expansion term, and the particles decouple from the plasma and fall out of equilibrium. To find the number density at the end of this freezing out process, one typically has to solve a differential equation (or a coupled set of differential equations for multiple particle species) derived from the Boltzmann equation.

Two particularly important cases are big bang nucleosynthesis and the decoupling of radiation from matter. The Standard Model describes the synthesis of light elements as occurring during a burst of nuclear interactions that transpire as the universe falls from a temperature of roughly 10^9 K, at $t \approx 3$ minutes, to 10^8 K, at about 20 minutes.[17] Prior to this interval, any deuterium formed by combining protons and neutrons is photodissociated before heavier nuclei can build up, whereas after this interval, the temperature is too low to overcome the Coulumb barriers between the colliding nuclei. But during this interval the deuterium nuclei exist long enough to serve as seeds for formation of heavier nuclei because they can capture other nucleons. Calculating the primordial abundances of light elements starts from an initial "soup" at $t \approx 1$ second, including neutrons, protons, electrons, and photons in local thermal equilibrium.[18] Given experimentally measured values of the relevant reaction rates, one can calculate the change in relative abundances of these constituents and the appearance of nuclei of the light elements. The result of these calculations is a prediction of light-element abundances that depends on physical features of the universe at this time, such as the total density of baryonic matter and the baryon to photon ratio. Observations of primordial element abundances can then be taken as constraining the cosmological model's parameters. Although there are still discrepancies (notably regarding Lithium 7) whose significance is unclear, the values of the parameters inferred from primordial abundances in conjunction with nucleosynthesis calculations are in rough agreement with values determined from other types of observations.

As the temperature drops below $\approx 4,000K$, "re-combination" occurs as the electrons become bound in stable atoms.[19] As a result, the rate of one of the reactions keeping the photons and matter in equilibrium (Compton scattering of photons off electrons) drops below the expansion rate. The photons decouple from the matter with a black-body spectrum. After decoupling, the photons cool adiabatically with the expansion, and the temperature drops as $T \propto 1/R$, but the black-body spectrum is unaffected. This "cosmic background radiation" (CBR) carries an enormous amount of information regarding the universe at the time of decoupling. It is difficult to provide a natural, alternative explanation for the black-body spectrum of this radiation.[20]

---

[17] See Olive, Steigman, and Walker (2000) for a review of big bang nucleosynthesis.

[18] These are called "primordial" or "relic" abundances to emphasize that they are the abundances calculated to hold at $t \approx 20$ minutes. Inferring the values of these primordial abundances from observations requires an understanding of the impact of subsequent physical processes, and the details differ substantially for the various light elements.

[19] The term "re-combination" is misleading, as the electrons were not previously bound in stable atoms. See Weinberg (2008) and section 2.3 for a description of the intricate physics of recombination.

[20] The black-body nature of the spectrum was firmly established by the COBE (Cosmic Background Explorer) mission in 1992. The difficulty in finding an alternative stems from the fact that the present universe is almost entirely transparent to the CBR photons, and the matter that does absorb and emit radiation is not distributed uniformly. To produce a uniform sea of photons with a black-body spectrum, one would need to introduce an almost uniformly distributed type of matter that thermalizes radiation from other processes to produce the observed microwave background, yet is nearly transparent at other frequencies. Advocates of the quasi-steady state cosmology have argued that whiskers of iron ejected from supernovae could serve as just such a thermalizer of radiation in the far infrared. See, e.g., Li (2003) for a discussion of this proposal and persuasive objections to it.

The initial detection of the CBR and subsequent measurements of its properties played a crucial role in convincing physicists to trust the extrapolations of physics to these early times, and ever since its discovery, the CBR has been a target for increasingly sophisticated observational programs. These observations established that the CBR has a uniform temperature to within 1 part in 10^5, and the minute fluctuations in temperature provide empirical guidance for the development of early universe theories.

In closing, two aspects of the accounts of the thermal history deserve emphasis. First, the physics used in developing these ideas has independent empirical credentials. Although the very idea of early universe cosmology was regarded as speculative when calculations of this sort were first performed (Alpher, Bethe, and Gamow 1948), the basic nuclear physics was not. Second, treating the constituents of the early universe as being in local thermal equilibrium before things get interesting is justified provided that the reaction rates are higher than the expansion rate at earlier times. This is an appealing feature, since equilibrium has the effect of washing away dependence on earlier states of the universe. As a result processes such as nucleosynthesis are relatively insensitive to the state of the very early universe. The dynamical evolution through nucleosynthesis is based on well-understood nuclear physics, and equilibrium effaces the unknown physics at higher energies.

2.3 Structure Formation

By contrast with these successes, the Standard Model lacks a compelling account of how structures like galaxies formed. This reflects the difficulty of the subject, which requires integrating a broader array of physical ideas than those required for the study of nucleosynthesis or the FLRW models. It also requires more sophisticated mathematics and computer simulations to study dynamical evolution beyond simple linear perturbation theory.

Newtonian gravity enhances clumping of a nearly uniform distribution of matter, as matter is attracted more strongly to regions with above average density. Jeans (1902) studied the growth of fluctuations in Newtonian gravity and found that fluctuations with a mode greater than a critical length exhibit instability and their amplitude grows exponentially. The first study of a similar situation in general relativity (Lifshitz 1946) showed, by contrast, that expansion in the FLRW models counteracts this instability, leading to much slower growth of initial perturbations. Lifshitz (1946) concluded that the gravitational enhancement picture could not produce galaxies from plausible "seed" perturbations and rejected it. Two decades later the argument was reversed: given the gravitational enhancement account of structure formation (no viable alternative accounts had been discovered), the seed perturbations had to be much larger than Lifshitz expected. Many cosmologists adopted a more phenomenological approach, using observational data to constrain the initial perturbation spectrum and other parameters of the model.

Contemporary accounts of structure formation treat observed large-scale structures as evolving by gravitational enhancement from initial seed perturbations. The goal is to account for observed properties of structures at a variety of scales—from features of galaxies to statistical properties of the large-scale distribution of galaxies—by appeal to the dynamical evolution of the seed perturbations through different physical regimes. Harrison, Peebles, and Zel'dovich independently argued that the initial perturbations should be scale invariant, that is, lacking any characteristic length scale.[21] Assuming that these initial fluctuations are small (with a density contrast $\frac{\delta \rho}{\rho} \ll 1$), they can be treated as linear perturbations to a background cosmological model where the dynamical evolution of individual modes is specified by general relativity. As the perturbations grow in amplitude and reach $\frac{\delta \rho}{\rho} \approx 1$, perturbation theory no longer applies and the perturbation mode "separates" from cosmological expansion and begins to collapse. In current models, structure grows hierarchically with smaller length scales going nonlinear first. Models of evolution of structures at smaller length scales (e.g., the length scales of galaxies) as the perturbations go nonlinear incorporate physics in addition to general relativity, such as gas dynamics, to describe the collapsing clump of baryonic matter.

The current consensus regarding structure formation is called the ΛCDM model. The name indicates that the model includes a nonzero cosmological constant (Λ) and "cold" dark matter (CDM). (Cold dark matter is discussed in the next section.) The model has several free parameters that can be constrained by measurements of a wide variety of phenomena. The richness of these evidential constraints and their mutual compatibility provide some confidence that the ΛCDM model is at least partially correct. There are, however, ongoing debates regarding the status of the model. For example, arguably it does not capture various aspects of galaxy phenomenology. Although I do not have the space to review these debates here, it is clear that current accounts of structure formation face more unresolved challenges and problems than other aspects of the Standard Model.

3. Dark Matter and Dark Energy

The main support for the Standard Model comes from its successful accounts of big bang nucleosynthesis, the redshift-distance relation, and the CBR. But pushing these lines of evidence further reveals that, if the Standard Model is basically correct, the vast majority of the matter and energy filling the universe cannot be ordinary matter. According to the "concordance model," normal matter contributes $\approx 4\%$ of the total energy density, with $\approx 22\%$ in the form of non-baryonic dark matter and another $\approx 74\%$ in the form of dark energy.[22]

---

[21] More precisely, the different perturbation modes have the same density contrast when their wavelength equals the Hubble radius, H^{-1}.

[22] Cosmologists use "concordance model" to refer to the Standard Model of cosmology with the specified contributions of different types of matter. The case in favor of a model with roughly these contributions to

Dark matter was first proposed based on observations of galaxy clusters and galaxies.[23] Their dynamical behavior cannot be accounted for solely by luminous matter in conjunction with Newtonian gravity. More recently, it was discovered that the deuterium abundance, calculated from big bang nucleosynthesis, puts a tight bound on the total amount of baryonic matter. Combining this constraint from big bang nucleosynthesis with other estimates of cosmological parameters leads to the conclusion that there must be a substantial amount of non-baryonic dark matter. Accounts of structure formation via gravitational enhancement also seem to require non-baryonic cold dark matter. Adding "cold" dark matter to models of structure formation helps to reconcile the uniformity of the CBR with the subsequent formation of structure. The CBR indicates that any type of matter coupled to the radiation must have been very smooth, much too smooth to provide seeds for structure formation. Cold dark matter decouples from the baryonic matter and radiation early, leaving a minimal imprint on the CBR.[24] After recombination, however, the cold dark matter perturbations generate perturbations in the baryonic matter sufficiently large to seed structure formation.

The first hint of what is now called "dark energy" also came in studies of structure formation, which seemed to require a nonzero cosmological constant to fit observational constraints (the ΛCDM models). Subsequent observations of the redshift-distance relation, with supernovae (type Ia) used as a powerful new standard candle, led to the discovery in 1998 that the expansion of the universe is accelerating.[25] This further indicates the need for dark energy, namely a type of matter that contributes to Eq. (2) like a Λ term, such that $\ddot{R} > 0$.[26]

Most cosmologists treat these developments as akin to Le Verrier's discovery of Neptune. In both cases, unexpected results regarding the distribution of matter are inferred from observational discrepancies using the theory of gravity. Unlike the case of Le Verrier, however, this case involves the introduction of new types of matter rather than merely an additional planet. The two types of matter play very different roles in cosmology, despite the shared adjective. Dark energy affects cosmological expansion but is irrelevant on smaller scales, whereas dark matter dominates the dynamics of bound gravitational systems such as galaxies. There are important

---

the overall energy density was made well before the discovery of cosmic acceleration (see, e.g., Ostriker and Steinhardt (1995); Krauss and Turner (1999)). Coles and Ellis (1997) give a useful summary of the opposing arguments (in favor of a model without a dark energy component) as of 1997, and see Frieman, Turner, and Huterer (2008) for a more recent review.

[23] See Trimble (1987) for a discussion of the history of the subject and a systematic review of various lines of evidence for dark matter.

[24] "Hot" vs. "cold" refers to the thermal velocities of relic particles for different types of dark matter. Hot dark matter decouples while still "relativistic," in the sense that the momentum is much greater than the rest mass, and relics at late times would still have large quasi-thermal velocities. Cold dark matter is "non-relativistic" when it decouples, meaning that the momentum is negligible compared to the rest mass, and relics have effectively zero thermal velocities.

[25] Type Ia supernovae do not have the same intrinsic luminosity, but the shape of the light curve (the luminosity as a function of time after the initial explosion) is correlated with intrinsic luminosity. See Kirshner (2009) for an overview of the use of supernovae in cosmology.

[26] These brief remarks are not exhaustive; there are further lines of evidence for dark matter and dark energy; see, e.g., Bertone, Hooper, and Silk (2005) for a review of evidence for dark matter and Huterer (2010) on dark energy.

contrasts in the evidential cases in their favor and in their current statuses. Some cosmologists have called the concordance model "absurd" and "preposterous" because of the oddity of these new types of matter and their huge abundance relative to that of ordinary matter. There is also not yet an analog of Le Verrier's successful follow-up telescopic observations. Perhaps the appropriate historical analogy is instead the "zodiacal masses" introduced to account for Mercury's perihelion motion before GR. Why not modify the underlying gravitational theory rather than introduce one or both of these entirely new types of matter?

The ongoing debate between accepting dark matter and dark energy vs. pursuing alternative theories of gravity and cosmology turns on a number of issues familiar to philosophers of science. Does the evidence underdetermine the appropriate gravitational theory? At what stage should the need to introduce distinct types of matter with exotic properties cast doubt on the gravitational theory and qualify as anomalies in Kuhn's sense? How successful are alternative theories compared to GR and the Standard Model, relative to different accounts of what constitutes empirical success? What follows is meant to be a primer identifying the issues that seem most relevant to a more systematic treatment of these questions.[27]

Confidence that GR adequately captures the relevant physics supports the mainstream position, accepting dark matter and dark energy. The application of GR at cosmological scales involves a tremendous extrapolation, but this kind of extrapolation of presumed laws has been incredibly effective throughout the history of physics. This particular extrapolation, furthermore, does not extend beyond the expected domain of applicability of GR. No one trusts GR at sufficiently high energies, extreme curvatures, and short length scales. Presumably it will be superseded by a theory of quantum gravity. Discovering that GR fails at low energies, low curvature, and large length scales—the regime relevant to this issue—would, however, be extremely surprising. In fact, avoiding dark matter entirely would require the even more remarkable concession that Newtonian gravity fails at low accelerations. In addition to the confidence in our understanding of gravity in this regime, GR has proven to be an extremely rigid theory that cannot be easily changed or adjusted.[28] At present, there is no compelling way to modify GR so as to avoid the need for dark matter and dark energy, while at the same time preserving GR's other empirical successes and basic theoretical principles. (Admittedly this may reflect little more than a failure of imagination; it was also not obvious how to change Newtonian gravity to avoid the need for zodiacal masses.)

The independence of the different lines of evidence indicating the need for dark matter and dark energy provides a second powerful argument in favor of the

---

[27] See Vanderburgh (2003, 2005) for a philosopher's take on these debates.

[28] See Sotiriou and Faraoni (2010) for a review of one approach to modifying GR, namely by adding higher-order curvature invariants to the Einstein-Hilbert action. These so-called "$f(R)$" theories (the Ricci scalar R appearing in the action is replaced by a function $f(R)$) have been explored extensively within the last five years, but it has proven to be difficult to satisfy a number of seemingly reasonable constraints. Uzan (2010) gives a brief overview of other ways of modifying GR in light of the observed acceleration.

mainstream position. The sources of systematic error in estimates of dark matter from big bang nucleosynthesis and galaxy rotation curves (discussed below), for example, are quite different. Evidence for dark energy also comes from observations with very different systematics, although they all measure properties of dark energy through its impact on spacetime geometry and structure formation. Several apparently independent parts of the Standard Model would need to be mistaken in order for all these different lines of reasoning to fail.

The case for dark energy depends essentially on the Standard Model, but there is a line of evidence in favor of dark matter based on galactic dynamics rather than cosmology. Estimates of the total mass for galaxies (and clusters of galaxies), inferred from observed motions in conjunction with gravitational theory, differ dramatically from mass estimates based on observed luminous matter.[29] To take the most famous example, the orbital velocities of stars and gas in spiral galaxies would be expected to drop with the radius as $r^{-1/2}$ outside the bright central region; observations indicate instead that the velocities asymptotically approach a constant value as the radius increases.[30] There are several other properties of galaxies and clusters of galaxies that lead to similar conclusions. The mere existence of spiral galaxies seems to call for a dark matter halo, given that the luminous matter alone is not a stable configuration under Newtonian gravity.[31] The case for dark matter based on these features of galaxies and clusters draws on Newtonian gravity rather than GR. Relativistic effects are typically ignored in studying galactic dynamics, given the practical impossibility of modeling a full galactic mass distribution in GR. But it seems plausible to assume that the results of Newtonian gravity for this regime can be recovered as limiting cases of a more exact relativistic treatment.[32]

There is another way of determining the mass distribution in galaxies and clusters that does depend on GR, but not the full Standard Model. Even before he had reached the final version of GR, Einstein realized that light-bending in a gravitational field would lead to the magnification and distortion of images of

---

[29] The mass estimates differ both in total amount of mass present and its spatial distribution. Estimating the mass based on the amount of electromagnetic radiation received (photometric observations) requires a number of further assumptions regarding the nature of the objects emitting the radiation and the effects of intervening matter, such as scattering and absorption (extinction).

[30] This behavior is usually described using the rotation curve, a plot of orbital velocity as a function of the distance from the galactic center. The "expected" behavior (dropping as $r^{-1/2}$ after an initial maximum) follows from Newtonian gravity with the assumption that all the mass is concentrated in the central region, like the luminous matter. The discrepancy cannot be evaded by adding dark matter with the same distribution as the luminous matter; in order to produce the observed rotation curves, the dark matter has to be distributed as a halo around the galaxy.

[31] In a seminal paper, Ostriker and Peebles (1973) argued in favor of a dark matter halo based on an N-body simulation, extending earlier results regarding the stability of rotating systems in Newtonian gravity to galaxies. These earlier results established a criterion for the stability of rotating systems: if the rotational energy in the system is above a critical value, compared to the kinetic energy in random motions, then the system is unstable. The instability arises, roughly speaking, because the formation of an elongated bar shape leads to a larger moment of inertia and a lower rotational energy. Considering the luminous matter alone, spiral galaxies appear to satisfy this criterion for instability; Ostriker and Peebles (1973) argued that the addition of a large, spheroidal dark matter halo would stabilize the luminous matter.

[32] This assumption has been challenged; see Cooperstock and Tieu (2007) for a review of their controversial proposal that a relativistic effect important in galactic dynamics, yet absent from the Newtonian limit, eliminates the need for dark matter.

distant objects. This lensing effect can be used to estimate the total mass distribution of a foreground object based on the distorted images of a background object, which can then be contrasted with the visible matter in the foreground object.[33] Estimates of dark matter based on gravitational lensing are in rough agreement with those based on orbital velocities in spiral galaxies, yet they draw on different regimes of the underlying gravitational theory.

Critics of the mainstream position argue that introducing dark matter and dark energy with properties chosen precisely to resolve the mass discrepancy is ad hoc. Whatever the strength of this criticism, the mainstream position does convert an observational discrepancy in cosmology into a problem in fundamental physics, namely that of providing a believable physics for dark matter and dark energy.

In this regard the prospects for dark matter seem more promising. Theorists have turned to extensions of the Standard Model of particle physics in the search for dark matter candidates, in the form of weakly interacting massive particles. Although the resulting proposals for new types of particles are speculative, there is no shortage of candidates that are theoretically natural (according to the conventional wisdom) and as yet compatible with observations. There also do not appear to be any fundamental principles that rule out the possibility of appropriate dark matter candidates.

With respect to dark energy, by contrast, the discovery of accelerating expansion has exacerbated what many regard as a crisis in fundamental physics.[34] Dark energy can either take the form of a true Λ term or some field whose stress-energy tensor effectively mimics Λ. As such it violates an energy condition associated with "ordinary" matter, although few theorists now take this condition as inviolable.[35] A more fundamental problem arises in comparing the observed value of dark energy with a calculation of the vacuum energy density in quantum field theory (QFT). The vacuum energy of a quantum field diverges. It is given by integrating the zero-point contributions to the total energy, $\frac{1}{2}\hbar\omega(k)$ per oscillation mode, familiar from the quantum harmonic oscillator, over momentum (k). Evaluating this quartically divergent quantity by introducing a physical cutoff at the Planck scale, the result is 120 orders of magnitude larger than the observed value of the cosmological constant.[36] This is sometimes called the "old" cosmological constant problem: Why isn't

---

[33] Gravitational lensing occurs when light from a background object such as a quasar is deflected due to the spacetime curvature produced, according to GR, by a foreground object, leading to multiple images of a single object. The detailed pattern of these multiple images and their relative luminosity can be used to constrain the distribution of mass in the foreground object.

[34] See, in particular, Weinberg (1989) for an influential review of the cosmological constant problem prior to the discovery of dark energy, and, e.g., Polchinski (2006) for a more recent discussion.

[35] Energy conditions place restrictions on the stress-energy tensor appearing in EFE. They are useful in proving theorems for a range of different types of matter with some common properties, such as "having positive energy density" or "having energy-momentum flow on or within the light cone." In this case the strong energy condition is violated; for the case of an ideal fluid discussed above, the strong energy condition holds iff $\rho + 3p \geq 0$. Cf., for example, chapter 9 of Wald (1984) for definitions of other energy conditions.

[36] In more detail, the relevant integral is

$$\rho_v = \int_0^\ell \frac{d^3k}{(2\pi)^3} \frac{\sqrt{k^2 + m^2}}{2} \approx \frac{\ell^4}{16\pi^2}$$

there a cancellation mechanism that leads to $\Lambda = 0$? Post-1998, the "new" problem concerns understanding why the cosmological constant is quite small (relative to the vacuum energy density calculated in QFT) but not exactly zero, as indicated by the accelerating expansion.

Both problems rest on the crucial assumption that the vacuum energy density in QFT couples to gravity as an effective cosmological constant. Granting this assumption, the calculation of vacuum energy density qualifies as one of the worst theoretical predictions ever made. What turns this dramatic failure into a crisis is the difficulty of controlling the vacuum energy density, by, say, introducing a new symmetry. Recently, however, an anthropic response to the problem has drawn increasing support. On this approach, the value of Λ is assumed to vary across different regions of the universe, and the observed value is "explained" as an anthropic selection effect (we will return to this approach in section 7 below).

Whether abandoning the assumption that the vacuum energy is "real" and gravitates is a viable response to the crisis depends on two issues. First, what does the empirical success of QFT imply regarding the reality of vacuum energy? The treatment of the scaling behavior of the vacuum energy density above indicates that vacuum energy in QFT is not fully understood given current theoretical ideas. This is not particularly threatening in calculations that do not involve gravity, since one can typically ignore the vacuum fluctuations and calculate quantities that depend only on relative rather than absolute values of the total energy. This convenient feature also suggests, however, that the vacuum energy may be an artifact of the formalism that can be stripped away while preserving QFT's empirical content. Second, how should the standard treatment of the vacuum energy from flat-space QFT be extended to the context of the curved spacetimes of GR? The symmetries of flat spacetime so crucial to the technical framework of QFT no longer obtain, and there is not even a clear way of identifying a unique vacuum state in a generic curved spacetime. Reformulating the treatment of the scaling behavior of the vacuum energy density is thus a difficult problem. It is closely tied to the challenge of combining QFT and GR in a theory of quantum gravity. In QFT on curved spacetimes (one attempt at combining QFT and GR) different renormalization techniques are used that eliminate the vacuum energy. The question is whether this approach simply ignores the problem by fiat or reflects an appropriate generalization of renormalization techniques to curved spacetimes. These two issues are instances of familiar questions for philosophers—what parts of a theory are actually supported by its empirical success,

---

For a Planck scale cutoff, $\ell_p \approx 1.6 \times 10^{-35} m$, the resulting vacuum energy density is given by $\rho_V \approx 2 \times 10^{110} erg/cm^3$, compared to observational constraints on the cosmological constant—$\rho_\Lambda \approx 2 \times 10^{-10} erg/cm^3$. Choosing a much lower cutoff scale, such as the electroweak scale $\ell_{ew} \approx 10^{-18} m$, is not enough to eliminate the huge discrepancy (still 55 orders of magnitude). Reformulated in terms of the effective field theory approach, the cosmological constant violates the technical condition of "naturalness." Defining an effective theory for a given domain requires integrating out higher energy modes, leading to a rescaling of the constants appearing in the theory. This rescaling would be expected to drive the value of terms like the cosmological constant up to the scale of the cutoff; a smaller value, such as what is observed, apparently requires an exquisitely fine-tuned choice of the bare value to compensate for this scaling behavior, given that there are no symmetry principles or other mechanisms to preserve a low value.

and what parts should be preserved or abandoned in combining it with another theory? Philosophers have offered critical evaluations of the conventional wisdom in physics regarding the cosmological constant problem, and there are opportunities for further work.[37]

Returning to the main line of argument, the prospects for an analog of Le Verrier's telescopic observations differ for dark matter and dark energy. There are several experimental groups currently searching for dark matter candidates, using a wide range of different detector designs and searching through different parts of the parameter space (see, e.g., Sumner 2002). Successful detection by one of these experiments would provide evidence for dark matter that does not depend directly on gravitational theory. The properties of dark energy, by way of contrast, insure that any attempt at a noncosmological detection would be futile. The energy density introduced to account for accelerated expansion is so low, and uniform, that any local experimental study of its properties is practically impossible given current technology.

There are different routes open for those hoping to avoid dark energy and dark matter. Dark energy is detected by the observed departures from the spacetime geometry that one would expect in a matter-dominated FLRW model. Taking this departure to indicate the presence of an unexpected contribution to the universe's overall matter and energy content thus depends on assuming that the FLRW models hold. There are then two paths open to those exploring alternatives to dark energy. The first is to change the underlying gravitational theory and to base cosmology on an alternative to GR that does not support this inference. A second would be to retain GR but reject the FLRW models. For example, models that describe the observable universe as having a lower density than surrounding regions can account for the accelerated expansion without dark energy. Cosmologists have often assumed that we are not in a "special" location in the universe. This claim is often called the "Copernican principle," to which we will return in section 5 below. This principle obviously fails in these models, as our observable patch would be located in an unusual region—a large void.[38] It has also been proposed that the accelerated expansion may be accounted for by GR effects that come into view in the study of inhomogeneous models without dark energy. Buchert (2008) reviews the idea that the back-reaction of inhomogeneities on the background spacetime leads to an effective acceleration. These proposals both face the challenge of accounting for the various observations that are regarded, in the concordance model, as manifestations of dark energy.

On the other hand, dark matter can only be avoided by modifying gravity—*including Newtonian gravity*—as applied to galaxies. Milgrom (1983) argued that a modification of Newtonian dynamics (called MOND) successfully captures several aspects of galaxy phenomenology. According to Milgrom's proposal, below an

---

[37] See, in particular, Rugh and Zinkernagel (2002) for a thorough critical evaluation of the cosmological constant problem, as well as Earman (2001), Saunders (2002), and Bianchi and Rovelli (2010).
[38] See ? for an overview of the use of inhomogeneous models as an alternative to dark energy.

acceleration threshold ($a_0 \approx 10^{-10}$ m/s^2) Newton's second law should be modified to $F = m\frac{a^2}{a_0}$. This modification accounts for observed galaxy rotation curves without dark matter. But it also accounts for a wide variety of other properties of galaxies, many of which Milgrom successfully predicted based on MOND (see, e.g., Sanders and McGaugh 2002, Bekenstein 2010 for reviews). Despite these successes, MOND has not won widespread support. Even advocates of MOND admit that at first blush it looks like an extremely odd modification of Newtonian gravity. Yet it fares remarkably well in accounting for various features of galaxies—too well, according to its advocates, to be dismissed as a simple curve fit. MOND does not fare as well for clusters of galaxies and may have problems in accounting for structure formation. In addition to these potential empirical problems, it is quite difficult to embed MOND within a compelling alternative to GR.

In sum, it is reasonable to hope that the situation with regard to dark matter and dark energy will be clarified in the coming years by various lines of empirical investigation that are currently underway. The apparent underdetermination of different alternatives may prove transient, with empirical work eventually forcing a consensus. Whether or not this occurs, there is also a possibility for contributions to the debate from philosophers concerned with underdetermination and evidential reasoning. The considerations above indicate that even in a case where competing theories are (arguably) compatible with all the evidence that is currently available, scientists certainly do not assign equal credence to the truth of the competitors. Philosophers could contribute to this debate by helping to articulate a richer notion of empirical support that sheds light on these judgments (cf. the closing chapter of Harper 2012).

4. Uniqueness of the Universe

The uniqueness of the universe is the main contrast between cosmology and other areas of physics. The alleged methodological challenge posed by uniqueness was one of the main motivations for the steady-state theory. The claim that a generalization of the cosmological principle, the "perfect cosmological principle," is a precondition for scientific cosmology, is no longer accepted.[39] It is, however, often asserted that cosmology cannot discover new laws of physics as a direct consequence of the uniqueness of its object of study.[40] Munitz (1962) gives a concise formulation of this common argument:

---

[39] The pronouncements of the steady state theorists drew a number of philosophers into debates regarding cosmology in the 1960s. See Kragh (1996) for a historical account of the steady-state theory, and the rejection of it in favor of the big bang theory by the scientific community, and Balashov (2002) for a discussion of their views regarding laws.

[40] See Pauri (1991); Scheibe (1991); Torretti (2000) for discussions of the implications of uniqueness and the status of laws in cosmology.

> With respect to these familiar laws [of physics] ... we should also mark it as a prerequisite of the very meaning and use of such laws that we be able to refer to an actual or at least possible *plurality* of instances to which the law applies. For unless there were a plurality of instances there would be neither interest nor sense in speaking of a law at all. If we knew that there were only one actual or possible instance of some phenomenon it would hardly make sense to speak of finding a law for this unique occurrence *qua* unique. This last situation however is precisely what we encounter in cosmology. For the fact that there is at least but not more than one universe to be investigated makes the search for *laws* in cosmology inappropriate. (Munitz 1962, 37)

Ellis (2007) reaches a similar conclusion:

> The concept of "Laws of Physics" that apply to only one object is questionable.
> *We cannot scientifically establish "laws of the universe" that might apply to the class of all such objects, for we cannot test any such proposed law except in terms of being consistent with one object (the observed universe).* (Ellis 2007, 1217, emphasis in the original)

His argument for this claim emphasizes that we cannot perform experiments on the universe by creating particular initial conditions. In many observational sciences (such as astronomy) the systems under study also cannot be manipulated, but it is still possible to do without experiments by studying an ensemble of instances of a given type of system. However, this is also impossible in cosmology.

If these arguments are correct, then cosmology should be treated as a merely descriptive or historical science that cannot discover novel physical laws. Both arguments rest on problematic assumptions regarding laws of nature and scientific method. Here I will sketch an alternative account that allows for the possibility of testing cosmological laws despite the uniqueness of the universe.

Before turning to that task, I should mention a different source of skepticism regarding the possibility of scientific cosmology based on distinctive laws. Kant argued that attempts at scientific cosmology inevitably lead to antinomies because no object corresponds to the idea of the "universe." Relativistic cosmology circumvents this argument insofar as cosmological models have global properties that are well-defined, albeit empirically inaccessible. (This is discussed further in section 5.) Yet contemporary worries resonant with Kant concerning how to arrive at the appropriate concepts for cosmological theorizing. Smolin (2000) criticizes relativistic cosmology for admitting such global properties and proposes instead that: "Every quantity in a cosmological theory that is formally an observable should in fact be measurable by some observer inside the universe."[41] A different question arises, for example, in extrapolating concepts to domains such as the early universe. Rugh and Zinkernagel (2009) argue that there is no physical footing for spacetime concepts in the very early universe due to the lack of physical processes that can be used to determine spacetime scales.

Munitz's formulation makes his assumptions about the relationship between laws and phenomena clear: the phenomena are instances of the law, just as $Fa \wedge Ga$ would be an instance of the "law" $\forall x(Fx \to Gx)$. Even if we grant this conception of laws, Munitz's argument would only apply to a specific kind of cosmological law. If we take EFE as an example of a "cosmological law," then it has multiple instantiations in the straightforward sense that every subregion of a solution of EFE is also a solution.[42] The same holds for other local dynamical laws applicable in cosmology, such as those of QFT. A single universe has world enough for multiple instantiations of the local dynamics. This is true as well of laws whose effects may have, coincidentally, only been important within some finite subregion of the universe. For example, consider a theory, such as inflation (see section 6 below), whose implications are only manifest in the early universe. The laws of this theory would be "instantiated" again if we were ever able to reach sufficiently high energy levels in an experimental setting. Although the theory may in practice only have testable implications "once," it has further counterfactual implications. Munitz's argument would apply, however, to cosmological laws that are formulated directly in terms of global properties, as opposed to local dynamical laws extrapolated to apply to the universe as a whole. Subregions of the universe would not count as instantiations of a "global law" in the same sense that they are instantiations of the local dynamical laws. Penrose's Weyl curvature hypothesis (proposed in Penrose 1979) is an example of such a law.[43] This law is formulated as a constraint on initial conditions and it does differ strikingly in character from local dynamical laws.

Phenomena are not, however, "instantiations" of laws of nature in Munitz's straightforward logical sense. Treating them as such attributes to the laws empirical content properly attributed only to equations derived from the laws with the help of supplementary conditions.[44] A simple example should help to make this contrast clear. Newton's three laws of motion must be combined with other assumptions regarding the relevant forces and distribution of matter to derive a set of equations of motion, describing, say, the motion of Mars in response to the Sun's gravitational field. It is this derived equation describing Mars's motion that is compared to

---

[41] This is the first of two principles Smolin advocates as necessary to resolve the problem of time, and he further argues that they bring cosmological theorizing more in line with scientific practice.

[42] That is, for any open set O of the spacetime manifold M, if $\langle M, g_{ab}, T_{ab}\rangle$ is a solution of EFE, then so is $\langle O, g_{ab}|O, T_{ab}|O\rangle$ taken as a spacetime in its own right.

[43] The Weyl tensor represents, roughly speaking, the gravitational degrees of freedom in GR with the degrees of freedom for the source terms removed. Penrose's hypothesis holds that this tensor vanishes in the limit as one approaches the initial singularity.

[44] This mistake also underlies much of the discussion of "ceteris paribus" laws, and here I draw on the line of argument due to Smith (2002); Earman and Roberts (1999).

the phenomena and used to calculate the positions of Mars given some initial conditions. The motion of Mars is not an "instance" of Newton's laws; rather, the motion of Mars is well approximated by a solution to an equation derived from Newton's laws along with a number of other assumptions.

Ellis's argument does not explicitly rest on a conception of phenomena as instantiations of laws. But he and Munitz both overlook a crucial aspect of testing laws. Continuing with the same example, there is no expectation that at a given stage of inquiry one has completely captured the motion of Mars with a particular derived equation, even as further physical effects (such as the effects of other planets) are included. The success of Newton's theory (in this case) consists in the ability to give more and more refined descriptions of the motion of Mars, all based on the three laws of motion and the law of gravity. This assessment does not depend primarily on "multiplicity of instances," experimental manipulation, or observation of other members of an ensemble. Instead, the modal force of laws is reflected in their role in developing a richer account of the motions. Due to this role they can be subject to ongoing tests.

The standard arguments that it is not possible to discover laws in cosmology assume that the universe is not only unique, but in effect "given" to us entirely, all at once—leaving cosmologists with nothing further to discover, and no refinements to make and test. A novel law in cosmology could be supported by its success in providing successively more refined descriptions of some aspect of the universe's history, just as Newtonian mechanics is supported (in part) by its success in underwriting research related to the solar system. This line of argument, if successful, shows that cosmological laws are testable in much the same sense as Newton's laws. This suggests that "laws of the universe" should be just as amenable to an empiricist treatment of the laws of nature as are other laws of physics.[45]

None of this is to say that there are no distinctive obstacles to assessing cosmological laws. But we need to disentangle obstacles that arise due to specific features of our universe from those that follow from the uniqueness of the object of study. Consider (contrary to the Standard Model) a universe that reached some finite maximum temperature as $t \to 0$, and suppose (perhaps more absurdly) that physicists in this universe had sufficient funds to build accelerators to probe physics at this energy scale. Many of the challenges faced in early universe cosmology in our universe would not arise for cosmologists in this other possible universe. They would have independent lines of evidence (from accelerator experiments and observations of the early universe) to aid in reconstructing the history of the early universe, rather than basing their case in favor of novel physics solely on its role in the reconstruction. This suggests that obstacles facing cosmology have to do primarily with theoretical and observational accessibility, which may be exacerbated by uniqueness, rather than with uniqueness of the universe per se.

---

[45] There may be other philosophical requirements on an account of laws of nature that do draw a distinction between laws of physics and laws of the universe.

5. Global Structure

The Standard Model takes the universe to be well-approximated by an FLRW model at sufficiently large scales. To what extent can observations determine the spacetime geometry of the universe directly? The question can be posed more precisely in terms of the region visible to an observer at a location in spacetime p—the *causal past*, $J^-(p)$, of that point. This set includes all points from which signals traveling at or below the speed of light can reach p.[46] What can observations confined to $J^-(p)$ reveal about: (1) the spacetime geometry of $J^-(p)$ itself, and (2) the rest of spacetime outside of $J^-(p)$? Here we will consider these questions on the assumption that GR and our other physical theories apply universally, setting aside debates (such as those in section 3) about whether these are the correct theories. How much do these theories allow us to infer, granting their validity?

Spacetime geometry is reflected in the motion of astronomical objects and in effects on the radiation they emit, such as cosmological red-shift. To what extent would the spacetime geometry be fixed by observations of an "ideal data set," consisting of comprehensive observations of a collection of standard objects, with known intrinsic size, shape, mass, and luminosity, distributed throughout the universe? Of course astronomers cannot avail themselves of such a data set. Converting the actual data recorded by observatories into a map of the universe, filled with different kinds of astronomical objects with specified locations and states of motion, is an enormously difficult task. The difficulty of completing this task poses one kind of epistemic limitation to cosmology. Exploring this limitation would require delving into the detailed astrophysics used to draw conclusions regarding the nature, location, and motion of distant objects. This kind of limitation contrasts with one arising from a different source, namely that we have an observational window on $J^-(p)$ rather than the entire spacetime. Even if we had access to an ideal data set, what we can observe is not sufficient to answer questions regarding global spacetime geometry unless we accept further principles underwriting local-to-global inferences.

The modest goal of pinning down the geometry of $J^-(p)$ observationally can be realized by observers with the ideal data set mentioned above (see Ellis 1980; Ellis et al. 1985). The relevant evidence comes from two sources: the radiation emitted by distant objects reaching us along our null cone, and evidence, such as geophysical data, gathered from "along our world line," so to speak. Ellis et al. (1985) prove that the ideal data set is necessary and sufficient, in conjunction with EFE and a few other assumptions, to determine the spacetime geometry of $J^-(p)$. Considering the ideal data set helps to clarify the contrast between what we can in

---

[46] In Minkowski spacetime, this set is the past lobe of the light cone at p, including interior points and the point p itself. A point p *causally precedes* q ($p < q$), if there is a future-directed curve from p to q with tangent vectors that are timelike or null at every point. The sets $J^\pm(p)$ are defined in terms of this relation: $J^-(p) = \{q : q < p\}$, $J^+(p) = \{q : p < q\}$, the *causal past* and *future* of the point p, and the definition generalizes immediately to spacetime regions.

principle determine locally, namely the spacetime geometry of $J^-(p)$, and what we can determine globally.

For points p, q with nonintersecting causal pasts, we would not expect the physical state on $J^-(p)$ to fix that of $J^-(q)$.[47] Does the spacetime geometry of $J^-(p)$, or of a collection of such sets, nonetheless constrain the large-scale or global properties of spacetime? Global properties of spacetime vary in general relativity, because unlike earlier theories such as Newtonian mechanics, spacetime is treated as dynamical rather than as a fixed background. EFE impose a local constraint on the spacetime geometry, but this is compatible with a wide variety of global properties.[48] Various global properties have been defined as part of stating and proving theorems such as the singularity theorems, including "causality conditions" that specify the extent to which a spacetime deviates from the causal structure of Minkowski spacetime (see Geroch and Horowitz 1979 for a clear introduction). For example, a *globally hyperbolic* spacetime possesses a Cauchy surface, a null or spacelike surface Σ intersected exactly once by every inextendible timelike curve. In a spacetime with a Cauchy surface, EFE admit a well-posed initial value formulation: specifying appropriate initial data on a Cauchy surface Σ determines a unique solution to the field equations (up to diffeomorphism). This is properly understood as a global property of the entire spacetime. Although submanifolds of a given spacetime may be compatible or incompatible with global hyperbolicity, this property cannot be treated as a property ascribed to local regions and then "added up" to deliver a global property.

What does $J^-(p)$ reveal about the rest of spacetime? Suppose we do not impose any strong global assumptions such as isotropy and homogeneity. Fully specifying the physical state in the region $J^-(p)$ places few constraints on the global properties of spacetime. This is clear if we consider what is shared by all the spacetimes into which $J^-(p)$ can be isometrically embedded, where we allow p to be any point in a given spacetime.[49] (That is, we shift from considering the causal past of a single observer to the causal past of *all* possible observers in the spacetime.) Call this the set of spacetimes "observationally indistinguishable" (OI) from a given spacetime. Except for the exceptional case where there is a p' such that, like Borges's Aleph, $J^-(p')$ includes the entire spacetime, there is a technique (due to Malament 1977; Manchak 2009) for constructing OI counterparts that do not share all the global properties of the original spacetime. The property of having a Cauchy surface, for example, will not be shared by all the members of a set of OI spacetimes.[50] More

---

[47] The Gauss-Codacci constraint equations do impose some restrictions on spacelike separated regions, although these would not make it possible to determine the state of one region from the other; see Ellis and Sciama (1972).

[48] A *local* property of a spacetime is one that is shared by locally isometric spacetimes, whereas global properties are not. (Two spacetimes are locally isometric iff for any point p in the first spacetime, there is an open neighborhood of the point such that it can be mapped to an isometric open neighborhood of the second spacetime (and vice versa).)

[49] The underdetermination problem still arises if we consider the past of future-inextendible curves; see Glymour (1977); Malament (1977) for discussion.

[50] Malament (1977) reviews several different definitions of observational indistinguishability and gives a series of constructions of OI spacetimes lacking specific global properties. Note that Malament defines OI in terms of the *chronological* rather than *causal* sets, which include the interior of the light cone but not the cone itself. (The definition follows the one given in footnote 46, dropping the phrase "or null.") Manchak (2009)

generally, the only properties that will be held in common in all members of the set of OI spacetimes are those that can be conclusively established by a single observer somewhere in the spacetime.[51]

The scope of underdetermination can be reduced by imposing constraints that eliminate potential OI spacetimes. Consider, for example, restricting consideration to spacetimes that are spatially homogeneous. The isometries on Σ (implied by homogeneity), which carry any point on Σ into any other, apparently block the construction of an indistinguishable counterpart with different global properties.[52] Homogeneity is just one example of a global property that could be imposed. Whatever property is imposed to eliminate underdetermination, it must be *global* to be effective given that the technique for constructing indistinguishable counterparts preserves local properties.

This line of argument clarifies the "cosmological principle." The cosmological principle is the strongest of many possible "uniformity principles" or global stipulations that allow local-to-global inferences. If we require only that the $J^-(p)$ sets for all observers can be embedded in a cosmological model, then the global properties of spacetime are radically underdetermined. Introducing different constraints on the construction of the indistinguishable counterparts mitigates the degree of underdetermination. The cosmological principle is the strongest of these constraints—strong enough to eliminate the underdetermination: every observer can take their limited view on the universe as accurately reflecting its global properties.

However, this merely pushes the original question back one step: What grounds do we have for imposing such a global constraint on spacetime?[53] It is unappealing to simply assert that the cosmological principle holds a priori, or to treat it as a precondition for cosmological theorizing. But one may hope to justify the principle by appealing to a weaker general principle in conjunction with theorems relating homogeneity and isotropy. Global isotropy around every point implies global homogeneity, and it is natural to seek a similar theorem with a weaker antecedent formulated in terms of observable quantities. The Ehlers-Geren-Sachs (EGS) theorem (Ehlers, Geren, and Sachs, 1968) shows that if all fundamental observers in an expanding model find that freely propagating background radiation is exactly

---

proves that Malament's technique for constructing such spacetimes fails only in the exceptional case noted in the text. Cf. Norton (2011), who argues that the inductive generalizations from $J^-(p)$ to other regions of spacetime lack clear justification.

[51] As Malament emphasizes, this includes the failure of the causality conditions to hold.

[52] Pick a point in $p \in M$ such that p lies in Σ and its image $\phi(p) \in M'$ under the isometric imbedding map ϕ. If homogeneity holds, then M' must include an isometric "copy" Σ' of the *entire Cauchy surface* Σ along with its entire causal past. Take ξ to be an isometry of the spatial metric defined on Σ, and ξ' an isometry on Σ'. Since $\phi \circ \xi(p) = \xi' \circ \phi(p)$, and any point $q \in \Sigma$ can be reached via ξ, it follows that Σ is isometric to Σ'. Mapping points along an inextendible timelike curve from M into M' eventually leads to an isometric copy of our original spacetime, assuming that both spacetimes are inextendible. Turning this into a proof that OI counterparts are completely eliminated requires further assumptions about the topology of the solutions. (de Sitter spacetime and its unrolled covering space are both inextendible and homogeneous yet have distinct global topology, as John Manchak reminded me.)

[53] See Beisbart (2009) for a thorough discussion of different attempts to justify the cosmological principle.

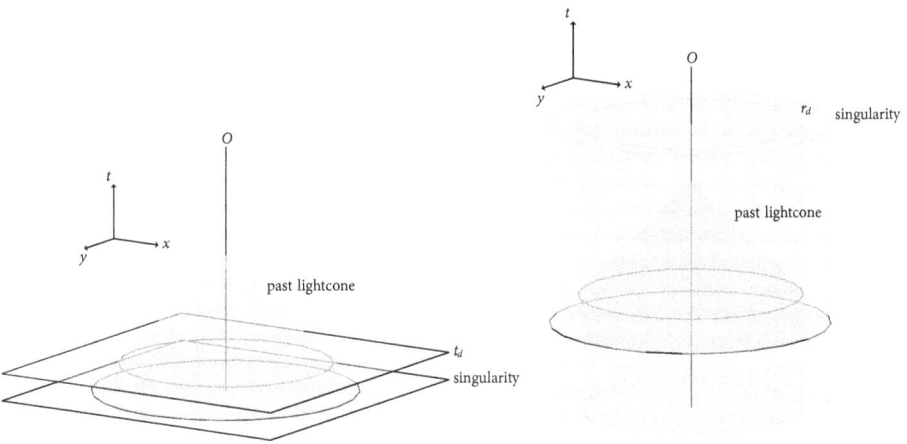

Figure 17.1 This figure contrasts the standard big bang model (a) and Ellis, Maartens, and Nel's (1978) model (b); in the latter, a cylindrical timelike singularity surrounds an observer O located near the axis of symmetry, and the constant time surface t_D from which the CBR is emitted in the standard model is replaced with a surface r_D at fixed distance from O.

isotropic, then their spacetime is an FLRW model.[54] If our causal past is "typical," observations along our worldline will constrain what other observers should see. This assumption is often called the "Copernican principle," which requires that no point p is distinguished from other points q by any spacetime symmetries. This principle rules out models such as Ellis, Maartens and Nel's (1978) example of a "cylindrical" counterpart to the observed universe (see figure 17.1).[55] (This example illustrates the tension between the Copernican principle and anthropic reasoning (see section 7 below). Ellis, Maartens, and Nel point out that in their model one would only expect to find observers near the axis of symmetry of the model, as that is the only region hospitable to life.) Combining the observed near isotropy of the CBR, the EGS theorem, and the Copernican principle yields an argument in favor of the approximate validity of the FLRW models.

Alternatively, one could dispense with the Copernican principle and its ilk by showing that an early phase of the universe's evolution leads to an approximately FLRW universe. This was the aim of Misner's "chaotic cosmology" program

---

[54] Recent work has clarified the extent to which this result depends on the various exact claims made in the antecedent. The fundamental observers do not need to measure exact isotropy for a version of the theorem to go through: Stoeger, Maarten, and Ellis have further shown that almost isotropic CMBR measurements imply that the spacetime is an almost FLRW model, in a sense that can be made precise; see Clarkson and Maartens (2010) for a review.

[55] Their model replaces *temporal* evolution in the Standard Model with *spatial* variation, with spherical symmetry around a preferred axis. They construct the model to recapture the observational results of the Standard Model for observers situated near the axis of symmetry. Such a preferred location is exactly what the Copernican principle rules out.

launched shortly after the discovery of the CBR, an aim taken up with greater acclaim by inflationary cosmology (see section 6 below). If this approach succeeds, then homogeneity and isotropy over some length scale would be a consequence of underlying physics, effectively replacing a priori principles regarding the uniformity of nature with factual claims about the universe's evolution. The warrant for an inductive inference regarding distant regions of the universe would then depend on the justification for this account. Note, however, that the account may not justify the conclusion that the universe is globally almost-FLRW. In the case of inflation, for example, homogeneity and isotropy hold in the interior of an inflationary bubble (which could be much larger than $J^-(p)$), but the universe at much larger scales has dramatic nonuniformities (bubble walls, colliding bubbles, regions between the bubbles, and so on).

The Copernican principle has come under increased scrutiny recently due to its role in the case for dark energy. Departures from an FLRW geometry could simply indicate the failure of the models rather than the presence of a new kind of matter. Recently there have been two suggestions for ways to test the Copernican principle on scales comparable to the observable universe. First, the Sunyaev-Zel'dovich effect[56] can be used to indirectly measure the isotropy of the CBR as observed from distant points. Any anisotropies in the CBR as seen at a distant point q will be reflected in a temperature difference in the scattered radiation; the distortion in the observed black-body spectrum in principle reveals the failure of isotropy from distant points not on our worldline (Caldwell and Stebbins, 2008). This allows one to prove that the local universe is almost-FLRW based on an EGS theorem and observations of the CBR without invoking the Copernican principle (Clifton, Clarkson, and Bull, 2011). A second test of the Copernican principle is based on a consistency relation between several observables that holds in the FLRW models (Uzan, Clarkson, and Ellis, 2008).

These discussions focus on whether $J^-(p)$ can be well approximated by an FLRW model. This question is closely tied to assessing the case for dark energy and in determining the parameters of the Standard Model. What are the further implications if the universe is almost-FLRW on much larger scales, or if the cosmological principle holds globally throughout all of spacetime? More generally, what are the empirical stakes of determining the global properties of spacetime? Some global spacetime properties are plausibly treated as preconditions for the possibility of formulating local dynamical laws.[57] And the global properties are obvious candidates for fundamental features of spacetime from a realist's point of view. Proofs of the singularity theorems require assumptions regarding global causal structure. Further, the origin and eventual fate of the universe are quite different in

---

[56] The Sunyaev-Zel'dovich effect refers to the distortion of the spectrum of CBR photons that results from scattering by hot gases in galaxy clusters. Due to the scattering by the hot gas the CBR spectrum will have an excess of high-energy photons and a deficit of low-energy photons; measurements of this distortion can in principle be used to measure the temperature and mass of the gas in the cluster.

[57] For example, topological properties such as temporal orientability, which allows for a globally consistent choice of the direction of time, seem to be presupposed in formulating local dynamical laws.

a globally almost-FLRW model and in an observationally indistinguishable counterpart to it. Yet despite all of this, there is a clear contrast between claiming that the observable universe is almost-FLRW and the extension of that to a global claim regarding all of spacetime. The former plays a fundamental role in evidential reasoning in contemporary cosmology, whereas the latter is disconnected from empirical research by its very nature. Thus, the status of the cosmological principle seems to differ significantly in practice from that of other principles supporting inductive generalizations—it does not lead, as in Newton's case of taking gravity to be truly universal, to a wide variety of further claims that can serve as the basis for a subsequent research program.

6. Early Universe Cosmology

Extrapolating the Standard Model backward in time leads to a singularity within a finite time, and as $t \to 0$ the temperature and energy scales increase without bound. Even if the singularity itself is somehow avoided, the early universe is expected to have reached energy scales far higher than anything produced at Fermilab or CERN. The early universe is thus a fruitful testing ground for high-energy physics, and since the early 1980s there has been an explosion of research in this area. Yet it is not clear whether observations of the early universe can play anything like the role that accelerator experiments did in guiding an earlier phase of research in particle physics. Other aspects of the Standard Model are based on extrapolating well-established physics, but the physics applied to the early universe often cannot be tested by other means. Instead the case in favor of new physical ideas is often based on their role in a plausible reconstruction of the universe's history. Here I will assess a common style of argument adopted in this literature, namely that a theory of early universe cosmology should be accepted because it renders the observed history of the universe probable rather than merely possible.

There is general agreement that the (cosmological) Standard Model should be supplemented with an account of physical processes in the very early universe. The early universe falls within the domains of applicability of both quantum field theory and general relativity, yet the two theories have yet to be combined successfully. The framework of the Standard Model is thus not expected to apply to the very early universe. Although research in quantum gravity is often motivated by calls for "theoretical unification" and the like, it can also be motivated by the more prosaic demand for a consistent theory applicable to phenomena such as the early universe and black holes (cf. Callender and Huggett, 2001). This "overlapping domains" argument does not imply anything in detail regarding what an early universe theory should look like, or how it would augment or contribute to the Standard Model.

The overlapping domains argument should not be confused with the common claim that general relativity is incomplete because it "breaks down" as $t \to 0$ and fails to provide a description of what happens at (or before) the singularity.[58] It is hard to see how general relativity can be convicted of incompleteness on its own terms. (Here I am following the line of argument in Earman (1995); Curiel (1999).) If general relativity proved to be the correct final theory, then there is nothing more to be said regarding singularities; the laws of general relativity apply throughout the entire spacetime, and there is no obvious incompleteness. On the other hand, there are good reasons to doubt that general relativity is the correct final theory, and further reasons to expect that the successor to general relativity will have novel implications for singularities. But then the argument for incompleteness is based on grounds other than the mere existence of singularities.

Cosmologists often give a very different reason for supplementing the Standard Model: it is *explanatorily deficient*, because it requires an "improbable" initial state. Guth (1981) gave an influential presentation of two aspects of the Standard Model as problematic:

> The standard model of hot big-bang cosmology requires initial conditions which are problematic in two ways: (1) The early universe is assumed to be highly homogeneous, in spite of the fact that separated regions were causally disconnected (horizon problem) and (2) the initial value of the Hubble constant must be fine tuned to extraordinary accuracy . . . (flatness problem). (Guth 1981, 347)

Horizons in cosmology measure the maximum distance light travels within a given time period; the horizon delimits the spacetime region from which signals emitted at some time t_e traveling at or below the speed of light could reach a given point. The existence of particle horizons in the FLRW models indicates that distant regions are not in causal contact.[59] There are observed points on the CBR separated by a distance greater than the particle horizon at that time (see figure 17.2). The Standard Model assumes that these regions have the same properties—e.g., the same temperature to within 1 part in 10^5—even though they were not in causal contact. In slightly

---

[58] Here I am adopting the usual way of describing the objection, although this language can be quite misleading as it implicitly assumes that the singularity can be "localized" in some sense. There are convincing arguments in favor of taking singular as an adjective describing spacetime as a whole; see Curiel (1999), Geroch, Can-bin, and Wald (1982).

[59] Following Rindler (1956), a horizon is the surface in a time slice t_0 separating particles moving along geodesics that could have been observed from a worldline γ by t_0 from those which could not. The distance to this surface, for signals emitted at a time t_e, is given by:

$$d = R(t_0) \int_{t_e}^{t_0} \frac{dt}{R(t)} \tag{3}$$

Different "horizons" correspond to different choices of limits of integration, with the "particle horizon" defined as the limit $t_e \to 0$. The integral converges for $R(t) \propto t^n$ with $n < 1$, which holds for matter or radiation-dominated expansion, leading to a finite horizon distance. See Ellis and Rothman (1993) for a clear introduction to horizons.

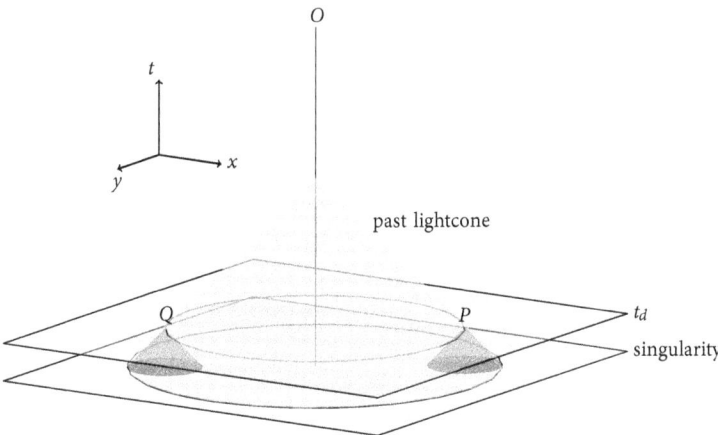

Figure 17.2 This figure illustrates the horizon problem. Lightcones are at 45° but distances are distorted, much like a Mercator projection. Two points P, Q on the surface of last scattering t_d, both falling within our past light cone, do not have overlapping light cones.

different terms, if one expects no correlations between the causally disjoint regions it is mysterious how the observable universe could be so well approximated by an FLRW model.

The flatness problem arises because the energy density at early times has to be very close to the value of the critical density $\Omega = 1$.[60] An FLRW model close to the "flat" $k = 0$ model, with nearly critical density, at some specified early time is driven rapidly away from critical density under FLRW dynamics; the flat model is an unstable fixed point under dynamical evolution.[61] This aspect of the dynamics makes it extremely puzzling to find that the universe is still close to the critical density—this requires an extremely finely-tuned choice of the energy density at the Planck time $\Omega(t_p)$, namely $|\Omega(t_p) - 1| \leq 10^{-59}$.

The horizon and flatness problems both reflect properties of the FLRW models. There are other similar "fine-tuning" problems related to other aspects of the Standard Model. The account of structure formation requires a set of "seed" perturbations that have two troubling features: first, the perturbations have to be coherent on super-horizon length scales, and, second, the amplitude of the perturbations was much smaller than one would expect for natural possibilities such as thermal fluctuations.[62] There are other puzzling features not related to the seed perturbations.

---

[60] $\Omega =: \frac{\rho}{\rho_c}$, where the critical density is the value of ρ for the flat FLRW model, $\rho_c = \frac{3}{8\pi}\left(H^2 - \frac{\Lambda}{3}\right)$.

[61] It follows from the FLRW dynamics that $\frac{|\Omega-1|}{\Omega} \propto R^{3\gamma-2}(t)$. $\gamma > 2/3$ if the strong energy condition holds, and in that case an initial value of Ω not equal to 1 is driven rapidly away from 1.

[62] One can evolve observed fluctuations backward to determine the amplitude of the fluctuation spectrum at a given "initial" time t_i. For t_i on the order of the Planck time, for example, Blau and Guth (1987) calculate that the fluctuations obtained by evolving backward from the time of recombination imply a density contrast of $\approx 10^{-49}$ at t_i, nine orders of magnitude smaller than thermal fluctuations. The comparison depends on the choice of the time t_i: if this is treated as a free variable, then there will be some time at which the fluctuations are comparable to thermal fluctuations.

It is not clear, for example, why the baryon-to-photon ratio, relevant to nucleosynthesis calculations, has the particular value it does. (This list could be extended.) The general complaint is that the Standard Model requires a variety of seemingly implausible assumptions regarding the initial state. Why did the universe start off with such a glorious pre-established harmony between causally disjoint regions? How was the initial energy density so delicately chosen that we are still close to the flat model? (And so on.) Although these features are all possible according to the Standard Model, the fact that they obtain seems, intuitively, to be incredibly improbable. The Standard Model treats these posits as brute facts not subject to further explanation.

By contrast, Guth proposed to supplement the Standard Model by modifying the very early expansion history of the universe, drawing on ideas in particle physics. Guth proposed that the universe underwent a transient phase of Λ-dominated, exponential expansion at roughly 10^{-35}s. Introducing this inflationary stage eases the conflict between a "natural" or "generic" initial state and the observed universe, in the following sense. Imagine choosing a cosmological model at random from among the space of solutions of EFE. Even without a good understanding of this space of solutions or how one's choice is to be "actualized," it seems clear that one of the maximally symmetric FLRW models must be an incredibly "improbable" choice.[63] New dynamics in the form of inflation makes it possible for "generic" pre-inflationary initial conditions to evolve into the uniform, flat state required by the Standard Model.[64] According to the Standard Model alone, what we observe is incredibly improbable; according to the Standard Model *plus inflation*, what we observe is to be expected.

This is an example of a general strategy, which I will call the "dynamical approach": given a theory that apparently requires special initial conditions, augment the theory with new dynamics such that the dependence on special initial conditions is reduced. McMullin (1993) describes a preference for this approach as accepting an "indifference principle," which states that a theory that is indifferent to the initial state, that is, robust under changes of it, is preferable to one that requires special initial conditions. Theorists who accept the indifference principle can identify fruitful problems by considering the contrast between a "natural" initial state and the observed universe and then seek new dynamics to reconcile the two.

This line of reasoning is frequently endorsed as a motivation for inflation in the huge literature on the topic following Guth's paper. However, a number of skeptics have challenged the dynamical approach as a general methodology and

---

[63] For any reasonable choice of measure over the space of solutions, these models are presumably a measure-zero subset.

[64] Inflation solves the horizon problem because the horizon distance increases exponentially during inflation; for a sufficiently long period of inflation, all the points on the surface of last scattering will have overlapping past light cones. The inflationary phase also reverses the dynamical feature of the FLRW models responsible for the flatness problem. Because $\gamma = 0$ (in the equation in f.n. 61) for most models of inflation, inflationary expansion drives Ω toward 1, enlarging the range of choices $\Omega(t_p)$ compatible with observations.

as a motivation for accepting inflation.⁶⁵ One line of criticism concerns whether inflation achieves the stated aim of eliminating the need for special initial conditions, as opposed to merely shifting it to a different aspect of the physics. In effect inflation exchanges the degrees of freedom associated with the spacetime geometry of the initial state for the properties of a field (or fields) driving an inflationary stage. This exchange has obvious advantages if physics can place tighter constraints on the relevant fields than on the initial state of the universe. What is gained, however, if the field (or fields) responsible for inflation has to be in a special state to trigger inflationary expansion, or to have other finely tuned properties, to be compatible with observations?

There are also direct challenges to the dynamical approach itself, sometimes presented in concert with advocacy of an alternative "theory of initial conditions" approach. First, why should we assume that the initial state of the universe is "generic"? Penrose, in particular, has argued that this proposal is not compatible with a neo-Boltzmannian account of the second law of thermodynamics (cf. Albrecht 2004). Penrose (1979) treats the second law as arising from a lawlike constraint on the initial state of the universe, requiring that it has low entropy. Rather than introducing a subsequent stage of dynamical evolution that erases the imprint of the initial state, we should aim to formulate a "theory of initial conditions" that accounts for its special features. Second, how should we make sense of the implicit probability judgments employed in these arguments? The assessment of an initial state as "generic," or, on the other hand, as "special," is based on a choice of measure over the allowed initial states of the system. But on what grounds is one measure to be chosen over another? Furthermore, how does a chosen measure relate to the probability assigned to the actualization of the initial state? It is clear that the usual way of rationalizing measures in statistical mechanics, such as appeals to ergodicity, do not apply in this case because the state of the universe does not "sample" the allowed phase space.⁶⁶

Assessing the dynamical approach depends on a number of central issues in philosophy of science. Philosophers steeped in debates regarding scientific explanation may find it exciting to discover a major scientific research program motivated by explanatory intuitions. Proponents of inflation often sound as though their main concern is to make the early universe safe for Reichenbach's principle of the common cause. Or, they emphasize the unification between particle physics and cosmology achieved in their models. While these connections are intriguing, they both must be treated with a grain of salt.⁶⁷ A more general question is whether the explanatory intuitions betray an overly strong rationalistic tendency to demand explanations of everything. Callender (2004a, b) argues in favor of accepting a posited initial state

---

⁶⁵ One of the main lines of criticism of inflation is due to Roger Penrose; see Penrose (2004, ch. 28) for a recent exposition. See Earman and Mosterin (1999) for a philosopher's take on inflation, Linde (2007), for example, for a recent review and Turok (2002) for a critical assessment.
⁶⁶ For further discussion, see, e.g., Callender (2004a); ?); Wald (2006); Wallace (2011).
⁶⁷ For further discussion of causality in relation to the horizon problem, see Earman (1995), and for a critical assessment of unification claims, see Zinkernagel (2002).

as a brute fact, in part by showing that purported "explanations" of it are mostly vacuous.[68]

A quite different approach purports to explain various features of the universe as necessary conditions for our presence as observers, to which we now turn.

7. Anthropic Reasoning

There has been a great deal of controversy regarding anthropic reasoning in cosmology in the last few decades.[69] Weinberg (2007) describes the acceptance of anthropic reasoning as a radical change for the better in how theories should be assessed, comparable to the introduction of symmetry principles. In assessing cosmological theories we need, on this view, to account for selection effects due to our presence as observers and to consider factors such as the number of observers predicted to exist by competing theories. How exactly this is to be done remains a matter of dispute. There is no widely accepted standard account of anthropic reasoning. Critics of this line of thought argue that insofar as anthropic reasoning introduces new aspects of theory assessment, as opposed to merely putting an anthropic gloss on some accepted inductive methodology, it is ill-motivated or even incoherent. A methodology that is itself controversial is not particularly useful in forging consensus, so the articulation and assessment of anthropic reasoning is clearly an essential task. Philosophers have already contributed to this effort and should continue to do so. My aim here is to provide a brief overview of the debate, with an emphasis on connections with the philosophical literature.

Two exemplary cases should suffice to introduce anthropic reasoning. Dirac (1937) noted that various "large numbers" defined in terms of the fundamental constants have the same order of magnitude. This coincidence (and others) inspired his "Large Number Hypothesis": dimensionless numbers constructed from the fundamental constants "are connected by a simple mathematical relation, in which the coefficients are of the order of magnitude unity" (Dirac, 1937, 323). Since one of these numbers includes the age of the universe t_0, so must they all. This implies time variation of the gravitational "constant" G. Dicke (1961) argued that attention to selection effects undermined the evidential value of this surprising coincidence. Surprise at the coincidence might be warranted if t_0 could be treated as "a random choice from a wide range of possible values" (Dicke, 1961, 440), but there can only be observers to wonder at the coincidence for some small range of t. Dicke (1961) argued that the value of t must fall within an interval such that Dirac's coincidence automatically holds given two necessary conditions for the existence of observers

---

[68] See Price (2004) for a defense of the opposing point of view, in an exchange with Callender (2004a,b).
[69] Barrow and Tipler (1986) is an influential early survey of the field; see Carr (2007) for a recent collection of essays.

like us.[70] The evidence allegedly provided by the large number coincidence bears no relation to the truth or falsity of Dirac's hypothesis or the Standard Model.[71] Taking the coincidence as evidence for the large number hypothesis would be as misguided as concluding (recycling Eddington's example) that there are no fish smaller than 6 inches in a pond based on the absence of such small fish in a fisherman's basket, even though the fisherman's net has gaps too large to hold these fish.

Attention has recently focused on a different kind of anthropic reasoning exemplified by Weinberg's (1987) prediction for Λ.[72] Just as in Dicke's arguments regarding t, within the Standard Model the value of Λ cannot be freely chosen. Because a Λ term does not dilute with expansion, a cosmological model with $\Lambda > 0$ will transition from matter-dominated to vacuum-dominated expansion. Weinberg showed that structure formation via gravitational enhancement stops in the vacuum-dominated stage. The existence of large gravitationally bound systems (large enough to lead to the formation of stars) then imposes an upper bound on possible values of Λ, keeping other aspects of the Standard Model fixed.[73] It is plausible to take the existence of gravitationally bound systems as a necessary precondition for the existence of observers. There is also a lower bound: a negative Λ term contributes to EFE like normal matter and energy, and adding a large negative Λ term leads to a model that recollapses before there is time for observers to arise.

So far the argument is similar to Dicke's elucidation of anthropic bounds on t. But Weinberg next predicted that Λ's observed value should be close to the mean of the values suitable for life. If we inhabit a "multiverse" in which the value of Λ varies in different regions,[74] the prediction is obtained by using the presence of observers as a selection effect. Weinberg assumed that the probability distribution for values of Λ in the multiverse is uniform within the anthropic bounds and that we are typical members of the reference class of observers in the universe Vilenkin (1995) calls this the "principle of mediocrity" (PM). In Bayesian terms, an initially flat probability distribution for the value of Λ is turned into a prediction—a sharply peaked distribution around a preferred value—by conditionalizing on the existence of large gravitationally bound systems, serving as a proxy for observers. Each of these assumptions is controversial. I will postpone more detailed discussion of the

---

[70] These necessary conditions are: (1) that main sequence stars are still burning, and (2) that an earlier generation of red giants had time to produce carbon in supernovae.

[71] Bayesians can account for this by explicitly conditionalizing on some claim characterizing the selection effect A: $P_s(\cdot) = P(\cdot|A)$. The selection effect may render an originally "informative" piece of evidence E useless, in that $P_s(E|H) \approx P_s(E|\neg H)$. In these terms, Dicke's argument shows that $P_s(LN|H_D) \approx P_s(LN|H_{SM}) \approx 1$, where LN is the large number coincidence, H_D is Dirac's cosmological theory, and H_{SM} is the Standard Model.

[72] This is not to say that Weinberg's paper is the first appearance of this kind of anthropic reasoning in contemporary cosmology; Collins and Hawking (1973) is an earlier influential example, in which similar reasoning is used to account for the isotropy of the universe.

[73] More precisely, the upper bound relates the Λ term to the total energy density in matter at the time when most galaxies formed; the upper bound on Λ is ≈ 200 times the present matter density. Considering variation of multiple parameters may undermine this bound; larger values of Λ can be tolerated if one increases the amplitude of the initial spectrum of density perturbations, for example. See Aguirre (2007) for a discussion of the problems associated with considering a single parameter.

[74] Weinberg (1987) did not base his suggestion on a particular multiverse proposal, instead listing four proposals that would provide a suitable setting for his argument.

multiverse until the next section and take up the PM shortly. The first assumption is often justified by appeals to simplicity or naturalness, but it is on unsure footing without further specification of how the multiverse is generated.[75] Nonetheless, Weinberg's prediction of a positive value of Λ within two orders of magnitude of currently accepted values has been widely cited as a striking success of anthropic reasoning.[76]

Different views regarding anthropic reasoning can be characterized in part by whether they take Weinberg's argument as a valid extension of Dicke's. Many anthropic skeptics accept Dicke's reasoning but see it as an illustration of how to take selection effects into account, without any truly anthropic elements (e.g., Earman, 1987; Smolin, 2007). Dicke simply follows through the consequences of the existence of main sequence stars and heavy elements. The nature of "observers" and whether they are typical members of a given reference class play no role. Furthermore, as Roush (2003) emphasizes, Dicke's argument devalues a particular body of evidence. The apparent coincidences that troubled Dirac reflect deep biases in the evidence available to us, and as a result have no value in assessing his hypothesis. Weinberg's argument, by contrast, takes the successful "prediction" of a surprising value for a particular parameter as evidence in favor of a multiverse. Thus it is more in line with Dirac's idea that such coincidences can be revealing rather than with Dicke's response. It also depends on assumptions regarding our "typicality" among members of a reference class, raising a number of issues that Dicke's argument avoids. Proponents of anthropic reasoning argue that these issues have to be dealt with in order to assess cosmological theories.

Some have argued that the PM must be assumed in order to extract any predictions at all from cosmological theories that describe an infinite universe.[77] Consider an observation O, for example that the CBR has an average temperature within the observer's Hubble volume of $T = 3.14159...K$, in agreement with the decimal expansion of π to some specified number of digits. Suppose we have a cosmological theory T that predicts the existence of an open FLRW model with infinite spatial slices Σ and also assigns a nonzero probability to O. Then there is an observer for whom O is true somewhere in the vast reaches of the infinite universe. The point generalizes to other observations and threatens to undermine the use of any observations to assess cosmological theories.[78] (This challenge arises even in the Standard Model, provided that the universe is not closed, and does not depend on more speculative multiverse proposals.) This skeptical conclusion can only be evaded by accepting the principle of mediocrity, according to this line of thought: we are interested not in

---

[75] There have been calculations for the prior probability distribution over Λ in different proposed multiverses; the assumption holds in some but not all of them (see, e.g., Garriga and Vilenkin 2000).

[76] This is a vast improvement on the estimates produced by particle physics, which are off by up to 120 orders of magnitude. In a later treatment, Weinberg argues for a lower anthropic bound, such that the probability assigned to current observations is either 5 or 12% (depending on other assumptions); see Weinberg (2007) for an overview and references.

[77] See, e.g., Vilenkin (1995); Bostrom (2002).

[78] Obviously this argument requires some assumptions regarding methodology; it is typically formulated within a Bayesian approach, and the conclusion need not follow on other accounts of inductive method. Shortly I will return to the question of whether this is a good argument even on a Bayesian approach.

the reports of such improbable "freak observers," but rather in our observations—where we regard ourselves as randomly selected from an appropriate reference class. Even "infinite universe" theories can make predictions by employing the PM, once the appropriate reference class has been specified.

The PM leads, unfortunately, to absurd results in other cases. These problems are arguably due to the explicit reliance on the choice of a reference class. This choice does not reflect a factual claim about the world, yet it can lead directly to striking empirical results, as illustrated in the Doomsday argument (e.g., Leslie 1992; Gott 1993; Bostrom 2002). The argument follows from applying the PM to one's place in human history, in particular by asserting that one should occupy a "typical" birth rank among the reference class consisting of all humans who have ever lived. This implies that there are roughly as many humans born before and after one's own birth. For this to be true, given the current rate of population growth, "doomsday"—a rapid drop in the growth rate of the human population—must be just around the corner.[79] The conclusion of the argument depends critically on the reference class. Starkman and Trotta (2006) argue that Weinberg's prediction of Λ is similarly sensitive to the reference class used in applying the PM.

Philosophers have discussed a number of other cases, from Sleeping Beauties to Presumptuous Philosophers, meant to test principles proposed for anthropic reasoning.[80] Stated more generally, these proposals regard how to incorporate indexical information (about, e.g., one's location in the history of mankind) in evidential reasoning. Straightforward modifications of the PM to avoid the Doomsday argument lead to counterintuitive results in these other cases. Bostrom (2002) advocates responding to the Doomsday argument by considering a different reference class when applying the PM, but his arguments that there is a unique reference class that resolves the problems are unconvincing. An alternative response is to take the number of observers in the reference class into account, by weighting the prior probability by this number.[81] For example, if a theory predicts that there will be 10^6 more observers (in the appropriate reference class) than a competing theory, then the prior probabilities should have this same ratio. This effectively blocks the Doomsday argument. It has unpalatable consequences of its own, however, if it is taken as a general methodological principle: it implies nearly unshakeable confidence in theories that predict large numbers of observers.[82]

---

[79] There are various different formulations of the argument (see Bostrom 2002 for an entry point into this literature). One formulation starts with the assumption that the probability of one's own birth rank being r is given by $Pr(r|N) = 1/N$, where N is the total number of humans ever born (assuming that $N \geq r$). If one further assigns a prior probability $Pr(N) = k/N$ (with a constant k), then the posterior probability obtained using Bayes's theorem is $Pr(N|r) = k/N^2$. It follows that there is a less than 5% probability that the total number of humans ever born will exceed $20r$. The argument is entirely general and results from invoking the PM in choosing a time within a process that extends over some finite duration.

[80] See Bostrom (2002) and Neal (2006) for discussions of the different versions of "anthropic reasoning" and the various puzzles they are meant to address.

[81] This was proposed by Dieks (1992) in response to the Doomsday argument; see Bostrom (2002) and Dieks (2007) for further discussion. The idea has also been discussed in light of Elga's (2000) Sleeping Beauty problem.

[82] Hence the Presumptuous Philosopher (see Bostrom 2002), whose posterior probability in the theory with more observers remains high despite receiving disconfirming evidence.

The combined effect of accepting PM and adjusting the priors to take account of the number of observers is to eliminate the dependence on a choice of a particular reference class, as Neal (2006) shows. Rather than introduce the reference class only to eliminate its impact, why not simply apply Bayesian conditionalization? Neal (2006) argues that standard Bayesian conditionalization on all nonindexical evidence available resolves the various puzzles associated with anthropic reasoning, with one caveat. On this approach anthropic reasoning is just a species of Bayesian conditionalization, and there is no need to introduce further methodological principles.[83] (It is crucial to conditionalize on everything because, as analyses of selection effects like Dicke's show, it is not always transparent which aspects of our evidence are relevant.)

This approach leads to the following assessment of anthropic predictions, such as Weinberg's prediction of Λ. Consider a multiverse theory T_M in which the value of Λ (and perhaps other parameters) takes on different values in different regions, contrasted with a theory T_1 in which the value of Λ is not fixed by theoretical principles, but does not vary in different regions. Suppose that Δ is the range of values of Λ compatible with all available evidence (including, for example, the existence of galaxies at high redshifts), and that according to T_M the fraction of regions with a value of Λ within Δ is given by f,[84] whereas T_1 assigns a probability of g to Δ. If one assigns equal priors to the two theories, the odds ratio for T_M to T_1 upon conditionalization will be given by f/g. The evaluation of the two theories depends on the probability they assign to a value of Λ within Δ. Whether the theory involves a "multiverse" with Λ varying in different regions is irrelevant to the comparison. The assessment also does not depend on considering how Δ compares to Δ', the range of parameter values of Λ compatible with "intelligent life" (or "advanced civilizations," etc.).

The caveat is that this analysis applies to universes in which the evidence is sufficiently rich to single out a unique observer. Neal acknowledges that in an infinite universe the argument above regarding "freak observers" poses a threat, given that there will be multiple observers with the same total body of evidence. He goes on to argue, however, that it is implausible that our evidential reasoning should depend on whether the universe is large enough to contain observers with exactly the same evidence. (This is, of course, exactly the context in which cosmologists feel the need to invoke the PM—see, e.g., Garriga and Vilenkin 2007.)

Philosophers have rejected the use of PM on other grounds. Norton (2010) has challenged the employment of probability distributions as a way of representing neutrality of evidential support, as part of a more general criticism of Bayesianism. He argues that the ability to get something from nothing—a striking empirical

---

[83] This is not to say that various considerations emphasized in the anthropic literature, such as the number of observers predicted to exist in a particular situation, are irrelevant. Rather, such factors can be accounted for in a Bayesian approach by paying careful attention to the details without adding further general principles.

[84] How to calculate this fraction depends upon the measure assigned over the multiverse, so that one can count regions. Here for the sake of illustration I will simply assume that such a fraction is well defined and that it yields a finite result.

result from innocuous assumptions, as in the Doomsday argument—reflects the extra representational baggage associated with describing ignorance using a probability measure. Probability measures are assumed to be countably additive, and this prevents them from expressing complete evidential neutrality. Assigning a uniform prior probability over the values of some parameter such as Λ implies that a value in a finite interval is disfavored by the evidence, rather than treating all of these values neutrally. One might hope that invoking a "random" choice among members of a reference class can underwrite ascriptions of probability. Norton counters that invocations of indifference principles such as PM actually support the ascription of neutral evidential warrant rather than uniform probability.

This brief survey has sketched three different lines of thought regarding anthropic reasoning. The most conservative option is to apply standard Bayesian methodology to cases where anthropic issues arise. The hope is that these cases can be treated by carefully attending to details without introducing new principles of general scope, and without invoking reference classes. One advantage of the conservative position is the availability of arguments in favor of the basic tenets of Bayesianism. It would be surprising if the validity of these methodological principles were in fact sensitive to whether we live in a vast, finite universe or a truly infinite universe. Against the conservatives, Norton directly attacks the use of probability to represent degrees of belief in cases of neutral support, such as undetermined parameters. This general criticism of Bayesianism has implications much broader than anthropic reasoning, but the conclusions it leads to in this case are similar to those of the conservative Bayesian: a rejection of the need to provide anthropic explanations of particular parameter values. Finally, a third position is that there are important and new methodological principles required to handle indexical information and selection effects. One goal of such an account would be to clarify this style of reasoning, which is widely employed within contemporary cosmology. What is lacking so far, in my view, is a compelling account of what these principles are and a motivation for accepting them.

8. Multiverse

Anthropic reasoning is often discussed in tandem with the multiverse (cf. Zinkernagel 2011). Weinberg's anthropic prediction for Λ is based on applying a selection effect to a multiverse in which the value of Λ varies in different regions. The multiverse idea has gained traction in part because Weinberg's approach is widely regarded as the only viable solution to the cosmological constant problem, and other similar problems may also admit only anthropic solutions.

Two different lines of thought in physics also support the introduction of the multiverse. First, within inflationary cosmology the same mechanism that produces a uniform, homogeneous universe on scales on the order of the Hubble radius leads to a dramatically different global structure of the universe. Inflation is said to be "generically eternal" in the sense that inflationary expansion continues in different regions of the universe, constantly creating bubbles such as our own universe, in which inflation is followed by reheating and a much slower expansion.[85] The individual bubbles are effectively causally isolated from other bubbles and are often called "pocket universes." The second line of thought relates to the proliferation of vacua in string theory. Many string theorists now expect that there will be a vast landscape of allowed vacua, with no way to fulfill the original hope of finding a unique compactification of extra dimensions to yield low-energy physics.

Both of these developments suggest treating the low-energy physics of the observed universe as partially fixed by parochial contingencies related to the history of a particular pocket universe. Other regions of the multiverse may have drastically different low-energy physics because, for example, the inflaton field tunneled into a local minima with different properties.[86] Here my main focus will be on a philosophical issue that is relatively independent of the details of implementation: In what sense does the multiverse offer satisfying explanations?

But, first, what do we mean by a "multiverse" in this setting?[87] These lines of thought lead to a multiverse with two important features. First, it consists of causally isolated pocket universes, and second, there is significant variation from one pocket universe to another. There are other ideas of a multiverse, such as an ensemble of distinct possible worlds, each in its own right a topologically connected, maximal spacetime, completely isolated from other elements of the ensemble. But in contemporary cosmology, the pocket universes are all taken to be effectively causally isolated parts of a single, topologically connected spacetime—the multiverse. Such regions also occur in some cosmological spacetimes in classical GR. In De Sitter spacetime, for example, there are inextendible timelike geodesics γ_1, γ_2 such that $J^-(\gamma_1)$ does not intersect $J^-(\gamma_2)$. In cases like this the definition of "effectively causally isolated" can be cashed out in terms of relativistic causal structure, but for a quantum multiverse the definition needs to be amended.

The example of pocket universes within De Sitter spacetime lacks the second feature, variation from one pocket universe to another. This can take several forms, from variation in the constants appearing in the Standard Models of cosmology and particle physics to variation of the laws themselves. Within the context of eternal inflation or the string theory landscape, what were previously regarded as "constants" may instead be fixed by the dynamics. For example, Λ is often treated as the consequence of the vacuum energy of a scalar field displaced from the minimum of its effective potential. The variation of Λ throughout the multiverse may then result from the scalar field settling into different minima. Greater diversity is suggested by

---

[85] Note that arguments to this effect usually involve a lot of hand-waving.
[86] The Everett interpretation of quantum mechanics attributes a branching structure to the universal wave function of the universe, and the individual branches can be regarded as something akin to pocket universes (see Wallace, this volume, for a discussion of the Everett interpretation). However, unlike the other accounts the laws of physics do not vary in the different branches. There is a clear distinction between the two cases, although recently there has been interest in exploring connections between these two lines of thought.
[87] See Tegmark (2009) for an influential classification of four different types or levels of the multiverse.

the string theory landscape, according to which the details of how extra dimensions are compactified and stabilized are reflected in different low-energy physics.

In the multiverse some laws will be demoted from universal to parochial regularities. But presumably there are still universal laws that govern the mechanism that generates pocket universes. This mechanism for generating a multiverse with varying features may be a direct consequence of an aspect of a theory that is independently well-tested. Rather than treating the nature of the ensemble as speculative or conjectural, one might then have a sufficiently clear view of the multiverse to calculate probability distributions of different observables, for example. In this case, there is a direct reply to multiverse critics who object that the idea is "unscientific" because it is "untestable": other regions of the multiverse would then have much the same status as other unobservable entities proposed by empirically successful theories.[88] Unfortunately for fans of the multiverse, the current state of affairs does not seem so straightforward. Although multiverse proposals are motivated by trends in fundamental physics, the detailed accounts of how the multiverse arises are typically beyond theoretical control. As long as this is the case, there is a risk that the claimed multiverse explanations are just-so stories where the mechanism of generating the multiverse is contrived to do the job. This strikes me as a legitimate worry regarding current multiverse proposals, but I will set this aside for the sake of discussion.

Suppose, then, that we are given a multiverse theory with an independently motivated dynamical account of the mechanism churning out pocket universes. What explanatory questions might this theory answer, and what is the relevance of the existence of the multiverse itself to its answers?[89] Here we can distinguish between two different kinds of questions. First, should we be surprised to measure a value of a particular parameter X (such as Λ) to fall within a particular range? Our surprise ought to be mitigated by a discussion of anthropic bounds on X, revealing various unsuspected connections between our presence and the range of allowed values for the parameter in question. But, as with Dicke's approach discussed above, this explanation can be taken to demystify the value of X without also providing evidence for a multiverse. The value of this discussion lies in tracing the connections between, e.g., the time-scale needed to produce carbon in the universe or the constraints on expansion rate imposed by the need to form galaxies. The existence of a multiverse is irrelevant to this line of reasoning.

A second question pertains to X, without reference to our observation of it: Why does the value of X fall within some range in a particular pocket universe? The answer to this question offered by a multiverse theory will apparently depend on contingent details regarding the mechanism that produced the pocket universe. This explanation will be historical in the sense that the observed values of the parameter will ultimately be traced back to the mechanism that produced the pocket

---

[88] This line of argument has appeared numerous times in the literature; see, e.g., Livio and Rees (2005) for a clear formulation.
[89] Here I am indebted to discussions with John Earman.

universe.[90] It may be surprising that various features of the universe are given this type of explanation rather than following as necessary consequences of fundamental laws. However, the success of historical explanations does not support the claim that other pocket universes must exist. Analogously, the success of historical explanations in evolutionary biology does not imply the existence of other worlds where pandas have more elegant thumbs.

To put the point in a slightly different form, the value of converting questions about modalities in cosmology into questions about location within a vastly enlarged ontology is not clear. Both types of questions can apparently be answered adequately without making the further ontological commitment to the actual existence of other pocket universes.

9. Conclusion

One theme running through the discussion above is the attempt to identify distinctive evidential challenges faced in cosmology. There is an echo of skepticism regarding the possibility of knowledge of the universe-as-a-whole in the discussion of global properties of the universe (section 5). Local observations are not sufficient to warrant conclusions regarding global properties without help from general principles like the cosmological principle, which is itself on unsure footing. This does not, however, support a general skepticism about cosmology. Most contemporary research in cosmology is compatible with agnosticism regarding the global properties of the universe. The challenges arise, not from the limits imposed by the causal structure of GR, but from the difficulty in gaining access to the relevant phenomena via independent routes. As the discussion in section 3 illustrates, assuming that the Standard Model is basically correct makes it possible to infer the presence of dark matter and dark energy. It is difficult to rule out the possibility that the same observations used as the basis for this inference instead reveal flaws in the Standard Model. Yet this does not mean that all the responses to the observations should be given equal credence. Philosophers of science ought to offer an account of empirical support that clarifies the assessment of different responses. Regarding early universe cosmology (section 6), the theory being used to describe the underlying physics is tested through its role in providing a reconstruction of the universe's history. The field has been partially driven by strong explanatory intuitions favoring a theory that renders the observed history probable or expected, although it is unclear how to move beyond intuitive discussions of probability. Cosmologists have to face the possibility that the data they use to assess theories is subject to unexpected anthropic selection effects (section 7). Whether these selection effects can be treated within standard approaches to confirmation theory or require new

---

[90] The explanation may also be path-dependent in the sense of depending not just on an initial state, but on various stochastic processes leading to the formation of the pocket universe.

principles of anthropic reasoning is currently being debated. Finally, cosmologists may also see their explanatory aims change, with various features of the universe traced to environmental features of our pocket universe rather than being derived from dynamical laws (section 8).

REFERENCES

Aguirre, A. (2007). Making predictions in a multiverse: Conundrums, dangers, coincidences. In *Universe or multiverse?* ed. B. Carr, 367–386.

Albrecht, A. (2004). Cosmic inflation and the arrow of time. In *Science and Ultimate Reality*, ed. J. D. Barrow, P. C. W. Davies, C. L. Harper. Cambridge: Cambridge University Press, 363–401.

Alpher, R. A., H. Bethe, and G. Gamow (1948). The origin of chemical elements. *Physical Review* 73: 803–804.

Balashov, Y. (2002). Laws of physics and the universe. In *Einstein Studies in Russia*, 107–148. Boston, MA/Basel/Berlin: Birkhäuser.

Barrow, J. D., and F. J. Tipler (1986). *The anthropic cosmological principle.* Oxford: Oxford University Press.

Beisbart, C. (2009). Can we justifiably assume the cosmological principle in order to break model underdetermination in cosmology? *Journal for General Philosophy of Science* 40: 175–205.

Bekenstein, J. D. (2010). Alternatives to dark matter: Modified gravity as an alternative to dark matter. *ArXiv e-prints* 1001.3876.

Bertone, G., D. Hooper, and J. Silk (2005). Particle dark matter: Evidence, candidates and constraints. *Physics Report* 405: 279–390. hep-ph/0404175.

Bianchi, E., and C. Rovelli (2010).Why all these prejudices against a constant? *ArXiv* preprint arXiv:1002.3966.

Blau, S. K., and A. Guth (1987). Inflationary cosmology. In *300 years of gravitation*, ed. S. W. Hawking and W. Israel. Cambridge: Cambridge University Press, 524–603.

Bostrom, N. (2002). *Anthropic bias: Observation selection effects in science and philosophy.* New York: Routledge.

Buchert, T. (2008). Dark energy from structure: a status report. *General Relativity and Gravitation* 40(2): 467–527.

Caldwell, R. R., and A. Stebbins (2008). A test of the Copernican principle. *Physical Review Letters* 100: 191302.

Callender, C. (2004a). Measures, explanations and the past: Should special initial conditions be explained? *British Journal for the Philosophy of Science* 55: 195–217.

———. (2004b). There is no puzzle about the low-entropy past. In *Contemporary debates in philosophy of science*, ed. Christopher Hitchcock. Oxford: Blackwell Publishing, 240–256.

Callender, C., and N. Huggett, eds. (2001). *Philosophy meets physics at the planck scale.* Cambridge: Cambridge University Press.

Carr, B., ed. (2007). *Universe or multiverse?* Cambridge: Cambridge University Press.

Carrera, M., and D. Giulini (2010). Influence of global cosmological expansion on local dynamics and kinematics. *Reviews of Modern Physics* 82: 169.

Clarkson, C., and R. Maartens (2010). Inhomogeneity and the foundations of concordance cosmology. *Classical and Quantum Gravity* 27: 124008.

Clifton, T., C. Clarkson, and P. Bull (2011). The isotropic blackbody cmb as evidence for a homogeneous universe. *ArXiv* preprint arXiv:1111.3794.

Coles, P., and G. F. R. Ellis (1997). *Is the universe open or closed?* Cambridge: Cambridge University Press.

Collins, C. B., and S. W. Hawking (1973).Why is the universe isotropic? *Astrophysical Journal* 180: 317–334.

Cooperstock, F. I., and S. Tieu (2007). Galactic dynamics via general relativity: A compilation and new developments. *International Journal of Modern Physics A* 22: 2293–2326.

Curiel, E. N. (1999). The analysis of singular spacetimes. *Philosophy of Science* 66: S119–S145.

Dicke, R. (1961). Dirac's cosmology and Mach's principle. *Nature* 192: 440–441.

Dieks, D. (1992). Doomsday, or: The dangers of statistics. *Philosophical Quarterly* 42: 78–84.

———. (2007). Reasoning about the future: Doom and beauty. *Synthese* 156: 427–439. ISSN 00397857. http://www.jstor.org/stable/27653528.

Dirac, P. A. M. (1937). The cosmological constants. *Nature* 139: 323.

Dodelson, S. (2003). *Modern cosmology*. New York: Academic Press.

Earman, J. (1987). The SAP also rises: A critical examination of the anthropic principle. *American Philosophical Quarterly* 24, 307–317.

———. (1995). *Bangs, crunches, whimpers, and shrieks*. Oxford: Oxford University Press.

———. (2001). Lambda: The constant that refuses to die. *Archive for History of Exact Sciences* 55: 189–220.

———. (2006). The "past hypothesis": Not even false. *Studies in History and Philosophy of Modern Physics* 37: 399–430.

Earman, J., and J. Mosterin (1999). A critical analysis of inflationary cosmology. *Philosophy of Science* 66: 1–49.

Earman, J., and J. Roberts (1999). *Ceteris Paribus*, there is no problem of provisos. *Synthese* 118: 439–478.

Ehlers, J., P. Geren, and R. K. Sachs (1968). Isotropic solutions of the Einstein–Liouville equations. *Journal of Mathematical Physics* 9: 1344–1349.

Einstein, A. (1917). Kosmologische Betrachtungen zur allgemeinen Relativitätstheorie. *Preussische Akademie derWissenschaften (Berlin). Sitzungsberichte*, 142–152.

Elga, A. (2000). Self-locating belief and the Sleeping Beauty problem. *Analysis* 60: 143–147.

Ellis, G. F. R. (1971). Topology and cosmology. *General Relativity and Gravitation* 2: 7–21.

———. (1980). Limits to verification in cosmology. In *9th Texas Symposium on Relativistic Astrophysics*, Vol. 336 of *Annals of the New York Academy of Sciences*, 130–160.

———. (1999). Before the beginning: Emerging questions and uncertainties. *Astrophysics and Space Science* 269: 691.

———. (2007). Issues in the philosophy of cosmology. In *Handbook for the philosophy of physics*, ed. J. Earman and J. Butterfield, vol. Part B. 1183–1286. Amsterdam: Elsevier.

———. (2011). Inhomogeneity effects in cosmology. *Classical and Quantum Gravity* 28: 164001.

Ellis, G. F. R., R. Maartens, and S. D. Nel (1978). The expansion of the universe. *Monthly Notices of the Royal Astronomical Society* 184: 439–465.

Ellis, G. F. R., S. D. Nel, R. Maartens, W. R. Stoeger, and A. P. Whitman (1985). Ideal observational cosmology. *Physics Reports* 124: 315–417.

Ellis, G. F. R., and T. Rothman (1993). Lost horizons. *American Journal of Physics* 61: 883–893.
Ellis, G. F. R., and D. W. Sciama (1972). Global and non-global problems in cosmology. In *General Relativity: Papers in Honour of J. L. Synge*, ed. L. O'Raifeartaigh. Oxford: Clarendon Press, 35–59.
Frieman, J., M. Turner, and D. Huterer (2008). Dark energy and the accelerating universe. *ArXiv* preprint arXiv:0803.0982.
———. (2008). Prediction and explanation in the multiverse. *Physical Review* D77: 043526.
Geroch, R., L. Can-bin, and R. Wald (1982). Singular boundaries of spacetimes. *Journal of Mathematical Physics* 23: 432–435.
Geroch, R., and G. Horowitz (1979). Global structure of spacetimes. In Hawking and Israel (1979), 212–293.
Glymour, C. (1977). Indistinguishable spacetimes and the fundamental group. In *Foundations of spacetime theories*, ed. J. Earman, C. Glymour and J. Statchel, Vol. VIII of Minnesota Studies in the Philosophy of Science. Minneapolis: University of Minnesota Press, 50–60.
Gott, J. R. (1993). Implications of the Copernican principle for our future prospects. *Nature* 363: 315–319.
Guth, A. (1981). Inflationary universe: A possible solution for the horizon and flatness problems. *Physical Review D* 23: 347–56.
Harper, W. L. (2012). *Isaac Newton's scientific method: Turning data into evidence about gravity and cosmology*. Oxford: Oxford University Press.
Hawking, S., and W. Israel, eds. (1979). *General relativity: An Einstein centenary survey*. Cambridge: Cambridge University Press.
Hubble, E. (1929). A relation between distance and radial velocity among extra-galactic nebulae. *Proceedings of the National Academy of Sciences* 15: 168–173.
Huterer, Dragan (2010). The accelerating universe. arXiv:1010.1162.
Jeans, J. H. (1902). The stability of a spherical nebula. *Philosophical Transactions of the Royal Society of London. Series A, Containing Papers of a Mathematical or Physical Character* 199: 1–53.
Kirshner, R. P. (2010). Foundations of supernova cosmology. In *Dark energy: Observational and theoretical approaches*, ed. P. Ruiz-Lapuente. Cambridge: Cambridge University Press, 151–176.
Kragh, H. (1996). *Cosmology and controversy*. Princeton: Princeton University Press.
Krauss, L. M., and M. S. Turner (1999). Geometry and destiny. *General Relativity and Gravitation* 31: 1453–1459. astro-ph/9904020.
Lachieze-Rey, M., and J. P. Luminet (1995). Cosmic topology. *Physics Reports* 254: 135–214.
Leslie, J. (1992). Doomsday revisited. *Philosophical Quarterly* 42: 85–89. http://www.jstor.org/stable/2220451.
Li, A. (2003). Cosmic needles versus cosmic microwave background radiation. *Astrophysical Journal* 584: 593.
Lifshitz, Y. M. (1946). On the gravitational stability of the expanding universe. *Journal of Physics USSR* 10: 116–129.
Linde, A. (2007). Inflationary cosmology. In Inflationary cosmology, volume 738 of *Lecture Notes in Physics*, ed. M. Lemoine, J. Martin, and P. Peter. Berlin: Springer, 1–54.
Livio, M., and M. Rees (2005). Anthropic reasoning. *Science* 309: 1022–1023.
Longair, M. (2006). *The cosmic century*. Cambridge: Cambridge University Press.
Malament, D. (1977). Observationally indistinguishable spacetimes. In *Foundations of spacetime theories*, ed. J. Earman, C. Glymour, and J. Statchel, Vol. 8 of Minnesota

Studies in the Philosophy of Science. Minneapolis: University of Minnesota Press, 61–80.

Manchak, J. B. (2009). Can we know the global structure of spacetime? *Studies in History and Philosophy of Modern Physics* 40: 53–56.

McMullin, E. (1993). Indifference principle and anthropic principle in cosmology. *Studies in the History and Philosophy of Science* 24: 359–389.

Milgrom, M. (1983). A modification of the Newtonian dynamics: Implications for galaxy systems. *Astrophysical Journal* 270: 384–389.

Munitz, M. K. (1962). The logic of cosmology. *British Journal for the Philosophy of Science* 13: 34–50.

Neal, R. M. (2006). Puzzles of anthropic reasoning resolved using full non-indexical conditioning. Unpublished, available at *ArXiv*: math/0608592.

North, J. D. (1965). *The measure of the universe*. Oxford: Oxford University Press.

Norton, J. D. (2010). Cosmic confusions: Not supporting versus supporting not. *Philosophy of Science* 77: 501–523.

———. (2011). Observationally indistinguishable spacetimes: A challenge for any inductivist. *Philosophy of science matters: The philosophy of Peter Achinstein*, 164.

Olive, K. A., G. Steigman, and T. P. Walker (2000). Primordial nucleosynthesis: Theory and observations. *Physics Reports* 333: 389–407.

Ostriker, J. P., and P. J. E. Peebles (1973). A numerical study of the stability of flattened galaxies: Or, can cold galaxies survive? *Astrophysical Journal* 186: 467–480.

Ostriker, J. P., and P. J. Steinhardt (1995). The observational case for a low-density universe with a non-zero cosmological constant. *Nature* 377: 600–602.

Pauri, M. (1991). The universe as a scientific object. In *Philosophy and the origin and evolution of the universe*, ed. E. Agazzi and A. Cordero. Dordrecht: Kluwer Academic Publishers, 291–339.

Peebles, P. J. E. (1993). *Principles of physical cosmology*. Princeton: Princeton University Press.

Penrose, R. (1979). Singularities and time-asymmetry. In Hawking and Israel (1979), 581–638.

———. (2004). *The road to reality*. London: Jonathan Cape.

Polchinski, J. (2006). The cosmological constant and the string landscape. *Arxiv* preprint hep-th/0603249.

Price, H. (2004). On the origins of the arrow of time: Why there is still a puzzle about the low entropy past. In *Contemporary debates in philosophy of science*, ed. Christopher Hitchcock. Oxford: Blackwell Publishing.

Rindler, W. (1956). Visual horizons in world models. *Monthly Notices of the Royal Astronomical Society* 116: 662–677.

Roush, S. (2003). Copernicus, Kant, and the anthropic cosmological principles. *Studies in History and Philosophy of Modern Physics* 34: 5–35.

Rugh, S. E., and H. Zinkernagel (2002). The quantum vacuum and the cosmological constant problem. *Studies in History and Philosophy of Modern Physics* 33: 663–705.

———. (2009). On the physical basis of cosmic time. *Studies in History and Philosophy of Modern Physics* 40: 1–19.

Sanders, R. H., and S. S. McGaugh (2002). Modified Newtonian dynamics as an alternative to dark matter. *Annual Reviews of Astronomy and Astrophysics* 40: 263–317. arXiv: astro-ph/0204521.

Saunders, S. (2002). Is the zero-point energy real? In *Ontological aspects of quantum field theory*, eds. M. Kuhlmann, H. Lyre, and A. Wayne. Singapore: World Scientific, 313–343.

Scheibe, E. (1991). General laws of nature and the uniqueness of the universe. In *Philosophy and the origin and evolution of the universe*, ed. E. Agazzi and A. Cordero. Dordrecht: Kluwer Academic Publishers, 341–360.

Smeenk, C. (2012). Einstein's role in the creation of relativistic cosmology. In *Cambridge companion to Einstein*, ed. M. Janssen and C. Lehner. Cambridge: Cambridge University Press.

Smith, S. (2002). Violated laws, *ceteris paribus* clauses, and capacities. *Synthese* 130: 235–264.

Smolin, L. (2000). The present moment in quantum cosmology: Challenges to the arguments for the elimination of time. In *Time and the Instant*, ed. R. Durie. Manchester: Clinamen Press, 112–143.

———. (2007). Scientific alternatives to the anthropic principle. In *Universe or multiverse*, ed. B. Carr. Cambridge: Cambridge University Press, 323–366.

Sotiriou, T. P., and V. Faraoni (2010). f (r) theories of gravity. *Reviews of Modern Physics* 82: 451.

Starkman, G. D., and R. Trotta (2006). Why anthropic reasoning cannot predict Lambda. *Phys. Rev. Lett.* 97: 201301. astro-ph/0607227.

Stoeger, W. R., R. Maartens, and G. F. R. Ellis (1995). Proving almost-homogeneity of the universe: An almost Ehlers-Geren-Sachs theorem. *Astrophysical Journal* 443: 1–5.

Sumner, T. (2002). Experimental searches for dark matter. *Living Reviews in Relativity* 5: 183.

Tegmark, M. (2009). The multiverse hierarchy. *Arxiv* preprint arXiv: 0905.1283.

Torretti, R. (2000). Spacetime models of the world. *Studies in the History and Philosophy of Modern Physics* 31: 171–186.

Trimble, V. (1987). Existence and nature of dark matter in the universe. *Annual Review of Astronomy and Astrophysics* 25: 425–472.

Turok, N. (2002). A critical review of inflation. *Classical and Quantum Gravity* 19: 3449.

Uzan, J. P. (2010). Dark energy, gravitation, and the Copernican principle. In *Dark energy: Observational and theoretical approaches*, ed. P. Ruiz-Lapuente. Cambridge: Cambridge University Press, 3–47.

Uzan, J. P., C. Clarkson, and G. F. R. Ellis (2008). Time drift of cosmological redshifts as a test of the Copernican principle. *Physical Review Letters* 100: 191303.

Vanderburgh, W. L. (2003). The dark matter double bind: Astrophysical aspects of the evidential warrant for general relativity. *Philosophy of Science* 70: 812–832.

———. (2005). The methodological value of coincidences: Further remarks on dark matter and the astrophysical warrant for general relativity. *Philosophy of Science* 72: 1324–1335.

Vilenkin, A. (1995). Predictions from quantum cosmology. *Physical Review Letters* 74: 846–849.

Wald, R. (1984). *General relativity*. Chicago: University of Chicago Press.

———. (2006). The arrow of time and the initial conditions of the universe. *Studies in History and Philosophy of Modern Physics* 37: 394–398.

Wallace, D. (2011). The logic of the past hypothesis. Unpublished. Available at http://philsci-archive.pitt.edu/id/eprint/8894.

Weinberg, S. (1972). *Gravitation and cosmology*. New York: John Wiley & Sons.

———. (1987). Anthropic bound on the cosmological constant. *Physical Review Letters* 59: 2607.
———. (1989). The cosmological constant problem. *Reviews of Modern Physics* 61: 1–23.
———. (2007). Living in the multiverse. In *Universe or multiverse*, ed. B. Carr. Cambridge: Cambridge University Press, 29–42.
———. (2008). *Cosmology*. New York: Oxford University Press.
Wolf, J. A. (2011). *Spaces of constant curvature*. 6th ed. Providence, RI: American Mathematical Society.
Zinkernagel, H. (2002). Cosmology, particles, and the unity of science. *Studies in the History and Philosophy of Modern Physics* 33: 493–516.
———. (2011). Some trends in the philosophy of physics. *Theoria* 71: 215–241.

Index

Ab initio strategy, the tyranny of scales, 263–64
Abraham-Lorentz equation, 119, 131, 137
ACDM model, 617, 618
Action-at-a-distance forces, 45
   rigid body mechanics, 73
Action principle, unitary equivalence, 491
Additivity, phase transitions, 201n2
Advanced Green's functions, classical mechanics, 109
   bounded spatial domain, wave equation in, 121–23
   Cauchy problem, 120
   Dirac delta function, 111–12
   Fourier transformations, 111
   overview, 109–13
   point source, 110–11
   spatial propagation, 120–21
   undamped harmonic oscillator, 110
   wave equation, 120–23
Airplane wings, boundary layers, 26
Airy, George Biddell, 17
"Algebraic Imperialism," unitary equivalence, 513
Analytic mechanics, 44
Anderson, Edward, 562
Anderson, Hans, 163
Anderson, Philip, 2, 142
Andrews, Thomas, 142, 147, 191
Anthropic reasoning, 638–43
   Bayesian conditionalization, 642, 643
   cosmic background radiation (CBR), 640
   Doomsday argument, 641, 643
   "Large Number Hypothesis," 638–39
   multiverse, 639
   "principle of mediocrity" (PM), 639–42
Anti-atomism, 55
Antifoundationalism, effective field theory (EFT), 240
Anti-particles, symmetry, 294n10
Antireductionism, effective field theory (EFT), 240
Approximations
   effective field theory (EFT), 243
   hydrodynamics, philosophy of, 31
Arc length along the slot, rigid body mechanics, 71–72, *72*
Asymmetry, indistinguishability, 365n33
Averaging and homogenization, differences, 277–78
Axiomatic presentation, 47–58
   choice of scale length, 50, *50*
   conceptual system, 57
   decompositional programs, 53, *53*
   degeneration, 53
   "escape hatches," 56
   force of rolling friction, 54, *54*
   foundational point of view, 48–49
   freezing to a scale level, 55
   homogenization theory, 51, 53, 55
   point-mass swarm, 51
   points, 49–50
   reduced variables, 51, *52,* 55n14, 56
   Riemann-Hugoniot approach to shock waves, 53n12
   rigid crystalline forms, 51
   rolling on a rigid track, 54–55
   rotation, 48
   scale sizes, relationships between, 51–52
   shock waves, Riemann-Hugoniot approach, 53n12
   Stieltjes-Lesbeque integration, 53n12
   theory facades, 48, 57
   viscosity of fluid, 54

Bach, Alexander, 355–56
Back-bending, infinite idealization, 206–8, *207*
Balance principles, continuum mechanics, 94
Barbour, Julian
   Hole Argument (Einstein), 578
   Mach's Principle: From Newton's Bucket to Quantum Gravity, 557
   rationality, failure of, 539–40
Batterman, R., emergence of EFT, 248–51
Bayesian conditionalization, 642, 643
Bead sliding on rigid wire, 71
Bélanger, Jean Baptiste, 22
Bell, John S., 481–82, 569
Bernoulli, Daniel. *See* Bernoulli-Euler beam; Bernoulli-Euler equation; Bernoulli's Law
Bernoulli-Euler beam, 99, *100*
Bernoulli-Euler equation, 90, *90,* 91
Bernoulli's Law, 15–16, 27
Best classical approximation, *46*
Best matching, Machian relationalism, *559,* 559–60
Best Systems prescription, have-it-all relationalism, 566
Big Bang, 609, 615
   electroweak theory, 303, 403
   global structure, cosmology, *631*
   singularity, 600

"Big dessert" assumption, 404
Billiard collisions, point-mass mechanics, 68–69, 69
Bit commitment problem, quantum mechanics, 429–30
Bjerknes, Vilhelm, 19
Black-body spectrum of radiation, 615
Black hole evaporation, 518
Black, Max, 374
Bloch sphere, 423
Blocks-and-cords construction, 91, 92
Block transforms, 170–72, 172
Blowup, classical mechanics, 45n3
Bohm-Aharonov effect, 507
Bohr, Niels, 9, 463
Boiling, 149, 163–64
   fluctuations, 151
   phase transitions, 147
Boltzmann equation, 614n16
Boltzmann-Gibbs formulation
   Ising model, 154
   matter, infinities and renormalization, 180
Boltzmann, Ludwig
   Boltzmann equation, 142
   indistinguishability
      entropy, 346–47
      haecceitism, 356
      Maxwell-Boltzmann statistics, 371n42
      quantum, indistinguishability and, 342
      thermodynamics, 343
   mechanics, principles of, 53
   statistical physics, theory of, 142
Born, Max, 431, 434
Born rule
   classic regime, 8, 417, 419, 431
   consistency, 442
   general phenomenology of measurements, 446, 448
   measurement, 451–52, 454
   proper mixtures, 424
   unsharp spin measurements, 445
Bose-Einstein, 206, 208
Bottom-up approach
   effective field theory (EFT), 224, 229–30
   the tyranny of scales, 257
Boundary conditions
   Dirichlet variety of, 122
   global spacetime structure, 600
   radiation theory, 123–34
Boundary layers, 13, 23–27, 53n12
   airplane wings, 26
   discontinuity surface, 24–25
   eddy resistance, 24, 25
   and modules, 37
   ships, 24
   skin resistance, 24
   wave resistance, 24
Bounded spatial domain, wave equation in, 121–23
Boussinesq, Joseph, 17
Branch count, Everett interpretation, 477–78
Broglie-Bohm theory, 431, 440, 482

Broken symmetries
   matter, infinities and renormalization, 144–45
Brown, Harvey R., 541–43, 544n46, 569, 570n92
Brownian particle, 437
BSW action, Machian relationalism, 562–64
Bubbles within bubbles, the tyranny of scales, 267, 284
Bulls-eyes, rigid body mechanics, 75–76
Burgess, C. P., 234–35
Butterfield, Jeremy, 219, 260–64

Cabibbo-Kobayashi-Maskawa framework, 402
Callen-Symanzik equation, 180
Caloric curve, back-bending of, 207
Canonical anticommutation relations (CARS), 489, 490, 501–2, 504
   "uniqueness results," 494, 497
Canonical ensemble, 147
Cao, T., 243, 244, 249–41, 407, 411
Carnap, Rudolf, 369, 372
CARS. See Canonical anticommutation relations (CARS)
Cartan distribution, 324
Cartesian coordinates, 536
Cartesian locations, 63
Cartesian motion, 524–26
Cassini space probe to Saturn, 61, 61
Cat example
   Everett interpretation, 464–65, 467
   Schrödinger, quantum mechanics, 435
Cauchy, Augustin
   continuum mechanics, "Cauchy's Law," 98–99
   point-mass mechanics, 67
   rigid body mechanics, "Cauchy's Law," 78n32
   surface waves, 17
   the tyranny of scales, Euler's recipes, 273–75
   vortex motion, 17
Cauchy integral, dispersion theory, 133
Cauchy problem
   advanced Green's functions, 120
   retarded Green's functions, 120
Cauchy's equation, the tyranny of scales, 270–75
"Cauchy's Law," 78n32, 98–99
Cauchy surface
   global structure, cosmology, 629
   spacetime properties, 597
Causal continuity condition, spacetime properties, 597
Causal curve, relativistic spacetime, 591
Causal directionality, retarded Green's functions, 116
Causal future, relativistic spacetime, 590–91
"Causality" as initial value problem, retarded Green's functions, 116–20
"Causality conditions," global structure, 629
Causal simplicity condition, spacetime properties, 597
Causal structure, spacetime properties, 596
Causation in classical mechanics, 107–40
   Abraham-Lorentz equation, 119, 131, 137
   advanced Green's functions, 4, 109

bounded spatial domain, wave equation in, 121–23
Cauchy problem, 120
Dirac delta function, 111–12
Fourier transformations, 111
overview, 109–13
point source, 110–11
spatial propagation, 120–21
undamped harmonic oscillator, 110
wave equation, 120–23
backward causation, 108n6
dispersion theory, 132–37
  Cauchy integral, 133
  description, 109
  Fourier transformations, 133, 135
  Hilbert transform pairs, 134, 136
  Kramers-Kronig relations, 135–36
  passivity and causality, 137
Eliezer's theorem, 132
Lorentz-Dirac equation
  description, 109
  point-particle electrodynamics, 132
overview, 4
point-particle electrodynamics, 129–32
radiation theory, 123–29
  boundary conditions, 123–34
  description, 109
  finiteness condition, 124
  Fourier transformation technique, 126–27
  Helmholtz equation, 123, 125–27, 129–30
  incoming waves, 127
  Laplace transform technique, 125
  outgoing waves, 127
  Principle of Limiting Amplitude, 128
  reasons for, 123–29
  reduced wave equation, 125. *See also* Helmholtz equation
  Sommerfeld radiation condition. *See* Sommerfeld radiation condition
  time dependent boundary condition, 124
  time-harmonic waves, 127–29
  waveguide, *124*
retarded Green's functions
  Abraham-Lorentz equation, 119
  bounded spatial domain, wave equation in, 121–23
  Cauchy problem, 120
  causal directionality, 116
  "causality" as initial value problem, 116–20
  contours for damped oscillator, *115*
  damping, added, 113–16, 135–36
  Dirac delta function, 111–12, 119
  "final condition," 119
  "final value problem," 119
  Fourier transformations, 111, 113, 114
  initial value problem, 116–20
  Lebesgue measure zero, 118
  overview, 4, 109
  physical motivation, 113
  point source, 110–11
  privileging, 113–20
  Residue Theorem, 114

spatial propagation, 120–21
undamped harmonic oscillator, 110
wave equation, 120–23
time-reversal invariance, 136n37
CBR. *See* Cosmic background radiation (CBR)
CCCRs. *See* Circular Canonical Commutation Relations (CCCRs)
CERN's Large Hadron Collider, 305
"Ceteris paribus" laws, 626n44
Chain reaction, unification in physics, 387–88
Chandler, David, 163
"Chaotic cosmology" program, 631–32
Charge conjugation, symmetry, 292–94
Charleton, Walter, 526
Chia Chiao Lin, 20
Choice of scale length, axiomatic presentation, 50, *50*
Christenson, Cronin, Fitch, and Turlay (CPT theorem), 294–95
Chronology condition, spacetime properties, 596
Chronology violating region of spacetime, relativistic spacetime, 591
Circular Canonical Commutation Relations (CCCRs), 506
Circular motion, 525
CKM. *See* Cabibbo-Kobayashi-Maskawa framework
Classical distillations of quantum processes, rigid body mechanics, 74–75
Classical dynamical behavior, quantum mechanics, 438
Classical electrodynamics, Everett interpretation, 466n6
Classical mechanics, 43–106
  action-at-a-distance forces, 45
  analytic mechanics, 44
  axiomatic presentation, 47–58
    choice of scale length, 50, *50*
    conceptual system, 57
    decompositional programs, 53, *53*
    degeneration, 53
    "escape hatches," 56
    force of rolling friction, 54, *54*
    foundational point of view, 48–49
    freezing to a scale level, 55
    homogenization theory, 51, 53, 55
    point-mass swarm, 51
    points, 49–50
    reduced variables, 51, *52*, 55n14, 56
    Riemann-Hugoniot approach to shock waves, 53n12
    rigid crystalline forms, 51
    rolling on a rigid track, 54–55
    rotation, 48
    scale sizes, relationships between, 51–52
    shock waves, Riemann-Hugoniot approach, 53n12
    Stieltjes-Lesbeque integration, 53n12
    theory facades, 48, 57
    viscosity of fluid, 54
  best classical approximation, 46
  blowup, 45n3

Classical mechanics (*Cont.*)
  causation in. *See* Causation in classical mechanics
  composite, 44
  "conceptually simple surrogate, 44
  connected rigid parts, *44*
  contact forces, 45
  continua, 44
  continuum mechanics, 83–104
    applied force, 84
    balance principles, 94
    Bernoulli-Euler beam, 99, *100*
    Bernoulli-Euler equation, 90, *90,* 91
    blocks-and-cords construction, 91, *92*
    "Cauchy's Law," 98–99
    compatibility, 85, *85*
    composite continua, 88
    constitutive assumptions, 98
    decorated points, *84,* 84–85
    dimensionless point cube, 95n42
    drumheads, 92
    Hooke's law, 99
    inertial force, 97n44
    infinitesimal cubes, 96
    infinitesimal traction vectors, 85
    labyrinth of the continuum, 90, 101
    lattice defects, 98
    loaded beam, 91n40
    material derivatives, 87
    mechanical elements, 91
    "method of extensive abstraction," 96n43
    physical infinitesimals, 84–85, *85*
    point-mass setting, 98
    Problem of the Physical Infinitesimal, 86, 90, 93
    reference plane, *96*
    response planes, 95
    shearing pattern, 85
    short-range forces, 97, *97,* 98
    special force laws, 99
    strain tensors, 85, 95, *95*
    stress and strain tensors, 96n43
    stress-energy tensors, 84
    stress tensors, 85, 95, *95,* 96
    surface forces, 87
    tensor fields, 98
    toothpaste, 89, *89*
    traction vectors, *83,* 83–84, 93
    vibrating strings, 92, *92,* 101
    wave movements, 88, *88*
  enriched relationalism, 545–53
    diffeomorphism group (DPM), 547, 550
    embedding of relational history, 546
    Galilei group, 549–50
    kinematic shift argument, 546–47
    Maxwell group, 550–51
    Maxwellian relationalist, 552
    Newton-Cartan theory, 551, 553
    nonsimultaneous events, 549
    Sklarations, 548–49
  "false facts," 47
  fields, 44
  flexible beam, *44*
  flexible bodies, 44
  mass point lattice, *44*
  mathematical complexity, 45
  new interest in, 2
  Newton's laws. *See* Newton, Isaac
  ontologically mixed circumstances, 44n2
  ordinary differential equations (ODEs)
    continuous variables, lift from, 55n14
    and foundational principles, 45n4
    PDE's distinguished, 45
    and Schrödinger equation, 46
    and spin, 45n5
    tasks governed by, 49
  overview, 3–4
  partial differential equations (PDEs)
    and foundational principles, 45n4
    ODE's distinguished, 45
  point-mass mechanics, 44–46, 57–69
    billiard collisions, 68–69, *69*
    Cartesian locations, 63
    Cassini space probe to Saturn, 61, *61*
    coefficient of restitution, 69
    constitutive modeling conditions, 66, 82
    inertial reaction, 65
    Lennard-Jones potential, 60
    magnitude F, 65, *65*
    matched asymptotics, 69
    methodologies of avoidance, 62
    "natural coordinates," 63
    purely elastic collision, 69n26
    representative center, 59
    rotating rigid objects, 58
    special force laws, 63
    steel ball pendulums, *64–65, 64–66*
  quantum mechanics, 45
  rigid body mechanics, 44, 67, 70–83
    action-at-a-distance forces, 73
    arc length along the slot, 71–72, *72*
    bulls-eyes, 75–76
    classical distillations of quantum processes, 74–75
    constraint relationships, 70
    contact forces, 73
    d'Alembert's principle, 79
    dead load, 73
    dimensional inharmonious quantities, 76
    dynamic loading, 73
    Eulerian cuts, 73, *74*
    forced closed, 70n27
    forces, 73–74
    free body diagrams, 73, *74*
    generalized coordinates, 71
    Greenwood's proofs, 80–82
    "higher or lower" pairs, 70n28
    independently variable, 72
    isolated particle, 81
    kinematics of mechanisms, 79
    Lagrange's principle, 78–82, *81,* 82
    meaningfully combined, 77
    mobility space, 79
    "pinned constraint," 71

point-mass perspective, 71
pressures, 74
principle of virtual work, 79n33
probability differences, 76
punctiform point of view, 79, *80*
redirection of thrust, 71
representative points, 81, 93
static load, 73
stress, 74
theory of measure, 76
torque r, 77, *78*
traction forces, 73
turning moment, 77, *78*
virtual displacement, *81*
virtual variations, 79n33
virtual-work reasoning, 72, *72*
Schrödinger equation, 46
small size sales, *46*
surrogate for classical doctrine, 44
Classical Particle Indistinguishability (Bach), 355–56
Classical regime, quantum mechanics. *See* Quantum mechanics
Clausius, Rudolf, 142, 191
Clocks, relativistic spacetimes, 536
Closed causal curve, relativistic spacetime, 591
Closing uniform topology, unitary equivalence, 510n23
Coarse-graining
  indistinguishability
    equilibrium entropy, 348
    statistical mechanics, 346
  renormalization group theory, phase transitions, 197
Coefficient of restitution, point-mass mechanics, 69
Cohen, E. G. D., 163
Coherent states, quantum mechanics, 432–34
Cold dark matter, 617
Collapse postulate, quantum mechanics, 8
  density operators, 424
  measurement, 417, 418, 454
Compactly generated Cauchy horizon, global spacetime structure, 602
Compact manifold, 588
Compatibility, 85, *85*
Complex inverse temperature, infinite idealization phase transitions, *209*
Composite, classical mechanics, 44
Composite continua, continuum mechanics, 88
Compton wavelength, 411, 413
Computational templates, hydrodynamics, 32–33n41
Conceptually simple surrogate, 44
Conceptual novelty, phase transitions, 199–200
  infinite idealization, 204–10
Conceptual system, axiomatic presentation, 57
Conciliatory approach, Everett interpretation, 475–77
"Concordance model," dark matter and dark energy, 617n22
Condensed matter physics, 3–4, 383–84

Conformal factor, relativistic spacetime, 592
Connected rigid parts, *44*
Constitutive assumptions, continuum mechanics, 98
Constitutive modeling conditions, point-mass mechanics, 66, 82
Constraint relationships, rigid body mechanics, 70
Constraint solutions, spacetime properties, 595
Contact forces, 45
  rigid body mechanics, 73
Continua, classical mechanics, 44
Continuous models, quantum mechanics, 438–39
Continuous phase transition, 147
Continuous symmetry, 295–97
  gauge theories, 299
Continuum EFTs, 237–39
Continuum mechanics, 83–104
  applied force, 84
  balance principles, 94
  Bernoulli-Euler beam, 99, *100*
  Bernoulli-Euler equation, 90, 91
  blocks-and-cords construction, 91, *92*
  "Cauchy's Law," 98–99
  compatibility, 85, *85*
  composite continua, 88
  constitutive assumptions, 98
  decorated points, *84,* 84–85
  dimensionless point cube, 95n42
  drumheads, 92
  "Euler's continuum recipe," 257
  Hooke's law, 99
  inertial force, 97n44
  infinitesimal cubes, 96
  infinitesimal traction vectors, 85
  labyrinth of the continuum, 90, 101
  lattice defects, 98
  loaded beam, 91n40
  material derivatives, 87
  mechanical elements, 91
  "method of extensive abstraction," 96n43
  point-mass setting, 98
  Problem of the Physical Infinitesimal, 86, 90, 93
  reference plane, *96*
  response planes, 95
  shearing pattern, 85
  short-range forces, 97, *97,* 98
  special force laws, 99
  strain tensors, 85, 95, *95*
  stress-energy tensors, 84
  stress tensors, 85, 95, *95,* 96
  surface forces, 87
  tensor fields, 98
  toothpaste, 89, *89*
  traction vectors, *83,* 83–84, 93
  the tyranny of scales
    Euler's recipes, 273–74
    "material particles," 270n18
  vibrating strings, 92, *92,* 101
  wave movements, 88, *88*
Continuum model equations, 256–57

Continuum versions, effective field theory (EFT), 251–52
Contours for damped oscillator, retarded Green's functions, *115*
Conventions, system of and indistinguishability, 246
Convex normal, spacetime properties, 593
Coordinate independent transformations, symmetries, 533
Copenhagen interpretation
  Everett interpretation, 463
  quantum mechanics, 420
Copernican principle, 631–32
Coriolis, Gaspard, 22
Correlation function calculations, 168–69
Correlation length, mean field theory, 162
Corresponding states, principle of, 162
Cosmic background radiation (CBR), 614–16, 632
  anthropic reasoning, 640
  dark matter and dark energy, 618
  early universe cosmology, 634
Cosmological principle, 610, 630
Cosmology, philosophy of, 10–11, 607–52
  anthropic reasoning, 638–43
    Bayesian conditionalization, 642, 643
    cosmic background radiation (CBR), 640
    Doomsday argument, 641, 643
    "Large Number Hypothesis," 638–39
    multiverse, 639
    "principle of mediocrity" (PM), 639–42
  dark matter and dark energy, 617–24, 646
    ACDM model, 618
    "concordance model," 617n22
    cosmic background radiation (CBR), 618
    dark energy, case for, 620–21
    Einstein-Hilbert actions, 619n28
    general relativity (GR), 619
    "hot" versus "cold," 618n24
    lensing effect, 621
    light-bending, 620–21
    modification of Newtonian dynamics (MOND), 623–24
    Newtonian gravity, 619–20
    "old" cosmological constant problem, 621–22
    Planck scale, 621
    quantum field theory (QFT), 621–22
    rotation curves, 620n30
    "standard candle," 618
    stress-energy tensors, 621n35
    systematic error, 620
    Type Ia supernovae, 618n25
    "zodiacal masses," 619
  early universe cosmology, 633–38
    cosmic background radiation (CBR), 634
    "dynamical approach," 636–37
    Einstein Field Equations (EFEs), 636
    flatness problem, 635
    Friedman-Lemaître-Robertson-Walker (FLRW) models, 634–36
    horizon problem, 634–35, *635*
    inflation, 636n64, 637
    "overlapping domains" argument, 633
    Standard Model, 633–36
    "theoretical unification," 633
    "theory of initial conditions," 637
    Friedman-Lemaître-Robertson-Walker (FLRW) models, 623
  global structure, 628–33
    Big Bang model, *631*
    Cauchy surface, 629
    "causality conditions," 629
    "chaotic cosmology" program, 631–32
    Copernican principle, 631–32
    cosmic background radiation (CBR), 632
    "cosmological principle," 630
    Ehlers-Geren-Sachs (EGS) theorem, 630–32
    Einstein Field Equations (EFEs), 628–29
    Friedman-Lemaître-Robertson-Walker (FLRW) models, 631–33
    Gauss-Codacci constraint equations, 629n47
    global hyperbolic spacetime, 10, 629
    homogeneity, 630
    local property of spacetime, 629n48
    Minkowski spacetime, 628n46, 629
    observationally indistinguishable (OI) spacetime, 629–30
    spacetime geometry, 628
    Sunyaev-Zel'dovich effect, 632
  multiverse, 643–47
    De Sitter universe, 644
    Everett interpretation, 644n86
    inflation, 644
    and string theory, 644
  "standard candle," 609n1
  Standard Model, 10–11
    ACDM model, 617
    assumption of, 628
    barriers, 615
    Big Bang, 609, 615
    black-body spectrum of radiation, 615
    Boltzmann equation, 614n16
    cold dark matter, 617
    cosmic background radiation (CBR), 614–16
    cosmological principle, 610
    Einstein Field Equations (EFEs), 610–11, 613
    expanding universe models, 609–14
    freeze out of particles, 614
    Friedman-Lemaître-Robertson-Walker (FLRW) models, 610, 612–13, 614n16, 617, 623
    galaxies and clusters of galaxies, length scale, 613
    global isotropy, reduction of, 611
    Mach's principle, 610
    overview of, 609–17
    parameters of, determining, 632
    Raychaudhuri equation, 611
    "re-combination," 615
    structure formation, 616–17
    thermal history, 614–16
    Type Ia supernovae, 609

uniqueness of universe, 624–27
    "ceteris paribus" laws, 626n44
    and "Laws of Physics," 625
    Mars's motion, 626–27
    and quantum field theory (QFT), 626
    and relativistic cosmology, 625–26
    Sun's gravitational field, 626–27
Coulomb force, infinite idealization, 205
Coulomb's law, 60, 401
Coulumb barriers, 615
CP symmetry, electroweak theory, 402–3
CPT theorem (Christenson, Cronin, Fitch, and Turlay), 294–95
Critical fixed points, 179, 180
Critical opalescence, fluctuations, 151–52
Critical point, 147
Crossover theory, infinite idealization, 220–21
Crystalline materials, 143–44
Crystals, Defects, and Microstructures (Phillips), 284–85
Curie, Pierre, 157, 290
"Curie Principle," 290
Cutoff, effective field theory (EFT), 241–43, 250n26
Cylindrical Minkowski spacetime, relativistic spacetime, *591*

Damping, retarded Green's functions, 113–16, 135–36
Dark matter and dark energy, 617–24, 646
    ΛCDM model, 618
    "concordance model," 617n22
    cosmic background radiation (CBR), 618
    dark energy, case for, 620–21
    Einstein-Hilbert actions, 619n28
    general relativity (GR), 619
    "hot" versus "cold," 618n24
    lensing effect, 621
    light-bending, 620–21
    modification of Newtonian dynamics (MOND), 623–24
    Newtonian gravity, 619–20
    "old" cosmological constant problem, 621–22
    Planck scale, 621
    quantum field theory (QFT), 621–22
    rotation curves, 620n30
    "standard candle," 618
    stress-energy tensors, 621n35
    systematic error, 620
    Type Ia supernovae, 618n25
    "zodiacal masses," 619
Darwin, Charles, 47
d'Alembert, Jean le Rond
    boundary layers, 23
    hydrodynamics, 13
    point-mass mechanics, 66
    rigid body mechanics, 79
    unification in physics, 389–90
    vortex motion, 19
Dead load, rigid body mechanics, 73
De Boer, Reint, 272
Debye, Peter, 163, 340

Decision-theoretic framework, construction of, 480
Decoherence
    Everett interpretation, 461, 468–70
    system-environment split, 468–69
    quantum mechanics, 437–41
    continuous models of, 438–39
    Everett interpretation. *See* Everett interpretation
    Newtonian behavior, 439
Decompositional programs, axiomatic presentation, 53, *53*
Decorated points, *84*
Decoupling, effective field theory (EFT), 238, 240–41
Degeneration, axiomatic presentation, 53
De Grav (Newton), 526
Dennett, Daniel, 472–73
Density operators, quantum mechanics, 420–23
    bit commitment problem, 429–30
    entangled states, 427
    Hilbert space, 425–26
    Hilbert-space vectors, 419n7, 420–21
    improper mixtures, 427
    no-go theorem for safe bit commitment protocols, 430
    no-hidden variables theorem, 422
    nontrivial spin properties, 428
    normalization, 422n10
    no-signaling theorem, 425–26
    proper mixtures, 423–28
    reduced states, 425
    simplex, 423–24
    spin-1/2 systems, 422–23, 425, 443
Dependence, relativistic spacetime, 590–93
Derivationally independent EFT, 246
Descartes, René, 70n27, 524, 525
De Sitter universe, 541n37
Deutsch, David, 478, 480–81
Diamond, Jared, 141
Dicke, G., 638, 640
Diffeomorphism group (DPM)
    enriched relationalism, classical mechanics, 547, 550
    Hole diffeomorphism, 575
    Machian relationalism, 557–58, 561
Differential equations, symmetries of, 323–26
    Cartan distribution, 324
    classical symmetries, 323–24, 334–37
    generalized symmetries, 323–26
    Kepler problem, 325
    Korteweg-de Vries vector, 325
    Lenz-Runge vector, 325
    Lie-Bäcklund transformations, 324–26
    local symmetries, 324–25
    Maxwell's theory, 325n21
    nonlocal symmetries, 323, 325
    and physical equivalence, 329–30
Dimensional inharmonious quantities, rigid body mechanics, 76
Dimensionless point cube, continuum mechanics, 95n42

Dirac delta function
  advanced Green's functions, 111–12
  retarded Green's functions, 113, 119
Dirac, Paul A.M., 640. *See also* Lorentz-Dirac equation
  on indistinguishability, 340
  "Large Number Hypothesis, 638–39
Dirac-von Neumann interpretation, 419
Dirichlet variety of boundary conditions, 122
Discrete symmetries, 292–94
Discretized position measurements, quantum mechanics, 443
Dispersion theory, 132–37
  Cauchy integral, 133
  description, 109
  Fourier transformations, 133, 135
  Hilbert transform pairs, 134, 136
  Kramers-Kronig relations, 135–36
Displacement, Maxwell's electrodynamics, 386–89
Distinguishability conditions, spacetime properties, 596
Distribution of zeros, infinite idealization, 208–10
Divergence symmetries, 327nn33 and 34
Domb, Cyril, 166
Dominant energy conditions, spacetime properties, 595
Donald, Matthew, 436
D1 and D2, symmetry and equivalence, 319–20
Doomsday argument, 641, 643
DPM. *See* Diffeomorphism group (DPM)
Drumheads, continuum mechanics, 92
D2, symmetry and equivalence, 329–30
Duhem, Pierre, 55, 67
Dynamical account, symmetry, 302n22
Dynamical approach
  early universe cosmology, 636–37
  have-it-all relationalism, 569–74
Dynamical-collapse theories, Everett interpretation, 463
Dynamical laws, have-it-all relationalism, 565
Dynamical symmetries
  Hole Argument (Einstein), 576
  spacetime and, 527–29
  spacetime, substantivalist and relationalist approaches to, 529
Dynamical systems theory, 180
"A Dynamical Theory of the Electromagnetic Field" (Maxwell), 390
Dynamic loading, rigid body mechanics, 73
Dynamics, 145–47
Dyson, Freeman, 288

Early universe cosmology, 633–38
  cosmic background radiation (CBR), 634
  "dynamical approach," 636–37
  Einstein Field Equations (EFEs), 636
  flatness problem, 635
  Friedman-Lemaitre-Robertson-Walker (FLRW) models, 634–36
  horizon problem, 634–35, 635
  inflation, 636n64, 637
  "overlapping domains" argument, 633
  Standard Model, 633–36
  "theoretical unification," 633
  "theory of initial conditions," 637
Earman, John
  have-it-all relationalism, 565
  Hole Argument (Einstein), 523, 575, 577
  Machian spacetime, defined, 557n75
  A Primer on Determinism, 1
  relationalism, 552
    enriched relationalism, 555n 2
  symmetry and equivalence, 318–19
Eddy resistance, 24, 25
Edinburgh *Festschrift*, 431, 434
E-expansion, renormalization group theory, 176–77
EFEs. *See* Einstein Field Equations (EFEs)
Effective field theory (EFT), 224–54
  antifoundationalism, 240
  antireductionism, 240
  approximations, EFTs and, 243
  autonomy, sense of, 247–48
    causal autonomy, 248
    explanatory autonomy, 248
    predictive autonomy, 247
    reductive autonomy, 247
  bottom-up approach, 224, 229–30
  "coarse" theory, 248
  continuum EFTs, 237–39
  continuum versions, 251–52
  cutoff, realistic interpretations of, 241–43, 250n26
    Wilsonian approach, 242
  decoupling, 238, 240–41
  definitional extension of T, 247
  derivationally independent EFT, 246
  duality transformation, 232
  emergence, EFTs, 243–51
    autonomy, sense of, 247–48
    Batterman's notion of emergence, 248–51
    causal autonomy, 248
    "coarse" theory, 248
    continuum versions, 251–52
    definitional extension of T, 247
    derivationally independent EFT, 246
    explanatory autonomy, 248
    Fermi EFT of weak force, 245
    intertheoretic relation, 245–47
    Lagrangian formalism, 247
    Lagrangian, initial and Lagrangian for superfluid Helium, 246
    limiting relations, emergence and, 248–51
    predictive autonomy, 247
    reductive autonomy, 247
    statistical mechanics, 249–50
    superfluid, EFT of, 245–46, *246*
    thermodynamics, 249–50
    top-down EFTs, 244–45
  Fermi EFT of weak force, 245
  Fourier transformations, 239
  Gauss's Law, 231–32
  Green's functions, 235, 251

INDEX 661

hierarchy problem, 228n6
Higgs term, 228n6
intertheoretic relation, 245–47
Lagrangian density, 226, 235, 251
Lagrangian formalism, 247
Lagrangian, initial and Lagrangian for superfluid Helium, *246*
limiting relations, emergence and, 248–51
low-energy superfluid helium-4 film, 230–32
    Lagrangian density, 230–32
mass-dependent schemes, 236–37
mass-independent schemes, 237–39
Minkowski spacetime, 231
nature of EFTs, 225–32
non-relativistic QCD, 229n7
ontological implications, 239–43
    antifoundationalism, 240
    antireductionism, 240
    approximations, EFTs and, 243
    cutoff, realistic interpretations of, 241–43
    decoupling, 240–41
    quantum field theory (QFT), 240–43
    quasi-autonomous domains, 240–41
    Wilsonian approach, 240
overview, 5
predictability, 232–35
quantum field theory (QFT), 240–43, 251
quasi-autonomous domains, 240–41, 243
    and emergence, 244
    Wilsonian approach, 241
relativistic quantum field theories (RQFTs)
    emergence, EFTs, 250–51
    overview, 224–25
renormalization group (RG), 232–35
    nonrenormalizable EFTs, 233
renormalization schemes, 235–39
    continuum EFTs, 237–39
    Green's function, 235
    Lagrangian density, 235
    mass-dependent schemes, 236–37
    mass-independent schemes, 237–39
    Wilsonian approach, 236–37
scalar field theory, 227n3
Standard Model, 228
statistical mechanics, 249–50
superfluid, EFT of, 245–46, *246*
thermodynamics, 249–50
top-down approach
    emergence of EFTs, 244–45
    naturalness, hypothesis of, 228
    overview, 225–29
    quantum chromodynamics (QCD), 228–29
    symmetry considerations, 228
types of EFTs, 235–39
unification in physics, 384, 406–13
    Gell-Mann/Low formulation, 408–9
    Hamiltonians, 410
    quantum electrodynamics (QED), 411
    reductionism, problems of, 411n25
    Wilson-Kadanoff model, 412
Wilsonian approach, 252
    cutoff, realistic interpretations of, 242

mass-independent schemes, 237–39
ontological implications, 240
and renormalization group (RG) techniques, 232
renormalization schemes, 236–37
top-down approach, 226, 228–29
Ehlers-Geren-Sachs (EGS) theorem, 630–32
Ehrenfest, P., 191, 350–51
Ehrenfest's theorem, quantum mechanics, 432–34
Ehrenfest-Trkal-van Kampen approach, indistinguishability, 350–51, 366
Eiffel, Gustav, 26
Eigenvectors, quantum mechanics, 418
"Eightfold Way," 302n22
Einstein, Albert
    corresponding states, principle of, 162
    critical opalescence, 151
    general relativity, 325, 537–39, 541, 543–44
    Hole Argument, 300n17, 523, 574–79
    indistinguishability
        quantum, indistinguishability and, 342–43
        quantum particles, 363
    on indistinguishability, 340
    light-bending, 620–21
    mean field theory, 158
    Noether, Emmy, obituary, 297
    Principle of Equivalence, 537
    quantum mechanics, 440
        classical regime, 431
        coherent states, 433–34
    substantivalist-relationalist debate, 522n1
    symmetry
        GTR principle, 292, 297
        STR, formulation of, 311–12
        theories of special and general relativity, 290
    unification in physics, 392
Einstein Field Equations (EFEs), 537, 543
    early universe cosmology, 636
    global structure, cosmology, 628–29
    relativistic spacetimes, 537
    Standard Model, 610–11, 613
Einstein-Hilbert actions, dark matter and dark energy, 619n28
Einstein's equation, spacetime properties, 595
Einstein tensor, 537, 594
Electroweak theory, 393–401
    Big Bang, 303, 403
    "big dessert" assumption, 404
    Cabibbo-Kobayashi-Maskawa framework, 402
    CP symmetry, 402–3
    gauge hierarchy problem, 404
    Glashow model, 397
    Higgs boson, 394–95, 403–6
    Higgs field, 398–401
    isospin, 396n14
    Lagrangian, 395
    Lagrangian invariance, 397–99
    Large Hadron Collider (LHC), 405–6
    Lie group, 395
    loop quantum gravity (LQG), 405n22
    from mathematics to physics, 397–401
    and Maxwell's theory, 396, 399

Electroweak theory (*Cont.*)
  multiplets, 396
  naturalness, 404n19
  Noether's theorem, 396–97
  non-Abelian case, 396, 401
  phase transformation, 396n13
  Planck scale, 403
  problems with, 403–6
  QCD vacuum, 394
  quantum chromodynamics (QCD), 402–3
  Schrödinger equation, 395
  Standard Model, 394, 402–6
  string theory, 405
  SU(2) and SU(3) color groups, 394, 396–97
  supersymmetry (SUSY), 405
  symmetry as tool for unification, 395–97
  Theory of Everything (TOE), 405
  Yukawa couplings, 402
  Yukawa interaction, 402n17
Electroweak unification, symmetry, 303–6
Elementary particles, renormalization group theory, 176
Elementary Principles in Statistical Mechanics (Gibbs), 352–53
Eliezer's theorem, 132
Eliminativism
  and Gibbs paradox, 378
  indistinguishability, 376–78
  "preferred basis problem," 377
  quantum fields, 377–78
Ellis, G. F. R., 625, 627
Embedding of relational history, enriched relationalism, 546
Emergence
  EFTs. *See* Effective field theory (EFT)
  of phase transitions, 197–204
Energy-momentum tensor, spacetime properties, 594
Enriched relationalism
  classical mechanics, 545–53
    diffeomorphism group (DPM), 547, 550
    embedding of relational history, 546
    Galilei group, 549–50
    kinematic shift argument, 546–47
    Maxwell group, 550–51
    Maxwellian relationalist, 552
    Newton-Cartan theory, 551, 553
    nonsimultaneous events, 549
    Sklarations, 548–49
  relativity, 553–57
    four-force, 554
    gravitational wave, 556
    kinematically possible models (KPMs), 555
    Minkowski distances, 553–56
Ensemble equivalent, 207
Ensemble generates averages, 180
Ensemble generates ensemble, 180
Entangled states, quantum mechanics, 427
Entropy, indistinguishability, 343–45
  Boltzmann definition, 346–47
Environment, entanglement with, 434–37
Epistemic puzzles, Everett interpretation, 479–81

Equal areas law, symmetry, 288n2
Equilibrium, 145–47
Equilibrium entropy, indistinguishability, 348–49
  coarse-graining, 348
Equivalence, symmetry and. *See* Symmetry and equivalence
Euclidean coordinate systems
  Machian relationalism, 557
  spacetime, substantivalist and relationalist approaches to, 528, 529, 557
Euclidean space, relativistic spacetimes, 536
Euclidean symmetries, Hole Argument (Einstein), 577
Eulerian flow, 26
Euler-Lagrange equations, symmetry, 296
Euler, Leonhard
  continuum mechanics, 257
    balance principles, 94
    Bernoulli-Euler beam, 99, *100*
    Bernoulli-Euler equation, 90, 91, 91n40
    laws of motion, 96
  Eulerian flow, 26
  explanatory progress, 29
  fluid motion, 13–14
    instabilities, 19
  point-mass mechanics, 59–60, 69
  rigid body mechanics
    Eulerian cuts, 73, *74*
    "First Law," 78n32
  vortex motion, 19
    vortex filaments, defined, 18
Euler's recipes, the tyranny of scales, 273–75
  and Cauchy, Augustin, 273–75
  continuum, 273–74, 278
  controversy, 274–75
  discrete, 273–74
  and Navier-Stokes equations, 275
  and Young's modulus, 275
Everett, Hugh I. *See also* Everett interpretation
  insight of, 463–66
Everett interpretation, 8–9, 460–88
  Bell's inequalities, 481–82
  branch count, 477–78
  "cat" example, 464–65, 467
  classical electrodynamics, 466n6
  conciliatory approach, 475–77
  Copenhagen interpretation, 463
  decision-theoretic framework, construction of, 480
  decoherence, 461, 468–70
    system-environment split, 468–69
  Dennett's criterion, 472–73
  dynamical-collapse theories, 463
  epistemic puzzles, 479–81
  higher-order ontology, 470–74
  Hilbert space, 464, 467n8
  indexical uncertainty, 476
  and many-exact worlds theories, 467–68
  and many-minds theories, 467–68
  measurement problem, 462–63
  mod-squared amplitude, 479n17, 480
  multiverse, 644n86

non-interacting particles, 463n5
overview, 460–61
parallel tradition, 478n16
pilot wave theory, 482
"preferred basis problem," 466–68
  and many-exact worlds theories, 467–68
  and many-minds theories, 467–68
  overview, 461
probability problem, 466, 474–77
  objective-probability role, 479n17
  parallel tradition, 478n16
  philosophical aspects of probability, 474–75
  and possibility, 475–77
  probability simpliciter, 475
  and uncertainty, 475–77
quantitative problem, 477–79
quantum mechanics, 440, 644n86
quasi particles, 473–74
semantics, 476
Standard Model, 471
structure, role of, 470–74
"superposition state," 462
wave packets, 461
Expanding universe models, 609–14
Explanatory irreducibility, phase transitions, 210–14
Explanatory progress, hydrodynamics, 27–30
  components of explanation, 28–29
  heterogeneous specializations, 30
  homogeneous specializations, 30
  pragmatic definition of, 29–30
  sources of, 27–28
Extended singularities
  Ising model, 152
  matter, infinities and renormalization, 183
Extended singularity theorem, Ising model, 154–55
Extensivity, phase transitions, 201n2

"False facts," classical mechanics, 47
Fecundity and generality, trade-off between, 335n13
Fermi EFT of weak force effective field theory (EFT), 245
Fermilab's Tevatron, 305
Fermions, indistinguishability, 364–65
Ferromagnetic phase transitions, 192
Ferromagnets, 145
  Ising model, *156*
  the tyranny of scales, 265–66, 278
    spontaneous magnetization, *265*
  unitary inequivalence, as example of, 508, *509*
Feynman, R., 119
  anti-particles, 294n10
  diagrams, indistinguishability, 367, *368*
  the tyranny of scales, 256
  unitary equivalence, 492
Fields, classical mechanics, 44
50-50 volume mixture, the tyranny of scales, *268*
Fine-graining, indistinguishability, 346
Finiteness condition, radiation theory, 124
First-order phase transition, 147

Fisher, Michael, 166, *170*, 208
Fixed points
  matter, infinities and renormalization, 178
  renormalization group theory, 177
Flatness problem, early universe cosmology, 635
Flexible beam, classical mechanics, *44*
Flexible bodies, classical mechanics, 44
Flows and flow diagrams
  matter, infinities and renormalization, 180–81
  one-dimensional Ising model, *180*
  two-dimensional Ising model, *181*
Fluctuations
  boiling, 151
  critical opalescence, 151–52
  liquid gas, 150n6, 151
Fluid mechanics. *See* Hydrodynamics, philosophy of
Fluid motion, hydrodynamics, 13–14, 22
Fluids
  mean field theory, 165
  velocity, hydrodynamics, 13
Forced closed, rigid body mechanics, 70n27
Force of rolling friction, axiomatic presentation, 54, *54*
Forces, rigid body mechanics, 73–74
Foundational point of view, axiomatic presentation, 48–49
Four-force, relativity, 554
Fourier space, renormalization group theory, 172–74
Fourier transformations
  advanced Green's functions, 111
  dispersion theory, 133, 135
  effective field theory (EFT), 239
  radiation theory, 126–27
  retarded Green's functions, 113, 114
Fraser, D., 251
Free body diagrams, rigid body mechanics, 73, *74*
Freeze out of particles, Standard Model, 614
Freezing to a scale level, 55
Friedman-Lemaître-Robertson-Walker (FLRW) models, 610, 612–13, 614n16, 617, 623
  early universe cosmology, 634–36
  global structure, cosmology, 631–33
Friedman, Michael, 548
Frisch, Mathias
  causation in classical mechanics, 108nn7 and 9, 109n10
  dispersion relations, 132
  dispersion theory, 135–37
  point-particle electrodynamics, 130–31
Froude, William
  boundary layers, 24–26
  surface waves, 16
  turbulence, 21
Fruitless definition, symmetry and equivalence, 322
Future Cauchy horizon, relativistic spacetime, 592
Future-directed curve, relativistic spacetime, 589
Future distinguishability condition, spacetime properties, 596

Future domain of dependence, relativistic spacetime, 592
Future endpoint, relativistic spacetime, 592
Future incomplete, spacetime properties, 598

Galaxies and clusters of galaxies, length scale, 613
Galilean boosts, symmetry and equivalence, 327, 331
Galilean covariance
　spacetime, substantivalist and relationalist approaches to, 523
Galilean idealization, 210n7
Galilean invariance, 527–31
　dynamical symmetries, spacetime and, 527–29
　dynamical symmetry group, 529
　Euclidean coordinate systems, 528, 529
　Galilei group, 529
　kinematically privileged systems, 527
　kinematic shift argument, 529–30
　Leibniz group of transformations, 528
　Leibnizian relationalism, 527–28, 530
　Newton group, 528
　Principle of Sufficient Reason (PSR), 530n14
　relative particle configurations, 527
　spacetime symmetry groups, 528
Galilean spacetime
　have-it-all relationalism, 570
　Hole Argument (Einstein), 576n104
　spacetime substantivalism, 531
Galilei group
　enriched relationalism, classical mechanics, 549–50
　physical equivalence, 330
　spacetime, substantivalist and relationalist approaches to, 529, 533
　symmetries, spacetime substantivalism, 534
Galileo, Church's condemnation of, 524n3
Games and gambling, the tyranny of scales, 275–76, 277
"Gauge argument," symmetry, 291, 298–300
　Lagrangian, 298–99
　　kinetic component, 300n17
Gauge hierarchy problem
　electroweak theory, 404
Gauge theories, symmetry, 300–303
　continuous symmetry, 299
　dynamical account, 302n22
　"Eightfold Way," 302n22
　Gell-Mann "strangeness," 302
　Glashow-Salam-Weinberg (GSW) model, 303
　"gluing" role, 301
　isospin idea, 301
　Lorentz invariant, 299
　"power of the gauge," 299
　quantum chromodynamics (QCD), 302
　weak neutral currents, 302n24
Gauss-Codacci constraint equations, 629n47
Gaussian bell curves, quantum mechanics, 432
Gaussian distribution, the tyranny of scales, 276, 279
Gaussian wave packets, quantum mechanics, 432–34

Gauss's Law, effective field theory (EFT), 231–32
Gell-Mann/Low formulation, effective field theory (EFT), 408–9
Gell-Mann, M.
　gauge theories, 302
　"Eightfold Way," 302n22
　omega minus hadron, prediction of, 306
　prediction, totalitarian principle, 309n39
　symmetry, predictive reasoning, 308–9
Gell-Mann "strangeness," 302
Generalized coordinates, rigid body mechanics, 71
Generally covariant equations, symmetries, 533
General relativity (GR), 325, 537–39, 541, 543–44
　dark matter and dark energy, 619
　have-it-all relationalism, 568–69
　Hole Argument (Einstein), 573–75, 578
　Machian relationalism, 563–64
　unification in physics, 383
Georgi, H., 238–39, 411n25
Geroch, Robert, 613
Gibbs, J. Willard. *See also* Boltzmann-Gibbs formulation
　Elementary Principles in Statistical Mechanics, 352–53
　on indistinguishability, 341–52. *See also* Indistinguishability
　phase transitions
　　infinite idealization, 207
　　statistical mechanical treatment, 195
　　thermodynamic treatment, 191
　statistical mechanics, 146
　thermodynamics, 142, 145
Gibbs paradox, indistinguishability, 378
Ginzburg, Vitaly, 167. *See also* Landau-Ginzburg-Wilson free energy
Glashow model, electroweak theory, 397
Glashow, S. *See* Glashow model; Glashow-Salam-Weinberg (GSW) model
Glashow-Salam-Weinberg (GSW) model, 303
Gleason, A. M., 422, 425, 449
Global continuous symmetry, 295–97
Global hyperbolic spacetime, 10, 597, 629
Global isotropy, reduction of, 611
Global spacetime structure, 10, 587–606
　Big Bang singularity, 600
　boundary conditions, 600
　compactly generated Cauchy horizon, 602
　hole-freeness, 603
　Klein-Gordon fields, 601
　manifold M, 588–90
　　compact manifold, 588
　　Hausdorff, 588
　metric, 588–90
　　Lorentzian metric, 588
　Minkowski spacetime
　　singularities, 601, *601*
　　time travel, 603
　Misner spacetime, *603*
　nakedly singular spacetime, 601
　reasonable properties, 598–603
　　Big Bang singularity, 600

boundary conditions, 600
compactly generated Cauchy horizon, 602
hole-freeness, 603
Klein-Gordon fields, 601
Minkowski spacetime, 601, 603
Misner spacetime, *603*
nakedly singular spacetime, 601
singularities, 598–601
time travel, 601–3
trapped surface, 600
relativistic spacetime, 587–93
causal curve, 591
causal future, 590–91
chronology violating region of spacetime, 591
closed causal curve, 591
conformal factor, 592
cylindrical Minkowski spacetime, *591*
dependence, 590–93
future Cauchy horizon, 592
future-directed curve, 589
future domain of dependence, 592
future endpoint, 592
influence, 590–93
isometric spacetimes, 589–90
locally isometric spacetime, 590
manifold M, 588–90
maximal spacetimes, 590
metric, 588–90
Minkowski spacetime, *591*, 591–93, *592*
Möbius strip, 589
null vectors, 589, *589*
past Cauchy horizon, 592
past-directed curve, 589
past domain of dependence, 592
past endpoint, 592
spacelike surface, 592
spacelike vectors, 589, *589*
temporarily orientable spacetime, 589
timelike future, 590–91
timelike vectors, 589, *589*
singularities, 598–601
spacetime properties, 593–98
Cauchy surface, 597
causal continuity condition, 597
causal simplicity condition, 597
causal structure, 596
chronology condition, 596
constraint solutions, 595
convex normal, 593
distinguishability conditions, 596
dominant energy conditions, 595
Einstein's equation, 595
Einstein tensor, 594
energy-momentum tensor, 594
future distinguishability condition, 596
future incomplete, 598
global hyporbolicity, 597
global properties, 596–98
local properties, 593–95
Minkowski spacetime, 597
null geodesic incompleteness, 598

past distinguishability, 596
past incomplete, 598
Ricci tensor, 594
singularities, 596
spacelike geodesic incompleteness, 598
stable causality condition, 596–97
strong causality condition, 596
strong energy conditions, 595
timelike geodesically incomplete, 598
time travel, 601–3
trapped surface, 600
Global structure, cosmology, 628–33
Big Bang model, *631*
Cauchy surface, 629
"causality conditions," 629
"chaotic cosmology" program, 631–32
Copernican principle, 631–32
cosmic background radiation (CBR), 632
"cosmological principle," 630
Ehlers-Geren-Sachs (EGS) theorem, 630–32
Einstein Field Equations (EFEs), 628–29
Friedman-Lemaître-Robertson-Walker (FLRW) models, 631–33
Gauss-Codacci constraint equations, 629n47
global hyperbolic spacetime, 10, 629
homogeneity, 630
local property of spacetime, 629n48
Minkowski spacetime, 628n46, 629
observationally indistinguishable (OI) spacetime, 629–30
spacetime geometry, 628
Sunyaey-Zel'dovich effect, 632
"Gluing" role, symmetry, 301
Glymour, Clark, 496, 499n9
Goldstone boson, symmetry, 304
GR. *See* General relativity (GR)
Gravitational wave, relativity, 556
Green, George, 269
Green, Melville, 168
Green's functions, point-mass mechanics, 67
Greenwood, Donald, 80–82
Greenwood's proofs, rigid body mechanics, 80–82
Group orbits, Machian relationalism, 559
Group theory, symmetry, 289
Guggenheim, E. A.
corresponding states, principle of, 162
mean field theory, 165
Guns, Germs and Steel (Diamond), 141
Guth, Alan, 634, 636

Haag, Rudolph, 492
Hacking, Ian, 13
Haecceitism, indistinguishability, 356–60
Hagen, Gotthilf, 20
Hamel, Georg, 48
Hamiltonian equations, symmetry, 289
Hamiltonian function
block transforms and scaling, 171
free energy, 173
Ising model, 153
matter, infinities and renormalization, 146
flows and flow diagrams, 180, 181

Haag, Rudolph (*Cont.*)
    mean field theory, 157
    naming of function, 146n3
    phase transitions, 190
        statistical mechanical treatment, 193
    renormalization group theory, 196–97
    and statistical mechanics, 146
    the tyranny of scales, 276
    universality classes, 178
Hamiltonians, effective field theory (EFT), 410
Hamiltonian symmetries
    outlook, 334
    and physical equivalence, 331–32
    symmetry and equivalence, 326–28
Hamiltonian systems, "uniqueness results," 493–94
Hamiltonian theory, unitary inequivalence, 507
Hamilton's "Principle of the Least Action," 295
Hamilton, William Rowan. *See* Hamiltonian equations; Hamiltonian function
Hausdorff, 588
Have-it-all relationalism
    Best Systems prescription, 566
    dynamical approach to relativity, 569–74
    dynamical laws, 565
    Galilean spacetime, 570
    general relativity (GR), 568–69
    homogeneous matter, 573–74
    Humean approach, 566
    Leibnizian relationalism, 567–68
    Lorentz contract, 570n92, 572
    Mill-Ramsey-Lewis's Best Systems prescription, 566
    Minkowski geometries, 569n91, 570–71
    regularity approach, 564–69
    rod, constituents of, 570
    rotation disks argument, 566
Hawking radiation exterior, 518
Hawking, Stephen, 599, 613
Heat capacity as measured, mean field theory, *166*
Heat transfer, indistinguishability, 344–45
Heisenberg model of ferromagnetism, 163
Heisenberg's "cut," quantum mechanics, 440–42
    Copenhagen interpretation, 441n30
Heisenberg, Werner, 20
    gauge theories, 301
    quantum mechanics
        classical regime, 430–31
        particle tracks, 439
Helmholtz equation, radiation theory, 123, 125–27, 129–30
Helmholtz, Hermann
    boundary layers, 23
    and explanatory progress, 28–29
    instabilities, 20
    thermodynamic calculations, 146
    turbulence, 21
    vortex motion, 18–19
Helmholtz-Kelvin instability, 19
Hermann, Grete, 422
Heterogeneous specializations, 30

Hierarchy problem, effective field theory (EFT), 228n6
Higgs boson
    electroweak theory, 394–95, 403–6
    symmetry, 291
Higgs field, electroweak theory, 398–401
Higgs mechanism, symmetry, 303–6
Higgs particle, unification in physics, 381–82
Higgs term, effective field theory (EFT), 228n6
High-energy particle acceleration, funding for, 2
Higher-order ontology, Everett interpretation, 470–74
"Higher or lower" pairs, rigid body mechanics, 70n28
Hilbert, David, 104
    axiomatic encapsulations, 56
    choice of length scale, 50
    decompositional programs, 53, *53*
    degeneration, 53
    dispersion theory, 134, 136
    homogenization, 53
    indistinguishability, 362, *363*
        ontology, 369
    list of problems, 47–48, 81–82
        sixth problem, 52–53
    symmetry, 289
Hilbert space
    Everett interpretation, 464, 467n8
    indistinguishability, 359
    quantum mechanics, 420–21, 425–26
        environment, 434
        ideal spin measurements, 443
        measurement, 417, 450n38
    unitary equivalence, physical equivalence and, 9, 489–90, 503n12, 510, 518
        competing criteria of equivalence, 513
        "uniqueness results," 494–95
    unitary inequivalence, as example of, 503
Hofstader, Douglas, 471
Hole Argument (Einstein), 300n17, 523, 574–79
    and dynamical symmetries, 576
    Euclidean symmetries, 577
    and Galilean spacetime, 576n104
    general relativity (GR), 573–75, 578
    hole diffeomorphism, 575
    individualistic facts, 577n108
    and kinematic shift argument, 576
    Leibniz and Clarke correspondence, 577n106
    Machian 3-space approach, 578
    and Maxwell group, 576n104, 578
    Newton-Cartan theory, 576–77
    pseudo-Reimannian metric field, 574n100, 578
    sophisticated substantivalism, 575
    structural realist interpretation of spacetime, 577
Hole-freeness, global spacetime structure, 603
Homogeneity
    global structure, 630
    have-it-all relationalism, 573–74
Homogeneous specializations, 30
Homogenization
    axiomatic presentation, 51, 53, 55

the tyranny of scales, 256, 280–83
   averaging and homogenization, differences, 277–78
   limit, *281*
Hooke's law
   continuum mechanics, 99
   steel beams, 258
Horizon problem, early universe cosmology, 634–35, *635*
"Hot" versus "cold," dark matter and dark energy, 618n24
Hubble, E., 613
Hubble radius, 617n21
Huggett, Nick
   effective field theory (EFT), 234
   enriched relationalism, classical mechanics, 548
   have-it-all relationalism, 564–69, 571–72, 573n97
   indistinguishability, haecceitism, 356
   regularity approach to relationalism, 564–69, 571–72
   relationalism, 545
   relational spacetime, 544
Hume, David, 56, 107
   have-it-all relationalism, 566
Hydrodynamics, philosophy of, 12–42
   approximation methods, 31
   articulation, 13
   Bernoulli's Law, 15–16, 27
   boundary layers, 13, 23–27
      airplane wings, 26
      discontinuity surface, *24*
      eddy resistance, 24, 25
      and modules, 37
      ships, 24
      skin resistance, 24
      wave resistance, 24
   calculation, 13
   computational templates, 32–33n41
   explanatory progress, 27–30
      components of explanation, 28–29
      heterogeneous specializations, 30
      homogeneous specializations, 30
      pragmatic definition of, 29–30
      sources of, 27–28
   fluid motion, 13–14, 22
   fluid velocity, 13
   history, 13–27
      Bernoulli's Law, 15–16, 27
      boundary layers, 13, 23–27
      fluid motion, 13–14, 22
      fluid velocity, 13
      instabilities, 19–21
      kinetic theory of gases, 22
      "losses of head," 16
      nonviscous fluid, 13
      pipe flow, 20
      plane parallel flow, 20
      plane Poiseuille flow, 21
      surface waves, 16–17
      turbulence, 21–23
      vortex motion, 17–19
   instabilities, 19–21
   interpretive schemes, 31
   kinetic theory of gases, 22
   "losses of head," 16
   modules, 33–38
      correspondence principle, 35–36
      defining, 33
      idealizing modules, 33
      and models, 36–38
      reducing modules, 34, 36n46
      shared module, 36
      specializing modules, 33
      structure, *35*
   nonviscous fluid, 13
   overview, 3
   physical theories, 31–32
   pipe flow, 20
   plane parallel flow, 20
   plane Poiseuille flow, 21
   surface waves, 16–17
   symbolic universe, 31
   theoretical laws, 31
   theory articulation, 13
   turbulence, 21–23
   the tyranny of scales, 279
   vortex motion, 17–19
   wing theory (Prandtl), 13, 30

Ideal spin measurements, quantum mechanics, 443–44
Identity conditions, indistinguishability, 372–76
Imprecision, Ising model, 155n10
Improper mixtures, quantum mechanics, 427
   indistinguishability from proper mixtures, 429
Incoming waves, radiation theory, 127
Indexical uncertainty, Everett interpretation, 476
Indistinguishability, 340–80
   "account of equality," 372
   asymmetry, 365n33
   classical indistinguishability, argument against, 354–56
   classical particle indistinguishability, 355–56
   coarse-graining
      equilibrium entropy, 348
      statistical mechanics, 346
   conventions, system of, 246
   Ehrenfest-Trkal-van Kampen approach, 350–51, 366
   eliminativism, 376–78
      and Gibbs paradox, 378
      Pauli exclusion principle, 377
      "preferred basis problem," 377
      quantum fields, 377–78
   entropy, 343–45
      Boltzmann definition of, 346–47
      equilibrium entropy, 348–49
      coarse-graining, 348
   fermions, 364–65
   Feynman diagrams, 367, *368*
   fine-graining, 346
   Gibbs paradox, 7, 341–52

Indistinguishability (*Cont.*)
  Ehrenfest-Trkal-van Kampen approach, 350–51, 366
  and eliminativism, 378
  equilibrium entropy, 348–49
  Feynman diagrams, 367, *368*
  haecceitism, 356
  N! puzzle, 349–52
  ontology, 366–69
  permutable coins, 366–67, *368*
  quantum, indistinguishability and, 341–43
  statistical mechanics, 346–47
  thermodynamics, 343–46
  uniform symmetry, 352–53
haecceitism, 356–60
Hilbert space, 359, 362, *363*
identity conditions, 372–76
Leibniz's principle, 373–74
Maxwell-Boltzmann statistics, 371n42
N! puzzle, 349–52
ontology, 365–78
  "account of equality," 372
  eliminativism, 376–78
  Gibbs paradox, 366–69
  identity conditions, 372–76
  Leibniz's principle, 373–74
  Maxwell-Boltzmann statistics, 371n42
  permutability of objects, 369–70
  philosophical logic, 369–72
  quantities invariant, 365
  quantum particles, 371–72
  spin, 375
  total symmetry, 369
overview, 6–7
parastatistics, 364n32
Pauli exclusion principle, 377
permutability of objects, 369–70
permutable coins, 366–67, *368*
phase space, reduced, 361–62, *362*
philosophical logic, 369–72
"preferred basis problem," 377
quantities invariant, 365
quantum fields, 377–78
quantum, indistinguishability and, 341–43
  "light quanta," 342
  microstate, 342n3
  and "permutable," 342n4
quantum particles, 371–72
quantum statistics, explanation of, 360–64
  Hilbert space, 362, *363*
  phase space dimension, 361–62
  subspace dimension, 362–63
  volume measures, 362–63
  weighting, 361, *362*
spin, 375
statistical mechanics, 346–47
  coarse-graining, 346
  entropy, 346–47
  fine-graining, 346
subspace dimension, 362–63
thermodynamics, 343–46
  conventions, system of, 246
  entropy, 343–45
  heat transfer, 344–45
  total symmetry, 369
as uniform symmetry, 352–53
  classical indistinguishability, argument against, 354–56
  fermions, 364–65
  Gibbs' solution, 352–53
  haecceitism, 356–60
  quantum statistics, explanation of, 360–64
  volume measures, 362–63
  weighting, 361, *362*
Individualistic facts, Hole Argument (Einstein), 577n108
Inertia
  continuum mechanics, 97n44
  Descartes, René, 524
  point-mass mechanics, 65
  spacetime explanation of, 541–44
Infinite idealization, 204–17
  crossover theory, 220–21
  and emergence of phase transitions, 202
  explanatory irreducibility, 210–14
  irreducibility, 221
  phase transitions, 202
  renormalization group, 216–23
    critical behavior of particular systems, 218
    crossover theory, 220–21
    irreducibility, 221
    universality, 217–18
Infinite limits, the tyranny of scales, 261–62
Infinitesimal cubes, continuum mechanics, 96
Infinitesimal traction vectors, 85
Infinite spin chain, as example of unitary inequivalence, 501–5, 508
Inflation
  early universe cosmology, 636n64, 637
  multiverse, 644
Influence, relativistic spacetime, 590–93
Instabilities, hydrodynamics, 19–21
Instantaneous relative distances, 540n35
Instantaneous states, symmetry and equivalence, 326n31
Intensive properties, phase transitions, 201
Interpretive schemes, hydrodynamics, 31
Intertheoretic relation, effective field theory (EFT), 245–47
Irreducibility
  infinite idealization, 221
  phase transitions, 199, 203
    explanatory irreducibility, 210–14
    infinite idealization, 221
Irrelevant scalings, 177–78
Irreversibility, 142
Ising, Ernst, 145
Ising model, 152–56
  Boltzmann-Gibbs formulation, 154
  defined, 152–53
  extended singularity theorem, 152, 154–55
  ferromagnets, *156*
  and Hamiltonian function, 153
  imprecision, 155n10

one-dimensional Ising model, *180*
Ornstein-Zernike infinity, 156
particle spin, 154
phase diagram, *156*
scaling, 171
singularity in phase transition, *155*
two-dimensional Ising model, *153, 181*
Isometric spacetimes, relativistic spacetime, 589–90
Isospin
  electroweak theory, 396n14
  symmetry, 301

Jacobi's principle, Machian relationalism, 558–59
Jeans, J., 289n3
Jet bundle, symmetry and equivalence, 324
Jordan-Wigner theorem, 9, 490–91, 501

Kadanoff, Leo P.
  the tyranny of scales, 266
  unification in physics, 408–9
Kant, Immanuel
  causation, 107
  classical mechanics, 89n39
  indistinguishability, haecceitism, 356
  and scientific cosmology, 625–26
  symmetry, incongruent counterparts, 293n9
Kármán, Theodore von, 22, 26
Kellers, L. F., 165, 168
Kelvin, Lord. *See* Thomson, William (Lord Kelvin)
Kepler, Johannes, 310
  Mysterium Cosmographicum, 287–88
Kepler problem, symmetry and equivalence
  differential equations, symmetries of, 325
  and physical equivalence, 332
Kinematically possible models (KPMs)
  diffeomorphism group (DPM), 533–34
  Machian relationalism, 558
  relativity, enriched relationalism, 555
  spacetime substantivalism, 531–33
Kinematically privileged systems, 527
Kinematics
  rigid body mechanics, 79
  unitary equivalence, kinematic pair, 495–96
  unitary equivalence, physical equivalence and, 492
Kinematic shift argument
  enriched relationalism, 546–47
  Hole Argument (Einstein), 576
  spacetime, 523, 544
    substantivalist and relationalist approaches to, 529–30
Kinetic theory of gases, 22
King's College (London) school, 166–67
Kirchhoff, G., 23–24
Klein, Felix, 289
Klein-Gordon fields, 601
Kolmogorov, Andrei, 167
Korteweg-de Vries vector, 325, 333
Koyré, Alexander, 524
KPMs. *See* Kinematically possible models (KPMs)

Kramers, Hendrik, 163
Kramers-Kronig relations, 135–36
Kretschmann, E., 537
Kuhn, Thomas
  criticisms of, 30n38
  dark matter and dark energy, 619
  matter, theories of, 150
  and "normal" phases of science, 12
  phase transitions, 168

Labyrinth of the continuum, 90, 101
Lagrange, Joseph Louis
  instabilities, 20
  rigid body mechanics, 78–80, 82
  surface waves, 16
  vortex motion, 17
Lagrangian
  density, effective field theory (EFT), 226, 235, 251
  formalism, effective field theory (EFT), 247
  initial and Lagrangian for superfluid Helium, effective field theory (EFT), *246*
  invariance, electroweak theory, 395, 397–99
  methodology
    symmetry, 295–96, 299
    symmetry and equivalence and physical equivalence, 331
  symmetry
    kinetic component, 300n17
    variational symmetries, 326–27
  unitary equivalence, physical equivalence and, 491–92
Lamb, Willis, 307n35
Lanczos, Cornelius, 82
Landau-Ginzburg-Wilson free energy, 173–74
Landau, Lev
  different phase transition problems, 160–61
  mean field theory, 158
    experimental facts, 164–65
    as model, 164
    summary of, 161–62
  phase transitions, 144, 152
    order parameter, 192
  universality, 161–62
Laplace transform technique, radiation theory, 125
Large Hadron Collider (LHC), 305, 381–82, 384
  electroweak theory, 405–6
  unification in physics, future output, 406
"Large Number Hypothesis," 638–39
Lattice defects, continuum mechanics, 98
Laws of motion. *See* Newton, Isaac
Lebesgue measure zero, 118
Lebowitz, J. L., 195, 198
Leeuwen, Hans van, 151n7
Leibniz, Gottfried Wilhelm
  axiomatic presentation, 48, 52
  Clarke, correspondence, 577n106
  continuum mechanics
    labyrinth of the continuum, 90
    loaded beam, 91n40
  indistinguishability

Leibniz, Gottfried Wilhelm (*Cont.*)
  haecceitism, 356
  Leibniz's principle, 373–74
Leibniz group of transformations, 528
  spacetime, substantivalist and relationalist approaches to, 333
Leibnizian relationalism
  have-it-all relationalism, 567–68
  Machian relationalism, 558
  spacetime, substantivalist and relationalist approaches to, 527–28, 530, 544–45, 548, 567–68
  symmetries, spacetime substantivalism, 535
Lennard-Jones potential, 60
Lensing effect, dark matter and dark energy, 621
Lenz-Runge vector, symmetry and equivalence, 325
Lenz, Wilhelm, 145, 153
Levanyuk, A. P., 167
Le Verrier, Urbain, 618, 619, 623
Lewis, David K., 482, 483, 566
LHC. *See* Large Hadron Collider (LHC)
Lie-Bäcklund transformations, 324–26
Lie group, electroweak theory, 395
Light-bending, dark matter and dark energy, 620–21
"Light quanta," 342
Linear operators, quantum mechanics, 417n4
Lipkin, H., 307
Liquid gas
  fluctuations, 150n6, 151
  infinite idealization, phase transitions, 215
Liu, C., 198
Loaded beam, continuum mechanics, 91n40
Locally isometric spacetime, relativistic spacetime, 590
Local properties, spacetime, 593–95, 629n48
Local symmetry, 323–25
London, Fritz, 297
Loop quantum gravity (LQG), electroweak theory, 405
Lorentz contract, have-it-all relationalism, 570n92, 572
Lorentz-Dirac equation
  causation in classical mechanics, 132
  description, 109
  point-particle electrodynamics, 132
Lorentzian metric, 588
Lorentz invariant
  gauge theories, 299
  Maxwell's equation, form of, 538
  symmetry, 292
Lorentz-Poincaré relativity, 292
Lorentz, quantum mechanics, 433
Lorentz transformations, relativistic spacetimes, 536
"Losses of head," hydrodynamics, 16
Low-energy superfluid helium-4 film, 230–32

Mach, Ernst, 55, 67, 541
Machian relationalism, 557–64
  best matching, 559, 559–60
  BSW action, 562–64
  diffeomorphism group (DPM), 557–58, 561
  Euclidean coordinate systems, 557
  general relativity (GR), 563–64
  group orbits, 559
  Jacobi's principle, 558–59
  kinematically possible models (KPMs), 558
  and Leibnizian relationalism, 558
  Reimannian 3-metrics, 561–62
  shape space, 558
  similarity groups, 559
  superspace, 562
  zero field strength, 557n75
Machian 3-space approach, 578
Mach's Principle: From Newton's Bucket to Quantum Gravity (Barbour), 557
Mach's principle, Standard Model, 610
Macromechanics, quantum mechanics, 430
Macroscopic scale behaviors, 256
Magnitude F, point-mass mechanics, 65, *65*
Magnitude of parameter, 144
Malament, David, 629n50
Manifold M
  global spacetime structure, 588–90
  compact manifold, 588
    Hausdorff, 588
  relativistic spacetime, 588–90
Manohar, A., 234, 246
Many-exact worlds theories, 467–68
Many-minds theories, 467–68
Many spins, mean field theory, 157
Many Worlds Theory. *See* Everett interpretation
Marginal scalings, 177–78
Mars's motion, 626–27
Mass-dependent schemes, effective field theory (EFT), 236–37
Mass-independent schemes, effective field theory (EFT), 237–39
Mass point lattice, classical mechanics, 44
Matched asymptotics, point-mass mechanics, 69
Material derivatives, continuum mechanics, 87
"Material particles," continuum mechanics, 270n18
Materials, discovery and invention of, 141–43
Mathematical complexity, classical mechanics, 45
Matter, infinities and renormalization, 141–88
  Boltzmann-Gibbs formulation, 180
  broken symmetries, 144–45
  Callen-Symanzik equation, 180
  canonical ensemble, 147
    use of term, 147n4
  continuous phase transition, 147
  correlation function calculations, 168–69
  corresponding states, principle of, 162
  critical fixed points, 179, 180
  critical point, 147
  crystalline materials, 143–44
  different phases of matter, 143–51
    broken symmetries, 144–45
    canonical ensemble, 147
    continuous phase transition, 147
    critical point, 147

crystalline materials, 143–44
dynamics, 145–47
equilibrium, 145–47
first mean field theory, 147–49
first-order phase transition, 147
Hamiltonian function, 146
magnitude of parameter, 144
Maxwell-Boltzmann distribution, 147
Maxwell's improvement, 149–51
mean field theory, 147–49
order parameters, 144–45
orientation of order parameter, 144
phase space, 146
phase transitions, 147
and statistical mechanics, 146
water, cartoon PVT diagram for, *148*
different scalings, 177–78
dynamical systems theory, 180
dynamics, 145–47
ensemble generates averages, 180
ensemble generates ensemble, 180
equilibrium, 145–47
extended singularities, renormalization group, 183
first mean field theory, 147–49
first-order phase transition, 147
fixed points, 178
flows and flow diagrams, 180–81
fluctuations, 151
 boiling, 151
 critical opalescence, 151–52
 liquid gas, 150n6, 151
Hamiltonian function, 146
irrelevant scalings, 177–78
irreversibility, 142
Ising model, 152–56
 Boltzmann-Gibbs formulation, 154
 defined, 152–53
 extended singularity theorem, 152, 154–55
 and Hamiltonian function, 153
 imprecision, 155n10
 Ornstein-Zernike infinity, 156
 particle spin, 154
 phase diagram, *156*
 singularity in phase transition, *155*
 two-dimensional Ising model, *153*
Landau's generalization, 160–61
magnitude of parameter, 144
marginal scalings, 177–78
materials, discovery and invention of, 141–43
Maxwell-Boltzmann distribution, 147
Maxwell's improvement, 149–51
mean field theory, 147–49, 156–60
 correlation length, 162
 experimental facts, 164–65
 first mean field theory, 147–49
 fluids, behavior of, 165
 and Hamiltonian function, 157
 heat capacity as measured, *166*
 many spins, 157
 mean field results, 157–58
 one spin, 157

order parameter jump, 162
 power laws, representing critical behavior by, 159–60
 renormalization calculations, 159–60
 scale transformation, 159–60
 scaling, 161
 spatial structures, 167
 symmetry, 160n12, 161
 theoretical facts, 165–67
 turbulence, 167
 universality, 161–62
mean field theory, summary of, 161–62
1937, events of, 160–64
 corresponding states, principle of, 162
 Landau's generalization, 160–61
 mean field theory, summary of, 161–62
 Netherlands meeting, 162–64
 universality, 161–62
order parameters, 144–45
orientation of order parameter, 144
phases of matter
 different phases, 143–51, 160–61
 fluctuations, 151
 thermodynamics, 145–46
 thermodynamics phases, 142
phase space, 146
phase transitions, 141–42, 147, 168
 word "phase," use of, 141n1
relevant scalings, 177–78
renormalization group as not a group, 181–82
scalings, different, 170–72
short-distance expansion, 182
snowflake, *143*, 143–44
splash, *143*
statistical equilibrium, 142
statistical mechanics, 142, 146
statistical physics, theory of, 141
sub-universe, 182
thermodynamic phases, 142
universality, 161–62
 classes, 178–79
U.S. National Bureau of Standards conference, 168
Virasoro algebra, 182
water, cartoon PVT diagram for, *148*
Widom scaling, 168–70
Wilson on renormalization group theory, 172–77
 calculational method, 174–75
 e-expansion, 176–77
 elementary particles, 176
 fixed-points, 177
 Fourier space, 172–74
 Landau-Ginzburg-Wilson free energy, 173–74
 physical space, 172–74
 running coupling constants, 175–76
 weak coupling fixed points, 177
Maudlin, Tim, 545n49
Maximal spacetimes, relativistic spacetime, 590
Maxwell-Boltzmann distribution, 147

Maxwell-Boltzmann statistics,
  indistinguishability, 371n42
Maxwell equations, quantum mechanics, 446
Maxwell group
  enriched relationalism, 550–52
  Hole Argument (Einstein), 576n104, 578
Maxwell, James Clerk
  different phases of matter
    boiling, 150
    improvement, 149–51
  "A Dynamical Theory of the Electromagnetic Field," 390
  Green's functions, 113
  modules, 34
  "On Physical Lines of Force," 387–93
  phase transitions, 147
  statistical physics, theory of, 141
  Treatise on Electricity and Magnesium, 391
  turbulence, 22
  vortex motion, 17
Maxwell's electrodynamics, 385–93
  Ampère law, 389
  chain reaction, 387–88
  currents, induction of, 386n5
  d'Alembert's principle, 389–90
  displacement, 391n9
  displacement current, 386–89
  electrodynamics, 386–87
  and Faraday's account of electromagnetism, 385, 387
  and fictional models, 387–93
  Lagrangian mechanics, 386–87, 390
Maxwell's equation, relativistic spacetimes, 538
Maxwell's improvement, 149–51
Maxwell's theory
  electroweak theory, 395, 399
  symmetry and equivalence, 325n21
  unification in physics, 7, 383
McMullin, Ernan, 636
Mean field calculations, the tyranny of scales, 266
Mean field theory, 147–49, 156–60
  correlation length, 162
  experimental facts, 164–65
  first mean field theory, 147–49
  fluids, behavior of, 165
  and Hamiltonian function, 157
  heat capacity as measured, *166*
  and long-range forces, 167n14
  many spins, 157–58
  mean field results, 157–58
  one spin, 157
  order parameter jump, 162
  power laws, representing critical behavior by, 159–60
  renormalization calculations, 159–60
  scale transformation, 159–60
  scaling, 161
  spatial structures, 167
  summary of, 161–62
  symmetry, 160n12, 161
  theoretical facts, 165–67
  turbulence, 167
  universality, 161–62
Meaningfully combined, rigid body mechanics, 77
Melnyk, Andrew, 198
Mercury's perihelion motion, 619
Metaphysical Foundations of Natural Science (Kant), 89n39
"Method of extensive abstraction," 96n43
Metric, global spacetime structure, 588–90
  Lorentzian metric, 588
Micromechanics, quantum mechanics, 430
Milgrom, M., 624
Mill-Ramsey-Lewis's Best Systems prescription, 566
Minkowski distances, 553–56
Minkowski geometries, have-it-all relationalism, 569n91, 570–71
Minkowski metric structure, relativistic spacetimes, 536
Minkowski spacetime
  effective field theory (EFT), 231
  global spacetime structure, 628n46, 629
    singularities, 601
    time travel, 603
  relativistic spacetime, 538, *591*, 591–93, *592*
  spacetime properties, 597, *597*
  substantivalist-relationalist debate, 522n1
  unification in physics, 393
  unitary equivalence, physical equivalence and, 518
Misner spacetime
  "chaotic cosmology" program, 631–32
  global spacetime structure, *603*
Mobility space, rigid body mechanics, 79
Möbius strip, relativistic spacetime, 589
Modification of Newtonian dynamics (MOND), 623–24
Mod-squared amplitude, Everett interpretation, 479n17, 480
Modules, hydrodynamics, 33–36
  correspondence principle, 35–36
  defining, 33
  idealizing modules, 33
  and models, 36–38
  reducing modules, 34, 36n46
  shared module, 36
  specializing modules, 33
  structure, *35*
Monte Carlo method. infinite idealization, 211–12
"More is Different" (Anderson), 2
Motion as change of place, 524
Motte, Andrew, 58
Multiple realization, phase transitions, 198n1
Multiplets
  electroweak theory, 396
  symmetry, 307–9
Multiverse, 639, 643–47
  De Sitter universe, 644
  Everett interpretation, 644n86
  inflation, 644
  and string theory, 644

Munitz. Milton K., 624, 626–27
Mysterium Cosmographicum (Kepler), 287–88

Nagel, Ernest, 1, 198, 260
Naimark dilation, quantum mechanics, 448
Nakedly singular spacetime, global spacetime structure, 601, *601*
"Natural coordinates," point-mass mechanics, 63
Naturalness, electroweak theory, 404n19
Navier-Cauchy equation, 6
Navier, Claude Louis
  instabilities, 19–20
  internal fluid forces, 14
  point-mass model, 67
Navier-Stokes equation, 256
  boundary layers, 25–27
  described, 14
  and explanatory progress, 27–28
  instabilities, 19–20
  low-density gas specialization, 34–35n44
  modules, 34–35
  turbulence, 22
  the tyranny of scales, 269–70, 275, 278–79
Ne'eman, Y., 306, 308–9
Neo-Newtonian spacetime, 531–33
Neptune
  discovery of, 618
  prediction of, 310
Netherlands meeting, 162–64
Neutrino, postulation of, 310
Newton-Cartan theory
  enriched relationalism, 551, 553
  Hole Argument (Einstein), 576–77
Newton group, 528
Newtonian behavior, quantum mechanics, 439
Newtonian boosts, symmetry and equivalence, 327n33
Newtonian gravity
  dark matter and dark energy, 619–20
  uniqueness of universe, 627, 632
Newtonian mechanics, unification in physics, 406
Newtonian scheme, symmetry, 295
Newtonian theory, symmetry and equivalence, 322n12, 324, 325, 327–28n36
  particles, 324, 325, 330
  physical equivalence, 328
Newton, Isaac
  billiard collisions, 68, *69*
  boundary layers, 23
  De Grav, 526
  law of gravitation, 60, 98
  laws of motion, 43, 58–59, 311
    Second Law, 77–78, 433
    third law, 63–64, 66, 81, 552
  "natural coordinates," 63
  planets, treatment of, 55
  Principia, 523
    Scholium, 523, 525n4, 526
  rotation, 48
  spacetime, substantivalist and relationalist approaches to
    diametrical symmetries, 527–29
    dynamical symmetries, spacetime and, 527–31
    Galilean invariance, 527–31
    have-it-all relationalism, 564–74
    Hole Argument (Einstein), 574–79
    inertia, spacetime explanation of, 541–44
    kinematic shift argument, 529–30
    Machian relationalism, 557–64
    neo-Newtonian spacetime, 531–33
    Newton's bucket, 523–27
    rationality, failure of, 539–40
    relationalism, varieties of, 544–74
Newton's bucket, 523–27
Noether, Emmy, 289, 291
  electroweak theory, 396–97
  "gauge argument," 291
  obituary (by Einstein), 297
  symmetry
    continuous symmetry, 295–97
    global continuous symmetry, 295–97
    local symmetry, 324–25
  symmetry and equivalence
    differential equations, symmetries of, 324–25, 328
    Hamiltonian symmetries, 328
No-go theorem for safe bit commitment protocols, 430
No-hidden variables theorem, 422
Non-Abelian case, electroweak theory, 396, 401
Noninertial motion, 542
Non-interacting particles, Everett interpretation, 463n5
Non-relativistic QCD, effective field theory (EFT), 229n7
Nonsimultaneous events, enriched relationalism, 549
Nontrivial spin properties, quantum mechanics, 428
Nonviscous fluid, hydrodynamics, 13
Normalization, quantum mechanics, 422n10
Normal science, hydrodynamics, 30n38
Norton, John D.
  anthropic reasoning, 642–43
  have-it-all relationalism, 572–73
  Hole Argument (Einstein), 523, 575
No-signaling theorem, quantum mechanics, 425
N! puzzle, indistinguishability, 349–52
NRQCD. *See* Non-relativistic QCD
Null geodesic incompleteness, spacetime properties, 598
Null vectors, relativistic spacetime, 589, *589*

Objective-probability role, Everett interpretation, 479n17
Observationally indistinguishable (OI) spacetime, 629–30
Ockham's razor, 523, 539
ODEs. *See* Ordinary differential equations (ODEs)
"Old" cosmological constant problem, dark matter and dark energy, 621–22
Oldenburg group, infinite idealization, 209

Olver, P., 322n10
Omega minus hadron, symmetry, 307
One-dimensional Ising model, 180
One spin, mean field theory, 157
Onnes, Heike Kamerlingh, 165
"On Physical Lines of Force" (Maxwell), 387–93
Onsager, Lars, 165–67, *166*, 195
Ontology and ontological issues
    classical mechanics, 44n2
        ontologically mixed circumstances, 44n2
    effective field theory (EFT), 239–43
        antifoundationalism, 240
        antireductionism, 240
        approximations, EFTs and, 243
        cutoff, realistic interpretations of, 241–43
        decoupling, 240–41
        quantum field theory (QFT), 240–43
        quasi-autonomous domains, 240–41
        Wilsonian approach, 240
    Everett interpretation, higher-order ontology, 470–74
    indistinguishability, 365–78, 374
        "account of equality," 372
        eliminativism, 376–78
        Gibbs paradox, 366–69
        identity conditions, 372–76
        Leibniz's principle, 373–74
        Maxwell-Boltzmann statistics, 371n42
        permutability of objects, 369–70
        philosophical logic, 369–72
        quantities invariant, 365
        quantum particles, 371–72
        spin, 375
        total symmetry, 369
    infinite idealization, phase transitions, 214–17
    retarded Green's functions, contours for damped oscillator, *115*
    symmetry, 306–7
Open channel flow, 28
Order parameters
    matter, infinities and renormalization, 144–45
    mean field theory, order parameter jump, 162
    orientation of order parameter, 144
Ordinary differential equations (ODEs)
    continuous variables, lift from, 55n14
    and foundational principles, 45n4
    PDE's distinguished, 45
    and Schrödinger equation, 46
    and spin, 45n5
    tasks governed by, *49*
Ordinary QM
    unitary equivalence, physical equivalence and, 495, 496, 500–502, 504, 510–13
    unitary inequivalence, as example of, 508
Ornstein, Leonard, 152, 158
Ornstein-Zernike infinity, Ising model, 156
Ostriker, P., 620n31
Our Knowledge of the "External World" (Russell), 96n43
Outgoing waves, radiation theory, 127
"Overlapping domains" argument, early universe cosmology, 633

Pairs of solutions, symmetry and equivalence, 333n55
Parallelly transported vector, spacetime properties, 593
Parallel tradition, Everett interpretation, 478n16
Paramagnetism, 192
Parastatistics, indistinguishability, 364n32
Parity, symmetry, 293n9
Partial differential equations (PDEs)
    and foundational principles, 45n4
    ODE's distinguished, 45
Particle spin, Ising model, 154
Passivity and causality, dispersion theory, 137
Past Cauchy horizon, relativistic spacetime, 592
Past-directed curve, relativistic spacetime, 589
Past distinguishability, spacetime properties, 596
Past domain of dependence, relativistic spacetime, 592
Past endpoint, relativistic spacetime, 592
Past incomplete, spacetime properties, 598
PAS, unitary equivalence, 504, 508, 514
Patashinskii, Alexander, 168–69
Patrizi, Francesco, 526
Pauli relations, unitary equivalence, 494
Pauli spins, as example of unitary inequivalence, 494, 501–4
Pauli, Wolfgang, 310, 377
Pearson, Karl, 271
Peebles, Phillip James Edward, 620n31
Penrose, Roger, 601, 613, 626, 637
Permutability of objects, indistinguishability, 369–70
Permutable coins, indistinguishability, 366–67, 368
PEV, unitary equivalence, 496–500, 496n6, 504, 508, 514
Phase diagram, Ising model, *156*
Phase space
    matter, infinities and renormalization, 146
    quantum statistics, explanation of, 361–62
    reduced phase space, 361–62, *362*
    symmetry and equivalence, 327
Phase transformation, electroweak theory, 396n13
Phase transitions, 189–223
    additivity, 201n2
    boiling, 147
    conceptual novelty, 199–200
    defined, 189
    emergence of, 197–204
    ensemble equivalent, 207
    experimental studies, 147
    explanatory irreducibility, 210–14
    extensivity, 201n2
    infinite idealization, 204–17
        back-bending, 206–8, *207*
        Bose-Einstein, 206, 208
        caloric curve, back-bending of, *207*
        complex inverse temperature, *209*
        conceptual novelty, 204–10
        distribution of zeros, 208–10

and emergence of phase transitions, 202
explanatory irreducibility, 210–14
Galilean idealization, 210n7
liquid-gas system, *215*
Monte Carlo method, 211–12
Oldenburg group, 209
ontological irreducibility, 214–17
renormalization group, 216–23
smooth phase transitions, 206
Yang-Lee theorem, 208
intensive properties, 201
irreducibility, 199, 203
explanatory irreducibility, 210–14
infinite idealization, 221
materials, discovery and invention of, 141–42
multiple realization, 198n1
Nagel's theory, 198
ontological irreducibility, 214–17
paramagnetism, 192
and reduction, 198–201
as reductionism in the core sense of, 198
renormalization group, infinite idealization in, 216–23
critical behavior of particular systems, 218
crossover theory, 220–21
irreducibility, 221
universality, 217–18
renormalization group theory, 195–97
coarse-graining, 197
and Hamiltonian function, 196–97
spin, 193–94
statistical mechanical treatment, 193–95
and Hamiltonian function, 193
spin, 193–94
Yang-Lee theorem, 195
and thermodynamic properties, 201
thermodynamic treatment, 191–93
continuous phase transitions, 191–92
critical exponent, 193
ferromagnetic transitions, 192
first-order phase transitions, 191
Helmholtz free energy, 192
order-disorder transitions, 192
order parameter, 192–93
U.S. National Bureau of Standards conference, 168
word "phase," use of, 141n1
Phenomenological arrow of time, quantum mechanics, 433n22
Phenomenological theories, generally, 2
Phillips, Rob, 272–73, 284–85
Physical equivalence
symmetry and equivalence, 328–33
differential equation, 329–30
D2, 329–30
Galilei group, 330
Hamiltonian symmetries, 331–32
Kepler problem, 332
Lagrangian treatment, 331
and Newtonian theory, 328
pairs of solutions, 333n55
spacetime symmetries, 331–32n49, 332

and unitary equivalence. *See* Unitary equivalence, physical equivalence and
Physical motivation, retarded Green's functions, 113
Physical space, renormalization group theory, 172–74
Pickering, Andy, 302n24
Pilot-wave theories, quantum mechanics, 440, 482
"Pinned constraint," rigid body mechanics, 71
Pipe flow, hydrodynamics, 20
Planck, M.
on indistinguishability, 7, 340
N! puzzle, 350
quantum, indistinguishability and, 341–43
quantum statistics, explanation of, 360
unitary equivalence, 494
Planck scale
dark matter and dark energy, 621
electroweak theory, 403
Plane parallel flow, hydrodynamics, 20
Plane Poiseuille flow, hydrodynamics, 21
Plücker, Julius, 49
Poincaré relativity, 292, 423
Poincaré sphere, 423
effective field theory (EFT), 238
quantum mechanics, 423
rationality, failure of, 539
Point-mass mechanics, 44–46, 57–69
billiard collisions, 68–69, *69*
Cartesian locations, 63
Cassini space probe to Saturn, 61, *61*
coefficient of restitution, 69
constitutive modeling conditions, 66, 82
inertial reaction, 65
isolated point mass, 59
Lennard-Jones potential, 60
magnitude F, 65, *65*
matched asymptotics, 69
methodologies of avoidance, 62
"natural coordinates," 63
purely elastic collision, 69n26
representative center, 59
rigid body mechanics, 71
rotating rigid objects, 58
special force laws, 63
steel ball pendulums, *64–65*, 64–66
Point-mass swarm, axiomatic presentation, 51
Point-particle electrodynamics, 129–32
Points, axiomatic presentation, 49–50
Poisson bracket
unitary equivalence, physical equivalence and, 493–94
unitary inequivalence, as example of, 505–6
Poisson, Siméon Denis
point-mass mechanics, 67–68
rigid body mechanics, 83
surface waves, 17
"two constant," 67
Pokrovsky, Valery, 168–69
Polarization of a system, 502

Polchinski, J., 227
Poncelet, Jean Victor, 22, 23
Possibility and Everett interpretation, 475–77
Post facto strategy, the tyranny of scales, 263
Post, H., 357
POV measure, quantum mechanics, 8, 448–49, 454
Power laws, representing critical behavior by, 159–60
Prandtl, Ludwig
  boundary-layer theory, 13, 25–27, 37, 53n12
  explanatory progress, 28–29
  instabilities, 20
  turbulence, 22
  wing theory, 13, 30
Prediction and predictability
  effective field theory (EFT), 232–35
  symmetry
    from multiplet scheme, 307–9
    Neptune, prediction of, 310
    of omega minus hadron, 306
    spin-3/2 baryon decuplet, 307–8, *309*
"Preferred basis problem"
  eliminativism, 377
  Everett interpretation, 466–68
    and many-exact worlds theories, 467–68
    and many-minds theories, 467–68
  overview, 461
Pressures, rigid body mechanics, 74
A Primer on Determinism (Earman), 1
Principia (Newton), 523
  Scholium, 523, 525n4, 526
Principle of Equivalence (Einstein), 537
Principle of Limiting Amplitude, 128
"Principle of mediocrity" (PM), 639–42
Principle of Sufficient Reason (PSR), 530n14
Principle of virtual work, rigid body mechanics, 79
Privileging, retarded Green's functions, 113–20
Probabilities
  Everett interpretation, 466, 474–77
    objective-probability role, 479n17
    parallel tradition, 478n16
    philosophical aspects of probability, 474–75
    and possibility, 475–77
    probability simpliciter, 475
    and uncertainty, 475–77
  quantum mechanics, 446
  rigid body mechanics, 76
Probability simpliciter, 475
Problem of the Physical Infinitesimal, 86, 90, 93
Projection postulate, quantum mechanics, 417
Proper mixtures, quantum mechanics, 423–28
  indistinguishability from proper mixtures, 429
Pseudo-Reimannian metric field, Hole Argument (Einstein), 574n100, 578
Punctiform point of view, rigid body mechanics, 79, *80*
Purely elastic collision, point-mass mechanics, 69n26
Puzzle solving, hydrodynamics, 30n38
PV measures, quantum mechanics, 446n34

Quantitative problem, Everett interpretation, 477–79
Quantizing, unitary equivalence, 492–94
Quantum chromodynamics (QCD)
  electroweak theory, 394, 402–3
  symmetry, 302
Quantum electrodynamics (QED)
  effective field theory (EFT), 411
  unification in physics, 396–97
Quantum fields, eliminativism, 377–78
Quantum field theory (QFT), 240–43
  dark matter and dark energy, 621–22
  effective field theory (EFT), 240–43, 251
  unification in physics, 382, 408–10, 412–13
  uniqueness of universe, 626
  unitary equivalence, physical equivalence and, 491, 518
Quantum, indistinguishability and, 341–43
  "light quanta," 342
  microstate, 342n3
  and "permutable," 342n4
Quantum locality, 1
Quantum mechanics
  bit commitment problem, 429–30
  Bloch sphere, 423
  Born rule, 8, 417, 419, 431
    consistency, 442
    general phenomenology of measurements, 446, 448
    measurement, 451–52, 454
    proper mixtures, 424
    unsharp spin measurements, 445
  Broglie-Bohm theory, 431, 440, 482
  Brownian particle, 437
  cat example (Schrödinger), 435, 440
  classical dynamical behavior, 438
  classical regime, 8, 416–59
    Born rule, 8, 417, 419, 431
    Brownian particle, 437
    cat example (Schrödinger), 435, 440
    classical dynamical behavior, 438
    coherent states, 432–34
    Ehrenfest's theorem, 432–34
    environment, entanglement with, 434–37
    Gaussian bell curves, 432
    Gaussian wave packets, 432–34
    Heisenberg's "cut," 440–42
    macromechanics, 430
    micromechanics, 430
    overview, 8
    phenomenological arrow of time, 433n22
    pilot-wave theories, 440
    uniqueness of, 439n28
  coherent states, 432–34
  collapse postulate, 8, 417, 418, 424, 454
  density operators, 425
  Copenhagen interpretation, 420
  decoherence, 437–41
    continuous models of, 438–39
    Everett interpretation. *See* Everett interpretation

Newtonian behavior, 439
density operators, 420–23
  bit commitment problem, 429–30
  collapse postulate, 425
  entangled states, 427
  Hilbert space, 425–26
  Hilbert-space vectors, 420–21
  improper mixtures, 427
  no-go theorem for safe bit commitment protocols, 430
  no-hidden variables theorem, 422
  nontrivial spin properties, 428
  normalization, 422n10
  no-signaling theorem, 425–26
  proper mixtures, 423–28
  reduced states, 424
  simplex, 423–24
  spin-1/2 systems, 422–23, 425, 443
Dirac-von Neumann interpretation, 419
discretized position measurements, 443
Ehrenfest's theorem, 432–34
entangled states, 427
environment, entanglement with, 434–37
Everett interpretation. *See* Everett interpretation
Gaussian bell curves, 432
Gaussian wave packets, 432–34
generally, 1, 45
Gleason's theorem, 449
Heisenberg's "cut," 440
  Copenhagen interpretation, 441n30
Hilbert space, 425–26, 450n38
  environment, 434
  ideal spin measurements, 443
Hilbert-space vectors, 419n7, 420–21
ideal spin measurements, 443–44
improper mixtures, 427
  indistinguishability from proper mixtures, 429
linear operators, 417n4
macromechanics, 430
Many Worlds Theory. *See* Everett interpretation
Maxwell equations, 446
measurement, 8, 416–59
  Born rule, 451–52, 454
  Broglie-Bohm theory, 482
  collapse postulate, 417, 418, 454
  discretized position measurements, 443
  eigenvectors, 418
  Everett interpretation. *See* Everett interpretation
  Gleason's theorem, 449
  Hilbert space, 417, 450n38
  ideal spin measurements, 443–44
  linear operators, 417n4
  Maxwell equations, 446
  Naimark dilation, 448
  phenomenology of, 417–19, 446–49
  POV measure, 8, 448–49, 454
  probabilities, 446

  problem, 451–55
  projection postulate, 417
  PV measures, 446n34
  Schrödinger equation, 420, 451–52
  self-adjoint operators, 417–18
  statistical algorithm of quantum mechanics, 417
  Stern-Gerlach magnetic field, 418, 419n6, 442–44, 446, 453
  subspace, 417–18
  transformation, 446
  "unsharp" spin measurements, 444–446
  up and down spin states, 418–19
micromechanics, 430
minimal interpretation, 419–20
Naimark dilation, 448
no-go theorem for safe bit commitment protocols, 430
no-hidden variables theorem, 422
nontrivial spin properties, 428
normalization, 422n10
no-signaling theorem, 425
phenomenological arrow of time, 433n22
pilot-wave theories, 440, 482
Poincaré sphere, 423
POV measure, 8, 448–49, 454
probabilities, 446
projection postulate, 417
proper mixtures, 423–28
  indistinguishability from proper mixtures, 429
PV measures, 446n34
radioactive decay, 437
reduced states, 420–30
resolution of the identity, 446
Schrödinger equation, 420, 451–52
  ideal spin measurements, 444
Schrödinger evolution, 418
self-adjoint operators, 417–18
simplex, 423–24
spin-1/2 systems, 422–23, 425, 443
standard interpretation, 419–20
statistical algorithm of quantum mechanics, 417
Stern-Gerlach magnetic field, 418, 419n6, 442–44, 446, 453
subspace, 417–18
transformation, 446
uniqueness of, 439n28
"unsharp" spin measurements, 444–446
up and down spin states, 418–19
Quantum particles, indistinguishability, 371–72
Quantum statistical mechanics (QSM), 491
Quantum statistics
  explanation of
    Hilbert space, 362, *363*
    phase space dimension, 361–62
    subspace dimension, 362–63
    volume measures, 362–63
    weighting, 361, *362*
  indistinguishability and, 360–64

Quasi-autonomous domains, effective field
    theory (EFT), 240–41, 243
  and emergence, 244
  Wilsonian approach, 241
Quasi particles, Everett interpretation, 473–74
Quine, W. van, 340–41

Radiation theory, 123–29
  boundary conditions, 123–24
  description, 109
  finiteness condition, 124
  Fourier transformation technique, 126–27
  Helmholtz equation, 123, 125–27, 129–30
  incoming waves, 127
  Laplace transform technique, 125
  outgoing waves, 127
  Principle of Limiting Amplitude, 128
  reasons for, 123–29
  reduced wave equation, 125. See also
      Helmholtz equation
  Sommerfeld radiation condition. See
      Sommerfeld radiation condition
  time dependent boundary condition, 124
  time-harmonic waves, 127–29
  waveguide, *124*
Radioactive decay, 437
Radioactive scattering, 52
Ramsey, Jeffrey, 13, 30
Random variables, the tyranny of scales, 278
Rankine, William John Macquorn, 24–25
"Rari-constancy" theorists, 270
Raychaudhuri equation, 611
Rayleigh, Lord
  boundary layers, 23–24
  instabilities, 20
  surface waves, 17
Reasonable properties, global spacetime
    structure. See Global spacetime structure
"Re-combination," Standard Model, 615
Redirection of thrust, rigid body mechanics, 71
Reduced phase space, 361–62, *362*
Reduced states, quantum mechanics, 420–30
Reduced wave equation, radiation theory, 125.
    See also Helmholtz equation
Reduction
  phase transitions, 198–201
  the tyranny of scales, 260
Reductionism
  effective field theory (EFT), 411n25
  phase transitions, 198
Reductive unity, unification in physics, 385–93
Regularity approach, 564–69
Reichenbach, H., 358–59, 637
Reimannian 3-metrics, 561–62
Relationalism
  Barbour's Machian relationalism, 557–64
  dynamical approach to relativity, 569–74
  enriched relationalism, 545–57
    classical mechanics, 545–53
    relativity, 553–57
  have-it-all relationalism
    Best Systems prescription, 566
  dynamical approach to relativity, 569–74
  dynamical laws, 565
  Galilean spacetime, 570
  general relativity (GR), 568–69
  homogeneous matter, 573–74
  Humean approach, 566
  Leibnizian relationalism, 567–68
  Lorentz contract, 570n92, 572
  Mill-Ramsey-Lewis's Best Systems
      prescription, 566
  Minkowski geometries, 569n91, 570–71
  regularity approach, 564–69
  rod, constituents of, 570
  rotation disks argument, 566
Machian relationalism, 557–64
  best matching, *559*, 559–60
  BSW action, 562–64
  diffeomorphism group (DPM), 557–58, 561
  Euclidean coordinate systems, 557
  general relativity (GR), 563–64
  group orbits, 559
  Jacobi's principle, 558–59
  kinematically possible models (KPMs), 558
  and Leibnizian relationalism, 558
  Reimannian 3-metrics, 561–62
  shape space, 558
  similarity groups, 559
  superspace, 562
  zero field strength, 557n75
regularity approach, 564–69
Relationalist approaches to spacetime. See
    Spacetime, substantivalist and
    relationalist approaches to
Relationalist, reasons for being, 539–40
Relative particle configurations
  spacetime, substantivalist and relationalist
      approaches to, 527
Relativistic cosmology, 625–26
Relativistic quantum field theories (RQFTs)
  emergence, EFTs, 250–51
  overview, 224–25
Relativistic spacetime, 587–93
  Cartesian coordinates, 536
  causal curve, 591
  causal future, 590–91
  chronology violating region of spacetime, 591
  clocks, 536
  closed causal curve, 591
  conformal factor, 592
  cylindrical Minkowski spacetime, *591*
  dependence, 590–93
  Einstein Field Equations (EFEs), 537
  Euclidean space, 536
  future Cauchy horizon, 592
  future-directed curve, 589
  future domain of dependence, 592
  future endpoint, 592
  influence, 590–93
  isometric spacetimes, 589–90
  locally isometric spacetime, 590
  Lorentz-invariant form of Maxwell's equation,
      538

Lorentz transformations, 536
manifold M, 588–90
maximal spacetimes, 590
Maxwell's equation, 538
metric, 588–90
Minkowski metric structure, 536
Minkowski spacetime, 538, *591,* 591–93, *592*
Möbius strip, 589
null vectors, 589, *589*
past Cauchy horizon, 592
past-directed curve, 589
past domain of dependence, 592
past endpoint, 592
spacelike surface, 592
spacelike vectors, 589, *589*
strong equivalence principle, 538
temporarily orientable spacetime, 589
timelike future, 590–91
timelike vectors, 589, *589*
"twin paradox" scenario, 536
Relativity, enriched relationalism, 553–57
 four-force, 554
 gravitational wave, 556
 kinematically possible models (KPMs), 555
 Minkowski distances, 553–56
Relevant scalings
 matter, infinities and renormalization, 177–78
Renormalization calculations, mean field theory, 159–60
Renormalization group (RG)
 e-expansion, 176–77
 effective field theory (EFT), 232–35
  nonrenormalizable EFTs, 233
 elementary particles, 176
 fixed-points, 177
 Fourier space, 172–74
 Landau-Ginzburg-Wilson free energy, 173–74
 matter, infinities and renormalization, 181–82
 overview, 5–6
 phase transitions, 195–97
  coarse-graining, 197
  and Hamiltonian function, 196–97
 phase transitions, infinite idealization, 216–23
  critical behavior of particular systems, 218
  crossover theory, 220–21
  irreducibility, 221
  universality, 217–18
 physical space, 172–74
 running coupling constants, 175–76
 the tyranny of scales, 264–66, 269, 275, 280
 unification in physics, 382, 385, 407, 410
 weak coupling fixed points, 177
 Wilson, Kenneth on, 172–77
  calculational method, 174–75
  e-expansion, 176–77
  elementary particles, 176
  fixed-points, 177
  Fourier space, 172–74
  Landau-Ginzburg-Wilson free energy, 173–74
  physical space, 172–74
  running coupling constants, 175–76

 weak coupling fixed points, 177
Renormalization schemes, effective field theory (EFT), 235–39
 continuum EFTs, 237–39
 Green's function, 235
 Lagrangian density, 235
 mass-dependent schemes, 236–37
 mass-independent schemes, 237–39
 Wilsonian approach, 236–37
Renormalization, unification in physics, 406–13
Representative center, point-mass mechanics, 59
Representative points, rigid body mechanics, 81, 93
Representative volume element (REV), 264, 267–68, 280
Resolution of the identity, quantum mechanics, 446
Response planes, continuum mechanics, 95
Retarded Green's functions
 Abraham-Lorentz equation, 119
 bounded spatial domain, wave equation in, 121–23
 Cauchy problem, 120
 causal directionality, 116
 "causality" as initial value problem, 116–20
 contours for damped oscillator, *115*
 damping, added, 113–16, 135–36
 Dirac delta function, 113, 119
 "final condition," 119
 "final value problem," 119
 Fourier transformations, 113, 114
 initial value problem, 116–20
 Lebesgue measure zero, 118
 overview, 109
 physical motivation, 113
 privileging, 113–20
 Residue Theorem, 114
 spatial propagation, 120–21
 undamped harmonic oscillator, 110
 wave equation, 120–23
Reynolds, Osborne
 boundary layers, 26
 instabilities, 20
 turbulence, 21, 22
RG. *See* Renormalization group (RG)
Ricci tensor, spacetime properties, 594
Riemann-Hugoniot approach to shock waves, 53n12
Riemann tensor, 539n34
Rigid body mechanics, 44, 67, 70–83
 action-at-a-distance forces, 73
 arc length along the slot, 71–72, *72*
 bead sliding on rigid wire, *71*
 bulls-eyes, *75,* 75–76
 classical distillations of quantum processes, 74–75
 constraint relationships, 70
 contact forces, 73
 d'Alembert's principle, 79
 dead load, 73
 dimensional inharmonious quantities, 76
 dynamic loading, 73

Rigid body mechanics (*Cont.*)
   Eulerian cuts, 73, *74*
   forced closed, 70n27
   forces, 73–74
   free body diagrams, 73, *74*
   generalized coordinates, 71
   Greenwood's proofs, 80–82
   "higher or lower" pairs, 70n28
   independently variable, 72
   kinematics of mechanisms, 79
   Lagrange's principle, 78–80, 82
   meaningfully combined, 77
   mobility space, 79
   "pinned constraint," 71
   point-mass perspective, 71
   pressures, 74
   principle of virtual work, 79
   probability differences, 76
   punctiform point of view, 79, *80*
   redirection of thrust, 71
   representative points, 81, 93
   sewing machine mechanism, *71*
   static load, 73
   stress, 74
   theory of measure, 76
   torque r, 77, *78*
   traction forces, 73
   turning moment, 77, *78*
   virtual displacement, 80, *81*
   virtual variations, 79n33
   virtual-work reasoning, 72, *72*
Rigid crystalline forms, axiomatic presentation, 51
Robin and Marian analogy, *75*, 75–76
Rod, constituents of, 570
Rolling on a rigid track, axiomatic presentation, 54–55
Rosenfeld, L., 356
Rotating rigid objects, point-mass mechanics, 58
Rotation curves, dark matter and dark energy, 620n30
Rotation disks argument, have-it-all relationalism, 566
Rotations
   axiomatic presentation, 48
   symmetry, 288
RQFTs. *See* Relativistic quantum field theories (RQFTs)
Running coupling constants, renormalization group theory, 175–76
Russell, Bertrand
   on causation, 4, 107–8, 137–38
   dispersion relations, 137
   indistinguishability, 369
   ontology, 372
   Our Knowledge of the "External World, 96n43
Russell, Scott, 17

Saint-Venant, Barré, Adhémar de, 22, 23
Saunders, Simon, 478, 480
Scalar field theory, 227n3
Scale sizes, relationships between, 51–52

Scale transformation, mean field theory, 159–60
Scaling, 170–72
   block transforms, 170–72
   mean field theory, 161
Schrödinger equation
   classical mechanics, 46
   electroweak theory, 395
   ordinary differential equations (ODEs), 46
   quantum mechanics, 438–39
      ideal spin measurements, 444
      measurement, 420, 451–52
   unitary equivalence, physical equivalence and, 494
Schrödinger, Erwin, 46
   electroweak theory, 395
   indistinguishability
      N! puzzle, 349–50
      quantum, indistinguishability and, 343
      as uniform symmetry, 354
   and ordinary differential equations (ODEs), 46
   quantum mechanics
      cat example, 435, 440
      classical regime, 430–31
      failure of, possible, 437
      measurement, 420
      wave function, 432n19
      wave functions, 433–34
   unitary equivalence, 494
Schrödinger evolution, 418
Schweber, S., 243, 244, 249–41, 407, 411
Schwinger, J., 397, 491, 492
Segal, Irving E., 516–17
Self-adjoint operators, quantum mechanics, 417–18
Semantics, Everett interpretation, 476
Sewing machine mechanism, *71*
Shape space, Machian relationalism, 558
Shearing pattern, 85
Ships, boundary layers, 24
Shock waves, Riemann-Hugoniot approach to, 53n12
Short-distance expansion, 182
Short-range forces, 97, *97*, 98
Similarity groups, Machian relationalism, 559
Simplex, quantum mechanics, 423–24
Singularities
   global spacetime structure, 598–601
   phase transition, Ising model, *155*
   spacetime properties, 596
Size of group, symmetry and equivalence, 322n10
Skin resistance, 24
Sklarations, 548–49
Sklar, Lawrence, 548
Smolin, Lee, 625–26
Smoluchowski, Marian, 151
Smooth phase transitions, 206
Snowflake, *143*, 143–44
   symmetry, 288
Sommerfeld radiation condition
   application of, 123
   description, 109
   time harmonic waves, 127, 129

Sophisticated substantivalism, Hole Argument (Einstein), 575
Spacelike geodesic incompleteness, spacetime properties, 598
Spacelike surface, relativistic spacetime, 592
Spacelike vectors, relativistic spacetime, 589, *589*
Spacetime geometry, global structure, 628
Spacetime properties, global spacetime structure, 593–98
 Cauchy surface, 597
 causal continuity condition, 597
 causal simplicity condition, 597
 causal structure, 596
 chronology condition, 596
 constraint solutions, 595
 convex normal, 593
 distinguishability conditions, 596
 dominant energy conditions, 595
 Einstein's equation, 595
 Einstein tensor, 594
 energy-momentum tensor, 594
 future distinguishability condition, 596
 future incomplete, 598
 global hyporbolicity, 597
 global properties, 596–98
 local properties, 593–95
 Minkowski spacetime, 597, *597*
 null geodesic incompleteness, 598
 parallelly transported vector, 593
 past distinguishability, 596
 past incomplete, 598
 Ricci tensor, 594
 singularities, 596
 spacelike geodesic incompleteness, 598
 stable causality condition, 596–97
 strong causality condition, 596
 strong energy conditions, 595
 timelike geodesically incomplete, 598
Spacetime, substantivalist and relationalist approaches to, 9–10, 522–86
 Cartesian coordinates, 536
 Cartesian motion, 524–26
 circular motion, 525
 clocks, 536
 De Sitter universe, 541n37
 diffeomorphism group (DPM), 533–34, 547, 557–58, 561
 dynamical symmetry group, 529
 Einstein Field Equations (EFEs), 537, 543
 Euclidean coordinate system, 529
 Euclidean coordinate systems, 528
 Euclidean space, 536
 Galilean covariance, 523
 Galilean invariance, 527–31
  dynamical symmetries, spacetime and, 527–29
  dynamical symmetry group, 529
  Euclidean coordinate system, 529
  Euclidean coordinate systems, 528
  Galilei group, 529
  kinematically privileged systems, 527
  kinematic shift argument, 529–30

 Leibniz group of transformations, 528
 Leibnizian relationalism, 527–28
 Leibnizian relationalist, 530
 Newton group, 528
 Principle of Sufficient Reason (PSR), 530n14
 relative particle configurations, 527
 spacetime symmetry groups, 528
 Galilean spacetime, 531–33
 Galilei group, 529, 533
 and general relativity, 537–39, 541
 gravitational field, 539n34
 Hole Argument (Einstein), 300n17, 523, 574–79
  and dynamical symmetries, 576
  Euclidean symmetries, 577
  and Galilean spacetime, 576n104
  general relativity (GR), 573–75, 578
  hole diffeomorphism, 575
  individualistic facts, 577n108
  and kinematic shift argument, 576
  Leibniz and Clarke correspondence, 577n106
  Machian 3-space approach, 578
  and Maxwell group, 576n104, 578
  Newton-Cartan theory, 576–77
  pseudo-Reimannian metric field, 574n100, 578
  sophisticated substantivalism, 575
  structural realist interpretation of spacetime, 577
 inertia, spacetime explanation of, 541–44
 instantaneous relative distances, 540n35
 kinematically possible models (KPMs), 531–33
 kinematically privileged systems, 527
 kinematic shift argument, 523, 529–30, 544, 546–47
 Leibniz group of transformations, 333, 528
 Leibnizian relationalism, 527–28, 544–45, 548, 567–68, 572
 Leibnizian relationalist, 530
 Lorentz-invariant form of Maxwell's equation, 538
 Lorentz transformations, 536
 Maxwell's equation, 538
 Minkowski metric structure, 536
 Minkowski's space time, 538
 motion as change of place, 524
 neo-Newtonian spacetime, 531–33
 Newton group, 528
 Newton's bucket, 523–27
 Ockham's razor, 523, 539
 Principle of Sufficient Reason (PSR), 530n14
 rationality, failure of, 539–40
 relationalism, varieties of, 544–74
  Barbour's Machian relationalism, 557–64
  enriched relationalism, 545–57
  have-it-all relationalism, 564–74
  Leibnizian relationalism, 527–28, 544–45, 548
 relationalist, reasons for being, 539–44
  inertia, spacetime explanation of, 541–44
  rationality, failure of, 539–40

Spacetime, substantivalist and relationalist approaches to (*Cont.*)
  relative particle configurations, 527
  relativistic spacetimes, 536–39
    Cartesian coordinates, 536
    clocks, 536
    Einstein Field Equations (EFEs), 537
    Euclidean space, 536
    Lorentz-invariant form of Maxwell's equation, 538
    Lorentz transformations, 536
    Maxwell's equation, 538
    Minkowski metric structure, 536
    Minkowski's space time, 538
    strong equivalence principle, 538
    "twin paradox" scenario, 536
  Riemann tensor, 539n34
  spacetime substantivalism
    diffeomorphism group (DPM), 533–34
    Galilean spacetime, 531–33
    kinematically possible models (KPMs), 531–32, 558
    neo-Newtonian spacetime, 531–33
    symmetries, 533–36
    transtemporal structure, 531
  spacetime symmetry groups, 528
  strong equivalence principle, 538
  symmetries
    coordinate independent transformations, 533
    Galilei group, 534
    generally covariant equations, 533
    Leibniz group of transformations, 533
    Leibniz relationalist, 535
    trajectories of force-free bodies, spacetime as, 542n41
  transtemporal structure, 531
  "twin paradox" scenario, 536
Spacetime symmetries, 292, 321–22
  and physical equivalence, 331–32n49, 332
  structure of, 319
Spacetime symmetry groups, 528
Spatial propagation
  advanced Green's functions, 120–21
  retarded Green's functions, 120–21
Spatial structures, mean field theory, 167
Special force laws
  continuum mechanics, 99
  point-mass mechanics, 63
Spin
  indistinguishability, 375
  and ordinary differential equations (ODEs), 45n5
  phase transitions, 193–94
  "unsharp" spin measurements, 444–446
Spin-1/2 systems, 422–23, 425, 443
Spin-3/2 baryon decuplet, 307–8, *309*
Splash, *143*
SSB insight, 304
Stable causality condition, 596–97
Stage-setting, symmetry and equivalence, 320–21
"Standard candle"
  cosmology, philosophy of, 609n1
  dark matter and dark energy, 618
Standard Model, 10–11
  ACDM model, 617
  barriers, 615
  Big Bang, 609, 615
  black-body spectrum of radiation, 615
  Boltzmann equation, 614n16
  cold dark matter, 617
  cosmic background radiation (CBR), 614–16
  cosmological principle, 610
  early universe cosmology, 633–36
  effective field theory (EFT), 228
  Einstein Field Equations (EFEs), 610–11, 613
  electroweak theory, 394, 402–6
  Everett interpretation, 471
  expanding universe models, 609–14
  freeze out of particles, 614
  Friedman-Lemaitre-Robertson-Walker (FLRW) models, 610, 612–13, 614n16, 617
  galaxies and clusters of galaxies, length scale, 613
  global isotropy, reduction of, 611
  Mach's principle, 610
  overview of, 609–17
  parameters of, determining, 632
  Raychaudhuri equation, 611
  "re-combination," 615
  structure formation, 616–17
  thermal history, 614–16
  unification in physics, 381, 383
Static load, rigid body mechanics, 73
Statistical algorithm of quantum mechanics, measurement, 417
Statistical equilibrium, 142
Statistical mechanical treatment, phase transitions, 193–95
  and Hamiltonian function, 193
  spin, 193–94
  Yang-Lee theorem, 195
Statistical mechanics, 142, 146
  effective field theory (EFT), 249–50
  indistinguishability, 346–47
    coarse-graining, 346
    entropy, 346–47
    fine-graining, 346
Statistical physics, theory of, 141
Statistical Thermodynamics (Schrödinger), 349–50
Steel ball pendulums, point-mass mechanics, 64–65, 64–66
Steel beams, the tyranny of scales, 258–59, *264*, 278, *279*
Steel, Gaussian and, *279*
Stein, H., 524, 526n5
Stern-Gerlach magnetic field, quantum mechanics, 418, 419n6, 442–44, 446, 453
Stern, Otto, 350
Stieltjes-Lesbeque integration, axiomatic presentation, 53n12
Stoker, J. J., 127–29

Stokes, George Gabriel
  boundary layers, 23
  fluid velocity, 14
  Navier-Stokes equation
    boundary layers, 25–27
    instabilities, 19–20
    low-density gas specialization, 34–35n44
    modules, 34–35
    point-mass mechanics, 67
    turbulence, 22
  point-mass mechanics, 67
  surface waves, 17
  turbulence, 22
  the tyranny of scales, 271
  vortex motion, 17
Stone-von Neumann theorem. *See also*
    "Uniqueness" results
  unitary equivalence, physical equivalence and,
    490–91, 501, 505–6
Strain tensors, 85, 95, *95*
Stress-energy tensors
  continuum mechanics, 84
  dark matter and dark energy, 621n35
Stress, rigid body mechanics, 74
Stress tensors, 85, 95, *95*, 96
String theory
  electroweak theory, 405
  and multiverse, 644
  unification in physics, 383n3, 406
Strong causality condition, spacetime properties,
    596
Strong energy conditions, spacetime properties,
    595
Strong equivalence principle, relativistic
    spacetimes, 538
Structural realist interpretation of spacetime, 577
Structure
  formation, Standard Model, 616–17
  role of, 470–74
Strutt, John William. *See* Rayleigh, Lord
SU(2) and SU(3) color groups, 394, 396–97
Subspace
  dimension, indistinguishability, 362–63
  quantum mechanics, measurement, 417–18
Substantivalist approaches to spacetime. *See*
    Spacetime, substantivalist and
    relationalist approaches to
Substantivalist-relationalist debate. *See*
    Spacetime, substantivalist and
    relationalist approaches to
Sub-universe, 182
Sun's gravitational field, 626–27
Sunyaey-Zel'dovich effect, 632
Superconductivity, unification in physics, 384
Superfluid, EFT of, 245–46, *246*
Superfluidity, unification in physics, 384
"Superposition state," Everett interpretation, 462
Superspace, Machian relationalism, 562
Supersymmetry (SUSY), 383, 405
Surface forces, continuum mechanics, 87
Surface waves, hydrodynamics, 16–17
Sykes, Martin, 166

Symbolic universe, 31
Symmetry, 287–317. *See also* Symmetry and
    equivalence
  anti-particles, 294n10
  CERN's Large Hadron Collider, 305
  charge conjugation, 292–94
  classification of, 292–95
    charge conjugation, 292–94
    CPT theorem (Christenson, Cronin, Fitch,
      and Turlay), 294–95
    discrete symmetries, 292–94
    Lorentz invariant, 292
    parity, 292–94
    spacetime symmetries, 292
    time-reversal, 292–94
  continuous symmetry, 295–97
    gauge theories, 299
  CPT theorem (Christenson, Cronin, Fitch, and
    Turlay), 294–95
  "Curie Principle," 290
  discrete symmetries, 292–94
  dynamical account, 302n22
  "Eightfold Way," 302n22
  Einstein's theories of special and general
    relativity, 290
  electroweak theory, 395–97
  electroweak unification, 303–6
  equal areas law, 288n2
  Euler-Lagrange equations, 296
  Fermilab's Tevatron, 305
  "gauge argument," 291, 298–300
    Lagrangian, 288–99
    Lagrangian, kinetic component, 300n17
  gauge theories, 300–303
    continuous symmetry, 299
    dynamical account, 302n22
    "Eightfold Way," 302n22
    Gell-Mann "strangeness," 302
    Glashow-Salam-Weinberg (GSW) model,
      303
    "gluing" role, 301
    isospin idea, 301
    Lorentz invariant, 299
    "power of the gauge," 299
    quantum chromodynamics (QCD), 302
    weak neutral currents, 302n24
  Gell-Mann "strangeness," 302
  Glashow-Salam-Weinberg (GSW) model, 303
  global continuous symmetry, 295–97
  "gluing" role, 301
  Goldstone boson, 304
  group theory, 289
  Hamiltonian equations, 289
  Hamilton's "Principle of the Least Action,"
    295
  Higgs boson, 291
  Higgs mechanism, 303–6
  Higgs particle, 6, 381–82
  isospin idea, 301
  Lagrangian, 295–96, 299
    kinetic component, 300n17
    variational symmetries, 326–27

Symmetry (*Cont.*)
  Lorentz invariant, 292
  mean field theory, 160n12, 161
  multiplet scheme, prediction from, 307–9
  Newtonian scheme, 295
  Noether's theorem, 289, 291, 295–97
  omega minus hadron, 307
  ontological issues, 306–7
  overview, 6
  parity, 292–94
  permutation symmetry, 6–7
  prediction
    from multiplet scheme, 307–9
    Neptune, prediction of, 310
    of omega minus hadron, 306
    spin-3/2 baryon decuplet, 307–8, *309*
  quantum chromodynamics (QCD), 302
  reversal of a trend, 310–12
  rotations, 288
  snowflake, 288
  spacetime substantivalism, 533–36
    coordinate independent transformations, 533
    Galilei group, 533, 534
    generally covariant equations, 533
    Leibniz group of transformations, 533
    Leibniz relationalisms, 535
  spacetime symmetries, 292
  square invariant, 288
  SSB insight, 304
  STR, formulation of, 311–12
  symmetry group of the square, 289
  time-reversal, 292–94
  totalitarian principle, 309n39
  total symmetry, indistinguishability, 369
  unification, 303–6
  "variational problem," 297
  weak neutral currents, 302n24
Symmetry and equivalence, 318–39
  Cartan distribution, 324
  classical symmetries, 323–24, 334–37
  differential equations, symmetries of, 323–26
    Cartan distribution, 324
    classical symmetries, 323–24, 334–37
    generalized symmetries, 323–26
    Kepler problem, 325
    Korteweg-de Vries vector, 325
    Lenz-Runge vector, 325
    Lie-Bäcklund transformations, 324–26
    local symmetries, 324–25
    Maxwell's theory, 325n21
    nonlocal symmetries, 323, 325
    and physical equivalence, 329–30
  disaster, recipe for, 321–22
  divergence symmetries, 327nn33 and 34
  D1 and D2, 319–20
  D2, physical equivalence, 329–30
  fecundity and generality, trade-off between, 335n13
  Fruitless definition, 322
  Galilean boosts, 327, 331
  Galilei group and physical equivalence, 330
  generalization, 321–22
  generalized symmetries, 323–26
  Hamiltonian symmetries, 326–27
    outlook, 334
    and physical equivalence, 331–32
  instantaneous states, 326n31
  jet bundle, 324
  Kepler problem
    differential equations, symmetries of, 325
    and physical equivalence, 332
  Korteweg-de Vries vector, 325, 333
  Lagrangian treatment and physical equivalence, 331
  Lenz-Runge vector, 325
  Lie-Bäcklund transformations, 324–26
  local symmetries, 324–25
  Maxwell's theory, differential equations, 325n21
  Newtonian boosts, 327n33
  Newtonian theory, 322n12, 324, 327–28n36
    particles, 324, 325, 330
    physical equivalence, 328
  Noether's theorem, 324–25, 328
  nonlocal symmetries, 323, 325
  outlook, 333–34
  overview, 6
  pairs of solutions and physical equivalence, 333n55
  phase space, 327
  and physical equivalence, 328–33
    differential equation, 329–30
    D2, 329–30
    Galilei group, 330
    Hamiltonian symmetries, 331–32
    Kepler problem, 332
    Lagrangian treatment, 331
    and Newtonian theory, 328
    pairs of solutions, 333n55
    spacetime symmetries, 331–32n49, 332
  size of group, 322n10
  spacetime symmetries, 321–22
    and physical equivalence, 331–32n49, 332
    structure of, 319
  stage-setting, 320–21
  symmetries abstractly speaking, 321
  symplectic form, 327n35
  variational symmetries, 326–27
Symmetry group of the square, 289
Symplectic form, 327n35
Synthetic unity, unification in physics, 393–401
Systematic error, dark matter and dark energy, 620

Tait, Peter, 59, 66, 105
Taylor, Geoffrey, 22
Temporarily orientable spacetime, relativistic spacetime, 589
Tensor fields, continuum mechanics, 98
Theoretical laws, hydrodynamics, 31
"Theoretical unification," early universe cosmology, 633
Theory facades, axiomatic presentation, 48, 57

Theory of Everything (TOE), 381, 383, 405
"Theory of initial conditions," 637
Thermal history, Standard Model, 614–16
Thermodynamic phases, 142
Thermodynamic properties, phase transitions, 201
Thermodynamics, 145–46
  effective field theory (EFT), 249–50
  indistinguishability, 343–46
    conventions, system of, 246
    entropy, 343–45
    heat transfer, 344–45
  limits, 259–69
  phases, 142
  reduction of, 260
Thermodynamic treatment, phase transitions, 191–93
  continuous phase transitions, 191–92
  critical exponent, 193
  ferromagnetic transitions, 192
  first-order phase transitions, 191
  Helmholtz free energy, 192
  order-disorder transitions, 192
  order parameter, 192–93
Thomson, J. J., 30
Thomson, William (Lord Kelvin)
  explanatory progress, 28
  Helmholtz-Kelvin instability, 19
  instabilities, 20
  surface waves, 17
  Treatise on Natural Philosophy, 59, 66
  unified theory, 387
  vortex motion, 17
Tidal forces, 539n34
Time dependent boundary condition, radiation theory, 124
Time-harmonic waves, radiation theory, 127–29
Timelike future, relativistic spacetime, 590–91
Timelike geodesically incomplete, spacetime properties, 598
Timelike vectors, relativistic spacetime, 589, *589*
Time-reversal
  causation, 136n37
  symmetry, 292–94
Time travel, 601–3
Todhunter, Isaac, 271
Tollmien, Walter, 20
Toothpaste, continuum mechanics, 89, *89*
Top-down approach
  effective field theory (EFT)
    naturalness, hypothesis of, 228
    overview, 225–29
    quantum chromodynamics (QCD), 228–29
    symmetry considerations, 228
    Wilsonian approach, 228–29
  "rari-constancy" theorists, 270
  the tyranny of scales, 257
Torque r, 77, *78*
Totalitarian principle, symmetry, 309n39
Traction forces, rigid body mechanics, 73
Traction vectors, 93
  continuum mechanics, *83*, 83–84

Trajectories of force-free bodies, spacetime as, 542n41
Transformation, quantum mechanics, 446
Transtemporal structure, spacetime substantivalism, 531
Trapped surface, global spacetime structure, 600
Treatise on Electricity and Magnesium (Maxwell), 391
Treatise on Natural Philosophy (Tait and Thomson), 59, 66, 105
Truesdell, Clifford, 93
Turbulence
  hydrodynamics, 21–23
  mean field theory, 167
Turning moment, rigid body mechanics, 77, *78*
"Twin paradox" scenario, relativistic spacetimes, 536
Two-dimensional Ising model, *153, 181*
Tyndall, John, 20
Type Ia supernovae, 609, 618n25
The tyranny of scales, 255–86
  ab initio strategy, 263–64
  averaging and homogenization, differences, 277–78
  "between" scale structures, 256, 284
  bottom-up approach, 257
  bubbles within bubbles, *267*, 284
  Cauchy's equation, 270–75
  continuum mechanics, "material particles," 270n18
  continuum model equations, 256–57
  controversy, 269–73
    Cauchy's equation, 270–75
    Navier-Stokes equations, 269–70
  empirical investigation of means, 276
  Euler's recipes, 273–75
    and Cauchy, Augustin, 273–75
    continuum, 273–74, 278
    discrete, 273–74
    and Navier-Stokes equations, 275
    and Young's modulus, 275
  ferromagnet model, 265–66, 278
    spontaneous magnetization, *265*
  50-50 volume mixture, *268*
  games and gambling, 275–76, *277*
  Gaussian distribution, 276, *279*
  Hamiltonian function, 276
  homogenization, 256, 280–83
    averaging and homogenization, differences, 277–78
    limit, *281*
  hydrodynamic theory, 279
  infinite limits, 261–62
  "material particles," 270n18
  mean field calculations, 266
  Navier-Stokes equations, 269–70, 275
  Navier-Stokes theory, 278–79
  overview, 5–6
  post facto strategy, 263
  random variables, 277
  reduction, 260

The tyranny of scales (*Cont.*)
   renormalization group (RG), 264–66, 269, 275, 280
   representative volume element (REV), 264, 267–68, 280, 284
   resolution to, 275–83
      averaging and homogenization, differences, 277–78
      "between" scale structures, 284
      empirical investigation of means, 276
      Euler's continuum recipe, 278
      games and gambling, 275–76, *277*
      Gaussian distribution, 276, *279*
      Hamiltonian function, 276
      homogenization, 280–83
      hydrodynamic theory, 279
      Navier-Stokes theory, 278–79
      random variables, 277
      renormalization group (RG), 280
      representative volume element (REV), 280, 284
      steel, Gaussian and, *279*
   steel beams, 258–59, *264*, 278, *279*
   steel, Gaussian and, *279*
   thermodynamic limits, 259–69
   top-down approach, 257
   and Young's modulus, 272, 275

Uhlenbeck, George, 162, 163
Uncertainty, Everett interpretation and, 475–77
Undamped harmonic oscillator
   advanced Green's functions, 110
   retarded Green's functions, 110
Unification in physics, 381–415
   condensed matter physics, 383–84
   effective field theory (EFT), 384, 406–13
      Gell-Mann/Low formulation, 408–9
      Hamiltonians, 410
      quantum electrodynamics (QED), 411
      reductionism, problems of, 411n25
      Wilson-Kadanoff model, 412
   electroweak theory, 393–401
      Big Bang, 303, 403
      "big dessert" assumption, 404
      Cabibbo-Kobayashi-Maskawa framework, 402
      CP symmetry, 402–3
      gauge hierarchy problem, 404
      Glashow model, 397
      Higgs boson, 394–95, 403–6
      Higgs field, 398–401
      isospin, 396n14
      Lagrangian, 395
      Lagrangian invariance, 397–99
      Large Hadron Collider (LHC), 405–6
      Lie group, 395
      loop quantum gravity (LQG), 405
      from mathematics to physics, 397–401
      and Maxwell's theory, 395, 399
      multiplets, 396
      naturalness, 404n19
      Noether's theorem, 396–97
      non-Abelian case, 396, 401
      phase transformation, 396n13
      Planck scale, 403
      problems with, 403–6
      QCD vacuum, 394
      quantum chromodynamics (QCD), 402–3
      Schrödinger equation, 395
      Standard Model, 394, 402–6
      string theory, 405
      SU(2) and SU(3) color groups, 394, 396–97
      supersymmetry (SUSY), 405
      symmetry as tool for unification, 395–97
      Theory of Everything (TOE), 405
      Yukawa couplings, 402
      Yukawa interaction, 402n17
   General Relativity, 383
   Higgs particle, 6, 381–82
   Large Hadron Collider (LHC), 381–82, 384
      future output, 406
   Maxwell's electrodynamics, 385–93
      Ampère law, 389
      chain reaction, 387–88
      currents, induction of, 386n5
      d'Alembert's principle, 389–90
      displacement, 391n9
      displacement current, 386–89
      electrodynamics, 386–87
      and Faraday's account of electromagnetism, 385, 387
      and fictional models, 387–93
      Lagrangian mechanics, 386–87, 390
   Maxwell's theory, 383
   Newtonian mechanics, 406
   overview, 7–8
   physical problem, analysis of, 384n3
   possibility of reduction, 384
   quantum electrodynamics (QED), 396–97
   quantum field theory (QFT), 382, 408–10, 412–13
   reductive unity, 385–93
   renormalization, 406–13
   renormalization groups (RGs), 382, 385, 407, 410
   Standard Model, 381, 383
   string theory, 406
   superconductivity, 384
   superfluidity, 384
   supersymmetry (SUSY), 383
   symmetry, 303–6
   synthetic unity, 393–401
   Theory of Everything (TOE), 381, 383
   universal behavior, 384
   universality class, 412
"Unified theory." *See* Unification in physics
Uniform symmetry, indistinguishability as, 352–65
   classical indistinguishability, argument against, 354–56
   fermions, 364–65
   Gibbs' solution, 352–53
   haecceitism, 356–60
   quantum statistics, explanation of, 360–64

Uniqueness of universe, 624–27
  "ceteris paribus" laws, 626n44
  and "Laws of Physics," 625
  Mars's motion, 626–27
  and quantum field theory (QFT), 626
  and relativistic cosmology, 625–26
  Sun's gravitational field, 626–27
Uniqueness results, unitary equivalence, 491–501
  action principle, 491
  analyzing physical equivalence, 496–501
  canonical anticommutation relations (CARS), 494, 497
  Hamiltonian system, 493–94
  Hilbert space, 494–95
  kinematic pair, 495–96
  kinematics, 492
  Lagrangian, 492
  ordinary QM, 495, 496, 500–502, 504
  Pauli relations, 494
  PEV, 496–500, 496n6, 504, 508
  physical equivalence, unitary equivalence as, 494–95
  Poisson bracket, 493–94
  preliminaries, 491–92
  quantizing, 492–94
  Schrödinger equation, 494
  weak operator topology, 495
Unitarily inequivalent representations, 491
Unitary equivalence, physical equivalence and, 9, 489–521
  action principle, 491
  "Algebraic Imperialism," 513
  analyzing physical equivalence, 496–501
  black hole evaporation, 518
  Bohm-Aharonov effect, 507
  canonical anticommutation relations (CARS), 489, 490, 494, 497, 501, 502, 504
  closing uniform topology, 510n23
  competing criteria of equivalence, 513–14
  GNS representation, 511–12
  Hadamard condition, 517–18
  Hamiltonian system, 493–94
  Hawking radiation exterior, 518
  Hilbert space, 494–95, 503n12, 510, 518
    competing criteria of equivalence, 513
    representations, 9, 489–90
  Jordan-Wigner theorem, 9, 490–91, 501
  kinematic pair, 495–96
  kinematics, 492
  Lagrangian, 492
  Minkowski spacetime, 518
  ordinary QM, 495, 496, 500–502, 504, 510–16
  PAS, 504, 508, 514
  Pauli relations, 494
  PEV, 496–500, 496n6, 504, 508, 514
  physical equivalence, unitary equivalence as, 494–96
  Poisson bracket, 493–94
  preliminaries, 491–92
  principles, 514–19
    Hadamard condition, 517–18
    Hawking radiation exterior, 518
    Minkowski spacetime, 518
    ordinary QM, 514–16
    quantum field theory (QFT), 518
  quantizing, 492–94
  quantum field theory (QFT), 491, 518
  quantum statistical mechanics (QSM), 491
  reasons for unitary equivalence, 513–14
  Schrödinger equation, 494
  Stone-von Neumann theorem, 490–91, 501, 505–6
  technical interlude, 509–13
  uniqueness results, 491–501
    action principle, 491
    analyzing physical equivalence, 496–501
    canonical anticommutation relations (CARS), 494, 497
    Hamiltonian system, 493–94
    Hilbert space, 494–95
    kinematic pair, 495–96
    kinematics, 492
    Lagrangian, 492
    ordinary QM, 495, 496, 500–502, 504
    Pauli relations, 494
    PEV, 496–500, 496n6, 504, 508
    physical equivalence, unitary equivalence as, 494–96
    Poisson bracket, 493–94
    preliminaries, 491–92
    quantizing, 492–94
    Schrödinger equation, 494
    weak operator topology, 495
  unitary inequivalence, examples of, 501–9
    bead on a circle, 505–7
    Bohm-Aharonov effect, 507
    Circular Canonical Commutation Relations (CCCRs), 506
    ferromagnet, 508, *509*
    Hamiltonian theory, 507
    Hilbert space, 503
    infinite spin chain, 501–5, 508
    ordinary QM, 508
    Pauli spins, 494, 501–4
    Poisson bracket, 505–6
    polarization of a system, 502
    Yang-Mills theory, 507
  weak operator topology, 495
Unitary inequivalence
  bead on a circle as example of unitary inequivalence, 505–7
  examples of. *See* Unitary equivalence, physical equivalence and
Universal behavior, unification in physics, 384
Universality, 161–62
  phase transitions, 217–18
Universality classes
  matter, infinities and renormalization, 178–79
  unification in physics, 412
Universe, uniqueness of. *See* Uniqueness of universe
"Unsharp" spin measurements, quantum mechanics, 444–446

Up and down spin states, quantum mechanics, 418–19
U.S. National Bureau of Standards conference, 168

Vaidman, Lev, 476
van der Waals, Johannes
  corresponding states, principle of, 162
  first mean field theory, 147–49, *149*
  memorial meeting, Netherlands meeting (1937), 162–64
  phase transitions, 142, 147
van Fraassen, Bas C., 565
van Kampen, N.
  Gibbs paradox, 368
  haecceitism, 356
  N! puzzle, 350–51
  thermodynamics, 345
  as uniform symmetry, 354–55
"Variational problem," symmetry, 297
Variational symmetries, 326–27
Veblen, O., 289n3
Verschaffelt, J. E., 164
Vibrating strings, continuum mechanics, 92, *92*, 101
Virasoro algebra, 182
Virtual displacement, rigid body mechanics, 80, *81*
Virtual variations, rigid body mechanics, 79n33
Virtual-work reasoning, rigid body mechanics, 72, *72*
Viscosity of fluid, axiomatic presentation, 54
Volume measures, indistinguishability, 362–63
von Neumann, John. *See also* Dirac-von Neumann interpretation; Stone-von Neumann theorem+
  discretized position measurements, 443
  and Heisenberg's "cut," 441, 442n31
  no-hidden variables theorem, 422
  quantum mechanics measurement, 452, 454
Voronel, Alexander, 165
Vortex motion
  Helmholtz-Kelvin instability, 19
  hydrodynamics, 17–19
  vortex filaments, defined, 18
Vorticity, 17

Water, cartoon PVT diagram for, *148*
Water Waves (Stoker), 127–28
Wave equation
  advanced Green's functions, 120–23
  retarded Green's functions, 120–23
Waveguide, radiation theory, *124*
Wave movements, continuum mechanics, 88, *88*
Wave packets, Everett interpretation, 461
Wave resistance, 24
Weak coupling fixed points, 177
Weak neutral currents, symmetry, 302n24
Weak operator topology, unitary equivalence, 495
Weeks, John, 163
Weighting, indistinguishability, 361, *362*

Weinberg, Steven
  anthropic reasoning, 638–40, 642, 643
  unification in physics, 383, 398, 400–401
Weiss, Pierre, 157
Weyl, Hermann, 289, 297, 626
Whitehead, A. N., 96n43
Widom, Benjamin, 168–70, *170*, 173
Widom scaling, 168–70
Wien's law, 342
Wigner, Eugene, 6, 291
  classifying, 292
  discussion, 310–12
  group theoretic approach, 306n33
  representations, forms of, 307n34
  "superprinciples," 291
Wilcox, C., 128–29
Wilsonian approach, effective field theory (EFT), 252
  cutoff, realistic interpretations of, 242
  mass-independent schemes, 237–39
  ontological implications, 240
  quasi-autonomous domains, 241
  and renormalization group (RG) techniques, 232
  renormalization schemes, 236–37
  top-down approach, 226, 228–29
Wilson-Kadanoff model, 412
Wilson, Kenneth, *175*. *See also* Landau-Ginzburg-Wilson free energy
  mean field theory, 152
    replacement theory, 164
  on renormalization group theory, 172–77
    calculational method, 174–75
    e-expansion, 176–77
    elementary particles, 176
    fixed-points, 177
    Fourier space, 172–74
    Landau-Ginzburg-Wilson free energy, 173–74
    physical space, 172–74
    running coupling constants, 175–76
Wing theory, 13, 30
  airplane wings, 26

Xia, Zhihong, 62, 68

Yang, C. N., 166
Yang-Lee theorem, 195, 208
Yang-Mills theory
  "gauge argument, 299, 301, 302n22
  unification in physics, 397
  unitary inequivalence, as example of, 507
Young's modulus, 272, 275
Yukawa couplings, electroweak theory, 402
Yukawa interaction, electroweak theory, 402n17

Zemanian, A., 137
Zernike, Frederik, 152, 158
Zero field strength, Machian relationalism, 557n75
"Zodiacal masses," 619